S0-BLS-183

Handbook of
Analytical Instruments

About the Author

Dr. R. S. Khandpur is currently Director General, Pushpa Gujral Science City, Kapurthala, Punjab—a joint project of the Government of India and Punjab Government. Prior to this, he was Director General, Centre for Electronics Design and Technology of India (CEDTI), an autonomous scientific society of the Ministry of Communication and Information Technology, Government of India. Dr. Khandpur was the Founder Director of CEDTI, which is the first ISO-9002 certified organisation of the Ministry of Information Technology.

Dr. Khandpur has served as a scientist for 24 years in CSIO, Chandigarh—a constituent laboratory of the Council of Scientific and Industrial Research (CSIR). He was Head of the medical instruments division (1975–1989) and Head of the electronics division (1986–1989).

Dr. Khandpur is a member of board of governors, Punjab Technical University; and director, board of directors, Punjab Information and Communication Technology Corporation Ltd. He is a AICTE distinguished visiting professor, and member of the vision group on IT, set up by the Punjab Government. He is a member of IEEE (Institution of Electronics and Electrical Engineers), USA, and fellow of IETE (Institution of Electronics and Telecommunication Engineers).

Dr. Khandpur has over four decades of experience in R&D, technology transfer, education and training, consultancy and management at national and international levels. He is the recipient of the 1989 Independence Day Award of the National Research and Development Corporation and IETE (Institute of Electronics and Telecommunication Engineers), for outstanding contribution to the development of the electronics industry.

Dr. Khandpur holds six patents of innovative designs, has authored seven books and published over 60 research and review papers.

Handbook of Analytical Instruments

R. S. Khandpur

Director General
Pushpa Gujral Science City, Kapurthala. Punjab, India

Formerly

Director General
Centre for Electronics Design and Technology of India (CEDTI)
Department of Information Technology, New Delhi, India
and
Director
CEDTI, Mohali (Chandigarh), Punjab, India

New York Chicago San Francisco Lisbon London Madrid
Mexico City Milan New Delhi San Juan Seoul
Singapore Sydney Toronto

The McGraw·Hill Companies

1 2 3 4 5 6 7 8 9 0 DOC/DOC 0 1 3 2 1 0 9 8 7 6

ISBN-13: 978-0-07-148746-7
ISBN-10: 0-07-148746-8

This book was first published in India in 2006 by Tata McGraw-Hill.

The sponsoring editor for this book was Stephehen S. Chapman and the production supervisor was Richard C. Ruzycka. The art director for the cover was Anthony Landi.

Printed and bound by RR Donnelley.

This book was printed on acid-free paper.

McGraw-Hill books are available at special quantity discounts to use as premiums and sales promotions, or for use in corporate training programs. For more information, please write to the Director of Special Sales, Professional Publishing, McGraw-Hill, Two Penn Plaza, New York, NY 10121-2298. Or contact your local bookstore.

Contents

Preface

Instruments used for analysis constitute the largest number of instruments in use today. Their range is spectacular and variety baffling. It is difficult to imagine a field of activity where analytical instruments are not required and used. They are used in hospitals for routine clinical analysis, drugs and pharmaceutical laboratories, oil refineries, food processing laboratories and above all for environmental pollution monitoring and control. This book has been designed to cater to the needs of a wide variety of readers working in these areas. The postgraduate students of chemistry and physics undergoing courses on instrumental analysis will find the book a useful text. It is also intended as a guide for the scientific investigator who wishes to acquire knowledge about the more widely-used analytical instruments. The availability of a bewildering array of instrumental techniques and a large variety of commercially available equipment have provided several courses of action to the investigator. The student who undergoes this course will be in a position to select instruments for a particular problem with some idea of their merits, demerits and limitations. The treatment of the subject is designed to be sufficient to give the reader the necessary background to fruitfully discuss the more esoteric details of an instrument with an expert in this area.

With the widespread use of analytical instruments, it has now become essential to have qualified and sufficiently knowledgeable service and maintenance engineers. Besides having a basic knowledge of the principle of operation, it is important for them to know the details of commercial instruments from different manufacturers. A concise description of instruments of each type from leading manufacturers has, therefore, been provided. Since rapid changes and improvements in instruments usually make some of the description obsolete, an attempt has been made to describe the principles that are basic to the various types of analytical instruments. The principles learned would enable the service engineers to carry over to new equipment as it appears in the market.

The area of analytical instrumentation involves a multidisciplinary approach, with electronics and optics forming the major disciplines. Highly populated printed circuit boards often have application-specific integrated circuits mounted on them. This has necessitated a new approach to repair and servicing involving replacement of the printed circuit card without having to repair the instrument at the component level. Work of this nature requires a knowledge of the various building blocks of an instrument, and this has been the approach in the preparation of the text. This approach will also be advantageous to students of chemistry, physics, chemical engineering, instrumentation and electronics, etc.

The book starts with an explanation of basic concepts in electronics and optics, with special reference to operational amplifiers, digital integrated circuits and microprocessors. The chapter also covers various types of display systems and laboratory recorders. The information in this chapter would serve as a base for the description in the rest of the book.

Colorimetry and spectrophotometry have become the central techniques of analytical chemistry. Spectrophotometry, particularly in the ultraviolet region, offers a sensitive, precise and non-destructive method of analysis of biologically important substances such as proteins and nucleic acids. The introduction of spectrophotometers, especially in the visible region of the spectrum, led directly to the explosion of analytical methodology, characterized by the use of an ever decreasing amount of sample size and multi sample analytical methods. The use of the double beam principle helped in the development of infrared spectroscopy, where direct recording of spectra is now routine. Similarly, flame photometers and atomic absorption spectrophotometers have become almost indispensable tools in clinical and research laboratories. These topics, along with fluorimeters and phosphorimeters, photoacoustic and photothermal spectrometers, mass spectrometers and Raman spectrometers, are covered in the next part of the book.

The most important spectroscopic techniques for structural determination are nuclear magnetic resonance and electron spin resonance spectrometry. These topics are covered in two chapters. Fourior transform NMR and spin decouplers are also concisely dealt with. A separate chapter on electron and ion spectroscopy, important analytical tools for surface analysis, has been included.

Qualitative and quantitative analyses have been greatly facilitated by the increasing availability of isotopically labelled compounds. The measurement of these has required the development of increasingly complex equipment, usually capable of measuring and printing out the radioactive counts from 100 or more samples wholly automatically. Multichannel counters, scanners and gamma cameras are representative of the highly sophisticated equipment and richly deserve their place in the book. X-ray spectrometers based on diffraction, absorption and fluorescence are also covered in detail. Analysis procedures have been automated, with the result that a very large number of samples can be handled per hour for a multiplicity of tests. Systems like autoanalysers have brought about tremendous procedural and conceptual innovations. A complete description of the automated analysis systems has been provided in a separate chapter.

Chromatography offers a unique method of separation of closely similar substances. Various methods like liquid-liquid partition, gas-liquid partition, ion-exchange, molecular sieve, etc. have all provided a basis for the same. Gas-liquid chromatography is perhaps the most widely used than any other single analytical method for small- to medium-sized molecules. However, a liquid mobile phase possesses certain advantages over a gaseous mobile phase since it can contribute to the separation achieved through the specificity of its interaction with solutes. This advantage has resulted in the development of high pressure, high performance liquid chromatography, using specialized column packing of very small size. Two chapters are devoted to the different types of chromatograph instruments. Perhaps the most sophisticated ultra-micro scale analytical instrument is the combination of the gas chromatograph with the mass spectrometer, and the same has been well illustrated in the chapter on mass spectrometry. A separate chapter on thermoanalytical methods gives a brief introduction to these techniques.

Electrophoresis, as a medium-scale method for the separation of proteins and its development into an ultra-micro method with the introduction of zone electrophoresis, is covered in the next chapter. The continuous improvements in staining and optical visualization methods have led to this technique being applicable to the microgram level or even less. The chapter also covers spectrodensitometers including those based on microprocessors.

Electrochemical instruments, based on a variety of principles of operation, have undergone considerable improvements in terms of accuracy, speed and automation. While the chapter on electrochemical instruments covers various types of instruments like conductivity meters and polarographs, the chapter on pH meters includes the latest in electronic circuitry, ion-selective electrodes and biosensors. An exhaustive treatment has been given to blood gas analyzers and industrial gas analyzers.

Awareness and concern about the deteriorating environment is increasing the world over. It is necessary to monitor changes taking place in the quality of the environment for initiating efforts to accomplish environmental pollution control. Many of the analytical tools used elsewhere in other areas of applications could be profitably utilized for air and water pollution monitoring purposes. However, a special chapter on this subject illustrates specific techniques employed for monitoring different pollutants in air and water.

The modern laboratory has a large number of analytical instruments churning out information. Electronic procedures for handling the huge amount of data have become imperative. The marriage of computers and instruments has offered tremendous possibilities of easing the burden on the scientists, as well as optimizing the performance of analytical instruments. Many of the leading instrument manufacturers are producing systems for use in the laboratory, both for data acquisition and for control purposes. The personal computer has been virtually responsible for a revolution in the methodology, quantity, economy and quality of experiments performed. Taking into consideration the impact of computers on analytic instruments, the topic is covered in the final chapter.

It is a matter of great pleasure to present this book, which is comprehensive but is yet sufficiently detailed for specialists and scholars who need not refer to specialised books on the various topics.

Some developments that increased the need for this book include:

- Integration of microcontrollers and personal computers, which have replaced long established recording techniques and display systems with ones that are intelligent, automated, and more precise.
- Spectrophotometry, which has led to an explosion in analytical methodology, characterised by the use of an ever decreasing sample size and multi-sample analytical techniques.
- Systems like auto-analysers, which have brought about tremendous procedural and conceptual innovations.
- Measurement of isotopically labelled compounds, which has greatly facilitated qualitative and quantitative analyses. This has necessitated the development of complex equipment, capable of automatically measuring and reporting radioactive counts from a very large number of samples per hour for a large number of tests.

- Development of high-pressure, high-performance liquid chromatography, using specialised column packing of a very small size.
- Continuous improvements in staining and optical visualisation methods that have led to electrophoresis being used at the microgram level.
- Electrochemical instruments, based on a variety of principles, which have undergone considerable improvements in terms of speed, automation and accuracy.

Rapid changes and improvements in analytical instrumentation render existing techniques and equipment obsolete, very fast. Hence the focus, in this book, is on describing the founding principles of analytical instrumentation. This will enable personnel to transition quickly to new and improved techniques, and equipment, as they become available.

The range of analytical instruments is spectacular and their variety is baffling! This book besides discussing technological advancements, and new methods of analysis, takes into account a bewildering array of commercially available equipment. To enable readers to make an informed decision about the instrument that best meets their requirement, the book covers a range of products. I would, however, like to mention that the discussion of these instruments does not imply they are superior to competing ones! My motive is to describe instruments typical of their class, or possessing features of special interest, as an illustration of the indicated principles and intended application.

I am indebted to the Director, CSIO, for his kind permission to publish the book and to Tata McGraw-Hill Publishing Company Limited for excellent editing and printing. I am thankful to TAB Books, USA, for their kind permission to reproduce some parts of my earlier book published by them. I am also grateful to my wife Mrs. Ramesh Khandpur and children Vimal, Gurdial and Popila for the help they extended me during preparation of the manuscript and proofreading.

R. S. KHANDPUR

Chapter 1

Fundamentals of Analytical Instruments

1.1 Elements of an Analytical Instrument

Analytical instruments provide information on the composition of a sample of matter. They are employed, in some instances, to obtain qualitative information about the presence or absence of one or more components of the sample, whereas in other instances they provide quantitative data. In the broadest sense, an analytical instrument (Figure 1.1) consists of the following four basic units:

- *Chemical information source*, which generates a set of signals containing necessary information. The source may be in the sample itself. For example, the yellow radiation emitted by heated sodium atoms constitutes the source of the signal in a flame photometer.

Chemical Information Source

- *Transducer*, which converts the nature of the signal. Because of the familiar advantages of electric and electronic methods of measurement, it is the usual practice to convert all non-electrical phenomena associated with the analysis of a sample into electrical quantities. For example, a photocell and a photomultiplier tube are transducers that convert radiant energy into electrical signals.

Transducer

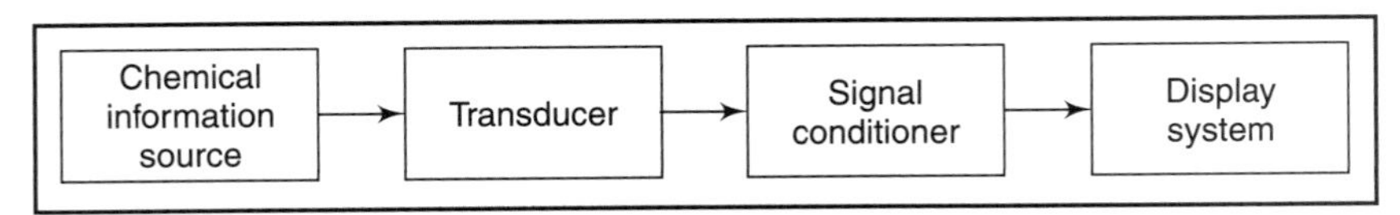

Figure 1.1 :: Elements of an analytical instrument

Signal conditioner

- *Signal conditioner,* which converts the output of the transducer into an electrical quantity suitable for operation of the display system. Signal conditioners may vary in complexity from a simple resistance network, or impedance matching device, to multi-stage amplifiers and other complex electronic circuitry. They help increase the sensitivity of instruments by amplifying the original signal or its transduced form.

Display system

- *Display system,* which provides a visible representation of the quantity as a displacement on a scale, or on the chart of a recorder, or on the screen of a cathode ray tube, or in numerical form.

Thus, the instrument can be considered in terms of flow of information, where the operation of all parts is essentially simultaneous.

Characteristic module

The first two blocks constitute the *characteristic* module, whereas the last two form the *processing* module of an instrument (Strobel, 1984a). The characteristic module in a pH meter consists of the glass membrane pH electrode and reference electrode immersed in the cell solution. Similarly, the constituents of the characteristic module in a UV-Vis spectrophotometer are a source of radiant energy, a monochromator, the sample holder and photodetector. Each one of the components of the characteristic module contributes to the performance specifications of an instrument. For example, by choosing a photomultiplier tube with the broadest spectral response available, good sensitivity is ensured for detection from 190 nm to beyond 800 nm. The higher its gain and the lower its noise, the better the possibility of working at trace concentration levels. The transducer also determines the limits of detection of measurement. Similarly, the monochromator fixes the resolution, signal-to-noise ratio, and level of stray light.

Processing module

The signal amplitude produced by a transducer is processed in the processing module. After adequate amplification, the processing can be carried out with the signal still in analog form (a signal of varying amplitude) or converted into digital form (a series of pulses whose number indicates the signal amplitude) by the use of an analog-to-digital (A/D) converter. This conversion ordinarily gives higher precision and accommodates the use of a microprocessor or computer for processing steps and instrument control.

Analog Meters or Digital Displays

The results of a measurement in analytical instruments are usually displayed on either analog meters or digital displays. Digital displays present the values of the measured quantities in numerical form. Instruments with such a facility are directly readable and slight changes in the parameter being measured are easily discernible in such displays, as compared to their analog counterparts. Because of their higher resolution, accuracy and ruggedness, they are preferred for display over conventional analog moving coil indicating meters. Different types of devices such as light emitting diodes and liquid crystal displays are available for display in the digital or numerical form.

Visual Display Units (VDUs)

Since computers are being increasingly used to control the equipment and to implement the man–machine interface, there is a growing appearance of high resolution colour graphic screens to display the course of analytical variables, laboratory values, machine settings or the results of image processing methods. The analog and digital displays have been largely replaced by video display units, which present information not only as a list of numbers but as elegant character and graphic displays, and sometimes as a three-dimensional colour display. Visual display units (VDUs) are usually monochrome as the CRTs in these units are coated with either white or green phosphors. Coloured video display units are also becoming popular.

Keyboard

A keyboard is the most common device connected into almost all forms of data acquisition, processing and controlling functions in analytical instruments. A keyboard can be as simple as a numeric pad with function keys as in a calculator, or it can be a complete alphanumeric and typewriter keyboard with an associated group of control keys suitable for computer data entry equipment. Most available keyboards have single contact switches, which are followed by an encoder to convert the key closures into ASCII (American Standard Code for Information Interchange) code for interfacing with the microprocessor.

All analytical instruments can thus be split (Strobel, 1984b) into the above indicated four sub-systems. Some examples of the instruments along with their sub-systems are given in Table 1.1.

Table 1.1 :: Examples of Instrument Sub-systems

Instrument	Characteristic Module		Processing Module	
	Analytical Signal	Input Transducer and its Output	Signal Processor	Readout
Photometer	Attenuated light beam	Photocell (electrical current)	None	Current meter
Atomic emission spectrometer	UV or visible radiation	Photomultiplier tube (electrical potential)	Amplifier, demodulator	Chart recorder
Coulometer	Cell current	Electrodes (electrical current)	Amplifier	Chart recorder
pH meter	Hydrogen ion activity	Glass-calomel electrodes (electrical potential)	Amplifier, A/D converter	Digital Display
X-ray powder diffractometer	Diffracted x-radiation	Photographic film (latent image)	Chemical developer	Black images on film
Colour comparator	Colour	Eye (optic nerve signal)	Brain	Visual colour response

The progress made in instrumental methods of analysis has closely paralleled developments in the field of electronics, because the generation, transduction, amplification and display of signals can be conveniently accomplished with electronic circuitry. All electronic circuits are constructed with the help of some basic components and circuit blocks, which are broadly described in Chapters 25 and 26 of this book.

1.2 Sensors and Transducers

Transducers are devices which convert one form of energy into another. Numerous methods have since been developed for this purpose and the basic principles of physics have been extensively employed. Variation in electric circuit parameters like resistance, capacitance and inductance in accordance with the events to be measured, is the simplest of such methods. Peizoelectric and photoelectric transducers are also very common. Chemical events are detected by the measurement of current flow through the electrolyte or by the potential changes developed across the membrane electrodes. A number of factors decide the choice of a particular transducer to be used for the study of a specific phenomenon. These factors include:

- The magnitude of quantity to be measured,
- The order of accuracy required,
- The static or dynamic character of the process to be studied,
- The type of application: *in vitro* or *in vivo*, and
- Economic considerations.

1.2.1 Classification of Transducers

Many physical, chemical and optical properties, and principles can be applied to construct transducers for applications in the analytical field. The transducers can be classified in many ways, such as:

(i) By the process used to convert the signal energy into an electrical signal. For this, transducers can be categorized as:
  - *Active Transducers*—Transducers that convert one form of energy directly into another. For example: a photovoltaic cell in which light energy is converted into electrical energy.
  - *Passive Transducers*—Transducers that require energy to be put into them in order to translate changes due to the measurand. They utilize the principle of controlling a dc excitation voltage or an ac carrier signal. For example: a variable resistance (such as a thermistor) placed in a Wheatstone bridge in which the voltage at the output of the circuit reflects the physical variable (temperature). Here, the actual transducer is a passive circuit element but needs to be powered by an ac or dc excitation signal.

(ii) By the physical or chemical principles used. For example: variable resistance devices, Hall effect devices and optical fibre transducers.

(iii) By the application for measuring a specific analyte variable. For example: flow transducers, pressure transducers, temperature transducers, etc.

1.2.2 Performance Characteristics of Transducers

A transducer is normally placed at the input of a measurement system, and therefore, its characteristics play an important role in determining the performance of the system. Ideally, a transducer should have a high level output, zero source impedance, and low noise and should be relatively linear. However, in practice, the ideal conditions are never met. Some of the important parameters which characterize transducers are:

- Input and output impedance
- Common mode rejection ratio
- Overload range
- Recovery time after overload
- Excitation voltage
- Shelf life
- Reliability
- Size and weight
- Response time: to a step change in the input (measured) and includes rise time, decay time and time constant.

Noise-equivalent power

However, manufacturers typically specify detectors by the noise-equivalent power (NEP), detectivity (D), responsivity (R) and time constant (t). These factors are described below.

Noise Equivalent Power is the rms (root mean square) value of sinusoidally modulated radiant power falling upon a detector that gives an rms signal voltage equal to the rms noise voltage from the detector.

Detectivity is defined as the inverse of noise-equivalent power.

Responsivity of a detector specifies its response to a unit change in the input.

Time constant of a detector is the measure of a detector's ability to respond to a rising or falling optical signal.

A majority of the transducers used in analytical instrumentation are analog devices. They measure continuous physical or chemical properties, and normally give continuous electrical outputs. For example, changes in the ion activity in a solution are measured with a specific ion electrode (transducer) whose output is available in terms of changes in voltages. Similarly, a polarographic cell measures the concentration of electroactive species and gives current as the output.

Detectors, transducers and sensors

The three terms, detectors, transducers and sensors, are often used interchangeably. However, there is a fine difference in their definitions and scope, which is explained below.

Detector: This is a device that indicates a change in environment. For example, a smoke detector or gas detectors.

Transducer: This is a device that converts one form of energy into another, specifically from non-electrical to electrical data. For example, a thermocouple which converts temperature into voltage.

Sensor: This is a device that converts chemical into electrical data. For example, a biosensor which converts biological activity into an electrical signal.

The action of transducers is based on some property of the analyte on the basis of which the measurements are done and instruments are constructed. Some of the properties of the analytes and the techniques employed are given in Table 1.2.

Table 1.2 :: Properties of Analytes and Techniques used in Analytical Instruments

Category	Property	Technique used
Radiation	Radiation absorption	Absorption spectroscopy • Photometry • Spectrophotometry • Nuclear Magnetic Resonance • Electron Spin Resonance
	Radiation emission	Emission Spectroscopy • Fluorescence • Phosphorescence • Luminescence
	Radiation scattering	• Raman Spectrometry • Turbidity Measurement
	Radiation refraction	• Refractometry • Interferometry
	Radiation diffraction	• X-ray Spectroscopy • Electron Spectroscopy
	Radioactivity	• Activation Analysis • Isotope Dilution
Electrical	Electrical potential Electrical current Electrical charge Electrical resistance	Potentiometry Amperometry Polarography • Coulometry • Conductometry
Mass	Mass Mass-to-charge ratio	Gravimetry Mass Spectrometry
Thermal	Temperature change	• Thermal Gravimetry • Calorimetry

The choice of an instrumental method and the equipment to be used for the same would depend upon the following performance requirements:

- Accuracy: How close to the true value?
- Precision: How reproducible?

- Sensitivity: How small a difference can be measured?
- Selectivity: How much interference?
- Dynamic Range: What range of amounts?

Section 1.7 details the performance requirements of instruments and the methods employed for purposes of analysis.

1.2.3 Smart Sensors

Any analytical measurement primarily involves two steps: the *reaction step* which involves sampling, sample transport and processing, separation and reaction, and the *measurement step*, involving transduction, signal acquisition and processing. An area of intense research and study at present concerns how these two steps can be combined to get integrated analytical systems. This is possible by having a *chemical sensor* with two integrated parts which are:

Chemical sensor

- Receptor, which is a highly selective recognition material, and
- Transducer, which is a material that converts the recognition signal (be it optical, electrochemical, mass or thermal) into a signal that is usually electrical in nature.

The recent designs of such chemical sensors which combine recognition and transduction have led to the development of new analytical instrumentation resulting in simplification, miniaturization, robustness, speed, mobility and cost reduction (Alegret, 2003).

The working of these chemical sensors depends, to a large extent, on the availability of highly selective recognition materials, which are of two types:

- Synthetic materials such as cyclic or macrocyclic ligands, cyclodextrines, and molecularly imprinted polymers; and
- Natural compounds such as enzymes, micro-organisms, animal or plant tissues, antibodies, DNA, etc.

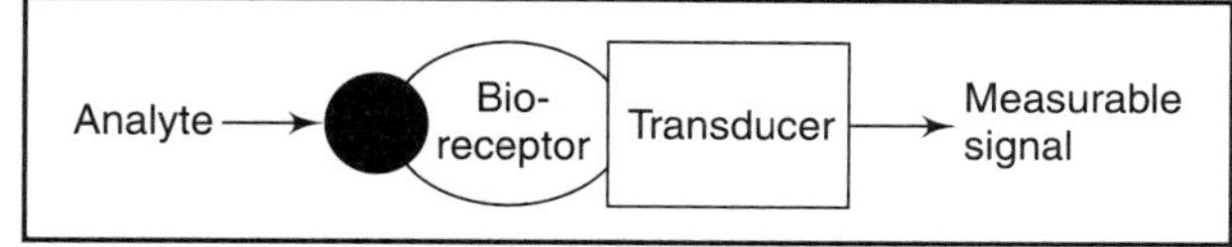

Figure 1.2 :: Principle of a biosensor

The chemical sensors that integrate synthetic recognition materials are known as *chemosensors* and when they integrate biomaterials, they are known as *biosensors* (Diamond, 1998). Figure 1.2 shows the constituent elements of a biosensor.

Chemosensors

Biosensors

In most cases, sensors are used to detect a single analyte. By integrating several sensing devices within the same platform, it is possible to have an array of micro-fabricated sensors with different selectivity for multi-component analysis, simultaneously. For example, gas-sensor arrays enable us to measure the concentrations of different gases in multi-component gas samples.

Rapid and continuous developments in the microelectronics industry have also had an influence in different fields such as chemistry and life sciences. The objective is to bring down the common laboratory instruments, devices and machines to the global size of a credit card. Examples of

Micro-electromechanical systems

current research in this area include the development of microsensors and micro-electromechanical systems (MEMS). These are integrated devices that combine both electrical and mechanical components within a micrometer dimension, and are fabricated in glass, quartz or plastic (Figeys and Pinto, 2000).

Lab-on-chip

The integration of fluidic microstructures with other analytical microstructures such as micro-reactors, microsensors and microactuators, has led to the development of the 'lab-on-chip' concept on a commercial scale. The technology includes any type of laboratory operation related to liquid handling, chemistry reactor technology and biotechnology, amplification, etc. The main advantages of these chips (micro-laboratories) are improved analytical performance in respect of higher throughputs, increased mechanical stability and reduced resource consumption (Manz and Becker, 1998).

Smart sensors

The sensors which have a tight coupling between the sensing and computing elements are also known as smart sensors. Their characteristics would normally include temperature compensation, calibration, amplification, some level of decision-making capability, self-diagnostic and testing capability and the ability to communicate interactively with external digital circuitry. The currently available smart sensors are actually hybrid assemblies of semiconductor sensors plus other semiconductor devices. In some cases, the coupling between the sensor and the computing element is done at the chip level on a single piece of silicon in what is referred to as an integrated smart sensor. In other cases, the term is applied at the system level. The important roles of smart sensors are as follows.

Signal Conditioning: The smart sensor converts a time-dependent analog variable into a digital output. Functions such as linearization, temperature compensation and signal processing are included in the package.

Tightening Feedback Loops: Communication delays can cause problems for systems which rely on feedback or which must react/adapt to their environment. By reducing the distance between the sensor and the processor, smart sensors facilitate significant advantages in these types of applications.

Monitoring/Diagnosis: Smart sensors which incorporate pattern recognition and statistical techniques can be used to provide data reduction, and to change detection and compilation of information for monitoring and control purposes.

Smart sensors divert much of the signal processing work load away from the general purpose external signal conditioning and comsputing. They offer a reduction in the overall package size and improved reliability, both of which are critical for *in situ* and sample return applications. Achieving a smart sensor depends on integrating the technical resources required to design the sensor and the circuitry, developing a manufacturable process and choosing the right technology. A typical example of a smart sensor is the ion-sensitive field-effect transistor.

1.3 Signal Conditioning in Analytical Instruments

The information from the transducer is often obtained in terms of current intensity, voltage level, frequency or signal phase relative to a standard. Voltage measurements are the easiest to make, as the signal from the transducer can be directly applied to an amplifier having a high input impedance. Usually, the input device in the amplifiers in modern equipment is FET or MOSFET as they inherently offer a high input impedance. The voltage signal source can be connected directly to a sensitive current measuring device (moving coil meter), but this is usually not done because of inadequate sensitivity and low input impedance of the meter.

Input impedance

Most of the transducers produce signals in terms of current, which, in many cases, is too small, thereby making direct connection between the transducer and the display device difficult. However, some of the earlier colorimeters and flame photometers used this arrangement. In modern instruments, the most widely used current-measuring arrangement is that of placing a resistor in series with the current source and measuring the voltage drop across it, which is subsequently measured by observing the current that it will drive into an amplifier. When impedance-matching problems preclude the use of dropping resistors, low input-impedance current-to-voltage conversion can be achieved by using operational amplifiers with appropriate feedback.

In order to make an accurate measurement of voltage, it is necessary to ensure that the input impedance of the measuring device is large as compared to the output impedance of the signal source. This is done to minimize the error that would occur, if an appreciable fraction of the signal source were dropped across the source impedance. Conversely, the need for an accurate measurement of current source signals necessitates the source output impedance to be larger than the receiver input impedance. Ideally, a receiver that exhibits a zero input impedance would not cause any perturbation of the current source. Therefore, high-impedance current sources are more easily handled than low-impedance current sources.

Output impedance

In general, the frequency response of the system should be compatible with the operating range of the signal being measured. A signal conditioner accepts information from a detector and presents it in a convenient form to the user. In order to process the signal waveform without distortion, the bandpass of the system must encompass all the frequency components of the signal that contribute significantly to the signal of interest. The range can be determined quantitatively by obtaining a Fourier analysis of the signal. The bandpass of an electronic instrument is usually defined as the range between the upper and lower half-power frequencies.

Frequency response

Electrical signals are invariably accompanied by components that are unrelated to the phenomenon being studied. Spurious signal components, which may occur at any frequency within the bandpass of the system, are known as noise. The instruments are designed in such a way that the noise is minimized to facilitate accurate and sensitive measurement. An important factor which characterizes a signal conditioner is the signal-power to noise-power (S/N). For extraction of information from noisy signals, it is essential to

Noise

Bandwidth

enhance S/N, for which several techniques have been put in practice. The simplest method is that of bandwidth reduction. The ratio S/N is improved by eliminating the unused segments of the spectrum. If it is possible to exclude especially noisy regions, such as 50Hz, by the bandwidth reduction technique, the result will be appreciable. However, if the reduced bandwidth extends to low frequencies, the results may not be satisfactory because of the presence of drift and flicker noise.

Signal averaging

Signal averaging is another noise reduction technique, in which the signal is averaged over a period that is long as compared to the periods of interfering noise components. The average of the noise signal will tend toward zero, while that of non-random information will be its actual value, even though signal and noise may have the same frequencies.

The limitation of using long observing time as compared to the period of the lowest frequency signal components of interest in the signal averaging technique can be avoided by the use of signal modulation and phase-sensitive detection. In modulation, the information-bearing signal is used to modify a second signal, which is termed as the carrier. The modification may involve encoding of information in terms of carrier amplitude (amplitude modulation) or of carrier frequency (frequency modulation). Certain frequency ranges are generally very troublesome, particularly those below 1Hz, in which drift and flicker noise cause difficulties, because ac-coupled amplifiers cannot be used in this case. Frequency modulation of the signal with a higher frequency conveniently solves this problem.

Phase-sensitive detection

Phase-sensitive detection is a demodulation technique, by which only signals that are coherent (signals which have a fixed phase relationship) with a reference signal are extracted from a modulated signal. This technique makes it possible to eliminate even those noise components which occur at signal frequency, as noise is not coherent with the other signals.

Digital Processing

The recent progress of digital technology, in terms of both hardware and software, makes digital processing more efficient and flexible than analog processing. Digital techniques have several advantages. Their performance is powerful as they are able to easily implement even complex algorithms. Their performance is not affected by unpredictable variables such as component aging and temperature, which can normally degrade the performance of analog devices. Moreover, design parameters can be more easily changed through digital techniques because they mostly involve software rather than hardware modifications.

The designs of analytical instruments have undergone tremendous changes with the use of microcomputers, resulting in the replacement of a large number of hardware components with software. Many circuit functions previously performed by hard-wired electronics are elegantly executed by the software. This has resulted in not only greater reliability with reduced maintenance costs but also improved performance and extended applications.

1.4 Read-out (Display) Systems

The results of a measurement in modern analytical instruments are usually displayed in the following forms:

- Analog meters
- Digital displays
- Laboratory recorders
- Cathode Ray Tube displays
- Video display units

1.4.1 Analog Meters

Moving coil meters are used in instruments which employ barrier-layer type photocells. Meters used in such instruments, when connected directly, should have very good current sensitivity. This is achieved by using strong permanent magnets and a phosphor bronze suspension wire. Current sensitivity of the order of 0.5 μA (full-scale deflection) has been employed in several direct reading commercially available colorimeters and flame photometers. Analytical instruments which incorporate electronic amplifiers and use analog meters for display, do not require such sensitive meters. Generally meters used in such instruments have a sensitivity of 50 μA, 100 μA or 1 mA. They often use shunt or series resistors to suit the amplifier output.

1.4.2 Digital Displays

Different types of devices are available for display of the measured quantities in numerical form. However, the seven-segment display arrangement commonly seen on digital watches has become quite popular in digital displays. The numerical indicator consists of seven bars positioned in such a way that the required figure can be displayed by selection of the appropriate bars. A special decoder is required for driving such an indicator from a BCD (Binary Coded Decimal) code. The seven segments can be illuminated by incandescent lamps, neon lamps, LEDs or directly heated filaments.

Light emitting diodes

Light emitting diodes (LEDs) arranged in seven-segment format are used in small-sized displays. These semiconductor diodes are made of gallium aresenide phosphide and are directly compatible with 5 V supplies typically encountered in digital circuitry. LEDs are very rugged and can withstand large variations in temperature. They are available in deep red, green and yellow colours.

Liquid crystal displays

Liquid crystal displays (LCDs) are the currently preferred devices for displays as they require very low current for their operation. LCDs with large screen sizes and full colour display capabilities are available commercially and are finding extensive and preferable applications in laptop computers and many portable analytical instruments.

The outputs from most of the transducers are analog signals. For digital operations which may include computations and display, these analog signals are converted into a digital representation.

Aanalog-to-digital converters

Several types of analog-to-digital converters (A/D) have been developed. They are commercially available in the form of a single monolithic chip. They even incorporate features like auto-zero and auto-polarity, so that they can be readily used for constructing digital panel meters. The three-digit A/D converter LD 130 from M/S Siliconix, USA is a typical example.

1.4.3 Laboratory Recorders

Every modern laboratory possesses recorders of different varieties for recording various types of analog signals generated by analytical instruments. Recorders are available in a large range of sizes, speeds, sensitivities and prices.

The most elementary electronic recording system consists of three important components, namely the signal conditioner, the writing part and the chart drive mechanism. The signal conditioners usually consist of a pre-amplifier and the main amplifier. Both these amplifiers have to satisfy specific operating requirements such as input impedance, gain and frequency response characteristics. For recording analog signals, the writing part may be of the direct writing galvanometric type, or the null balancing potentiometric type. The dot matrix printer and the laser printer are normally used with PC-based instruments.

1.4.3.1 Direct Writing Recorders

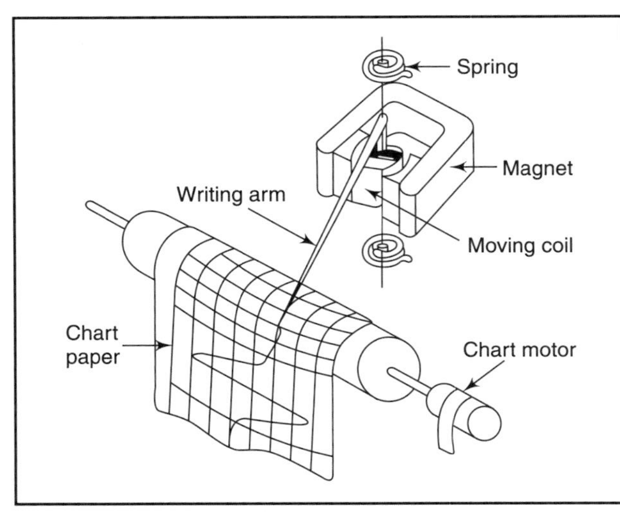

Figure 1.3 :: Principle of direct writing recorder

D' Arsonval meter movement

In the most commonly used direct writing recorders, a galvanometer activates the writing arm called the pen or the stylus. The mechanism is a modified form of the D' Arsonval meter movement. This arrangement owes its popularity to its versatility combined with its ruggedness, accuracy and simplicity.

A coil of thin wire, wound on a rectangular aluminium frame, is mounted in the air space between the poles of a permanent magnet (Figure 1.3). Hardened steel pivots attached to the coil frame fit into jewelled bearings, so that the coil rotates with a minimum of friction. Most often, the pivot and jewel is replaced by a taut-band system. A lightweight pen is attached to the coil. Springs attached to the frame always return the pen and coil to a fixed reference point.

When current flows through the coil, a magnetic field is developed, which interacts with the magnetic field of the permanent magnet. It causes the coil to change its angular position, as in an electric motor. The direction of rotation depends upon the directison of flow of current in the coil. The magnitude of pen deflection is proportional to the current flowing through the coil. The stylus can have an ink tip, or it can have a tip that is the contact for an electro-sensitive, pressure-sensitive or heat-sensitive paper. If a writing arm of fixed length is used, the ordinate will be curved. In order to convert the curvilinear motion of the writing tip into rectilinear motion, various writing mechanisms have been devised to change the effective length of the writing arm, as it moves across the recording chart.

The usual paper drive is by a synchronous motor and a gear box. The speed of the paper through the recorder is determined by the gear ratio. In order to change the speed of the paper, one or more gears must be changed.

1.4.3.2 Potentiometric Recorder

Strip-chart recorders

For recording of low frequency phenomenon, strip-chart recorders based on the potentiometric null-balance principle are generally used. The operating principle of a potentiometric recorder is shown in Figure 1.4.

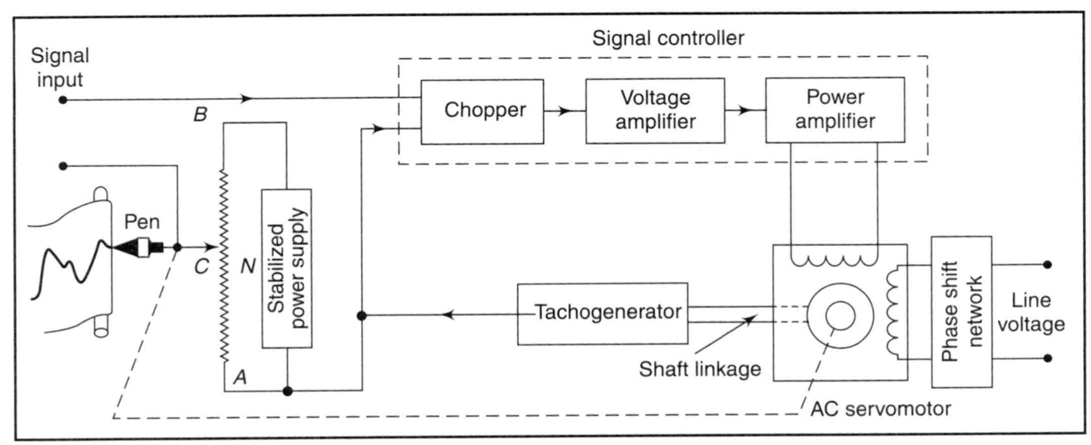

Figure 1.4 :: Principle of working of potentiometric recorder

A slide-wire AB is supplied with a constant current from a stabilized power supply. The slide-wire is constructed from a length of resistance wire of high stability and uniform cross-section, such that the resistance per unit length is constant. The unknown dc voltage is fed between the moving contact *C* and *A* of the slide-wire. The moving contact is adjusted so that the current flowing through the galvanometer placed across *AC* is zero. At that moment, the unknown input voltage is proportioned to the length of the wire *AC*. In practice, the slide-wire is calibrated in terms of span voltage, the typical spans being 100 mV, 10 mV or 1 mV.

For converting the simple potentiometer into a recording unit, the moving contact of the slide-wire is made to carry a pen which writes on a calibrated chart moving underneath it. For obtaining the null-balance, a self-balancing type potentiometer is generally used. The balancing of the input unknown voltage against the reference voltage is achieved by using a servo-system. The potential difference between the sliding contact *C* and the input dc voltage is given to a chopper type dc amplifier in place of a galvanometer. The chopper is driven at the mains frequency and converts this voltage difference into a square-wave signal. This is amplified by the servo-amplifier and then applied to the control winding of a servo-motor.

Most servo-motors are conventional rotary motors and are connected to the slide-wire and pen-through mechanical linkages. In order to simplify the design, linear motors have been developed, in which the prime mover is in a straight line instead of a circle.

The tacho-generator used to provide damping by electrical feedback is usually a two-phase device with a squirrel cage rotor. One phase is supplied from the ac line voltage to induce eddy currents in the rotor. When the rotor moves, it causes a shift in the space alignment of the eddy current field so that voltage is induced in the other phase. This induced voltage is directly proportional to the rotor velocity.

T–Y recorders

X–Y recorders

The chart is driven by a constant speed motor to provide a time axis. Therefore, the input signal is plotted against time. Recorders of this type are called T–Y recorders. If the chart is made to move according to another variable, then the pen would move under the control of the second variable in the X-direction. This type of recorders are called X–Y recorders.

Most analytical instruments provide output terminals for connecting a chart recorder. The commercially available recorders usually have multiple input ranges (1 to 10 mV), achievable by switching in various voltage dividers on the input. Single range recorders can be used, provided that the amplifier's full-scale output is equal to or greater than the recorder's span. If it is greater, a suitable voltage divider network is incorporated between the amplifier output and the recorder input.

Recorders offer several advantages over meter read-outs. They give a continuous record of the measurement, and can accurately measure the magnitude of any noise pattern. Any sudden or subtle changes in the signal can be determined. A record can give an accurate measurement of signal-to-noise ratio, which is an extremely important factor to determine the detection limit in some instruments, like flame spectrophotometry.

1.4.3.3 Dot Matrix Printers

With the increasing use of personal computers with analytical instruments for data processing, display and record, various types of printers are being used. The most common and low cost printer used with computerized systems is the dot matrix printer. This is an impact type of printer which produces dot matrix characters. These printers carry seven independent, solenoid-driven print wires or needles in a 7×5 dot matrix. A point-head containing the solenoids and needles travels across the fixed carriage at a constant speed. Appropriate solenoids are triggered by electronic circuitry to produce the desired characters. These printers are usually microprocessor-controlled. The character codes can also be stored in an ROM that commands the print-head's seven print needles. The ROM's output fed through power amplifiers actuates solenoids that cause the print wires to impact the ribbon and paper. The microprocessor evaluates the next line of data while a line is being printed. The processor decides whether printing the upcoming line is faster in the forward or reverse direction and issues instructions accordingly. Figure 1.5 shows a schematic of a typical serial dot matrix printer mechanism.

Impact printer

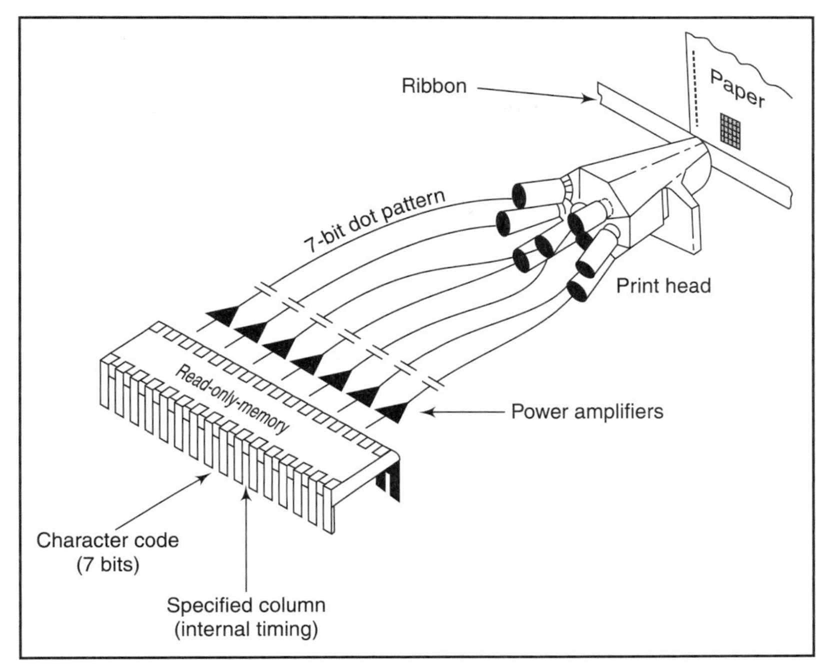

Figure 1.5 :: Schematic diagram of a Dot Matrix Printer

Dot matrix printers are almost universally bidirectional printing units as they print on both passes across the paper. They usually feature logic-seeking print-heads which seek the shortest path between two printing points, thus cutting down printing time. Since they are logic-seeking and microprocessor-controlled, dot matrix printers are usually capable of graphics output. Typically, most of them have a special programmable graphics mode that is capable of laying down nearly 80×80 dots/inch and even more, if special graphics software is used.

Dot matrix printers tend to be fairly reliable as there are only a few moving parts, other than the line feed motor and the motor used to move the print-head. Print-heads last a long time, of the order of 200 million characters.

The major drawbacks of dot matrix printers are their noise and the quick wear-out of ribbons. Most dot matrix printers emit noise in the 65 to 85 db range, though newer machines tend to be a little quieter. Due to the wearing out of the ribbon, early print-outs are of good quality, but later print-outs tend to be lighter. Also, if the ribbons are not used quickly, they tend to dry out and produce lighter print-outs when used.

Most dot matrix printers produce 80 or more columns and generate fairly good graphics. Problems of uneven pin wear have been noticed which produce weak or unbalanced characters. The most important development is the multi-mode impact matrix printer with the versatility and local intelligence to handle a wide range of output needs. Such printers typically offer a choice of printing modes: high speed coupled with data processing quality, medium speed draft output and low speed near letter quality printing. These printers can accommodate virtually any type of character font.

1.4.3.4 Laser Page Printers

Laser printers depend on photocopier technology for their action. A drum has a positively charged photoconductive layer. The drum acts as an insulator in the dark. The positive charges dissipate when illuminated by the laser beam. The laser printers (Figure 1.6) write an image by exposing the background area and turning off the laser beam to leave positively charged dots.

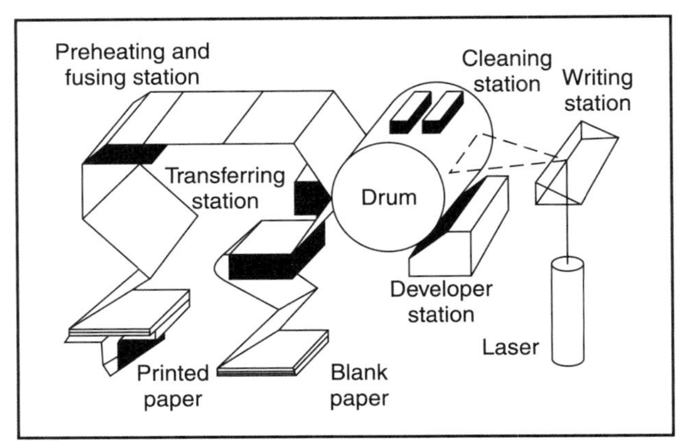

Figure 1.6 :: Schematic diagram of a laser printer

A negatively charged toner is then spread over the drum, developing the image by sticking to the positive charges. The toned image is then transferred to the paper and fused. Finally, the excess toner material is removed from the drum and the printing process is ready to be repeated for a new page. The modern laser page printers employ gallium arsenide laser diodes.

Most common low cost laser printers provide a resolution of 600 dpi × 600 dpi whereas high resolution laser printers give a resolution of 1200 dpi × 1200 dpi. Monochrome printers have a printing speed of 35 pages per minute whereas colour printers give 16 pages per minute of A-4 size paper.

1.4.4 The Oscilloscope

Among the indicating and display devices, the cathode ray oscilloscope occupies an important place in analytical instrumentation. Like a pen recorder, the oscilloscope presents a two-dimensional graphical display. In recorders, a mechanical device does the writing, while in the oscilloscope, an electronic beam does the same job, with speed and elegance that are unthinkable in other types of display devices.

A block diagram showing the various stages required for a complete oscilloscope and location of various important controls is shown in Figure 1.7. It will be seen that in effect, there are two separate sections in the oscilloscope. These are the horizontal and vertical sections. Each section has its own amplifiers, controls and deflection plates, and so on. Basically, whatever is done on the vertical section does not affect the horizontal section and vice versa. But by working together, they will be able to display the incoming signal on the face of the cathode ray tube.

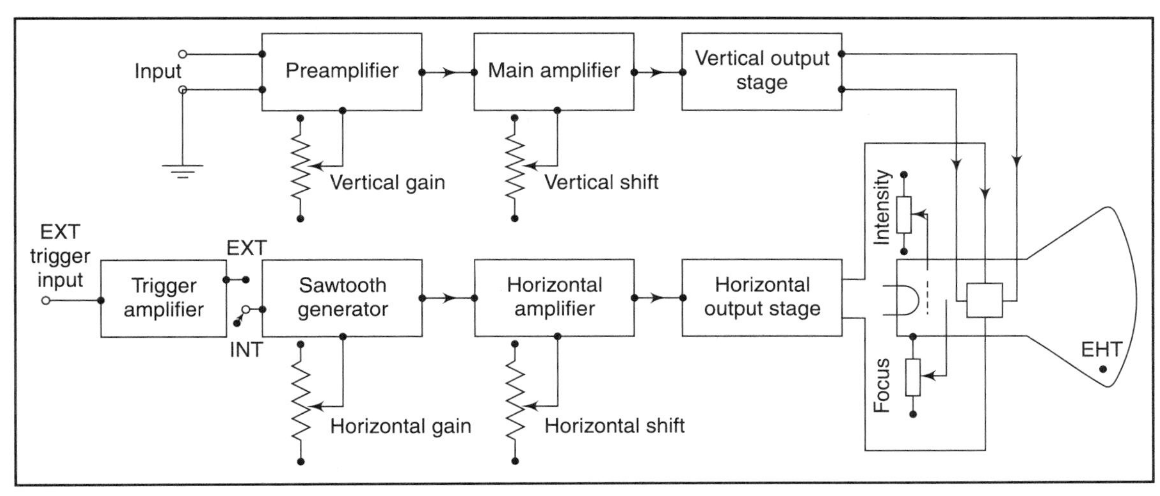

Figure 1.7 :: Block diagram of a cathode ray oscilloscope

The oscilloscope produces a visual display of electrical phenomenon. Its chief advantage is that it produces visual representation directly, with extremely high speed. This is made possible because of the very high velocity with which the electrons can move.

1.4.5 Video Display Units

With the incorporation of microcomputers in analytical instruments, the display systems for results and processed data in these instruments have undergone a complete change. The analog and digital displays have been replaced by video display units, which present information not only as a list of numbers but as elegant character and graphic displays and sometimes as a three-dimensional colour display.

Visual display units (VDUs) are usually monochrome as the CRTs in these units are coated with either white or green phosphors. They have only a single electron gun. Coloured display on a CRT is created by adding together varying amounts of three-coloured light sources, with each of them covering a selected passband in the wavelength region between 380 and 780 nm. Classically, the primary additive colours are red, blue and green. The CRT screen is coated with phosphor triads of these colours placed as dots at the corners of an equilateral triangle. A shadow mask located in front of the phosphor screen has a small hole located over each triad to help direct the electron beams. The CRT display is a rectangular array of visual points created by selectively exciting phosphor areas. The minimum picture element is called a *pixel*. There are typically 2.5 triads/pixel in a high resolution display.

Pixel

The most common technique used to build displays is the raster scan technique similar to commercial television. In a display, a row of pixels is called a scan line and a collection of scan lines is a raster frame. The raster frame is driven from a high-speed frame buffer memory, which can be considered as a matrix whose planar array dimensions, in bits, are determined by the number of pixels on the screen; for example 512 × 512 or 1024 × 1024 bits. Low resolution displays (256 × 256) appear with very jagged edges. Therefore, acceptable quality graphics require at least 512 × 512 to 1024 × 1024 bits displays, whereas for high resolution picture quality, 2048 × 2048 bits are preferred.

Sometimes, the scientist would like to use colour texture to describe surfaces, wherein a three-dimensional character must be expressed, or use related colours to express chemical concentrations.

Bit-planes

The third dimension depth is obtained by adding Z to the matrix and if expressed in bit-planes, is determined by the colour characteristics required. If it is one-bit deep (one-bit-plane) only monochrome information can be displayed, as the single electron gun is turned on or off by the binary bit pattern stored in this bit-plane. A monochrome display having three-bit-planes would provide eight levels of gray scale, as the beam current in the electron gun can be set at eight levels. In a colour CRT, the simple eight colour display would also have three bit-planes, having one bit for each colour gun. As many as 64 different colours would be obtained if two bits are allocated to each colour gun, allowing four energy levels for each (six-bit-planes) colour. If we allocate eight bits to each colour gun, it would provide 16 million colours, but requires 24-bit-planes and 3 Mbyte of memory.

Character display is usually done in a 5×7 or 7×9 matrix. Alphanumeric display is done with

ASCII Code

ASCII Code (American Standard Code for Information Interchange) which occupies 7 bits (128 possible characters). Alphanumeric terminals can draw graphs only by using symbols in the 24 line×80 character/line positions. Bit-mapped addressing is also possible in these display terminals. Good bit-mapped displays can superimpose alphanumeric information over graphs and drawings.

1.5 Intelligent Analytical Instrumentation Systems

Intelligent technology is pervading every area of modern society, from satellite communications to washing machines. The field of analytical instrumentation is no exception. This has become possible due to the availability of high performance microprocessors, microcontrollers and personal computers.

Microprocessors

The application of microprocessors in analytical instrumentation has matured following a series of stages. In the first stage, the microprocessors simply replaced conventional electronic systems that were used for processing data. This resulted in more reliable and faster data. This was soon followed by the use of the microprocessor to control logic sequences required in instrumentation. Thus, the microprocessor replaced programming devices as well as manual programming, making possible digital control of all the functions of the analytical instruments. With the availability of more powerful microprocessors and large data storage capacity, it has become possible to optimize the analytical conditions as the analysis is proceeding. They can be used to replace the complicated instructional procedures that are now required in several analytical techniques.

Extensive use has been made of microprocessors in analytical instruments designed to perform routine measurements, particularly in situations wherein data computing and processing can be considered as a part of the measurement and diagnostic procedure. The incorporation of microprocessors into instruments facilitates a certain amount of intelligence or decision-making capability. This decision-making capability increases the degree of automation of the instrument and reduces the complexity of the man–machine interface. Systems have been designed with numerous safety back-up features and real-time self-diagnostics and self-repair facilities. The reliability of many transducers has been improved and many measurements can now be made *in-situ* because of the added computational ability of microprocessors. This computational capability facilitates features such as automatic calibration, operator guidance, trend displays, alarm priority and automated record keeping. The use of microprocessors in various instruments and systems has been explained at various places in the text.

Microprocessor-based instrumentation is facilitating the ability to make intelligent judgment and provide diagnostic signals in case of potential errors, provide warnings or preferably make appropriate corrections. Already, the microprocessors are assisting in instruction-based servicing of equipment. This is possible by incorporating monitoring circuits that will provide valuable diagnostic information on potential instrumentation failure modes and guide the operator in their correction. The instrument diagnostic microprocessor programs would sense such a potential failure of the unit and switch on the stand-by unit. The operator is told to remove and service the defective part while the analytical work proceeds uninterrupted. Modern analytical instruments have been greatly affected by the use of microprocessors and have provided various degrees of control. The types of controls that are exercised for the operation of various instruments are described below.

Manual Operation: In this case, the operator is provided with sufficient information in the operating or instruction manual to permit him to operate the instrument. The operator is expected to set the operating conditions to achieve optimum performance. For example, in order to operate a spectrophotometer, the settings relating to source intensity, monochromator slit width, scanning speed and range, amplifier gain and system time constant, etc. are made by the operator. If performance varies, the values are required to be reset appropriately by the operator.

Automatic Operation: In automatic operation, after the device is started, stable operating conditions and an appropriate signal-to-noise ratio are ensured, even though conditions may change during a measurement or from sample to sample. Here, the system constituents are self-standardizing and/or operate automatically during unattended operation.

Automatic Control: For complete control and automation, one or more microprocessors (microcomputers) are wired into an instrument. Here, programs for operation are entered by the manufacturer into the read-only memory incorporated into the microcomputer (a microcomputer has an in-built microprocessor, read-only memory and random-access memory). The programs also include provision for the user to interact with the computer. The system is menu-driven and the computer prompts the user to supply information about the kind of measurements to be made. The programs indicated by these data provide for appropriate operation of sub-systems and data acquisition. The computer processes the data and extracts chemical information.

The incorporation of computer control necessitates some redesign of the instrument. For example, sensing devices such as position encoders must be inserted to furnish data to the computer. Devices that the computer can actuate, such as stepping motors are required to be added to allow program control of scanning and other operations.

1.6 PC-Based Analytical Instruments

Hardware

Software

The propularity of so-called personal computers (PCs) or home computers has fuelled intense commercial activity in the field of computers. Hardware is a comprehensive term for all the physical parts of a computer, as distinguished from the data it contains or operates on, which is the software that provides instructions for the hardware to accomplish tasks. Typically, the personal computer contains in a desktop or tower case the following parts:

- *Motherboard* which holds the *CPU*, *main memory* and other parts, and has slots for expansion cards;
- *Power supply*, which is a case that holds a transformer, voltage control and fan;
- *Storage controllers*, of *IDE*, *SCSI* or other type, that control *hard disk*, *floppy disk*, *CD-ROM* and other drives; the controllers sit directly on the motherboard (on board) or on expansion cards;
- *Graphics controller* that produces the output for the *monitor*;
- The *hard disk*, floppy disk and other drives for mass storage; and
- Interface controllers (*parallel*, *serial*, *USB*, *Firewire*) to connect the computer to external *peripheral* devices such as *printers* or *scanners*.

The low cost and increasing power of the personal computers are making them popular in the analytical field (Chu and Zilora, 1986). Also, software for personal computers is largely commercially available and the users can purchase and use it conveniently. Personal computers are now well-established and widely accepted in the analytical field for data collection, manipulation and processing, and are emerging as complete workstations for a variety of applications. A personal computer becomes a workstation with the simple installation of one or more 'instruments-on-a-board' in its accessory slots, and with the loading of the driver software that comes with each board. The concept has proven to be an ideal instrument, providing a low cost yet highly versatile computing platform for the measurement, capture, analysis, display and storage of data derived from a variety of sources.

Workstation

Figure 1.8 illustrates the typical configuration of a PC-based workstation. It is obvious that the system is highly flexible and can accommodate a variety of inputs, which can be connected to a PC for analysis, graphics and control. The basic elements in the system include sensors or

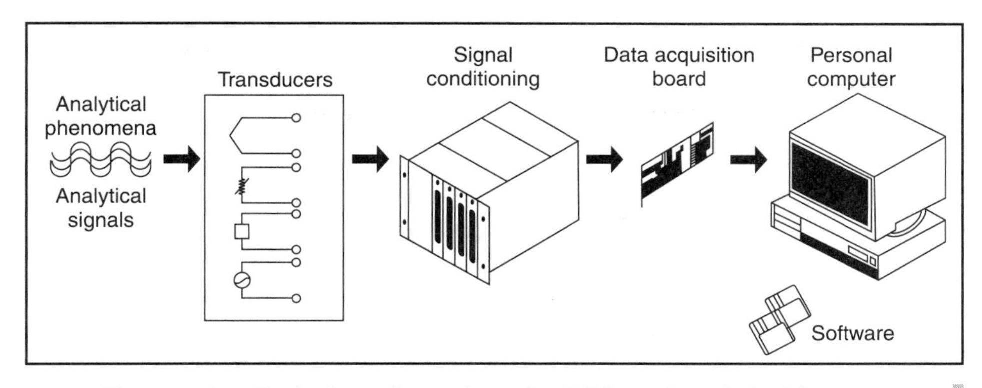

Figure 1.8 :: Typical configuration of a PC-based analytical instrument

transducers that convert physical or chemical phenomena into a measurable signal, a data acquisition system (a plug-in instrument/acquisition board), an acquisition/analysis software package or program and a computing platform. The system works totally under the control of software. It may operate from the PC's floppy and/or hard disk drive. Permanent loading or unloading of driver files can be accomplished easily. However, for complex applications, some programming in one or several of the higher level programming languages such as 'C language' may be needed. Data received from the measurements can be stored in a file or output to a printer, plotter or other device via one of the ports on the computer.

Programming

Voltage can be digitized into computer memory through plug-in data acquisition (DAQ) boards that contain analog-to-digital converters or stand-alone interfaces that transfer data to the computer through a serial port. Many of the commercial DAQ boards also support digital and trigger inputs, as well as analog and digital outputs for integrated data acquisition and instrument control. A variety of software packages support the commercially available DAQ boards for data acquisition and instrument control. National Instruments, Inc. (*http://www.natinst.com*) sells a wide range of DAQ boards and LabView software for sophisticated instrument control, data acquisition, and data analysis. Vernier Software, Inc. (*http://www.vernier.com*) sells a variety of probes for biological, chemical and physical measurements, computer and calculator interfaces, and data logging software.

Data acquisition

PC-based analytical instruments are gaining popularity for several reasons including price, programmability and the performance specifications they offer. Software development, rather than hardware development, increasingly dominates new product design cycles. Therefore, one of the most common reasons why system designers are increasingly choosing PC and PC architecture is the PC's rich and cost-effective software tool set. This includes operating systems, device drivers, libraries, languages and debugging tools. Several examples of PC-based analytical instruments can be found at various places in this book.

1.7 Performance Requirements of Analytical Instruments

1.7.1 Errors in Chemical Analysis

Many times, measurement systems consist of several elements and total control over a number of experimental variables is usually difficult. These variables include sampling methods, analytical techniques and instrument-related errors. In cases where a measurement consists of a single reading on a simple equipment, like taking a reading of temperature on a thermometer, the number of variables contributing to uncertainties in that measurement is fewer than a measurement involving a multi-step process, like the use of a variety of reagents. It is thus important to be able to estimate the uncertainty in any measurement because not doing so would mean lack of knowledge about the variability of the measurement, which would lead to erroneous results. Statistical methods are often employed as a means of objectively evaluating the source and amount of error in analytical methods.

Uncertainty

An 'error' by definition is the difference between the measured value and the true value.

Error $e = V_{measured} - V_{true}$

However, we normally do not know the true value, otherwise there would be no reason to make the measurement. Therefore, we estimate the likely upper bound of an error, expressed as the uncertainty. Therefore,

$$-u \leq e \leq u$$

or

$$-u \leq V_{true} \leq V_{meas} + u,$$

where u = uncertainty.

Errors arise from many sources and therefore, it is necessary to determine their likely dominant sources. Armed with this knowledge, anyone involved in measurement would be in a position to control the errors wherever possible and consider the effects of the error on the results.

1.7.1.1 Types of Errors

Errors are broadly classified into three types—random, systematic or gross.

Random errors: These are the result of the intrinsically uncertain nature of the measurement technique and are invariably present in every measurement. For example, thermal, shot or flicker noise are sources of random errors. Their magnitude is usually small. They can usually be minimized by filtering techniques through hardware or software implementation.

Systematic errors: They are procedural errors and arise from incorrect calibration of the instrument, its improper operation and the use of impure reagents. These are identifiable errors and can mostly be eliminated by modification of the analytical procedure.

Another type of error is *bias*, which represents a systematic distortion in a measurement. It is a non-compensating error and arises due to:

- Flaw in measurement instrument,
- Flaw in the method of selecting the sample,
- Flaw in the technique of estimating a parameter, or
- Subjectivity of operators.

Bias errors

The only practical way to minimize systematic or bias errors is through a continual check on instruments and assumptions, care in the use of instruments and application of methods, and meticulous training of the instrument operator. It is essential to check all instruments before commencement of any important measuring session and to re-check them periodically during the course of the project.

Gross errors: The common sources of gross errors include mistakes made in analytical techniques, improper reading or recording of results and data, and errors in calculations. Gross errors are usually irregular in nature and are characterized by large amplitude.

Mistakes are caused by human carelessess, a casual approach or by fallibility. Even though there is no excuse for mistakes, we all make them. The advice, therefore is to never be satisfied with a single reading, no matter what is being measured. Repeating the measurement will also improve the precision of the final result.

In addition, *accidental errors* take place due to uncontrolled environmental conditions, and limitations or deficiencies of instruments, assumptions and methods. Accidental errors can be reduced by using more accurate and precise equipment, but this can be expensive. Therefore, it is advisable to make an appropriate choice from the available equipment so that one can obtain results of an accuracy that is sufficient for the task in hand.

Total error

For many systems, the error in a measurement is not isolated to one source but is rather due to several different individual errors. A simplistic method of calculating the total error is simply to add the errors from each source together. The key is to make certain that the error is expressed in terms of the final measurement quantity.

Net error

In many instances, a measurement is taken many times to improve the overall quality of the measurement. This averaging improves error levels if the noise is random. For example, if a measurement M is taken three times, the average value for M is found by taking the sum of the three measurements and dividing it by three. Similarly, the net error in the average is found by taking the total error for the three measurements and dividing it by three.

Hysteresis: This is a specific type of error that is very common in mechanical systems. This error appears as a different output value for the same input value, with the difference occurring due to the manner in which the input value is set. Specifically, if the input value is set by increasing the physical parameter, it will have a different output value than if the input value is set by decreasing the input value. This is illustrated in Figure 1.9.

Repeatability Error: Repeatability is similar to hysteresis except that it refers to the difference in output values for the same input values when the input value is produced in the same fashion. For example, a repeatability error would be the difference in output voltage for a pressure

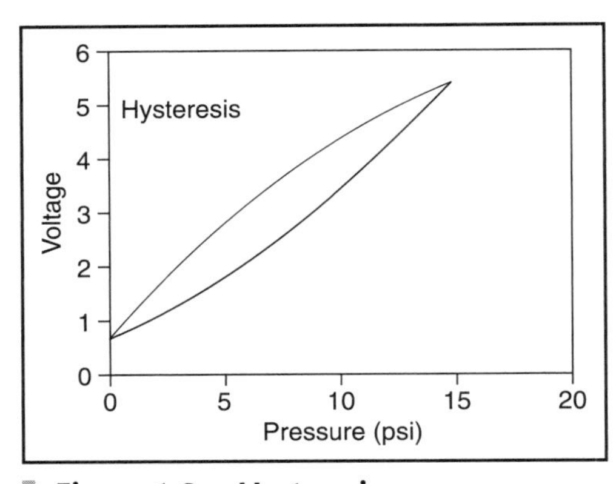

Figure 1.9 :: Hysteresis error

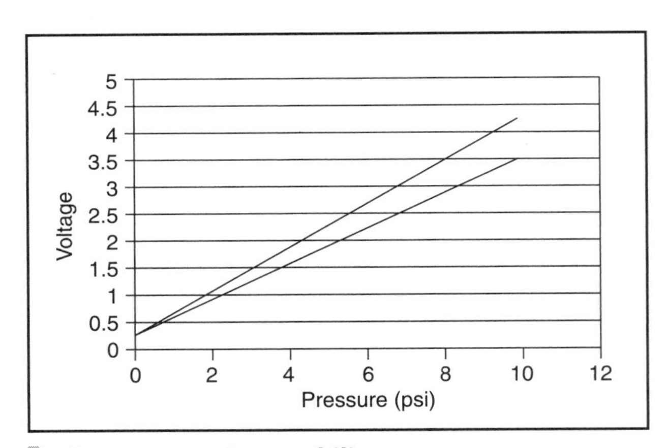

Figure 1.10 :: Repeatability error

transducer when an identical pressure is placed on the sensor with the pressure being either increased or decreased each time. Figure 1.10 illustrates the concept of repeatability error.

As one can see, hysteresis and repeatability are both errors that quantify the difference in output for identical input; the difference between the two is how the input quantity is applied to the transducer or sensor.

1.7.2 Accuracy and Precision

The accuracy of a measurement pertains to how close a result comes to the true value. Broadly speaking, accuracy is the degree to which a statement or quantitative result approaches the true value. Accuracy refers to the size of the total error and this includes the effects of systematic errors. An estimated value may be inaccurate because of the occurrence of one or more kinds of error. Determining the accuracy of a measurement usually requires calibration of the analytical method with a known standard.

Accuracy includes the effects of repeatability, hysteresis, and linearity in a single term. It is a statement of the relationship between the input parameter and the output signal. Accuracy is typically specified as ± a fraction of either the full scale reading of the instrument or of the value that is being measured. For example, if a pressure transducer is specified as ±1 per cent accurate of its full scale value and the transducer is used at pressures from zero to 50 psi, the error in any measurement will be ±0.5 psi.

Precision refers to the reproducibility of the multiple measurements. Precision has several variants depending on use, such as:

- The fitness of a single measurement, which refers to the resolving power of the measuring device and is ordinarily indicated by the number of decimal places in the measurements made with the device,

- The degree of agreement in a series of measurements,
- The clustering of sample values about their own average, and
- The reproducibility of an estimate in repeated sampling.

Accuracy and precision are not synonymous. Accuracy signifies freedom from all types of errors whereas precision implies freedom from variation—one does not imply the other, i.e. precise measurements may not be accurate, as illustrated in Figure 1.11. An accurate measurement is one in which the systematic and random errors are small whereas precise measurement is one in which the random errors are small.

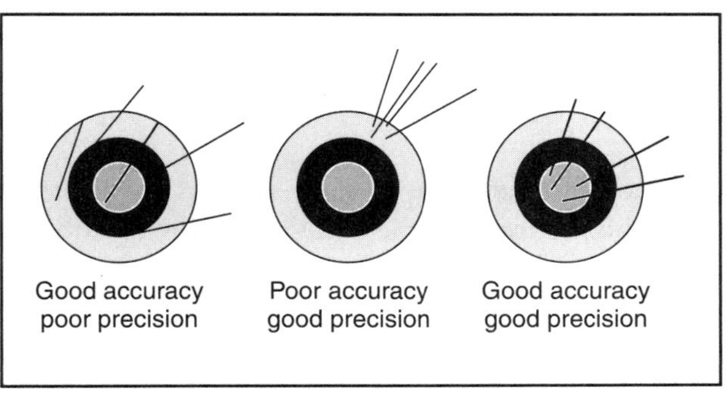

Figure 1.11 :: Concept of accuracy and precision (after Tissue, 2004a)

Bias

The uncertainties in measurement generally emanate from two common causes, which are bias and precision as shown in Figure 1.12. An example of a bias (measure of systematic or determinate error of a method) error is poor calibration. Because of the poor calibration, all measurements are incorrect by the same amount, say 5 per cent, whereas precision errors are variable and random. An example of precision error is, when repeated measurements are supposed to be the same, but they are not. Figure 1.12 shows that a measurement with a small range of precision errors can be called precise. However, the average value of measurements might have a large bias error, so the measurement would not be accurate, i.e. it does not represent the true value.

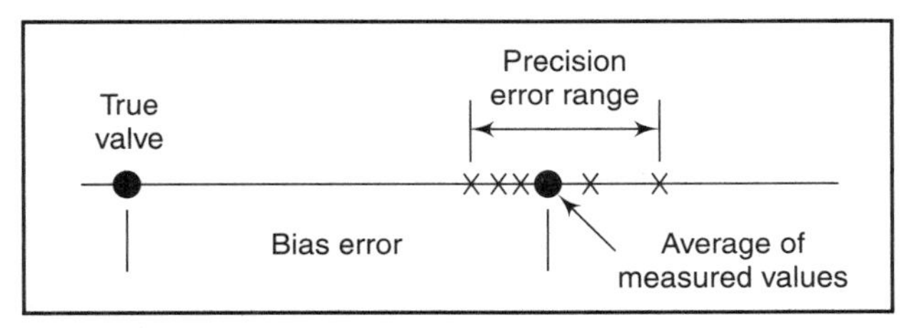

Figure 1.12 :: Bias and precision error (after Wheeler and Ganji, 1996)

Bias errors are usually caused by:

- Incorrect calibration—a system may give values that err by a fixed percentage.
- Systematic human errors—not synchronizing readings between two experimenters.
- Defective equipment—incorrect scale graduations.
- Loading—the heated thermistor inserted into an ice water bath, which might change the temperature of the bath.

When plotting data on a graph, bias errors usually shift the total data set away from the true line, or change its slope.

Precision errors reflect the fact that instrumentation or measurement devices are both imperfect and of finite resolution. They cause scatter about the true line. While it is easier to spot precision errors, bias errors can be identified only through comparison with an independent model or data set.

1.7.3 Significant Figures

All measurements carry some degree of uncertainty. The degree of uncertainty depends upon both the accuracy of the measuring device and the skill of its operator. For example, in a digital balance, the mass of a sample substance can be measured to the nearest 0.1 g, i.e. mass differences less than this cannot be detected on this balance. We might therefore, indicate the mass of a substance measured on this balance as 3.2 ± 0.1 g; the ±0.1 (read plus or minus 0.1) is a measure of the accuracy of the measurement. It is important to have some indication of the accuracy of any measurement. The ± notation denotes the understanding that there is uncertainty of at least one unit in the last digit of the measured quantity. In general, the measured quantities are reported in such a way that only the last digit is uncertain. All the digits, including the uncertain one, are called significant digits or, more commonly, significant figures. The number 3.2 has two significant figures, while the number 3.2405 has five significant figures.

Significant digits

In order to determine the number of significant digits in a measurement or number, the following rules may be applied:

- If there is no decimal point, the right-most non-zero digit is the least significant digit.
- In case of numbers that include a decimal point, the right-most digit is the least significant digit regardless of its value.

Significant digits should not be considered the same as uncertainty in measurement systems. They are used for manipulating experimental data, but not for expressing uncertainty. When calculating results from the measurement, if your calculator displays too many significant digits, round off to the proper number of significant digits.

Significant figures have economic significance simply because the more the significant numbers expected from the measurement, the more expensive will be the equipment and higher the quality of reagents required. Care should therefore, be taken to properly define the optimum number of significant figures when attempting an analysis.

1.7.4 Application of Statistical Methods

Statistical methods of data analysis are used to provide the analyst with a systematic and objective approach to the problems involved in error analysis. We know that the precision of an analytical method is usually indicated by the range of values about the mean that includes a specified percentage of the total observation. When many independent random factors act in an additive manner to create variability, data will follow a bell-shaped distribution called the 'Gaussian

Distribution'. This distribution is also called 'Normal Distribution' which is shown in Figure 1.13. Gaussian distribution has some special mathematical properties that form the basis of many statistical tests. However, the catch is that the sample number has to be reasonably large, normally more than 30.

Gaussian distribution

For computing, normal probabilities for any normal distribution can be characterized by

$$Z = x - \mu/\sigma$$

where

μ = the population mean of the distribution. Because of symmetry, this occurs at the peak of the distribution.

σ = the population standard deviation of the distribution

Z = standard normal random variable

x = Value of a given measurement.

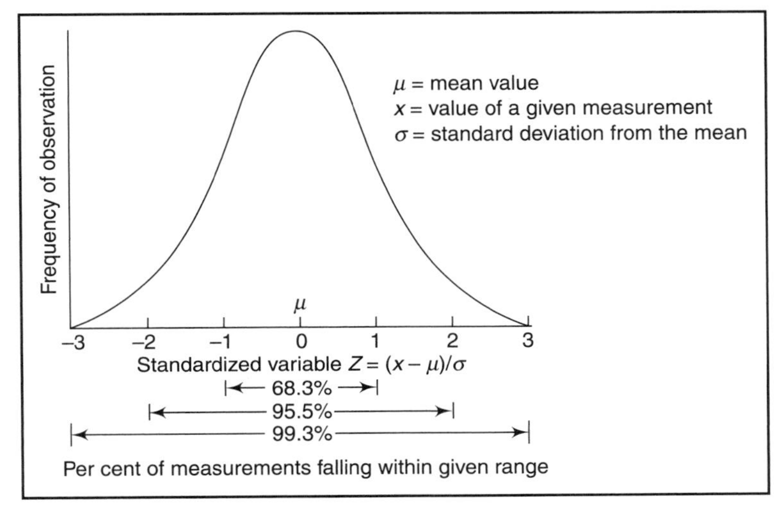

Figure 1.13 :: Normal Gaussian Distribution (adapted from Willard *et al.*, 1988)

The Gaussian distribution, as shown in the Figure 1.13, is a continuous function, and is normalized so that the sum over all values of x gives a probability of 1. The nature of Gaussian gives a probability of 0.683 of being within one standard deviation of the mean. The mean value is $a = np$ where n is the number of events and p, the probability of any integer value of x. Similarly, the percentage measurements falling within two standard deviations is 95.5 per cent and it is 99.3 per cent for three standard deviations.

Probability

Confidence Interval is defined as the probability that the data point is found in a certain neighbourhood of the mean. Three different confidence intervals are commonly used. They are usually called one-sigma, two-sigma and three-sigma intervals. The probability that the data point is within a one-sigma interval around the mean is 0.6828. In other words, approximately 68.28 per cent of data will be found within a one-sigma confidence interval around the average. The two-sigma and the three-sigma confidence intervals are interpreted in exactly the same way, wherein two-sigma has 95.5 per cent and three-sigma has 99.3 per cent confidence intervals. One-sigma also indicates one standard deviation from the mean value.

Confidence Interval

As is evident from the Figure 1.13, precision increases with the decreasing value of standard deviation. It also indicates that the distribution of measurement results carries a wealth of information which can be used to improve the measurements and the results. For example, the results of individual samples falling outside the desired confidence levels may be rejected. It also helps to establish trends in results to facilitate corrective measures.

1.7.5 Signal-to-Noise Ratio

The term signal-to-noise ratio, often abbreviated as SNR or S/N, represents the ratio between the maximum possible signal (meaningful information) and the background noise,

$$\text{S/N} = \frac{\text{Average signal amplitude}}{\text{Average noise amplitude}}$$

In other words, S/N expresses the ability of an instrument to discriminate between signals and noise. Since many signals have a very wide dynamic range, S/N is often expressed in terms of the logarithmic decibel scale. Due to the definition of decibel, the S/N gives the same result, independent of the type of signal which is evaluated (i.e. power, current, voltage). Often the signals being compared are electromagnetic in nature, though it is also possible to apply the term to sound and light stimuli. The S/N in decibels is twenty times the base-10 logarithm of the amplitude ratio, or ten times the logarithm of the power ratio.

Decibel

It may be observed that the higher the signal-to-noise ratio, the better is the measurement result. A highesr S/N is possible only if the noise value is reduced. Higher levels of amplification do not improve S/N because an increase in the magnitude of the signal will be accompanied by a proportional increase in the value of noise.

1.7.6 Other Performance Parameters

1.7.6.1 Sensitivity

The sensitivity of an instrument is the change in output signal while a change in the physical parameter is being measured. For linear systems, this is simply the proportionality constant used to relate output to input.

The sensitivity of a measurement system is determined with the help of the calibration curves. The sensitivity is constant over the entire range in case of a linear response system as illustrated in Figure 1.14(a). However, from the slopes of curves A and B, it is evident that the sensitivity of the method is much greater in case of substance B than substance A but constant for the same substance. In case, the response curve is non-linear Figure 1.14(b), the sensitivity is found to be changing with respect to the concentration. Evidentally, the sensitivity for substance C decreases with an increase in concentration.

In case of analytical instruments, sensitivity is usually expressed as the concentration of analyte required to cause a given instrument response.

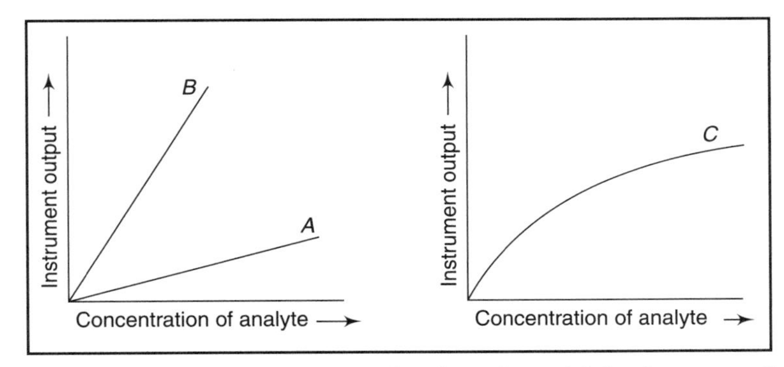

Figure 1.14 :: Determination of sensitivity from calibration curve (a) linear response (b) non-linear response

1.7.6.2 Selectivity

The selectivity of an analytical method is defined as its ability to accurately measure an analyte in the presence of interference that may be expected to be present in the sample matrix.

1.7.6.3 Specificity

The term 'specificity' generally refers to a method that produces a response for a single analyte only.

1.7.6.4 Resolution

Resolution is the smallest amount of input signal change that the instrument can detect reliably. This term is determined by the instrument noise which could be either circuit or quantization noise.

1.7.6.5 Range

The range of an analytical method is the interval between the upper and lower levels (including these levels) that have been demonstrated to be determined with precision, accuracy and linearity by using the specified method. The range is normally expressed in the same units as the test results (e.g. percentage, parts per million) obtained by the analytical method.

1.7.6.6 Limit of Detection

The limit of detection is the point at which a measured value is larger than the uncertainty associated with it. It is the lowest concentration of analyte in a sample that can be detected but not necessarily quantified. For example, in chromatography, the detection limit is the injected amount that results in a peak with a height which is at least twice or three times as high as the baseline noise level.

Detection limits are generally defined at 95 per cent confidence level, which means that in 95 out of 100 measurements, the measurements can be done reliably. However, concentration values that are smaller than the detection limit may be qualitatively detected but will be quantitatively unreliable.

1.7.6.7 Linearity

The linearity of an analytical method is its ability to elicit test results that directly, or by means of well-defined mathematical transformations, are proportional to the concentration of analytes in samples within a given range. Frequently, the linearity is evaluated graphically in addition or alternatively to mathematical evaluation. The evaluation is made by visual inspection of a plot of signal height or peak area as a function of analyte concentration. Since deviations from linearity are sometimes difficult to detect, two additional graphical procedures can be used.

The first procedure involves plotting the deviations from the regression line versus the concentration, or versus the logarithm of the concentration, if the concentration range covers several decades. Regression analysis, a statistical technique, provides the means for objectively obtaining a linear line with minimum deviation between the plot and data. Many computer spreadsheet programs have in-built functions which can perform linear regression analysis. For linear ranges, the deviations should be equally distributed between the positive and negative values.

Regression analysis

The second method is to divide signal data by their respective concentrations, thereby yielding the relative responses. A graph is plotted with the relative responses on the Y-axis and the corresponding concentrations on the X-axis on a log scale. The resultant plot should be horizontal over the full linear range. At higher concentrations, there will typically be a negative deviation from linearity. Parallel horizontal lines are drawn in the graph corresponding to, for example, 95 per cent and 105 per cent of the horizontal line. The method is linear up to the point where the plotted relative response line intersects the 95 per cent line. Figure 1.15 shows a comparison of the two graphical evaluations of a caffeine sample using high-performance liquid chromatography.

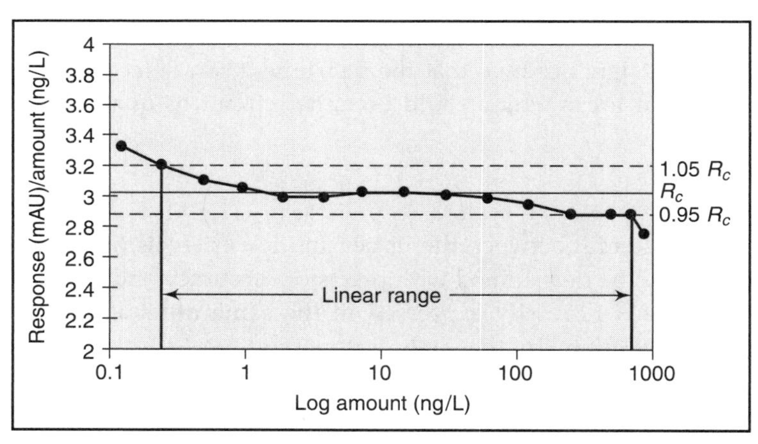

Figure 1.15 :: Graphical presentation of linearity plot of a caffeine sample using HPLC. Plotting the amount (concentration) on a logarithmic scale has a significant advantage for a wide linear ranges *Rc* = line of constant response (*www.labcompliance.com/methods/methval.html/linearity* and calibration curve)

The range within which a linear relationship exists between a measured value and the corresponding concentration is called the working range. The minimum concentration for the working range is known as limit of detection whereas the maximum concentration is called the Limit of Linearity. It is generally attempted to adjust the concentration by dilution or volume reductions to bring it within the working range.

Working range

It is not always possible to obtain a linear graph due to interferences, signal noise, sample matrix or deviations from Beer's Law. However, several programs are commercially available for computing the best 'fit' curve for the data set.

1.8 Instrument Calibration Techniques

Taking measurements with any analytical method or instrument necessitates calibration to ensure the accuracy of the measurement. The process of quantitatively defining the system responses to known, controlled signal inputs is known as 'calibration'. The three commonly used methods to calibrate the instruments are the:

Calibration

- Calibration Curve Method,
- Standard Addition Method, and
- Internal Standard Method.

The choice of the calibration technique used is determined by the instrumental method, instrument response, number of samples to be analyzed and the interferences present in the sample. A brief description of each of these methods is given below.

Working curve method

1.8.1 Calibration Curve Method

In the calibration or working curve method, a series of standard solutions containing known concentrations of the analyte, is prepared. These solutions should cover the concentration range of interest. A blank solution containing only the solvent matrix is also analyzed. The net readings—standard solution minus blank (background)—are plotted versus the concentrations of the standard solutions to obtain the working calibration curve. Figure 1.16 shows a typical calibration curve. The non-linearity in the curve is usually corrected by computer software techniques.

The chief advantage of the working curve method is its speed as in it a single set of standards can be used for the measurement of multiple samples.

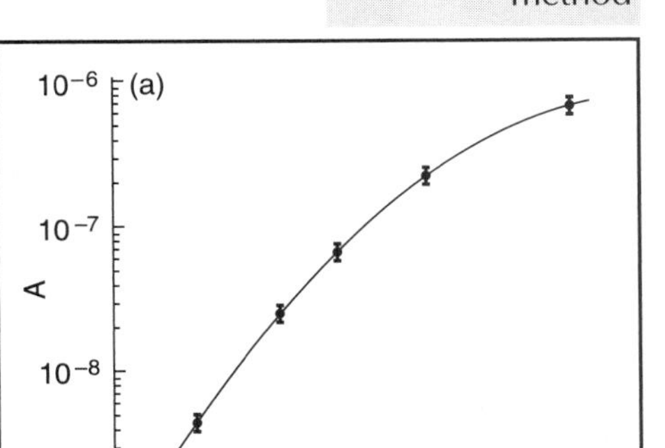

Figure 1.16 :: Illustration of analytical (working) curve method for calibration

1.8.2 Standard Addition Method

Due to matrix effects, the analytical response for an analyte in a complex sample may not be the same as that for the analyte in a simple standard. In such a case, calibration with a working curve would require standards that closely match the composition of the sample. For routine analyses, it is feasible to prepare or purchase realistic standards, e.g. NIST standard reference materials. However, for diverse and one-of-a-kind samples, this procedure is time-consuming and often impossible.

An alternative calibration procedure is the standard addition method. An analyst usually divides the unknown sample into two portions so that a known amount of the analyte (a spike) can be added to one portion. These two samples, the original and the original plus spike, are then analyzed. The sample with the spike will show a larger analytical response than the original sample due to the additional amount of analyte added to it.

The difference in analytical response between the spiked and unspiked samples is due to the amount of analyte in the spike. This provides a calibration point to determine the analyte concentration in the original sample.

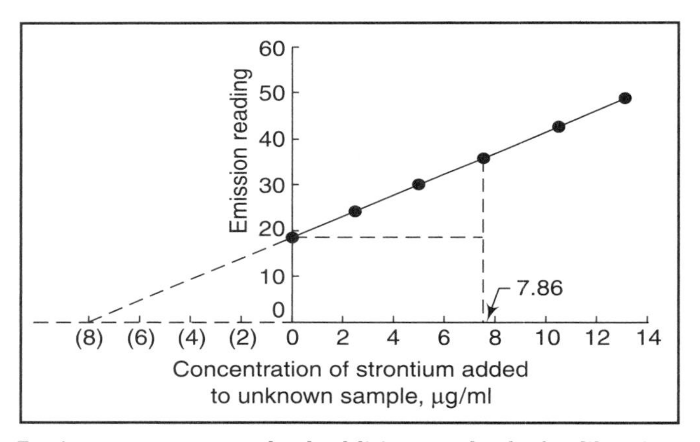

Figure 1.17 :: Standard addition method of calibration

Let us assume that with an unknown concentration X, the instrument gives a reading R_x. If a known concentration Y is added to the sample solution to give a reading R_Y, then X can be calculated from the equation:

$$X/X + y = R_X/R_Y$$

The principle of standard addition is shown in Figure 1.17.

The concentration scale is on the X-axis. When the analyte is added to the sample solution, the unknown concentration can be determined by the point at which the extrapolated line intersects the concentration axis.

The method of standard addition gives more accurate results than the calibration curve method. It is particularly used in electroanalytical chemistry wherein the unknown and standard solutions are measured under identical conditions. These conditions exist in matrix-sensitive volumetric techniques such as anode stripping voltametery. The method also provides a systematic means of identifying sources of error in analyses which could be due to a defective instrument, incorrect internal standard or depletion of test reagents.

1.8.3 Method of Internal Standard

An internal standard is a known amount of a compound, different from an analyte that is added to the unknown. The signal from the analyte is compared with the signal from the internal standard to find out how much analyte is present. This known substance must not be present in the original unknown substance.

Thereafter, the response of the analyte and internal standard are determined and the ratio of the two responses is calculated. The response ratio of analyte to internal standard would depend only on the analyte concentration because the internal standard and analyte are generally affected by a variation in one or more of the parameters affecting the measured response. The calibration curve is generated by plotting the response ratio as a function of analyte concentration. This method is used in emission spectroscopy, gas chromatographic analysis and improved spectroscopy. While using the internal standard method, the following points should be taken into consideration:

- The internal standard should be a substance similar to the analyte.
- The internal standard should have an easily measured signal that does not interfere with the response of the analyte.
- The response of the internal standard should not be affected by other components of the sample.
- The concentration of the internal standard should be of the same order of magnitude as that of the analyte in order to minimize error in calculating the ratio.

When used in spectroscopy, the internal standard method requires instruments with a special capability of determining intensity or absorbance simultaneously at two wavelengths one for the standard and the other for the unknown. The output of two detectors (known versus standard) is read out as a ratio, which is then plotted against concentration and the analyte in the unknown is determined from the graph.

In order to illustrate the method for internal standard when determining sodium concentration in a solution by using a flame photometer, *Li* is added to each of the series of known and standard solutions. Figure 1.18 shows a plot of intensity ratio vs. concentration of *Na* from which the concentration of *Na* in unknown solutions can be calculated. The graph is non-linear and therefore the determination is made graphically. For example, the unknown *'A'* has a concentration of 3.2 ppm Na whereas unknown *'B'* has a concentration of 6.9 ppm Na. If the graph were linear, then a linear equation could be developed from the calculation of the slope and Y-intercept.

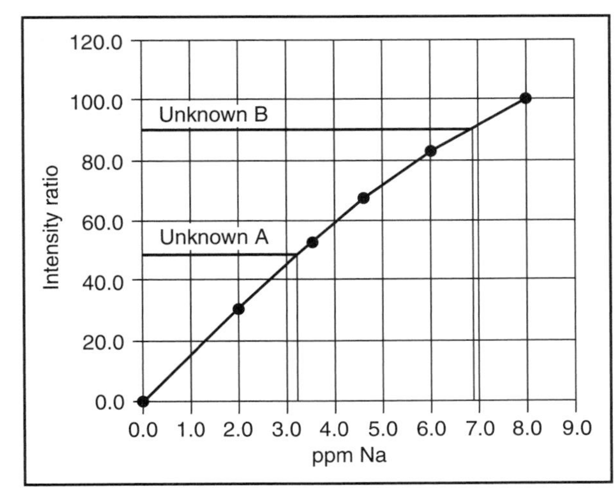

Figure 1.18 :: Internal standard method of calibration (after Moore, 2000)

Internal standards are especially useful for analyses in which the quantity of sample analysed or the instrument response varies slightly from run to run. Therefore internal standards, are widely used in chromatography in which a small quantity of sample solution is injected into the chromatograph is not very reproducible in some experiments.

1.9 Validation

The process of using independent means to assess the quality of the data products derived from the system outputs is termed as 'validation'. As regards analytical instruments/equipment, method validation is usually resorted to in order to confirm that the analytical procedure employed for a specific test is suitable for its intended use. Every analytical method needs to be validated/revalidated:

Method validation

- Before its introduction into routine use,
- Whenever the conditions change for which the method has been validated i.e. instrument with different characteristics, or
- Whenever the method is changed.

The validity of a specific method is closely related to the type of equipment and its location where the method will be run. Before an instrument is used to validate a method, its performance should be verified by using generic standards. Satisfactory results for a method can only be obtained if the equipment used is performing well. Special care should be taken to assess the characteristics that are critical for the method. For example, if the detection limit is critical for a specific method, it is advisable to verify the instruments' specification for baseline noise and the response of the detectors to specified compounds.

Method validation has received considerable attention from several organizations and regulatory agencies. For example, the Guidance on the Interpretation of the EN45000 Series of Standards and ISO/IEC (International Standards Organization/International Electrotechnical Commission) Eurachem Guide 25 (1993) includes a chapter on the validation of analytical methods.

Chapter 2

Colorimeters and Spectrophotometers (Visible–Ultraviolet)

The most important among all the instrumental methods of analysis are the methods based on the absorption of electromagnetic radiation in the visible, ultraviolet and infrared ranges. According to the quantum theory, the energy states of an atom or molecule are defined and for any change from one state to another, they would, therefore, require a definite amount of energy. If this energy is supplied from an external source of radiation, the exact quantity of energy required to bring about a change from one given state to another will be provided by photons of one particular frequency, which may thus be selectively absorbed. The study of the frequencies of the photons that are absorbed would thus indicate a lot about the nature of the material. Also, the number of photons absorbed may provide information about the number of atoms or molecules of the material present in a particular state. It thus provides us with a method to carry out the qualitative and quantitative analysis of a substance.

Photons

Molecules possess three types of internal energy—electronic, vibrational and rotational. When a molecule absorbs radiant energy, it can increase its internal energy in a variety of ways. The various molecular energy states are quantized and the amount of energy required to cause any change in any one of the above energy states would generally correspond to specific regions of the electromagnetic spectrum. Electronic transitions correspond to the ultraviolet and visible regions, vibrational transitions to the near-infrared and infrared regions, and rotational transitions to the infrared and far-infrared or even microwave regions (Figure 2.1).

The method based on the absorption of radiation of a substance is known as 'absorption spectroscopy'. The main advantages of spectrometric methods are speed, sensitivity to very small amounts and a relatively simple operational methodology. The time required for the actual measurement is very short and most of the analysis time, in fact, goes into preparation of the samples. Absorption spectroscopy has a tremendously wide range of analytical applications and is proving to be extremely useful for analysis even at trace levels.

Absorption spectroscopy

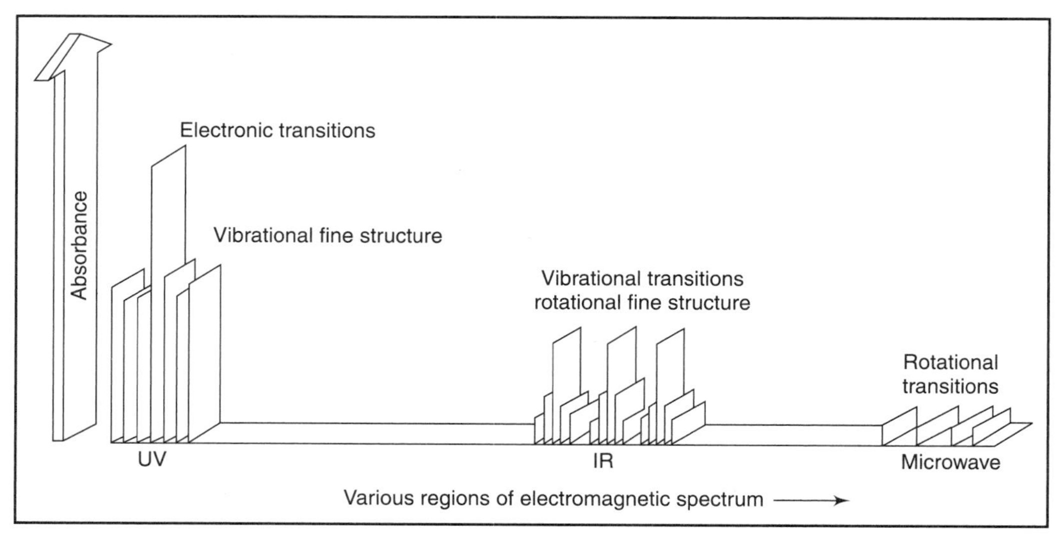

Figure 2.1 :: Relationship of wavelength and energy-induced transitions (after Human, 1985)

2.1 Electromagnetic Radiation

Electromagnetic radiation is a type of energy that is transmitted through space at a speed of approximately 3×10^{10} cm/s. Such a radiation does not require a medium for propagation and can readily travel through vacuum. Electromagnetic radiation may be considered as an amalgam of discrete packets of energy called photons. A photon consists of an oscillating electric field component (E) and an oscillating magnetic field component (M). The electric and magnetic fields are perpendicular to each other (Tissue, 1996). These fields are, in turn, perpendicular (orthogonal) to the direction of propagation of the photon. The electric and magnetic fields of a photon flip direction as the photon travels. The number of flips or oscillations of a photon that occur in one second represents the frequency, expressed in Hertz or units of oscillations per second. This is shown in Figure 2.2. The relation between the energy of a photon and the frequency of its propagation is given by:

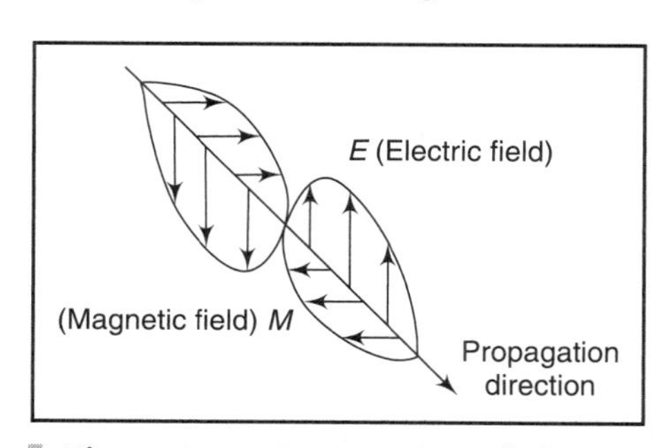

Figure 2.2 :: Propagation of electromagnetic radiation

$$E = h\upsilon$$

where E = energy in ergs

υ = frequency in cycles

h = Planck's constant (6.625×10^{-27} ergs-s.)

Wave motion

For some purposes, electromagnetic energy can be conveniently considered as a continuous wave motion in the form of alternating electric field in space. The electric field produces a magnetic field at right angles to its own direction. These fields, in turn, are mutually perpendicular to the direction of propagation. If λ is the wavelength (interval between successive maxima or minima of the wave), then:

$$C = \upsilon\lambda,$$

where C is the velocity of propagation of radiant energy in vacuum and λ is the frequency in cycles. The practical units employed to express wavelength are:

Wavelength

nm (nanometers) or mμ (millimicron) = 10^{-9} meters

μm (micrometer) or μ (micron) = 10^{-4} cm

Å (Angstrom) = 0.1nm = 10^{-8} cm.

The wavelength is now commonly expressed in nanometers. The use of other units like Angstrom (Å) and millimicron (mμ), is discouraged.

1 nanometer = 1 nm = 1 mμ = 10 Å.

Wave number

Wave number: Wave number is defined as the number of waves per centimeter.

2.1.1 The Electromagnetic Spectrum

Figure 2.3 shows the various regions in the electromagnetic spectrum which are normally used in spectroscopic work. For convenience, the photons in all these regions have the same electromagnetic nature, but because of their vastly different energies, they interact with matter very differently. However, some of the boundaries between spectral regions are not as well-defined as those between ultraviolet and visible radiation. Visible light represents only a

Visible light

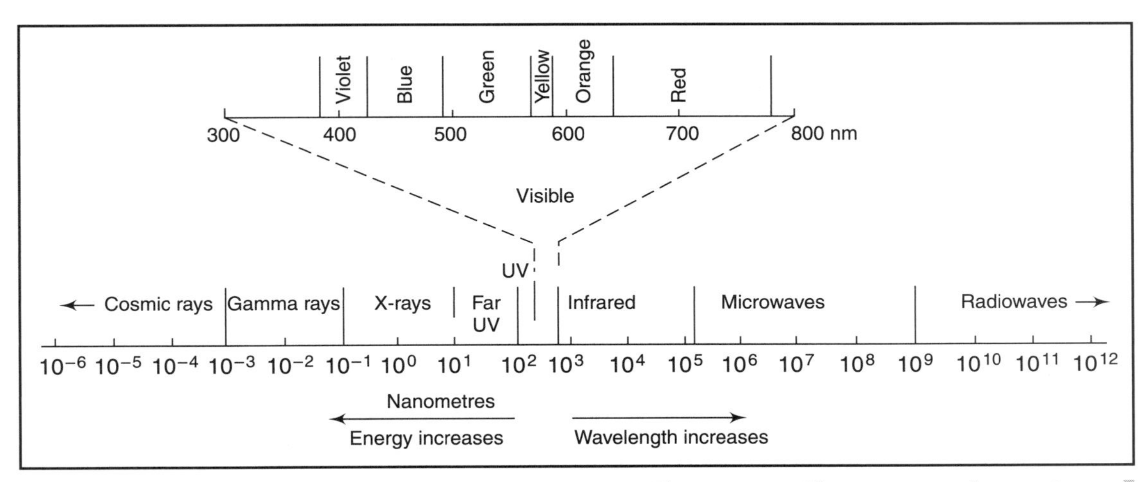

Figure 2.3 :: Electromagnetic spectrum

very small portion of the electromagnetic spectrum and generally covers a range from 380 to 780 nm. The short wavelength cut-off (380 nm) is due to absorption by the lens of the eye and the long wavelength (780 nm) cut-off, due to the decrease in sensitivity of the photoreceptors in the retina for a longer wavelength. The ultraviolet region extends from 185 nm to the visible range. Shorter wavelengths lie in the far ultraviolet region, which overlaps the soft X-ray part of the spectrum. The infrared region covers wavelengths above the visible range. Table 2.1 shows the frequencies and wavelengths of different spectral regions.

Ultraviolet region

Iinfrared region

Table 2.1 :: Electromagnetic Spectrum

Type of Radiation	Wavelength Range	Frequency Range (Hz)
Gamma rays	< 1 pm	$10^{20} - 10^{24}$
X-rays	1 nm-1 pm	$10^{17} - 10^{20}$
Ultraviolet	400 nm-1 nm	$10^{15} - 10^{17}$
Visible	750 nm-400 nm	$4 - 7.5 \times 10^{14}$
Near-infrared	2.5 μm-750 nm	$1 \times 10^{14} - 4 \times 10^{14}$
Infrared	25 μm-2.5 μm	$10^{13} - 10^{14}$
Microwaves	1 mm-25 μm	$3 \times 10^{11} - 10^{13}$
Radio waves	> 1 mm	$< 3 \times 10^{11}$

2.1.2 Interaction of Radiation with Matter

When a beam of radiant energy strikes the surface of a substance, the radiation interacts with the atoms and molecules of the substance or molecular ions or solids. The radiation may be transmitted, absorbed, scattered or reflected, or it may excite fluorescence depending upon the properties of the substance. The interaction, however, does not involve the permanent transfer of energy.

The velocity at which radiation is propagated through a medium is less than its velocity in vacuum. It depends upon the kind and concentration of atoms, ions or molecules present in the medium. Figure 2.4 shows various possibilities which might result when a beam of radiation strikes a substance. These are:

(a) The radiation may be transmitted with little absorption taking place, and therefore, without much energy loss.
(b) The direction of propagation of the beam may be altered by reflection, refraction, diffraction or scattering.

(c) The radiant energy may be absorbed in part or entirely by the substance.

Absorption spectrophotometry

In absorption spectrophotometry, we are usually concerned with absorption and transmission. Generally, the conditions under which the sample is examined are selected to keep reflection and scattering to a minimum.

Absorption spectrophotometry is based on the principle that the amount of absorption which occurs is dependent upon the number of molecules present in the absorbing material. Therefore, the intensity of the radiation leaving the substance may be used as an indication of the concentration of the material. The sample is usually examined in solution.

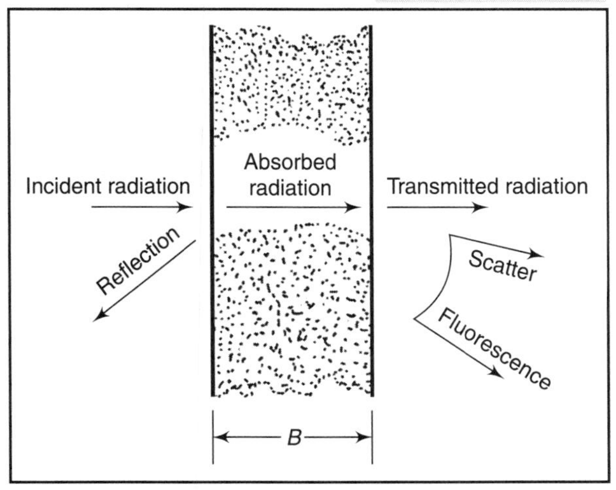

Figure 2.4 :: Interaction of radiation with matter

2.2 Laws Relating to Absorption of Radiation

2.2.1 Lambert's Law

Lambert's Law states that each layer of equal thickness of an absorbing medium absorbs an equal fraction of the radiant energy that traverses it.

The proportion of incident light absorbed by a given thickness of the absorbing medium is independent of the intensity of the incident light, provided that there is no other physical or chemical change in the medium.

Let us suppose that I_o is the incident radiant energy and I is the energy which is transmitted. The ratio of the radiant power transmitted by a sample to the radiant power incident on the sample is known as the transmittance. Lambert's Law is expressed as:

Transmittance

Transmittance $T = I/I_o$.

It is customary to express transmittance as a percentage

% Transmittance $= I/I_o \times 100$

The logarithm to the base of the reciprocal of the transmittance is known as absorbance.

Absorbance

Absorbance $= \log_{10} (1/T) = \log_{10} (I_o/I)$

Optical density $= \log_{10} (100/T)$.

It can be immediately seen that in order to determine the concentration of an unknown sample, the percentage transmittance of a series of solutions of known concentration or 'standards' can be plotted and the concentration of the 'unknown' can be read from the graph. It will be found that the graph is an exponential function, which is obviously inconvenient for easy interpolation.

2.2.2 Beer's Law

This law states that the absorption of light is directly proportional to both the concentration of the absorbing medium and the thickness of the medium in the light path.

2.2.3 The Beer-Lambert Law

A combination of the two laws, known jointly as the Beer Lambert Law, defines the relationship between absorbance (A) and transmittance (T). It states that the concentration of a substance in solution is directly proportional to the 'absorbance', A, of the solution.

Absorbance $A = \varepsilon\, cb$,

where

A = absorbance (no unit of measurement)
ε = molar absorptivity ($dm^3\ mol^{-1}\ cm^{-1}$)
c = molar concentration ($mol\ dm^{-3}$)
b = path length (cm).

It may be noted that ε is a function of wavelength. So, the Beer Lambert Law is true only for light of a single wavelength or monochromatic light. Absorptivity is a constant, depending upon the wavelength of the radiation and nature of the absorbing material. Absorptivity is also sometimes referred to as specific extinction and absorbance as 'Optical Density'. Absorbance is the property of a sample, whereas absorptivity is the property of a substance and is a constant.

Absorptivity

Mathematically, absorbance is related to percentage transmittance T by the expression:

$$A = \log_{10} (I_o/I) = \log_{10} (100/T) = \varepsilon\, bc$$

Monochromatic

The relationship between energy absorption and concentration is of great importance for the purpose of analysis. The amount of monochromatic radiant energy absorbed or transmitted by a solution is an exponential function of concentration of the absorbing substance present in the path of radiant energy. This means that successive equal thickness of a homogenous absorbing medium will reduce the intensity by successive equal fraction and, therefore, radiant energy will diminish in geometric or exponential progression. In other words, if a particular thickness absorbs half the radiant energy, the thickness which follows the first and is equal to it, will not absorb the entire second half, but only a half of this half and will consequently reduce it to one quarter.

Consider a condition when three samples (standard solutions) having identical absorption are introduced in a beam of monochromatic light. Each of the samples is chosen so that precisely one half of the intensity of the incident radiation is transmitted (T = 50%). If the intensity of the incident radiation is 100% T, then their intensity after each sample will be:

After sample, $S_1 = 1 \times 0.5 = 50\%\ T$
After sample, $S_2 = 50\% \times 0.5 = 25\%\ T$
After sample, $S_3 = 25\% \times 0.5 = 12.5\ \%\ T$.

If we plot the concentration against transmission for the above samples, it will be found that the resultant graph is exponential (Figure 2.5).

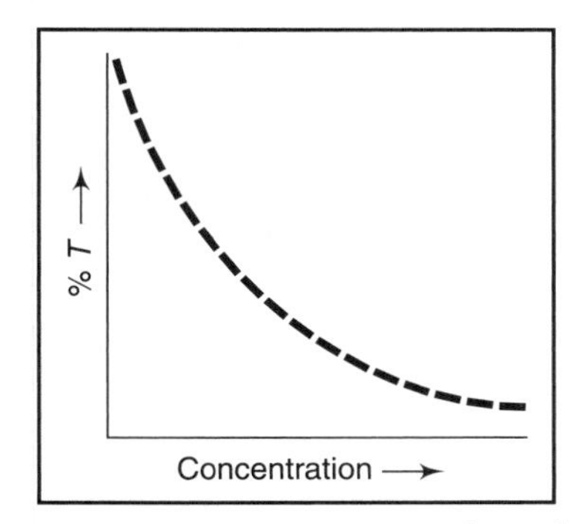

Figure 2.5 :: Transmission plotted against concentration

However, the expression relating to absorbance 'A' to transmittance 'T' ($A = \log 100/T$) shows that the absorbance after each sample will be

After sample, $S_1 = 0.301$
After sample, $S_2 = 0.602$
After sample, $S_3 = 0.903$.

By plotting the absorbance against concentration, it will be seen that the plot is linear (Figure 2.6). It is, therefore, more convenient to express results in terms of absorbance rather than transmission when measuring unknown concentrations since linear calibration plots are available.

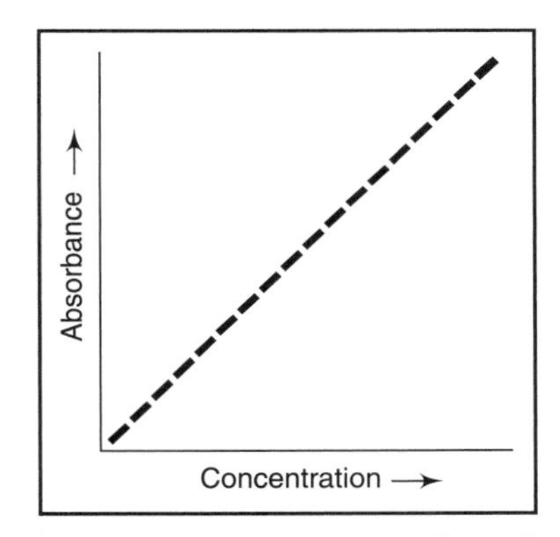

Figure 2.6 :: Absorbance plotted against concentration

An alternative to plotting calibration curves is to make use of the following relationship:

$$C = kA,$$

where C = the concentration of the unknown
A = The measured absorbance of the unknown
k = is factor derived from the reference or standard solution.

In order to determine k, the absorbance of a standard solution of known concentration is measured and the concentration is divided by the absorbance.

$$K = \frac{\text{concentration (standard)}}{\text{Absorbance (standard)}}$$

The factor k may be applied to a series of absorbance measurements on a similar solution measured in the same conditions to give results directly in concentration.

Photometers

In modern photometers, the output electronics provide the means of entering the concentration value of the standard or the factor to the calculation so that instrument readings are expressed directly in concentration units.

$E^{1\%}{}_{1\text{cm}}$ = The value of ε for a 1 per cent sample concentration of 1 cm thickness.
ε = The molar extinction coefficient, i.e. the value of ε for a sample concentration of 1 g molecule per liter and 1 cm thickness.

The relationship between transmittance and absorbance as marked on the scales of analog indicating meters, is shown in Figure 2.7.

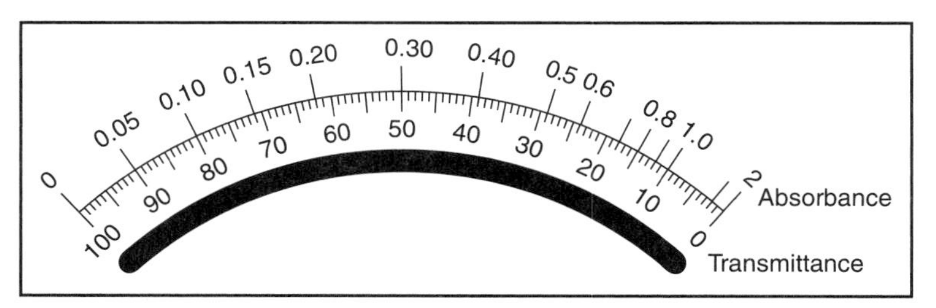

Figure 2.7 :: Absorbance and transmittance scale

2.2.4 Deviations from Beer's Law

Beer's Law only describes the relationship among absorbance, thickness and concentration. It does not imply that these are the only factors that affect absorbance. The direct, linear relationship between absorbance and concentration is used as the fundamental test of a system's conformity to the combined laws. A straight line passing through the origin indicates conformity to Beer's Law. Discrepancies are usually found when the absorbing solute dissociates or associates in solution.

Bandwidth

Beer's Law is derived by assuming that the beam of radiation is monochromatic. However, in all photometers and in most of the spectrophotometers, a finite bandwidth of frequencies is always present. The wider the bandwidth of radiation passed by the filters or dispersing device, the greater is the deviation of a system from adherence to Beer's Law.

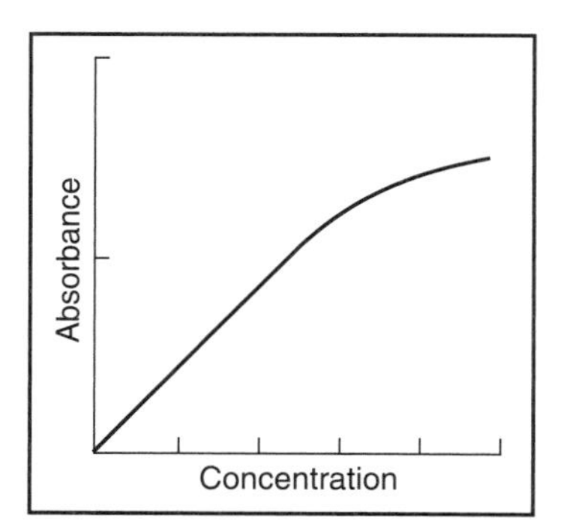

Figure 2.8 :: Representation of Beer's Law and deviations

Beer's Law is a limiting law and should be expected to apply only at low concentrations. It has been observed that the deviation becomes evident at higher concentrations (Figure 2.8) on an absorbance versus concentration plot, when the curve bends towards the concentration axis.

The linearity of the Beer Lambert Law is limited by chemical and instrumental factors. The causes of non linearity include:

- Deviations in absorptivity coefficients at high concentrations (>0.01 M) due to electrostatic interactions between molecules in close proximity,
- Scattering of light due to particulates in the sample,
- Fluorescence or phosphorescence of the sample,
- Changes in the refractive index at high analyte concentration,

- Shifts in chemical equilibrium as a function of concentration,
- Non-monochromatic radiation, wherein deviations can be minimized by using a relatively flat part of the absorption spectrum such as the maximum of an absorption band, and
- Stray light.

2.2.5 Quantitative Analysis

The most commonly employed quantitative method consists of comparing the extent of absorption or transmittance of radiant energy at a particular wavelength by a solution of the test material and a series of standard solutions. It can be done with visual colour comparators, photometers or spectrophotometers.

For quantitative analysis, normally the radiation used is of a wavelength at which k, the extinction coefficient, is a maximum, i.e. at the peak of the absorption band, for the following reasons:

- The change in absorbance for a given concentration change is greater, leading to greater sensitivity and accuracy in measurement.
- The relative effect of other substances or impurities is smaller.
- The rate of change of absorbance with the wavelength is smaller, and the measurement is not as severely affected by small errors in the wavelength setting.

In order to carry out an analysis of a unknown material, it should be possible in theory to measure the absorption of a sample at its absorption maximum, measure the thickness, obtain values of ε from tables, and calculate the concentration. In practice, the values of ε depend upon instrumental characteristics, which is why published values from tables are not accurate enough, and a reliable analyst will usually plot the absorbance values of a series of standards of known concentration; the concentrations of actual samples can then be read directly from the calibration graph.

Concentration

Using today's modern software-driven instruments, the analyst will usually only require one standard which can be used to calibrate the instrument so that it displays the concentration of subsequent samples directly in the required units. If the calibration curve is non-linear due to either instrumental or chemistry-related considerations, then some instruments will automatically construct a 'best fit' calibration when presented with a number of standards.

2.2.6 Choice of Wavelength

The selection of a suitable wavelength in the spectrum for the quantitative analysis of a sample can be made during the course of preparation of the calibration curve for the unknown material. The calibration curve is plotted to show the absorbance values of a series of standards of known concentration and after which the concentrations of actual samples can then be read directly

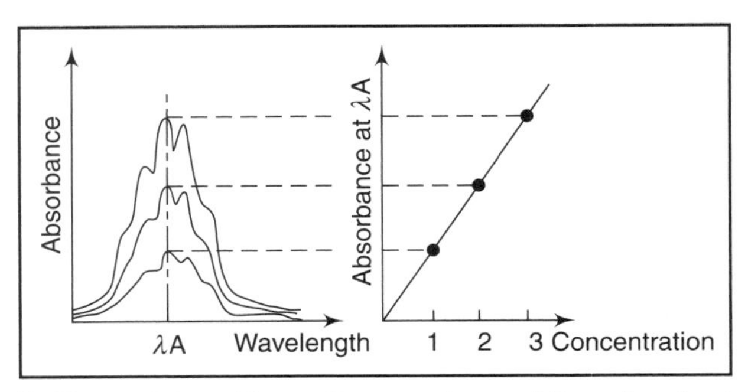

Figure 2.9 :: Absorption spectra and Beer-Lambert calibration

(Figure 2.9). A series of standard solutions is prepared along with a blank. Using one filter at a time, calibration curves are plotted in terms of absorbance versus concentration. The filter which provides the closest adherence to linearity over the widest absorbance interval and which yields the largest slope with a zero intercept, will constitute the best choice for analysis. In a spectrophotometer, the wavelength of maximum absorbance is readily ascertained from the absorbance wavelength curve for the material.

2.2.7 Simultaneous Spectrophotometric Determination

It is often found in practice that when there are several components which absorb radiation of the same wavelength, their absorbances add together as a result of which the absorbance of the sample is no longer proportional to the concentration of one component. If there is no reaction or interaction between the different solutes, the absorbances are additive of all the components at a given frequency. The Beer Lambert Law can then be written as:

$$A = a_1bc_1 + a_2bc_2 + a_3bc_3 + \cdots + a_nbc_n,$$

where the subscripts refer to the respective components. While determining the various values for a, if the cell thickness b is held constant, the b may be included in the a. Hence this is an equation with n c's, when n compounds are present. If n such equations were determined from values of the absorbance at n different frequencies, n simultaneous linear equations would result, which could be solved to find the required concentrations. Certainly, the procedure is difficult, and due to the presence of many components, one would really need a computer to solve the equations. However, much of the classical work in the analysis of petroleum fractions was carried out in this way. The method is not preferred when there are more than two or three components absorbing radiation of the same wavelength.

Multi-components analysis

The multi-wavelength or multi-components analysis technique has seen a resurgence of interest over the last few years. This has been due to the advent of data processing techniques. Various algorithms are available and the analyst is generally required to input the number of components, measurement wavelengths and concentration values of the standards. Once the standards are measured, the samples can be processed and the results presented appropriately. Programs of this type, including measurement parameters, can then be stored on disk, and recalled and run with minimum operator intervention, providing results for complex mixtures containing up to ten components.

A simplified and more common method is to convert the component under analysis, by adding a chemical reagent, which specifically reacts with it to form a highly absorbing compound. The addition of this reagent to the mixture would result in a change of the wavelength of the absorption maxima, so that interference no longer occurs among the components. The analysis then becomes very simple.

2.3 Absorption Instruments

Figure 2.10 shows an arrangement of components of an absorption instrument. These essential components are:

- A source of radiant energy, which may be a tungsten lamp, xenon-mercury arc, hydrogen or deuterium discharge lamp, etc.;
- Filtering arrangement for selection of a narrow band of radiant energy; it could be a single wavelength absorption filter, interference filter, a prism or a diffraction grating;
- An optical system for producing a parallel beam of filtered light for passage through an absorption cell (cuvette); the system may include lenses, mirrors, slits, diaphragm, etc.;
- A detecting system for measurement of unabsorbed radiant energy, which could be the human eye, barrier layer cell, phototube or photomultiplier tube; and
- A read-out system or display, which may be an indicating meter or numerical display.

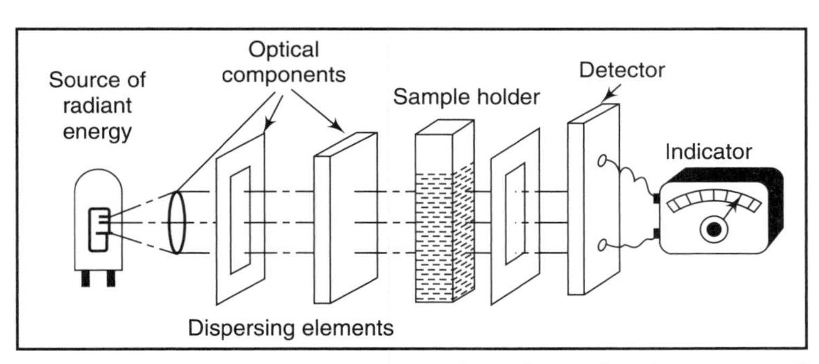

Figure 2.10 :: Various components of an absorption instrument

Generally, the components are selected in consonance with their intended use. Figure 2.11 shows the optical characteristics of various optical components and their range of suitability in the electromagnetic spectrum.

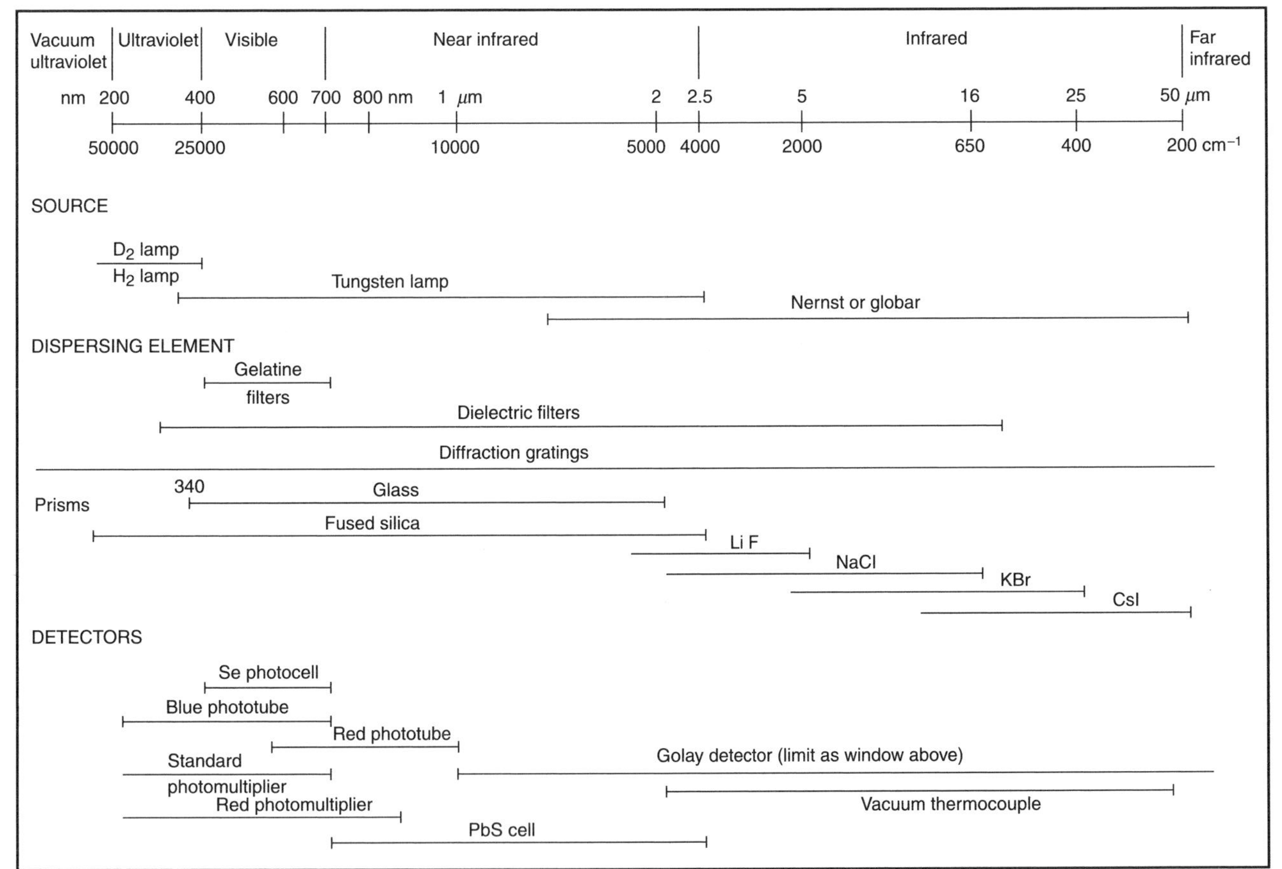

Figure 2.11 :: Spectral characteristics of various optical components and their range of suitability in the electromagnetic spectrum

2.3.1 Radiation Sources

The function of the radiation source is to provide sufficient intensity of light for making a measurement. The most common and convenient source of light is the tungsten lamp. Lamps convert electrical energy into radiation. Different designs and materials are needed to produce light in different parts of the electromagnetic spectrum. The following sections describe several different types of lamps which are useful in spectroscopy.

2.3.1.1 Blackbody Sources

A hot material, such as an electrically heated filament in a light bulb, emits a continuum spectrum of light. The spectrum is approximated by Planck's radiation law for blackbody radiations:

$$B = \{2hv^3/c^2\}\ \{1/exp\ (hv/kT) - 1\},$$

Planck's radiation

where h is Planck's constant, v is frequency, c is the speed of light, k is the Boltzmann constant, and T is temperature in k.

The most common incandescent lamps and their wavelength ranges are:

Tungsten filament lamps: 350 nm – 2.5 μm

Glowbar: 1 – 40 μm

Nernst glower: 400 nm – 20 μm.

Tungsten lamps are used in visible and near-infrared (NIR) absorption spectroscopy, while the glowbar and Nernst glower are used for infrared spectroscopy.

This lamp consists of a tungsten filament enclosed in a glass envelope. It is cheap, intense and reliable. A major portion of the energy emitted by a tungsten lamp is in the visible region and only about 15 to 20 per cent is in the infrared region. When a tungsten lamp is being used, it is desirable to use a heat absorbing filter between the lamp and the sample holder in order to absorb most of the infrared radiation without seriously diminishing energy at the desired wavelength. For work in the ultraviolet (UV) region, a hydrogen or deuterium discharge lamp is used.

Tungsten

Deuterium discharge lamp

In these lamps, the envelope material of the lamp puts a limit on the smallest wavelength, which can be transmitted. For example, quartz is suitable only up to 200 nm and fused silica up to 185 nm. The radiation from the discharge lamps is concentrated into narrow wavelength regions of emission lines. Practically, there is no emission beyond 400 nm in these lamps. For this reason, spectrophotometers for both the visible and ultraviolet regions always have two light sources, which can be manually selected for an appropriate wavelength.

For work in the infrared region, a tungsten lamp may be used. However, due to their high absorption of the glass envelope and due to the presence of unwanted emission in the visible range, tungsten lamps are not preferred. In such cases, nernst filaments or other sources of similar type are preferred. They are operated at lower temperatures but still radiate sufficient energy.

Nernst filaments

The radiation from hot solids is made up of many wavelengths and the energy emitted at any particular wavelength depends largely upon the temperature of the solid and is predictable from the probability theory. The curves in Figure 2.12 show the energy distribution for a tungsten filament at three different temperatures. Such radiation is known as 'blackbody radiation'. Note how the emitted energy increases with the temperature and how the wavelength of maximum energy shifts to shorter wavelengths. More recently, it has become common practice to use a variant of this—the tungsten–halogen lamp. The quartz envelope transmits radiation well into the UV region. For the UV region itself, the most common source is the deuterium lamp and a UV-visible spectrophotometer will usually have both lamp types to cover the entire wavelength range.

Tungsten–halogen lamp

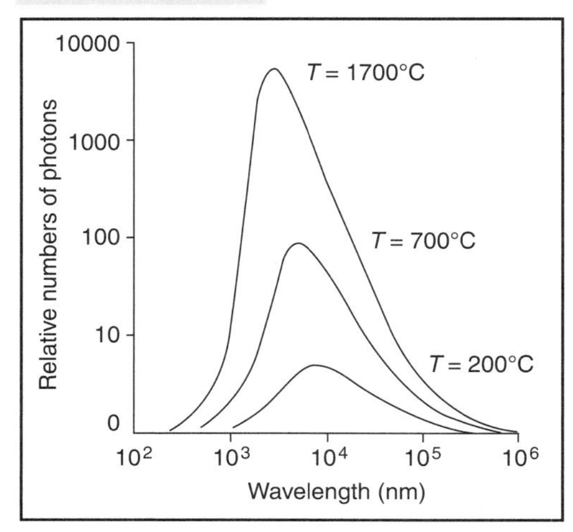

Figure 2.12 :: Tungsten filament radiation

The introduction of the tungsten–halogen light source has a higher intensity output than the normal tungsten lamp in the changeover region of 320-380 nm used in colorimetry and spectrophotometry. It also has a larger life and does not suffer from blackening of the bulb glass envelope. In the ultraviolet region of the spectrum, the deuterium lamp has superseded the hydrogen discharge lamp as a UV source. The radiation sources should be highly stable and preferably emit a continuous spectrum.

Lamp Regulator

A simplified circuit diagram of a lamp regulator that delivers constant power to a tungsten lamp is shown in Figure 2.13. A part of the voltage across the lamp is summed up with a voltage proportional to the current through the lamp. This total voltage V_f is compared with a reference voltage V_{ref} by the operational amplifier A_1. A_1 controls the voltage at the emitter of transistor Q, such that V_f will always be equal to V_{ref}.

$$V_f = V_{ref} = V_1 + V_2$$

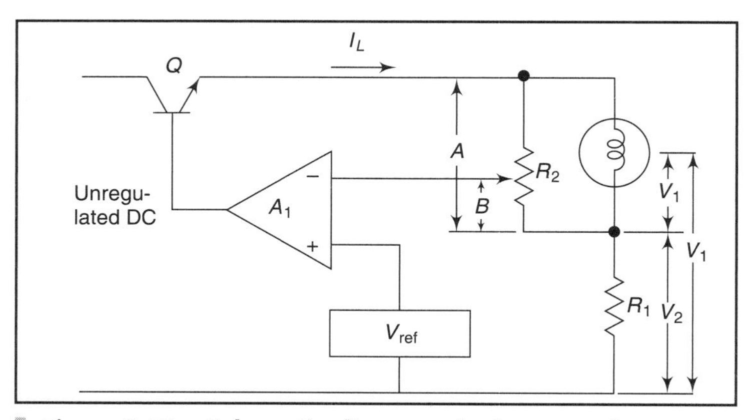

Figure 2.13 :: Schematic diagram of a lamp regulator

The voltage V_f can be written in terms of lamp current (I_L) and filament resistance R_f.

$$V_f = b/a.\ I_L.\ R_f + I_L.\ R_1,$$

where b/a is the ratio corresponding to the setting of potentiometer R_2. R_1 is the current sensing resistor.

The ratio b/a can be selected, so that constant power is delivered for a particular type of lamp. The power delivered to the lamp can be varied by adjusting the reference voltage.

2.3.1.2 Discharge Lamps

Discharge lamps such as neon signs pass an electric current through a rare gas or metal vapour to produce light. The electrons collide with gas atoms, exciting them to higher energy levels which then decay to lower levels by emitting light. Low pressure lamps have a sharp line emission characteristic of the atoms in the lamp, and high pressure lamps have broadened lines superimposed on a continuum spectrum.

Discharge lamps

The common discharge lamps and their wavelength ranges are:

Hydrogen or deuterium: 160 – 360 nm
Mercury: 253.7 nm, and weaker lines in the near-UV and visible
Ne, Ar, Kr, Xe discharge lamps: many sharp lines throughout the near-UV to near-lR
Xenon arc: 300 – 13nm.

Deuterium lamps are presently used as the UV source in UV-vis absorption spectrophotometers. The sharp lines of the mercury and rare gas discharge lamps are useful for wavelength calibration of optical instrumentation. For fluorescent work, an intense beam of ultraviolet light is required. This requirement is met by a xenon arc or a mercury vapour lamp. A cooling arrangement is absolutely necessary when these types of lamps are used.

Mercury lamps are usually run direct from the ac power line via a series ballast choke. This method gives some inherent lamp power stabilization and automatically provides the necessary ionizing voltage. The ballast choke is physically small and a fast warm-up to the lamp operating temperature is obtained.

A deuterium arc lamp provides emission of high intensity and adequate continuity in the 190-380 nm range. A quartz or silica envelope is necessary not only to provide a heat shield, but also to transmit the shorter wavelengths of the ultraviolet radiation. The limiting factor is normally the lower limit of atmospheric transmission at about 190 nm. Figure 2.14 shows the energy output as a function of wavelength in case of the deuterium arc lamp and tungsten–halogen lamp. Wilson (1985) describes a circuit for a stable constant current supply for a deuterium lamp.

In the modern spectrophotometers, the power supply arrangements including any necessary start-up sequences for arc lamps, as well as the changeover between sources, at the appropriate wavelength are automatic mechanical sequences. Lamps are generally supplied on pre-set focus mounts or they incorporate simple adjustment mechanisms for easy replacement.

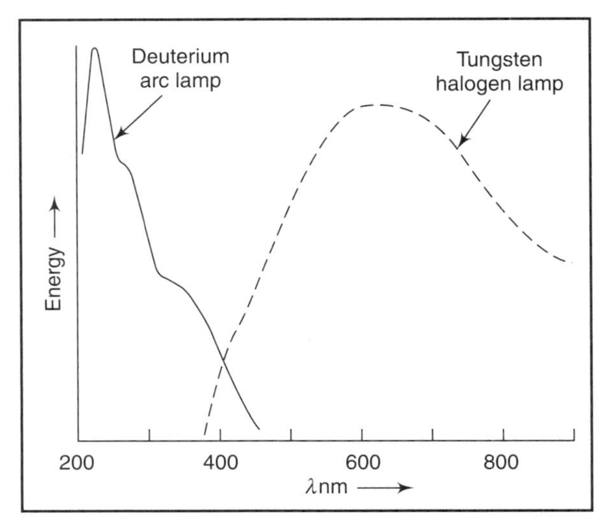

Figure 2.14 :: Energy output as a function of wavelength for deuterium arc lamp and tungsten halogen lamp

2.3.1.3 Lasers

The term 'laser' has been coined by taking the first letters of the expression 'Light amplification by simulated emission of radiation'. Although it is an amplifier, as suggested by the abbreviation, the laser is invariably used as a generator of light. But its light is quite unlike the output of conventional sources of light. The laser beam has spatial and temporal coherence, and is monochromatic (pure wavelength). The beam is highly directional and exhibits high density energy which can be finely focused.

The laser is acquiring an increasingly important role in analytical chemistry and it is important for the practising analyst to understand both its capabilities and its problems (Wright and Wirth, 1980a).

Laser beam

Lasers have a number of advantages over conventional sources such as glow bars, continuous discharges, pulsed discharges and X-ray tubes, which make them useful for analytical chemistry applications. These advantages include higher intensity, monochromaticity, low beam divergence, the availability of short and ultra-short pulses for studies of transient phenomena, and coherence (well-defined phase).

Lasers have been used in many spectroscopic applications such as Raman Spectroscopy (West, 1984). However, there is little that the laser can offer for UV–visible spectrophotometry. Perhaps, a high peak power will allow a measurable amount of light to get through an optically dense sample so that high absorbance can be measured.

Another advantage of the highly collimated beam of a laser is its ability to measure absorption over a very long path length. Both these situations are specialized. Wright and Wirth (1980b) explain the principles of working of different types of lasers such as solid, gas and liquid lasers.

2.3.2 Optical Filters

A filter may be considered as any transparent medium which, by its structure, composition or colour, enables the isolation of radiation of a particular wavelength. For this purpose, ideal filters should be monochromatic, i.e. they must isolate radiation of only one wavelength. A filter must meet the following two requirements:

(a) high transmittance at the desired wavelength, and
(b) low transmittance at other wavelengths.

However, in practice, the filters transmit a broad region of the spectrum. Referring to Figure 2.15, they are characterized by the relative light transmission at the maximum of the curve T_{λ}, the width of the spectral region transmitted (the half-width—the range of wavelength, between the two points on the transmission curve at which the transmission value equal 1/2 T_{λ}) and T_{res}, (the residual value of the transmission in the remaining part of the spectrum). The ideal filter would have the highest value of T_{λ} and the lowest values for the transmission half-width and T_{res}.

Ideal filter

Filters can be broadly classified as absorption filters and interference filters.

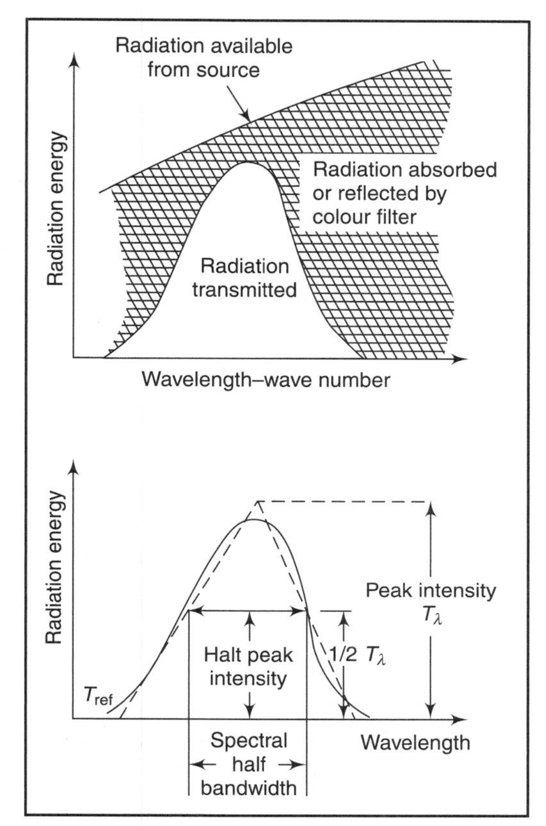

Figure 2.15 :: Optical properties of a light filter

2.3.2.1 Absorption Filters

The absorption type optical filter usually consists of colour media including colour glasses, coloured films (gelatin, etc.), and solutions of the coloured substances. This type of filter has a wide spectral bandwidth, which may be 40 to 50μ in width at one-half the maximum transmittance. Their efficiency of transmission is very poor and is of the order of 5 to 25 per cent.

It is possible to obtain more selective light filters from coloured media by increasing their thickness by two, three or more times. Here the transmission of the filter for the light of the wavelength isolated is decreased, but there is a simultaneous increase in the selectivity. By using this technique, it is theoretically possible to achieve a very good selectivity, but the fall in transmission efficiency would have to be compensated by suitable amplification of the photocurrent. As absorption type filters do not provide a high degree of monochromaticity required for isolating complex systems, their use is restricted to only the very simple type of photometers.

Selectivity

Composite filters consisting of sets of unit filters are often used. In the combination, one set consists of long wavelength, sharp cut-off filters and the other of short wavelength, cut-off filters; combinations are available from a range of about 360 nm to 700 nm.

The glass filter consists of a solid sheet of glass that has been coloured with a pigment, which is either dissolved or dispersed in glass, whereas the gelatin filter consists of a layer of gelatin impregnated with suitable organic dyes and sandwiched between two sheets of glass. Gelatin filters are not suitable for use over long periods. With the absorption of heat, they tend to deteriorate due to changes in the gelatin and bleaching of the dye.

2.3.2.2 Interference Filters

Interference filters usually consist of two semi-transparent layers of silver, deposited on glass by evaporation in vacuum and separated by a layer of dielectric (ZnS or MgF_2). In this arrangement, the semi-transparent layers are held very close together. The spacer layer is made of a substance which is of a low refractive index. The thickness of the dielectric layer determines the wavelength transmitted. Figure 2.16 shows the path of light rays through an interference filter. Some part of light that is transmitted by the first film is reflected by the second film and is again reflected on the inner face of the first film, as the thickness of the intermediate layer is one-half the wavelength of a desired peak wavelength. Only light which is reflected twice will be in-phase and will come out of the filter, while other wavelengths with phase differences would cause destructive interference. Constructive interference between different pairs in superposed light rays occurs only when the path difference is exactly one wavelength or some multiple thereof. The relationship expressing a maximum for the transmission of a spectral band is given by

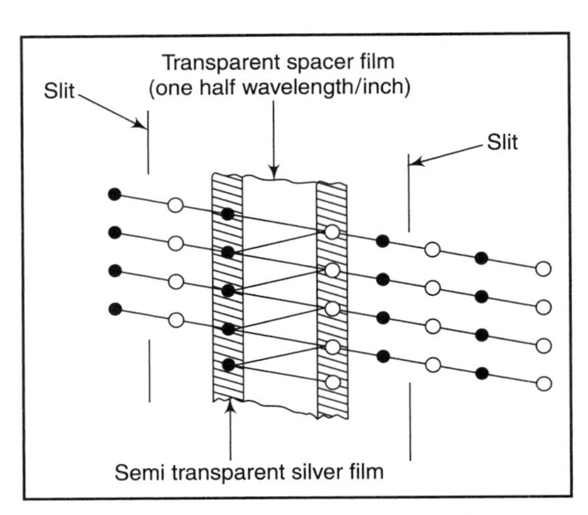

Figure 2.16 :: Path of light rays through an interference filter

$$m\lambda = 2\ d(n)\ \sin\theta,$$

when light is incident normally, $\sin\theta = 1$

$$m\lambda = 2d(n),$$

where d is the thickness of the dielectric spacer, whose refractive index is n. The multiple of frequencies harmonically related to the wavelength of the first order rays is the order (m) of the interference.

Interference filters allow a much narrower band of wavelengths to pass and are similar to monochromators in selectivity. They are simpler and less expensive. However, as the selectivity increases, the transmittance decreases. The transmittance of these filters varies between 15 to 60 per cent with a spectral bandwidth of 10 to 15 nm.

One type of interference filters is the continuous wedge filter, which permits a continuous selection of different wavelengths. The continuity of an interference filter is achieved by using a spacer film of graded thickness between the two semi-transparent layers of silver. They usually have a working interval of 400 to 700 nm and a dispersion of 5.5 nm/mm. The transmittance is usually not more than 35 per cent. With less transmittance, the sensitivity gets lower, which may be compensated by using electronic amplifiers after the photodetectors.

Wedge filter

For efficient transmission, multi-layer transmission filters are often used. They are characterized by a bandpass width of 8 nm or less and a peak transmittance of 60-95 per cent. Interference filters can be used with high intensity light sources, since they remove unwanted radiation by transmission and reflection rather than by absorption.

2.3.3 Monochromators

Monochromators are optical systems, which provide better isolation of spectral energy than optical filters, and are therefore preferred in cases where it is required to isolate narrow bands of radiant energy. Monochromators usually incorporate a small glass of quartz prism or a diffraction grating system as dispersing media. The radiation from a light source is passed either directly or by means of a lens or mirror into the narrow slit of the monochromator and allowed to fall on the dispersing medium, where it gets isolated. The efficiency of such monochromators is much better than that of filters, and spectral half-bandwidths of 1 nm or less are obtainable in the ultraviolet and visible regions of the spectrum.

2.3.3.1 Prism Monochromators

The isolation of different wavelengths in a prism monochromator is based upon the fact that the refractive index of materials is different for radiation of different wavelengths. If a parallel beam of radiation falls on a prism, the radiation of two different wavelengths will be bent through different angles. The greater the difference between these angles, the easier it is to isolate the two wavelengths. This becomes an important consideration for selection of material for the prisms, because only those materials are selected whose refractive index changes sharply with the wavelength.

Figure 2.17 shows the use of a prism as a monochromator. Light from the source S is made into a parallel beam and made to fall on a prism after it is passed through entrance slit S_1 and mirror M_1. The entrance silt is at the focus of mirror M_1. The prism disperses the light and photons of different wavelengths are deflected at different angles. If the dispersed beam is refocused, the focal point for photons of one wavelength will be displaced from that for photons of a different wavelength. The light of any one wavelength can be selected by moving a slit across the focal plane. The required wavelength passes through the slit, while the other wavelengths are blocked. The optical arrangement used in practice may differ from that illustrated in Figure 2.17, but the principle is the same. In most of the cases, the prism is moved to shift the spectrum across the exit slit, rather than the slit being moved across the focal

Slit

plane. The same collimating mirror is used for both M_1 and M_2 to save the cost of high grade optical components. The instru-ment manuals can be consulted to learn about the alternative systems.

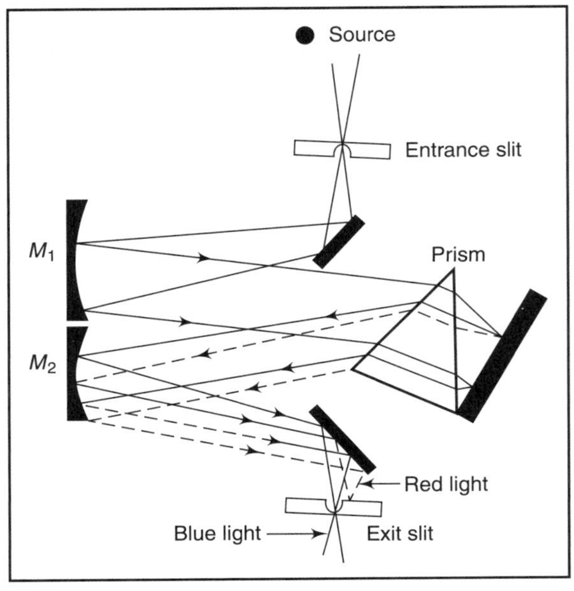

Figure 2.17 :: Prism monochromator

The prism may be made of glass or quartz. The glass prisms are suitable for radiations essentially in the visible range whereas the quartz prism can cover the ultraviolet spectrum also. It is found that the dispersion given by glass is about three times that of quartz. However, quartz shows the property of double refraction. Therefore, two pieces of quartz, one right-handed and the other left-handed, are taken and cemented back-to-back in the construction of a 60° prism (Cornu mounting), or the energy must be reflected and returned through a single 30° prism, so that it passes through the prism in both directions (Littrow mounting). The two surfaces of the prism must be carefully polished and optically flat. Prism spectrometers are usually expensive because of exacting requirements and difficulty in obtaining quartz of suitable dimensions.

Cornu mounting

Littrow mounting

There are several ways of selecting a particular wavelength in prism monochromators. It may be chosen by local selection with movable exit slits, or by local selection with fixed slits, behind which are placed the same number of photosensitive elements as that of slits. The selection can also be achieved by prism rotation, in which all the lines of the spectrum are passed through a fixed slit one after the other. The wavelength scale in this case is non-linear.

2.3.3.2 Diffraction Gratings

Monochromators may also make use of diffraction gratings as a dispersing medium. A diffraction grating consists of a series of parallel grooves ruled on a highly polished reflecting surface. When the grating is put into a parallel radiation beam so that one surface of the grating is illuminated, this surface acts as a very narrow mirror. The reflected radiation from this grooved mirror overlaps the radiation from neighbouring grooves (Figure 2.18).

The waves would, therefore, interfere with each other. On the other hand, it could be that the wavelength of radiation is such that the separation of the grooves in the direction of the radiation is a whole number of wavelengths. Then the waves would be in-phase and the radiation would be reflected undisturbed. When this is not a whole number of wavelengths, there would be destructive interference, the waves would cancel out and no radiation would be reflected. By changing the angle at which the radiation strikes the grating, it is possible to alter the wavelength reflected.

The expression relating the wavelength of the radiation and the angle (θ) at which it is reflected is given by:

$$m\lambda = 2d \sin \theta,$$

where d is the distance separating the grooves and is known as the grating constant and m is the order of interference.

When $m = 1$, the spectrum is known as first order, and when $m = 2$, the spectrum is known as second order.

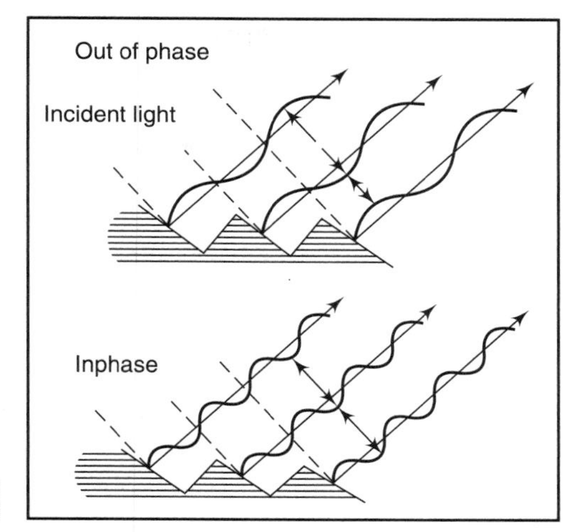

Figure 2.18 :: Dispersion phenomenon in diffraction gratings

Resolving power

The resolving power of a grating is determined by the product mN, where N is the total number of grooves or lines on the grating. Higher dispersion in the first order is possible when there are a larger numbers of lines. When compared with prisms, the gratings provide much higher resolving powers and can be used in all spectral regions. Gratings would reflect, at any given angle, radiation of wavelength λ and also $\lambda/2$, $\lambda/3$, etc. This unwanted radiation must be removed with filters or pre-monochromators, otherwise it will appear as stray light.

Figure 2.19 shows a typical diffraction grating monochromator. The entrance slit is illuminated by the light source, and light from the slit is focused to a parallel beam by the collimating mirror, with this beam being incident on the grating. The grating is rotated to diffract light of the required wavelength on to the focusing mirror, which, in turn, focuses it on to the exit slit.

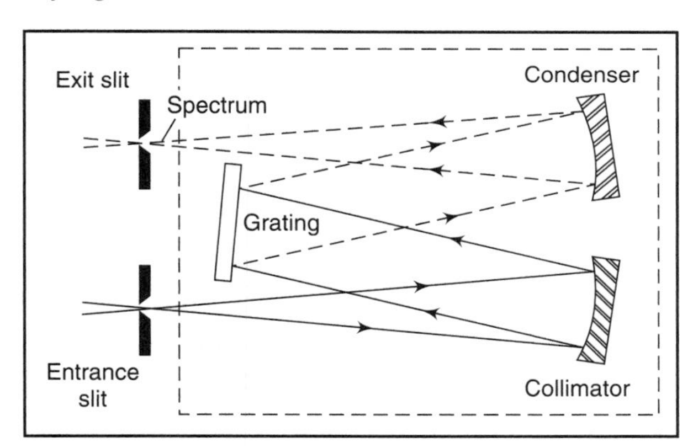

Figure 2.19 :: Use of diffraction grating in a monochromator

A good monochromator design can be obtained by ensuring that the mirrors and dispersing elements are of high quality with little scatter and that the scattered light is minimized by baffling, so that it cannot reach the exit slit. In addition, the light can be diffracted twice or more by the grating on reflection from the mirrors. Higher order spectra are usually removed by suitable filters.

Most modern instruments now use a diffraction grating as a dispersing element in the monochromator, as prisms, in general, have a poorer stray light performance and require complex precision cams to give a linear wavelength scale. Replica gratings can also be produced more cheaply than prisms and require only a simple sine bar mechanism for the wavelength scale.

A typical reflection grating may have 1200 grooves/mm, which means the grooves are spaced at about 800 nm intervals. The grating may have a width of 20 mm or more, giving a total of at least 24,000 grooves. In order to obtain constructive interference across this number of grooves with little light scattering, the spacing and form of the grooves must be accurate to within a few nanometers to give a high quality grating. To obtain high quality grating as near as possible to this accuracy with mechanical diamond ruling engines is a difficult technological problem.

2.3.3.3 Holographic Gratings

Precision spectrophotometers use holographic or interference gratings, which have superior performance in reducing stray light as compared to diffraction gratings. Holographic gratings are made by first coating a glass substrate with a layer of photo-resist, which is then exposed to interference fringes generated by the intersection of two collimated beams of laser light. When the photo-resist is developed, it gives a surface pattern of parallel grooves. When coated with aluminium, this becomes diffraction gratings.

A compared to ruled gratings, the grooves of a holographic grating are more uniformly spaced, and smoothly and uniformly shaped. These characteristics result in much lower stray light levels. Moreover, the holographic gratings can be produced in much less time than the ruled grating. Holographic gratings used in commercial spectrophotometers are either original master gratings produced directly by an interferometer or replica gratings. Replica gratings are reproduced from a master holographic grating by moulding its grooves onto a resin surface on a glass or silica substrate. Both types of gratings are coated with an aluminium reflecting surface and finally with a protective layer of silica or magnesium fluoride. Replica gratings give performance, which is as good as master gratings. The holographic process is capable of producing gratings that almost reach the theoretical stray-light minimum.

Replica gratings

2.3.4 Optical Components

Several different types of optical components are used in the construction of analytical instruments based on the radiation absorption principle. They could be windows, mirrors and simple condensers. The material used in the construction of these components is a critical factor and depends largely upon the range of wavelength of interest. Normally, the absorbance of any material should be less than 0.2 at the wavelength of use. The following factors need to be considered while selecting optical components:

- Ordinary silicate glasses are satisfactory from 350 to 3000 nm.
- From 300 to 350 nm, special corex glass can be used.
- Below 300 nm, quartz or fused silica is utilized, and the limit for quartz is 210 nm.
- From 180 to 210 nm, fused silica can be used, provided the monochromator is flushed with nitrogen or argon to eliminate absorption by atmospheric oxygen.

Reflections from glass surfaces are reduced by coating these with magnesium fluoride, which is one-quarter wavelength in optical thickness. With this, the scattering effects are also greatly reduced. However, using a layer of magnesium fluoride over aluminium coating does not offer a satisfactory solution, as the layer is soft and has poor chemical resistance (Sharpe, 1984). It cannot be easily cleaned. A better solution is to use a silica or synthetic quartz coating, which is hard and chemically resistant. If these coatings become dirty, they can be washed with a mild detergent and distilled water to restore the original high reflectance. The use of silica-coated aluminium mirrors ensures long mirror life with enhanced reflectance in the UV region and minimum deterioration of stray light performance. Table 2.2 provides data on materials that are used for optical components in different parts of the electromagnetic spectrum.

Table 2.2 :: Materials for Optical Components

Region of the Spectrum	Mirrors	Lenses	Windows
X-ray	—	—	Beryllium
Ultraviolet	Aluminium	Fused silica (synthetic quartz), Sapphire	Fused silica, Sapphire
Visible	Aluminium	Glass, Sapphire	Glass, Sapphire
Near-infrared	Gold	Glass, Sapphire	Glass, Sapphire
Infrared	Copper, Gold	CaF_2, ZnSe	NaCl, BaF_2, CaF_2, SnZe

In order to reduce the beam size or render the beam parallel, condensers are used. These condensers operate as simple microscopes. In order to minimize light losses, lenses are sometimes replaced by front-surfaced mirrors to focus or collimate light beam in absorption instruments. Mirrors are aluminized on their front surfaces. With the use of mirrors, chromatic aberrations and other imperfections of the lenses are minimized.

Beam splitters

Beam splitters are used in double-beam instruments. These are made by giving a suitable multi-layer coating on an optical flat. The two beams must retain the spectral properties of the incident beam. Half-silvered mirrors are often used for splitting the beam. However, they absorb some of the light in the thin metallic coating. Beam splitting can also be achieved by using a prismatic mirror or a stack of thin horizontal glass plates, silvered on their edges and alternatively oriented to the incident beam.

Fibre Optics

Optical fibres are now being extensively used in spectroscopic applications to transmit light for quantitative measurements. The fibres used in most of the analyzers have core diameters of 200 to 600 microns. They come in two types based upon their hydroxyl content.

Fibre attenuation

Figure 2.20 shows fibre attenuation with respect to wavelength. Fibre attenuation is typically specified in decibels (*dB*) per kilometer. Most on line applications use fibres of less than 50 m, and 10 m is more typical.

Low-OH fibres (shown by the solid transmission spectrum) are used for visible and near-infrared (NIR) applications from 400 to 2500 nm. The transmittance of a typical 10 m Low-OH fibre is 99.77 per cent. Thus, for most applications, loss of light in the fibre optic cables is not a limiting factor. Other transmission values for 10 m fibres are also shown in Figure 2.20.

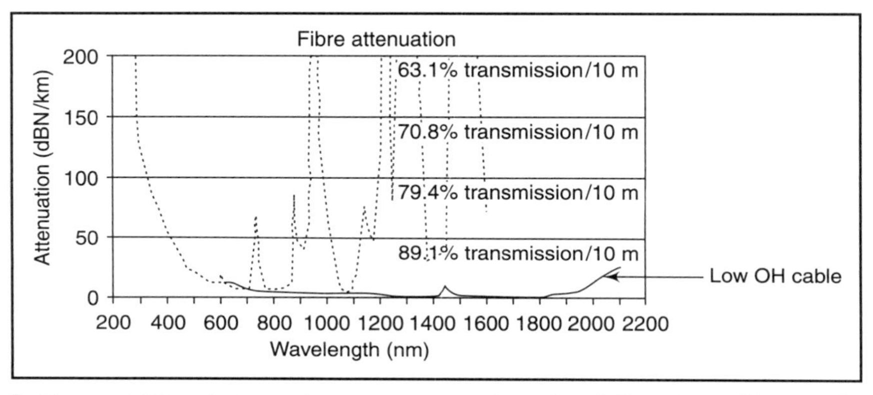

Figure 2.20 :: Attenuation versus wavelength of light in a fibre optic cable

High-OH fibres are primarily used in the ultraviolet (UV) region, but can also be used in the visible region of the spectrum and up to about 800 nm in the NIR. The high-OH content produces too much absorption of light at longer NIR wavelengths.

The fibre is protected inside the cable with polypropylene or Teflon tubing, which, in turn, is surrounded by a braided Aramid fiber mesh for strength. The protected fiber is often placed in PVC-covered monocoil jacket, which is a strip-wound metal tube that adds protection for the assembly.

2.3.5 Photosensitive Detectors

After isolation of radiation of a particular wavelength in a filter or a monochromator, it is essential to have a quantitative measure of their intensities. This is done by causing the radiation to fall on a photosensitive element, in which the light energy is converted into electrical energy. The electric current produced by this element can be measured with a sensitive galvanometer directly or after suitable amplification.

Any type of photosensitive detector may be used for the detection and measurement of radiant energy, provided it has a linear response in the spectral band of interest and has a sensitivity that is good enough for the particular application. There are two types of photoelectric cells: photovoltaic cells and photo-emissive cells.

2.3.5.1 Photovoltaic or Barrier Layer Cells

Photovoltaic or barrier layer cells usually consist of a semi-conducting substance, which is generally selenium deposited on a metal base that may be iron and that acts as one of the electrodes. The semi-conducting substance is covered with a thin layer of silver or gold deposited by cathodic deposition in vacuum. This layer acts as a collecting electrode. Figure 2.21 shows the construction of the barrier layer cell. When radiant energy falls upon the semiconductor surface, it excites the electrons at the silver–selenium interface. The electrons are thus released and collected at the collector electrode.

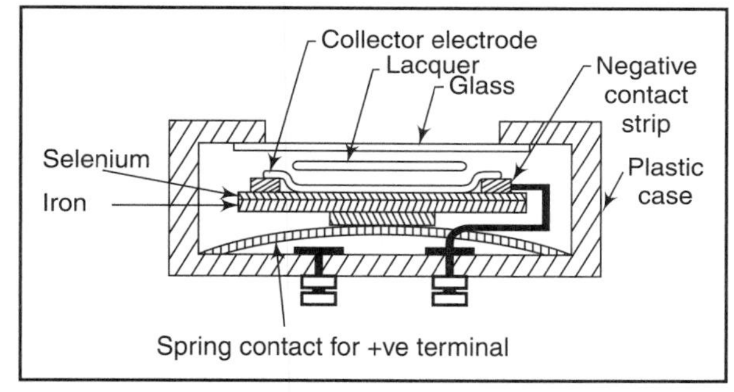

Figure 2.21 :: Construction of a barrier layer cell

The cell is enclosed in a housing of insulating material and covered with a sheet of glass. The two electrodes are connected to two terminals for connecting the cell with other parts of the electrical circuit. Photovoltaic cells are very robust in construction, need no external electrical supply and produce a photocurrent which is sometimes stronger than other photosensitive elements. The typical photocurrents produced by these cells are as high as 120 μA/lumen. At constant temperature, the current generated in the cell usually shows a linear relationship with the incident light intensity. Since selenium photocells have very low internal resistance, it is difficult to amplify the current they produce by using dc amplifiers. The currents are usually measured directly by connecting the terminals of the cell to a very sensitive galvanometer.

Photocurrents

Selenium cells are sensitive to almost the entire range of wavelengths of the spectrum. However, their sensitivity is greater within the visible spectrum and highest in the zones near the yellow wavelengths. Figure 2.22 shows the spectral response of the selenium photocell and the human eye. The human eye can only detect radiation that is in the visible region of the spectrum, and hence the name. These photons are transmitted by the lens of the human eye as well as absorbed by the photoreceptors in the retina.

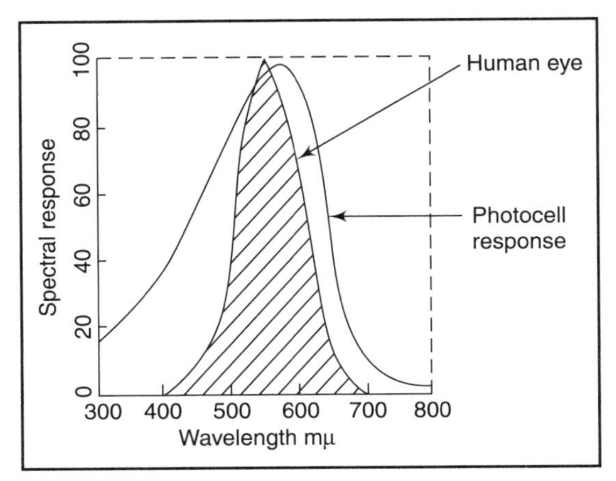

Figure 2.22 :: Spectral response of a selenium photocell and the human eye

Since selenium cells have a high temperature coefficient, it is necessary to allow the instrument to warm up before the readings are commenced.

Fatigue effects

They also show fatigue effects. When illuminated, the photocurrent rises to a value several percentage points above the equilibrium value and then falls off gradually. When it is connected in the optical path of the light rays, care should be taken to block all external light and to see that only the light from the source reaches the cell. Apart from selenium, photocells may be made of some other materials. The spectral sensitivity is different for different types of cells and should be chosen in accordance with the wavelength of the radiation to be measured.

Selenium cells are not suitable for operations in instruments where the levels of illumination change rapidly, because they fail to respond immediately to those changes. They are thus not suitable where mechanical choppers are used to interrupt light 15–60 times a second.

2.3.5.2 Photo-emissive Cells

Photo-emissive cells are of three types: (a) high vacuum photocells, (b) gas-filled photocells and (c) photomultiplier tubes. All these types differ from selenium cells in that they require an external power supply to provide a sufficient potential difference between the electrodes to facilitate the flow of electrons generated at the photosensitive cathode surface. Also, amplifier circuits are invariably employed for the amplification of this current.

2.3.5.3 High Vacuum Photo-emissive Cells

The vacuum photocell consists of two electrodes, a cathode having a photosensitive layer of metallic cesium deposited on a base of silver oxide and an anode, which is either an axially centred wire or a rectangular wire that frames the cathode. The construction of the anode is such that no shadow falls on the cathode. The two electrodes are sealed within an evacuated glass envelope.

When a beam of light falls on the surface of the cathode, electrons are released from it, which are drawn towards the anode that is maintained at a certain positive potential. This gives rise to photocurrents, which can be measured in the external circuit. The spectral response of a photo-emissive tube depends upon the nature of the substance coating the cathode, and can be varied by using different metals or by variation in the method of preparation of the cathode surface.

Cesium–silver oxide cells are sensitive to the near-infrared wavelengths. Similarly, potassium–silver oxide and cesium–antimony cells have maxima of sensitivity in the visible and ultraviolet regions. The spectral response also depends partly upon the transparency to different wavelengths of the medium to be traversed by the light before reaching the cathode. For example, the sensitivity of the cell in the ultraviolet region is limited by the transparency of the wall of the envelope. For this region, the use of quartz material can be avoided by using a fluorescent material like sodium salicylate, which when applied to the outside of the photocell, transforms the ultraviolet into visible radiations.

Figure 2.23 shows the current–voltage characteristics of vacuum photo-emissive tube at different levels of light flux. They show that as the voltage is increased, the point is reached where all the photoelectrons are swept to the anode as soon as they are released and result in saturation photocurrent. It is not desirable to apply very high voltages, as they would result in excessive dark current without any gain in response.

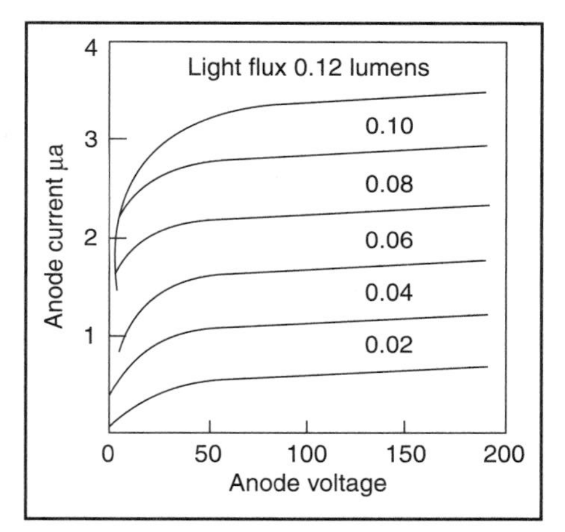

Figure 2.23 :: Current–voltage characteristics of vacuum photo-emissive tube

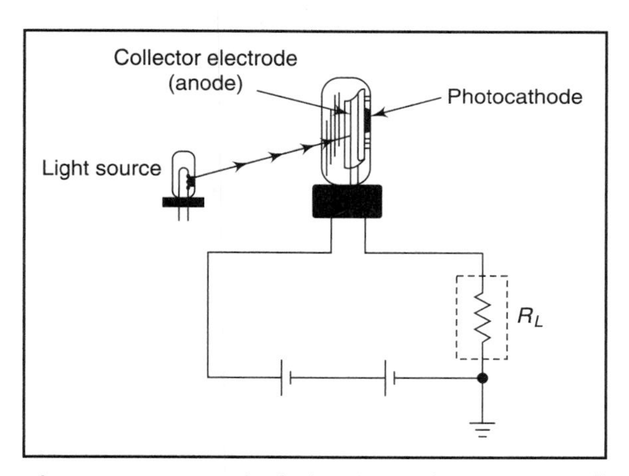

Figure 2.24 :: Typical circuit configuration employed with photo-emissive tubes

Figure 2.24 shows a typical circuit configuration usually employed with photo-emissive tubes. Large values of phototube load resistor are employed to increase the sensitivity up to the practical limit. Load resistances as high as 10,000 MΩ have been used. This, however, almost puts a limit, as any further increase in sensitivity causes difficulties in the form of noise, non-linearity and slow response. At these high values of load resistors, it is essential to shield the circuit from moisture and electrostatic effects. Therefore, a special type of electrometer tubes, carefully shielded and with a grid cap input, are employed during the first stage of the amplifier.

2.3.5.4 Gas-filled Photo-emissive Cells

This type of cell contains small quantities of inert gas like argon, whose molecules can be ionized when the electrons present in the cell possess sufficient energy. The presence of the small quantities of this gas prevent the phenomenon of saturation current, when higher potential differences are applied between the cathode and anode. Due to repeated collisions of electrons in the gas-filled tubes, the photoelectric current produced is greater even at low potentials.

2.3.5.5 Photomultiplier Tubes

Photomultiplier tubes (PMTs) convert photons into an electrical signal. They have a high internal gain and are sensitive detectors for low-intensity applications such as fluorescence spectroscopy.

Photomultiplier tubes are used as detectors when it is required to detect very weak light intensities. The tube consists of a photosensitive cathode and has multiple cascade stages of electron amplification in order to achieve a large amplification of primary photocurrent within the envelope of the phototube itself. When a photon of sufficient energy strikes the photocathode, it ejects a photoelectron due to the photoelectric effect. The photocathode material is usually a mixture of alkali metals, which make the PMT sensitive to photons throughout the visible region of the electromagnetic spectrum. The photocathode is at a high negative voltage, typically -500 to -1500 volts. The photoelectron is accelerated towards a series of additional electrodes called dynodes. Each of these electrodes is maintained at successively less negative potentials. Additional electrons are generated at each dynode. There may be 9-16 dynodes (Figure 2.25). This cascading effect creates 10^5 to 10^7 electrons for each photoelectron that is ejected from the photocathode. The amplification depends upon the number of dynodes and the accelerating voltage. This amplified electrical signal is collected at an anode at ground potential, which can be measured.

Dynodes

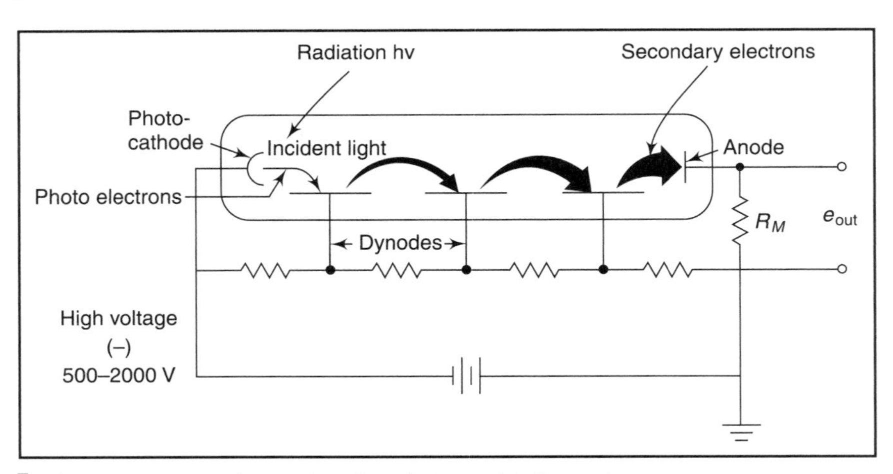

Figure 2.25 :: Schematic of a photomultiplier tube

The sensitivity of the photomultiplier tube can be varied by regulating the voltage of the first amplifying stage. Because of the relatively small potential difference between the two electrodes, the response is linear. The output of the photomultiplier tube is limited to an anode current of a few milliamperes. Consequently, only low intensity radiant energy can be measured without causing any appreciable heating effect on the electrode surface. They can measure light intensities about ten times weaker than those measurable with an ordinary phototube. For this reason, they should be carefully shielded from stray light. The tube is fairly fast in response to the extent that they are used in scintillation counters, where light pulses as brief as 10^{-9} duration are encountered. A direct current (dc) power supply is required to operate a photomultiplier, the stability of which must be at least one order of magnitude better than the desired precision of measurement; for example, to attain precision of 1 per cent, fluctuation of the stabilized voltage must not exceed 0.1 per cent.

Fatigue and saturation can occur at high illumination levels. The devices are sensitive to electromagnetic interference, and are also costlier than other photoelectric sensors. Photomultipliers are not uniformly sensitive over the whole spectrum and in practice, manufacturers incorporate units best suited for the frequency range for which the instrument is designed. In the case of spectrophotometers, the photomultipliers normally supplied cover the range of 185 to 650 nm. For measurements at longer wavelengths, special red sensitive tubes are offered. They cover a spectral range from 185 to 850 nm, but are noticeably less sensitive at wavelengths below 450 nm, than the standard photomultipliers. Photomultiplier tubes show a quantum efficiency of 1-10 per cent, with a response time of 1-20 ns.

Photomultiplier tubes may be damaged if excessive current is drawn from the final anode. Since accidental overload may easily occur in a laboratory and the tubes are too expensive to replace, it is advisable to adopt some means of protection from overloads. Generally, the circuits are so designed that they automatically cut off the EHT supply to a photomultiplier tube if accidental overload of the tube should occur. Once cut off, the EHT has to be reset manually.

2.3.5.6 Silicon Diode Detectors

Photodiode

The photomultiplier which is large and expensive, and requires a source of stabilized high voltage can be replaced by a photodiode. This diode is useable within a spectral range of 0.4–1.05 μm, in analytical instruments such as spectrophotometers and flame photometers. The photodiode can be powered from low voltage source. The signal is amplified by a low noise op-amp.

When a photon strikes a semiconductor, it can promote an electron from the valence band (filled orbital) to the conduction band (unfilled orbital) creating an electron (–)–hole (+) pair. The concentration of these electron–hole pairs is dependent upon the amount of light striking the semiconductor, making the semiconductor suitable as an optical detector. There are two ways to monitor the concentration of electron–hole pairs. In photodiodes, a voltage bias is present and the concentration of light-induced electron–hole pairs determines the current through a semiconductor.

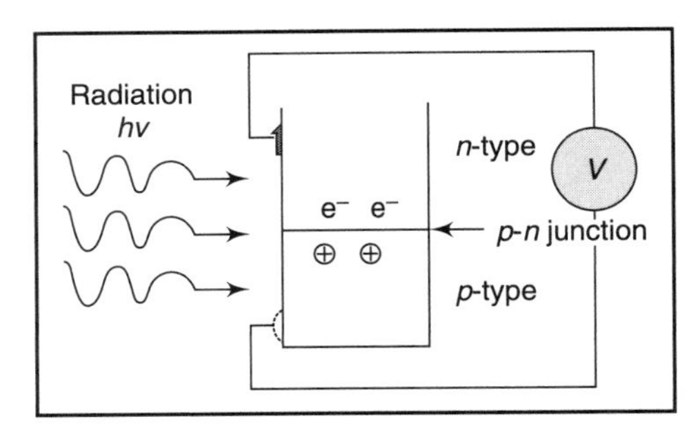

Figure 2.26 :: Schematic of a semi-conductor detector

Table 2.3 :: Wavelength Range

Detector type	λ range (μ)
Si	0.2 – 1.1
Ge	0.4 – 1.8
InAs	1.0 – 3.8
InSb	1.0 – 7.0
InSb (77K)	1.0 – 5.6
HgCdTe (77K)	1.0 – 25.0

Photovoltaic detectors contain a p–n junction that causes the electron–hole pairs to separate to produce a voltage that can be measured. Figure 2.26 shows a schematic diagram of a semi-conductor photovoltaic detector.

However, photodiode detectors are not as sensitive as PMTs but are small and robust. The wavelength range for different solid state detectors is given in Table 2.3.

Silicon diode detectors (when integrated with an operational amplifier) have performance characteristics which compare with those of a photomultiplier over a similar wavelength range. Figure 2.27 shows the spectral response of silicon diode detectors. Being solid state, the devices are mechanically robust and consume much less power. Their dark current output and noise levels are such that they can be used over a much greater dynamic range.

2.3.5.7 Photodiode Arrays (PDAs)

Diode arrays are assemblies of individual detector elements in linear or matrix form, which in a spectrophotometer, can be mounted so that the complete spectrum is focused on to an array of

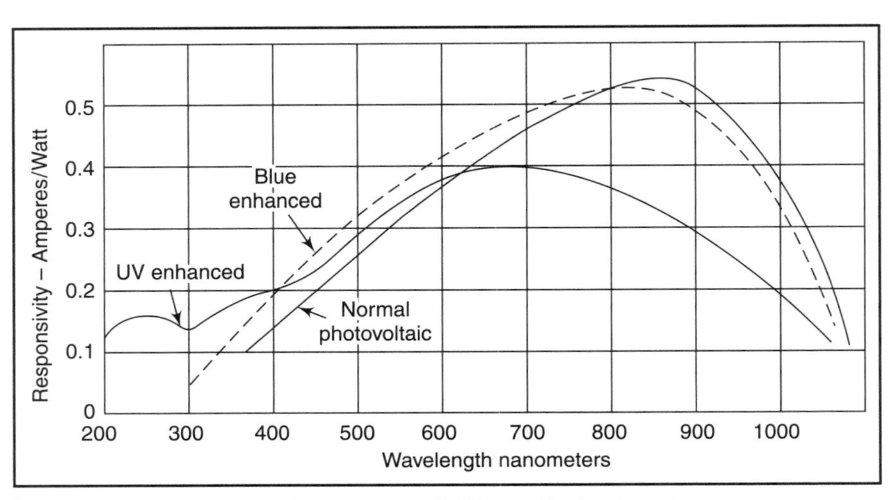

Figure 2.27 :: Spectral response of silicon diode detectors

appropriate size. The arrangement does not require any wavelength selection mechanism and the output is instantaneously available. However, resolution in diode arrays is limited by the physical size of individual detector elements, which at present is about 2 nm.

A photodiode array (PDA), consisting of discrete photodiodes, is available on an integrated circuit (IC) chip. For spectroscopy, it is placed at the image plane of a spectrometer to allow a range of wavelengths to be detected simultaneously. In this regard, it can be thought of as an electronic version of a photographic film. Array detectors are especially useful for recording the full UV–vis absorption spectra of samples that are rapidly passing through a sample flow cell, such as in HPLC detectors. PDAs are available with 512, 1024, or 2048 elements with typical dimensions of 25 μm wide and 1–2 mm high. A schematic diagram of a photodiode array is shown in Figure 2.28.

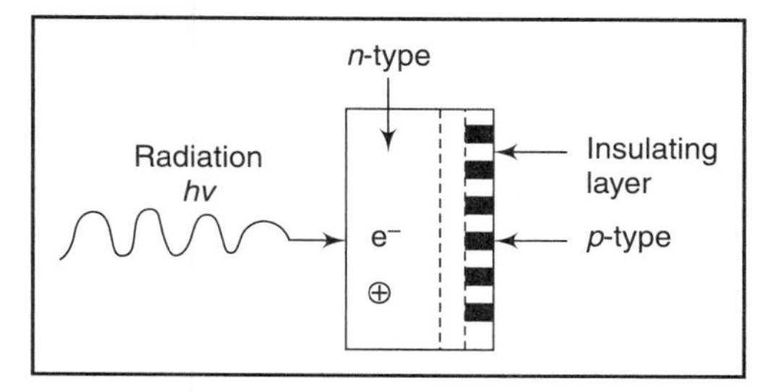

Figure 2.28 :: Schematic of a photodiode array

Light creates electron–hole pairs and the electrons migrate to the nearest PIN junction. After a fixed integration time the charge at each element is read with solid-state circuitry to generate the detector response as a function of linear distance along the array.

Knud *et al.* (1980) describes the details of diode array photodetectors used in the Hewlett Packard spectrophotometer Model 8450A. The detector consists of two silicon integrated circuits, each containing 211 photosensitive diodes and 211 storage capacitors. The photodiode array is PMOS (p-channel metal-oxide semiconductor) integrated circuit that is over 1.25 cm long. Each photosensitive diode in the array is 0.05 by 0.50 mm and has a spectral response that extends well beyond the 200-800 nm range.

Diode array chip

A functional block diagram of the diode array chip is shown in Figure 2.29. In parallel with each of the 211 photodiodes is a 10 pF storage capacitor. These photodiode capacitor pairs are sequentially connected to a common output signal line through individual MOSFET switches. When a FET switch is closed, the pre-amplifier connected to this signal line forces a potential of –5 V on to the capacitor-diode pair. When the FET switch is opened again, the photocurrent causes the capacitor to discharge towards zero potential. The serial read-out of the diode array is accomplished by means of a digital shift register designed into the photodiode array chip.

The diode arrays typically exhibit a leakage current of less than 0.1 pA. This error term increases exponentially with temperature, but because the initial leakage value is so low, there is no need to cool the array at high ambient temperatures.

The obvious advantage of using a photodiode array is the short amount of time required for making a measurement. The measurement time in the spectrophotometers using a photodiode array is hardly 5 seconds as compared to more than 5 minutes in the usual scanning monochromator-based instruments.

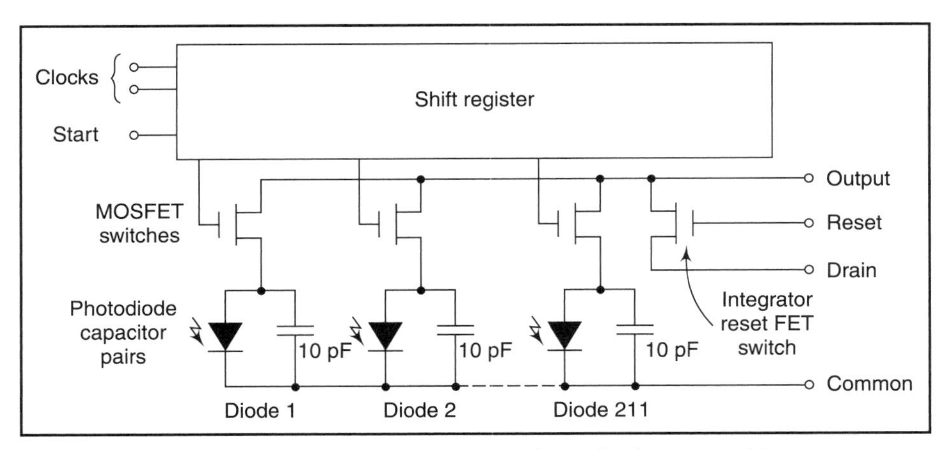

Figure 2.29 :: Functional diagram of the photodiode array chip

The major disadvantage of using a diode array instrument is the limited resolution. Scanning instruments, depending upon the resolution of the excitation monochromator, are able to easily achieve resolutions of the order of 0.1 nm, whereas in photodiode array instruments, the resolution is 1 nm, and it cannot be changed.

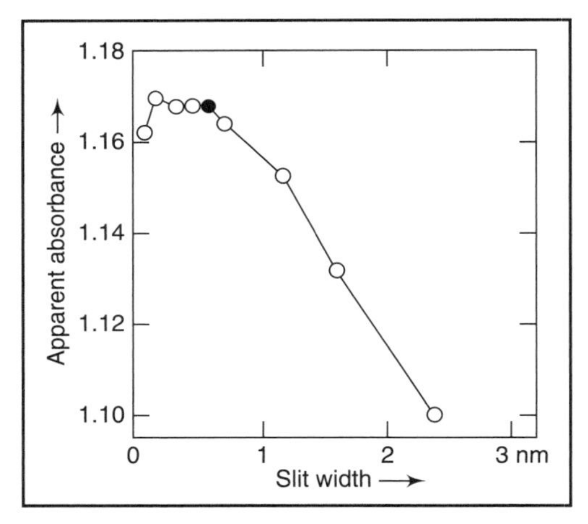

Figure 2.30 :: Effect of slit width on absorbance

2.3.6 Slit Width

The resolution of a spectrophotometer is usually limited by the spectral purity and intensity of the monochromator light output and the detector sensitivity at that wavelength. In some instruments, the control of the energy level reaching the detector is obtained by adjusting the aperture of the slit at the monochromator exit. When more than one slit width is available, it gives the user a means of trading energy against spectral sensitivity. However, most instruments using diffraction gratings take advantage of the linear dispersion and provide fixed slit widths to give a known and controlled band-width at the exit slit of the monochromator.

Where it is necessary to accurately determine the absorbance at λ_{max}, it is better to first plot apparent absorbance against slit width. Figure 2.30 shows that a slit width greater than about 0.75 mm may result in significant error in the measurement of the absorbance concerned.

2.3.7 Sample Holders

Liquids may be contained in a cell or cuvette made of transparent material such as silica, glass or Perspex. The faces of these cells through which the radiation passes are highly polished to keep reflection and scatter losses to a minimum. Solid samples are generally unsuitable for direct spectrophotometry. It is usual to dissolve the solid in a transparent liquid. Gases may be contained in cells which are scaled or stoppered to make them air-tight. A sample holder is generally inserted somewhere in the interval between the light source and the detector.

For the majority of analyses, a 10 mm path–length rectangular cell is usually satisfactory. For the far ultraviolet region below 210 nm, a 10 mm path–length rectangular cell made from special silica that has better transmission characteristics at shorter wavelengths than does the standard cell is recommended. However, some samples, such as turbid or densely coloured solutions, may absorb so strongly that shorter path–lengths are necessary for the sample to transmit sufficiently. Rectangular liquid cells are commercially available in both 5 and 1 mm path–lengths, while the standard 10 mm path–length cell can be reduced to 1 mm path–length by using a silica spacer. These shorter path–length cells have lesser volumes—of the order of 0.43 ml for a 1 mm path–length, which are necessary for some studies.

Liquid cells

In analyses where only minimal volumes of liquid samples are practical, microcells, which have volumes that are as small as 50 μl can be employed. Most of the rectangular liquid cells have caps and, for analyses of extremely volatile liquids, some of the cells have ground-glass stoppers to prevent the escape of vapour. Studies of dilute or weakly absorbing liquid samples or of samples where trace components must be detected, require a cell with a long path-length. For such applications, a 50 cm path–length with about 300 ml volume cell is employed.

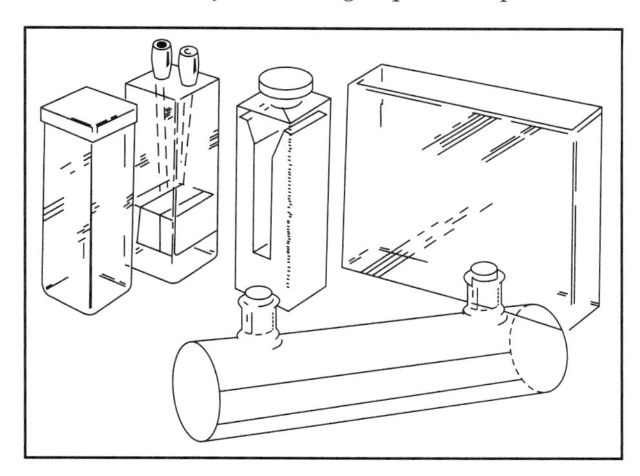

Figure 2.31 :: Selection of sample cuvettes

Cylindrical liquid cells offer higher volume to path–length ratios than do rectangular cells, being available in path–lengths of 20, 50 and 100 mm and in volumes of 4, 8, 20 and 40 ml, respectively.

Demountable cells

Similar to these cylindrical cells are the demountable cells that have easily removable silica windows. This demoun-table feature is especially useful for containment of samples that are difficult to remove and clean from conventional cylindrical cells. Demountable cells are equipped with ground-glass stoppers. Figure 2.31 shows a selection of typical sample cuvettes.

Gas cells

Gas cells are available in both rectangular and cylindrical configurations. Both configurations incorporate glass stopcocks. Two stopcocks are installed on the cylindrical cells to permit its connection into a flow system for dynamic measurements. Rectangular gas cells usually have path lengths of 2 and 10 mm, while cylindrical gas cells have a path–length of 100 mm.

Specially designed cells are required for making precise absorbance measurements of dilute aqueous solutions over the temperature range 25-250°C. The cell must take care of the major problem, which stems from the reactivity of the hot aqueous solution, when measurements are made at high temperatures.

The measurement of absorption spectra of liquids at low temperatures leads to practical problems like the misting of cell windows and accommodation of attachment for accurate control of the sample temperature in the small space normally available in the cell compartment. Problems of misting are eliminated by immersing the cells in liquid and placing long silica rods in the light path. For this purpose, an attachment is inserted between the monochromator and the photocell compartment in place of the cell compartment generally attached in the instrument.

2.4 Ultraviolet and Visible Absorption Spectroscopy (UV–Vis)

Ultraviolet and visible (UV-Vis) absorption spectroscopy is the measurement of the attenuation of a beam of light after it passes through a sample or after reflection from a sample surface in the ultraviolet and visible range of the electromagnetic spectrum. Absorption measurements can be done at a single wavelength or over an extended spectral range. Ultraviolet and visible light are energetic enough to promote outer electrons to higher energy levels. UV-Vis spectroscopy is usually applied to molecules or inorganic complexes in solution. The UV-Vis spectra have broad features that are of limited use for sample identification but are very useful for quantitative measurements. The concentration of an analyte in solution can be determined by measuring the absorbance at some wavelength and applying the Beer Lambert Law.

UV-Vis spectra

Since the UV-Vis range spans the range of human visual acuity of approximately 400–750 nm, UV-Vis spectroscopy is useful for characterizing the absorption, transmission and reflectivity of a variety of technologically important materials such as pigments, coatings, windows and filters. This more qualitative application usually required recording at least a portion of the UV–Vis spectrum for characterization of the optical or electronic properties of materials (Manning, 1969).

The UV-Vis spectral range is approximately 190 to 900 nm, as defined by the working range of typical commercial UV-Vis spectrophotometers. The short-wavelength limit for simple UV-Vis spectrometers is the absorption of ultraviolet wavelengths of less than 180 nm by atmospheric gases. Purging a spectrometer with nitrogen gas extends this limit to 175 nm. For working beyond 175 nm requires a vacuum spectrometer and a suitable UV light source. The long-wavelength limit is usually determined by the wavelength response of the detector in the spectrometer.

High-end commercial UV-Vis spectrophotometers extend the measurable spectral range into the NIR region to as far as 3300 nm.

The light source is usually a deuterium discharge lamp for UV measurements and a tungsten–halogen lamp for visible and NIR measurements. The instruments automatically swap lamps when scanning between the UV and visible regions. The wavelengths of these continuous light sources are typically dispersed by a diffraction/holographic grating in a single or double monochromator or spectrograph. The spectral bandpass is then determined by the monochromator slit width or by the array–element width in array–detector spectrometers. Spectrometer designs and optical components are optimized to reject stray light, which is one of the limiting factors in quantitative absorbance measurements. The detector in single–detector instruments is a photodiode, phototube, or photomultiplier tube (PMT). UV–Vis–NIR spectrometers utilize a combination of a PMT and a Peltier-cooled PbS IR detector. The light beam is redirected automatically to the appropriate detector when scanning between the visible and NIR regions. The diffraction grating and instrument parameters such as slit width can also be changed.

Most commercial UV-Vis absorption spectrometers use one of the following three overall optical designs:

- a fixed or scanning spectrometer with a single light beam and sample holder,
- a scanning spectrometer with dual light beams and dual sample holders for simultaneous measurement of I and I_0, or
- a non-scanning spectrometer with an array detector for simultaneous measurement of multiple wavelengths.

In single-beam and dual-beam spectrometers, the light from a lamp is dispersed before reaching the sample cell. In an array–detector instrument, all wavelengths pass through the sample and the dispersing element is between the sample and the array detector.

2.5 Colorimeters/Photometers

A colorimetric method in its simplest form uses only the human eye as a measuring instrument. This involves comparison by visual means of the colour of an unknown solution, with the colour produced by a single standard or a series of standards. The comparison is made by obtaining a match between the colour of the unknown and that of a particular standard by comparison with a series of standards prepared in a similar manner, to the unknown. Errors of 5 to 20 per cent are not uncommon because of the relative inability of the eye to compare light intensities.

In the earlier days, visual methods were commonly employed for all colorimetric measurements, but now photoelectric methods have largely replaced them and are used almost exclusively for quantitative colorimetric measurements. These methods are more precise and eliminate the necessity of preparing a series of standards every time a series of unknowns is run.

Strictly speaking, a *colorimetric determination* is one that involves visual measurement of colour; and a method employing photoelectric measurement is referred to as a photometric or spectrophotometric method. However, usually any method involving the measurement of colour in the visual region of the electromagnetic spectrum (400-700 mμ) is referred to as the colorimetric method.

In a colorimeter, the sample is normally a liquid. The sample compartment of a colorimeter is provided with a holder to contain the cuvette, in which the liquid is examined. Usually this holder is mounted on a slide with positions for at least two cuvettes so that sample and reference cuvettes are measured first and a shutter is moved into or out of the light beam until the microammeter gives a full-scale deflection (100 per cent T-scale reading). The sample is then moved into the beam and the light passing through it is measured as a percentage of the reference value.

$$\text{Sample concentration} = \text{Standard concentration} \times \frac{\text{Sample reading}}{\text{Reference reading}}$$

Colorimeters are extremely simple in construction and operation. They are used for a great deal of analytical work, where high accuracy is not required. The disadvantage is that a range of filters is required to cover different wavelength regions. Also the spectral bandwidth of these filters is large in comparison with that of the absorption band being measured.

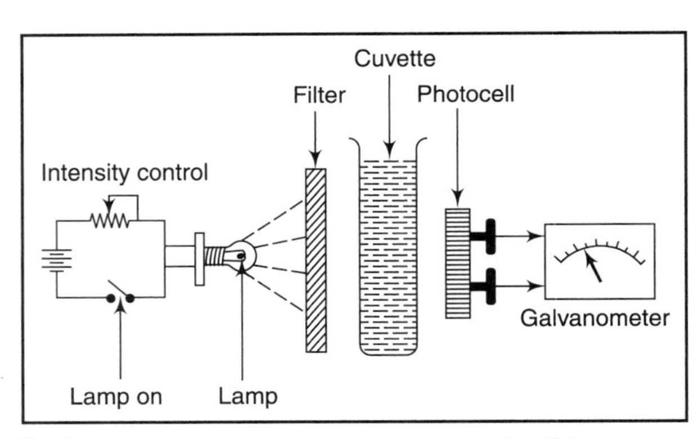

Figure 2.32 :: Basic components of a filter photometer

2.5.1 Single-beam Filter Photometers

Figure 2.32 illustrates the basic components of a filter photometer. The source of light is a tungsten filament lamp, which is held in a reflector and which throws light on the sample holder through a filter. The filter may be either of absorption or interference type. The sample holder is a cuvette with parallel walls or may be a test tube. The light, after passing through the sample holder, falls on the surface of the photocell. The output of the cell is measured on a microammeter.

The lamps must be energized from a highly stabilized dc source, or by the output of a constant voltage transformer. In order to operate the instrument, the following steps are taken:

(i) With the photocell darkened, the meter is adjusted mechanically to read zero.

(ii) The blank or a pure solvent or a reference solution is inserted in the path of the light beam and the incident light intensity is regulated. This can be done either by adjusting the rheostats in series with the lamp or rotating the photocell about an axis perpendicular to the light beam or adjusting a diaphragm in the light beam. With the help of any one of these adjustments, the meter reading is brought to 100 scale divisions.

(iii) Solutions of both standards and unknowns are inserted in place of the blank, and the reading of the specimen relative to the blank is recorded. The meter scale is calibrated in linear transmittance units (0-100 per cent). Such types of instruments are easy to operate and are also inexpensive. Errors of 1 to 5 per cent are quite common, but are nevertheless acceptable in many applications.

2.5.2 Double-beam Filter Photometer

In double-beam filter photometers, two photocells are normally employed. The two photocells are connected to two potentiometers P_1 and P_2, as shown in Figure 2.33. Each of the two potentiometers is of low resistance and is wound linearly. Light from the source lamp is made to pass through the filter F and is then divided into two parts, with one part passing through the solution in the cuvette before falling on the measuring photocell, and the other part passing directly on to the reference photocell. The galvanometer G receives opposing currents through it. The potentiometer P_1 is graduated in transmittance and absorbance units.

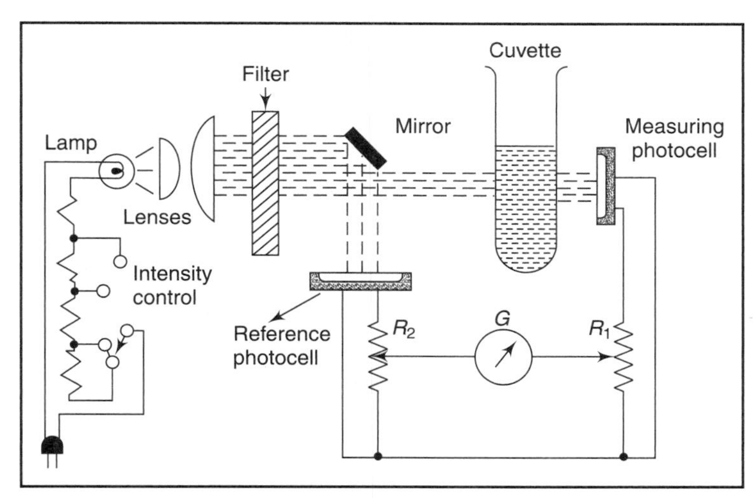

Figure 2.33 :: Schematic of a double-beam filter photometer

The operation of the instrument is very simple. With the lamp off, the galvanometer zero is adjusted mechanically. The potentiometer R_1 is set to $T = 1$, or $A = 0$. Then with the lamp on the blank solution is placed in the light path of the measuring cell. Potentiometer R_2 is adjusted until the galvanometer G reads zero. The solution to be analyzed is then substituted for the blank and P_1 is adjusted until the current through the galvanometer is zero, with the setting of R_1 remaining unchanged. The absorbance or transmittance can then be read directly on the scale of potentiometer R_1.

In two-cell photometers, the errors resulting from the fluctuations of the lamp intensity are minimized. The scale of potentiometer R_1 (transmittance scale) can be made much larger in size than the scale of the meter in single-cell instruments.

Different manufacturers adopt different means for restoring the balance of current through the galvanometer, e.g. in one arrangement, it is done by adjusting the intensity of the reference light beam by means of a diaphragm (Figure 2.34). Alternatively, some photoelectric colorimeters make use of the rotation of the reference photocell about an axis perpendicular to the light beam through an angle of 90°, plus a series of fixed apertures for coarse adjustment. These adjustments remain unchanged while standards and unknowns are inserted. The potentiometer in series with the reference photocell is adjusted to obtain the scale reading.

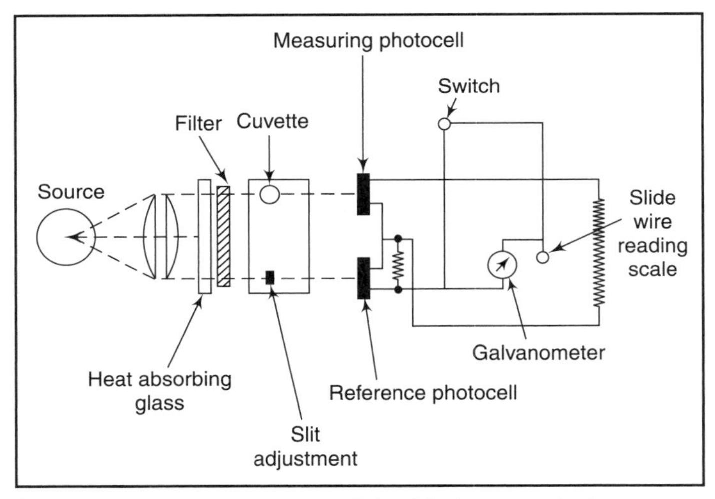

Figure 2.34 :: Schematic of double-beam colorimeter

2.5.3 Multi-channel Photometer

An increasing number of chemical analyses are carried out in the laboratories of industry and hospitals, and in most of these, the final measurement is performed by a photometer. Obviously, it is possible to increase the capacity of the laboratory by using photometers, which have a large measuring capacity. One of the limitations for rapid analyses is the speed at which the samples can be transferred in the light path.

In a multi-channel photometer, instead of introducing one sample at a time into a single light path, a batch of samples is introduced and measurements are carried out simultaneously, using a multiplicity of fibre optic light paths (Figure 2.35) and detectors, and then scanned electronically instead of mechanically. The 24 sample cuvettes are arranged in a rack in a three by eight matrix. The 25th channel serves as a reference beam and eliminates possible source and detector drifts. The time required to place the cuvette rack into the measuring position corresponds to the amount of time necessary to put one sample into a sample changer.

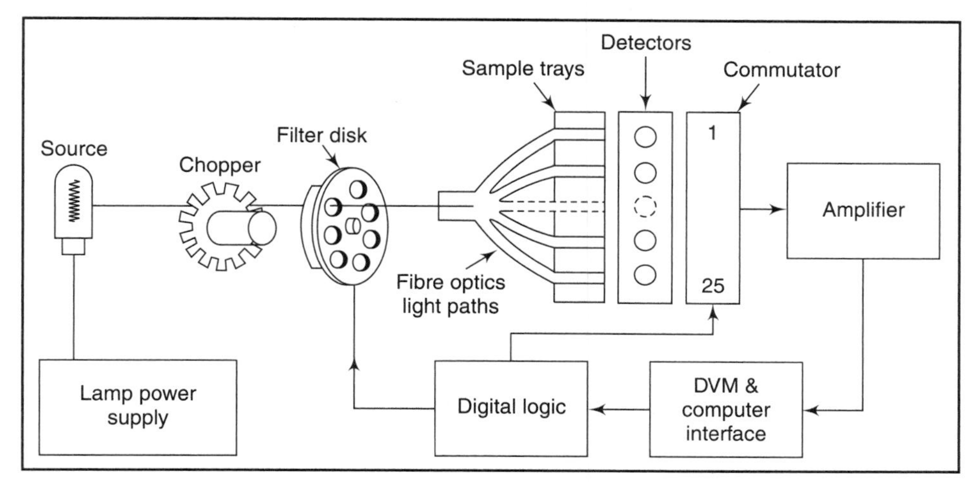

Figure 2.35 :: Schematic of a multi-channel photometer

The light source is a 50 W tungsten–halogen lamp, driven from a precisely controlled voltage source. The light is chopped by means of a mechanical rotating chopper. A lens focuses the light on the end of a bundle of fibre optic elements. The output of the detectors is amplified and displayed on digital voltmeter. The whole operation is synchronized with digital logic circuits.

2.5.4 Process Photometers

For on-line applications, M/s Optical Solutions, USA have introduced process photometers, known as ChemView, which allow measurement from 250 to 2150 nm without using any moving parts. This results in lower maintenance problems. The photometer uses a low-power tungsten lamp in the visible and NIR region. By continuously adjusting its brightness with a detector in a feedback circuit, very low optical drift is achieved. The company calls this arrangement 'StabLamp'. In the UV, a pulsed neon source is used.

Figure 2.36 shows the block diagram of a single probe instrument. Light is sent to the fibre optic probe and returned to the analyzer. Beam splitters divide the light and a portion passes through narrow-band wavelength filters into each detector. Up to six detectors, one or more of

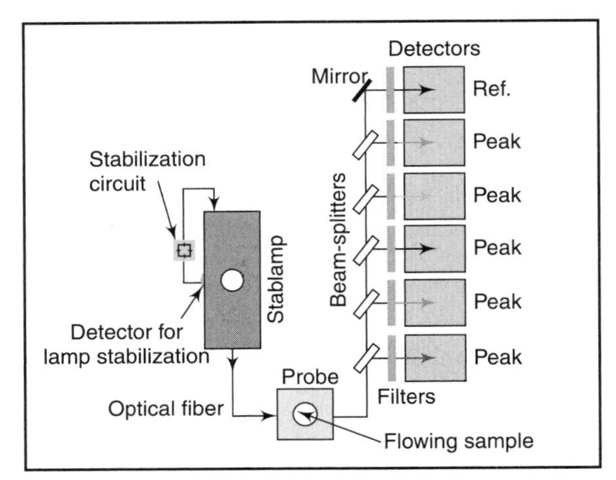

Figure 2.36 :: Optical diagram of a single channel UV-Vis and NIR process photometer

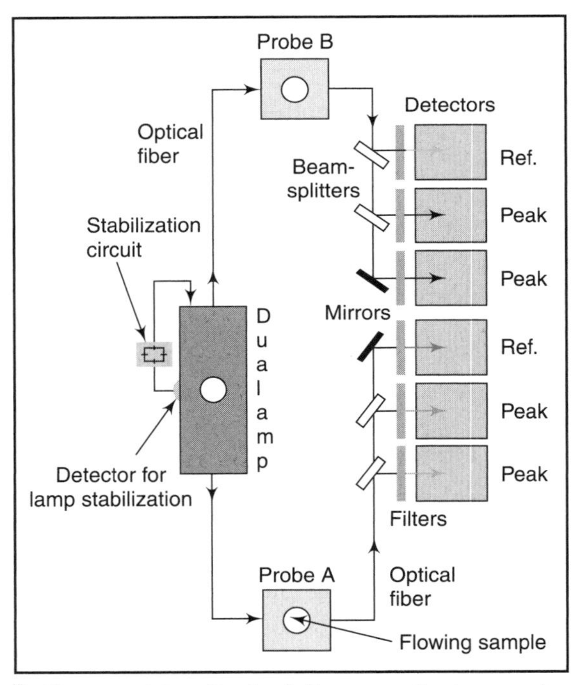

Figure 2.37 :: Optical diagram of a two probe process photometer

which are references, may be used. Any combination of NIR (In GaAs) and UV-Vis (Si) detectors can be placed in any position, providing flexibility in operation.

Voltage from each detector is logarithmically converted, and then the signals from each peak and its reference are subtracted, producing a voltage equivalent to optical absorbance units (Au). A micro-processor converts the signals into chemical units for display and transmits them as 4–20 mA signals to the process computer.

Two probe photometers have also been developed that can measure two sample probes simultaneously. A separate reference wavelength is used for each probe and upto three analyzing wavelengths may be used with one of the probes, as shown if Figure 2.37. It uses a dual lamp light source. This lamp has optics on either side of the tungsten lamp and couples light into two optical fibres, each connected to a different fibre optic probe. With the pulsed xenon source in the UV, however, light is branched into two fibres mounted side by side in the same connector. The peak and reference wavelengths are calculated independently and simultaneously for each probe by a microprocessor and displayed.

The instrument uses reference voltage (i.e. the signal strength at the reference wavelength) to identify fibre breakage, probe fouling or intermittent cloudiness due to filter breakthrough. It additionally uses an internal circuit to indicate if the lamp is about to burn out or if it has already burnt out.

Multi-probe analyzers for applications in UV-Visible (200–800 nm) and near infrared (1100–2150) using diode array detectors are also available from the same company. The arrangement is shown in Figure 2.38. Compatible with PC and using popular GRAMS spectral software to operate the system, the instrument facilitates control of the diode array settings, screen displays and referencing. The software also enables one to perform virtually any manipulation of the spectra, including derivatives, peak areas, base line corrections, smoothing, subtraction, etc.

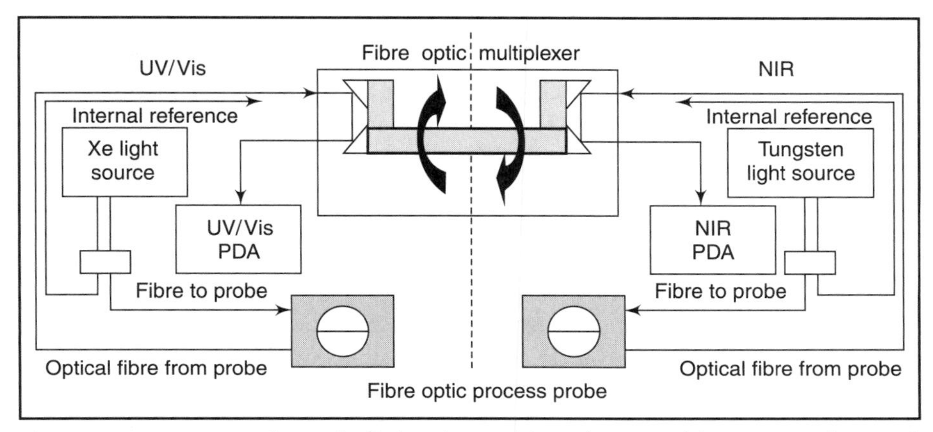

Figure 2.38 :: Operating principle of a multi-probe portable spectrophotometer

2.6 Spectrophotometers

A spectrophotometer is an instrument which isolates monochromatic radiation in a more efficient and versatile manner than colour filters used in filter photometers. In these instruments, light from the source is made into a parallel beam and passed on to a prism or diffraction grating, where the light of different wavelengths is dispersed at different angles.

The amount of light reaching the detector of a spectrophotometer is generally much smaller (Figure 2.39) than

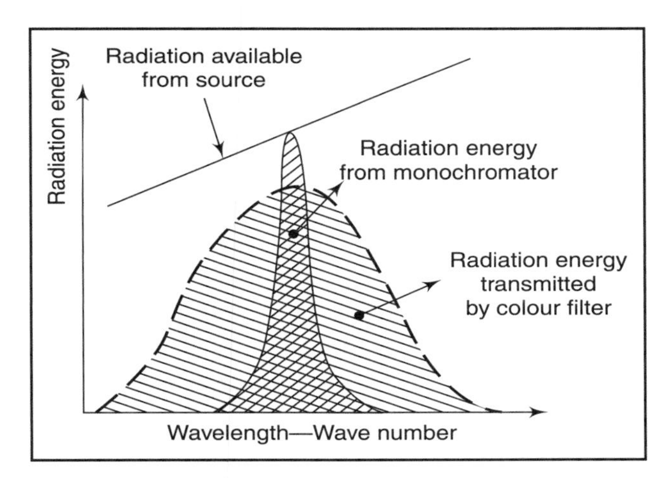

Figure 2.39 :: Comparison of radiation energy from a colour filter and monochromator

that available for a colorimeter, because of the small spectral bandwidth. Therefore, a more sensitive detector is required. The electrical signal from the photoelectric detector can be measured by using a sensitive microammeter. However, it is difficult and expensive to manufacture a meter of the required range and accuracy. In order to overcome this problem, either of the following two approaches is generally adopted:

(a) The detector signal may be measured by means of an accurate potentiometric bridge. A reverse signal is controlled by a precision potentiometer, until a sensitive galvanometer shows that it exactly balances the detector signal and no current flows through the galvanometer. This principle was adopted in the earlier Beckman Model DU single-beam spectrophotometer.

(b) The detector current is amplified electronically and displayed directly on an indicating meter or in digital form. These instruments have an advantage in terms of the speed of measurement. As in the case of colorimeters, the instrument is adjusted to give a 100 per cent transmission reading, with the reference sample in the path of the light beam. The sample is then moved into the beam and the percentage transmission is observed.

Modern commercial instruments are usually double-beam, digital reading and/or recording instruments, which can provide absorbance, concentration, per cent transmission and differential absorbance readings. It is also possible to make reaction rate studies. They can be used to include specialized techniques such as automatic sampling and batch sampling, with the addition of certain accessories. The measurements can be made generally with light at wavelengths from 340 to 700 nm, and from 190 to 700 nm with a deuterium source. In variable-slit type of instruments, the slit can be made to vary from 0.05 to 2.0 mm. The wavelength accuracy is ± 0.5 nm. The recorders are usually single-channel, strip-chart potentiometric recorders. They are calibrated from 0.1 to 2.0 A or 10 to 200 per cent T full-scale. The recorders used with spectrophotometers have adjustable wavelength scanning speeds (100, 50, 20 and 5 nm/mm) and chart speeds (10, 5,2, 1, 0.5, 0.2 and 0.1 inch/min). They have a sensitivity of 100 mV absorbance units or 100 mV/ 100 per cent T.

When scanning a narrow wavelength range, it may be adequate to use a fixed slit width. This is usually kept at 0.8 mm. In case of adjustable slit width instruments, they should be so selected that the resultant spectral slit width is approximately one-tenth of the observed bandwidth of the sample, i.e. if the absorption band is 25 nm wide at half its height, the spectral slit width should be 2.5 nm. This means that the slit width set on the instrument should be 1.0 mm. This is calculated from the dispersion data, as the actual dispersion in grating instruments is approximately 2.5 nm/mm slit width.

Spectrophotometers generally employ a 6 V tungsten lamp, which emits radiation in the wavelength region of visible light. Typically, it is 32 candle power. These lamps should preferably be operated at a potential of say 5.4 V, when its useful life is estimated at 1200 h. The life is markedly decreased by an increase in the operating voltage. With time, the evaporation of tungsten produces a deposit on the inner surface of the tungsten lamp and reduces the emission of energy. Dark areas on the bulb indicate this condition. It should then be replaced.

The useful operating life of the deuterium lamp normally exceeds 500 hours under normal conditions. The end of the useful life of this lamp is indicated by its failure to start or by a rapidly decreasing energy output. Iionization may occur inside the anode rather than in a concentrated path in front of the window. Generally, this occurs when the lamp is turned on while still hot from a previous operation. If this occurs, the lamp must be turned off and allowed to cool before restarting.

Spectrophotometers should be placed in an area which is reasonably free of dust and excessive moisture, and not subject to significant temperature variations. As they are sensitive instruments, their performance is likely to be affected by strong electromagnetic fields, as would exist in proximity to diathermic machines or large electric motors. Disturbances of this nature should be avoided when determining location. The surface on which the instrument is to be placed must be stable and free from vibrations.

Harris (1982) states that instrumental design for quantitative analysis by solution spectrophotometry has remained conceptually stable since the early part of the last century. However, the advent of stable electronics has spawned a group of techniques for measuring absorbance that provides a quantum leap in spectrophotometric sensitivity.

2.6.1 Single-beam Null-type Spectrophotometers

A typical example of the single-beam optical null-type spectrophotometer is that of the Beckman Model DU. Although this is an old instrume ... explained to illustrate the principle of this once popular instrument.

The schematic diagram of the optical s
Light from the source *S* is focused on the c
in a beam to the 45° slit-entrance mirror *C*
slit *D* on to the collimating mirror *E*. Lig
and reflected on the prism *F*, where
undergoes refraction. The back surface
the prism is aluminized. The light reflec
at the first surface is reflected back thro
the prism, undergoing further refractic
it emerges. This is called Littrow mou
and employs only one piece of quartz
the back surface of the prism meta
Since the light passes back and
through the same prism and le
polarization effects are eliminated.
mounting results in a compact ins

The desired wavelength of
selected by rotating the prism m
the table on which the prism is m

slowly rotated by the wavelength drum, the prism provides a series of images of the entrance slit at the exit slit, as the spectrum is swept past. In order to achieve perfect correspondence of slit widths, the upper and lower parts of the same slits are generally used as entrance and exit slits.

The spectrum from the prism is directed back to the collimating mirror. The light passes through the sample and finally reaches the photodetector *D*. The resulting current passing through the load resistor develops the voltage, which is amplified in a dc amplifier using high input impedance electronic circuit. The amplifier circuit is of null-type, to provide absorbance or transmittance measurement. The electrometer plate current is indicated on a 1 mA milliammeter.

The slide wire of potentiometer, which is used to obtain 'null' reading, is calibrated in absorbance and transmittance units.

In such type of instruments, two interchangeable phototubes are employed, which are mounted on a sliding carriage. For the visible and near ultraviolet range (220–625 nm), a blue sensitive tube is used. It has a cesium–antimony photosensitive surface and an insert of fused silica in the envelope. In the range 625–1000 nm, a red sensitive phototube with cesium–oxide coated photocathode, is utilized. When photomultipliers are used for increased sensitivity or for operation at smaller slit widths, they usually replace the blue sensitive tube. The slit width is often marked in millimeters to serve as a performance check and to allow resetting of a previous slit opening.

The load resistance in such type of circuits is very high (20 to 200 MΩ), so that the circuit is sensitive to extremely small currents of the order of 10^{-13} A. Therefore, the load resistance and detectors are housed in a compartment, which is kept moisture-free by putting silica gel crystals in a bag. In the presence of moisture, leakage current may develop and result in a drift of the null-balancing circuit. Silica gel crystals should be dried frequently and replaced to overcome this problem.

2.6.2 Direct Reading Spectrocolorimeters/Spectrophotometers

-balance
truments

Direct reading instruments offer greater speed in operation and convenience. However, they have lower accuracy and precision than the null-balance type instruments. Figure 2.41 (a) shows the schematic of a direct reading absorption meter, (Model SP 1400 of Unicam). The prism, in this instrument, is of ith an aluminized rear surface. Wavelength selection is done by rotating the The entrance and exit slits are each of 0.25 mm. The detector is a standard output is developed across a 250 MΩ resistor and applied to a single r. The output from the photocell is applied to a sensitive galvanometer. are provided for direct readings in percentage transmission or

irect reading instrument of using a reflection grating is makes use of a grating monochromator and can be used rcury vapour lamp. A colour filter is used to block e grating.

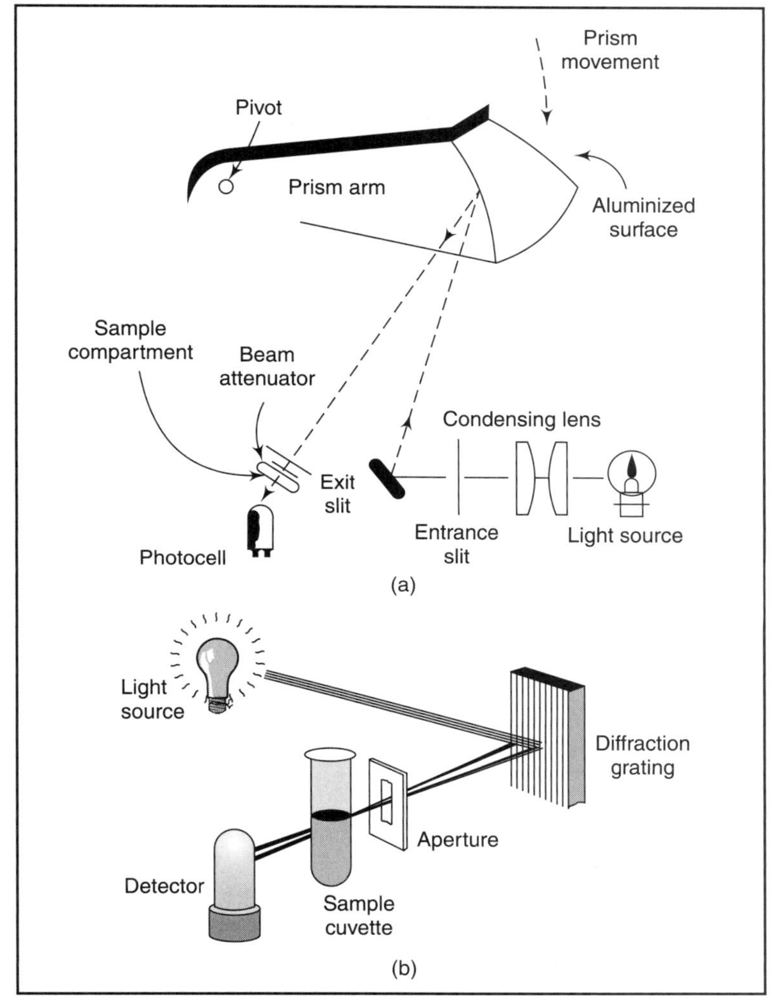

Figure 2.41 :: (a) Typical optical diagram of spectrocolorimeter using prism monochromator (b) Optical diagram of a spectrocolorimeter using diffraction grating as a monochromator

Spectronic 20 Spectrocolometer/Spectrophotometer

Bausch and Lomb Spectronic 20 is a direct reading grating spectrophotometer/spectrocolorimeter. In this instrument, the normal range of operation is 350 to 650 nm and can be extended up to 900 nm by the use of a red sensitive phototube. The monochromator comprises a reflection grating, lenses and a pair of fixed slits (Figure 2.42).

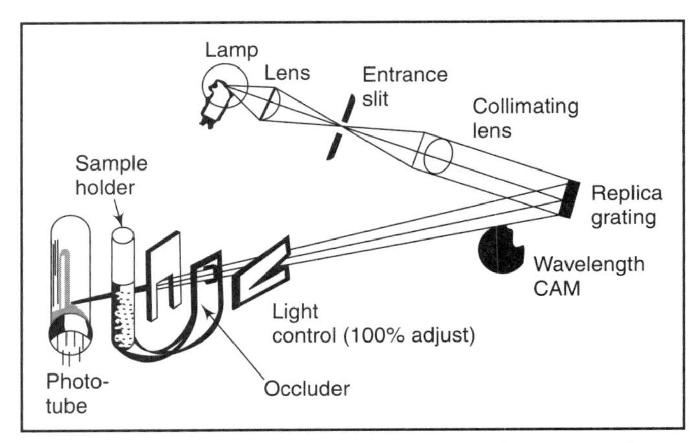

Figure 2.42 :: Optical arrangement of Spectronic-20 spectrocolorimeter

The standard Spectronic 20 phototube is a type S-4 cesium–antimony tube, which is most sensitive to light in the blue region. A red sensitive phototube is used above 625 nm.

The curved arm of metal at the bottom of the sample chamber is the occluder, which when mounted on a pivot, opens and closes the shutter gate which allows light into the sample. It is automatically pushed aside by the cuvette when a reading is taken.

Continuous light over the entire visible spectrum is emitted from the tungsten lamp. This light is collected by the field lens and collimated (made parallel) by objective lens. The white light next falls upon the diffraction grating, which disperses it horizontally into the familiar sequence of spectral colours. The grating is ruled at 600 lines per millimeter. The dispersed light next passes through the light control, the occluder and the exit slit. Only when a cuvette is inserted into the sample chamber is the occluder moved out of the way. The width of the exit slit and the spread or dispersion of the light from the diffraction grating together determine the spectral bandwidth, which is 20 nm. Thus, for example, if the instrument is set at 600 nm, light with wavelengths in the range 590 to 610 nm will be passed to the sample and phototube.

Only that portion of the light dispersed by the grating, which falls on the exit slit, is passed to the sample. The selection of the desired portion of the spectrum to be passed is accomplished by adjusting the angle of incidence between the source ray and the diffraction grating. The control on the instrument performs this task by physically turning the grating and is known as the wavelength control knob. Attached to this control is a dial calibrated in nanometers, which indicates the setting.

As the grating produces a dispersion that is independent of the wavelength, a constant bandwidth of the order of 20 μm can be obtained throughout the operating range. The electrical components include the power supply, the amplifier which strengthens signals coming from the phototube, and the read-out device which, in this case, is a voltmeter calibrated in both absorbance and per cent transmittance units. In the earlier versions of this instrument, a differential amplifier was used to amplify the photocurrent from the detector. Since the amplifier current is proportional to the radiant power, the scale of the meter can be calibrated to read transmittance and absorbance. The amplifier is so constructed that the current through the meter is zero under no signal (dark current) conditions. With the detector input fed to the amplifier, the unbalance current as indicated in the meter, is proportional to the radiant energy falling on the detector tube. The design of the amplifier is such that electrical fluctuations get cancelled out. The wavelength scale is linear and is coupled to the grating with a sine-bar drive.

Quantitative spectrophotometry

In quantitative spectrophotometry, one is interested not in the absolute intensity of light passing through a sample, but rather in the relative intensity of such light with respect to the intensity of light passing through a reference or blank solution. It is thus necessary to set the 0 and 100 per cent transmittance limits between which the transmittance of samples will be measured. The 0 per cent T is set in such a way that no light reaches the phototube, that is, with no cuvette in the sample chamber and with all light blocked by the occluder. This setting is performed with the amplifier control on the instrument. This control adjusts the gain, or sensitivity of the amplifier, and the offset. This determines how much needle deflection is caused by a specific intensity of light striking the phototube. The 100 per cent T is set with the light control, usually with a cuvette of water or another blank solution in the sample chamber. Adjusting the light control knob causes a V-shaped slit behind the knob to move into or out of the light beam. It is thus a purely mechanical control, which simply physically blocks out more or less of the diffracted light.

Spectronic 21

Spectronic 21 is an advanced version of a single-beam spectrophotometer from Bausch and Lomb. The instrument is designed to provide a wide wavelength range (200–1000 nm), a narrow spectral width (10 nm) and more accuracy due to very low stray radiant energy. Figure 2.43 shows the optical diagram of the Spectronic 21. The system has two lamps mounted in an exterior lamp house. Light from the proper lamp is selected by appropriate positioning of the lamp interchange mirror, either to pass the deuterium light or to reflect the white light from the tungsten lamp. A silica relay lens system focuses on the light near the chopper. The light is chopped and focused by a condenser lens onto the entrance slit of the monochromator after passing through the appropriate filter. The optical arrangement used in the monochromator is called a folded crossover Czerny–Turner configuration, which provides higher quality optical characteristics over a wider wavelength range. After passing through the slit, the light is incident on a collimating mirror, which collimates the light onto the diffraction grating. The light dispersed from the grating is picked up by the focusing mirror, which focuses the light onto the exit slit. The exiting light then passes through the sample and is incident on the photodiode.

An image of the radiation source is produced on the entrance slit via the rigidly built-in condenser and the mirror. The light converted into parallel rays by a collimator objective strikes

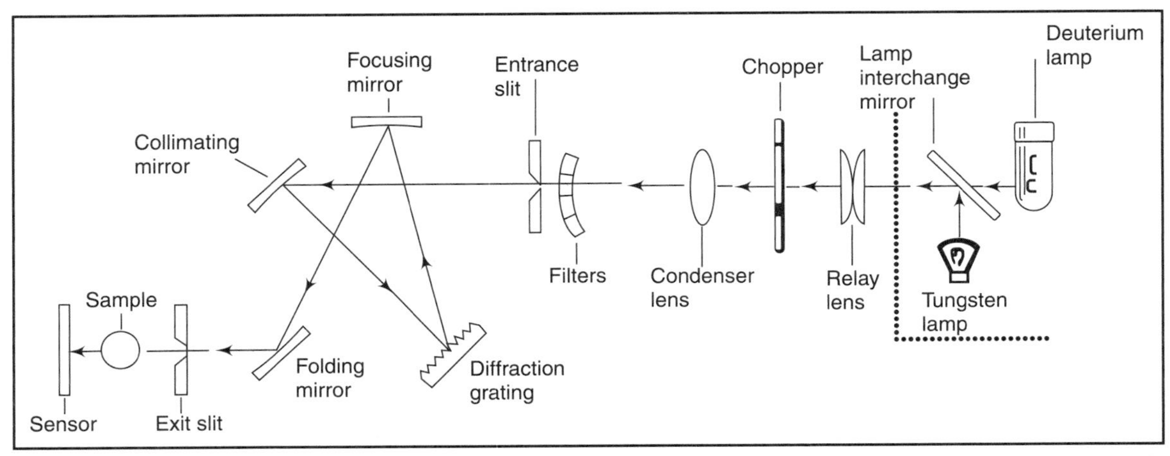

Figure 2.43 :: Optical diagram of the SPECTRONIC 21 Spectrophotometer (Courtesy Bausch and Lomb, USA)

the diffraction grating and is spectrally dispersed by the achromat. The diffracted light is focused in the plane of the exit slit. Monochromatic radiation separated from the spectrum through the slit passes the measurement sample and strikes the radiation detector. The photocurrent produced is amplified in the amplifier and fed to the indicating instrument. The wavelength is set by turning the grating by means of the drum provided with 1 nm divisions in the range from 330 to 850 nm.

Signal Processing in Direct Reading Spectrophotometers

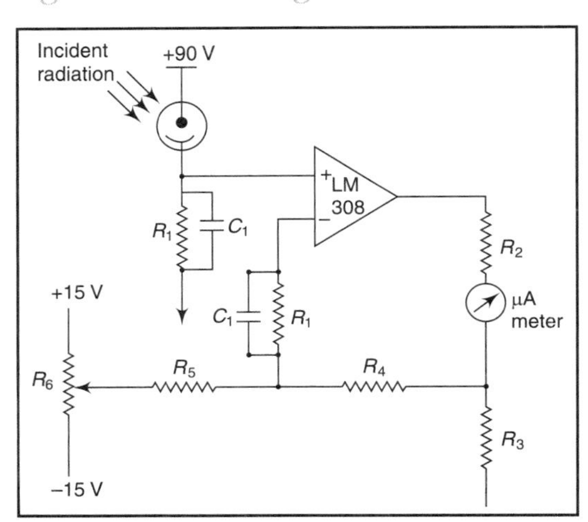

Figure 2.44 :: Typical circuit used with direct reading spectrocolorimeter/spectrophotometer

In single-beam direct reading instruments, operational amplifiers are used for handling signals from the detectors, which could be a vacuum phototube, photodiode, photomultiplier or photocell.

Figure 2.44 shows a typical circuit diagram for amplification and measurement of current from the phototube. The phototube anode is given 90 V from a stabilized power supply and its cathode is connected to ground through a high resistance R_1. Operational amplifier A_1 is of very high input impedance, low drift and low noise. The op-amp offers a very high input impedance, as the input is connected to the non-inverting input. The inverting input terminal is connected to a variable supply voltage (±15 V). This control is provided on the front panel to set zero, initially when light is blocked from falling on the phototube. This control not only balances the dark current, but also nullifies the offset voltage of the op-amp.

The feedback resistance uses a T-resistor network instead of a single resistor. This arrangement gives a very high effective feedback resistance and at the same time, facilitates the use of resistors of lower values, which at the required precision of 0.1 per cent are substantially cheaper and stable. The output of the amplifier gives linear readings on the transmittance scale. If absorbance is to be displayed, the output of the transmittance stage is given to a logarithmic converter, whose output is equal to the minus log of transmittance.

$$A = -\log_{10} T$$

The logarithmic modules are available in encapsulated form and can be directly used.

Currents from photomultipliers are usually very small. Therefore, considerable amplification is necessary before the current can be suitably recorded. This is done by using an operational amplifier having a field effect transistor in the input stage and using a high value of the feedback resistance R_f (Figure 2.45). As the input to the amplifier is connected by a shielded cable, there exists a stray or parasitic capacitance at the input terminals. It necessitates the use of capacitance C_f in the feedback to ensure amplifier stability. The capacitance also makes the bandwidth narrower, which significantly improves the signal-to-noise ratio. The non-inverting amplifier input terminal is connected to a suitable voltage to compensate for the dark current.

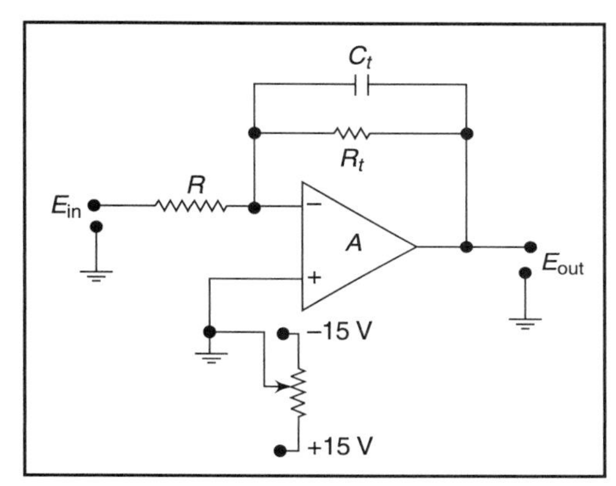

Figure 2.45 :: Amplifier circuit for use with a photomultiplier

Spectrophotometer Using Diode Array Detector

The optical system of the spectrophotometer Model 8453 from M/s Agilent is shown in Figure 2.46. Its radiation source is a combination of a deuterium discharge lamp for the ultraviolet (UV) wavelength range and a tungsten lamp for the visible and short wave near-infrared wavelength range. The image of the filament of the tungsten lamp is focused on the discharge aperture of the deuterium lamp by means of a special rear-access lamp design, which allows both light sources to be optically combined and to share a common axis to the source lens. The source lens forms a single, collimated beam of light. The beam passes through the shutter/stray-light correction filter area, then through the sample of the spectrograph lens and slit. In the spectrograph, light is dispersed onto the diode array by a holographic grating. This allows simultaneous access to all wavelength information. The result is a fundamental increase in the rate at which spectra can be acquired.

Spectrograph

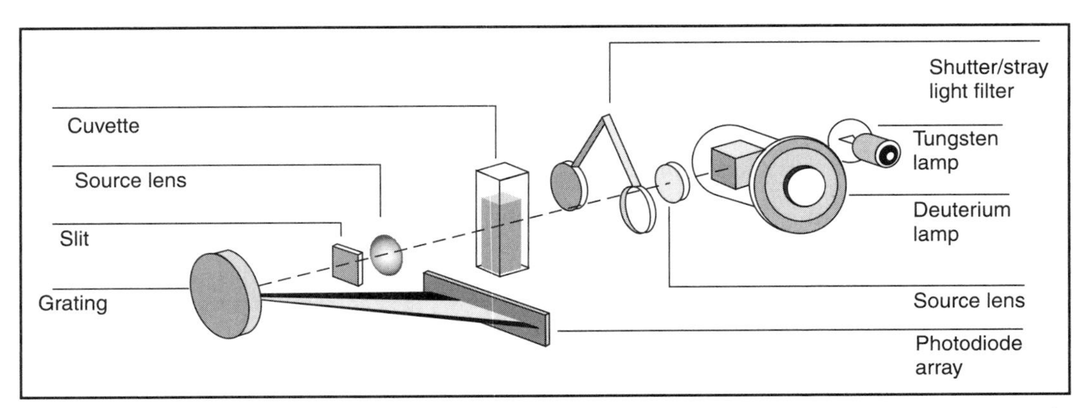

Figure 2.46 :: Optical system of a spectrophotometer using diode array (Courtesy M/s Agilent)

As a result of plasma discharge in a low pressure deuterium gas, the deuterium lamp emits light over the wavelength range of 190 nm to approximately 800 nm. Similarly, the low-noise tungsten lamp emits light over the 370 nm to 1100 nm wavelength range. The source lens receives the light from both the lamps and collimates it. The collimated beam passes through the sample in the sample compartment.

The shutter is electromechanically actuated. It opens and allows light to pass through the sample for measurements. Between sample measurements, it closes to limit exposure of the sample to light. If the measurement rate is very fast, the shutter can be commanded to remain open (Agilent ChemStation software) or it stays open automatically.

In a standard measurement sequence, reference or sample intensity spectra are measured without and then with the stray-light filter in the light beam. Without the filter, the intensity spectrum over the whole wavelength range from 190-1100 nm is measured. The stray-light filter is a blocking filter with 50 per cent blocking at 420 nm. With this filter in place, any light measured below 400 nm is stray light. This stray-light intensity is then subtracted from the first spectrum to give a stray-light corrected spectrum. Depending upon the software, you can either switch off the stray-light correction in case you want to do very fast repetitive scans or it is switched off automatically.

The spectrophotometer has an open sample compartment for easier access to sample cells. Because of the optical design, a cover for the sample area is not required. The spectrophotometer is supplied with a single-cell holder already installed in the sample compartment. This can be replaced with the Peltier temperature control accessory, the thermostattable cell holder, the adjustable cell holder, the long path cell holder or the multi-cell transport. All these optional cell holders mount in the sample compartment using the same quick, simple mounting system. An optical filter wheel is also available for use with the spectrophotometer and most of the accessories.

The spectrograph housing material is of ceramic in order to reduce thermal effects to a minimum. The main components of the spectrograph are the lens, the slit, the grating and the photodiode array with front-end electronics. The mean sampling interval of the diode array is 0.9 nm over the wavelength range 190 nm to 1100 nm. The nominal spectral slit width is 1 nm. The spectrograph lens refocuses the collimated light beam after it has passed through the sample.

The slit is a narrow aperture in a plate located at the focus of the spectrograph lens. It is exactly the size of one of the photodiodes in the photodiode array. By limiting the size of the incoming light beam, the slit makes sure that each band of wavelengths is projected only on to the appropriate photodiode. The combination of dispersion and spectral imaging is accomplished by using a concave holographic grating. The grating disperses the light onto the diode array at an angle proportional to the wavelength.

The photodiode array is the heart of the spectrograph. It is a series of 1024 individual photodiodes and control circuits etched onto a semiconductor chip. With a wavelength range from 190 nm to 1100 nm, the sampling interval is a nominal 0.9 nm. The electrical signal from the diode arrays is then amplified and displayed by the system electronics.

Absorbance is a ratiometric measurement, which requires the measurement of the ratio of the intensity of light transmitted through the sample to the intensity of light transmitted through a 'blank'. With a dual beam instrument, this ratio is measured directly as the user instals both a blank and a sample into the instrument. Diode array instruments are single-beam instruments, and the blank and sample must be measured sequentially. In case of the HP 8453 photodiode array spectrophotometer, when the blank is measured, it is stored in a register which identifies it as the 'blank'. The most recently measured 'blank' is always used to calculate the absorbance spectrum of all subsequently measured samples.

Blank

2.6.3 Double-beam Ratio-recording Spectrophotometers

Most modern spectrophotometers are double-beam instruments. These instruments are configured to allow the automatic and simultaneous (or near simultaneous) measurement of the sample and reference beam intensities. This can be accomplished by using a beam splitter to divide the beam into the reference beam and sample beam. In the double-beam-in-space configuration, the intensities of the split beams are measured simultaneously after passing through the sample and reference cells. Alternately, in a double-beam-in-time spectrophotometer, the beams pass through a modulator, which allows the detector to see either the reference beam or the sample beam. Since the source beam is directed through a reference cell part of the time and through the sample cell the rest of the time, the term 'double-beam-in-time' is used.

The main disadvantage of single-beam spectrophotometers is that the instrument settings have to be adjusted to give a 100 per cent reading with the reference in the beam, before the sample is examined. This drawback is overcome by using a double-beam instrument, wherein the arrangement is such that the radiation beam is shifted automatically to pass alternately through the sample and reference cuvettes. The cuvettes themselves are not shifted and remain in their fixed position. There are several ways in which this can be achieved. Most commonly, a single rotating sector mirror is used. Monochromators used in these instruments are similar than those used in single-beam instruments.

Double-beam instruments are generally of the recording type. This facility allows for the rapid and accurate reproduction of spectrograms. The instrument automatically compares the sample beam energy with reference beam energy. The ratio of the two would be the transmittance of the sample. This procedure is followed over a sequence of wavelengths. A graph is plotted with transmittance (absorbance) as ordinate and wavelength as abscissa, giving the absorption spectrum of the sample under analysis.

In the prism type instruments, the Littrow monochromator disperses the radiation of the deuterium or tungsten lamp. The monochromatic light produced in this manner is modulated at 400 Hz and alternately led via a rotary mirror through the measuring and reference cells into a photomultiplier, with multi-alkali cathode. The photomultiplier produces a 400 Hz ac voltage, the amplitude of which periodically oscillates between the measuring and reference value. After amplification, the signal is rectified in a phase-sensitive detector, so that two dc voltages are produced that are proportional to the measuring and reference radiation. The quotient of these voltages is measured with an automatically balancing potentiometer and transferred to the recorder. A linear or logarithmical potentiometer permits recording of transmission or extinction. A slit program controls the monochromator slits, so that there is approximately the same signal at the photomultiplier.

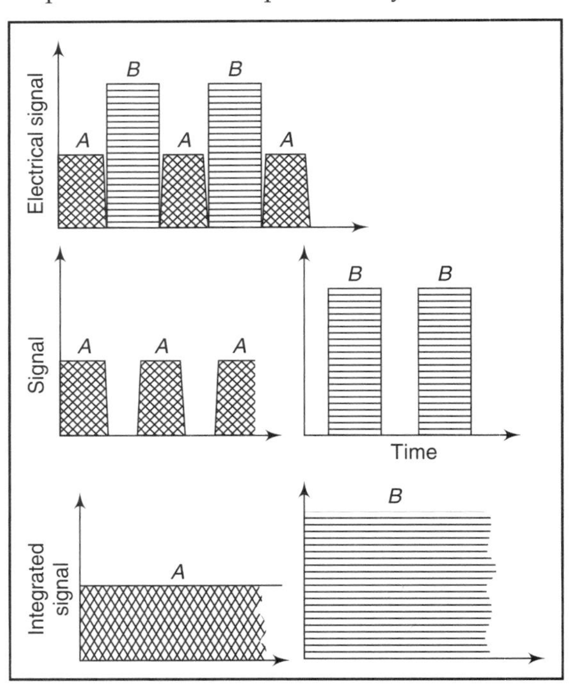

Figure 2.47 :: Principle of ratio-recording spectrophotometric system

Figure 2.47 shows the signals in a ratio recording double-beam spectrophotometer. The signals are in the form of pulses, which are directly proportional to intensities of radiation passing through the sample and reference. These two signals are resolved electronically into two dc voltages corresponding to the sample and reference beams, and the latter is used as the standardizing potential on a potentiometric recorder. Double-beam instruments are also constructed on the optical-null photometer system (Figure 2.48). In these instruments, the

Optical-null photometer

alternating signal from the reference and sample is not separated into its components, but is used directly to drive a servo-motor. The motor drives an optical attenuator or wedge into or out of the reference beam of the spectrophotometer, until it has the same intensity as the sample beam. The electrical signals produced at the detector from the reference and sample sides are equal and there is no signal to drive the servo-motor and it stops. The optical attenuator arrangement is such that the distance from which it is driven into the reference beam is proportional to either the transmittance or absorbance. Therefore, the distance it has moved when the signals become equal and the motor stops, is proportional to the relative intensities of the sample arid reference beams. The absorbance or transmittance can be recorded by mechanically coupling the pen of a recorder to the attenuator.

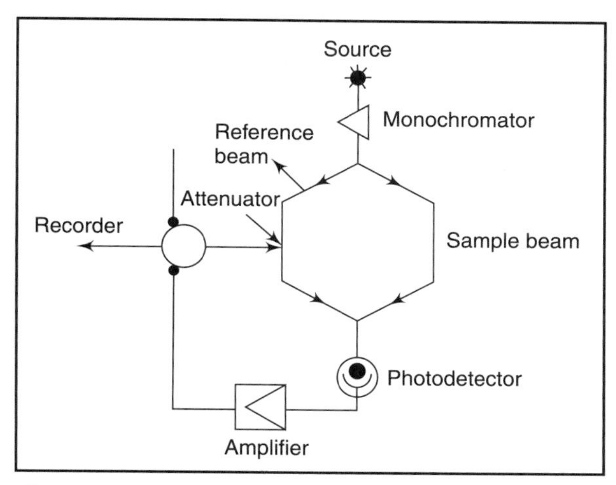

Figure 2.48 :: Principle of optical-null spectrophotometric system

Figure 2.49 shows the optical arrangement of the double-beam spectrophotometer from M/a Varian. Either tungsten or a deuterium lamp can be switched in the system to cover an operating range from 200 to 700 nm, to cover the ultraviolet and visible regions. The monochromator comprises a diffraction grating and two narrow slits, one of which serves as an entrance, and the other as an exit. It is possible to set the spectral bandwidth at 2, 4, 8 or 16 nm. The larger bandwidths are used for increased light output, while the smaller bandwidth is used where spectral resolution is of importance. A four-segment filter automatically performs order sorting and rejects stray light from the monochromator, holding it to 0.1 per cent at 340 nm, giving low noise over the entire wavelength range.

The instrument chopper is so synchronized that light passes alternately through the sample and the reference cells. In this arrangement, the entire light passes through a given cell when the transmission of that cell is being measured. This offers a distinct advantage over beam splitter arrangements, where only 50 per cent of the available light passes through each cell during measurement.

Figure 2.50 shows the arrangement employed in measurement of transmittance in a double-beam instrument. The measuring process basically consists of comparing the photocurrents from two photocells or phototubes. The input circuits are current followers and their output signals are compared in a differential amplifier. If it is required to measure absorbance A, the output signals

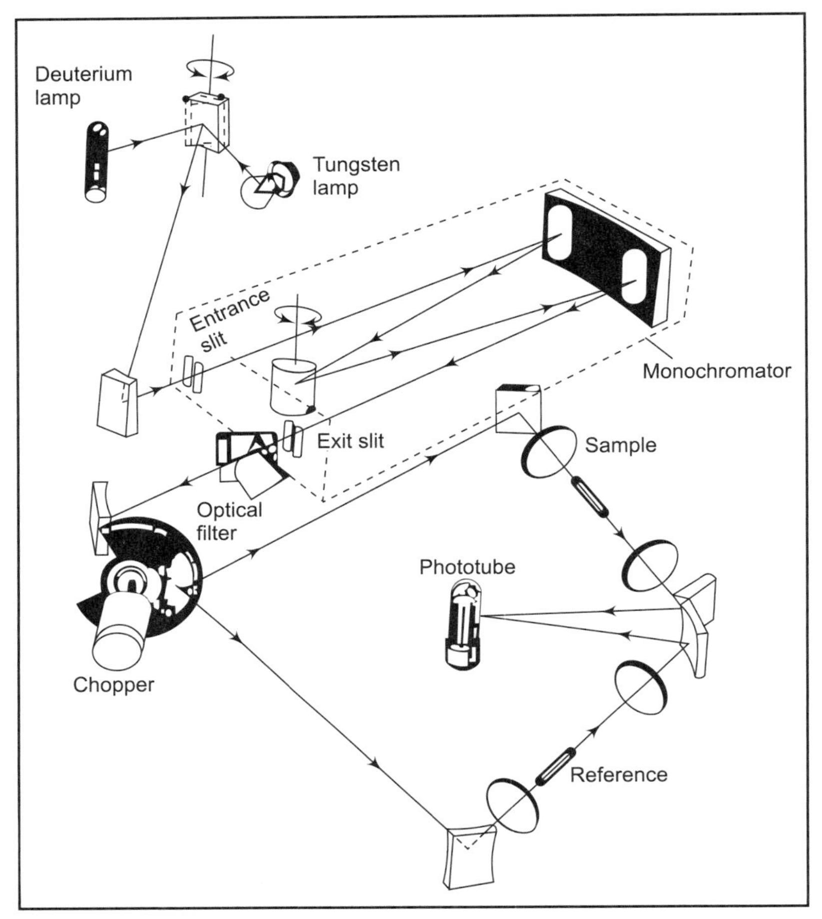

Figure 2.49 :: Optical arrangement of a double-beam spectrophotometer (Courtesy M/S Varian, USA)

Logarithmic amplifiers

from a current follower are given to logarithmic amplifiers. The logarithms of the signals are subtracted in a differential amplifier. The output of the amplifier is given by

$$E_0 = K \cdot \log I_o/I = KA,$$

where K is constant and I_0 and I are the incident and transmitted light intensities respectively. The circuit used for this purpose is shown in Figure 2.51.

This circuit enables us to take the logarithm of a ratio of the two voltages or currents, equalling the difference of the logarithms of the input signals. The outputs of the logarithmic amplifiers are brought to the inputs of a differential amplifier. This circuit is used mainly in spectrophotometric measurements, when the logarithm of the ratio of the radiant power of the incident radiation to that of the radiation after passage through the absorbing layer is determined. In the circuits

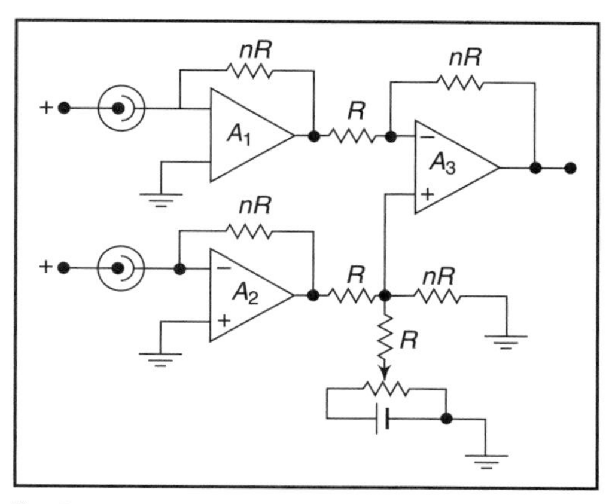

Figure 2.50 :: Basic measuring circuit in double-beam spectrophotometer

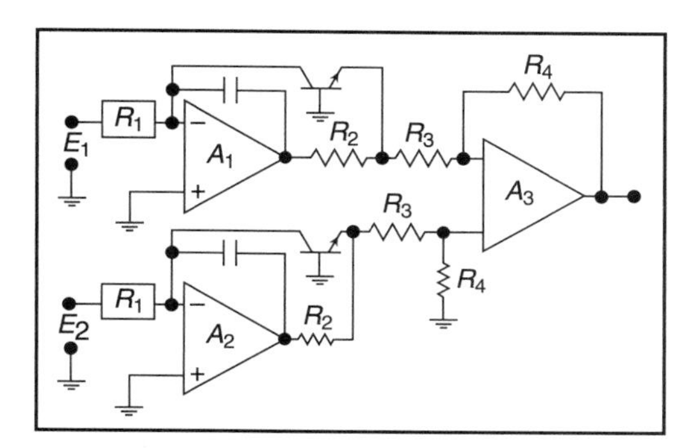

Figure 2.51 :: Circuit for measuring logarithm of a ratio of two currents or voltages used in double-beam spectrophotometers

described above, both transistors should have identical properties, and matched temperature-dependent resistors should be employed. Integrated circuit logarithmic amplifiers are used in most of the instruments.

The optical system as shown in Figure 2.52 is that of Model 24-25 from M/s Beckman Instruments double-beam instruments. Energy of the appropriate wavelengths is produced by the appropriate source lamp. This energy is converted to monochromatic light by using a filter-grating optical system. The grating has 1200 lines/mm and is blazed at 250 nm. The filters are necessary to

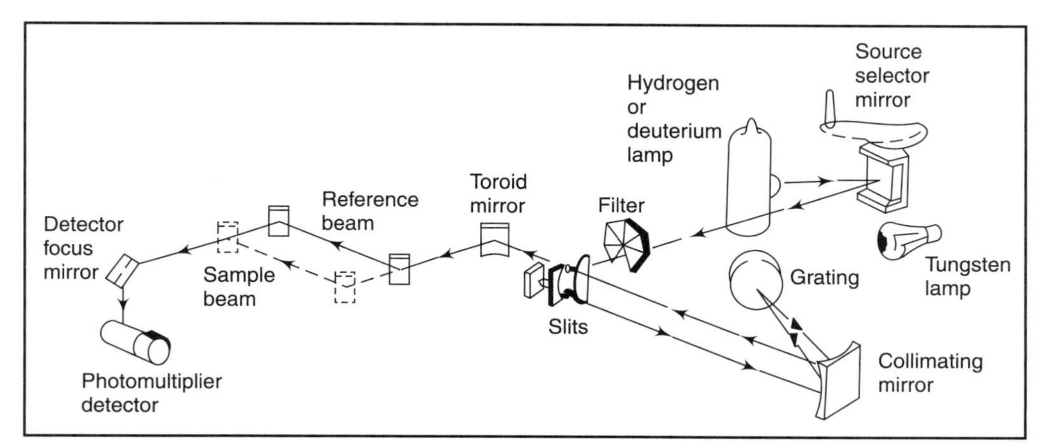

Figure 2.52 :: Optical arrangement of Model 24-25 Beckman double-beam spectrophotometer

eliminate unwanted orders from the grating. Six different filters cover the wavelength regions from 300 to 900 nm.

The filter wheel is driven by a dc motor, which is synchronized with the wavelength cam. As the wavelength cam moves, it causes the filter motor to drive till the correct filter comes into position. The wavelength cam drives the wavelength arm, which, in turn, causes the grating to pivot on its own axis, thereby causing the wavelength of light coming out of the monochromator to change. The monochromatic light is then directed to the sample and reference via a vibrating mirror bridge, which vibrates horizontally at a certain frequency. This bridge allows light to pass into the sample and reference cell holders alternately, with a frequency equal to the displacement frequency of the bridge. The vibrating bridge is controlled by the bridge drive circuitry. The reference and sample pulse train is then passed to the photomultiplier tube, which converts the monochromatic light pulses to current pulses.

Vibrating bridge

Figure 2.53 is a block diagram of the electronic part of the instrument. As the vibrating bridge chops light energy coming from the monochromator at 35 Hz, the output of the PMT is an ac signal. This signal is passed on to the pre-amplifier where it is amplified. The input of the pre-amplifier is a FET, which offers a high input impedance to the signal. The amplified signal is then given to a demodulator, which separates reference pulses from sample pulses and converts sample and reference pulses to a dc potential. However, the demodulation process requires synchronization, so that when the bridge directs light through the sample path, the electronics demodulates the sample pulse, and the same holds true for the reference pulse. In order to achieve synchronization, the same signal that is used to drive the coil for the vibrating bridge is tapped off and used as the input to the demodulator.

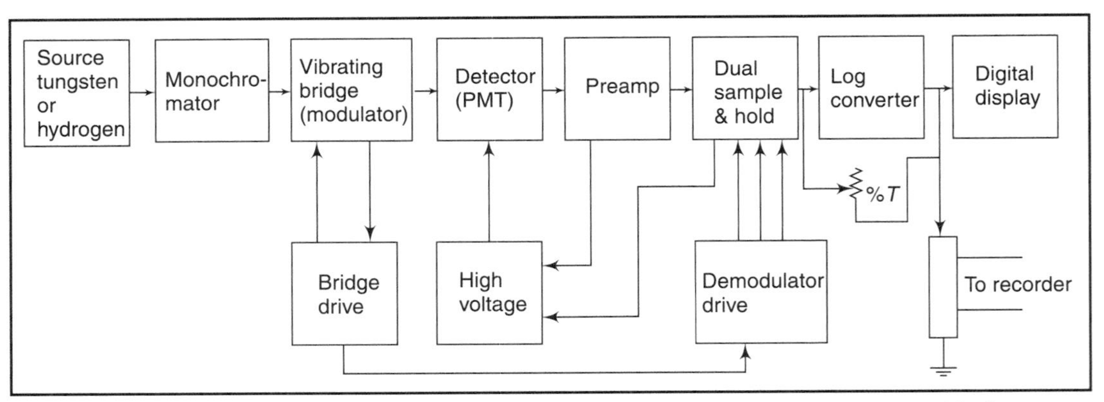

Figure 2.53 :: Block diagram of electronic part of Model 24-25 Beckman double-beam spectrophotometer

The dc output of the reference side is also fed back to the high voltage power supply. If the reference signal should decrease, the high voltage would increase, thus restoring the reference channel to a constant potential. This would ensure constant energy through the reference channel during the scan.

For making absorbance measurements, the dc potential from the sample and hold amplifier is passed on to the log converter. Differential absorbance can be measured between – 0.3 A to +0.7 A by switching in the bucking potential at the output stage. The analog dc potential is converted into binary coded decimal and this information is then displayed as an absorbance or concentration value on the digital display. The output of the log converter is applied to a divider network, which, in turn, drives an external recorder.

2.6.4 Microprocessor-based Spectrophotometers

Computers have long been used in spectrophotometry, especially for on-line or off-line data processing. Since the advent of the microprocessor, their application has not been limited to processing of data from analytical instruments, but has been extended to the control of instrument functions and digital signal processing, which had been performed conventionally by analog circuits. This has resulted in improved performance, operability and reliability over purely analog instruments.

In a spectrophotometer, a microprocessor can be used for the following functions:

- *Control functions:* Wavelength scanning, automatic light source selection, control of slit width and detector sensitivity, etc.
- *Signal processing functions:* Baseline correction, signal smoothing, calculation of % *T*, absorbance and concentration, derivative, etc.
- *Communication functions:* Keyboard entry, menu-driven operation, data presentation, warning display, communication with external systems. etc.

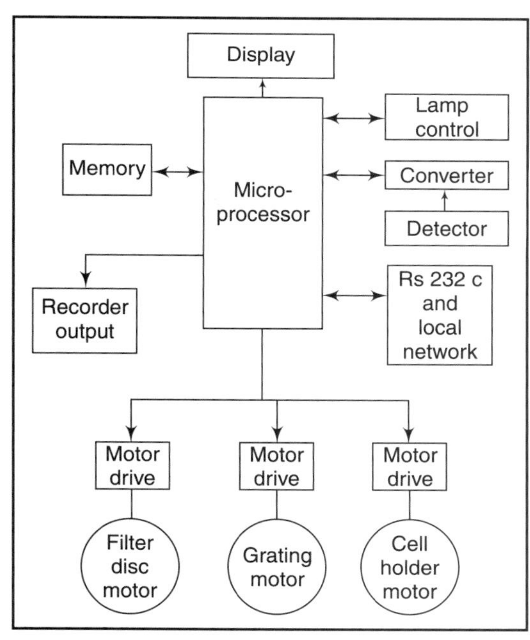

Figure 2.54 :: Block diagram of a microprocessor-controlled spectrophotometer

Figure 2.54 shows the block diagram of a microprocessor-controlled spectrophotometer. The diagram shows only the post-detector electronic handling and drive systems, all controlled via a single microprocessor. Once the operator defines such parameters as wavelength, output mode and relevant computing factors, the system automatically ensures the correct and optimum combination of all the system variables. Selection of the source and detector are automatically determined; any filters introduced at appropriate points, and sample and reference cells are correctly managed in the sample area. Output in the desired form (transmittance, absorbance, concentration, etc.) is presented along with the sample identification. Secondary routines such as wavelength calibration and self-tests are available on demand.

For wavelength scanning, a stepper motor is used, which ensures accurate and fast scanning. The automatic selection of samples is also made with a motor-driven system under the control of a microprocessor.

The signal from the photodetector is amplified in a pre-amplifier and converted into digital form in an A-D converter. The signals are differentiated into sample signal S, reference signal R and zero signal Z and stored in the memory. From these values, the microprocessor calculates the transmittance $T = (S - Z)/(R - Z)$ and absorbance $= -\log T$. In order to obtain R or S values within a specified range, the microprocessor provides control signals for slit width and high voltage for photomultiplier.

Baseline compensation

Baseline compensation due to solvent and optical unmatching of cells, which is difficult to achieve with conventional instruments, is conveniently possible in microprocessor-based systems. Improvements have also been achieved in such functions as auto-zero, expanding and contracting of the photometric scale, automatic setting of wavelength as well as in ensuring repeatable and more accurate results.

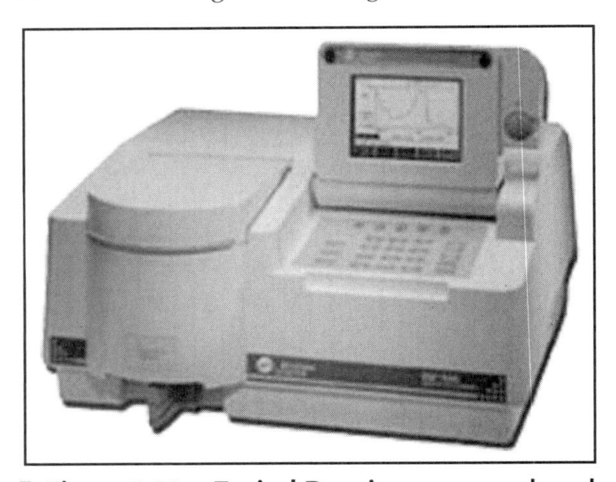

Figure 2.55 :: Typical Du microprocessor based spectrophotometer (Courtesy Beckman Coulter Instruments)

The digital output from the microprocessor is converted into analog form with a D-A converter and given to an X-Y recorder as a Y-axis signal, whereas the wavelength forms the X-axis, to obtain absorption or reflected spectra. Microprocessors have also helped in making such measurements as higher order derivative spectra and high speed sampling and storage of fast reaction processes, and for presenting processed data during and after the completion of the reaction. Figure 2.55 shows a Du series UV/VIS spectrophotometer from M/s Beckman Coulter.

Perkin-Elmer LAMBDA 9 Double-beam Spectrophotometer

Perkin Elmer LAMBDA 9 is a microcomputer-based double-beam UV-Vis–NIR (near-infrared) spectrophotometer. It incorporates a video display, a keyboard and a printer/plotter for recording of spectra and data. This instrument covers the wavelength range from 3200 nm to 185 nm. A photomultiplier is used as a detector in the UV-Vis range, whereas a PbS (lead sulphide) cell takes over in the NIR range. The detector changes at 860.8 nm. An aligned deuterium lamp for the UV range and a halogen lamp for the visible and the NIR range serves as light source. The source change takes place automatically at 319.2 nm. The double monochromators in the Littrow arrangement are each provided with one holographic grating with 1440 lines for the UV-Vis range and one holographic grating with 360 lines for the NIR range. In the UV-Vis range, the spectral bandwidth can be adjusted from 0.05 to 5 nm and on the NIR range from 0.1 to 20 nm.

Figure 2.56 shows the optical arrangement of the instrument. For operation in the near-infrared and visible ranges, the source mirror *M*1 reflects the radiation from the halogen lamp onto mirror *M*2. At the same time, it blocks radiation from the deuterium lamp. For operation in the ultraviolet range, the source mirror *M*1 is raised to permit radiation from the deuterium lamp to the strike mirror *M*2. The source change is automatic during monochromator scanning.

From mirror *M*4, the radiation is reflected through the entrance slit of monochromator 1. The radiation is collimated at mirror *M*5 and reflected to one of the gratings depending upon the wavelength range. After the appropriate segment of the spectrum is selected, it is reflected to Mirror *M*5 and from there to the exit slit. The exit slit of monochromator 1 serves as the entrance slit of monochromator 2. The radiation is reflected via mirror *M*6 to the appropriate grating and then back via mirror *M*6 through the exit slit to mirror *M*7.

A choice is provided between a fixed slit width and a servo-slit programme in the UV-Vis range. During scanning, the slit widths change automatically to maintain constant energy to the detector. A servo-slit program is provided for the NIR range.

From mirror *M*7, the radiation beam is reflected via toroid mirror *M*8 to the chopper assembly. When the chopper rotates, a mirror segment, a window segment and dark segments are brought alternately into the radiation beam. The resulting beams emerge as follows:

- When a mirror segment enters the beam, radiation is reflected via mirror *M*9, creating the sample beam *S*.
- When a window segment enters the beam, radiation is reflected via mirror *M*10, creating the reference beam *R*.
- With a dark beam, no radiation reaches the detector, thereby creating the dark signal.

The radiation passing alternately through the sample and reference beams falls on the appropriate detector, a photomultiplier for the UV-Vis range while Pbs detector for the NIR range. Detector

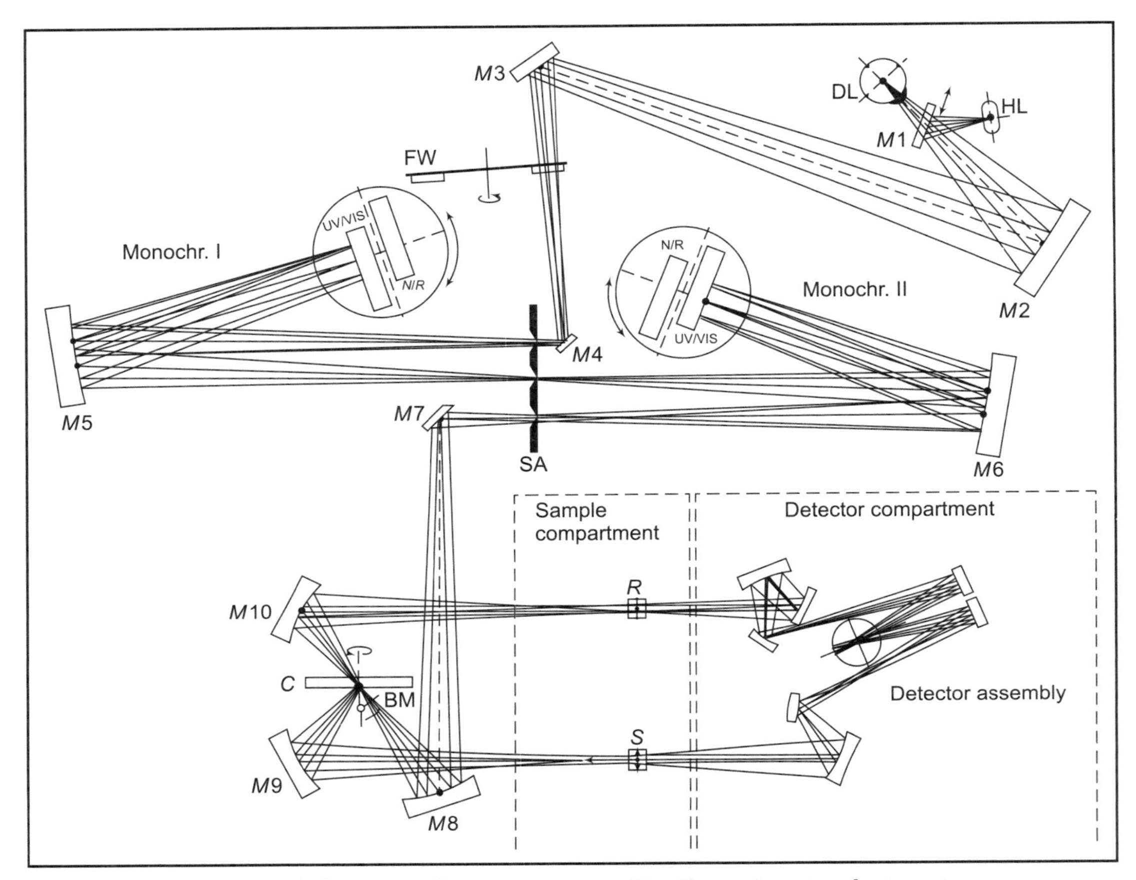

Figure 2.56 :: Optical diagram of LAMBDA 9 Perkin-Elmer Spectrophotometer

change is automatic during monochromator scanning. During all scanning operations, the monochromators stop slewing and the plotter/printer chart advance is stopped until the respective filter, source or detector change is completed.

The electronic part of the instrument is based around the Motorola MC 6808 microprocessor, which performs the control and data processing functions. The processor has 8-bit data and 16-bit address buses. The programme is stored in 3 EPROMS type 2764. Four RAMs with 2k × 8 bit each are used for data storage. In case of power failure, the RAMs and their control obtain their operating voltage from a battery.

The photomultiplier, which is used as a detector in the UV-Vis range, converts the light received into current. The supplied current of approximately 0.1 μA is converted to approximately 2 V in a pre-amplifier. Similarly, the resistance changes detected in the Pbs detector are converted into a voltage signal of approximately 2 V for 100 per cent transmittance. Following amplification of the signal is an A-D converter, which converts the analog signal to digital information for processing by the microcomputer.

Conversion is effected according to the single slope mode with automatic calibration. The resolution is 12 to 20 bits depending upon the input signal, high resolution available for low input voltage. The conversions carried out continuously one after the other are the sample signal, reference signal and the calibration signal.

The wavelength drive, slit drive, filter wheel drive, source selector mirror drive, detector change and grating change are all carried out by stepper motors. Stepping control is effected by the microcomputer, with the pulse frequency depending on the individual scan speed.

A computer unit processes the data supplied by the microprocessor and transmits them in the suitable format to the video display screen and to the keyboard. It also supplies the data to the printer/plotter.

2.6.5 High Performance Spectrophotometers

The model 8450A spectrophotometer from Hewlett Packard is based on a different concept, in which the sequence of optical components is reversed compared to the more traditional design approach. The principle is shown in Figure 2.57. The source is now focused on the sample instead of the monochromator. After passing through the sample, the remaining light enters a spectrograph, rather than a monochromator. This permits access to all wavelength information simultaneously, rather than serially, as with a monochromatic system. Thus, we can acquire the same information on all resolution elements simultaneously, which offers a tremendous speed advantage. Also, since there are no moving parts in the system, the uncertainty caused by wavelength reproducibility due to mechanical linkages is avoided (Knud and Widmayer, 1980).

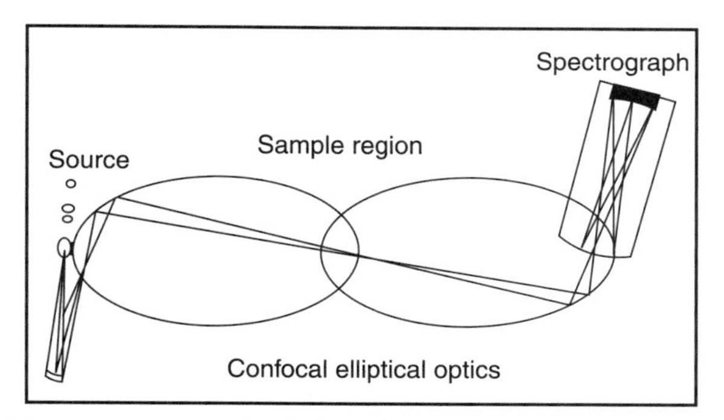

Figure 2.57 :: Principle of HP spectrophotometer Model 8450A. Here, the traditional optics are reversed. The light passing through the sample is wideband instead of a single wavelength. The detector is a spectrograph that measures all wavelength simultaneously, instead of one at a time (Courtesy Hewlett Packard, USA).

Adaptive optical system

Baseline stability in a reversed optical system can be a serious problem, which is taken care of by the adaptive optical system shown in Figure 2.58. The system is basically divided into three principal sections, which are detailed below.

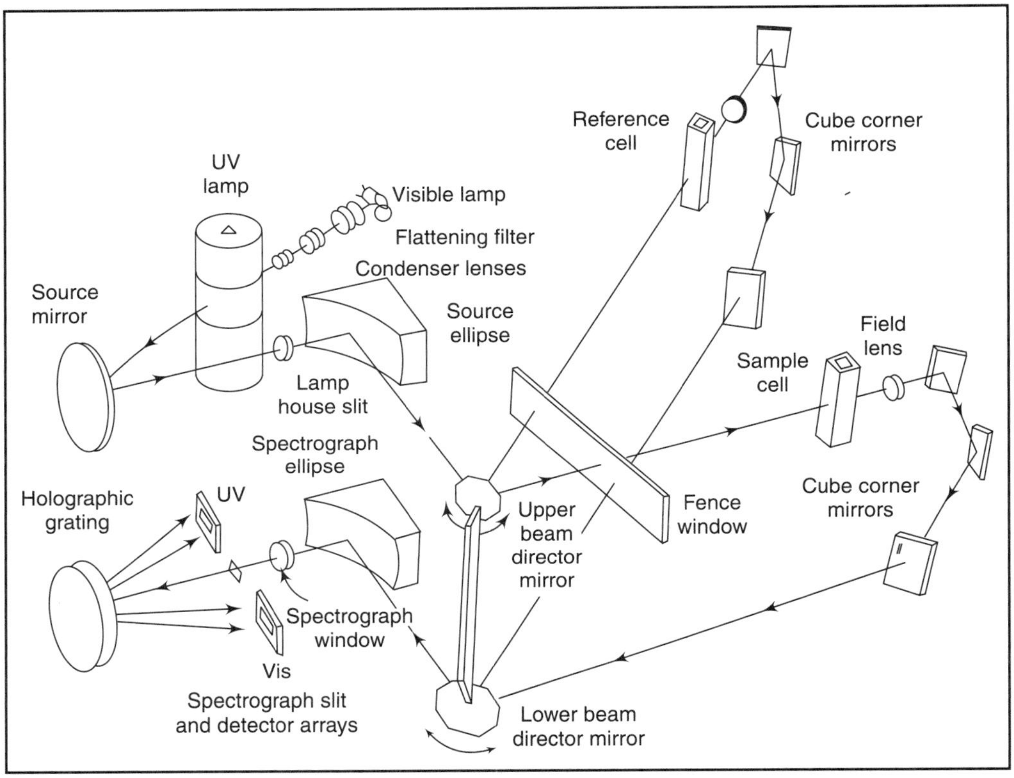

Figure 2.58 :: The folded optical system of the H.P 8450A spectrophotometer (Courtesy Hewlett Packard, USA)

The *source section* contains two light sources which are combined and focused on a source slit. The first element is a 20 W tungsten–halogen lamp. Its filament is imaged through a spectral flattening filter onto the aperture of a see-through deuterium lamp. The source slit is 0.120 mm by 0.600 mm. A replicated ellipsoid with the lamp at one focus and the slit at the other focus serves as the condenser mirror.

The *sample section* contains two flat mirrors on a common shaft, which move the light beam to different sample positions. The shaft is under the control of the computer and a servo mechanism. The light diverging from the common ellipsoid focus at the sample cell position strikes a field lens, which is located as close as is mechanically feasible to the sample position. This field lens images the first sample section ellipsoid onto the second sample section ellipsoid, to reduce clear aperture requirements at the second ellipsoid and to increase flux through the system.

Next, there are three flat mirrors. The mirrors form a cube corner, whose diagonal intersects the beam director shaft midway between the centres of the two-beam director mirrors. The cube corner returns the entering beam back, along the direction it entered. After reflection from the lower beam director mirror, light is focused onto a 0.05 mm by 0.50 mm slit, by the second sample section ellipsoid. The image of the 0.120 mm by 0.600 mm source slit is formed on top of the spectrograph slit. The slit jaws are chisel mirrors tilted at 15° to direct light from the sides of the image on to photodiodes, on each side of the entering beam. These diodes provide a signal for the servo mechanism that controls the beam director shaft location, by balancing signals from the two diodes.

The *spectrograph section* contains two holographically recorded diffraction gratings, which receive light passing through the slit. The gratings are formed on a common substrate and the grating lines are tilted at 3.5° to the vertical to separate their spectra. A photodiode array is positioned at the first order spectrum of each grating (Hopkins and Schwartz, 1980). Each of these arrays uses 200 elements. One grating (UV) covers 200 nm to 400 nm, while the other grating (the Visible) covers over 400 nm to 800 nm. Second order spectra signals are eliminated by using absorption filters placed over appropriate portions of the diode arrays.

Figure 2.59 gives a block diagram of the overall control and communication system of the spectrophotometer Model 8450 A. It incorporates the HP MC-5805 (Silicon-on-Sapphire) 16-bit microprocessor. The interactive control program occupies 57,344 bytes of ROM, which makes the control of the instrument highly user-oriented. Besides the ROM control memory, the processor makes use of 32 kbytes of RAM, which is expandable to 64 kbytes. The signal from the photodiode arrays is connected to the input of the first amplifier, which is configured as a low-noise charge integrator. This results in a voltage proportional to light intensity, at the output of the integrator stage. This voltage is stored by the sample-and-hold circuit and the integrator is reset, and ready to access the next diode.

Programmable-gain amplifier

The output of the integrator is fed into the input of a programmable-gain amplifier, which normalizes the level of signal from each diode to the ADC (analog-to-digital converter) input voltage range. This is necessary because the photocurrent generated in the diodes varies considerably over the spectrum, primarily because of variations in the output of the lamps. The ADC converts the analog signals at the sample-and-hold stage into 14-bit digital words that are read by the microprocessor system. The processor enables the system to select the appropriate service routines and carries out the necessary data processing and calculations.

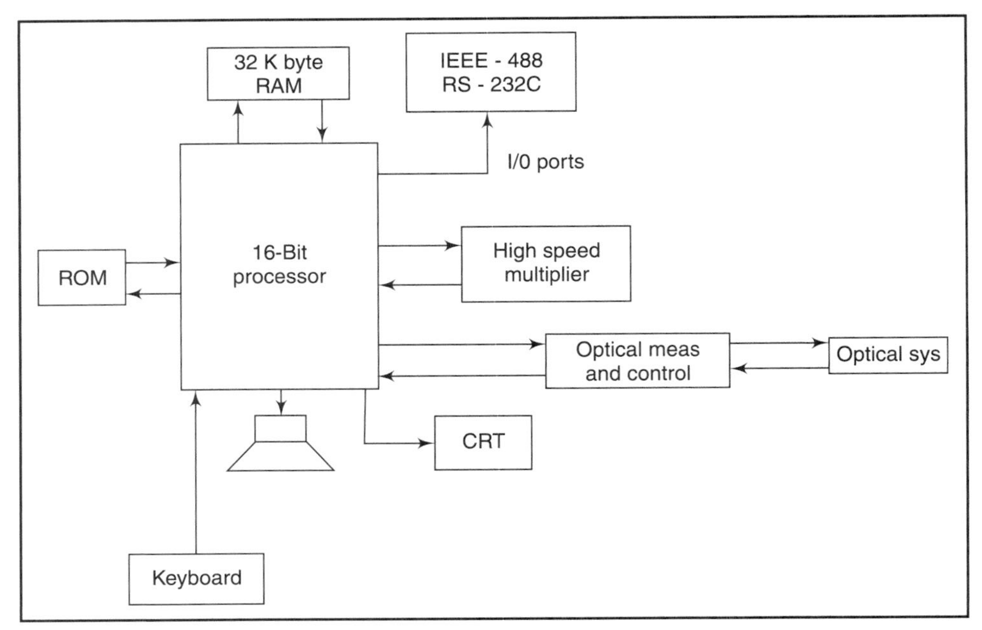

Figure 2.59 :: Block diagram of the microprocessor control of the Model 8450A Spectrophotometer (Courtesy Hewlett Packard, USA)

The operation of the spectrophotometer and the measurement sequence is controlled from a keyboard. For a one-second measurement, the beam director starts from the resting dark position. It is commanded to move to the sample position, and when the system verifies that it is there, the sample integration begins. When this is complete, the beam moves to the reference position and then the integrator cycle for the reference path is initiated. After the sample and reference measurements, the beam is returned to the dark position and one additional dark measurement is made for the running average.

When all the measurements have been taken, the final calculation of absorbance is initiated:

$$A = -\log_{10}\frac{S-D}{R-D} - B,$$

where D is the dark current measurement, R is the reference path measurement, S is the sample path measurement, and B is the optical balance, which is given by

$$B = \log\frac{S'-D}{R-D},$$

where S' is the sample path measurement, with sample path in a reference state.

A fast table-look-up algorithm for the calculation of logarithms with interpolation for the necessary decision is used to complete the calculation for all 401 wavelength values and then statistics within one second.

Like all modern microprocessor-based instruments, the display of results and information is on the CRT, which can show alphanumeric and graphic plots simultaneously. The user can communicate with a printer for the hard copy of the results and also with a remote computer. Communication is established by connecting these devices to either IEEE-488 or the RS-232C connectors.

2.6.6 Dual Wavelength Spectrophotometer

The dual wavelength spectrophotometer permits the recording of absorbance changes in the same sample to be made at two different wavelengths alternately, and virtually simultaneously. This function is performed by means of a control and reversible motor, which automatically adjusts the monochromator wavelength control to alternate between the two wavelengths selected. Following completion of the measuring cycle of the cuvette positioned at the first wavelength, the direction of the drive motor is reversed and the monochromator is adjusted to the second wavelength. The two absorbance measurements are registered on the chart almost simultaneously.

There are three modes of operation for dual wavelength spectrophotometers, all of which depend upon chopping the light source in such a way as to time-share two signal sources on the detector output. The first is the dual wavelength mode. In this arrangement, two wavelengths of light are alternately passed through a single sample. The difference between transmittance or absorbance at the two wavelengths is measured as a function of time. This mode is primarily used to monitor the kinetics of reactions in the sample. In the second method, the two wavelengths of light are alternately passed through a single sample. In this case, one wavelength is scanned over some small range, while the other wavelength is held fixed at some reference point. The third mode is called the split-beam mode. Here a single wavelength is alternately passed through two separate samples. The wavelength is scanned over any region and a transmittance or absorbance difference between spectra of the two samples is the desired output.

Dual wavelength mode

Figure 2.60 illustrates the optical system of a dual wavelength spectrophotometer. Light from the source is focused on the entrance slit of the duochromator by quartz condensing systems L_1 and L_2. The lens L_3 causes an image of the mask in back of L_2 to be projected on the gratings G_1 and G_2. This mask has two adjacent rectangular windows. Light going through one window illuminates grating G_1 while the light going through the other illuminates grating G_2. The rotating shutter blade mounted on the motor allows the alternate opening and closing of the two windows

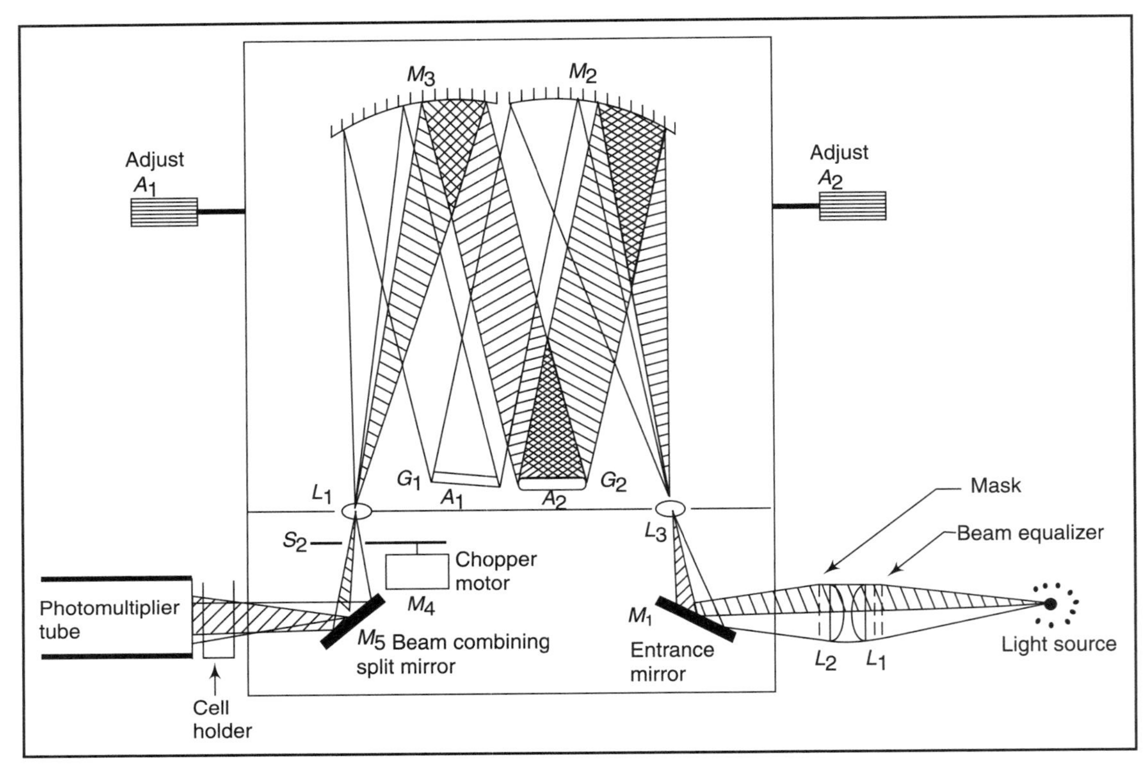

Figure 2.60 :: Optical system of dual wavelength spectrophotometer

in a 60 Hz sequence, and the two monochromatic beams leaving the exit slit S_2 are time-shared and bear a fixed frequency and phase relationship with the driving line voltage.

In the split-beam mode of operation, since both beams must be of the same wavelength, the gratings are driven angularly in synchronism by the scanning motor. This is achieved by setting the reference grating at any wavelength below the region to be scanned, for example 300 nm. When this is done, the measure grating drives the reference grating at an equal angle in a captive fashion. The wavelength driving motor scan is continuously indicated by the measure side. The voltage developed across a potentiometer, which is coupled to the lead screw that determines the grating position, is used to drive the X-axis of the X-Y recorder. Upon leaving the duochromator, both beams are merged through reference and measure cuvettes by mirrors M_4 and M_5.

In the dual wavelength mode, each of the gratings is set by dialling the appropriate wavelengths by means of the two controls on the panel of the instrument. Thus, the radiant energy of these selected wavelengths passes alternately through the sample, where it is absorbed by or transmitted through the material under study. When a difference of transmittance along the two optical paths occurs, there is an alternating error signal, which is amplified, demodulated and read out as the difference between the absorbance readings at the two wavelengths.

Monitoring at alternate wavelengths can result in considerable time savings on lengthy reactions. In addition, the comparison of absorbance measurements at two different wavelengths can result in a more effective analysis of the effluent from chromatographic columns.

Figure 2.61 shows the schematics of the electronic system. It consists of two basic circuits: the measure circuit and the reference circuit. The measure circuit provides a signal to the recorder, which is the difference of the intensities of the measure beam from the reference beam in optical density units. In a dual wavelength operation, the difference in intensity of the reference–monochromatic wavelength beam from the measure–monochromatic wavelength beam is plotted on the recorder in terms of an optical density difference. In a split-beam operation, a difference or a ratio spectrum can be obtained. The reference circuit serves to internally standardize the instrument electronically at the frequency of the mains supply. It does this through the dynode feedback control technique. In this method, the signal from the photomultiplier is electronically sampled during the time when the reference cell or reference wavelength is being seen. It is then compared to a reference voltage and an error signal is generated, which is fed back to raise or lower the high voltage applied to the photomultiplier, so that the signal coming from it during the reference portion of the operation is exactly equal to the reference voltage. In a dual wavelength

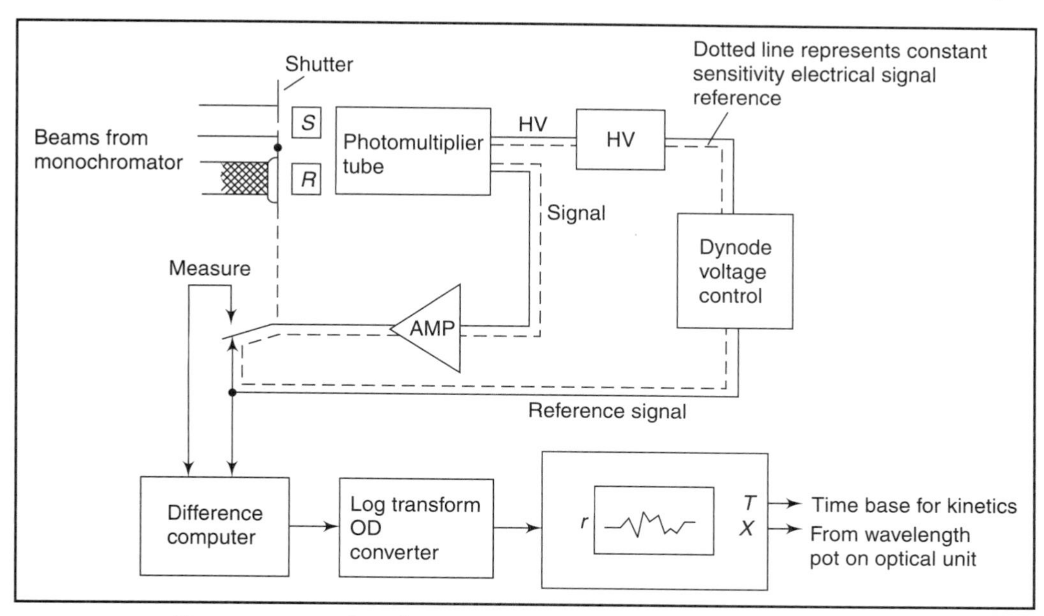

Figure 2.61 :: Signal processor for dual wavelength spectrophotometer

operation, the reference circuit assures that the basic sensitivity of the instrument does not change with time, since there is continuous re-standardization. This is important because changes in lamp intensity and photomultiplier instabilities would cause the instrument to drift. The signals are processed in a computer and given to the recorder.

2.6.7 The Derivative Technique

The absorption spectrum of solids consists of a large number of overlapping absorption bands. In order to isolate the bands, it is necessary to use a system having a high spectral resolution. However, in most of the cases, it is only possible to resolve the first few absorption bands, and the higher quantum number bands appear as points of inflection on the sloping recorder trace of transmitted light intensity due to overlap. In such cases, $dT/d\lambda$ can give a clearer indication of small spectral details (Chopra, 1986).

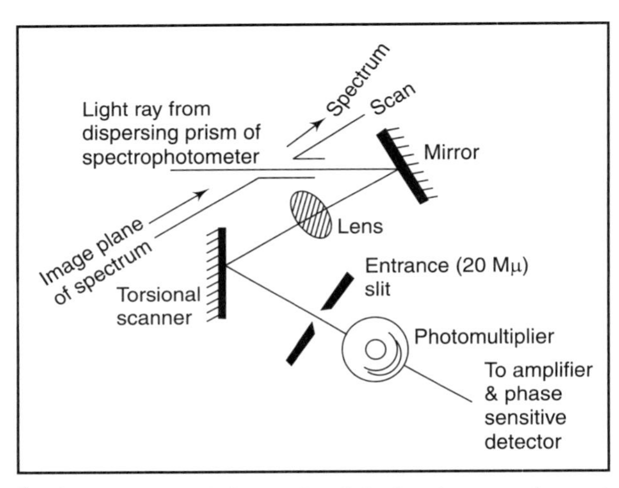

Figure 2.62 :: Schematic of derivative attachment

A derivative attachment for a recording prism spectrophotometer, which gives a direct measurement of the first or second wavelength derivative of the light intensity transmitted by a sample, is shown in Figure 2.62. The arrangement essentially employs a torsional scanner, which consists of a small mirror mounted on one prong of a tuning fork, which can be set into torsional oscillations at its resonant frequency. The amplitude of oscillation can be varied by the use of an electronic drive circuit, which provides a pure sine wave reference signal. The oscillating mirror moves a selected portion of the spectrum sinusoidally across the photomultiplier entrance slit. The original paper shows mathematically that the signal transferred to the recorder would be proportional to either the first or the second wavelength derivative of transmittance. The calibration method is also illustrated.

Torsional scanner

Some spectrophotometers are provided with a plug-in derivative mode module. This module enables the spectrophotometer to measure the instantaneous slope of the absorbance curve at any wavelength. In other words, the spectrophotometer can measure $dA/d\lambda$. The module utilizes a passive resistor-capacitor electrical network to create the value $dA/d\lambda$, at each wavelength for the corresponding sample absorption curve. Due to a slight lag in the signal response with the electrical network, an absolute value of $dA/d\lambda$ is obtained only if the scan speed is very low. The derivative curve can be used as a qualitative analysis tool to ascertain impurities, whose absorption peaks occur near or under the absorption peaks of the sample of interest.

The wavelength scanning speed influences the amplitude of the derivative spectrum and the background noise. Increasing the scanning speed generally improves the signal-to-noise ratio.

Modern spectrophotometers and associated software are frequently configured to operate in the first to fourth derivative modes. The use of even higher derivative orders is possible and they can be expected to produce even greater improvements in resolution by further band sharpening. However, the apparent usefulness of high derivative orders may be compromised by the increasingly complex effect of side band satellites on the overall profile.

2.7 Sources of Error in Spectrophotometric Measurements

The sources of error in spectrophotometric measurements can be divided into two categories: instrument-related errors and non-instrumental errors.

2.7.1 Instrument-related Errors

The major source of instrument-related errors is stray light, which is the unwanted component of radiant energy outside the spectral bandwidth. Stray light causes serious measurement errors, and its primary effect particularly, is to reduce the observed peak height. Where absorbance is high (at an absorption peak), or where instrument sensitivity is low (at the wavelength limits or near 190 nm), the errors introduced by stray light are relatively larger.

Stray light

The main source of stray light in most spectrophotometers is usually the dispersing element in the monochromator, which is either a prism or a diffraction grating. The scattering of light and unwanted reflections from other optical elements can also add significantly to the stray light. It is also possible for stray light to arise outside the monochromator, such as from light leaks in the instrument, allowing some light directly to the sample or detector from outside the instrument or directly from the light source. However, in a well-designed and well-constructed instrument, the latter source of stray light would be negligible.

Stray light causes negative deviations from Beer's Law and a level of 0.1 per cent stray light at any wavelength will prevent accurate absorption measurements of greater than 3 A.

Electronic noise in the detector and the noise element associated with the random fluctuations of the photon beam reaching the detector are also sources of problems in spectrophotometers. The noise is usually apparent in the amplifier output, especially where the beam energy is low. Noise problems may be reduced by integration with respect to time, or by digital signal processing by using microprocessors.

2.7.2 Non-instrumental Errors

These errors may originate from the nature of the solution to be analyzed. Multi-component mixtures usually create difficulty as more than one constituent absorbs light at a wavelength of interest. The absorbance in these conditions is additive and a Beer's Law plot for one component may no longer be valid. However, it is normally possible to take readings at several wavelengths, to construct a set of simultaneous equations and to solve them with a computer interfaced to a spectrophotometer.

The total attenuation of beam radiation may be due to the following factors:

- Reflection of air/cell and solution/cell interfaces,
- Scattering by any suspended particles, or
- Absorption by the solution.

In practice, the effects of reflection and scattering are restricted to less than significant levels by the use of quality sample cells, matched where possible, and by careful sample handling practice. There may be errors due to additional fluorescence component, as a result of absorbed energy being re-emitted at a longer wavelength from that of the incident radiation. Fluorescence effects may be reduced by chemical inhibition, or by appropriate cut-off filters.

2.8 Calibration

Holmium oxide filter

The wavelength calibration of a spectrophotometer can be checked by using a holmium oxide filter, as a wavelength standard. Holmium oxide glass has a number of sharp absorption bands, which occur at precisely known wavelengths in the visible and ultraviolet regions of the spectrum. The holmium oxide filter wavelength peaks are given below:

279.3 ultraviolet range with deuterium lamp
287.6
360.8 visible range with tungsten lamp
418.5
453.4
536.4
637.5

In double-beam spectrophotometers, zero is adjusted with the sample and reference beam. Then a holmium oxide filter is placed in the sample beam. The wavelength control is manually scanned through each wavelength, until the absorption peak is found, always approaching each point from the longer to the shorter wavelength. The spectrum is then recorded.

The wavelength calibration can also be checked in the visible region by plotting the absorption spectrum of a didymium glass, which has, in turn, been calibrated at the National Bureau of Standards, USA.

Chapter 3

Infrared Spectrophotometers

3.1 Infrared Spectroscopy

The infrared (IR) spectrophotometer has become almost indispensable in the chemistry laboratory, as it is ideally suited for carrying out qualitative and quantitative analyses, particularly of organic compounds. Its use in the applications to inorganic compounds is limited, because of the strong absorption of infrared radiation by water. This constitutes a serious limitation in practical applications, since it necessitates the study of inorganic materials in the solid state.

Infrared region

The infrared region extends from 0.8 to 200 μ in the electromagnetic spectrum. However, most of the commercial instruments are available in the region from 0.8 to 50 μ. The position of absorp-tion bands in the infrared spectrum is expressed in both wavelength as well as wave numbers. The wavelength λ is generally measured in microns (μ). The wavenumber (υ) is the number of wavelengths per centimetre and is given by

$$\upsilon \text{ (in cm}^{-1}) = 10^4/\lambda \text{ (in } \mu)$$

Infrared spectrophotometers using prisms produce spectra which are spread linearly with wavelength, whereas instruments fitted with gratings generally deliver spectra that are spread linearly with wave number. However, the results are preferably reported in wave numbers in either case, since these are proportional to molecular properties like frequency and energy, whereas the wavelength is a property of the radiation only.

Each of the vibrational motions of a molecule occurs with a certain frequency, which is characteristic of the molecule and of the particular vibration. The energy involved in a particular vibration is characterized by the amplitude of the vibration, so that the higher the vibrational energy, the larger is the amplitude of the motion. A series of energy levels or states is associated with each of the vibrational motions of the molecule. The molecule may be made to ascend from one energy level to a higher one by absorption of a quantum of electromagnetic radiation. While undergoing such a transition, the molecule gains vibrational energy, which is manifested in an increase in the amplitude of the vibration. The frequency of light required to cause a transition for a particular vibration is equal to the frequency of that vibration, so that we may measure the vibrational frequencies by measuring the frequencies of light which are absorbed by the molecule.

Since most vibrational motions in molecules occur at frequencies of about 10^{14} sec^{-1}, then the light of the wavelength $\lambda = c/f = 3 \times 10^{10}$ cm/sec/10^{14} sec^{-1} = 3×10^{-4} cm = 3 microns will be required to cause transitions. As it happens, the light of this wavelength lies in the so-called infrared region of the spectrum. IR spectroscopy, then, deals with transitions between vibratinoal energy levels in molecules, and is therefore also called vibrational spectroscopy. An IR spectrum is generally displayed as a plot of the energy of the infrared radiation (expressed either in microns or wave numbers) versus the percentage of light transmitted by the compound. This is indicated schematically in Figure 3.1. Here, sample transmittance is usually presented linearly on the vertical axis in an IR spectrum and absorption bands are generally presented pointing downwards. Transmittance, in this case, is defined as the radiant power of the radiation, which is incident on the sample, divided by the radiant power transmitted by the sample.

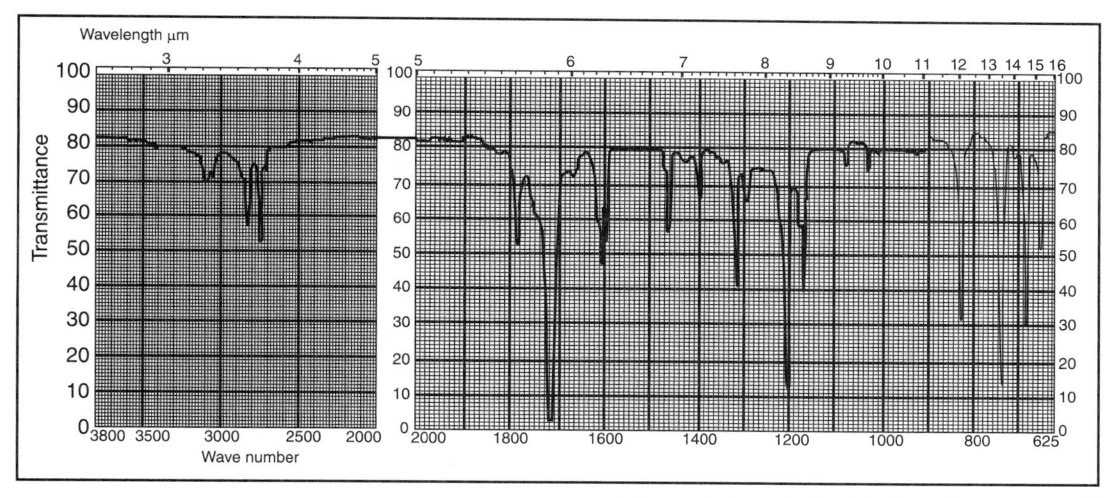

Figure 3.1 :: Typical infrared spectrum

The energy acquired by a molecule can be utilized in three ways, namely electronic excitation, vibrational change and rotational change, and the various molecular energy transformations that may take place are quantized. The amount of energy required to cause the various types of transitions generally corresponds to definite regions of the electromagnetic spectrum. For a molecule, the average energy involved in electronic excitation is 5 eV. For molecules in a particular electronic state, the average energy involved in a vibrational excitation is 0.1 eV and a rotational excitation involves about 0.005 eV.

By definition:

$$1 \text{ eV} = 1.602 \times 10^{-19} \text{ J}$$

$$h = 6.626 \times 10^{-34} \text{ JS}$$

and $E = h\,\upsilon$

(i) 5 eV corresponds to $\upsilon = 1.2 \times 10^{15}$ Hz
so $\upsilon = 40{,}000\ cm^{-1}$
or $\lambda = 2500$ Å $= 250$ nm

(ii) 0.1 eV corresponds to $\upsilon = 2.4 \times 10^{13}$ Hz
so $\upsilon = 833\ cm^{-1}$
or $\lambda = 12\ \mu m$

(iii) 0.005 eV corresponds to $v = 1.2 \times 10^{12}$ Hz
so $\upsilon = 40\ cm^{-1}$
or $\lambda = 250\ \mu m$.

This shows that light absorption at 250 nm (ultraviolet range) produces electronic change. Light absorption at 12 μm (infrared) produces vibrational change. Light absorption at 250 μm (far-infrared) produces rotational change.

In infrared spectroscopy, we are interested mainly in the vibrations and rotations induced in a molecule by the absorption of radiation. The infrared spectrum based on these absorption properties constitutes a powerful tool for the study of molecular structures and identification. Chemical identification is based on the empirical correlations of vibrating groups, with specific absorption bands and the quantitative estimations are dependent on the intensity measurements.

Normally, the data obtained from an IR spectrum should be used in conjunction with all other available information like physical properties, elemental analysis, NMR, ultraviolet, etc., as infrared data taken in isolation can at times be misleading.

Infrared spectrophotometry is applicable to solids, liquids and gases. It is fast and requires small sample sizes. It can also differentiate between subtle structural differences. On the other hand, interpretation of infrared spectra is highly empirical and requires huge libraries of reference spectra. The interpretation of spectra is a skilled art; nevertheless, it is now a part of the repertoire of many chemists. Infrared spectrophotometry provides the means for monitoring many common atmospheric pollutants such as ozone, oxides of nitrogen, carbon monoxide, sulphur dioxide and others.

Interpretation of spectra

The region of the infrared spectrum which is of greatest interest to organic chemists is the wavelength range 2.5 to ≈ 15 micrometers (μ). In practice, units proportional to frequency, (wave-number in units of cm^{-1}) rather than wavelength, are commonly used and the region 2.5 to ≈ 15 μ corresponds to approximately 4000 to 600 cm^{-1}.

Molecular asymmetry is a requirement for excitation by infrared radiation and fully symmetric molecules do not display absorbances in this region unless asymmetric stretching or bending transitions are possible.

Since most organic molecules have single bonds, the region below 1500 cm^{-1} can become quite complex and is often referred to as the fingerprint region, that is, if you are dealing with an unknown molecule which has the same fingerprint in this region, that would be considered as evidence that the two molecules may be identical.

Modern infrared instruments employ microprocessors to facilitate the control of spectrometer functions. This provides better control of the spectrometer and in many cases, additional features such as peak sensing or quantitative analytical capability. The internal representation of data in such systems is, of course, digital. Therefore, it would be quite feasible to produce data, which can be accepted by more powerful computer systems.

It is useful to divide the infrared region into three sections: near-, mid- and far- infrared as given in Table 3.1.

Table 3.1 :: Various Regions of the Infrared Range of the Spectrum

Region	Wavelength range (μm)	Wave number range (cm^{-1})
Near	0.78–2.5	12800–4000
Middle	2.5–50	4000–200
Far	50–1000	200–10

The most useful I.R. region lies between 4000 and 600 cm^{-1}.

Most commercial infrared spectroscopy instruments now offer data handling facilities for a variety of purposes: signal to-noise ratio improvement by computer averaging, difference spectroscopy by subtraction of spectra to reveal minor differences in related samples, spectral deconvolution to provide resolution enhancement, and spectral searching to identify unknown samples and mixtures. However, the most significant change that has resulted from decrease in cost and increase in sophistication of small computer systems is the availability of Fourier transform infrared (FTIR) spectrometers for routine applications at a price that is competitive with that of grating instruments (Morrisson, 1984).

3.2 Basic Components of Infrared Spectrophotometers

Spectrophotometers for the infrared range are composed of the same basic elements as instruments in the visible and ultraviolet range, namely a:

- Source of radiation,
- Monochromator for dispersing the radiation, and
- Detector which registers the residual intensity after selective absorption by the sample.

However, the materials used in the construction of the optical parts are quite different and they are suitably described in the following sections.

3.2.1 Radiation Sources

Blackbody radiator

The source of radiation in infrared spectrophotometers is ideally a blackbody radiator. All practical sources fall short of this to a lesser or greater extent. The energy emitted by a blackbody radiator varies with wavelength and with temperature. In particular, increasing the temperature of the source raises the energy of emission enormously in the short wavelength region, but has a relatively small effect at long wavelengths.

The optimum infrared source is an inert solid, heated electrically to temperatures between 1500 and 2000° K. The maximum radiant intensity at these temperatures occurs at 1.7 to 2 μ (5000 to 6000 cm^{-1}). At larger wavelengths, the intensity reduces continuously, until it is about 1 per cent of the maximum at 15 μ (667 cm^{-1}). On the short wavelength side also, the decrease is much more rapid and a similar reduction is noticed at about 1 μ (10,000 cm^{-1}).

There are three common practical infrared sources: the *Globar rod*, the *Nernst filament* and the *nichrome wire*. The spectral characteristics of these IR sources are shown in Figure 3.2.

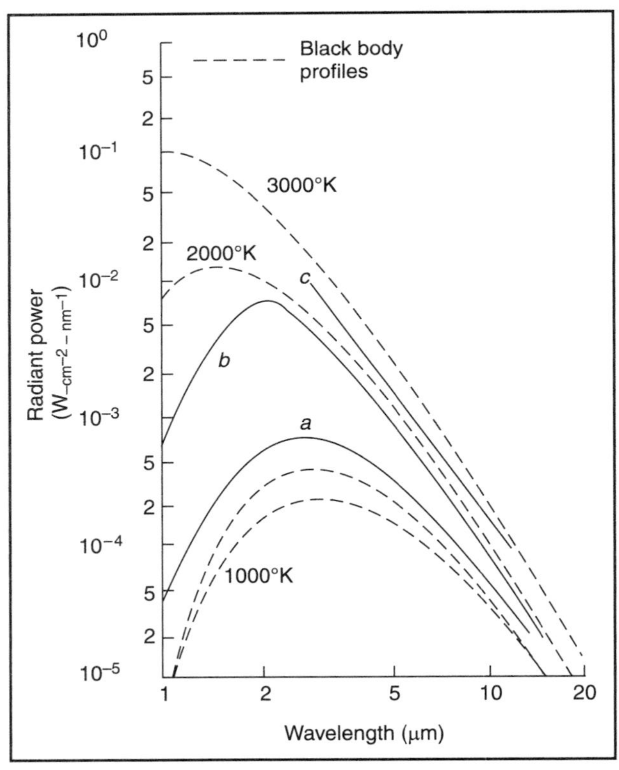

Figure 3.2 :: Typical spectral characteristics of some IR sources (a) Globar, (b) Nernst glower, (c) tungsten glower

Globar

The Globar is a silicon carbide rod, which has a positive temperature coefficient of resistance. The rod is about 5 cm in length and 0.5 cm in diameter. It is electrically heated and run at a current to produce a temperature of about 1300° K. The heat dissipation at the ends of the rod is high, as the latter are the coolest parts. This sometimes leads to arcing and burnout. Therefore, water cooling of electrical parts is required to prevent arcing. The Globar finds applications for work at wavelengths longer than 15 μ (650 cm^{-1}), because its radiant energy output decreases less rapidly. Since the resistance of the rod increases with the length of time, provision must be made for increasing the voltage across the element. It can be conveniently done with a variable transformer.

Nernst filament

The Nernst filament is a small rod composed of fused rare earth oxides of zirconium and yttrium. The filament is of cylindrical shape with a diameter of about 1–2 mm and a length of 20–30 mm.

Platinum leads are sealed to the ends of the cylinder to allow passage of current through it. The filament is heated to temperatures between 1500 and 2000°C and produces maximum radiation at about 7100 cm^{-1}. The device has a negative temperature coefficient of electrical resistance and is therefore, operated in series with ballast resistance in constant voltage circuit. It must be externally heated to a dull-red hot, because it is non-conducting when cold, and has a tendency to crack or separate from its connections on cooling. Therefore, it should be run continuously whenever possible.

Nichrome strip

The nichrome strip, though, gives less energy than the Globar or Nernst, and is extremely simple and reliable in operation. In construction, it could be a tightly wound spiral of nichrome wire, heated by a passage of current. Its temperature is about 800–900°C. Some IR spectrophotometers employ a ceramic rod source, in which is embedded a platinum–rhodium filament producing temperatures up to 1200°C. This radiator consumes only 45 W power.

Several workers have reported the use of tunable lasers as energy sources. Their major advantages are high energy concentrated in a narrow spectral band, a highly directional beam and a very short pulse time (5 ns).

Light from any one of these sources is concentrated by a mirror and is focused onto the entrance slit of the monochromator.

Gagliardi *et al.* (2002) have demonstrated the generation of continuous wave tunable far-infrared radiation by mixing a quantum cascade laser and a CO_2 laser in a metal–insulator–metal (MIM) diode. The system provides full spectral coverage of the FIR (far-infrared radiation) region upto 6 THz (Tera Cycles), making a fundamental contribution to coherent spectroscopy.

3.2.2 Monochromators

Light from the entrance slit is rendered parallel after reflection from a collimating mirror and falls on the dispersing element. The dispersed light is subsequently focused on to the exit slit of the monochromator and passes into the detector section. The sample is usually placed at the focus of the beam, just before the entrance slit to the monochromator. Infrared monochromators generally employ several mirrors for reflecting and focusing the beam of radiation in preference over lenses to avoid problems with chromatic aberrations.

Both prisms and gratings are used for dispersing infrared radiation. However, the use of gratings is a relatively recent development. As a general rule, instruments operating below 25-40 μ have prism monochromators, whereas reflection gratings are utilized above 40 μ, because transparent materials are not easily available in that range.

Materials for prism construction, which are found to be most suitable in the infrared region are listed in Table 3.2.

An ideal prism instrument would contain a large number of prisms made from different optical materials, so that each could be used in sequence in its most effective region. Such an instrument would, of course, be extremely expensive. High resolution prism instruments contain a combination of SiO_2, NaCl and KBr prisms. Low-cost instruments use an NaCl prism over the

Table 3.2 :: Materials for Prism with Their Range of Operation

Material	Optimum Range U Prisms
Glass (SiO_2)	300 mμ to 2 μ (5000 cm^{-1})
Quartz	800 mμ to 3 μ (3300 cm^{-1})
Lithium fluoride	600 mμ to 6.0 μ (1670 cm^{-1})
Calcium fluoride	200 mμ to 9.0 μ (1100 cm^{-1})
Sodium chloride	200 mμ to 14.5 μ (625 cm^{-1})
Potassium bromide	10 to 25 μ (400 cm^{-1})
Cesium iodide	10 to 38 μ (260 cm^{-1})

full range. They give the highest resolution in the vital fingerprint region. Some of these materials are very hygroscopic. Therefore, a good control of the humidity in the room, in which an infrared monochromator is installed, is essential. A prism monochromator is sensitive to temperature changes and must be thermostated to maintain constant wavelength calibration.

Prism monochromators

Prism monochromators employ the Littrow mount, which reflects the beam from a plane mirror behind the prism and returns it through the prism a second time (Figure 3.3), thereby doubling the dispersion produced. In a double-pass system, the beam returns through the prism again, producing a total of four passes through the prism, which brings about an improvement in resolution.

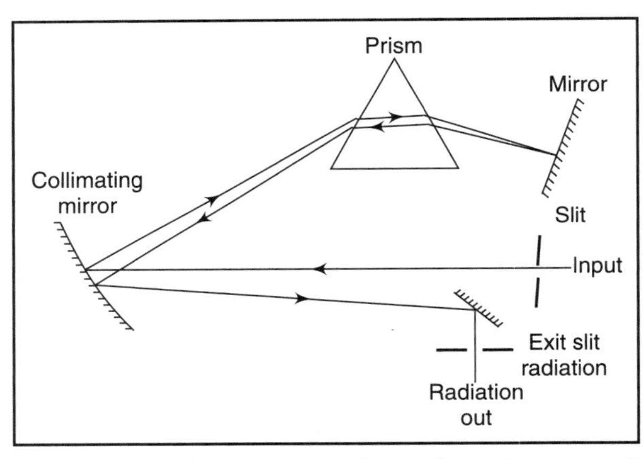

Figure 3.3 :: Littrow mounting infrared monochromator

Figure 3.4 shows the optical arrangement of the SPECORD IR spectrophotometer from Carl Zeiss. Emanating from the radiation source (1), the reference beam is reflected by spherical mirrors (2) and (3) through the attenuator diaphragm (4) and passes the reference cell (5). A plane mirror (6) and a concave mirror (7) project it on to the rotating sector mirror (8). The sample beam takes an almost analogous course via mirrors (10) and (9), through the sample cell (11) and the 100 per cent adjusting diaphragm (12), and via a concave mirror (13). The rotating sector mirror (8) alternatingly allows one of the two light beams to reach the entrance slit of the monochromator, via a toroidal mirror (14). The monochromator used is according to the Ebert principle and is equipped with a prism in Littrow arrangement with 67° prisms of either NaCl or kBr.

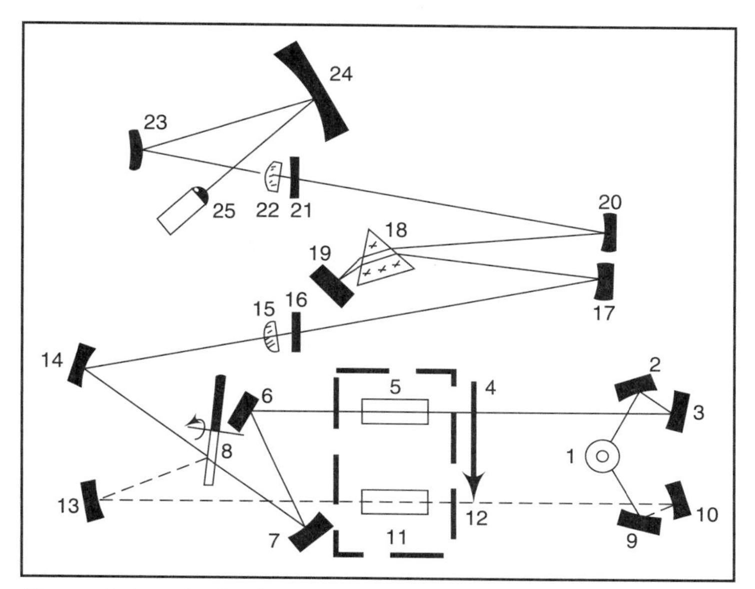

Figure 3.4 :: Optical arrangement of SPECORD IR spectrophotometer (Courtesy Carl Zeiss, Jena)

The instrument provides high spectral resolution achieved by long, curved entrance and exit slits. The scanning conditions within the entire wavenumber region are improved by flushing desiccated air through the path of rays. All hygroscopic elements are protected from the ambient air. In order to obtain wavenumber stability, the NaCl prism is temperature-regulated to ± 1°C. An optical imaging system consisting of a field lens (22), two mirrors (23 and 24), and the lens cemented on the detector (25) projects the beam, leaving through the monochromator's exit slit on to the receiver surface. The detector is a thermocouple of high sensitivity and short time constant.

Efficient balancing of the intensities of the two beams within the entire wavenumber region is achieved by the 100 per cent adjusting diaphragm (12). It is operated by means of a cam, which allows balancing of the 100 per cent line at any part of the wavenumber scale.

The use of gratings as dispersing elements in the monochromators offers a number of advantages for the infrared region. Better resolution is possible, because there is less loss of radiant energy than in a prism system. This facilitates the employment of narrower slits. Also, gratings offer nearly linear dispersion. Because of these advantages, gratings have almost replaced prisms.

The standard grating used in infrared spectrophotometry is an echelon reflection grating. These gratings are usually constructed from glass or plastic that is coated with aluminium. Grating instruments incorporate a sine-bar mechanism to drive the grating mount, when a wavelength read-out is desired, and a cosecant-bar drive when wavenumbers are desired.

One disadvantage of using gratings is that they disperse radiant energy into more than one order. The higher order reflected rays would emerge from the monochromator at the same angle and act as unwanted radiation passing through the sample giving rise to high transmission values. This necessitates the use of an additional separation method. There are two methods of removing these higher unwanted orders of wavelength, using either optical interference filters or a low-dispersion prism. Some instruments make use of two gratings, each covering a part of the range, along with filters for removing unwanted orders. The two gratings have different construction, one having 60 lines/mm for the range 400-133 cm^{-1} and 180 lines/mm for 1200-1400 cm^{-1}. It is possible to maintain comparatively high resolution at a much lower cost by using an Ebert monochromator. It employs a single grating in its first and second orders. Figure 3.5 shows the optical diagram of IR spectrophotometer SP 1000 of Pye Unicam, making use of Ebert monochromator. The spectrum is scanned by rotating the grating about a vertical axis.

Ebert monochromator

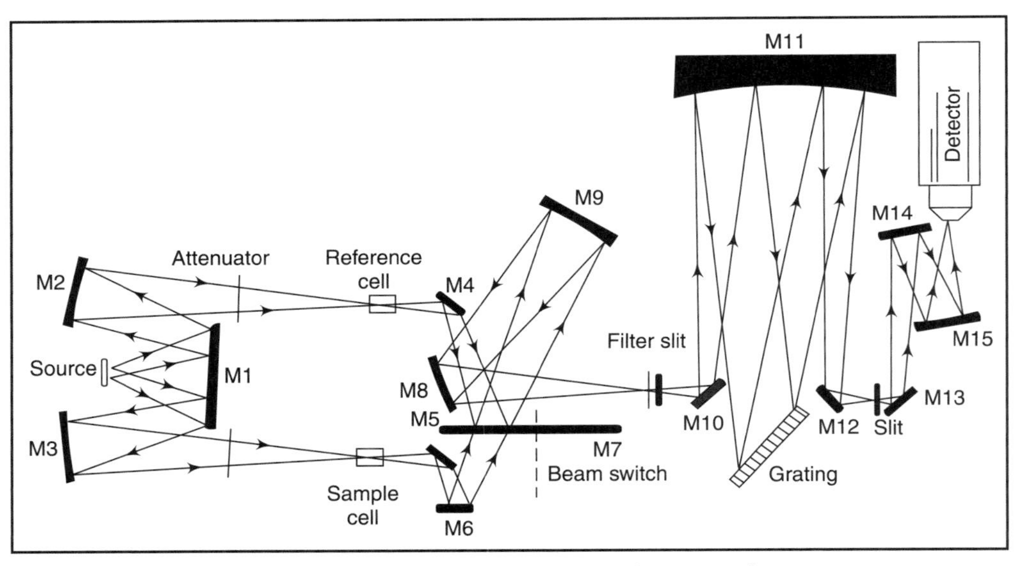

Figure 3.5 :: Ebert Monochromator used in infrared spectrophotometer

In the Beckman 4200 series IR spectrophotometers, monochromaticity is provided by a grating and a set of circular variable filters (Figure 3.6). Two annular filter segments are mounted on a rotating wheel to select the desired wavelengths from the continuous output of a nichrome wire energy source. This is designated as a wedge filter system and is coupled to a grating wavelength selector. The annular filter segments consist of IR transparent material, coated to transmit a continuously varying band of wavelengths, when rotated in front of a V-shaped slit. The slit is positioned on the optical centre line of the filter system.

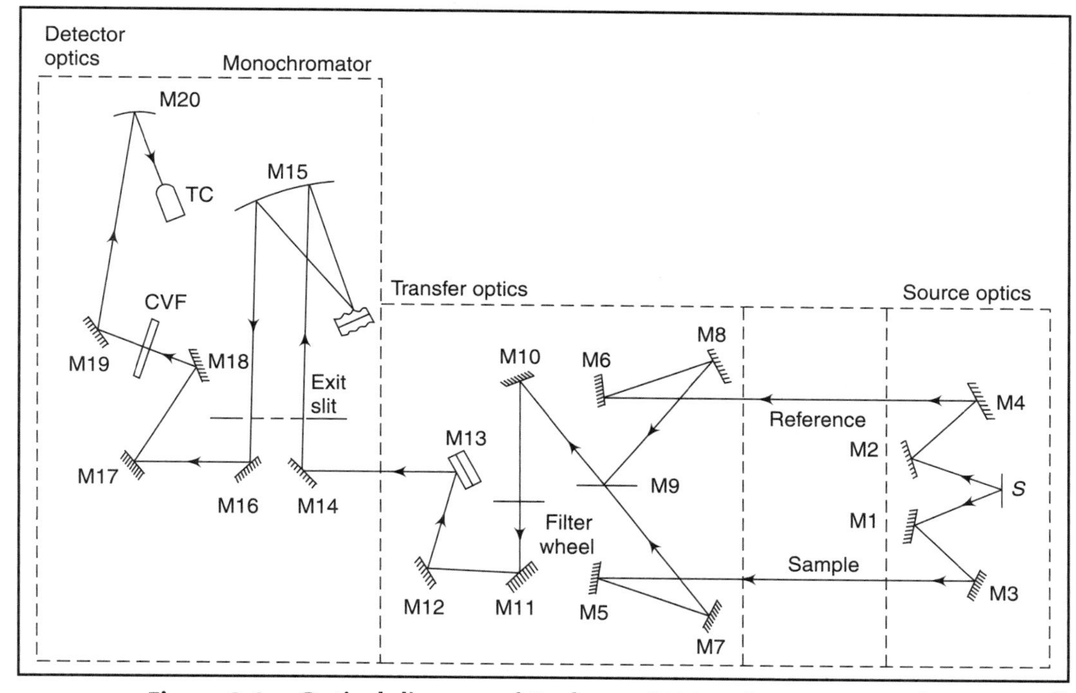

Figure 3.6 :: Optical diagram of Beckman 4200 series IR spectrophotometer

A single element thermocouple receives energy transmitted by the wedge filter and generates voltage for amplification by the electronics system. In a double-beam operation, the amplified thermocouple output, which is proportional to the difference of the reference and sample beam intensities, drives an optical attenuator into the reference beam until the beam intensities become equal, i.e. until the optical null is achieved. The attenuator position in the reference beam is thus directly proportional to the sample transmission at any given wavelength. Since the recorder pen is coupled mechanically to the attenuator, the pen records the sample transmission as a function of wavelength. In single-beam operation, with the reference beam physically blocked by the operator, sample beam transmission produces a thermocouple voltage that is proportional to the sample beam intensity. The recorder functions potentiometrically. Amplified thermocouple voltage is compared with the output of the pen potentiometer, which receives reference voltage from a regulated supply. The difference between the two voltages is then directed to the pen servo-motor, wherein it results in linear representation of energy in the sample beam.

The optical arrangement incorporates a number of mirrors, and entrance and exit slits in the optical system. The slit width is adjustable manually from 0.005 to 7.0 mm. It also permits slit selection in a programmed condition and keeps energy within a dynamic range of the servo system.

Diffraction gratings should not be touched and unless absolutely essential, cleaning should be done only by blowing with air. In no case should these be washed with corrosive solvents. Prisms and lenses are generally hygroscopic and can get easily scratched. In case they become foggy, the only remedy is to get them polished, but in most cases, they are beyond repair and hence extreme care should be taken to prevent exhalation of vapours near instruments that are opened.

3.2.3 Entrance and Exit Slits

In all continuous comparison techniques in spectrophotometry, part of the optical radiation emerging from the monochromator is directed on to the reference detector and the rest is allowed to fall into the sample. As the wavelength of the monochromator is scanned, it is arranged such that the signal from the reference detector is maintained at an approximately constant level. It is possible to adjust the intensity of a tungsten lamp to obtain this condition, but when a wide range of wavelengths in the infrared region are to be covered, the only practical method is to control the monochromator slits. Most of the monochromators use a specially shaped cam to control the slit width, as the wavelength is scanned. Besides this, a servo system can also be employed to achieve the same purpose as the slit-servo system, which maintains the output of an optical monochromator at a constant level, as the wavelength is scanned. In this system, after amplification, the signal from the detector is compared with a reference voltage and the error signal is converted into a frequency. The series of pulses drive a dc stepper motor coupled to the monochromator slits, and as balance is reached, the motor slows down, thus preventing overshoot.

The slit width control has the same importance as it has in ultraviolet and visible spectrophotometers, and a compromise of different factors needs to be made while selecting the slit width to be used. Narrow slits produce smaller bandwidth, which consequently result in better spectra definition, whereas wider slits permit a larger amount of radiant energy to reach the detector and consequently greater photometric accuracy.

3.2.4 Mirrors

As the materials that are used for lenses are not transparent to IR radiation over the entire wavelength range, lenses are generally not preferred. Front surfaced mirrors are usually used in the IR instruments, of which plane, spherical, parabolic and toroidal types are the most common. Although highly reflecting aluminium coated mirrors with a protective coating are usually employed, after long periods of operation, the mirrors are bound to become somewhat dusty. Under these conditions, the recording accuracy may not be affected much; the reflective power will be reduced. As fingerprints or other contaminations may give their own absorption bands, mirrors should be cleaned, if required, only by blowing hot air over them. In case the mirrors are too dirty, these can be cleaned by washing with detergents and rinsing with distilled water. Under no conditions should corrosive solvents be used.

3.2.5 Detectors

Detectors used in infrared spectrometers usually convert the thermal radiant energy into electrical energy, which can subsequently be plotted on a chart recorder. The detectors range in format from single element, uncooled detectors to specialized multi-spectral array detectors. The selection of a specific detector depends upon the wave band of interest, the sensitivity required and the cost constraints. The detectors used may be divided into thermal and quantum types. Quantum detectors are useful in the near-infrared region, whereas thermal detectors can be used much beyond this range.

Thermal detectors

The main features of thermal type detectors include responsivity with little dependence on wavelength and operation at room temperature. However, the response speed and detectivity are lower than in the quantum type. The commonly used thermal detectors are thermopiles, bolometers, pneumatic detectors and pyroelectric detectors.

Quantum detectors

Quantum type detectors feature high detectivity and fast response speed. Responsivity is wavelength-dependent and except for detectors in the near-infrared range, cooling is normally used with these detectors. Quantum type detectors are classified into intrinsic types and extrinsic types.

The wavelength limits of intrinsic type detectors are determined by their inherent energy gap, and the responsivity drops drastically when the wavelength limit is exceeded. Typical examples of this type of detectors are photoconductive and photovoltaic detectors such as HgCdTe or PbSnTe.

Extrinsic types of detectors are photoconductive detectors whose wavelength limits are determined by the level of impurities doped in high concentrations to the Ge or Si semiconductors.

The major difference between intrinsic type detectors and extrinsic type ones lies in the operating temperature. Extrinsic type detectors must be cooled down to the temperature of liquid helium.

3.2.5.1 Quantum Type Detector

Photoconductive cells: These are essentially electrical resistors, which decrease in resistance in relation to the intensity of light striking their surface and are characterized by greater sensitivity and rapidity of response. They are constructed from a thin layer (0.1 μ) of semiconductor, like lead sulphide or lead telluride supported on a backing medium like glass and sealed into an evacuated glass envelope. These detectors are sensitive up to 3.5 and 6 μ for the lead sulphide and lead telluride cells, respectively. The typical response time of these cells is 0.5 ms.

Solid state photodetectors: Certain semiconductor materials exhibit a photoelectric effect, which can be used for the detection of IR radiation. Detectors of PbS were the earliest example of this type of detectors. However, they do not respond to radiation at wavelength greater than 3 μm. Indium antimonide (In Sb) is another detector which is useful in the 2-6 μm range. But they are highly sensitive and are capable of detecting small temperature variations as compared to thermistors. Another detector making use of an alloy of cadmium, mercury and telluride (CMT) and cooled with liquid nitrogen, though showing a peak response at 10–12 μm, has a moderate sensitivity. The spectral response of various photodetectors is shown in Figure 3.7.

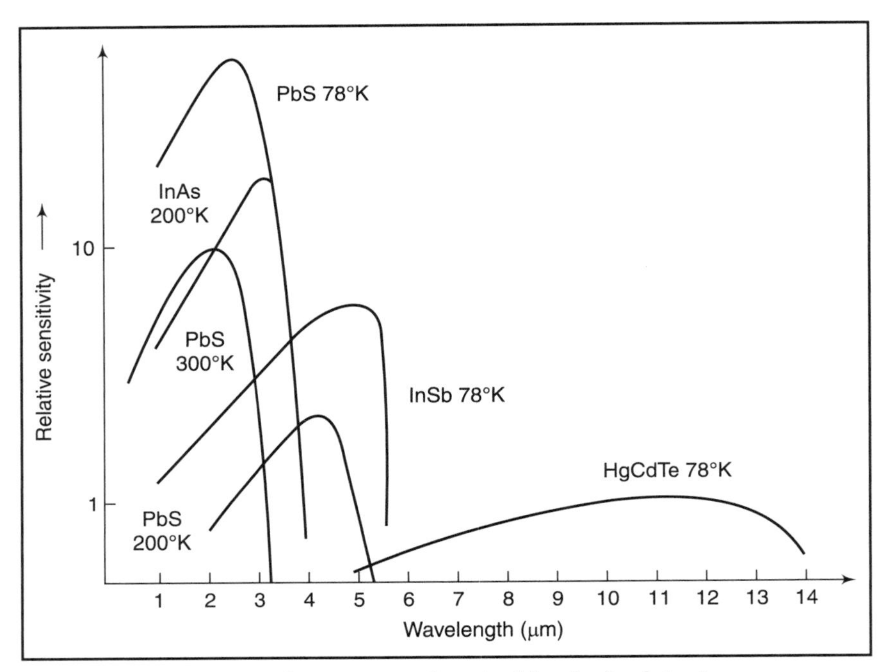

Figure 3.7 :: Spectral response of typical intrinsic detectors

It may be noted that the excitation energy required is very small in the long wavelength devices, and the detectors must be cooled to prevent background thermal excitation, thereby obscuring measurement of the incident IR radiation. Figure 3.7 shows the high sensitivity at peak response, and the temperature required for proper operation of the detectors.

Photodiode

InGaAs photodiode arrays are specifically suited for near-infrared spectroscopy using reflection measurements, and for line scan imaging applications. The linear image sensor is a hybrid assembly of an array of InGaAs photodiodes connected to a charge amplifying multiplexer.

The sensors come in vacuum-sealed packages with anti-reflection coated windows. A proper cooling arrangement reduces the dark current and facilitates longer exposure times, ranging from 1 μsec up to 20 msec. The sensors are delivered with a flat bottom package with fixation holes for optimal thermal coupling. The typical operating temperature is between 210 K and 225 K when the base plate temperature is kept at 293 K. The array is useful between 1.1–2.5 μm wavelength range and can be organized with 128, 256 and 512 pixels linear array format. The linear array infrared sensor from M/s XenIcs (*www.XenIcs.com*) is shown in Figure 3.8 along with its typical spectral response.

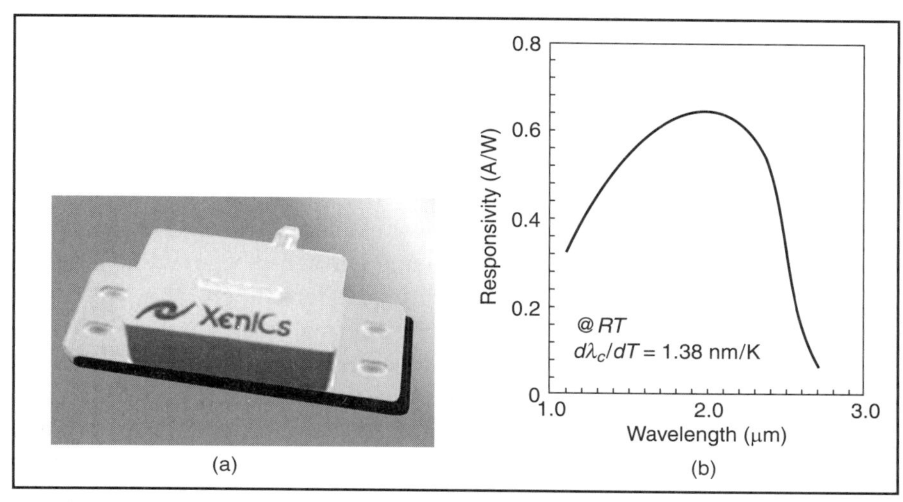

Figure 3.8 :: (a) Photodiode (InGaAs) array for infrared spectroscopy (b) Typical spectral response

3.2.5.2 Thermal Detectors

These detectors depend upon the heating effect of the radiation for their response. They are useful for detection of all but the shorter infrared wavelengths. The infrared radiations are absorbed by a small blackbody and the resultant temperature increase is measured. Even under the best of circumstances, the temperature changes are extremely small and are confined to a few thousandths of a degree centigrade. Thermocouples, bolometers, pneumatic and pyroelectric detectors, are the commonly used thermal detectors.

Thermocouples

Thermocouples are the most widely used detectors and are employed in infrared spectrophotometers. In these detectors, the signal originates from a potential difference caused by heating a junction of unlike metals. They are made by welding together two wires of metals in such a manner that they form two junctions. One junction between the two metals is heated by the infrared beam while the other junction is kept at a constant temperature. Due to a difference in the work functions of the metals with temperature, a small voltage develops across the thermocouple. The receiver element is generally blackened gold or platinum foil, to which are welded the fine wires comprising the thermoelectric junction. The other junction is shielded from the incident radiation. Changes in temperature of the order of 10^{-6} °C can be detected. It is possible to increase the output voltage by connecting several thermocouples in series. This arrangement is called the thermopile. Thermopiles are made from both metals and semiconductors.

The average electrical output is about I μV. Amplification of such low signals is difficult, because of their low resistance (10–20 Ω) and the slow response of the average thermocouple. The thermocouple is enclosed in magnetically shielded housing to reject spurious signals. Stray light can often be troublesome and should be avoided. Some of the combinations of metals for thermocouples that have been used are Ag-Pd, Sb-Bi and Bi-Te.

Bolometers

Bolometers give an electrical signal as a result of the variation in resistance of a conductor with temperature. It consists of a thin platinum strip in an evacuated glass vessel, with a transparent window in the infrared range. Irradiation by the infrared beam produces an increase in resistance of the metal strip, which is measured with a Wheatstone bridge. Usually, two identical elements are used in the opposite arms of the bridge. One of the elements is placed in the path of the infrared beam while the other is used to compensate for the changes in the ambient temperature. Alternatively, the platinum strips may be replaced with thermistors, which show a negative thermal coefficient of electrical resistance. Bolometer arrays have become the focus of most uncooled detector development.

Pneumatic detector

The pneumatic detector described by Golay (1947) essentially measures the intensity of infrared radiation by following the expansion of a gas upon heating. The Golay cell (Figure 3.9) comprises a chamber containing xenon, a gas of low thermal conductivity. It is sealed at its front end by a blackened receiver. The rear wall is a flexible membrane with a mirrored surface on its rear side. Infrared energy falling on the receiver warms up the gas in the chamber. A rise in temperature of the gas in the chamber produces a corresponding rise in pressure and therefore, a distortion of the mirror diaphragm. Light from a lamp inside the detector housing can be focused on the diaphragm, which reflects the light on to a photocell. Movements of the diaphragm corresponding to the amount of incident infrared energy change the incident light energy on the photocell surface and cause a change in the photocell output. By periodically interrupting the incident radiation with a chopper, an ac signal is produced by the photocell which can be amplified.

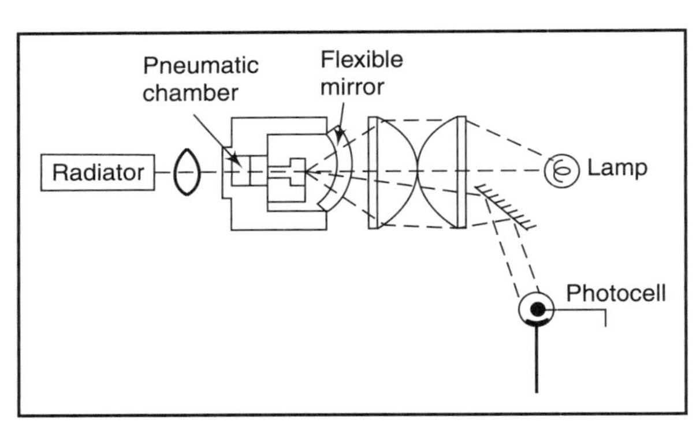

Figure 3.9 :: Golay pneumatic cell

Another form of pneumatic detector is the capacitor microphone type in which the varying expansion of the gas affects the capacitor film, which, in turn, produces the variation in the electrostatic capacity.

Pneumatic detectors with large receiver areas are suitable for instruments in which wide slits are necessary. They function at all wavelengths throughout the infrared region (as far as 400 μ using a diamond window). They have a low response time. The whole receiver is sensitive to the radiations, thereby eliminating the need for optical alignment.

Pyroelectric detectors

In the case of pyroelectric detectors, certain types of crystals (ferroelectric) get polarized in a well-defined direction, known as the polar axis. Since the degree of polarization is temperature-dependent, heating or cooling a slice of such a crystal will create an accumulation of charge (on the faces normal to the polar axis), that is proportional to the variation in polarization caused by the temperature change. This is the pyroelectric effect. A pair of electrodes normal to the polar axis of the crystal may be used to measure the voltage generated within the crystal due to temperature changes.

Pyroelectric detectors are mostly made from single crystal triglycine sulphate (TGS). The detector construction is similar to that of capacitor. Two electrodes, one of them transparent, are formed on opposite sides of a TGS slice. The transparent electrode allows the radiation to fall on the slice. The voltage generated is usually applied to a field effect transistor, which is an integral part of the detector package (Figure 3.10). Pyroelectric detectors are current sources with an output that is proportional to the rate of change of their temperatures.

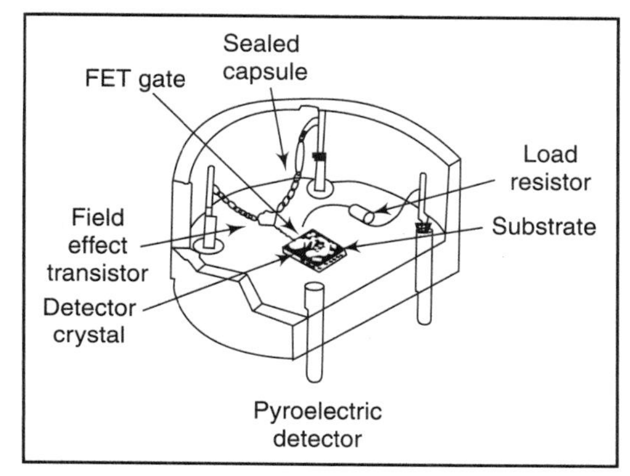

Figure 3.10 :: Pyroelectric detector

Figure 3.11 shows a circuit arrangement to process the electrical signal with an FET in the input stage. The FET source terminal pin 2 connects through a pull down resistor of 100 K to ground

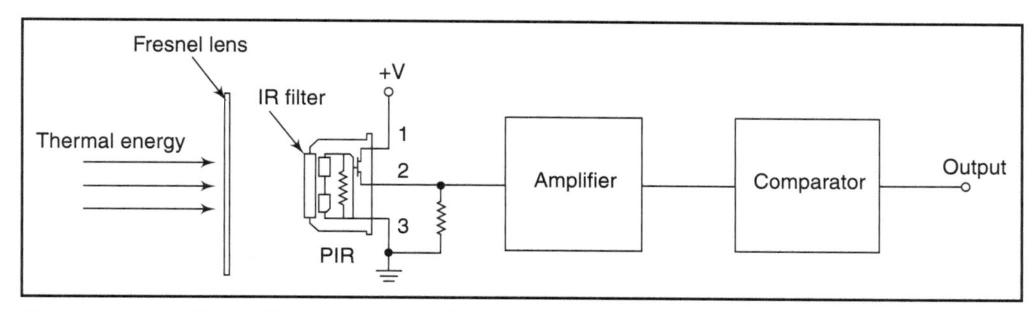

Figure 3.11 :: Block diagram of signal processing circuit using pyroelectric detector

and feeds into a two-stage amplifier having signal conditioning circuits. The amplifier is typically bandwidth limited to below 10 Hz in order to reject high frequency noise, and is followed by a window comparator that responds to both the positive and negative transitions of the sensor output signal. A well-filtered power source from 3 to 15 volts is connected to the FET drain terminal pin 1.

The sensor has two sensing elements connected in a voltage bucking configuration. This arrangement cancels the signals caused by vibration, temperature changes and sunlight. A body passing in front of the sensor will activate first one and then the other element whereas other sources will affect both elements simultaneously and then be cancelled. The radiation source must pass across the sensor in a horizontal direction when sensor pins 1 and 2 are on a horizontal plane so that the elements are sequentially exposed to the IR source. A focusing device is usually used in front of the sensor.

Fresnel lens

The device used for focusing is a Fresnel lens which is a Plano convex lens that has been collapsed on itself to form a flat lens that retains its optical characteristics but is much smaller in thickness and therefore has less absorption losses.

The FL65 fresnel lens from M/s Glolab (*www.glolab.com*) is made of an infrared transmitting material that has an IR transmission range of 8 to 14 μm which is most sensitive to human body radiation. It is designed to have its grooves facing the IR sensing element so that a smooth surface is presented to the subject side of the lens, which is usually the outside of an enclosure that houses the sensor.

The lens element is round with a diameter of 1 inch and has a flange that is 1.5 inches square. This flange is used for mounting the lens in a suitable frame or enclosure.

The lens has a focal length of 0.65 inches from the lens to the sensing element. It has been determined by experiments to have a field of view of approximately 10 degrees when used with a Pyroelectric sensor.

Pyroelectric materials absorb quite strongly in the far-infrared region and have an essentially flat wavelength response from the near-infrared through the far-infrared. Although the pyroelectric detectors have low sensitivity as compared to the Golay detector, they are preferred in the far-infrared range due to their faster response.

General Considerations Regarding Detectors

Due to the low intensity of available sources and the low energy of the infrared photon, the task of measurement of infrared radiation is particularly difficult. The electrical signal produced by the various detectors is quite small and requires large amplification before it can be put to a recorder. In order to prevent the very small signals being lost in the noise signals, which might be picked up by the connecting wires, the pre-amplifier is located as close to the detector as possible.

In addition to this, the radiation beam is chopped with a low frequency light interrupter. By using narrow bandwidth electronic amplifiers, the alternating signal, corresponding only to the chopping frequency, is measured. This arrangement minimizes stray light signals.

The response times of different types of detectors are typically as follows:

Photoconductive cells	0.5 ms
Golay cell	4 ms
Bolometer	4 ms
Thermocouple	15 to 60 ms

The sensitivity of photoconductive detectors is generally higher than that of thermal detectors. Nevertheless, there is no strong choice among thermal detectors on sensitivity considerations.

When using thermal detectors, it is essential to shield the detector to reduce its heating by nearby extraneous objects. Therefore, the absorbing element is placed in a vacuum and is carefully shielded from thermal radiation emitted by other bodies in the area.

3.2.5.3 Pre-amplifier for use with Photoconductive Infrared Detectors

When photoconductive infrared detectors such as cadmium mercury telluride are used, it is necessary to employ a low-noise pre-amplifier, such as shown in Figure 3.12. Basically, the circuit consists of an FET input stage, which has sufficient gain to make the noise contribution of the next stage negligible. This is followed by a low-noise integrated amplifier. A common source stage Q_1 is used with a bootstrapped drain load Q_2. In order to obtain the low noise, the FETs are used with drain currents near to I_{DSS} (drain current of FET with a zero source gate voltage). The feedback resistor is made as small as possible, so that the minimum bias is applied to Q_1. The current in Q_2 is determined by the current in Q_1. Therefore, in order to ensure that the gate junction of Q_2 is not forward-biased, the FET with the higher I_{DSS} is used in this position. The amplifier, at room temperature, with a 51 Ω source impedance and a bandwidth from 8 Hz to 10 kHz, has a noise level of 2 μV peak-to-peak (Gore and Smith, 1974).

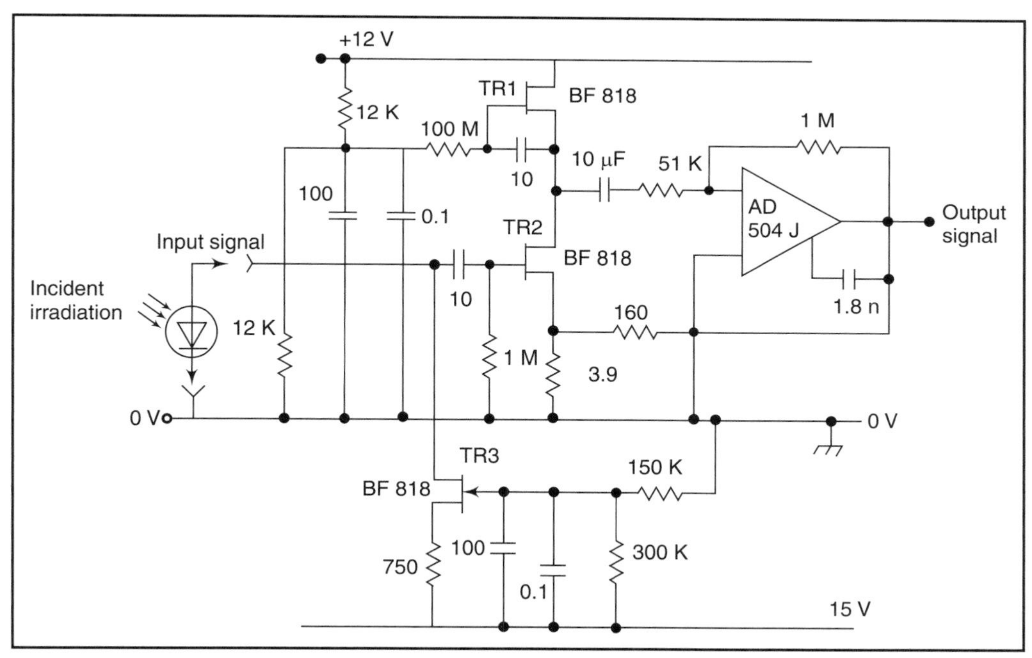

Figure 3.12 :: Pre-amplifier for use with photoconductive infrared detectors

The detector requires a 10 mA bias current supply, which must have a low noise level, if the system performance is not to be degraded. This current is obtained from Q_3, which is also a low-noise FET. The noise voltage across the detector is about 19 mV rms, which is small as compared to the theoretical amplifier noise.

3.3 Types of Infrared Spectrophotometers

Infrared spectrophotometers are produced by several instrument manufacturers. Almost all the commercial designs employ a double-beam system, wherein the radiant energy passes alternately through the sample and then through a reference to a single detector. They incorporate a low-frequency chopper (5 to 13 Hz) to modulate the output radiation from the source. The various sources and detectors described above are incorporated in one or more of the commercial instruments.

Double-beam system

The double-beam system offers the advantage that scans are not disturbed by absorption bands produced by atmospheric water vapour and carbon dioxide. Also, fluctuations in source output detector sensitivity and gain have no influence on the record.

There are two types of arrangements for recording the infrared spectra. These are the ratio recording and the optical null methods.

3.3.1 Optical Null Method

In this method, the infrared radiation is passed simultaneously through two separate channels, one containing the sample, and the other, the reference. The two beams are recombined into a common axis and are alternately focused on the detector. If the intensities of the sample and reference beams are exactly equal, then no alternating intensity radiation goes through the slit. If the sample absorbs some radiation, an alternating intensity radiation is observed by the detector, which produces an ac signal. This ac signal can be selectively amplified in a tuned amplifier. As the amplifier is tuned to the chopping frequency of one light beam, the signals of frequency that are different from the chopping frequency are not amplified. The alternating signal from the detector is used to drive a servo-motor, which is mechanically coupled to an optical wedge or a fine-toothed comb attenuator. The movement of the comb occurs when a difference in power of the two beams is sensed by the detector. The motor will stop when the reference and the sample beam intensities are exactly equal. The movement of the motor is synchronized with the recorder pen so that its position gives a measure of the relative power of the two beams and thus the transmittance of the sample. The record obtained therefore is of the sample absorption as a function of spectral frequency. The teeth of the attenuator comb are accurately cut so that a linear relationship exists between the lateral movement of the comb and the decrease in power of the beam.

Comb attenuator

Figure 3.13 shows the diagram of a typical double-beam optical-null principle infrared spectrophotometer. Emanating from the source, the two light beams pass through the sample and reference cells. With a switching frequency of 12.5 Hz, the rotary sector mirror alternately conducts the radiation of the sample and reference channels into the monochromator. Here, the light of the respective wavenumber is reflected and passed on to the thermocouple, which delivers a 12.5 Hz ac voltage signal, when the intensities of the two optical paths are not equal. After amplification

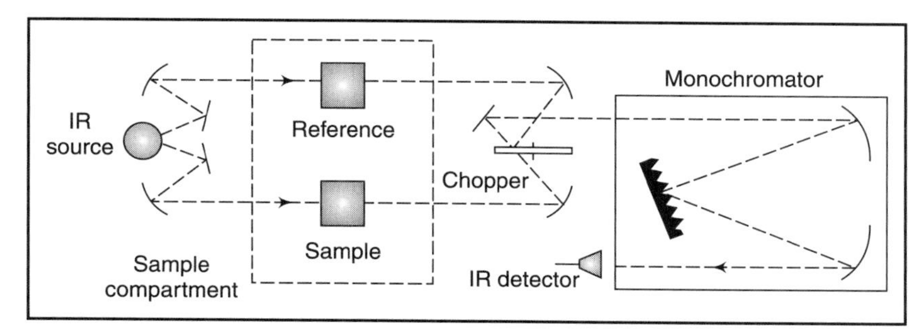

Figure 3.13 :: Block diagram of a double-beam infrared spectrophotometer

in the pre-amplifier and main amplifier, and after phase-sensitive rectification in the demodulator, this ac voltage signal controls the servo-motor, which adjusts the attenuator diaphragm in the reference channel, until the intensities of both beams are equal and the signal disappears. The servo-motor also drives the recording pen, which records the sample's transmittance corresponding to the position of the attenuator diaphragm.

The Beckman Model IR 4200 series infrared spectrophotometers are automatic recording double-beam optical null instruments, which cover a range of 4000-200 cm^{-1}. They employ the Nichrome/Nernst source and a thermocouple detector, and are available with scanning speeds ranging from 2 cm^{-1}/mm to 1000 cm^{-1}/mm. The monochromator is a filter/grating combination and provides a resolution of 0.5 cm^{-1} at 900 cmn^{-1} and 1.4 cm^{-1} at 3000 cm^{-1} in the better models.

3.3.2 Ratio Recording Method

In the direct ratio recording system, radiation from the source is passed alternately through two separate channels, one passing through the sample cell and the other through the reference cell. The two signals from the detector are amplified, rectified and recorded separately. The sample signal is displayed as a proportion of the reference signal. In this method, attempts are not made to achieve a physical equalization of the sample and the reference beam intensities. The reference signal is used to drive the slit width control so that the energy level reaching the detector is constant. This system is, however, not favoured in infrared instruments, as it requires a very stable amplifier system and chopping system.

Perkin Elmer 780 Series IR Spectrophotometers

Modern IR spectrophotometers are microprocessor-based and provide a continuous record of the infrared transmittance or absorbance of a sample as a function of frequency, expressed in wavenumber units. The chart is driven in synchronism with the monochromator so that a pen moving laterally across chart records the sample transmittance or absorbance as a function of the wavenumber.

In the Perkin-Elmer 780 series IR spectrophotometers, two microprocessors have been used (Figure 3.14). The wavenumber scan motor drives the grating monochromator scan mechanism, which is synchronized with the recorder drive by the abscissa microprocessor, in order to accurately reproduce the wavenumber settings on the chart. The instrument is a double-beam system (Figure 3.15) in which the energy radiated by the source is split into the sample and reference beams. In the photometer section, the two beams are combined by a rotating sector mirror to form a single beam containing pulses of radiation from the sample and reference beams. This combined pulsed beam passes into the monochromator where it is dispersed by the grating into its spectral components. As the grating is rotated, the dispersed spectrum is scanned across the monochromator exit slit. The mechanical width of the monochromator slit determines the width of the wavenumber band emerging from the monochromator.

After leaving the monochromator, the radiation passes through one of a set of optical filters. The appropriate filter is automatically selected for the spectral region being scanned. The optical filter rejects unwanted radiation diffracted from the grating. Finally, the transmitted radiation is focused onto a thermocouple detector.

The electrical signal from the thermocouple detector is applied via a pre-amplifier to an analog-to-digital converter. The digital data produced is applied to the ordinate microprocessor, which controls the ordinate function of the instrument. The microprocessor output is converted back to an analog form, which is applied to the servo system.

Although a smaller slit width decreases the bandwidth and provides improved resolution, it results in decreased intensity of the emerging radiation, thereby decreasing the signal-to-noise ratio. In order to maintain a uniform signal-to-noise response, the detector output is generally maintained at an approximately constant level over the range of the instrument by programming the widths of the monochromator slits by means of a cam drive.

A ceramic tube heated by an internal metallic element acts as a source of radiant energy. The ceramic tube, when heated to about 1100°C, produces a continuous spectrum of electromagnetic energy, most of which is in the infrared region. The incandescent portion of the ceramic tubing is approximately 15 mm long and 3 mm in diameter.

The radiant energy is split into the reference and sample beams by the plane mirror *M1* and toroidal mirrors *M2* and *M3*. The toroidal mirror *M2* focuses the sample beam onto the sample aperture in the sample panel and *M3* focuses the reference beam onto the reference aperture in the sample panel. The use of toroidal mirrors eliminates astigmatism in the source images.

The sample and reference beams are combined by the sector mirror, which rotates at ten revolutions per second and gives a modulation frequency of 20 Hz. Plane mirrors *M4*, *M5* and *M6* are so oriented that the sector mirror will alternately pass the sample beam and reflect the reference beam through the aperture stop which is imaged on the grating in the monochromator and ensures that both beams are the same size. The toroidal mirror *M7* then focuses this combined beam on the monochromator entrance slit *SI*, with the beam being directed by the plane

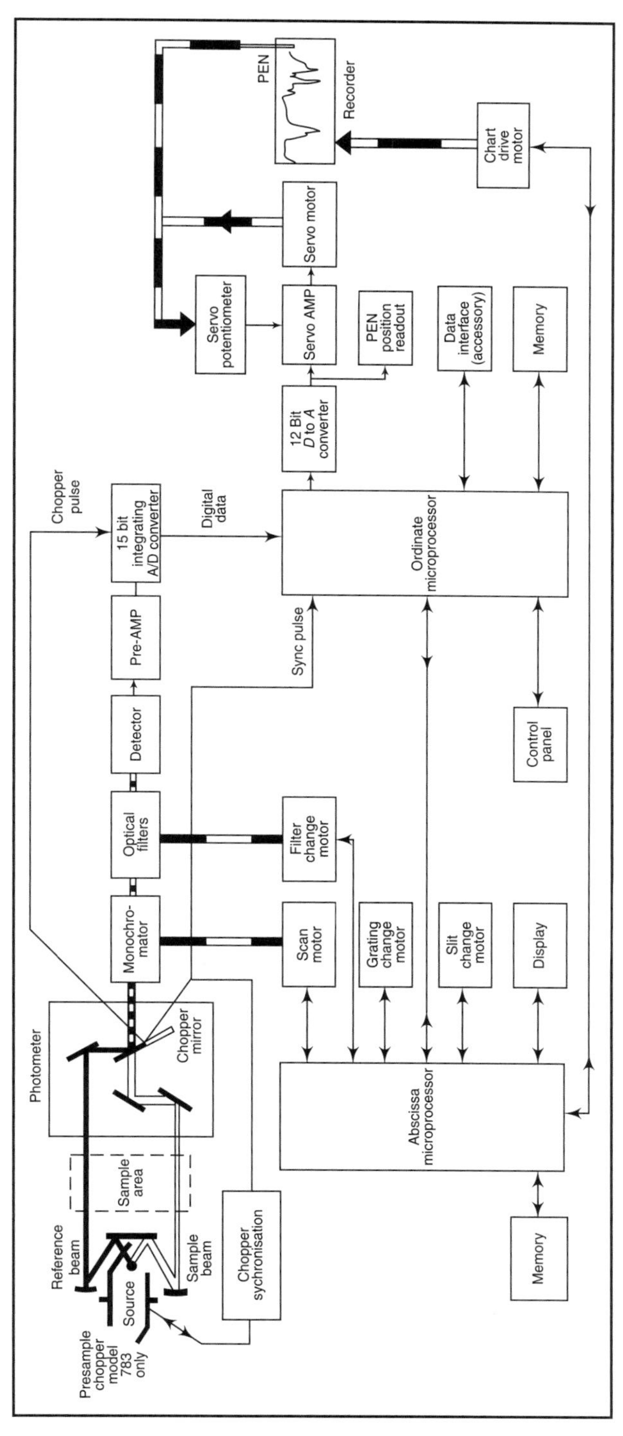

Figure 3.14 :: Block diagram of a scanning infrared ratio recording infrared spectrophotometer

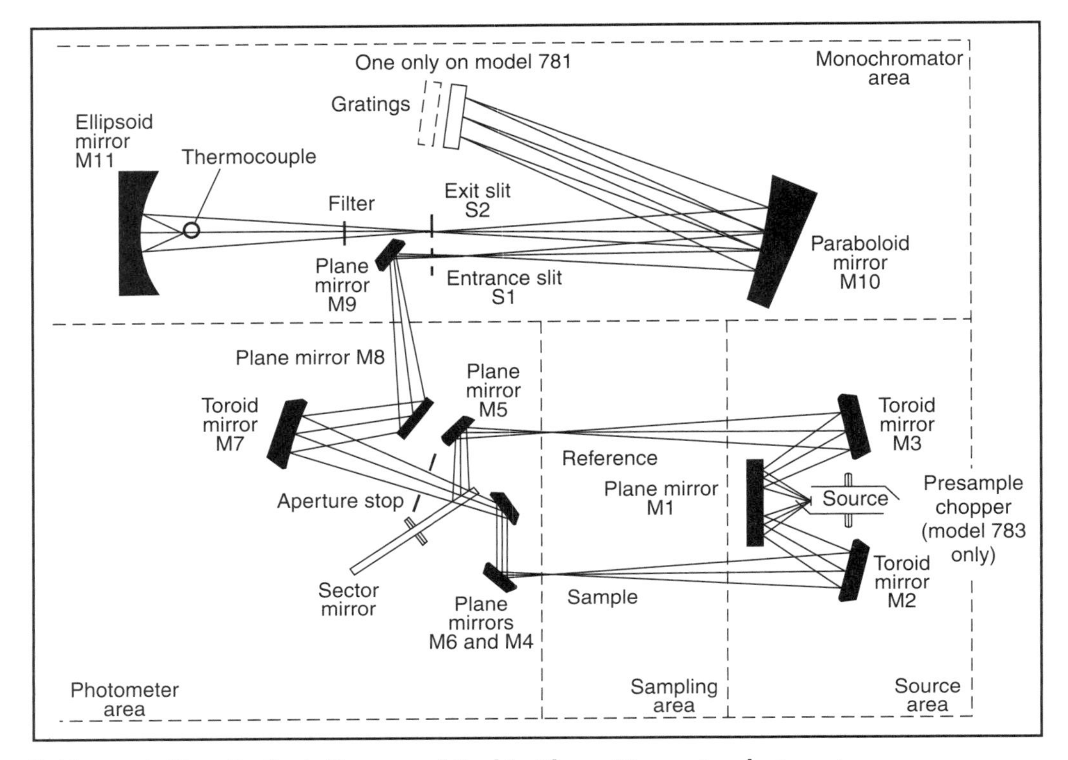

Figure 3.15 :: Optical diagram of Perkin Elmer IR spectrophotometer

mirrors *M8* and *M9*. The beam diverges from *S1* until the 19° off-axis parabolic mirror *M10* reflects it as collimated radiation onto the grating. The radiation, after being diffracted by the grating is focused in the plane of the exit slit *S2* by *M10*.

In order to make measurements over a wide spectral range, two gratings are used, which are mounted back to back. One grating has 100 while the other has 25 lines per millimeter. The grating is selected automatically depending upon the wavenumber setting of the instrument. In order to cover the full range of the instrument, the gratings are selected as follows:

Range	Grating	Order
4000 to 2000 cm^{-1}	100 lines/mm.	2nd
2000 to 600 cm^{-1}	100 lines/mm.	1st
600 to 200 cm^{-1}	25 lines/mm.	1st

The radiation of different wavenumbers corresponding to different orders of diffraction from the grating emerges from the exit slit and impinges on the optical filter assembly. The optical filter rejects radiation from all but the desired order of diffraction. The optical filters are mounted on an eight-position wheel. The first three positions of the wheel correspond to the second order of the 100 lines/mm grating. In order to obtain second order diffraction radiation to pass while blocking first order radiation, two transmission filters are used in each of the three positions. In the other five filter positions, the gratings are operated in their first orders only for which single long wavelength pass filters are employed. Filter changes occur at 3120, 2500, 2000, 1150, 700, 400 and 250 cm^{-1}. Variation of the slit width over the range of the instrument is achieved by the action of a cam, whose radius is sensed to control the separation of the slit jaws with the help of the slit motor.

The detector (thermocouple) and the monochromator exit slit are located at the foci of the ellipsoid M11, with the ratio of its focal lengths being 1:8. The radiant energy leaving the exit slit is therefore, in linear dimensions by a factor of eight. A cesium iodide lens on the thermocouple assembly further reduces the linear dimensions of the slit, with the image falling on the thermocouple target by a factor of 1.4.

The signal from the thermocouple is approximately 0.5 μV with the medium slit width. The signal is amplified in a low noise-high gain pre-amplifier before it is given to an A-D converter. The digital data from the A-D converter is applied to the ordinate microprocessor. The function of the instrument is controlled through two stepper motors; the monochromator scan and chart drive motors. The speeds of these motors are determined by the repetition rate of the pulses applied to their stator coils. These pulses are derived from a voltage controlled oscillator which is phase locked to the mains frequency of 50 Hz.

The Perkin-Elmer instrument is a double-beam ratio recording instrument which covers a range of 4000 to 200 cm^{-1}, has three scan speeds and provides for a slit selection from four ranges to give resolution from 1.2 cm^{-1} to 5.5 cm^{-1}. The recommended environmental conditions for the instrument are 15°C to 35°C temperature and 75% maximum relative humidity.

3.4 Sample Handling Techniques

A wide variety of samples can be analyzed by using IR spectroscopy. Solids, liquids or gases can all be handled. However, sample handling presents a number of problems, since no rugged window material for cells exists, which is transparent over the entire IR range and is also inert. The most commonly used window material is NaCl, which transmits down to about 650 cm^{-1}. KBr transmits down to about 400 cm^{-1}, CaBr to about 250 cm^{-1} and CsI to about 200 cm^{-1}. Since these materials are all water-soluble, the surfaces of the windows made from them are easily fogged by exposure to atmospheric water vapour or moist samples. Therefore, they require frequent polishing when used under such conditions.

3.4.1 Gas Cells

Gas cells usually have a path length of 100 mm, since this thickness gives a reasonable absorbance level for the majority of gases and vapours at the normally encountered partial pressures. The inner cell diameter is typically 40 mm and the volume, about 125 cm^3. The ends of the cells are ground square, to which infrared transparent windows are glued with sealing gaskets. Two tubes with a stopcock are attached to the cell for connecting the cell to the gas handling system. A typical glass cell may have a body of uniform cross-section, but it is preferable to use a tapered construction to conform closely to the section of the radiation beam and requires less sample volume for a given thickness. When weak bands are to be studied, it is necessary to use very long cell paths, which it is inconvenient fit into the instrument. In such cases, multiple reflections using a combination of a mirror system, with a beam condenser, enable the spectra of minor components to be recorded.

3.4.2 Liquid Cells

The extinction coefficient of most liquid hydrocarbons in the infrared region is such that a pure sample of thickness between 0.01 and 0.05 mm gives an absorption spectrum that is quite suitable for analysis. The transmittance lies between 15 and 70 per cent. Other liquid materials are not markedly different.

Two types of cells are generally available for the examination of liquid samples: sealed cells for liquids of high vapour pressure and demountable ones for all other liquids. Both types are with path lengths of 0.02, 0.04, 0.06, 0.10, 0.16, 0.25, 0.4, 0.6 and 1.0 mm, and with the earlier stated window materials. The external cell diameter is 30 mm. About 100 μl sample volume per 0.1 mm path length is required.

Liquid cells of this thickness consist of IR transparent windows separated by thin gaskets of copper and lead, which have been amalgamated with mercury before assembly is securely clamped together. As the mercury penetrates the metal, the gasket expands providing a tight seal. The filling is done with hypodermic syringe. The demountable type cells can be assembled for different path lengths, in which case circular spacing foils are placed between the windows.

3.4.3 Variable Path Length Cells

These are employed for determining the absorption coefficient and concentration of the test substance, as well as for establishing calibration curves. When using the cell in the reference beam, the solvent's absorption bands may be completely compensated by continuously changing the path length. The cell path length is usually adjustable (Figure 3.16), in the range 0–5 mm via a differential thread with an accuracy of ± 1. The path length can be read off the setting drum to an accuracy of 1 μ. In precisely fitted piston guideways, the cell windows are displaced against each other, thus achieving an extreme tightness of the cell. This permits long-time scans to be carried out with highly volatile solvents. This type of cell is normally supplied with KBr windows. CaF_2 or NaCl windows are also available. The length of the cell is typically 120 mm and the diameter, 75 mm.

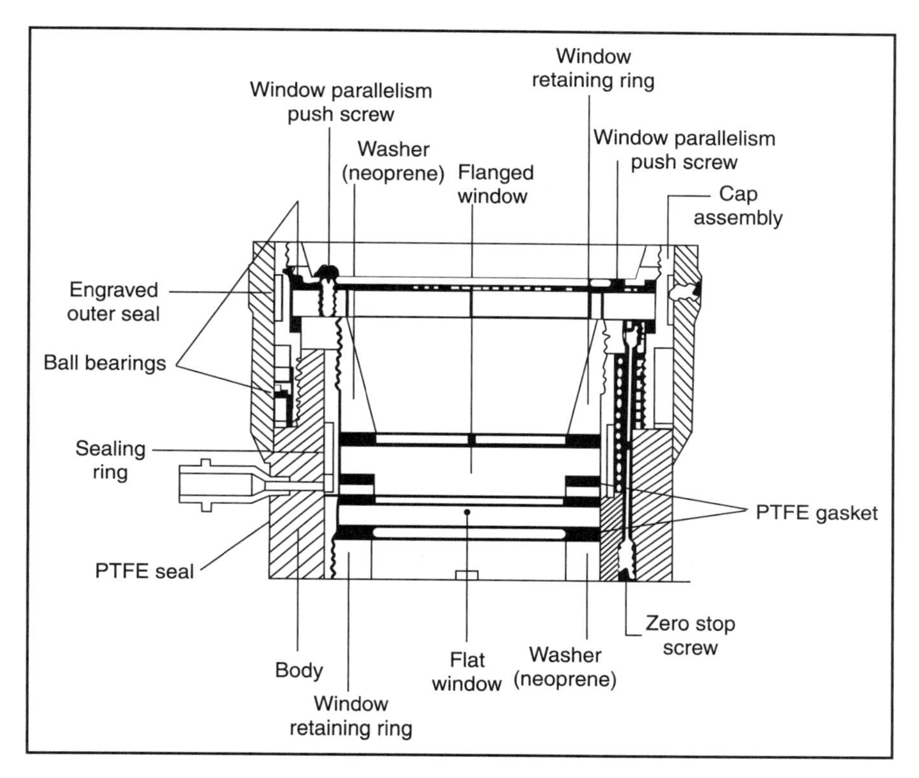

Figure 3.16 :: Variable path–length cell

In order to achieve repeatable cell thickness, it is important to be able to measure the path length. The path length can be conveniently measured by inserting the empty cell into the spectrometer, and observing the interference pattern created by the reflection of part of the radiation beam from the internal faces of the cell. The pattern is a regular sine wave, when recorded on a linear wave number scale and is of amplitude between 2 and 15 per cent T. The path length of the cell is given by

$$\text{Path length (in cm)} = \frac{n}{2(v_1 - v_2)},$$

where n is the number of fringes between the wavenumbers v_1 and v_2. The method is not practical in case the cells are in a bad condition. Alternative methods like the use of a travelling microscope, may then be employed.

3.4.4 Sampling of Solids

Different methods are available for sampling solids for examination in the infrared spectrometer. These methods are detailed below.

Solids Dissolved in Solutions

In case it is possible to obtain suitable solutions of the solids, the solids are dissolved and examined as dilute solutions by running in one of the cells for liquids. However, there is no single solvent which is transparent through the entire infrared range. In order to cover the main spectral range between 4000 and 650 cm^{-1} (2.5–15.4 μ), a combination of carbon tetrachloride and carbon disulphide are employed.

Pressed Pellet Technique

In the pressed pellet technique, the solid sample is finely ground and mixed with an alkali halide like KBr, and then pressed into the form of a disc for examination in the instrument. Pressing is done in an evacuable die, under high pressure. The resulting disc or pellet to be directly run is transparent. Other alkali halides like CsI and CsBr are used for measurement at longer wavelengths.

Mull Technique

The infrared spectrum of finely powdered solids may be obtained by dispersing the powder within a liquid medium, so that a thick slurry is produced. The thick slurry is spread between IR transmitting windows. The liquid is so chosen that it has the same refractive index as the sample, so that energy losses due to scattering of light are minimized. The most commonly used liquid mulling agent is a mixture of liquid paraffins, known as Nujol (mineral oil). Although Nujol is transparent throughout the IR spectrum, it cannot be used if C-H stretching and bending frequencies are to be observed. In this case, a second mull like perfluorokerosene of hexchlorobutadiene may be used.

3.4.5 Micro-sampling

When the quantity of sample available for infrared spectrophotometric examination is very small, standard sampling techniques cannot generally be used. This is because the absorbance of a given sample is proportional to the number of grams of absorbing material and inversely proportional to the area of the sample. Thus, in order to increase the absorbance for a given weight of sample, it is necessary to decrease the area of the sample. The identification of gas chromatographic fractions, new compounds and materials isolated from natural products, frequently involves microgram quantities of samples and necessitates the use of micro-sampling techniques for the measurement of their infrared spectra.

The simplest method of coping with small quantities of a sample is to use cells having reduced apertures, so that the spectrophotometer beam falls on the entire sample. These cells (for example, the M-OX series of Beckman-RIIC microcells) are generally individually tailored to the beam dimensions in the sample compartment of a particular instrument in order to facilitate the most efficient use of the limited amount of sample available.

In cases where the quantity of available sample is very small, or where the spectrophotometer design does not permit small sample areas to be used, it is necessary to employ a beam condenser. Basically, a beam condenser consists of an appropriate combination of lenses and mirrors, which reduce the size of the spectrophotometer beam, focus it on the sample, and then return it to its normal size before entering the monochromator. Beam condensers are usually designed to fit directly into the sample compartment and to hold a variety of micro- liquid, solid, and gas cells. Although most beam condensers waste some energy because of reflection losses and optical aberrations, most are highly efficient and conserve sufficient energy to permit extremely small samples to be examined in a nearly routine fashion. Typically, beam condensers have condensing factors of about four or six-to-one and are available either with all-reflecting optics or with refracting optics, employing alkali halide lenses.

The normal sealed liquid cell has a nominal volume of about 0.1 ml with a 0.1 mm spacer. By using a cell with a specially shaped spacer for the size of the beam, this volume can typically be reduced to about 0.01 ml. Ultra-micro cells for use in beam condensers have volumes of around 2 μl, with a 0.1 mm spacer. Powder sample techniques are probably the most widely used in the infrared analysis of micro-solids. In micro-sample applications, a small quantity of finely ground powder can be suspended in mineral oil and smeared on a demountable cell window; or it can be ground with KBr, pressed into a micro-disk with a micro die, and mounted on a micro solid sample holder.

Of all the techniques, the micro KBr disk is by far the most popular, because of its ability to deal with even the most minute samples. Samples, as small as 1–100 μg, can be examined routinely, depending upon the disk size.

3.5 Fourier Transform Infrared Spectroscopy (FTIR)

Fourier transform infrared spectroscopy (FTIR) is rapidly becoming a common feature in modem spectroscopy laboratories. A wide range of commercial FTIR spectrometers with very different specifications are now available. This has become possible with the availability of inexpensive microcomputers.

The fourier transform technique depends upon the basic principle, that any wave function could be represented as a series of sine and cosine functions with different frequencies (Davies, *et al.* 1985). This is illustrated in Figure 3.17 which shows the simple ease of adding a series of sine waves. Figure 3.17 (a) shows two sine waves, one having half the amplitude and double the

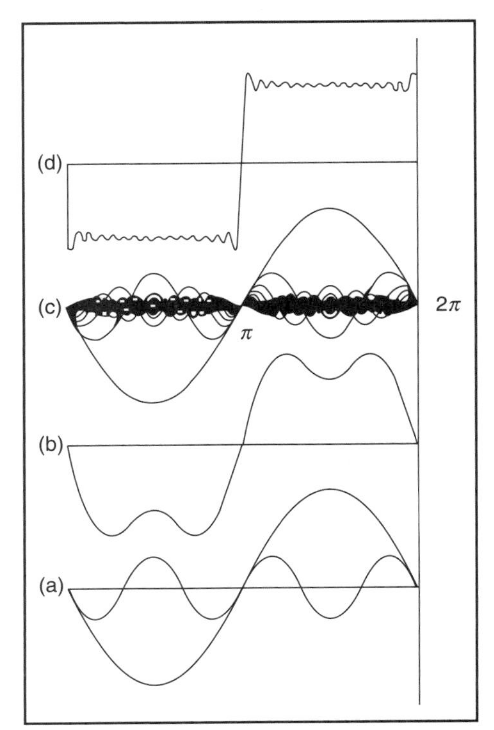

Figure 3.17 :: Summation of sine waves to illustrate the idea of the Fourier series

frequency of the other. The sum of these two sine waves is shown in Figure 3.17 (b). Curves of higher and higher frequency can be added and, in fact, Figure 3.17 (d) shows a series of 15 sine waves and their addition. It can be seen that in the limit, the curve will become a square wave. Logically, in order to fit more complicated waves, it becomes necessary to consider both the sine and cosine series. The determination of the sine and cosine components of a given wave function is known as Fourier transformation (FT).

FT facilitates two different presentations of the same experimental data, known as domains. These are most commonly the time domain, in which the data are recorded as a series of measurements at successive time intervals and the frequency domain, in which the data are represented by the amplitudes of its sine and cosine components at different frequencies. Putting the FT into practice can become tedious and time-consuming even when using computers, and especially when a large number of points have to be considered. The situation was considerably eased by the development of fast Fourier Transformation (FFT) algorithm by Cooley and Tukey (1965), which made it possible to carry out FT of complex data in a matter of few seconds. Most of the present day applications of FT in analytical techniques are dependent on Cooley–Tukey FFT.

There are many advantages of using the FT spectroscopic technique over dispersive or continuous wave (CW) instruments, to record signal intensities directly as a function of frequency. If we measure a signal as a function of time, we obtain information on all frequencies in the spectrum simultaneously, whereas a CW instrument is confined to measuring only one frequency, or a very narrow band of frequencies at a time. Further, a typical spectrum consists of a few sharp peaks, with long stretches of noisy base line, i.e. CW scan wastes most of the measuring time by recording this base line. A considerable gain in signal-to-noise for a given total measurement time is achieved by repeatedly measuring an interferogram and then with Fourier transforming the data, rather than by scanning through the frequencies directly.

3.5.1 FTIR Spectrophotometers

The heart of an FTIR spectrometer is a two-beam interferometer, most commonly of the Michelson type (Chalmers, 1983). The basic optical components of the interferometer are shown in Figure 3.18.

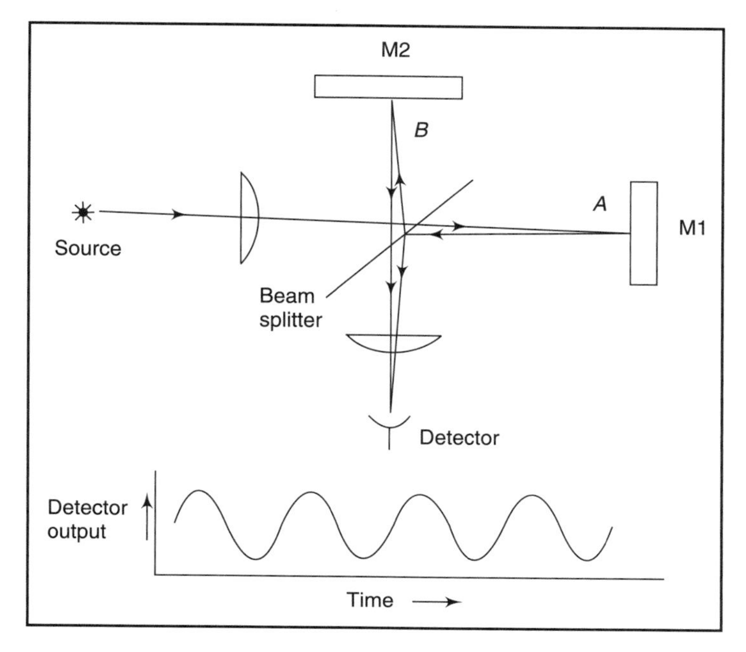

Figure 3.18 :: Principle of Michelson interferometer

Beam splitter

The radiation entering the interferometer is split into two beams by a beam splitter. Beam *A* follows the straight path before returning to the beam splitter, whereas the distance travelled by beam *B* can be varied, before it re-combines with beam *A*. When the beams *A* and *B* re-combine, an interference pattern is produced, which is incident on a detector. When the two beams are in phase at the beam splitter, the maximum intensity will reach the detector. The intensity will be minimum in case the beams are out-of-phase. If provision is made in such a way that mirror M_2 is displaced uniformly, the detector output will be a sine wave, whose frequency is determined by the translation velocity of M_2 and the wavelength of the monochromatic radiation, and the amplitude of the signal will depend upon the intensity of the incoming radiation.

If the incoming radiation is polychromatic, the detector output wave of unique frequency will be produced for each component. The overall detector output as a function of time (interferogram) will be the sum of the waves for each frequency component. It contains all the information associated with a conventional spectrum, but since it is the Fourier transform of the spectrum, the spectral information can be extracted by performing the appropriate inverse Fourier transformation, with the help of a computer.

Figure 3.19 shows the schematic of a Fourier transform spectrometer. The source and detector are of the conventional types that are used in IR spectrometry. A small computer is employed to control the scan system and carry out the mathematical transformation. The heart of the system is the mirror scan mechanism. For faithful reproduction of the spectrum from the interferogram, the

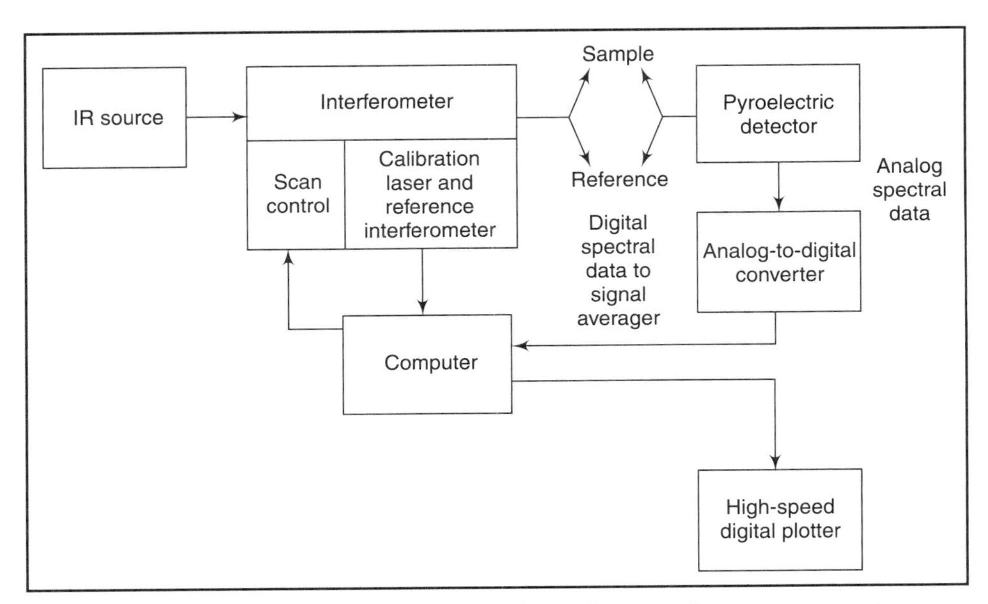

Figure 3.19 :: Block diagram of a typical Fourier transform spectrometer

detector output must be known as a function of the mirror displacement. The measurement of mirror displacement becomes increasingly difficult as the wavelength decreases. At shorter wavelengths, a special technique known as the 'fringe referencing technique' is used. Here, one transducer is used to simultaneously drive the movable mirrors of two interferometers through identical displacements. The sample radiation passes through one interferometer and produces the sample interferogram. The reference interferogram is recorded by using a He-Ne laser. This reference laser ensures that sampling of the detector signal occurs reproduceably from scan to scan. Since the interferogram of a monochromatic source (laser) is a sine wave, its zero crossings provide an accurate trigger for digitized data the collection of the main interferogram. Further, the use of He-Ne laser reference interferogram provides an internal frequency calibration, important to high resolution studies.

Figure 3.20 shows the double-beam Model 1800 FTIR spectrophotometer from Perkin–Elmer, which is based on the scanning Michelson interferometer. The instrument gives a resolution of better than 0.2 cm^{-1} in the wavelength range of 4500-450 cm^{-1}. The instrument uses the TGS (triglycine sulphate) detector and Ge/KBr beam splitter. The instrument employs three different microprocessor systems. The first is the system controller, a Perkin-Elmer PE 7700 professional computer. The PE 7700 is a 16-bit microcomputer, which comes equipped with 1 Mb RAM, two double-sided, double density floppy disk drives, a 20 or 40 M byte disk drive, colour CRT with eight soft keys, and four RS-232C and IEEE-488 communications ports. The PE 7700 processor performs all the data acquisition and subsequent calculations. A number of software packages are available for different data processing routines. The other two microprocessor systems are coupled

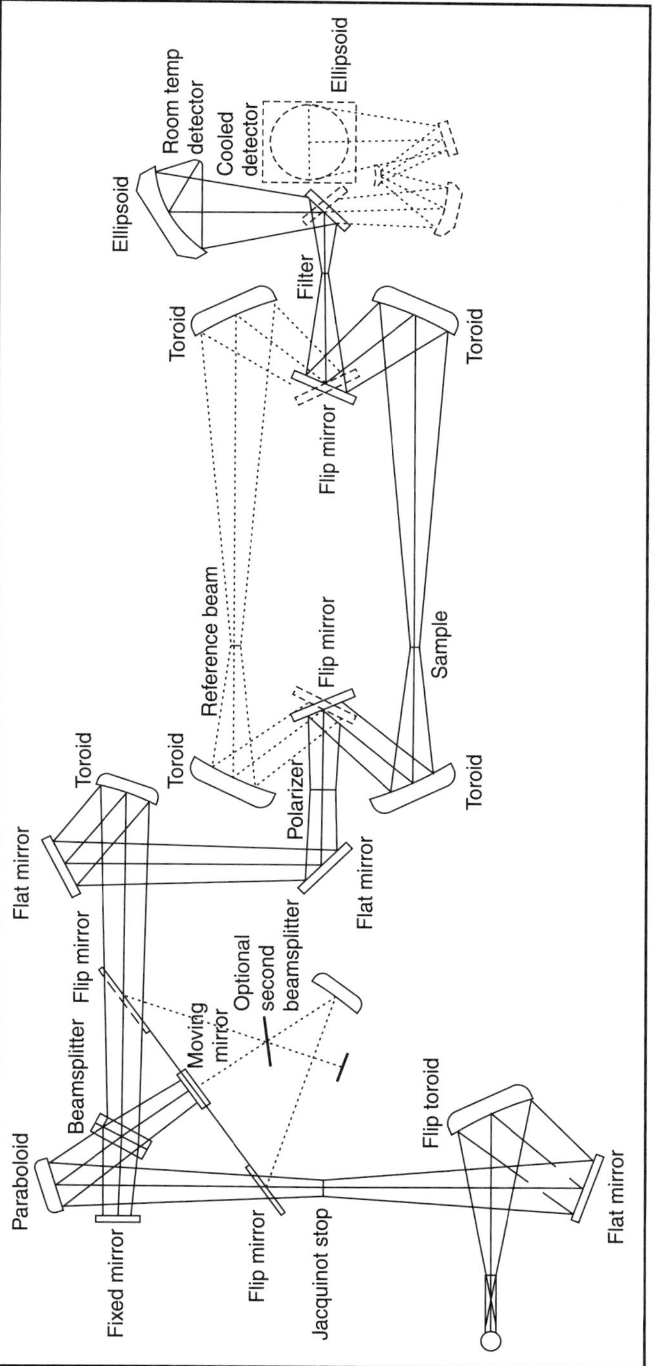

Figure 3.20 :: The optical layout of the Model 1800 Perkin Elmer Fourier Transform infrared spectrophotometer

with the optical unit. The instrument supervisor is based on the Motorola MC 68000. The actual control of the optical hardware is performed by a coprocessor based upon the Motorola 6802. The instrument processor seems to acquire and perform a wide range of calculations using 32-bit precision.

The recording spectrum by the Fourier transform technique requires more effort than conventional dispersion technique. Therefore, at present, the chief areas of application of FTIR are in the far-infrared absorption studies with remote sources, in IR emission studies of weak emitters, and in rapid scanning measurements.

The application of Fourier transform spectrometry has been extended to the UV/Visible region in the entire spectral range from 160 nm–700 nm (Snook, 1986). The resolution achieved is much greater than in the case of instruments based on grating monochromators.

3.6 Calibration

In order to check the IR spectrophotometer, the spectrum of a polystyrene film is run and the absorption spectrum is compared with that provided by the company. In case any difference in the absorption spectrum is detected, the instrument has to be thoroughly checked and recalibrated.

Broad-band wavelength calibrators

In order to facilitate speedy recalibration of low resolution far-infrared instruments, special broad-band wavelength calibrators are often used. They usually employ polythene powder as the matrix. However, their fabrication is a difficult procedure. A broad-band wavelength calibrator for use with low resolution far-infrared monochromators, which makes use of thin polythene sheets rather than polythene powder as the substrate, can also be used. The method of construction involves taking a clear polythene sheet and stretching it 12" below a 250 W infrared lamp. Using an electric spray, a suspension of mercuric oxide powder in carbon tetrachloride is sprayed on to the polythene sheet, in sprays of a few seconds each. When a uniform layer of mercuric oxide has been deposited, another polythene sheet is pressed on top, and the sandwich is left under the infrared lamp for a further 15 minutes. A filter of approximately 1.5 inch is cut from the uniform part of the sandwich and is mounted in a metal holder.

The transmittance of the filter obtained between 800 and 350 cm^{-1} on a Perkin-Elmer 457 spectrophotometer shows that it provides three accurate calibration points: one due to the polythene matrix itself (at $722 \pm 1\ cm^{-1}$) and the other two (at $590 \pm 2\ cm^{-1}$ and $500 \pm 1\ cm^{-1}$) due to mercuric oxide (Siddiqui and Stewart, 1974).

3.7 Attenuated Total Reflectance (ATR) Technique

Infrared methods can be used to study materials which are light scattering or opaque, or in the form of coatings on opaque materials. The technique usually employed is based on the obtaining of the reflectance spectrum. However, a spectrum obtained by reflection of the radiation from the surface of a chemical material is generally very poor. In order to overcome this difficulty, a technique known as 'attenuated total reflectance', which enables one to obtain reflection spectra of satisfactory quality, is used.

In this technique, energy from the source enters into a prism and is reflected almost entirely from face *C* when the beam strikes it at less than a critical angle. The prism consists of a crystal of material of high refractive index and transparent to infrared radiation. Silver chloride and KRS-5 satisfy these conditions.

If the sample is placed in contact with the face *C*, the internal reflection will be attenuated or reduced at the sample–crystal surface. The attenuation will take place at those wavelengths, where the material absorbs infrared energy. This facilitates achievement of absorption spectra of coatings and liquids even if they are opaque. The information conveyed by the ATR spectra is essentially the same as that conveyed by the transmission spectra. The optical arrangement of a simple ATR system is shown in Figure 3.21.

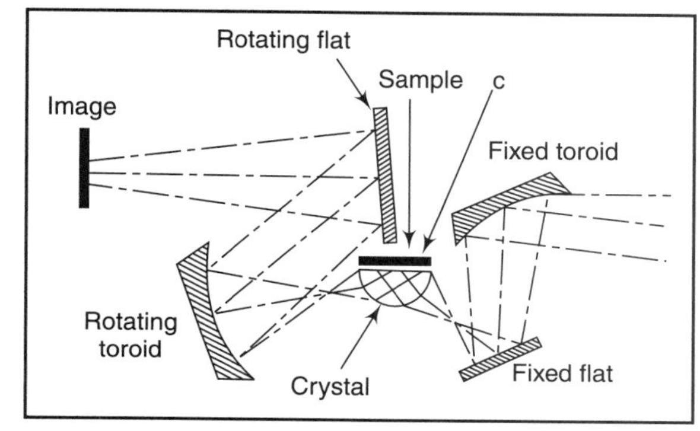

Figure 3.21 :: Optical arrangement of a simple attenuated total reflectance system

Single reflection ATR is not always adequate because of the low sensitivity observed. Improvement in sensitivity can be obtained by the use of multiple internal reflection systems, in which internal reflections take place 25 to 50 times before emerging out of the crystal (Figure 3.22). This strengthens the absorption pattern of a material placed on one surface of the optical flat. This is best achieved by reducing the width of the entrance face of the crystal and by increasing its length. However, there is a practical limitation, as the entrance face cannot be made smaller than the monochromator slit, otherwise it would result in serious loss of energy.

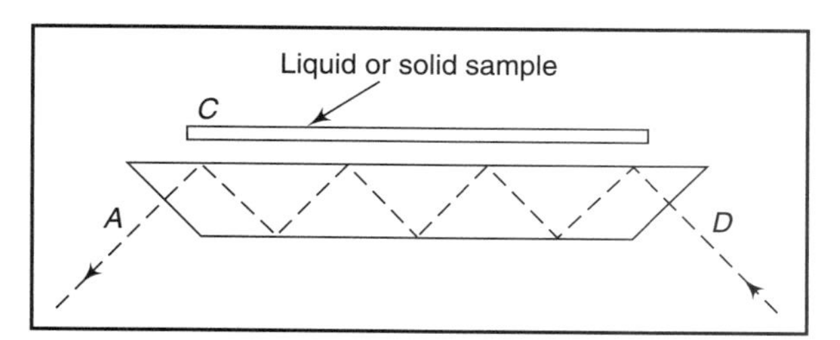

Figure 3.22 :: Multiple internal reflections

One of the main advantages of the ATR technique is that the spectrum obtained is independent of the sample thickness. Typically, the reflected radiation penetrates the sample to a depth of only a few microns. Consequently, the method has been found to be particularly useful for establishing the surface characteristics of many materials. However, because of its small penetration depth, ATR spectra is inherently weak.

ATR analysis

The prime objective in preparing samples for ATR analysis is to obtain intimate sample–prism contact. Since most of the commonly used ATR prism materials are fairly soft and easily deformed, only a limited amount of direct pressure can be applied to the sample, in the attempt to obtain the accessory contact. Too much sample pressure can seriously distort the optical surface of the prism and result in lost energy and poor spectra. Films can, of course, be cast directly upon the optically active surface of the prism, provided all traces of solvent can be removed without heat. Flexible, malleable samples, e.g. rubber and plastics, can be simply pressed against the prism surface. Rigid, irregularly shaped samples are, by far, the most difficult to handle, since they frequently require grinding, sanding and/or polishing to achieve the flat surface required for this type of work.

Chapter 4

Flame Photometers

4.1 Principle of Flame Photometry

Quantum theory

According to the quantum theory, the energy states of an atom or molecule are sharply defined and any change from one state to another therefore requires a sharply defined quantity of energy. When radiation falls on a material, or the material is supplied with extra energy in some form, some part of the energy is taken up by the material and alters the state of the atoms or molecules of which it is composed. The atoms or molecules of the material are promoted to higher energy states. However, the higher energy states are rather unstable. The particles at the higher energy levels tend to lose the extra energy and return to the original level or ground state, either by undergoing a chemical reaction, or by dissipating the energy as heat, or by emitting the energy as radiation. If it loses all or part of the energy as radiation, it will emit photons of an energy corresponding to a difference between two energy levels. Since the levels are clearly defined for a given atom, the radiation will be emitted at clearly defined frequencies only. The frequencies are shown up as bright lines if the emitted light is dispersed as a spectrum. By measuring the wavelength of the emission, one can find out the atoms that are present. Also, by measuring the intensity of the emission, one can compute the concentration of the element.

In short, the principle of flame photometry is based on the fact that if an atom is excited in a flame to a high energy level, it will emit light as it returns to its former energy level. By measuring the amount of light emitted, we can measure the number of atoms excited by the flame.

The method used for flame photometric determinations is simple. A solution of the sample to be analysed is prepared. A special sprayer operated by compressed air or oxygen is used to introduce this solution in the form of a fine spray (aerosol) into the flame of a burner operating on some fuel gas, like acetylene or hydrogen. The conversion of sample solution into an aerosol by atomizer does not bring about any chemical change in the sample. However, the heat of the flame which vaporizes sample constituents, molecules and ions of the sample species are decomposed and reduced to give atoms. The heat of the flame causes excitation of some atoms into higher electronic states. Excited atoms revert to the ground state by emission of high energy of characteristic wavelength.

The radiation of the element produced in the flame is separated from the emission of other elements by means of light filters or a monochromator. The intensity of the isolated radiation is measured from the current it produces when it falls on a photocell. The measurement of current is done with the help of a galvanometer, whose readings will be proportional to the concentration of the element. After carefully calibrating the galvanometer with solutions of known composition and concentration, it is possible to correlate the intensity of a given spectral line of the unknown sample, with the amount of the same element present in a standard solution.

Flame photometry is characterized by a high degree of constancy and reproducibility. The spectrum of an element as produced in a flame is relatively simple, consisting normally of only a few lines. Identification of the line is simple and spectral interference is less frequent. The most usual application of flame photometry is for the analysis of sodium and potassium in body fluids and these analyses constitute the bulk of the determinations generally performed. However, the method is slowly replacing the more troublesome methods for other elements also.

Flame photometry is mostly concerned with atoms. Molecules cannot normally survive the high temperatures employed in flame photometry.

The number of elements to which flame photometric methods can be applied depend mainly upon the source temperature developed by the fuel mixture employed. This is because not all atoms are easily excited in the flame. The energy levels of the atoms of some elements are so spaced that they emit a large number of different lines, while for some others, almost all the light they emit is concentrated in one spectra line. The upper energy levels or atoms of some elements are so high above the ground state that they are very difficult to excite. Also, their emission lines could be of a wavelength that may not be directly usable. All these factors determine the lowest concentration of atoms which can be detected in the sample and thus varies very widely for different atoms.

Emission intensity versus concentration

A typical plot of emission intensity versus concentration of ionic species in the solution being measured is linear over a wide range, but with a deviation at both low and high concentrations. This is shown in Figure 4.1. The deviation from linearity at very low concentrations is because the emission falls below the expected value due to ionization where some atoms get converted back to ions. The ionization is insignificant at higher concentrations. On the other hand, the negative deviation observed at high concentration is due to self-absorption. This involves partial absorption of the photons emitted by excited atoms by the ground state atoms in flame.

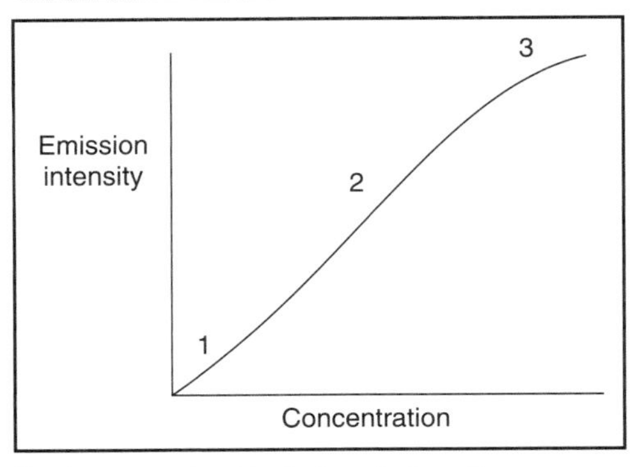

Figure 4.1 :: Emission intensity vs. concentration

Flame photometry offers the following advantages:

- The technique is very rapid. It does not require any chemical preparation except preparation of a solution of suitable concentration.
- The method is highly useful for the analysis of some elements, which are difficult to measure by other methods.
- The technique is most suited to analytical problems, in which a large number of samples of similar types have to be measured.
- The method is quite cheap as it does not require any other expensive reagents.

4.2 Constructional Details of Flame Photometers

A flame photometer has three essential parts (Figure 4.2).

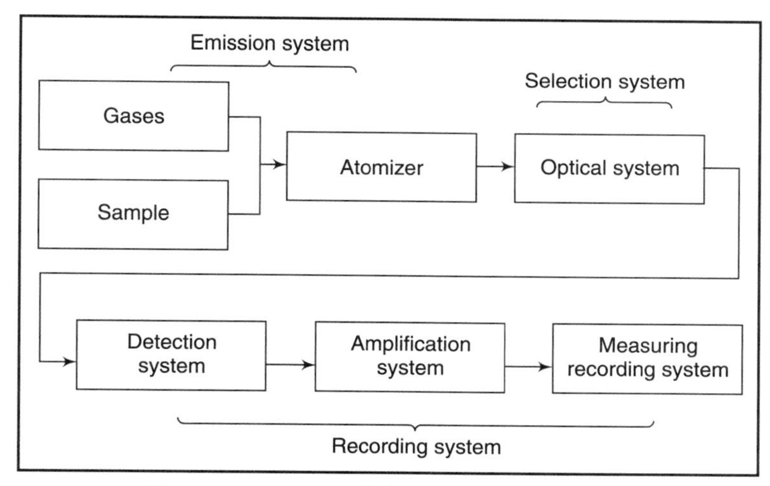

Figure 4.2 :: Essential parts of a flame photometer

Emission System: It consists of the following:

(i) Fuel gases and their regulation, comprising the fuel reservoir, compressors, pressure regulators and pressure gauges;
(ii) Atomizer, consisting, in turn, of the sprayer and the atomization chamber, where the aerosol is produced and fed into the flame.
(iii) Burner, which receives a mixture of the combustion gases.
(iv) Flame which is the true source of emission.

Optical System: It consists of the optical system for wavelength selection (filters or monochromators), lenses, diaphragms, slits, etc.

Recording System: It includes detectors like photocells, phototubes, photomultipliers, photodiodes, etc., and the electronic means of amplification, measuring and recording.

Figure 4.3 shows a typical block diagram of a typical flame photometer.

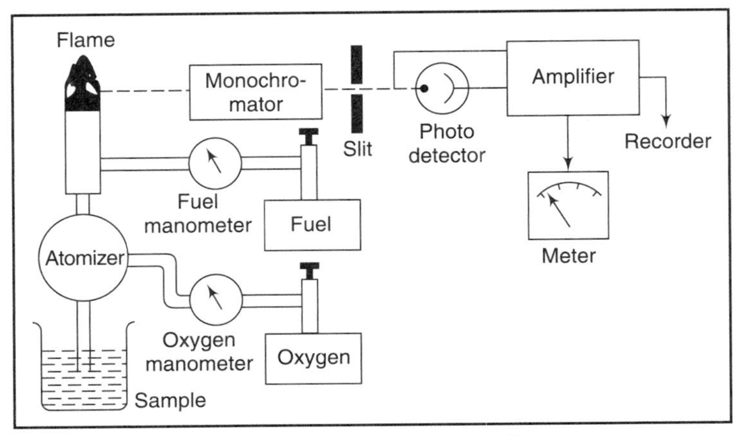

Figure 4.3 :: Block diagram of a flame photometer

4.2.1 Emission System

(a) Fuel Gases and their Regulation

Pressure Regulators: In order to obtain a steady emission reading, it is imperative to have a flame that is perfectly steady and free from flickers. In order to achieve this, the air or oxygen and fuel pressure has to be maintained constant during the operation of the instrument. Suitable pressure gauges are therefore provided in the instrument to indicate the pressure that is actually present in the line. A 25-lb gauge for the oxygen or air supply and a 10-lb gauge for the fuel, are generally used. Pressure regulators are usually followed by needle valves for controlling the flow. Gauges provided with regulators are often not sensitive enough to detect small changes of pressure, which have a profound effect on the flame photometer operation. Therefore, narrow range pressure regulators and manometers are installed in the line, in order to observe small changes in pressure or gas flow.

Flowmeters: A flowmeter may be inserted in the line from the gas reservoir to the atomizer in order to detect any clogging of the orifice. For the same flow rate, an appreciable change in the gauge pressure indicates a partially clogged orifice. By controlling the individual flow rates of the fuel and oxygen, the operator can choose various fuel–oxygen mixtures ranging from lean flame mixtures to fuel-rich types of flames. The flow rates usually vary from 2 to 10 cu. ft/h.

Fuel Supply: The fuel gas normally used in flame photometry is acetylene gas, which is commercially available in cylinders of various sizes. Cylinder acetylene consists of acetylene gas dissolved in acetone, which, in turn, is absorbed on a porous filling material. Consequently, after the flame is lit, it should be allowed to burn for several minutes before attempting adjustments in order to vent the excess acetone initially present in the vapour phase. The consumption of acetylene ranges from 1 to 5 cu. ft/h. The other fuels used in flame photometry are propane, butane and hydrogen. When the available gas pressure is less or is variable, a booster pump is necessary. The pump generally used is a motor-driven diaphragm pump, which delivers the gas to the burner at the required volume and pressure.

Oxygen Supply: Oxygen from cylinders should be supplied to the burner through a regulator capable of delivering approximately 12 cu. ft/h. at a pressure of 12–15 lb per sq. inch.

Air Supply: The air supply can be provided from a cylinder of compressed air or from an air compressor through a tank held at about 10 lb/per sq. inch. A pressure compensation valve is placed between the reservoir and the burner. When compressors are used, the air should be filtered through glass wool. Consumption of air is approximately 10 cu. ft/h for an acetylene air flame.

(b) Atomizer

One of the most exacting problems in the flame photometer design is the manner in which the sample is fed to the flame at a uniform rate. The usual method is to prepare solutions of known concentrations and to spray these into the flame, by using some form of aerosol production. The use of an aerosol permits its distribution throughout the body of the flame, rather than its introduction at a single point. In flame photometry, the name 'atomizer' is given to a system which is used to form aerosol by breaking a mass of liquid into small drops. This little device is responsible for introducing the liquid sample into the flame at a stable and reproducible rate. The atomizer must not be attacked by corrosive solutions.

Two types of atomizers which are in common use are: (i) Those, which introduce the spray into a condensing chamber and into the flame by the air of the combustible gas air mixture. Large droplets are removed in the condensing chamber. (ii) Those in which the sample is introduced directly into the flame, i.e. the atomizer and the burner are an integral unit.

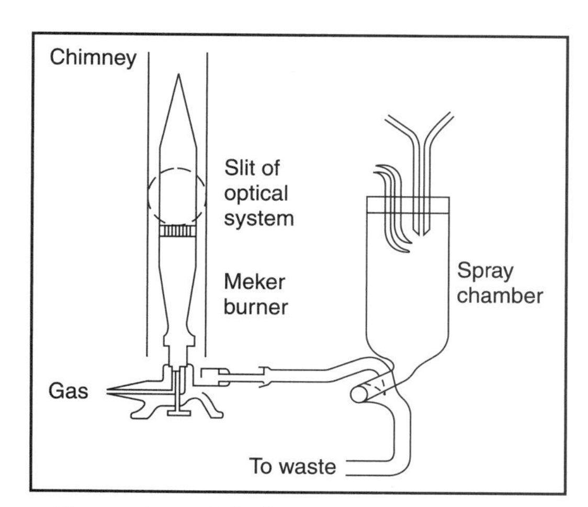

Figure 4.4 :: Discharge type atomizer with condensing chamber and burner

Figure 4.4 shows the construction of the first type of atomizer. This is called the discharge-type atomizer and consists of two capillary tubes sealed into the walls of a glass chamber in such a manner that their bores are perpendicular to each other. The sample solution is poured into a funnel or drawn up from a container and is atomized by the blast of air from the tip of the other capillary. However, the atomized stream is composed of coarse spray with large droplets, which condense on the walls of the chamber and helical tube leading to the burner. The condensate flows down to the waste drain. The smaller droplets, in the form of a virtual fog, are carried by the air stream into the burner, where the condensate is mixed with the burner gases and carried into the region of active combination. Two removable hypodermic needles of stainless steel or glass are commonly used. With this type of atomizer, the consumption of the sample is comparatively high and ranges between 4 and 25 ml of solution per minute. Of this amount, only 5 per cent actually reaches the flame. The sensitivity of this type of atomizer can be markedly increased by using a chamber which is heated by an electric heater placed around its walls. This hastens the process of vaporization of the solvent and produces an aerosol of very fine particles, all of which are swept into the burner. In this case, sample consumption is only 0.2 to 0.6 ml/min and a substantial portion is carried directly into the burner to yield a much higher sensitivity.

Integral burner atomizer

The second type of atomizer is the integral burner atomizer. The unit made of glass or metal is constructed of two concentric tubes. The sample solution is drawn through the innermost tube by the passage of oxygen through the orifice of the middle annulus. At the tip of the inner sample capillary, the liquid is sheared off and dispersed into droplets. The outer annulus supplies the combustible gas to the flame. The body of the unit is machined from brass and the capillary for the solution intake is of palladium. All droplets, both large and small, are introduced directly into the flame. The sample consumption is between 0.8 and 2 ml/min. Each atomizer requires separate calibration and is not strictly interchangeable with another of the same type.

Pneumatic nebulization, though one of the most common methods of aerosol generation, produces a relatively wide droplet size distribution and limited aerosol density. Other techniques such as the spinning disk aerosol generator, atomizer impactor and isolated droplet generation techniques have also been used, but they produce very limited concentrations of aerosol. Ultrasonic techniques, when used for aerosol generation, offer a unique combination of high solution to aerosol conversion rates, relatively independent of carrier flow gas and have the ability to generate varying sized droplet populations. Denton and Swartz (1974) describe the design of an ultrasonic nebulizer system for the generation of high density aerosol dispersions.

Periodic cleaning of the atomizer and capillary tube is necessary as the blocked atomizer will give an unstable or intermittent reading and low sensitivity. Cleaning is usually done by flushing with a copious amount of distilled water. If blockage occurs, the atomizer is removed from its reading and flushed with dry air or cleaned by using a thin wire.

(c) Burner

The burner brings the fuel, oxidant and sample aerosol together, so that they may react safely and produce a good flame. Burners used in flame photometry must fulfil the following conditions:

(i) Supply fuel and oxygen at constant pressure to enable the shape, size and temperature of the flame to remain constant,

(ii) Assist in a perfect distribution of the mixture of gases and the aerosol, which carries the atomized solution under analysis,

(iii) Have a tip of suitable shape to produce a symmetrical flame and ensure a homogeneous flow and distribution of the gases, in order to avoid a strike-back in the burner from accidental fluctuations in the feed system, and

(iv) Prevent condensation of the aerosol in the stem of the burner, which would reduce the effective quantity of the sample brought into the flame.

The most commonly used burner for low temperature flames is the Meker type. Here, the fuel gas issues from a small orifice and passes through a venture throat, where considerable air is entrained. The mixture of gas and the entrained air passes up the burner tube and burns at the top of the burner, where the combustion is assisted by the surrounding air.

A deep grid of metal across the mouth of the burner prevents the flame from striking back down the tube. In order to screen the flame from air drafts, it is surrounded by a glass chimney, which also protects the operator.

In the integral burner–atomizer unit, the sample solution is introduced directly into the flame. A cut action of this type of burner is shown in Figure 4.5.

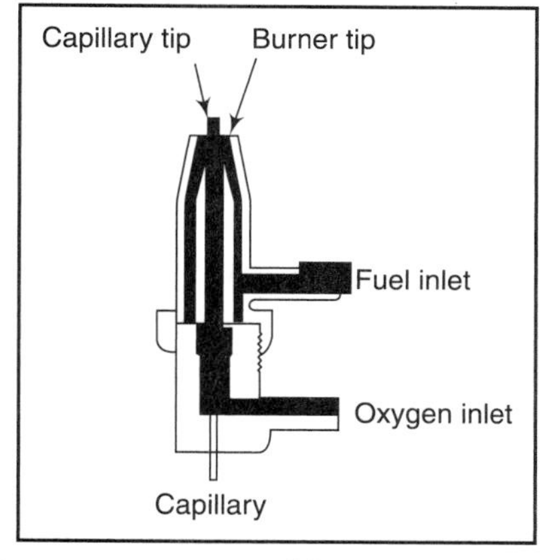

Figure 4.5 :: Integral burner–atomizer

(d) Flame

The flame is the most important part of the flame photometer, since it forms the source in which the light radiations characteristic of the elements under analysis are produced. The flame performs the following functions:

(i) It converts the constituents of the sample to be analysed from the liquid or solid state into the gaseous state.

(ii) It decomposes these constituents into atoms or simpler molecules.

(iii) It excites the resulting atomic or molecular species to emit light radiations.

In order to produce accurate results, the flame must be very stable. If its temperature or structure shows a change over a period of time, the emission produced by a given sample will also change.

The flame temperature must be high enough to excite the atoms to higher energy levels, so that emission may take place. In general, the flames are produced by burning the gases given in Table 4.1:

Table 4.1 :: Flame Temperature (°C)

Fuel	In Air	In Oxygen
Propane	1925	2800
Butane	1900	2900
Hydrogen	2100	2780
Acetylene	2200	3050

All the fuels employed in flame emission spectroscopy produce both continuous background as well as certain band spectra. The nature and intensity of the spectrum for a given fuel are strongly dependent upon the fuel-to-oxidant ratio and the flame temperature.

4.2.2 Optical System

The complex light emission produced in the flame by a solution containing one or more elements makes it imperative to select the radiations of different wavelengths for each of the elements present in the solution. This job is accomplished by the optical system, which collects the light from the steadiest part of the flame, renders it monochromatic and then focuses it on to the surface of the photodetector. A concave mirror is often placed behind the flame, with its centre of curvature in the flame. In this way, the intensity of the emitted light is nearly doubled.

4.2.2.1 Filters

Atoms in the vapour state give line spectra and not band spectra. As there are no covalent bonds on the atoms, there would be no vibrational sub-levels to cause broadening. Therefore, wavelength selection may be done by optical filters or monochromators. Less expensive instruments make use of filters, which may be of the absorption type or of the interference type. The usual glass or gelatin absorption filters have wide spectral bandwidths, of 35 to 45 nm, in width at one-half the maximum transmittance. Their overall transmittance is only 5–25 per cent, decreasing with improved spectral isolation. Absorption filters do not give the degree of monochromaticity required for analysing complex systems and hence the flame photometers thus equipped are restricted in use to the determination of only sodium, potassium, calcium and lithium.

Absorption filters

Interference filters

Interference filters are often used in order to improve the resolution. They are the most suitable for flame photometry as compared to absorption filters, because they allow a much narrower band of wavelengths to pass. They are similar to monochromators in selectivity, but are less expensive. Standard interference filters having spectral bandwidths from 10 to 17 nm at one-half the maximum transmittance are commercially available. The same characteristics are shown by continuous wedge filters, which allow a continuous selection of different wavelengths. The continuity of an interference filter is achieved by using a spacer film of graded thickness between the two semi-transparent layers of silver. These filters usually have a working interval of 400 to 700 nm. They provide a dispersion of 5.5 nm/mm, with a transmittance which is usually not higher than 35 per cent. The filters must be kept clean as a dirty filter can result in unstable reading.

4.2.2.2 Monochromators

Monochromators are incorporated in the instruments, which are expected to provide better isolation of spectral energy. This requirement emanates when the spectrum lines are very close or very weak or both. By using narrow slit widths and sensitive detecting circuits, narrow bands of radiant energy can be isolated. There are two means of dispersing radiation—by refraction or by interference. Examples of these two types of radiation dispersal are prism monochromator and diffraction grating, respectively. In flame photometry, prism monochromators have been the most widely accepted. They may be of glass or quartz. Glass prisms are only suitable for emissions essentially in the visible region, whereas quartz prisms enable many atomic lines to be studied that occur in the far ultraviolet region of the spectrum. In prism monochromators, emission bands or lines are chosen by local selection with movable exit slits or by prism rotation, wherein all the lines of the spectrum are passed through a fixed slit one after the other. The wavelength scale is non-linear as in prism spectrographs.

Dispersal of light may also be obtained by means of diffraction in a grating. A grating is produced by ruling grooves at extremely close intervals, of the order of 15,000 grooves per inch, on a highly polished surface. The dispersion is linear, which permits a constant bandwidth to be used throughout the spectrum. The wavelength scale in this case is linear.

Resolving power

For an analysis of complex spectra, the resolving power should be very high. Resolving power is a measure of the ability of the monochromator to resolve two closely spaced lines of about equal intensity. Perhaps the straightest test of an instrument's resolving power concerns the separation of the manganese line at 403.3 nm, the potassium line at 404.4 nm, and the lead line at 405.8 nm from each other.

4.2.2.3 Other Optical Components

Reflectors and condensing lenses can be placed in the optical path to facilitate concentration of the light. A concave metal mirror with its focus in the flame itself is placed immediately behind the flame. Condensing lenses are used to render parallel beams of light before the flame falls on the interference filter. Some apparatus are fitted with an iris diaphragm in the optical path of the rays, which serves to reduce the sensitivity when very concentrated solutions are being used. Similarly, uniform metal grids with different size meshes may be used. The slit is of special importance in apparatus fitted with monochromators. Its width is more important than its length, because the sensitivity, the accuracy, the ratio of total emission–flame background and the resolving power, depend upon the width of the slit.

4.2.3 Photosensitive Detectors

After the isolation of the lines or bands emitted by the chemical elements under analysis in the optical system, the light intensities of these radiations have to be measured quantitatively. This is done by causing the selected radiation to fall on a photosensitive element. The electric current produced is then measured with a galvanometer, either directly or after amplification. The intensity of electric current produced is a function of the concentration of the solution under analysis. The photodetectors are either of photovoltaic or photo-emissive type.

Photovoltaic cells have very little internal resistance and the current they produce cannot be amplified by dc amplifiers. The current is measured directly by connecting the terminals of the cell to a highly sensitive galvanometer. The spectral response of a selenium cell adequately covers various wavelengths encountered in the analysis of common alkali and alkaline earth metals. The sensitivity that can be achieved with these cells is considerable, as they give currents of more than 5×10^{-10} A, when they receive a light flux of more than 10^{-6} lumen.

Selenium cells are highly sensitive to temperature variations. The cells have certain inertia and at least a quarter of an hour must be allowed for them to settle, with the burner lit before the readings are taken. A fall in efficiency due to aging is a well-known characteristic of these cells and they are also subject to fatigue upon exposure to prolonged illumination.

Dirty photocells usually result in intermittent readings, which necessitates their regular cleaning. Faulty photocells normally give low sensitivity, and need to be changed.

Phototubes of several types may also be used for detecting the radiations. They differ from barrier layer cells in that they need an external electrical supply to provide a sufficient potential between the electrodes. Phototubes are less sensitive than barrier layer cells and the current produced usually needs amplification. Since they exhibit large internal resistance, amplification of the cathode current by an external circuit can be conveniently done.

The preferred detector for weak light intensities is the photomultiplier tube. This is many times more sensitive than the simple phototube, which, in turn, permits greater spectral resolution without loss of signal. They are available with envelopes of ordinary glass or with quartz window inserts for use in the ultraviolet region below 350 nm.

4.2.4 Recording System

The final step in recording the light radiations produced in the source of emission involves transforming the electrical signal produced by the selected radiation in the photosensitive element into a reading which is related to the concentration of the solution under analysis.

4.2.4.1 Amplifiers for Phototube Circuits

The current from the phototube is generally so small that it is necessary to amplify it, quite often by a factor of 10^5 to 10^6. The amplification of direct current can be achieved with the help of integrated circuits with high input impedance.

In order to reduce the rapid and random fluctuations, a long enough time constant is incorporated in the amplifier circuit to aid in the integrating process, and thus the meter is rendered unresponsive to sudden flashes of light in the flame. Usually, the time constant is chosen to be approximately 1s and is obtained by connecting a capacitor in parallel with the coupling resistor. In order to overcome the problem of drift of dc amplifiers, the light from the flame can be interrupted, and the photoelectric current can be amplified by an ac amplifier. A synchronously rotating disc is placed before the flame and the photoelectric current is thus convened to a chopped dc signal, which can then be amplified by an ac amplifier.

4.2.4.2 Amplifiers for Use with Photomultiplier Tubes

The output current from the last amplifying stage in a photomultiplier tube for many solutions and concentration ranges normally employed, often lies between 10 and 100 μA and can be directly measured by a high sensitivity galvanometer (Figure 4.6), or simply a rugged pointer type micrometer. Photomultiplier tubes possess an appreciable output current even when the tube is dark, and therefore, for direct measurements, it is necessary to carry out a background correction before measurement. This is done by compensating with an auxiliary battery and adjusting R_2. The sensitivity of the photomultiplier tube can be adjusted by varying the voltage per dynode or preferably by variation of the voltage applied to the first dynode.

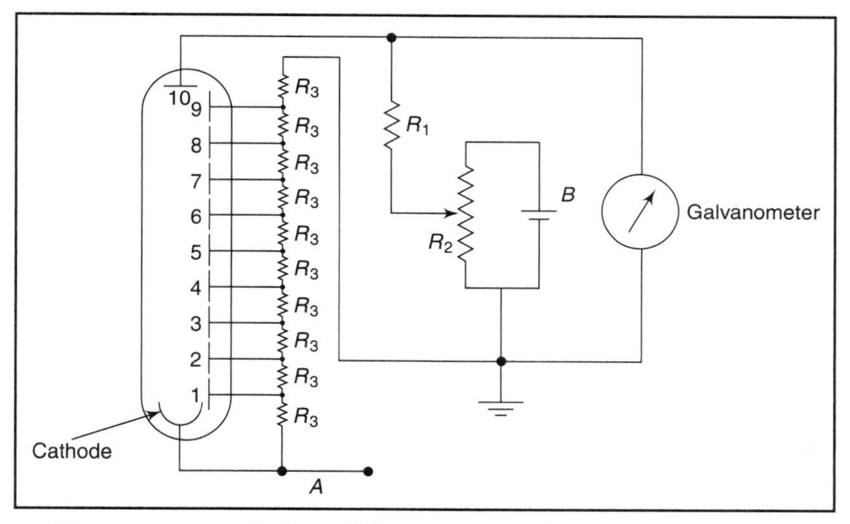

Figure 4.6 :: Typical amplifier circuit used in flame photometers

Although the photomultiplier tube itself provides a large amplification, this can be appreciably increased by an amplifying circuit. Figure 4.7 shows a circuit diagram, which can be used to amplify current obtained in a photomultiplier tube. The output of the PMT is applied to the gate of FET Q_1. Q_1 inverts this negative going signal to positive and applies this to the non-inverting

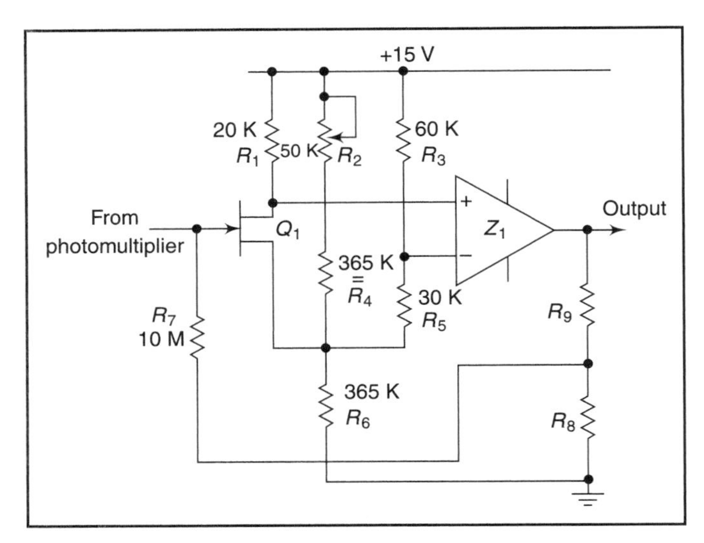

Figure 4.7 :: Circuit diagram for direct reading flame photometer using a photomultiplier tube

side of operational amplifier Z_1. Due to the divider networks R_9 and R_8, the gain of the operational amplifier Z_1 is held at 10. The output of the circuit can be directly connected to a moving coil meter.

Digital display type flame photometers are now available in which the analog voltage is converted into digital form and then given to LEDs for numerical display.

4.3 Types of Flame Photometers

4.3.1 Single-beam Instruments

Figure 4.8 shows a simple arrangement for directly measuring the output of a single barrier-layer cell. This type of circuit is suitable for a single-beam instrument, which contains only one set of optics. The photoelectric current generated in the photocell by the light is given to the galvanometer through the coarse and fine sensitivity controls at B and A, respectively. Selenium cells exhibit no dark current, which would require suppression, but a suppressor circuit might be useful in bucking out a reading resulting from the flame background.

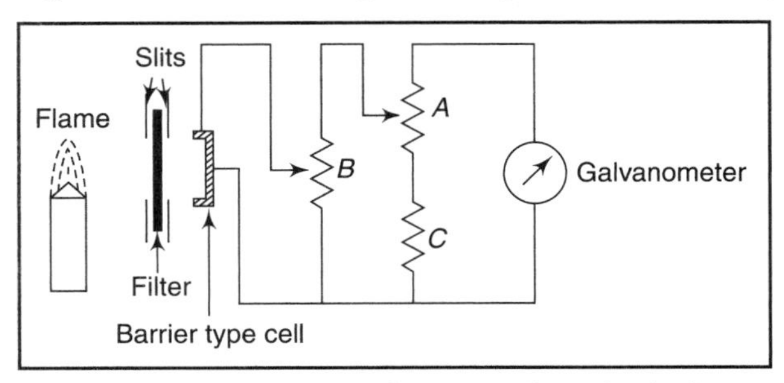

Figure 4.8 :: Circuit diagram for single-beam direct-reading flame photometer

4.3.2 Double-beam Instruments

In double-beam instruments, provision is made for a second light path, which is emitted by an internal standard element. The internal standard and unknown wavelengths are isolated by means of optical filters in a dual optical system, and each beam is focused on a photoelectric cell (Figure 4.9). The circuit is so arranged that the photoelectric currents produced by the emitted radiations of unknown and internal standard elements oppose each other through a galvanometer. By means of an accurate potentiometer P_1, the opposing photoelectric currents can be balanced. The settings of the potentiometer required to produce balance, are calibrated in terms of known solutions. Sensitivity adjustments are made with another potentiometer P_2 by shunting currents generated by the measuring cell. This makes it possible to read more concentrated solutions of the test element without changing the concentration of the internal standard used.

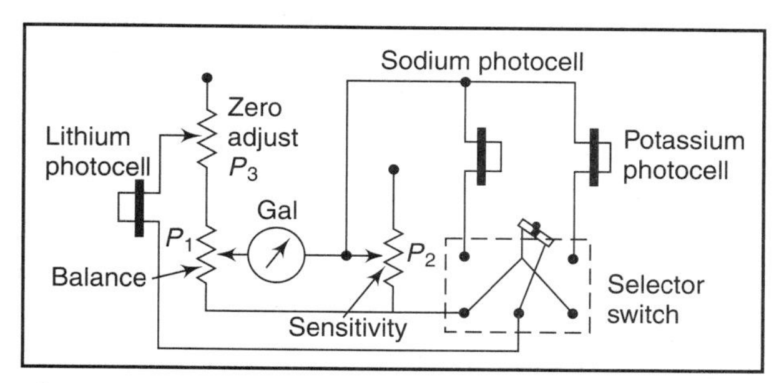

Figure 4.9 :: Circuit diagram for double-beam flame photometer

4.3.3 Recording Type Flame Photometers

Recording flame photometers are useful when one has to determine the background, and consequently the emission intensity of the line or band more accurately than is possible on a direct reading instrument. In instruments of this type, the measured emission intensity of an element in a flame is recorded by means of an electronic recorder, or a moving paper strip in the form of intercepts of various lengths. The wavelength selection is carried out by attaching a spectrum scanning mechanism. This mechanism consists of a synchronous motor, which rotates the shaft of the prism rotation drum of the monochromator through a reducing gear. The recorder normally used is of the potentiometric type, with a signal range of 0-50 mV.

Figure 4.10 shows the general arrangement of the EEL flame photometer. Air is introduced around the atomizer nozzle through the inlet *F* at a controllable pressure. The pressure is indicated by a pressure gauge provided on the front panel of the instrument. Gas enters the mixing chamber through the tube *M* and is likewise controllable. The air flow around the atomizer is applied to draw the sample *B* up the capillary tube *C* and through the atomizer, which injects it in a fine spray into the mixing chamber *D*. Larger droplets falling from the air-stream pass through the drain tube *E*. The baffles *G* produce an even mixture of gas, air and sample, which passes to the burner *L* and is ignited, to burn with a broad, flat flame within a well-ventilated chimney.

Light from the flame is collected by the reflector *H* and focused by the lens *I* on to the photocell *K*, the critical wavelength being isolated by the interchangeable filter *J*. The photocell output is then applied to deflect a taut suspension reflecting galvanometer, in proportion to the intensity of the emitted light.

In order to maintain a steady supply of clean, compressed air to the instrument, it is necessary to use an oil-free and portable compressor. The compressors are generally required to give an output of 6 litres/min at 0.7 kg/cm^2 (10 1b/in^2) and make use of an asynchronous motor 1/20 hp and running at 1400 rpm.

The instrument provides full-scale deflection with 2 ppm sodium, 3 ppm potassium, 15 ppm lithium and 40 ppm calcium. It needs only a 2 ml sample and is provided with a sampling attachment, which enables the operator to make a series of tests rapidity.

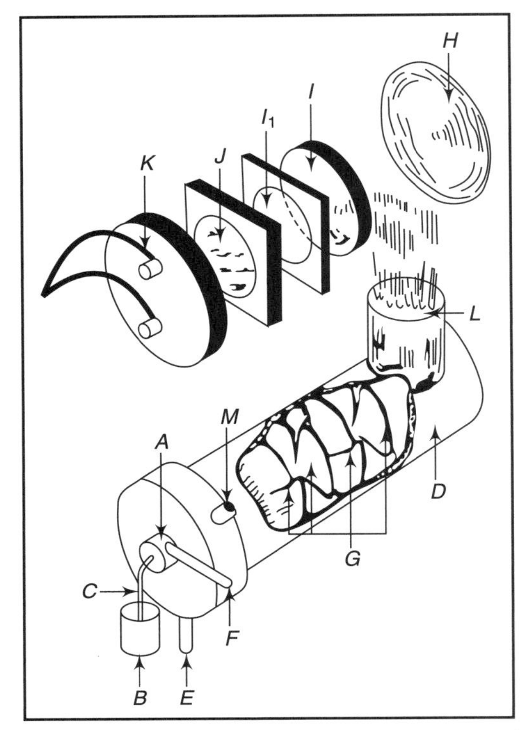

Figure 4.10 :: General arrangement of the atomizer, combustion chamber and optical system of EEL flame photometer

4.4 Clinical Flame Photometers

The flame photometer is one of the most useful instruments in clinical analyses. This is due to the suitability of the flame photometer for clinically analysing sodium, potassium and calcium, which are of immense importance in the development of the living being and indispensable to his physiological functions. In the clinical analysis of sodium and potassium, the flame photometer gives numerous differential data for normal and pathological values rapidly and accurately.

The normal plasma sodium concentration is about 140 mmol/L. A high plasma sodium concentration of more than 145 mmol/L is referred to as 'hypernatremia'. This can occur due to simple dehydration, excess sodium intake, steroid therapy as well as in diabetic insipidus. 'Hyponatremia', with a plasma sodium concentration of less than 130 mmol/L can occur due to diuretic medication, kidney disease, excessive sweating, congestive heart failure or gastrointestinal disorder.

Potassium is the major intracellular cation and influences muscular activity, cardiac function and nerve conduction process. Acute hyperklemia condition, in which the plasma potassium concentration exceeds 5.5 mmol/L, is a medical emergency. In hypoklemia, the potassium level is less than 3.5 mmol/L, which can occur due its excessive loss in gastrointestinal secretions and urine.

Many companies manufacture flame photometers. The KLiNa Flame System of M/S Beckman is a dedicated instrument used for the simultaneous analysis of sodium and potassium. In this instrument, sample handling is automatic, as the system has a turntable, which will hold up to 20 samples in cups and an automatic positive piston displacement dilutor that dilutes the sample prior to entering the spray chamber (Figure 4.11).

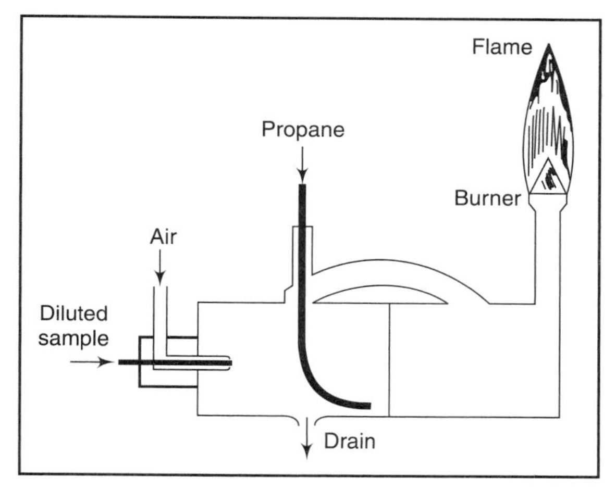

Figure 4.11 :: Flame spray chamber and burner of KLiNa flame photometer (Courtesy M/S Beckman Instruments, USA)

The ignition and shutdown of the flame are automatic. When the calibrate button is depressed, the flame is ignited and the circuits are energized. Passing a standby button extinguishes the flame, but maintains thermal equilibrium. Twenty-four microlitres of samples are aspirated for simultaneous sodiums and potassiums, thereby facilitating micro-sample analysis suitable for paediatric or geriatric work. After analysis, the results are displayed directly in millimoles per litre or milliequivalents per litre.

For the precise and accurate determination of sodium and potassium concentrations, use is made of the fact that lithium normally not present in significant concentrations in serum, exhibits about the same flame emission characteristics as Na and K. Lithium ions are added to the diluent used for samples, standards and controls. The lithium in the diluent is referred to as the internal standard.

The diluted sample containing the fixed known amount of lithium, in the form of a dissolved salt, is nebulized and carried by the air supply into the first of two compartments in the spray chamber (Figure 4.12). Heavier droplets fall out of the stream on to the chamber walls or separate from the stream upon striking a partition in the chamber and flow to a drain from the compartment. Propane enters the first chamber to mix with the air and sample stream, and carry it through a tubular glass bridge into a second compartment; the aerosol and propane mixture travels up from the chamber to the burner head, where the mixture is burned. Exhaust gases are vented to room air from a cover located on top of the instrument.

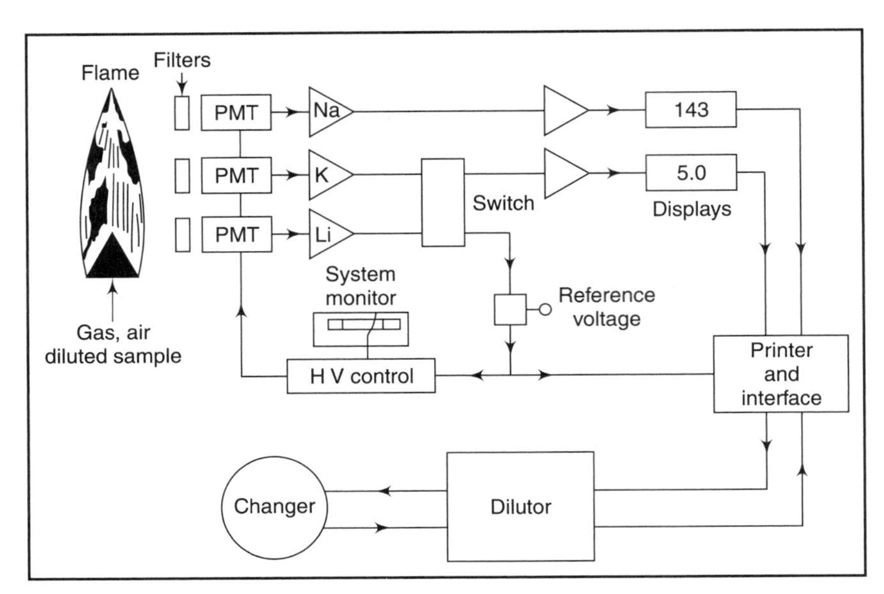

Figure 4.12 :: Schematic diagram of Beckman KLiNa flame photometer

In order to provide internal standardization, the response of the sodium and potassium detector is a ratio function of the response by the lithium detector. Thus, any change in the air flow rate or fuel pressure that may affect the sample would proportionately affect the lithium detector.

The modern flame photometer determines and provides digital display of sodium and potassium or lithium concentrations in a sample, by responding optically and electronically to the intensity of the principal emission lines that characterize each ion as it is excited in the propane and air flame.

The flame is monitored continuously by three photomultiplier detectors. Each detector views the flame through an optical filter that passes only the wavelength band of interest to the particular detector. The sodium detector therefore responds only to wavelengths in a narrow band centred at 589 nm, the potassium detector to wavelengths in a narrow band centred at 766 nm, and the lithium detector to wavelengths in a narrow band centred at 671 nm.

The specifications for the KLiNa Flame System are ± 0.2 mmol/litre for potassium and lithium, and ± 2.0 mmol/litre for sodium. Potassium and lithium both show linearity to 20 mmol/ litre, while sodium is linear to 200. In addition to the 0-20 scale, potassium may be rescaled to readout to 200 mmol/litre for convenient analysis of urine samples.

The fully automatic ignition and flame optimization sequences reduce the set-up and calibration time. An automatic gas shut-off mechanism activates if the flame is accidentally extinguished. If no flame is detected within a fixed time, about ten seconds, the system automatically shuts down the flow of fuel and air (Alpret, 1974). Warning lights indicate whether the problem has to do with either low air or low fuel pressure. The lithium internal standard signal reduces fluctuation in flame conditions, drift and dilution errors and ensures reproducible results. In the microprocessor-based instruments monitoring and control software allow measurements only after blanking and calibration.

By observing simple precautions and maintenance procedures, flame photometers have been found to give a long working life. The following procedures may be followed to correct any malfunctioning of the instrument:

(i) If the instrument shows low *sensitivity*, it may be due to a blocked atomizer, low gas pressure, or a faulty photocell. The necessary corrective steps should be taken to clean the blockage of the atomizer, such as restoring the gas pressure or changing the photocell.

(ii) The reason for varying or intermittent readings is usually a contaminated burner, faulty photocell, low gas pressure, blocked atomizer, improper air supply and excessive vibration. The instrument must be placed on a shock-proof base. Appropriate checks must be made for the above faults.

Similarly, the corning EEL Model 450 is a low temperature internal standard flame photometer, designed specifically for the determination of sodium, potassium and lithium in clinical samples. Results are read directly from two digital panel meter (DPM) displays. Lithium determinations are performed separately and the result is presented on the potassium DPM display, the sodium DPM display being blanked off. The model 449 Dilutor is available as an optional extra to facilitate Operation of the Model 450. The Model 449 is a peristaltic pump, which dilutes the sample and the internal standard with distilled water in the ratio 1:100.

The capability of flame photometers has been extended by including the measurement of barium (Ba) and calcium (Ca) in addition to sodium (Na) and potassium (K). This facility is available in the instrument Model Superspec F20-F (Figure 4.13) from M/s Spectro Lab Analytical Ltd, UK,

The instrument is capable of measuring all four elements from a single aspiration and applying all the necessary connections and linearizing functions with a microprocessor. It includes automatic filter selection and electronic ignition of the burner flame. The inbuilt memory facilitates storage of five point calibration curves for upto four elements. The instrument has a sensitivity of less than 0.2 ppm for Na, K, Li and less than 10 ppm for calcium.

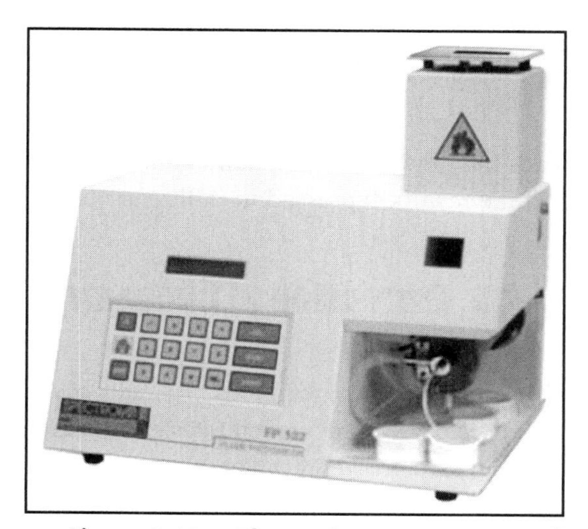

Figure 4.13 :: Flame photometer (Courtesy M/s Spectro Lab Analytical Ltd., UK)

4.5 Accessories for Flame Photometer

Modern flame photometers come with many useful accessories. For example, the KLiNa flame photometer of M/S Beckman comes with a diluter, sample changer and a printer.

Diluter

The diluter is a motor-driven cam-programmed system that functions through a cycle of operations. These operations involve sample pick-up and transport to an internal mixing cup, the addition of a measured volume of diluent, mixing to ensure a properly prepared sample aliquot, coupling of the mixing cup to spray chamber, so that sample aspiration can occur, and finally the washing and draining of the mixing cup. The diluter uses positive displacement pumps to ensure exact sample dilutions in the operator selectable ratios of 50:1, 100:1 or 200:1. The solutions are generally diluted to fall within a linear part of the emission curve.

The dilutor, based on a peristaltic pump, provides accurate dilute ratios independent of sample viscosity. Variation in dilution ratios is achieved by changing the bore size of the pump tubing.

Automatic changer

The automatic changer facilitates automatic presentation to the diluter sample probe of up to 20 successive samples. It is a turntable, which rotates step-wise to locate each sample cup under the extended and down position of the diluter sample probe. The probe is extended from the diluter, once each sample determination probe tip enters the sample and a measured volume is taken for transport to the diluter mixing cup. Individual sample trays are placed on the chamber turntable. Each tray can hold 20 sample cups. The sample cup could be 0.25 ml, 0.5 ml or 2.00 ml size. The 2.0 ml size is generally recommended.

Printer

The printer provides a paper tape print-out to record the result of each sample determination. Each print-out is a data line that from left to right, reads the sample position number in the Changer sample tray, the type of determination performed and the results expressed in milli-equivalents per litre.

4.6 Expression of Concentration

The concentration of solutions is usually expressed in parts per million (ppm) in flame photometry. This type of expression facilitates easy calculations on dilute solutions and the concentrations can be expressed in weight/weight, weight/volume and volume/volume ratios. The ratios can be expressed in volume, if the density of the liquid is near unity and hence 1 ml = 1 g. This is true especially in dilute solutions, whose density is approximately that of distilled water. The equivalents are as follows:

1 ppm = 1 mg/l = 1 mg/kg

$1 \text{ ppm} = 1 \times 10^{-3} \text{ g/l} = 1.\ 10^{-3} \text{ mg/ml}$
$1 \text{ ppm} = 1 \times 10^{-6} \text{ g/ml}$

4.7 Interferences in Flame Photometry

Flame photometry is subject to several types of interferences, which arise during the course of analysis through variations in any of the parts of the instrumental system or in the composition of the samples. These interferences give rise to experimental errors. Some of these interferences are discussed below.

4.7.1 Flame Background Emission

A flame is an extremely complex mixture of reacting gases. Since these gases are hot, they emit a certain amount of light, even when no sample atoms have been injected into the flame. This light is the flame background. The lines produced by a sample are superimposed on this background. Failure to correct properly for the background reading can be a source of serious error. It is desirable to eliminate, as far as possible, the flame background and other emission lines from the measurements. The narrower the bandwidth of the monochromator used, the more efficient is this elimination.

4.7.2 Direct Spectral Interference

This interference occurs when the emission lines are placed so close together that the monochromator cannot separate them. For example, the emission of the orange band of CaOH interferes with the sodium line at 589 nm. Interference is more serious when absorption filters are used in place of monochromators. In this case, the two lines overlap partially or completely, and will be read together in proportion to the degree of overlap. If the interference cannot be obviated by increased resolution, the difficulty must be overcome by prior removal of one element, perhaps by selective solvent extraction.

4.7.3 Self-absorption

Interference due to self-absorption occurs due to the absorption of radiant energy through collision with atoms of its own kind present in the ground energy level. If some of the radiant energy is self-absorbed, the strength of the spectral line is weakened. Self-absorption primarily depends upon the number of atoms present in the ground state, the concentration of atomized solution and by the probability that these atoms will be excited by the incident radiation from excited atoms of its own kind. Self-absorption is insignificant at very low concentrations of a test element.

4.7.4 Effect of Anions

Anionic disturbances result from a radiation in the number of free metallic ions by a stable combination in the flame with the anions. There is usually no interference with concentrations of less than 0.1 M. Above this concentration, sulphuric, nitric and phosphoric acids, in particular, show a marked effect in lowering metallic emission. For example, the strong depression of the calcium emission in the presence of phosphate, aluminate and other similar anions has been fairly well-known. In practice, over limited intervals of concentration, the depression is linear and this forms the basis for indirect determination of the depressant in the presence of a standard amount of calcium.

4.7.5 Effect of Ionization

The easily excitable alkali metals such as potassium show the phenomenon of ionization through excessive dissociation and consequently result in the depletion of the number of available neutral atoms that can be excited. This, in turn, weakens the intensity of the atomic spectrum, whereas any ionic spectrum is strengthened. When small quantities of easily ionized metals are added to a flame, the number of neutral atoms tend to increase more rapidly in comparison to the ionized atoms, in proportion to a concentration of metal sprayed into the flame. As a result, the curve of intensity versus concentration may initially be concave upward.

4.7.6 Solution Characteristics

The physical properties of the solutions strongly influence the atomization rate, which, in turn, influences the luminescence of the flame. The main factors responsible for this are the differences in viscosity, tension, density, volatility and temperature between standards and samples. For example, an increase in viscosity lowers the atomization rate, with a consequent weakening of the emission, thereby causing errors and loss of sensitivity. Similarly, variation in surface tension is considered as another source of error. They affect the emission in a similar manner to viscosity, through their effect on atomizing conditions. Added salts and acids hinder the evaporation of solvent, large droplets result in a diminished quantity of aerosol reaching the burner, because the larger droplets are more likely to settle upon the walls of the condensing chamber.

There are several standard ways of eliminating or avoiding the influence of the above-mentioned factors. The techniques of overcoming interferences demand high skill and experience. The development of atomic absorption spectrophotometry, in which the problems of interference and flame stability are not as severe, has largely limited flame emission analysis to the determination of sodium and potassium, and less frequently of lithium, calcium, manganese and copper in relatively simple mixtures.

4.8 Procedure For Determinations

4.8.1 Calibration Curve Method

In order to determine the concentration of an unknown element *X* in a given sample, or a series of samples, the instrument is first standardized by making a calibration curve. For this, it is desirable to know the probable range of concentration of *X* in the sample, so that a series of calibration standards can be made up; having a suitable spread over the range expected. The instrument is first adjusted to give a zero meter reading, when pure distilled water is aspirated into the flame. The instrument settings are then added to give the full-scale reading, with the most concentrated standard sample. Without altering this adjustment, emission intensities for the remaining standards and the unknown are measured. The instrument response is proportional to the concentration and hence a straight line is obtained. Calibration curves are thus plotted and the concentrations read off against this calibration thus, concentration of the unknown is obtained from the calibrated curve.

Since it is difficult to obtain a flame of high long-term stability, it is usual to check the calibration at frequent intervals by spraying a standard. This ensures that the drift can be measured and taken into account. In fact calibration curves are difficult to reproduce on the same day, virtually impossible on different days.

The concentrations to which the samples are diluted for measurement can often be selected, if the concentration in the original solution is much greater than the detection limit of the element concerned. Careful choice of the concentration is very important, as it has a considerable effect on the accuracy of analysis for the following reasons:

(i) Calibration curves are not necessarily linear. It is found that the curves for a given element are linear over a limited range (Figure 4.1). At stronger concentrations, the same element would give a curved calibration. This happens due to self-absorption. Photons emitted by excited atoms are partly absorbed by ground state atoms in flame, thus giving a negative deviation at high concentration.

(ii) In case of a high total concentration of solids in the flame, the flame background is greatly increased and the burner may be fouled.

(iii) If the viscosity of the sample is too high, it may not spray efficiently.

(iv) If the solution is diluted excessively, errors of dilution may occur and the signal-to-noise ratio may decrease. At very low concentrations, emission falls below expected due to ionization as some atoms get converted back to ions.

4.8.2 Standard Addition Method

A number of radiation interferences can be avoided by using the method of standard addition. In this method, net emission readings are obtained on two solutions, solution *A* containing an aliquot of the unknown solution and solution *B* containing the same quantity of unknown solution plus a measured amount of a standard solution of the element. The amount of test element in each of the two solutions is then determined from their measured emission intensities and the standard calibration curve. Subtracting the quantity of unknown found in solution *A* from that found in solution *B*, gives the amount of test element equal to that added, when there is no depression or enhancement. When one of these effects is present, the corrections are applied, which correct for the interference. It is necessary that the calibration curve be linear over the range of concentrations employed.

4.8.3 Internal Standard Method

The flame system is a critical part of the flame photometer and its long-term stability is dependent upon the: (i) fuel flow, (ii) air pressure, (iii) nebulization rate, and (iv) burner/atomizing chamber temperature. While the first two factors can be effectively controlled to within acceptable limits by suitable instrument design, the other two are less easily controlled in this way. In order to overcome variation from these sources, an internal standard procedure has been developed.

In this method, a fixed quantity of the internal standard element is added to the blank sample and standard solution. The radiation of the element is measured together with those of the elements under analysis by dual detectors, or by scanning successively the two emission lines. The ratio of the emission intensity of the analysis line to that of the internal standard line is plotted against the concentration of the analysis element on log-log paper, to prepare the calibration curve for a number of standards. Corrections for background radiations are made for each line. The plot of log (emission ratio) versus log (concentration of test element) will produce a straight line, whose slope ideally is 45° over limited concentration intervals. On most double-beam instruments, this ratio is given by the reading of the balancing potentiometer. Calibration curves may be drawn on a linear coordinate paper as a ratio rather than an absolute light intensity that is being measured, because of which fluctuations due to variable operation of the nebuliser-burner system, are greatly minimised. This procedure also reduces errors due to differences in the composition of the test and standard solutions.

The choice of a suitable element and spectral line depends upon the following analytical conditions:

(i) The element should give spectral lines that are close to those of the analysis element, so that it is not necessary to use a different type of photodetector. Thus, lithium 671 nm line is a suitable reference for the analysis of potassium at 761 nm and sodium at 589 nm, but it is less satisfactory for the determination of calcium at 422.7 nm.

(ii) The element must be one that is not present in the samples and must be available in a high state of purity.

(iii) It must be added in concentrations very near to those of the analysis element, so that radiation of approximately equal intensities may be obtained. For this reason, the concentration of lithium is normally adjusted to 100 μg/cm^3 (15 mEq), in solutions where the sodium and potassium concentration lie in the 0 to 100 μg/cm^3 range.

(iv) The reference element must produce radiations that are perfectly selectable.

(v) When added, the element must not alter the physical properties of the solution.

For an analysis of alkali metals and even of other elements, lithium has been preferred as an internal standard, as it satisfies most of the above-mentioned requirements.

Chapter 5

Atomic Absorption Spectrophotometers

5.1 Atomic Absorption Spectroscopy

Atomic absorption spectroscopy is an analytical technique based on the absorption of radiant energy by atoms. It was explained in the chapter on flame photometry, that when a dispersion of the atoms of a sample is produced in a flame, some of these atoms get thermally excited and emit characteristic radiation, as they return to the ground level. Most of them, however, remain in the ground state. When a beam of light is made to pass through the flame, a portion of it will be absorbed by dispersed atoms, in the same manner that a beam of light passing through a solution will be absorbed by the dispersed molecules of a solute. It is possible to find a series of absorption bands corresponding to the energy levels of the atoms sprayed into the flame. The wavelength of the bands is characteristic of the atoms of the element concerned and the absorbance of the band is proportional to the concentration of the atoms in the flame.

The potential of atomic absorption spectroscopy for the determination of metallic elements in chemical analysis was first realized by Alan Walsh, who, during the mid-1950s developed it into its modern form. Standard commercial equipment became available by about 1960. Since that time, the use of this technique in routine analysis has become widespread, and has replaced many traditional wet methods for the estimation of metals in solution.

The versatility of atomic absorption spectroscopy is demonstrated by the fact that it permits the estimation of between 60 and 70 elements at concentrations that range from trace to macro-quantities. It is applicable to the estimations of metals in organic and mixed organic-aqueous solvents, as well as to those in aqueous solution.

In atomic absorption spectrophotometry, the absorption lines are very narrow, and are approximately 0.02 Å wide. This is because being atomic bands, they are not broadened by the rotational or vibrational structure, as is the case in molecular absorption bands. However, the lines are very weak, because there would be relatively few atoms in the light path. Also, a narrow absorption band of the atoms in the flame necessitates the use of a very narrow source line, and therefore, a continuous radiation source like tungsten or hydrogen lamp cannot be used in atomic absorption spectrophotometers. Since a different hollow cathode lamp is required for each element to be determined, atomic absorption is practically useless for qualitative analysis. Quantitative analysis is easily performed by a measurement of the radiation absorbed by the sample. As little as 0.01 to 0.1 ppm of many elements can be determined.

In technique, atomic absorption spectrophotometry is quite similar to flame photometry. In principle, the two methods are complementary and therefore, in combination, they provide a powerful analytical tool. While the flame emission technique measures energy that is emitted by atoms, atomic absorption measures the energy that is absorbed by the atoms. Atomic absorption is able to make use of a much larger number of metal atoms in the flame that do not acquire sufficient energy to emit light. These are the atoms in the ground state.

Each element has its own unique absorption wavelengths. A source is chosen for the element to be tested, which emits the characteristic line spectrum of that element. The sample, normally in liquid solution for fine suspension, is sprayed into a flame. The flame vaporizes the sample and puts the atoms in a condition where they can absorb energy. The atoms must be chemically uncombined and in their minimum energy or ground state. A monochromator selects a characteristic sample line and illuminates adjacent lines. A photo-detector measures the light passing through the flame, both before and after introduction of the sample into the flame.

The amount of light absorbed by the atoms in the flame is proportional to the concentration of the element in the sample solution. The unknown concentration of the element in the sample may be determined by comparing the absorbance reading on a meter with a calibrated chart showing the relationship of absorbance versus concentration made from data of known concentrations of the element.

5.2 Atomic Absorption Instrumentation

Figure 5.1 shows a block diagram of the atomic absorption spectrophotometer.

The source of radiation is a hollow cathode lamp, whose cathode is made of the element to be investigated. This lamp emits the line spectrum of that element Figure 5.2 (A).

The sample is sprayed into the flame and the atoms of the element are dispersed in the gaseous phase. The atoms absorb the radiation only at a particular line called the resonance line [Figure 5.2 (B)], giving the result as shown in Figure 5.2 (C). The resonance line gets diminished after passing through the flame containing the sample vapour, while all other lines remain unaffected. The other lines are removed by a monochromator, which selects a band of wavelengths around the resonance line and rejects all other lines [Figure 5.2 (D)]. The photodetector measures the diminished resonance line (Figure 5.2 (E)) and displays it on a suitably calibrated scale. Atomic absorption obeys Beer's Law and calculations for concentration can be performed accordingly.

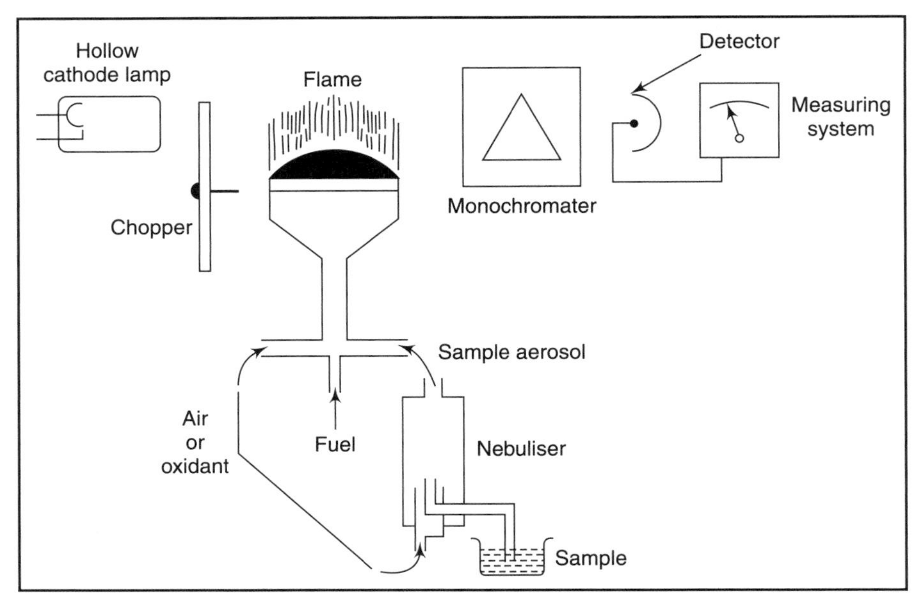

Figure 5.1 :: Block diagram of an atomic absorption spectrophotometer

Many elements emit light strongly in the flame at the same wavelength at which they absorb. This creates a serious error in measurement of the absorption. This error can be eliminated by modulating the light from the lamp, by using either an ac discharge or a mechanical or electronic chopper between the lamp and the flame. The detector circuit is so designed that only the modulated light is recorded and the unmodulated flame emission gets eliminated.

5.2.1 Radiation Sources

5.2.1.1 Hollow Cathode Lamps

The hollow cathode lamp is the most commonly used source of radiation for atomic absorption spectroscopy. It is a discharge lamp which emits the characteristic light of the element to be analyzed.

The lamp consists of a cylindrical thick walled glass envelope, which has a transparent window of glass or silica affixed to one end (Figure 5.3). It contains one anode and a cup-shaped cathode, which are both connected to tungsten wires, for taking out the electrical connections. The cathode is a hollow metal cylinder 10 to 20 mm in diameter, and is constructed of the metal whose spectrum is desired, or it may simply serve as a support for depositing a layer of the element. The tube is filled with a highly pure inert gas at low pressures of 1 or 2 mm. The gases generally used are helium, neon or argon. Mica sheets are placed inside the lamp to limit the radiation to within the cathode.

The choice of the window material depends upon the wavelengths of the resonance lines of the element concerned. For wavelengths that are shorter than 250 nm, a silica window is chosen, whereas for wavelengths longer than 250 nm, either silica or glass be employed.

Hollow cathode lamps have been constructed for almost every naturally available element. However, for atomic absorption spectroscopy, only lamps of the metallic elements are used. They usually have a long operating life, of the order of 1000 hours. The operating current requirements of a particular lamp would depend on the element concerned, and it is usually between 5 and 25 mA. The warm-up period of such type of lamps is normally between 5 to 30 minutes. A lamp which requires longer than 30 min may be regarded as unsatisfactory.

When a potential is applied between the electrodes, a discharge is struck and ionization of the gas takes place. Current would flow as a result of movement of the ions to the electrodes. If the potential is sufficiently large, the gaseous cations acquire sufficient kinetic energy to dislodge some of the metal atoms from the cathode surface and produce an atomic cloud. This process is known as sputtering. Some of the sputtered metal elements in the excited state give out light characteristic of the excited atoms. If we excite an atom of element X by an electrical discharge, into a higher energy state, and if it returns to the ground state and emits a photon, this photon would have exactly the correct energy to be absorbed by another atom of X and to raise it to the same excited state.

In atomic absorption spectrophotometers, a delay is encountered when one element is being changed into another, due to the warm-up time

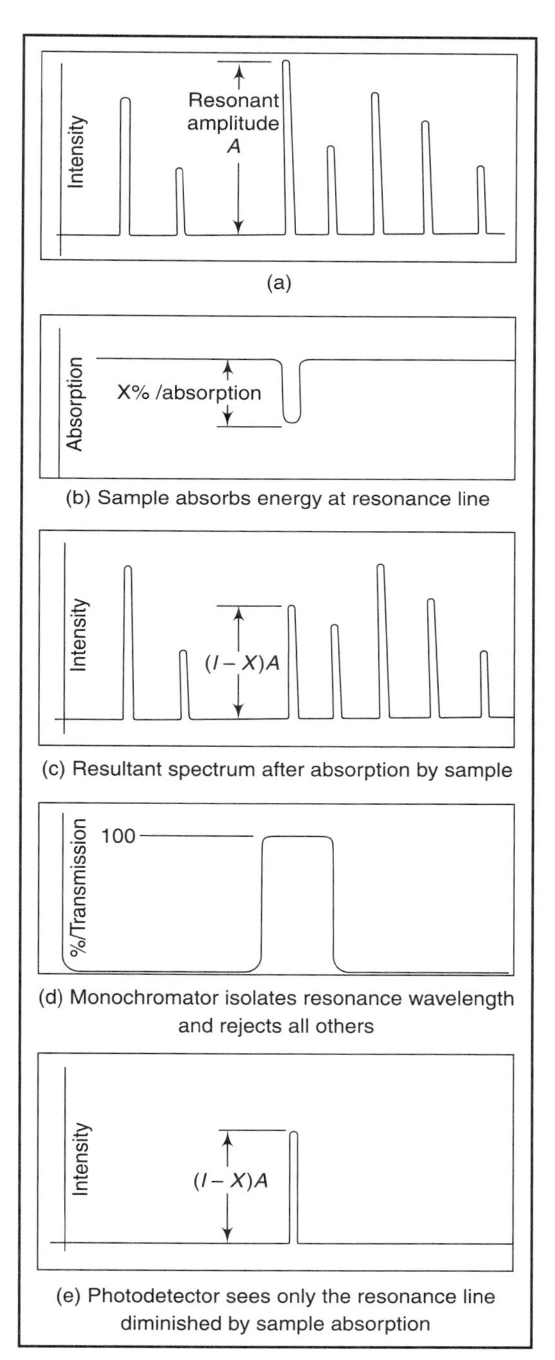

Figure 5.2 :: Principle of an atomic absorption spectrophotometer

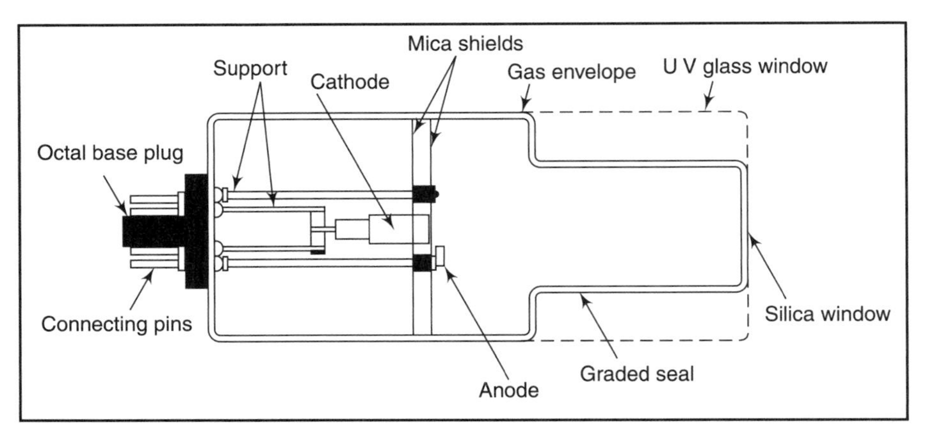

Figure 5.3 :: Constructional details of a hollow cathode lamp

for needed hollow cathode lamps. It may be possible to overcome this lag by warming up several lamps at once, but this way the lamp life gets wasted. It is possible to combine certain elements in the same hollow cathode in such a way as to provide very nearly equal performance to single element lamps. Sintered cathodes are used for some of these lamps. Highly pure powders of the elements are mixed in the correct proportions and press-formed. These lamps are called multi-element lamps. The instruments are provided with a switching system, which puts each lamp into its approximate current range. Multi-element lamps emit very complex spectra. Therefore, only dual or triple element lamps, specially for related elements such as calcium-magnesium, sodium-potassium, copper-nickel-cobalt are generally constructed. However, they have the disadvantage that with continued operation, the resonance line intensity of one or more elements may fall off more rapidly than others.

The hollow cathode lamps are operated from a regulated power supply, which provides chopped or ac power to ensure that the emission from the flame which is unchopped and, therefore results in dc signal, is ignored by the electronic system.

Several researchers have been using a continuum source for atomic absorption (AA) to avoid the need for element-specific light sources. An important advantage of such a technique is that, it provides the opportunity for multi-element analysis rather quickly. The instrument makes use of a high-resolution monochromator having a smaller spectral bandpass than the conventional AA instruments. Such a monochromator is echelle monochromator and the system provides separate detector channels for each of the elements to be determined. By modulating the wavelength at relatively high frequency over a narrow spectral range, a signal is generated that permits background measurements to be made. The chief limitation of this technique is that signal-to-noise ratio is diminished by the requirements of the continuum sources, especially if the system is to be used for very fast analytical signals. In practice, this is quite serious for the elements that are determined in the far UV, where the very bright continuum sources are not very useful (Galan, 1986).

5.2.1.2 Electrodeless Discharge Lamp

For most elements, the hollow cathode lamp is a completely satisfactory source for atomic absorption. In a few cases, however, the quality of the analysis is impaired by limitations of the hollow cathode lamp. These are primarily the volatile elements where low intensity and short lamp life are a problem. The atomic absorption determination of these elements can often be dramatically improved with the use of the 'electrodeless discharge lamp' (Beaty, 1978).

The electrodeless discharge lamp (EDL) design is shown in Figure 5.4. A small amount of the element or salt of the element for which the source is to be used is sealed inside a quartz bulb. This bulb is placed inside a ceramic cylinder on which the antenna from a radio frequency generator is coiled. When an RF field of sufficient power is applied, the coupled energy will vaporize and excite the atoms inside the bulb into emitting their characteristic spectrum. An accessory power supply is required to operate an EDL.

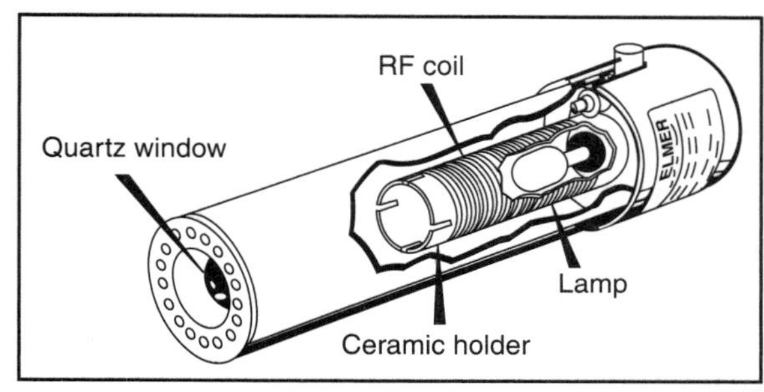

Figure 5.4 :: Electrodeless discharge lamp (Beaty, 1978)

Electrodeless discharge lamps are typically much more intense and, in a few cases, more sensitive than comparable hollow cathode lamps. They therefore offer the analytical advantages of better precision and lower detection limits. In addition to providing superior performance, the useful lifetime of an EDL will exceed that of a hollow cathode lamp for the same element. It should be noted, however, that the source of light in the EDL is considerably larger than that in the hollow cathode lamp. As a result, the performance benefits of the EDL can only be observed in instruments with optical systems designed to take advantage of these benefits.

Electrodeless discharge lamps are available for the 17 elements listed in Table 5.1. Hollow cathode lamps also exist for these elements. The EDLs, however, offer definite performance and lifetime benefits for instruments which are optically compatible with the EDL source.

5.2.2 Burners and Flames

The essential requirements of a burner for atomic absorption are unique. The flame must extend over a reasonable length to improve sensitivity. This is directly analogous to the use of longer path cells in normal spectrophotometry. This follows from the fact that if the flame can be spread out into a long, narrow band, a greater number of the atoms that it contains may be brought into the light beam. In this way, absorption bands become strong and improve the sensitivity. Also, in order to avoid problems caused by scattering of light by unburned droplets and large sample

Table 5.1 :: Electrodeless Discharge Lamps—Elements (Courtesy M/s Perkin-Elmer)

Antimony	Arsenic	Bismuth
Cadmium	Cesium	Germanium
Lead	Mercury	Phosphorus
Potassium	Rubidium	Selenium
Tellurium	Thallium	Tin
Titanium	Zinc	

droplets, which cannot be burned, they must be removed before reaching the flame. In addition to this, the burner must be able to take up solutions with a high concentration of dissolved solids, limit clogging problems and reduce requirements for sample dilution.

The burners employed for the flame emission instruments are of the total consumption type. In these burners, the sample solution, the fuel, and the oxidizing gas are carried through separate passages and meet at an opening at the base of the flame. However, for atomic absorption studies, a pre-mix type of burner (Figure 5.5) is mostly preferred. In this type of burner, the sample is aspirated through a thin capillary tube by the air flowing into the atomiser section. The air sample mixture emerges from the atomizer as a fine spray of droplets, which is then mixed with the fuel, usually acetylene. The mixture is rendered turbulent by the flow spoilers and is then forced up into the burner head. Larger drops of the sample collect at the bottom of the chamber and are drained off.

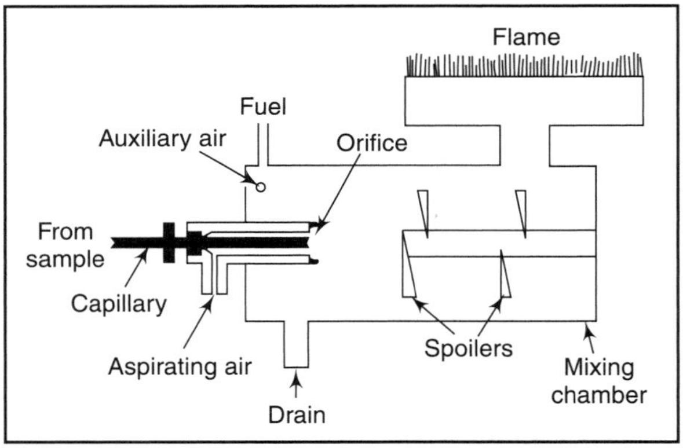

Pre-mix burner

Figure 5.5 :: Constructional details of a pre-mix burner

In another design, gas mixing and large drop removal are achieved by passing the aerosol through a constriction, which functions as a turbulent jet. This geometry leads to improved detection limits in atomic absorption spectroscopy. Because of the long path available, a slot burner is used in atomic absorption spectrometry. The slot dimensions are dependent upon the fuel and oxidant gases. Usually, a 10 cm slot is used for an air-acetylene flame and a 5 cm slot for a nitrous oxide-acetylene flame. The port is usually 0.4–0.6 mm in width. Very thin flames are used, so as to obtain high analyte concentrations in the flame for a longer slot. The burner head is fabricated from aluminium, stainless steel or titanium.

Another burner system, denoted as the Autolam-burner, consists of a nebulizer, spray chamber and burner head. It differs from the pre-mix burners in that the nebulizer and burner head are 180° out of the alignment; the analyte droplets must complete a semi-circle from the inlet to the flame. The design enables uniform mist to reach the flame and presents clogging of the slot.

The desirability of avoiding flashback has led to the development of capillary burners. They are more stable and for a number of elements, the emission signal is greater than that from a conventional burner. With a suitable adaptor, they can be made to fit any pre-mix–burner assembly.

Atom production

The mechanism of atom production is quite complex. Apparently, it is not the same for all solvents and flames. One of the important factors that needs consideration is the possible formation of oxides. Many metal oxides are very stable and once formed are very difficult to reduce, even in a high temperature flame. Therefore, the reducing nature of the flame (amount of fuel to oxidant) assumes critical importance.

For flame atomic absorption spectrophotometers, the design of nebulizers and spray chambers appears to have been empirically optimized to provide the best aerosol drop size in the flame for interference-free analyte vaporization. It is thus no surprise that nebulizers and spray chambers have undergone a steady progression during the past 20 years. However, the pneumatic nebulization approach is used in the vast majority of atomic spectroscopy determinations in case of liquid samples.

Pneumatic nebulizers

The main practical requirements (Browner and Boorn, 1984) for pneumatic nebulizers are the following:

- A high velocity gas stream,
- A reasonable pressure drop of the liquid injection capillary for venturi-effect natural aspiration,
- Maximum interaction between the gas and liquid streams for fine aerosol production, and
- Freedom from blockage resulting from either particles suspended in the solution or from salt build-up at the nebulizer top.

All these requirements mean that the construction of pneumatic nebulizers is a demanding engineering challenge, as tolerances must be kept precisely on annular spaces. These spaces may be as small as 10-20 μm for ICP nebulisers, compared to 150–250 μm for flame atomic absorption nebulizers.

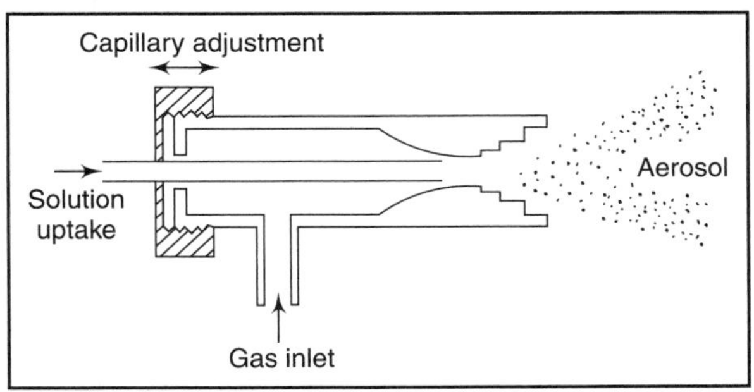

Figure 5.6 :: Adjustable concentric nebulizer (redrawn after Browner and Boorn, 1984)

A substantial control over the gas-liquid interaction can be obtained by varying the position of liquid uptake tube in a conically or parabolically converging gas tube (Figure 5.6). This need is avoided in

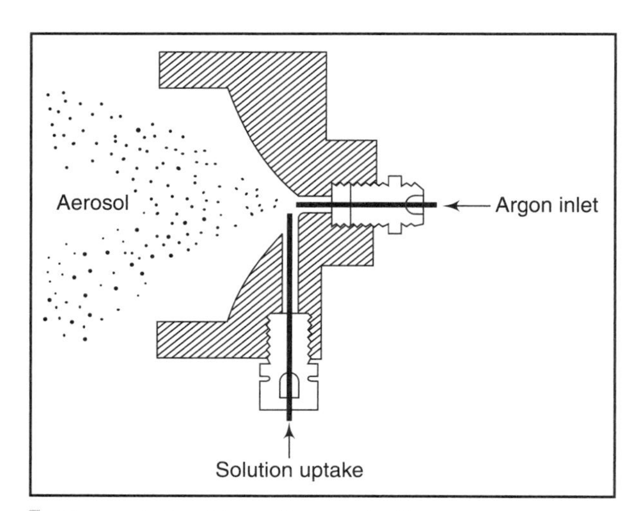

Figure 5.7 :: Cross-flow nebulizer

cross-flow nebulizers (Figure 5.7). However, it requires very precise and rigid positioning of the gas liquid tubes. Cross-flow nebulizers are much less fragile than concentric nebulizers, which are usually made of glass. The major limitation of air pneumatic nebulizers of conventional design is that they produce aerosols with a wide drop range. This means that high transport efficiency can be achieved only by allowing large drops to reach the atomizer. One device that produces a much finer aerosol is the ultrasonic nebulizer. In this device, up to 30 per cent efficient production of droplets in the size range from 1.5 to 2.5 μm has been found at a 0.3 ml/min solution flow rate.

Ultrasonic nebulizer

Figure 5.8 shows the constructional details of an ultrasonic nebulizer. In the case of this device, the principle of aerosol production is significantly different from pneumatic nebulization. In the ultrasonic nebulizer, instead of drops being stripped from a liquid cylinder by a high velocity gas jet, surface instability is generated in a pool of liquid by an ultrasound beam, which is generated by a piezoelectric transducer.

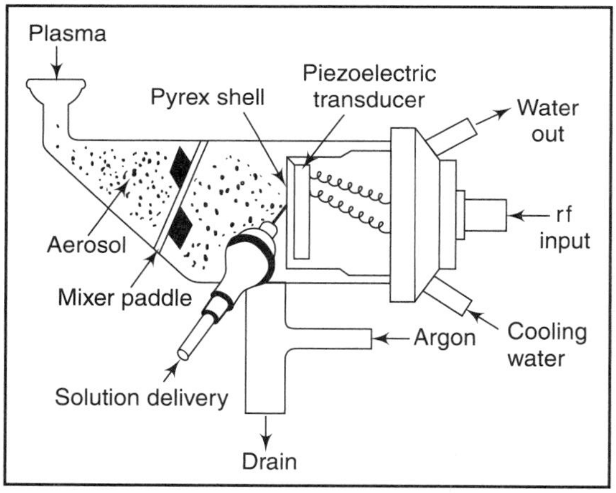

Figure 5.8 :: Ultrasonic nebulizer (redrawn after Browner and Boorn, 1984)

Ultrasonic nebulizers produce aerosols with mean drop diameters that appear to be a function of the exciting frequency. At frequencies below 50 kHz, cavitation is the main mode of droplet production. At the high frequencies commonly used in modern ultrasonic nebulizers, typically 1 MHz or greater, the mechanism of aerosol production changes from cavitation to geyser formation. With the geyser formation mechanism, it is the power density and not the operating frequency which determines the drop size.

Although ultrasonic nebulizers have the advantage of producing small-sized droplets, many unanswered questions remain about the general reliability and freedom from interference offered by these devices.

The problems of nebulizer blockage inherent in pneumatic nebulizers are effectively overcome with the Babington type pneumatic nebulizer. The principle of this device is based on the concept used for paint spraying. It involves a spherical surface with an array of holes around a circumference. The gas supply comes from within the sphere, and as the liquid flows over the outside of the sphere and passed over the gas stream, it is nebulised (Figure 5.9). Babington type nebulizers are ideal when solutions with suspended particles must be analyzed and when prior acid or other dissolution to dissolve the particles is not convenient. These nebulizers can be made either entirely of glass, metal or teflon, or by embedding a sapphire orifice in a teflon block.

Babington nebulizer

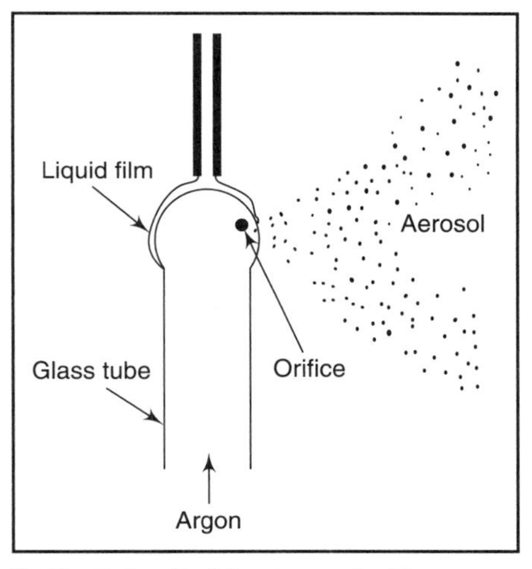

Fig 5.9 :: Babington nebulizer

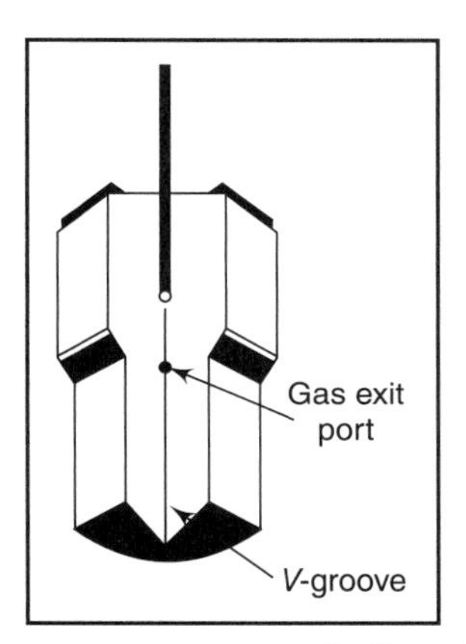

Figure 5.10 :: V-groove nebulizer (re-drawn after Browner and Boorn, 1984)

Another design of a V-groove Nebulizer is the type (Figure 5.10) in which a liquid stream is passed down a V-groove, with a small hole drilled in its centre for the gas stream. This type of nebulizer is quite popular and is available commercially.

V-groove Nebulizer

Nebulizers and spray chambers operate interactively. They must be optimized as a unit rather than individually. However, the spray chamber must meet the following specific requirements:

- Effective removal of aerosol droplets larger than the cut-off diameter; this is necessary for interference-free movement.
- Rapid wash-out characteristics, both to increase the possible rate of analysis and to avoid cross-contamination problems; and
- Smooth drainage of waste aerosol from the chamber, to avoid pressure pulses in the atomizer.

Since the thermal energy from the flame is responsible for producing the absorbing species, flame temperature is an important parameter in atomic absorption spectrometry. The temperatures for some flames used in atomic absorption are given in Table 5.2.

Table 5.2 :: Temperature of Pre-mix Flames

Gas	Temperature
Air-Methane	1875
Air-Hydrogen	2000
Air-Acetylene	2300
N_2O- Acetylene	2955
Air-Propane	1925

It may be noted that cooler flames are subject to more interference problems resulting from insufficient energy for complete atomization. While the air-acetylene flame is satisfactory for a majority of the elements determined by atomic absorption, the hotter nitrous oxide-acetylene flame is required for many refractory-forming elements. Moreover, nitrous oxide-acetylene flame is effective in interference control in other situations. Table 5.3 gives a list of elements which may be determined by atomic absorption spectroscopy with different flames.

Table 5.3 :: Elements which may be Determined by Atomic Absorption Spectroscopy with Different Flames

Air/propane or Air/acetylene	Air/acetylene	Air/acetylene or N_2O/acetylene	N_2O/acetylene
Arsenic	Antimony	Barium	Aluminium
Bismuth	Chromium	Calcium	Beryllium
Cadmium	Cobalt	Molybdenum	Germanium
Copper	Iron	Strontium	Silicon
Gold	Lithium	Tin	Tantalum
Lead	Magnesium		
Manganese	. Nickel		Titanium
Mercury	Platinum		Tungsten
Potassium			
Silver			
Sodium			
Zinc			Zirconium

If a high standard of analytical performance is to be expected from the instrument, the burner must be kept in a clean condition. The burner head can usually be disassembled for cleaning purposes. Inorganic deposits on the surface of the burners can be removed by soaking it in water. Layers of organic material may be removed with the help of a fine emery paper. Soaking in acid solutions is generally not recommended. The burner should be cleaned after every day of operation.

5.2.3 Plasma Excitation Sources

In order to produce strong atomic emission from all chemical elements, it is necessary to attain temperatures considerably above those of simple flames. A convenient means of obtaining temperatures in the range of 7000 K–14000 K is to generate an inert gas plasma. Plasma is a gaseous fourth state of matter, which is electrically conductive because of the major concentrations of essentially free electrons and highly charged cations present in it. Plasma is a very effective medium for volatilization, atomization and ionization.

Plasma

A significant development in atomic spectrochemical equipment has been the development of the inert gas, electrical discharge plasma excitation source. Unlike other electrical discharges that have been optimised principally for solid or powdered samples, the inert gas plasma has been developed specifically for solution sample types, as a replacement for the flame excitation source.

The inert gas electrical discharge plasma, commonly using argon, comes in the following principal versions:

- Direct current plasma (DCP),
- Inductivity coupled plasma (ICP), and
- Microwave induced plasma (MIP).

The DCP and ICP have been specifically developed for solution analysis. The MIP does not perform as well for solution analysis. This is because one of the limitations of the MIP is that the sample introduction is improperly matched to the material input. Organic liquids are handled well by DCP and ICP systems. All the plasmas do poorly with solid sample introduction. They are, however, not as robust as dc arcs and graphite atomizers.

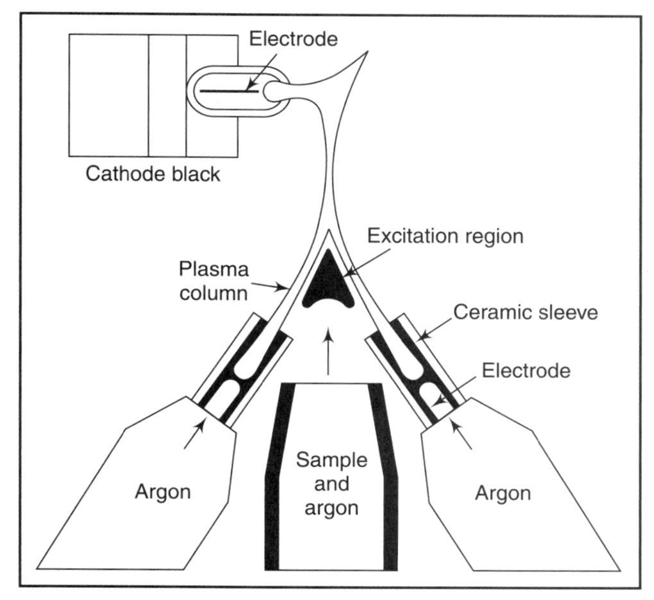

Figure 5.11 :: Multiple-electrode direct current plasma system (redrawn after Zander, 1986)

5.2.3.1 Direct Current Plasma (DCP)

The DCP operates at dc and uses a multiple electrode system as shown in Figure 5.11. Typically, it is a three electrode system having two pyrolytic graphite anodes and one thoriated tungsten cathode. It has four electrical poles, two anodes and two cathodes, and uses two power supplies. The single cathode connects to the negative pole of both power supplies. Each electrode is sufficiently cooled by flowing argon. The electrode set keeps the generated plasma volume in position, which provides much of the stability of the generated signal.

The DCP uses nearly 10 l/min of argon, with the bulk of that going to the single type of nebulizer developed for it, the argon carrier gas for the DCP should be at least 99.995 per cent pure. The generation of a DCP within a three electrode set requires two 7A arcs at 40–50 V. Consequently, about 0.75 kW (max.) can be applied to the discharge.

A result of the size of the generated plasma volume is the size of the viewing zone for spectroanalytical observation. The DCP viewing zone is a volume confined between two pinched arcs. The cross-sectional area imaged on the spectrometer slit block containing the most useful emission is about 8 mm^2. The gas temperature is about 4000–6000 K.

5.2.3.2 Inductively Coupled Plasma

The operating principle of inductively coupled plasma torch is simple in that an electric discharge (e.g. from a Tesla coil) is used to cause partial ionization in argon gas and this gas is then passed through an oscillating magnetic field from an induction coil operating at radiofrequencies. The ions initially created are accelerated and through collision with atoms, cause further ionization until a plasma results. Power consumption in the RF generator for the induction coil is typically around 10KW.

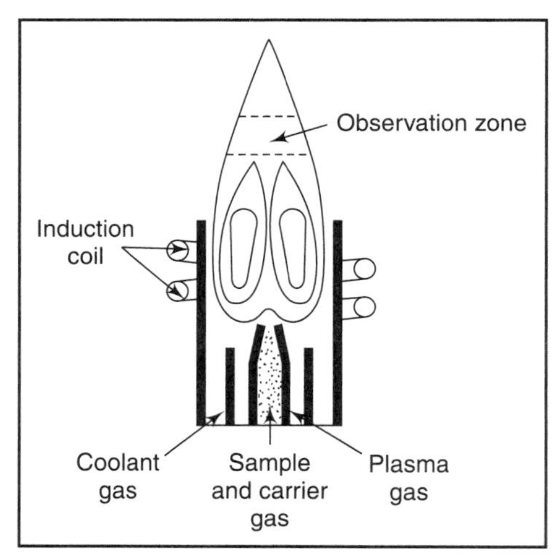

Figure 5.12 :: Schematic diagram of an inductively coupled plasma (redrawn after Zander, 1986)

Inductively coupled plasma (ICP) systems typically operate at 27.12 MHz or 40.68 MHz. The arrangement (Figure 5.12) uses a radiofrequency coil antenna surrounding a ceramic, flow controlling torch. Usually, the coil is made of copper, is protected from oxidation and corrosion and is water cooled. The torch body generally a concentric arrangement of quartz tubes, is usually made of silica and must be cooled by an argon flow. A separate argon flow is used in plasma generation and a third is necessary as the sample aerosol carrier, as sample introduction is that used in conventional AAS. The pneumatic arrangement tends to hold the generated plasma volume in position.

Most ICPs use 10–20 l/min of argon, the bulk of which cools the gas that cools the quartz torch. The ICP requires more power capacity, since it generally uses more power at 220 V, 20–30 A, single-phase. The ICP is generated from about 2.5 kW radiated power at a coil antenna. The viewing zone in ICP is 20–60 mm^2 and the temperature 4000–6000 K. The ICP is available commercially as a complete atomic emission analysis system from several companies. The wash-out-times necessary for a drop to 1 per cent and 0.1 per cent of peak for a typical ICP spray chamber are 25 s and 40 s compared to AA values of 1 s and 3 s. Therefore, larger wash-out times is a particular problem in ICP systems.

A major development in the area of atomic absorption spectroscopy has been the introduction of commercial inductively coupled plasma (ICP) torches in place of chemical flames. The ICP torch provides extremely high sensitivity, low matrix interferences and multi-element analysis capability for samples in solution. However, the ICP systems are extremely expensive and less convenient to operate than the typical flame spectrometers. Nevertheless, their preference cannot be overlooked in laboratories, where high sample throughput is imperative and where determination of at least five elements in each sample is required. In most other cases, the chemical flame will remain the choice for the majority of determinations. The attractive features of ICP are:

- The high efficiency of atomization and excitation gives high sensitivity for most elements. This facilitates working at low detection limits.
- The high temperature, relatively uniform structure and chemical mildness of a plasma minimizes both chemical and spectral interferences. This leads to a greater dynamic range.
- Only one optical system can be used for all elements. There is no need for a separate component for every element as with hollow cathode lamp-based AAs.
- Simultaneous determination of all elements in a sample is possible, leading to a very high throughput in analysis.

The main disadvantage of IPC AEs analysis is the extreme complexity of atomic emission spectra, i.e. the presence of lines due to different elements close to each other. This requires very high resolution and the use of an expensive monochromator.

5.2.3.3 Microwave Induced Plasma (MIP)

Resonant cavity

The MIP uses 2450 MHz resonant cavity in a quartz or ceramic tube, which is cooled by the flowing carrier gas. The resonant cavity can be made of pure copper. Brass or aluminium can also be used if they are silver-coated. Figure 5.13 shows a schematic diagram of the resonant MIP cavity. The MIP typically uses argon as the carrier gas, but newer versions permit atmospheric pressure operation with helium. It uses very little carrier gas, 50 ml/min to a few litres per minute. The popular versions of MIP use more expensive helium, which greatly offsets its reduced gas consumption. The MIP is generated from a few watts to a few hundred watts of radiated power.

Sample introduction in the MIP may usually cause an alteration of the power transfer efficiency to the plasma and consequent power density fluctuations. As a result, sample introduction into an MIP with retention of power suitability and thereby signal stability is quite difficult. Even if the MIP system is operated at relatively high power, the small size of the plasma and the viewing of the plasma itself make sample introduction a complex problem.

The details of various types of plasma sources have been given by Zander, 1986.

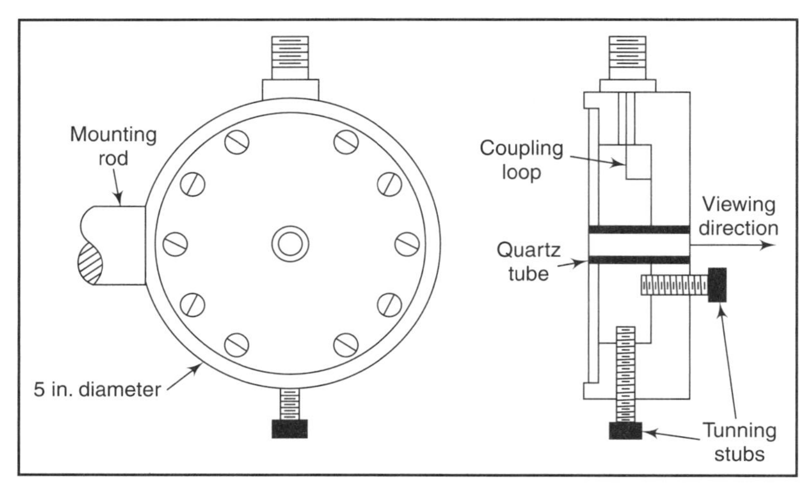

Figure 5.13 :: Schematic diagram of a 2450 MHz microwave induced plasma (redrawn after Zander, 1986)

5.2.4 Graphite Furnace for Atomization

In recent years, there has also been an increasing interest in the use of non-flame techniques like electrothermal and cold vapour atom formation devices. They primarily include carbon filament and furnaces, metal filament atomizers, cold vapour devices, etc.

The furnace AA system is probably the most sensitive analytical technique for the determination of trace metals. In this method, a very small sample is taken and completely converted into an atomic vapour. The subsequent integration of the absorbance pulse yields a useful signal about the element of interest. The instrument manufacturers have modified the earlier designs to carry out the furnace work due to the usefulness of the graphite furnace.

There are different methods by which the sample can be placed in the furnace (Slavin, 1982). The simplest technique is the one in which the sample is applied on the wall of the furnace. When the wall reaches a temperature at which the analyte will vaporize, the metal is driven from the surface into the gas phase. This temperature will vary depending upon the matrix constituents. Also, the rate at which the metal comes off will depend upon the quantity and the specific nature of the matrix constituents.

An improved technique is to add a graphite platform (plate) at the bottom of the graphite tube, on which the sample is deposited. Here, the graphite plate is heated by radiation from the walls, so that the temperature of the sample on the plate is delayed relative to the wall of the tube, and therefore to the gaseous vapour within the tube. In this situation, the atomic vapour is generated at a constant temperature and the absorbance signal is directly proportional to the number of atoms present in the sample and independent of the rate at which the atoms are generated.

In order to operate a stabilized temperature–platform furnace, the tube must be rapidly heated to ensure that the platform would have lagged the heating cycle sufficiently for the tube and the gas to come to steady-state temperature, prior to the evolution of the analyte of interest. A heating rate in excess of 1000°C seems necessary to achieve the necessary degree of interference freedom. Also, the electronic system used to record the integrated absorbance signal must be very fast, since these peaks are generated quite rapidly and appear quite rapidly. The older analog electronic circuits developed for flame AA, which have relatively long time constants, will yield analytical errors when used for the graphite furnace.

Probe furnace

Another variation of the AA furnace is called a probe furnace, which involves putting the sample on a surface outside the furnace. The sample is dried and charred just outside the furnace, thus avoiding contamination of the analytical furnace itself, by the char products. The furnace is then brought to the desired atomization temperature. When that temperature has been achieved, the probe with the charred sample on it is inserted rapidly into the furnace. The technique is far from practical due to some difficult engineering problems, but its physical considerations are correct. Figure 5.14 shows a cross-sectional view of the graphite furnace.

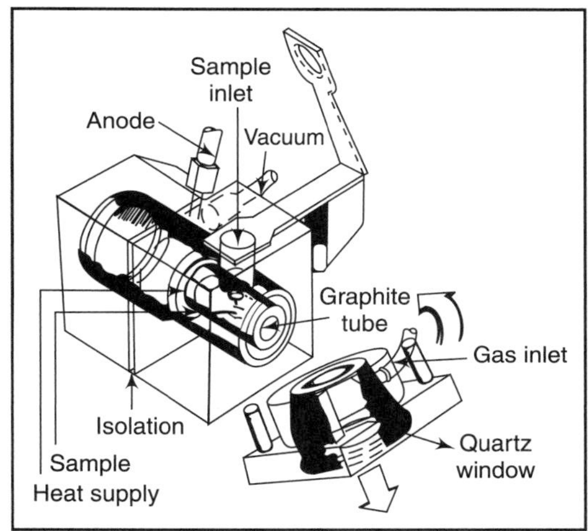

Figure 5.14 :: Cross-sectional view of the graphite furnace

Graphite tube

The graphite tube, which is aligned in the spectrophotometer optical path, is an open-ended cylinder of graphite with a small hole in the center through which the sample is introduced. The arrangement is shown in Figure 5.15. The tube is held between two large graphite rings, which provide electrical contact. The graphite tube acts as a resistor in an electrical circuit. When a voltage is applied to the tube, current flow will cause an increase in temperature of the tube. By controlling the amount of current flow, the temperature of the tube can be adjusted to any desired level up to a maximum of about 3000°C. The graphite system is held in place by a water-cooled housing. By circulating water through the housing, the outside temperature of the furnace is kept at a safe level, and after atomization the graphite tube is rapidly cooled to room temperature where it is ready to accept the next sample.

In order to prevent tube destruction at high temperatures from air oxidation, an inert gas purge of the graphite tube is provided. An external purge around the outside of the tube shields the tube from the oxidizing atmosphere. A separately controllable internal gas is extremely valuable in controlling matrix effects from major constituents of the sample. The internal gas is introduced directly into the inside of the tube from the open ends and escapes out the center hole of the tube. The internal gas flows only during furnace operation and interrupts after the completion of each measurement to allow the next sample to be introduced.

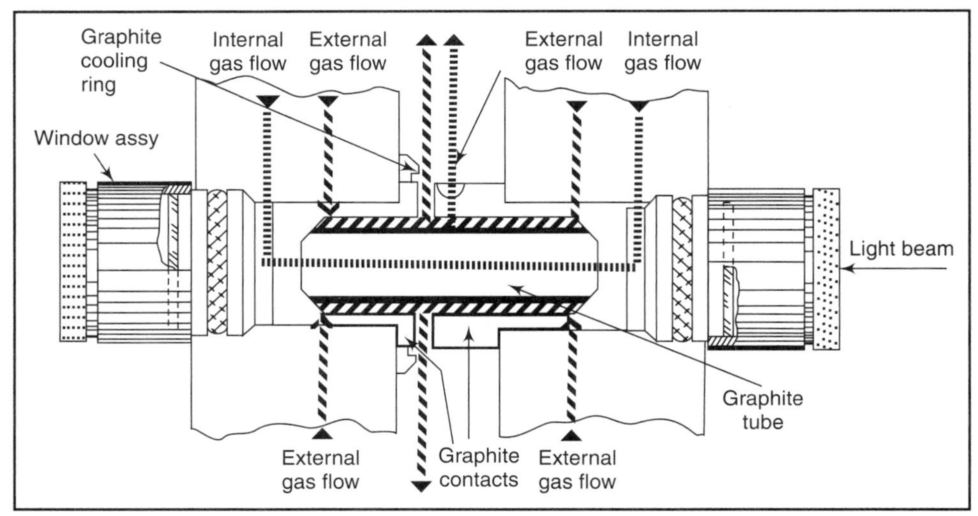

Figure 5.15 :: Longitudinally heated graphite furnace atomizer (after Beaty and Kerber, 1993)

The graphite tube may be constructed entirely of high density graphite, or it may be covered with a relatively thin coating of pyrolytic graphite. The sensitivity for some elements is enhanced with the pyrolytic graphite tube. The refractory forming elements (elements which form compounds of a very high thermal stability) usually benefit most from pyrolytic graphite. Figure 5.16 shows the Perkin Elmer Model 30303B atomic absorption spectrophotometer with a graphite furnace attachment for flameless analysis.

The prospect of release from the tedious and time-consuming sample dissolution step used in an analysis, has prompted further studies into the possibility of the direct introduction of solid samples. The fundamental advantages of vapour introduction, as compared to liquid sample introduction, are as follows:

- It allows pre-concentration of the sample from a relatively large volume of solution, into a relatively small volume of vapour. Therefore, sample transport can be achieved with an efficiency approaching 100 per cent, as compared to the 0-10 per cent typical of liquid sample introduction.
- The greater transport efficiency gives adequate detection limits, which can be critically important for several elements like arsenic, selenium and tellurium.

The disadvantage of electrothermal atomizers is that sample inhomogeneity and dispensing problems account for poor precision.

Carbon Rod Analyzer

This device can be used to convert a powder sample into atomic vapour. A current is applied to a very thin, heated carbon rod that contains the solid sample in order to vaporize it.

Figure 5.16 :: A typical atomic absorption spectrophotometer (Courtesy M/s Perkin Elmer)

Tantalum Boat Analyzer

This is another technique that produces an atomic vapour from a solid sample. A tantalum boat is electrically heated in a manner similar to the carbon rod system within an inert atmosphere.

5.2.5 Optical System

The optical system of an atomic absorption spectrophotometer is typically as shown in Figure 5.17.

The system employs a photometer which is of the single-beam type. The source lamp is a hollow cathode lamp, which is ac-operated and produces a modulated light beam. The light beam is passed through a limiting stop and focused at the mid-point of a slot burner. After travelling through the flame, the beam is re-focused on the entrance slit of the Littrow grating monochromator. The slit width is kept continuously variable and may be controlled by a cam and cam-follower mechanism.

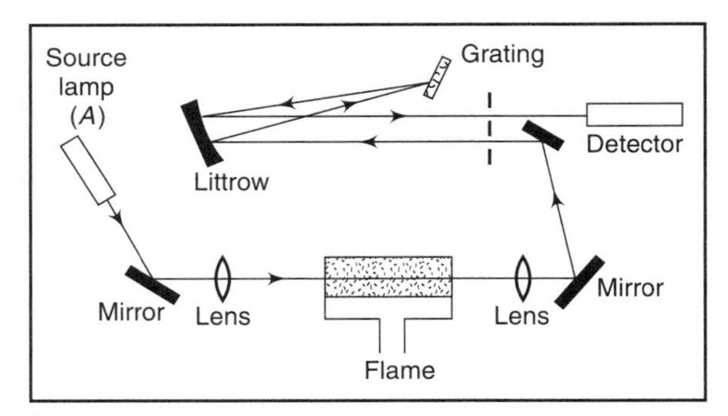

Figure 5.17 :: Optical system of an atomic ab-sorption spectrophotometer

Littrow mounting uses only one piece of quartz or silica with the back surface of the prism metallized. Since the light passes back and forth through the same prism and lens, the polarization effects are eliminated. This type of monochromator provides adequate resolution to separate the resonance line in the lamp from the other lines. Also with its high dispersion, it is possible to open the slits wide and admit high energy. Conventional prism systems do not meet these specifications. Figure 5.18 shows the optical diagram using a silica prism as a monochromator.

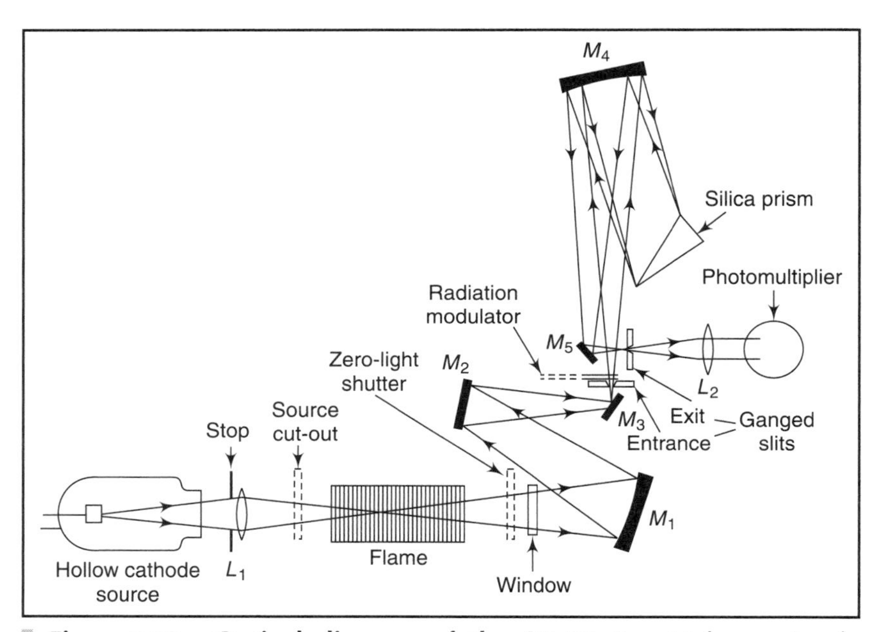

Figure 5.18 :: Optical diagram of the SP 90 Pye Unicam atomic absorption spectrophotometer

A Littrow grating monochromator with a dispersion of 16 Å/mm provides a bandpass as low as 2 nm at the standard slit width of 1 mm. Instruments using grating monochromators provide not only good uniform dispersion over the whole spectrum, but also high resolving power. They have a superior dispersion to a prism instrument at high wavelengths, and are at least as good at low wavelengths.

For atomic absorption determination of most of the common metals, a very moderate monochromator is quite suitable. For determinations of metals for which the lamps possess poor output and complex spectra, a better monochromator is necessary. Floyd *et al.*, (1980) illustrate a computer-controlled scanning monochromator system which can be combined with an inductively coupled plasma excitation source.

The radiation emerging from the exit slit falls on the cathode of a photomultiplier, sensitive over the range 210–800 nm. The lenses used in the system are of fused silica, whereas the mirrors are coated with MgF_2 to preserve reflectivity in the UV region.

5.2.6 Electronic System

The emission system not only produces light of a particular wavelength, originating from the lamp and falling upon the photodetector, but also light of the same wavelength arising from the flame. It is necessary to distinguish between these two sources of radiation, since it is the measurement of that from the lamp only, which is required. This requirement is attained by modulating the light from the lamp, with either a mechanical chopper or by using an ac power supply and tuning the electronics of the detector into this particular frequency. On the other hand, with a dc system, it would become very difficult to determine calcium in the presence of sodium, because the presence of sodium or potassium in a flame contributes a variable amount of continuous background radiation, which adds to the measured intensity of all spectral lines. Sodium also interferes with magnesium when the dc system is employed, because there is a sodium emission line at 2852.1 Å, only 0.7 Å away from the magnesium line at 2852.8 Å. On the other hand, the ac system is entirely free from spectral interference and the dc signals from the flame and other extraneous sources are eliminated.

The main advantage of electronic modulation over a mechanical chopper is that no moving parts are involved. However, in a double-beam system, in which the light from the measuring and reference beams is passed on to the same photomultiplier, the use of a mechanical chopper becomes essential.

The random background noise that originates from the flame is generally noticeable at low frequencies. Also the high frequency photomultiplier noise increases above at about 1000 Hz. The optimum modulation frequency of the chopper is therefore, selected around 350 to 400 Hz.

Zeeman effect

The most innovative method developed for background correction is the one involving the Zeeman effect. In this method, a magnetic field is applied to either source, or to the generated atoms to split the Zeeman components of each spectral line. Since these components are polarized, it is possible to isolate them by using stationary, or rotating optical polarizers. This method, known as Zeeman modulation, can thus be employed for high precision background correction. Accurate instrumental subtraction of the flame background can also be accomplished by directly measuring the background in a reference flame and subtracting it from the signal produced by an analytical flame. This dual flame approach has been employed in some systems.

The Zeeman background correction technique has been incorporated into several new AA spectrophotometers which utilize, in addition, a considerably improved form of the graphite furnace atomizer. The simple insertion of a small platform into the graphite tube furnace, instead of the use of a cup built into the furnace wall, ensures volatilization and atomization under nearly constant temperature conditions, giving a far more reliable and reproducible transient.

Hieftje method

Another form of background correction is the Hieftje method in which the hollow cathode lamp (HCL) depends on the so-called self-reversal of HCL emission at high current. The operation of an HCL at high current not only broadens the atomic emission line (because of the higher operating temperature), but because of the high concentration of unexcited atoms produced, it also leads to absorption within the HCL so that an apparent doublet, centred about the true emission line, is emitted.

Thus, by measuring absorption when the lamp is operating at low current and then repeating the operation while the lamp is briefly operated at high current, the true atomic absorption will be given by the difference between the two. The absorbance under the high current conditions is, of course, assumed to be due to a broad molecular band and could be an over-estimate of the background if this is not so.

Figure 5.19 shows a block diagram of the electronic system of an atomic absorption spectrophotometer.

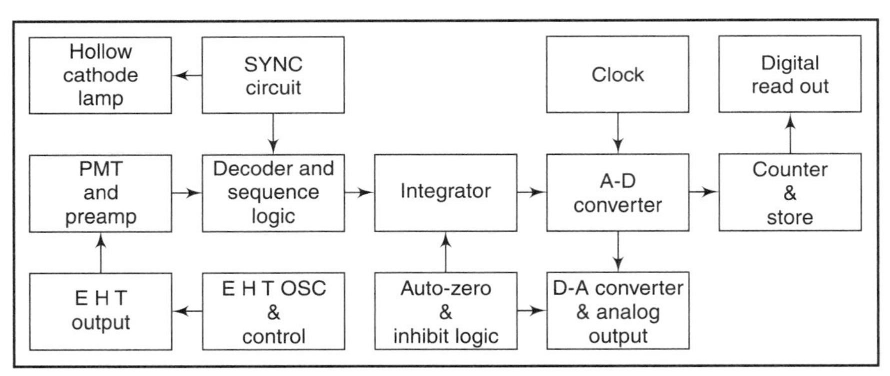

Figure 5.19 :: Block diagram of the electronic circuit of an atomic absorption spectrophotometer

The low level signal developed across a PMT anode resistor is either chopped by mechanical interruption of the light beam directed on to the PMT or is electronically chopped by a transistor switch. In most of the spectrophotometers, where a transistor switch is used, it is operated by the waveform supplied from the synchronizing circuits to ensure that the chopped amplified signals are correctly phased before being applied to the synchronized decoder circuits. In either mode of operation, the signal is usually chopped at the frequency used in the synchronizing system, say 50 Hz, and is amplified in a low-noise amplifier. In order to provide high input impedance, the input stage contains a bipolar transistor. The pre-amplifier circuit also contains signal limiting diodes, the circuit for eliminating the spikes produced by electronic chopping and suitable filters in the power supply line.

The circuit shown in Figure 5.20 has a gain of approximately 4. The signal applied via C_1 to the gate of field effect transistor T_1 originates at the anode of the photomultiplier, with R_1 being the anode resistor. The zener diode protects the gate emitter junction of T_1 by limiting the signal amplitude. The stage gain of the amplifying stage is linearized by a dc feedback from the junction of R_3 and R_4 to T_1 source. Capacitor C_2 provides ac feedback between the collector and base of T_2. R_5 and C_3 act as filter in the power supply line.

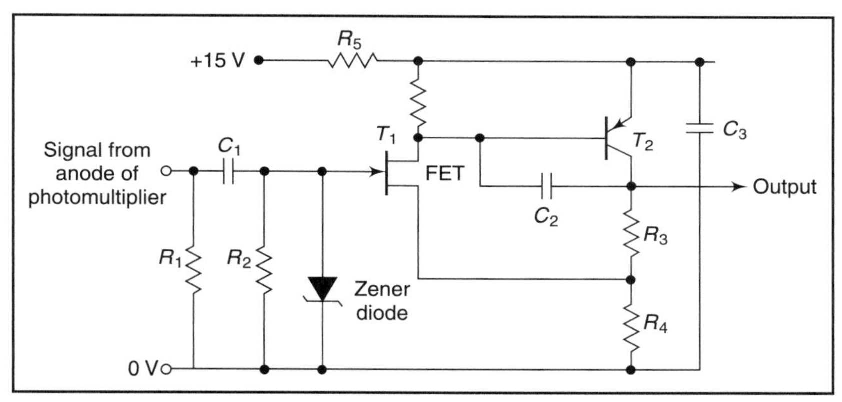

Figure 5.20 :: Pre-amplifier circuit used with photomultiplier detector

EHT

The EHT oscillator is a high frequency oscillator, operating in the range of 15–20 kHz. It is an astable multivibrator, whose output is connected to the primary winding of the EHT (extra high tension) transformer, through an output stage comprising a Darlington stage.

The EHT requirement for PMT is of the order of 2 kV at a maximum current of 200 μA. The transformer's primary windings may be bifilar wound to reduce losses and are connected in series to permit a push-pull signal. The secondary winding is connected to four EHT rectifying diodes (Figure 5.21) arranged in quadrupler rectifier circuit, which is followed by smoothing filters. In most of the spectrophotometers, the positive EHT line is earthed at the photomultiplier pre-amplifier and the polarity of the output is, therefore, negative.

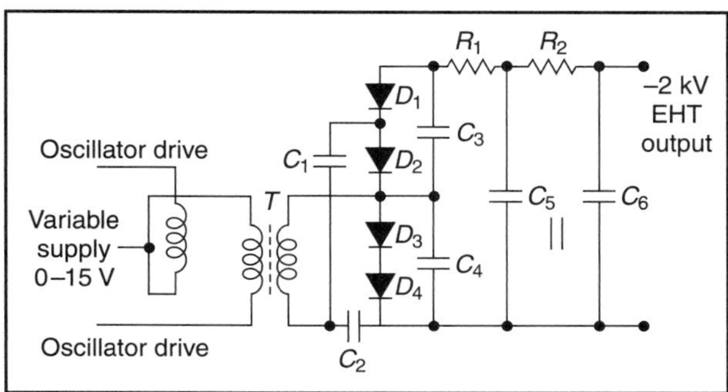

Figure 5.21 :: EHT supply circuit for a photomultiplier

The main synchronizing circuit derives its input from the mains supply sine wave signal, or 50 Hz frequency. The signal is squared in a Schmidt trigger circuit, to get lamp modulation waveform at 50 Hz pulse frequency. The same circuit also provides for sample decoder drive at 50 Hz and reference decoder voltage, which are connected to PMT signal processing circuit in decoder and sequence logic circuit

The amplified signal from the PMT pre-amplifier is given to the decoder circuit, where it is amplified, encoded and given to the integrator. The sequence logic circuit controls various timings of the cycle. The operation can be manual or remote-controlled.

The integrator provides dc voltage corresponding to the signal and reference signals. The integrator could be a conventional op-amplifier connected in integrator configuration, having a capacitor in the feedback circuit. Auto-zero potentiometer is connected in feedback of the amplifier following the integrator.

Basically, the analog-to-digital converter (ADC) employs single ramp principle. It consists of a comparator circuit, which evaluates the difference between the reference and sample signals. It produces a pulse of width proportional to the signal difference. The conversion is carried out according to linear or logarithmic law, the latter is required in the absorbance mode.

The comparator is fed with a ramp generator and a switch circuit. The ramp waveform is determined by the voltage charge on the ramp capacitor. When a linear ramp is required, the ramp voltage decreases at a constant rate, the capacitor getting discharged at a constant current. A logarithmic ramp is produced when the capacitor discharges towards a preset voltage and the discharge current is passed via a resister.

The output of the comparator is given to a circuit of logic gates, which produces two output signals in anti-phase. The output signals indicate the polarity of the signal differences evaluated in the comparator.

An additional circuit provides for timing pulses like 10 Hz reset pulse output, a 5–Hz set zero pulse output and a 5 Hz transfer data pulse output. These pulses are obtained by dividing the mains frequency in a divider.

The basic clock frequency for the ADC is obtained from an astable multivibrator. The clock pulses are responsible for causing operation in digital display. The clock frequency can be varied for scale expansion. The logic circuit provides gating of the clock output with the comparator output pulse. The resulting bursts of oscillator pulses provide the digital equivalent of the voltage difference of the dc analog signals.

The bursts of clock pulses are counted in decade counters and given to BCD to seven-segment decoder/drivers, which contain the decoding logic and drive transistors for each of the seven segments of each numeral of the display.

Modern instruments are computer-controlled in which one significant development is the use of multi-lamp turrets controlled by the same computer which tunes the desired wavelength. This makes the technique multi-element in nature. The computer also controls the flame composition and burner height. The instruments are also self-optimizing with the gases controlled via a binary flow system with feedback of absorbance signal to automatically set the optimum fuel flow (Ebdon, 1984).

5.2.7 Signal Integration

In atomic absorption spectroscopy, the signal-to-noise ratio limits the work at low concentrations. This can be greatly improved by signal integration. If the detector signal is integrated for a certain time, the output voltage of the integrator would be the sum of the signal and the noise. Since the noise is mostly random, its total component will be zero in the ideal case.

Integrated readings are obtained by feeding the amplified current from the photomultiplier to a capacitor, for a period of time precisely controlled by an electronic timer. At the end of this period, the voltage attained by the capacitor is applied to the readout meter. The reading obtained remains steady until a reset control is operated.

The integration time can be varied from 2 to 40 s. The accuracy and repeatability of measurement increases with the integration time. The random variations in the absorbance value of aspirated samples vary from instant to instant, due to variations in the number of excited neutral atoms in the flame. These instantaneous variations from the average level are summed up and presented at the DVM (Digital Voltmeter) or recorder. The integration starts as soon as the start button is actuated. Immediately, the rising signal can be observed on the recorder or DVM. At the end of the integration time, the memory holds the reading for a short duration and then returns the pen to zero.

A signal recorded on the integration mode will be about four to five times greater than that recorded on the conventional mode. The advantage of integration mode is clearly shown in Figure 5.22.

In the memory mode, the integrated value of absorbance is held on the digital display, until the beginning of the next analysis. But the potentiometric recorder registers zero until the measuring time, or until the integrating time elapses. At the end of the integrating time, the pen is given the signal to record the value held by the memory for the same period as memory time. The advantage of memory is that it is noise-free and control settings such as scale expansion can be adjusted, to bring the reading on scale. Such changes do not affect the measured value (Faithful, 1983).

5.2.8 Sampling System

A sampling system is used to supply a constant feed rate of sample into the burner. The most common method employed makes use of a solution which is introduced into the flame by an atomizer. The design and construction of the atomizer are similar to that used in emission spectroscopy. A complete sampling system comprises a nebulizer and a spray chamber (Figure 5.23).

The nebulizer reduces the sample solution to a spray of droplets of various diameters by a pneumatic action. The spray is directed into the spray chamber, where the larger droplets are precipitated and the air stream carries the remaining spray into the burner and thence to the flame. Normally, the sample uptake rate of the system is 3-4 ml/min. It has been experimentally established that an improvement in analytical sensitivity, when non-aqueous solvents are being

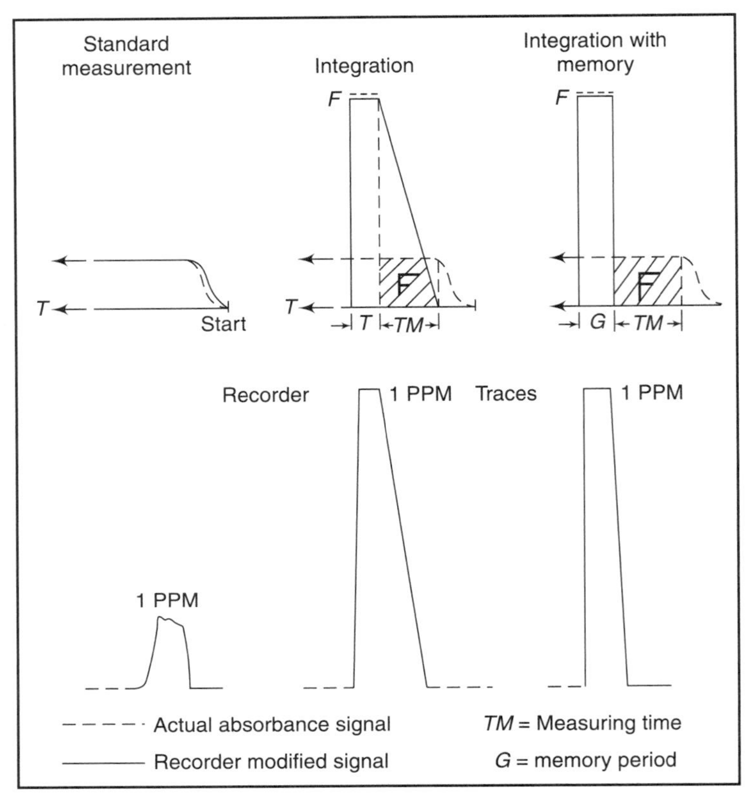

Figure 5.22 :: Principle of operation of different modes in atomic absorption spectrophotometers

used, occurs by employing a more efficient nebulization of the sample solution. Organic solvents are often used in place of water, when the sample may be an organic liquid or a solid, soluble only in organic solvents. The choice of a suitable organic solvent is very important, because the solvent can affect the physical properties of the flame and may influence the decision, and sensitivity of the analysis.

The sampling system may be coupled with an automatic sample changer, which may have a number of sample-holding polythene cups. Each cup passes the sampling point in turn, and its contents are sampled by an automatic head.

One way to reduce the wash-out time significantly is to use the flow injection technique, a block diagram of which is shown in Figure 5.24. The pump used is peristaltic and the system injects the sample in pulsed-form, similar to the one used in liquid chromatography. This technique offers the following advantages:

- Only a relatively small volume of sample is necessary to achieve a signal comparable to continuous nebulization.

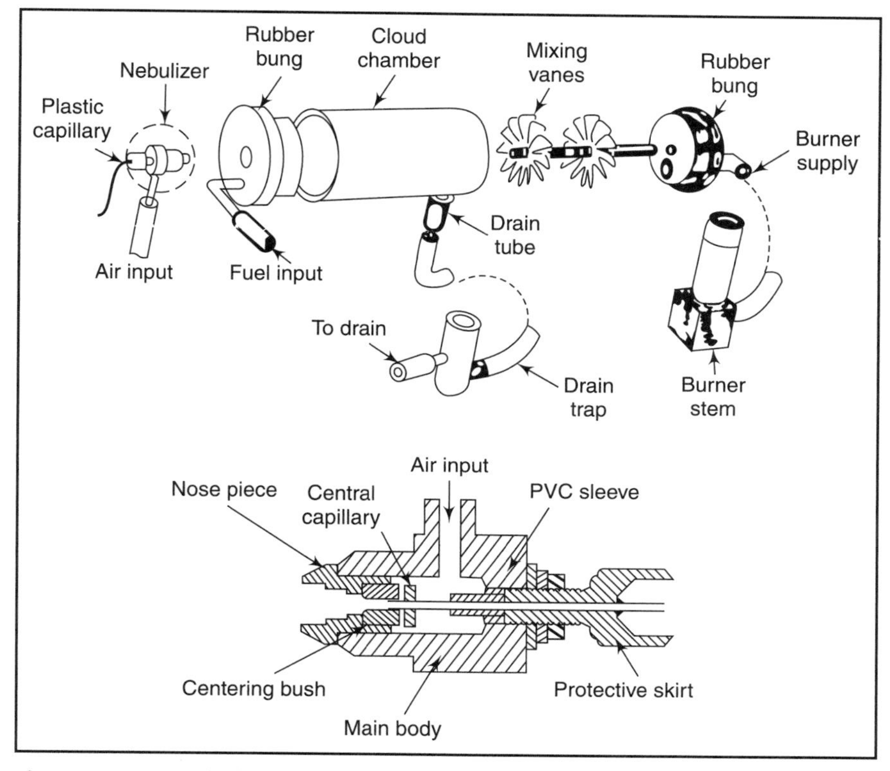

Figure 5.23 :: Nebulizer and spray chamber used in atomic absorption spectrophotometric measurements

- Because of the transient nature of the signal, exponential decay in the wash-out process starts much sooner than with continuous sample introduction. Consequently, the signal decays to baseline more quickly than with continuous sample introduction. It is, therefore, possible to inject samples at the rate of approximately 4/min, as against 1.5/min, with conventional sample introduction.

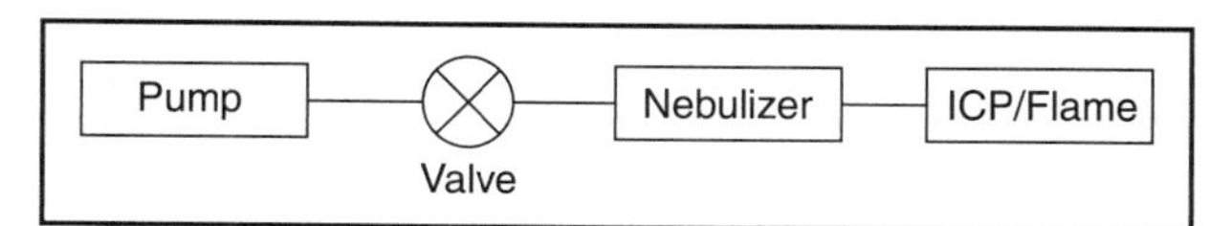

Figure 5.24 :: Flow-injection technique for reduction of wash-out time

The sample introduction procedure selection depends upon the:

- Type of sample—liquid, solid or gas,
- Levels and also the range of levels for the elements to be determined,
- Accuracy required,

- Precision required,
- Amount of material available, and
- Number of determinations required per hour.

In addition, all the measurement techniques available, regardless of whether they are flame atomic absorption spectroscopy, inductively coupled plasma atomic emission spectroscopy or dc plasma atomic emission spectroscopy, have a major effect on the choice of the procedure selected.

In ICP systems, the following three important considerations need to be followed in sample introductions:

- It is necessary to introduce the sample to the atomizer with drops that are strictly of a certain maximum size.
- The solvent introduction rate must fall within a certain permissible band of values.
- In order to maintain good system reproducibility, the temperature must be maintained constant in the plasma box, as a means to reduce the baseline drift caused by variable solvent vapour loading.

The automation of atomic absorption spectrophotometers is highly developed. For example, the Perkin Elmer Model 5000 coupled with its automated sample-handling system, can determine six elements in 50 samples in less than 35 minutes, complete with standardization, etc., which is equal to about 8.5 quantitated determinations per minute or about 500/h. Modern instruments can be easily coupled to computer systems for processing the massive amounts of information that the instruments generate. The use of microprocessors in AAs provides the optimized choice of experimental conditions, detection of errors and assurance of analytical reliability. Loon (1980) reviews the methods used for direct trace elemental analysis of solids by atomic spectrometry.

5.2.9 Performance Aspects

The performance of an atomic absorption spectrophotometer is usually described by sensitivity and detection limit.

Sensitivity

By definition, sensitivity is the concentration of an element in solution, which absorbs 1 per cent of the incident light. Its value depends upon atomizer efficiency, the design of the burner and optical system. It is independent of amplifier gain and noise from flame or lamp.

Detection limit

Detection limit is defined as that concentration of an element, in ppm, which gives a reading equal to twice the standard deviation of the background signal, the standard deviation being computed from at least ten readings. In practice, the detection limit is the minimum concentration of the element in solution that can just be detected. This parameter depends both upon sensitivity and noise generated by the flame and lamp. It is, therefore, more meaningful and a better criterion of spectrophotometric performance.

5.3 Sources of Interferences

5.3.1 Anionic Interference

The atomic absorption spectroscopic technique is generally recognized to be free from spectral interference, because the bandwidth or the monochromator is essentially the width or the source emission line. Thus, closely spaced lines such as Mg 285.21 and Na 285.28, would present no problem in absorption work. The wavelengths at which the individual elements absorb are well-defined and the possibility of two elements absorbing at exactly the same wavelength is quite remote.

However, the response of an element at its resonance wavelength may sometimes depend upon another component in the same solution. This effect is known as chemical interference. The major source of this effect is the presence of anions with the metal ions in the sample, which affect the stability of the metal compounds formed during atomization and hence, the efficiency of metal atom production. The effect is important when determining calcium and magnesium in biological fluids and aluminium alloys, etc. Sulphate, phosphate, silicate, aluminate and a number of other radicals all interfere with these metals, by tending to bind the calcium into a compound that is not associated, in the flame. Generally, all these interferences may be completely removed by the addition of excess lanthanum, which competes with the calcium for the interfering anions. Lanthanum is also useful for removing similar interferences from strontium and barium.

5.3.2 Viscosity Interference

Viscosity interference is caused by the fact that solutions of widely differing viscosities enter the burner at different rates. Fortunately, this effect is negligible in most of the analyses. However, if it is desirable to overcome this interference, it may be necessary to match the standards to the samples. The method usually adopted is that of additions. In this method, a crude determination of the concentration of the desired element is made, after which the sample is divided into three aliquots. Different additions are made to two of the aliquots, while the third is left blank. All three are then determined by atomic absorption, and a graph is plotted for absorbance versus ppm added. The intercept of the curve on the ppm axis gives the concentration in the unknown. With the method of additions, atomic absorption measurements are independent of sample characteristics, and the making up of standards is not necessary.

5.3.3 Ionization Interference

Ionization interference is usually observed at high flame temperature and manifests itself as an enhancement in the response of the element under determination. At any temperature in the flame, there is an equilibrium between neutral atoms and ions. For elements having relatively low ionization potentials, the proportion of ions will be high. If another easily ionizable element is added into the flame, the equilibrium for both elements is shifted towards the atomic state, producing an enhancement in analytical sensitivity. The interference is generally overcome by addition of the same quantity of the interference to the standard solutions, as is present in the samples, or by the addition of an easily ionizable metal like sodium or potassium to both standard and sample solutions.

Since the concentration of the analyte element is considered to be proportional to the ground state atom population in the flame, any factor that affects the ground state population of the analyte element can be classified as an interference. Factors that are likely to affect the ability of the instrument to read this parameter can also be classified as an interference.

5.3.4 Broadening of the Spectral Line

There are several factors which lead to broadening of the spectral line. These are:

- *Doppler Effect*: This effect arises because atoms will have different components of velocity along the line of observation.
- *Lorentz Effect:* This effect occurs as a result of the presence of foreign atoms in the environment of the emitting or absorbing atoms. The magnitude of broadening varies with the pressure, the foreign gases and their physical properties.
- *Quenching Effect:* In a low pressure spectral source, quenching collision can occur in flames as the result of the pressure of the foreign gas molecules with vibrational levels very close to the excited state of the resonance line.
- *Self-absorption Effect:* The atoms of the same kind as that emitting radiation will absorb maximum radiation at the centre of the line than at the wings, resulting in the change of shape of the line as well as its intensity. The effect is more predominant if the vapour which is absorbing radiation is considerably cooler than that which is emitting radiation.

5.4 Meter Scale

The linear relationship between concentration and absorbance (Beer's Law) is followed only approximately in atomic absorption spectroscopy. In general, calibration curves tend to approach an asymptote drawn parallel to the concentration axis. The calibration curve correlating concentration and per cent transmission can also be drawn, but it shows more pronounced curvature. The meter scale is usually calibrated in absorbance (0-1.0 A), which approximates to 0-100 per cent T.

A circuit which enables to continuously record the output from a Perkin Elmer 303 atomic absorption spectrophotometer, directly as optical density is shown in Figure 5.25. The circuit has a pair of operational amplifiers (A_1, A_2) as its first stage, which operate as unity gain buffers. These amplifiers receive the signal from sample and reference detectors and are followed by a linear-to-log converter. This is a circuit comprising of op-amps having transistors in their feedback circuits, There is a provision for offset adjustment and impedance matching output stage, before the circuit is connected to the recorder.

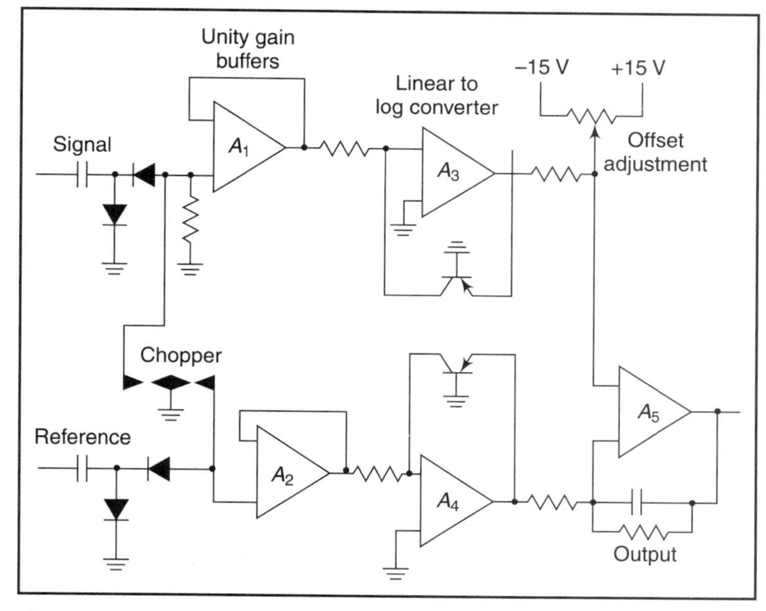

Figure 5.25 :: Linear-to-log recorder adapter for atomic absorption spectrophotometric measurements

5.4.1 Curve Correction for Linearization

When solutions with higher concentrations are being analyzed, or if the element being analyzed is a strong absorber, a linear plot of concentration or absorbance is not normally obtained. This is because, at higher concentration, the indicated absorbance is less than the expected value. Deviations from linearity usually occur, especially as the concentration of metallic analytes increases due to various reasons such as unabsorbed radiation, stray light or disproportionate decomposition of molecules at high concentrations. The lower concentrations of the same element, however, show a linear behaviour.

Curve correction

The earlier instruments did not incorporate the facility for curve correction. Due to the advent of modern techniques in electronics such as the use of integrated circuits, and now computer circuits (microprocessors), curve correction facility is introduced automatically in the programming of the microprocessor.

The idealized calibration or standard curve is stated by Beer's Law that the absorbance of an absorbing analyte is proportional to its concentration. Figure 5.26 (a) shows a deviation of plot from linearity. In order to linearize the plot, after zeroing with blank, aspirate a standard solution of highest concentration of interest and set the digital display to the required reading by the Auto-Cal control, such that the value is registered with memory for internal calibration, as shown in Figure 5.26 (b). The mid-scale standards will show a higher reading than the dotted values. The required correction can be calculated as:

$$\text{Per cent deviation} = \frac{\text{Read-out concentration deviation of mid-range standards}}{\text{Actual concentration of mid-range standards}} \times 100$$

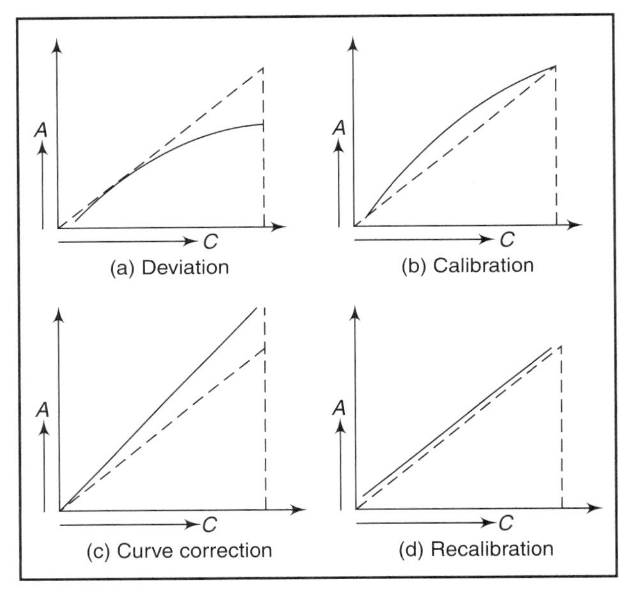

Figure 5.26 :: Curve correction for linearization

The curve correction knob must be set to the computed value, which makes the plot as shown in Figure 5.26 (c). The standards have to be aspirated once again to verify linearization.

Should any further deviation exist, the same method of computing and correcting must be carried out to establish the required linearity, as shown in Figure 5.26 (d).

The use of a an in-built microcomputer in an atomic absorption system provides the ability to conveniently calibrate and accurately compute concentrations from absorbance data, even for non-linear calibration curves. In the linear region, data on one standard and a blank are adequate for defining the relationship between concentration and absorbance. Where the relationship becomes non-linear, however, more standards are required. The accuracy of a calibration computed for a non-linear relationship depends upon the number of standards and the equations used for calibration. For the equation format which optimally fits atomic absorption data, it has been experimentally shown that accurate calibration can be achieved with three standards plus a blank, even in cases of severe curvature.

Chapter 6

Fluorimeters and Phosphorimeters

6.1 Fluorescence Spectroscopy

The analytical technique based on the absorption of infrared, visible and ultraviolet light has found extensive applications in chemistry. However, a limitation of this technique is the difficulty in differentiating between different substances with similar absorption spectra. Many solutions when irradiated with visible or ultraviolet light, not only absorb this light, but re-emit light of a different wavelength. This effect is known as fluorescence and its exploitation opens up possibilities in the discrimination and accurate determination of many substances in very dilute solutions. The difference in wavelength between incident and re-emitted light facilitates efficient discrimination between substances with similar absorption characteristics.

Fluorescence spectroscopy is thus a form of analysis that utilizes the emission properties of specific molecules rather than a tendency to absorb certain wavelength of light. This inherently makes it more selective than absorption spectroscopy due to the fact that all molecules that fluoresce must absorb, but not all molecules that absorb necessarily fluoresce. Therefore, only specific types of molecules can be studied using fluorescence spectroscopy.

The main advantage of fluorescence detection as compared to absorption measurements is the greater sensitivity achievable because the fluorescence signal has, in principle, a zero background. Its major analytical applications include quantitative measurements of molecules in solution and fluorescence detection in liquid chromatography.

Fluorescent analysis makes valuable contributions to many branches of research and industry. However, it has found the widest application in biomedical laboratories. Clinical laboratories employ it for studies such as adrenalin and noradrenalin in urine and blood, and in screening tests for tumours of adrenal glands. Tests for transminase, lactic dehydrogenase, DPN (diphosphoneucleotide), steroids, DNA (deoxyribonucleic acid), histamine and many others are routinely made with fluorimeters. The value of the fluorimetric technique, for studies such as air pollution and analysis of alkaloids, has tremendously increased. Industrial health laboratories use fluorimeters for the routine determination of beryllium. Agriculture and food chemists use this technique for vitamin and insecticide residue studies. Water, air pollution and sanitation engineers find it effective and convenient for studying flow and the diffusion of air and water.

6.2 Principle of Fluorescence

Normally, in the ground state of molecules, molecular energies are constant and present at minimum values. When a quanta of light impinges on a molecule, it is absorbed and an electronic transition to a higher electronic state takes place. This absorption of radiation is highly specific, and radiation of a particular wavelength and energy is absorbed only by a characteristic structure. In the ground state of most molecules, each electron in the lower energy levels is paired with another electron, whose spin is opposite to its own spin. This state is called a singlet state. But when the molecules absorb energy, they are raised to an energy level of an upper excited singlet state or S_1 or S_2 or S_3 (Figure 6.1), represented by the transition G-S_1 or G-S_2. These singlet transitions are unstable and the molecular energy tends to revert almost immediately to a lower level. While going to a lower energy level, the absorbed energy is lost by steps from the upper singlet state to the lowest single-state through the transition S_3-S_1 or S_2-S_1. These singlet transitions are responsible for the visible and ultraviolet absorption spectra.

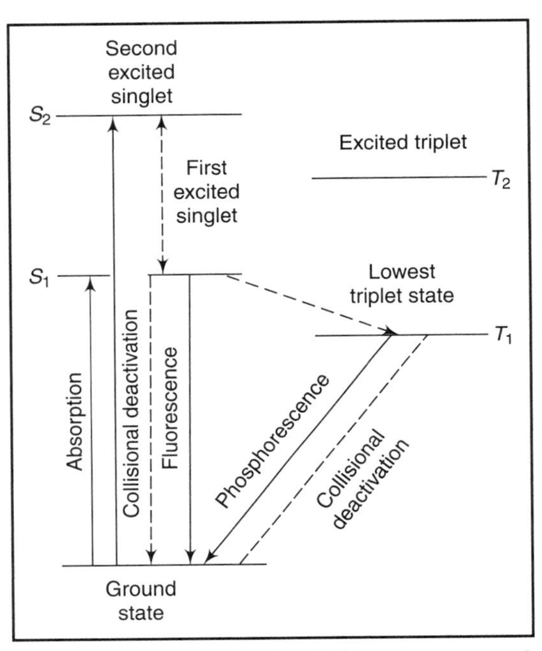

Figure 6.1 :: Principle of fluorescence and phosphorescence

The excited singlet state persists for a time, which is of the order of 10^{-8} to 10^{-4} s. During this time interval, some energy in excess of the lowest vibrational energy level is rapidly dissipated. In case all the excess energy is not further dissipated by collisions with other molecules, the electron would return to the ground state with the emission of energy. This phenomenon is called 'fluorescence'. Since this transition involves less energy than in the original absorption process, the fluorescence is emitted at wavelengths longer than those of the exciting source. This is shown in Figure 6.2. Many vibrational levels are actually involved, so molecular fluorescence spectra are generally observed as continuous bands, with one or more maxima. Filters are used to filter out the exciting wavelengths, so that only fluorescent energy reaches the detector.

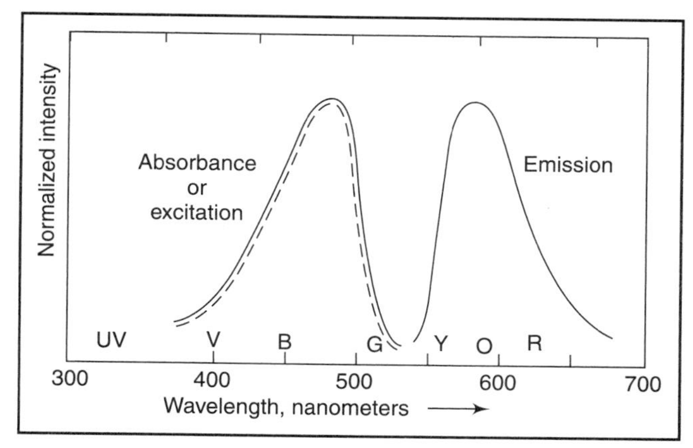

Figure 6.2 :: Representation of excitation (absorption) and emission spectrum (*www.turner-designs.com*)

Any fluorescent molecule has two characteristic spectra, the *excitation spectrum* and the *emission spectrum.* The excitation spectrum shows the relative efficiency of different wavelengths of exciting radiation to cause fluorescence, whereas the emission spectrum would indicate the relative intensity of radiation emitted at various wavelengths. Normally, the shape of the excitation spectrum should be identical to that of the absorption spectrum of the molecule, and independent of the wavelength at which fluorescence is measured. However, due to instrumental errors, this is seldom the case. Also, the shape of the emission spectrum is independent of the wavelength of exciting radiation. If the exciting radiation is at a wavelength different from the wavelength of the absorption peak, there will be less absorption of radiant energy, and consequently less energy will be emitted.

Luminescence

It may be noted that luminescence is the generic name used to describe all forms of light emission other than that arising from elevated temperature.

'Photoluminescence' relates to the emission of light through absorption of ultraviolet or visible energy. Photoluminescence can be further divided into:

'Fluorescence', which is the immediate release (10^{-8}s) of absorbed light energy as opposed to phosphorescence, which is the delayed release (10^{-6} to 10^{2} s) of absorbed light energy.

'Chemiluminescence' is the emission of light energy caused by chemical reactions.

'Bioluminescence' refers to the light emission through the use of enzymes in living systems.

6.2.1 Relationship between Concentration and Fluorescence Intensity

The total fluorescence intensity is equal to the rate of light absorption multiplied by the quantum efficiency of fluorescence (η), that is

$$F = \eta\,(P_0 - P) \tag{1}$$

where P_0 is the power of the beam incident upon the solution and P is its power after traversing a length b of the medium.

In order to relate F to the concentration c of the fluorescing particle, we can apply Beer's Law. In such a case, the fraction of light transmitted is

$$P/P_0 = e^{-\varepsilon bc}, \tag{2}$$

where τ is the molar absorptivity of the fluorescent molecules and τbc is its absorbance A.

$$1 - P/P_0 = 1 - e^{-\tau bc}P$$

$$(P_0 - P) = P_0\,(1 - e^{-\tau bc}) \quad (3)$$

substituting Eq.(3) in Eq.(1), we get

$$F = \eta P_0\,(1 - e^{-\tau bc}) \quad (4)$$

For very dilute solutions in which $A\ (\tau\, bc) < 0.05$ and only a small fraction of light is absorbed, the exponential term can be expanded, and Eq. (4) reduces to

$$F = K'.\ \eta.\ P_0\,(2.3\ \tau\, bc)$$

or $\quad F = K.\ C_M.$

This equation shows that for very dilute solutions, a plot of the fluorescent power of a solution versus the concentration of the emitting particles would be linear. However, at higher concentrations (Figure 6.3) linearity is lost and the curve bends downwards as the concentration increases.

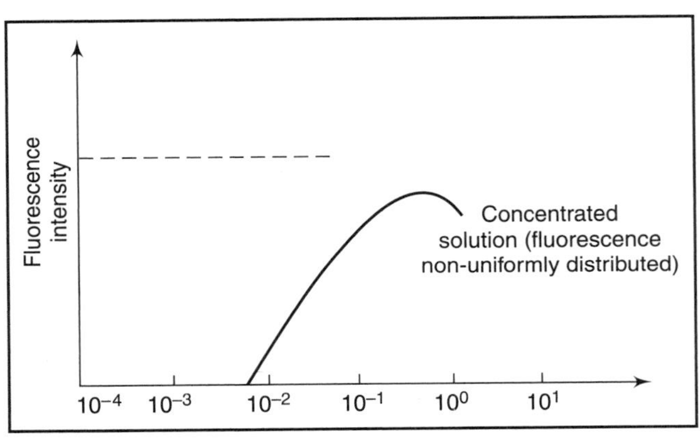

Figure 6.3 :: Plot between concentration and fluorescence intensity

The linearity of a sample depends upon many factors, including the chemical composition of the sample and the path length that the light must travel. Also, concentration has an effect on linearity. When concentration is too high, light cannot pass through the sample to cause excitation. This high concentration can have low fluorescence. Similarly, at intermediate concentrations, the surface portion of the sample nearest the light absorbs so much light that little is available for the rest of the sample. So, the readings will not be linear, though they will be within the range of a calibration curve.

While making fluorescent measurements, it is always advantageous to make the exciting energy as large as possible. However, photodecomposition should be avoided.

6.2.2 Factors Affecting Fluorescent Yield

Fluorescence intensity, which is important for quantitative work, depends upon numerous factors, some of which are characteristic of the fluorescent substance and some of which are dependent upon the measuring technique employed. The failure to realize the significance of these factors can seriously impair quantitative work.

Some substances are strongly fluorescent, while others have weak or undetectable fluorescence. The efficiencies are defined by the fluorescent yield, that is the ratio of radiation emitted to that absorbed. Some substances such as certain dyes, are highly efficient and provide yields approaching 1, whereas other substances have fluorescent yields less than 0.0001. The efficiency for a given substance is not constant, but depends upon environmental effects, some of which are described below.

Fluorescent yield

Fluorescence intensity depends upon both the intensity and the spectral distribution of the exciting radiation. Therefore, for reliable analyses, it is essential to employ stable sources, which provide intense radiation in the proper wavelength regions. Since intensities vary between sources and with time, reference materials are commonly utilized to maintain uniformity between readings.

The temperature of the sample has a considerable effect upon the fluorescence. As a result of the reduced probability of collisions and reactions, the fluorescent yield is increased by cooling the sample. Reduced temperatures also enhance the fine structure observed in many spectra.

Solvent effects similar to those noted in absorption work are common in fluorescence work, as changes in both the wavelength position and intensity of fluorescence peaks are observed. Quenching effects are greatly influenced by the solvent employed.

The pH of the sample solution is critical for many substances. In some instances, highly fluorescent materials becomes completely non-fluorescent when the pH is altered by only a few pH units.

Most fluorescent substances decompose in solution, particularly when subjected to ultraviolet radiation. This is one of the major disadvantages of utilizing powerful sources as a means of increasing sensitivity.

In some instances, the fluorescence spectrum of a substance overlaps the region in which it shows absorption. Consequently, part of the emitted radiation is re-absorbed before it leaves the solution. Since the absorption occurs at selected wavelengths, both the total intensity and the shape of the fluorescence spectrum are affected.

Various foreign substances present in the sample can quench the fluorescence radiation. Thus, oxygen diminishes the fluorescence of polyaromatics, and transition elements such as iron and manganese quench the uranyl fluorescence. Dilution techniques are commonly employed to detect impurity quenching, and in some cases to correct for its effect. In other cases, it is necessary to remove the interfering substance before reliable measurements can be obtained.

The observed spectral fluorescence spectrum is dependent upon the dispersion, stray radiation, and transmitting properties of the optical system. For a given instrument, these factors are essentially constant. So simple calibration curves can be utilized for quantitative work. However, since significant differences may exist between instruments, caution should be exercised in comparing inter-laboratory results.

The inherent advantages of the fluorescence technique are:

- *Sensitivity:* The sensitivity of a fluorometer refers to the minimum detectable quantity of a compound of interest under specified instrument condition. High sensitivity of fluorometry allows the reliable detection of fluorescent materials using small sample sizes. Fluorometers achieve 1000 to 500,000 times better limits of detection as compared to spectrophotometers detectability to parts per billion or even parts per trillion is common for most analytes. Therefore, picogram quantities of luminescent materials are frequently studied.

- *Selectivity:* Fluorometers are highly specific and less susceptible to interferences because fewer materials absorb and also emit high fluorosce. This is because two characteristic wavelengths are involved in these measurements. In contrast, spectrophotometric techniques are prone to interference problems because many materials absorb high, making difficult to isolate the targeted analyte in a complex matrix.
- *Variety of sampling methods:* This include dilute and concentrated samples, suspensions, solids, surfaces and combination with chromatographic systems.
- *Wide concentration range:* Since the fluorescence output is linear to sample concentration over a broad range, the technique can be used over three to six decades of concentration without sample dilution or modification of the sample cell.
- *Simplicity and speed:* Fluorometry is a relatively simple analytical technique. Its sensitivity and specificity reduce or eliminate the sample preparation procedures and thereby expedite analysis.

6.3 Measurement of Fluorescence

Fluorescence measurements are carried out in instruments called fluorometers or fluorimeters. The various components of these instruments are similar to those of photometers. These are analogous to photometers, as they too make use of filters to restrict the wavelengths of the excitation and emission beams. Fluorescence measuring instruments making use of monochromators like spectrophotometers are called spectrofluorimeters. These are of two types: in one type, a filter is employed to limit the excitation radiation and grating or prism monochromator to isolate a peak of the fluorescent emission spectrum. The other type of instruments are equipped with two monochromators. One of the monochromators restricts the excitation radiation to a narrow band, whereas the other permits the isolation of a particular fluorescent wavelength. The latter type are the true spectrofluorimeters.

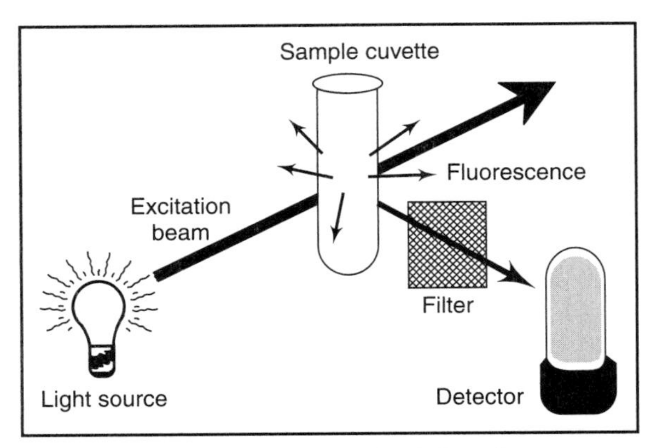

Figure 6.4 :: The basic principle of measurement of fluorescence (Tissue, 2004b)

The basic principle of measurement of fluorescence is shown in Figure 6.4. The arrangement contains an excitation source, sample cell and fluorescence detector. Molecules in solution are usually excited by UV light and the excitation source is a deuterium or xenon lamp. Broad band excitation light from a lamp passes through a monochromator, which passes only a selected wavelength. The fluorescence is dispersed by another monochromator and detected by a photomultiplier tube while the excitation monochromator gives an excitation spectrum, scanning the fluorescence monochromator gives the fluorescence spectrum. Simple instruments sometimes use only a bandpass filter to select the excitation wavelength.

6.3.1 Single-beam Filter Fluorimeter

Figure 6.5 shows a block diagram of a single-beam filter fluorimeter. The source of light is a high pressure mercury discharge lamp, with glass or fused silica envelope. Selection of the exciting wavelength is made by inserting a primary filter in the incident beam. This helps to isolate one of the principal lines at 365, 405, 436, 546 or 579 nm and to have intense line spectrum with good spectral purity. Low pressure lamps, equipped with a silica envelope, can produce intense radiation at 254 nm. The key components of a signal beam filter based fluorometer are shown in Figure 6.6. The lamp or light source which provides the energy that excites the compound of interest include xenon lamps, high pressure mercury vapour lamps, xenon–mercury arc lamps, lasers and LED's. Lamps emit a broad range of light, i.e. more wavelength than those required to excite the compound. LED's and lasers emit more specific wavelengths.

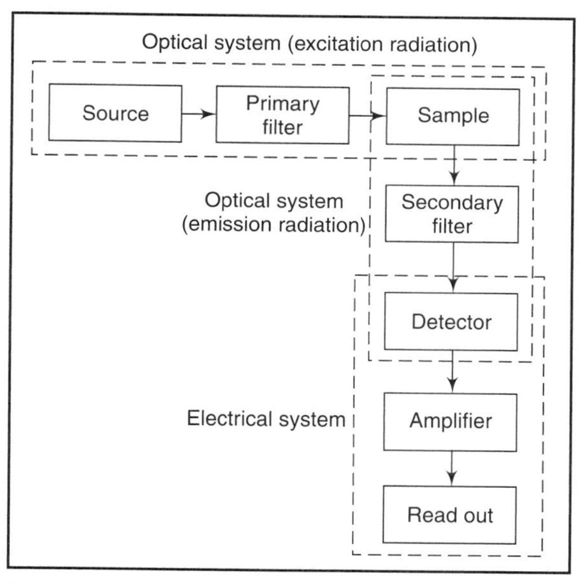

Figure 6.5 :: Schematic diagram of the components of typical fluorimeter

Both cylindrical and rectangular cells constructed of glass or silica are employed in fluorescence measurements. The cuvette material must allow the compound's absorption and emission light energy to pass through. Also, the size of the sample cell affects the measurement. The greater the path length (or diameter) of the cell, the lower is the concentration that can be read. Fluorometers commonly use 10 mm sq. cuvettes and/or 13 mm or 25 mm test tubes. For small volumes of solutions, adaptors are available for capillary tubes (9ml) and minicells (100 ml). Cuvettes are made from borosilicate or quartz glass that can pass the selected wavelengths. Adequate care is taken in the design of the cell compartment to reduce the amount of scattered

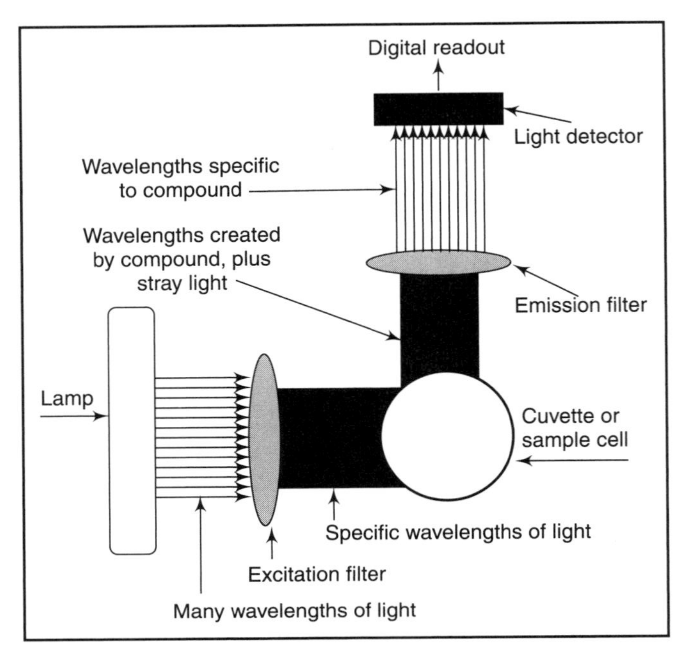

Figure 6.6 :: Key components of a filter fluorimeter, *www.turnerdesigns.com*

radiation reaching the detector. For continuous, on-line monitoring of samples for fluorescence studies, flow-cells are used in the instruments sample chamber, through which samples are pumped. Flow cells too, are available in various diameters. The fluorescence emitted by the sample is measured by selecting the wavelength of fluorescence radiant energy by a second optical filter, called the secondary filter. This is placed between the sample and a photodetector located at a 90° angle from the incident optical path. The fluorescent signal is generally of low intensity and requires large amplification for carrying out necessary measurements. The detectors used in fluorescence measurements are the sensitive photomultiplier tubes. After signal amplification, the signal is displayed on a duly calibrated microammeter or recorder.

It is always difficult to measure absolute fluorescence intensity. Therefore, measurements are made with reference to some standard substance, which may be chosen arbitrarily. The circuit is balanced at any chosen setting by placing the standard solution in the instrument. Without making any adjustment, the standard solution is replaced by additional standard solutions of lesser concentration. Lastly, the fluorescence of the solvent and cuvette alone is measured to establish a true zero concentration. With the help of these settings, a curve is plotted between concentration against scale readings, which gives the fluorescence–concentration curve. Unknown concentrations can be estimated from this curve, if the scale reading is known.

It is known that the dark current of a photomultiplier is primarily due to the emission of thermionic electrons from the cathode and it is generally reduced by cooling the photomultiplier. In the case of very weak light measurements, the reduction of the dark current is an important factor for improving the signal-to-noise ratio. Special arrangements are used to cool photomultipliers to a temperature, where the thermal component of the dark current is virtually eliminated.

The detection system used in the commercially available spectrofluorimeters is almost exclusively based on the use of the photomultiplier tube in a single-channel configuration. Yet, the parallel multi-channel optoelectronic image detectors should be clearly superior, because of its ability to record the entire spectrum simultaneously. Parallel detection can result in either a significant increase in signal-to-noise ratio, or a corresponding reduction in analysis time. The detector is a silicon intensified target vidicon, whose control is achieved with a computer.

In order to study the dynamic properties of excited molecules, one must know not only the decay time of the luminescence, but also its quantum yield, that is, the number of emitted photons divided by the number of absorbed photons. The number of absorbed photons can be determined easily by any absorption measurements, but the number of emitted quanta cannot be found in a straightforward manner. The recorded emission spectra for the spectral response of the detection system thus needs to be corrected. The spectral response of the detection system is not always known, and also it is considered too much of a trouble to redraw the spectra point by point. The latter can be conveniently done by interfacing the detection system to a computer. A simple and cheap solution to the problem is possible by the use of programmable read only memories.

Fluorescence Decay

For studying the Fluorescence of Decay on the nanosecond time scale, the impulse technique is generally preferred. In this method, the sample is excited by a short pulse of light of suitable wavelength and the emitted fluorescence is detected and recorded as a function of time. The excitation and detection of the fluorescence are repeated many times and the results are averaged in order to obtain acceptable signal-to-noise ratios. However, if any drift occurs in the excitation light source, or in the detection electronics, the measured profile of the light source will not truly represent the correct profile and may introduce serious errors. Automated luminescence spectrometers using a microprocessor for correcting and recording absorption, excitation and emission spectra are greatly extending the qualitative and quantitative analytical capabilities of the technique.

A single beam fluorometer is a good choice when sensitive quantitative measurements are desired for specific compounds. The comparative ease of handling and low cost make filter fluorometers ideal for dedicated and routine measurements.

Filter fluorometers are calibrated by compensating for blank (solution containing zero concentration of the substance to be read) and adjusting the instrument to reflect a known concentration of sample (standard). Calibration is required for each instrument from the same manufacturer and certainly from manufacturer to manufacturer. As fluorescence is subject to temperature and environmental effects, the fluorometer must be calibrated in conditions as close as possible to the actual conditions of study. Due to the extremely sensitive nature of measurement in fluorometry, it is important to be rigorous in laboratory procedures such as carefully preparing standards and cleaning labware.

6.3.2 Double-beam Filter Fluorimeter

Double-beam instruments are of the direct reading type and are generally preferred because of this obvious advantage. In these instruments, the circuit is so arranged that the current from the reference detector opposes that generated by the measuring photocell. The reading on the calibrated scale at which the two currents neutralize each other directly gives the fluorescence reading.

Figure 6.7 shows a schematic diagram of a double-beam fluorimeter. A mercury lamp is used as a source of radiation. A collimated beam is passed through the primary filter, which then falls on a cuvette containing the sample. The measuring photocell is mounted at right angles to the excitation beam. Fluorescent radiation passes through secondary filters, which remove scattered ultraviolet radiation, but pass visible light.

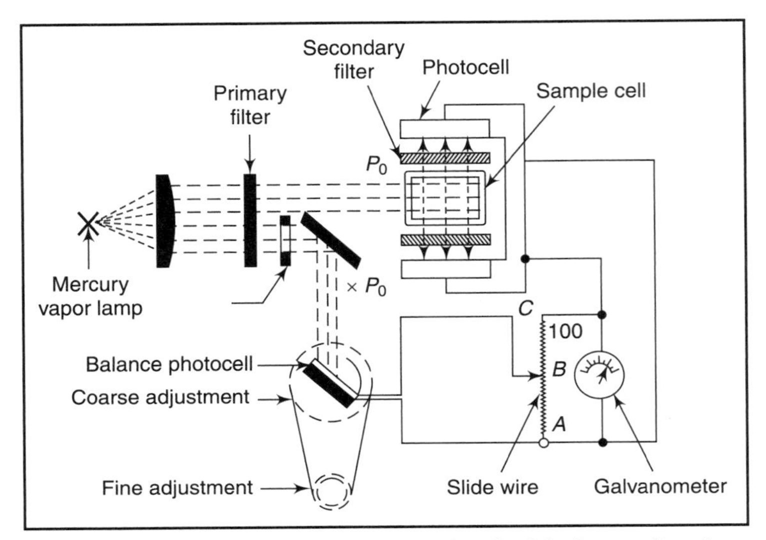

Figure 6.7 :: Schematic diagram of a double-beam fluorimeter

A part of the excitation beam is passed through a reduction plate for reading the intensity of the reference beam, in order that its power be of the same order of magnitude as the weak fluorescent beam. The attenuated reference beam is reflected on to the surface of a reference photocell, mounted on a rotating table. By adjusting the angle of the photocell, the amount of radiation falling on the photocell and thus the current output, can be adjusted.

The outputs of the reference and the measuring photocells is compared and a sensitive galvanometer is used to check the null-point. The circuit is arranged, so that the current from the reference cell opposes that generated by the measuring photocell, which responds to fluorescent radiation. In this arrangement, the concentration C is given by

$$C = K.\, R_{AB}/R_{AC},$$

where AC is the total resistance. The constant K is independent of P_0.

Quite often, two photocells, connected in parallel, are used to measure the weak fluorescent radiation. This arrangement increases the sensitivity of the instrument.

6.3.3 Ratio Fluorimeters

A ratio fluorimeter which enables direct readings for rapid analysis of multiple samples is shown in Figure 6.8.

In this instrument, the exciting energy is matched with the sample by the two primary filters located on both sides of a specially designed source lamp. The lamp cycles in-phase with the ac line voltage and has a dark time, as the lamp anodes alternately reach zero potential during the change in phase of the line voltage.

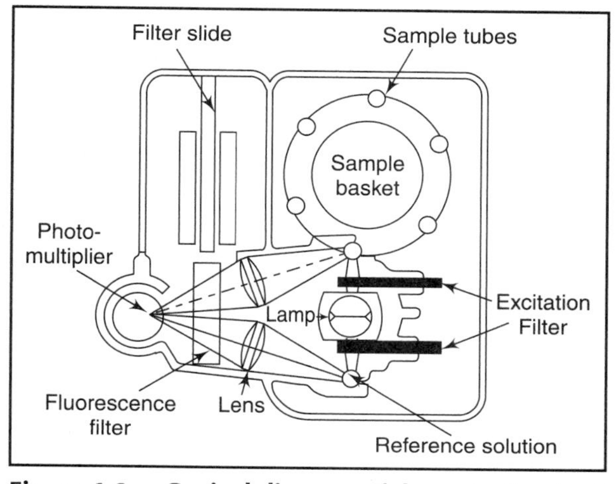

Figure 6.8 :: Optical diagram of the ratio fluorimeter

During one phase, light reaches the reference test tubes and produces fluorescence, which strikes the photomultiplier and produces the reference signal. During the other phase, light falls on the sample and causes it to fluoresce. When this fluorescent energy strikes the photomultiplier, it produces the sample signal. The reference and sample beams pass through the secondary filter. This filter isolates the fluorescence radiation from spurious radiation. By setting the reference signal equal to 100 per cent on the meter, the signal at the meter indicates the ratio of the sample to the reference. The ratio can also be recorded by using a high impedance recorder.

The source of excitation used in a typical instrument is divided into two sections by a light shield, with electrons flowing from a common cathode alternately to two anodes, one on the reference side, and the other on the sample side, as the lamp cycles in-phase with ac line voltage. When one beam is on, the other is completely off. Thus, the lamp supplies excitation radiation alternately to the reference and sample photocells. The lamp is surrounded by a phosphor-coated glass sleeve, which has a clear portion plus coatings with three different phosphors. By rotating the sleeve, mercury wavelengths about 237 nm through the clear portion, plus peaks at 310, 360 and 450 nm through the phosphor coated portions are achieved. This dual source lamp provides double-beam operation without light choppers, beam splitters or other similar devices. The sample beam and the reference beam consist of fluorescent energy of the same wavelength, thus making readings independent of lamp temperature variations. Two types of filters are employed: two primary (excitation) filters and one secondary (fluorescence) filter. Both types of filters can be easily changed for analysis of different samples. Glass test tubes can be used as sample containers for most of fluorescence that the photomultiplier receives. Since fluorescence is greater at lower temperatures, provision is made for water cooling of the sample, so that sample fluorescence is as strong as possible. The sample tube area may be cooled by a continuously flowing supply of water.

Reference bars are supplied with fluorimeters for establishing upscale meter values. These are glass rods fused with varying concentration of uranium salts. Each bar emits about three times more fluorescence intensity, successively from numbers 1 through 6. They cover an approximate range of 5–1000 ppb of quinine sulphate equivalent. In some cases, they may be more useful than standard solutions, because their fluorescence values are stable indefinitely, if not scratched or chipped.

Double-beam fluorimeters using two photomultipliers to compensate for fluctuations in light level from the exciting source, are generally used. Nevertheless, the stability of these instruments is not particularly good, and is generally not better than ±1 per cent over a ten-minute period. This poor performance has been found to be due to gain changes and drifts in dark current of the two photomultipliers, caused by variation in high voltage ambient temperature and fatigue effects. A simple solution to this problem is to use a single photomultiplier for both reference and sample beams.

Another version of the double-beam type fluorimeter makes use of an optical bridge that is analogous to the Wheatstone bridge, used in measuring electrical resistance. The bridge detects and measures the difference between the light emitted by a sample and that of the calibrated rear light path. This is the principle of the Turner Model 110 Fluorimeter (Figure 6.9).

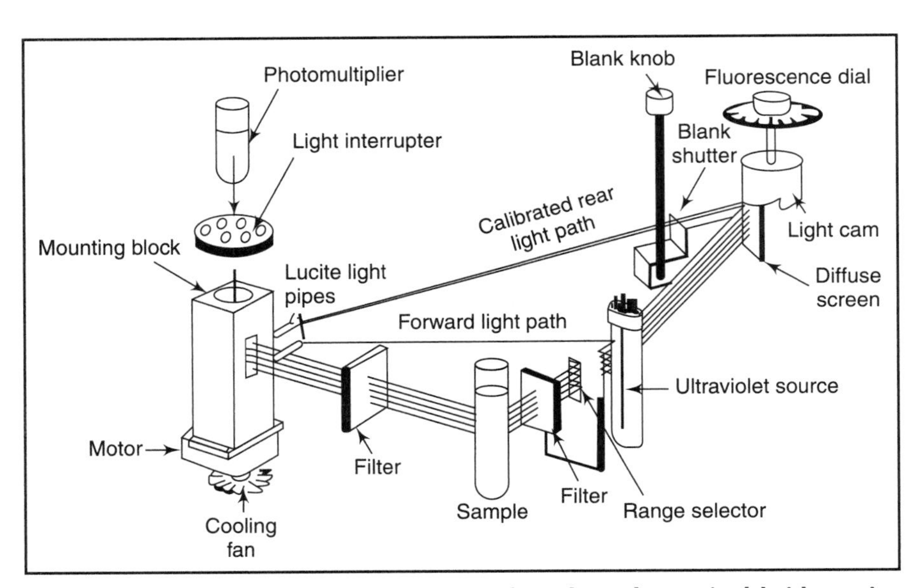

Figure 6.9 :: Double-beam fluorimeter based on the optical bridge principle (principle of Turner fluorimeter)

This instrument employs a single photomultiplier (PMT) tube as a detector. A rotating mechanical light interrupter surrounds the photomultiplier and causes the reference beam and the fluorescent beam to strike the PMT, alternately. The output of the PMT is thus an alternating current, which can be amplified in an ac amplifier. The next stage is a phase-sensitive detector, whose output will either be positive or negative depending upon whether the fluorescent beam or the reference beam is stronger. The power of the reference beam can be adjusted manually or automatically, by the rotation of the light cam which increases or decreases the intensity of reference beam reaching the detector. This is done until the amount of light reaching the PMT from the rear light path equals the amount from the sample. The light cam is attached to a linear dial, each small division of which corresponds to an equal fraction of light. For a totally non-fluorescent sample, no light would reach the detector in one of the phases and the null-point could then be obtained from one direction only. For accurately adjusting the null-point from both directions, a third (forward) light path of constant intensity is made to fall on the detector, so that under all conditions some radiation strikes the PMT. This permits a correct operation, both above and below zero with non-fluorescent blanks. Two lucite light pipes are employed in the optical system, for the rear light path and forward light path.

The PMT is mounted inside a rotating cylindrical interrupter, with two sets of light slots. The upper slots allow light to pass alternately to the PMT from the rear calibrated light path and the combination of the sample and forward light path. The lower slots interrupt the light path from the reference beam, which in turn generates the reference signal. This circuit arrangement cancels out variations in line voltage, light source and photomultiplier sensitivity.

The PMT produces an ac signal at about 400 Hz, that is proportional to the light unbalance. This signal is amplified and fed as signal to the phase-sensitive detector.

The reference signal is generated by a lamp with special glass filter around it, which passes only infrared light and directs it on to a red-sensitive phototube. The reference lamp and phototube generates an ac signal. This signal is amplified and is given as reference to the phase detector, its phase depending upon the position of the light interrupter. The phase detector generates a dc signal proportional to the signal received, which is applied to the chopper, where the dc signal is converted to ac. The ac signal is amplified. The ac signal is used to drive the servomotor, which in turn drives the light cam, so that a null-balance is obtained.

6.4 Spectrofluorimeters

Spectrofluorimeters are true fluorescence spectrometers, which normally make use of two monochromators, one to selectively supply the excitation radiation and the other to isolate and analyze the fluorescence emission. The dual monochromator system thus permits the determination and selective utilization of the peaks of excitation and fluorescence wavelength. The spectrum of the fluorescence emitted by the sample can be studied at any frequency of exciting light. Also, the variation of the intensity of fluorescence with the frequency of the exciting light used can be determined.

Figure 6.10 shows a schematic of the optical system used in a spectrofluorimeter. The source of light is a 150 W high pressure xenon arc lamp. The ellipsoidal mirror collects radiation from this lamp and focuses an enlarged image of the arc upon the entrance slit of the excitation monochromator, completely filling the f/3.5 mirrors.

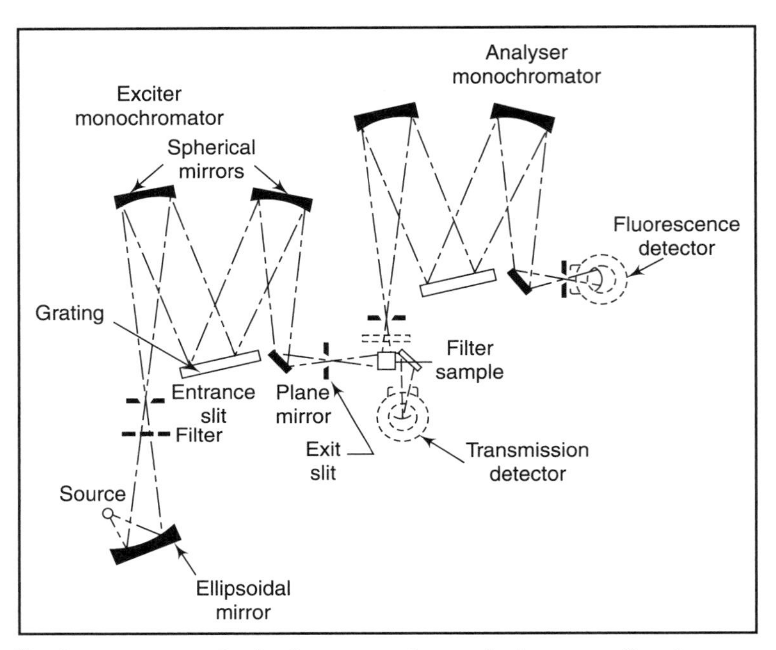

Figure 6.10 :: Optical system of a typical spectrofluorimeter

Monochromatic radiation of specific wavelength is selected by angular positioning of the grating. A desired bandpass is achieved by the selection of appropriate interchangeable slits for the entrance and exit positions of the monochromator. The radiant energy emerging from the exit slit of the excitation monochromator activates specific molecules within the sample, resulting in fluorescent emission. The emitted radiation is viewed by the analyzing monochromator, which permits only characteristic sample emission to reach the detector, blocking all undesired spectral regions.

When fluorescent energy falls on the photomultiplier, the registered photocurrent can be indicated on the meter, recorder or oscilloscope. The spectrum can be manually or automatically scanned.

Many compounds exhibit far greater fluorescence at reduced temperatures than at normal, ambient or higher temperatures. Therefore, spectrofluorimeters are usually accompanied by cryogenic accessories to cool the sample by means of an immersion probe, which is cooled by an easily replenished reservoir refrigerant. Using liquid nitrogen, for example, provides a sample temperature very close to 77 K.

The Baird Atomic Model SF 1OOE spectrofluorimeter (Figure 6.11) employs a unique optical system, comprised of four grating monochromators, two of which are used in tandem for dispersing excitation energy and the other two for observation of fluorescent spectra. This double-grating system in conjunction with folded optics offers a significant advantage in the reduction of physical dimensions. Each monochromator incorporates a 600 line/mm plane reflection grating. Wavelength control is provided by means of a cam, which may be operated manually or automatically. The calibrated operating range of each pair of monochromators is 220 to 700 nm. Wavelength settings are indicated by a scale, which is calibrated in 2 nm increment. The spectral widths of the entrance and exit slits to each pair of monochromators are each independently variable, in steps of 2, 8 and 32 nm.

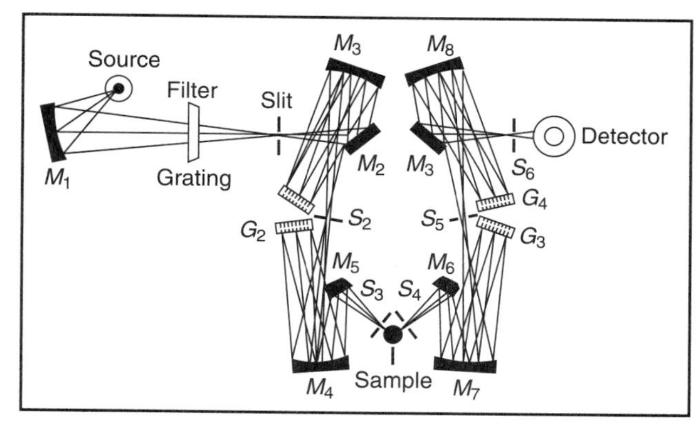

Figure 6.11 :: Optical diagram of a spectrofluorimeter (Courtesy M/S Baired Atomic, USA)

The excitation source is a specially selected 150 W dc xenon arc lamp, which gives a very stable, high intensity light output in the UV, visible and infrared regions. An off-axis ellipsoidal mirror forms an enlarged image of the source arc at the entrance slit of the excitation monochromator. The lamp can be very simply focused and laterally adjusted.

The detector used is a photomultiplier, which is supplied with a stabilized power supply. The detector signal is amplified in an operational amplifier.

Miller (1984) has reported the successful adoption of pulsed xenon lamps, in place of the continuous arc lamps used previously. The pulsed source have much longer life-time, and do not require bulky power supplies. They can be conveniently used with electronic gating systems, to distinguish between the virtually instantaneous emission of fluorescence and the delayed emission of phosphorescence.

6.5 Microprocessor-Based Spectrofluorometer

Iwata *et al.*, (1986) illustrate the details of a microprocessor-based spectrofluorometer (Figure 6.12). The instrument takes advantage of the microprocessor in both instrument control and data manipulation originating in the data station unit. The microprocessor facilitates a variable

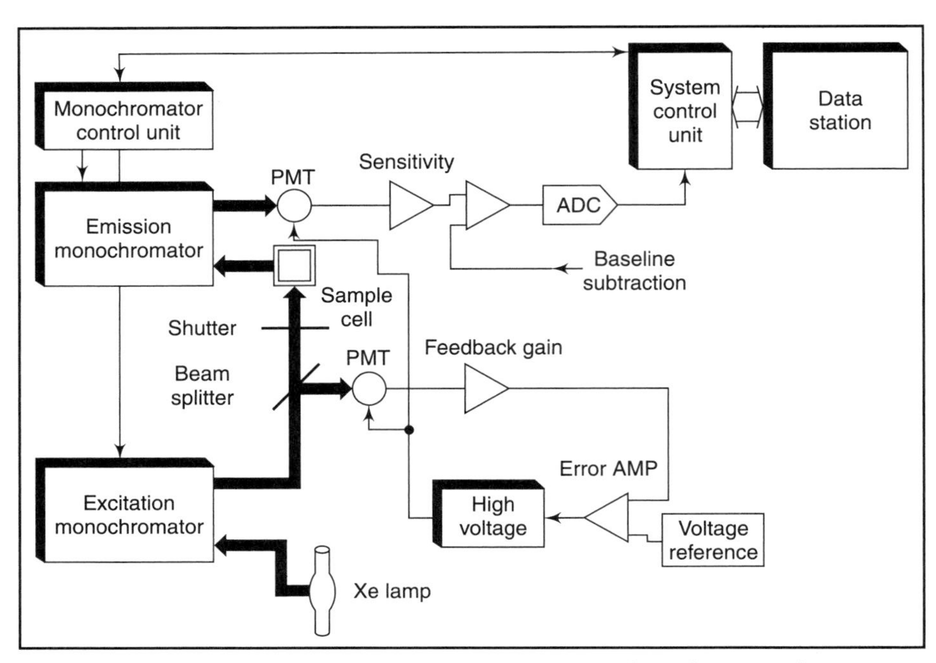

Figure 6.12 :: Block diagram of a microprocessor-based spectrofluorimeter (after Iwata *et al.*, 1986)

spectral bandwidth in steps from 1.5 to 20 nm, wavelength scanning steps selectable in steps from 1 to 1000 mm/min, a softwave-controlled shutter that closes the aperture of the excitation monochromator and an auto-zero function.

Sensitivity improvement in the instrument is achieved by: (i) using holographic gratings of 1800 grooves/mm. The stray light of this optical system is less than 3.5% at 220 nm excitation wavelength and 5-nm spectral bandwidth, (ii) rotating the slit image of the monochromator formed on the sample cell by 90°, in which the observed volume of the sample illumination is enlarged by the slit's length-to-width ratio, resulting in higher emission intensity and an increase in sensitivity.

A dynode-feedback technique is employed to compensate for characteristics of the light source and excitation monochromator, as well as light source fluctuation. The feedback electronically controls the photomultiplier high voltage. By changing the gain of the feedback loop in steps, the feedback technique can effectively cover a wide dynamic range of incident radiation.

The data acquisition hardware comprises three modules: (i) a high resolution, cathode ray tube for display, (ii) a thermal graphic plotter/printer, and (iii) processor module that houses microprocessors, memory, bus and associated electronics. In order to maximize the data throughput, two slave processors are employed to help the master processor by sharing relatively low priority jobs. The spectra produced by fluorescence spectrometers are affected by the solvent background and the optical characteristics of the instrument. Appropriate software permits one to obtain corrected emission spectra. This can be achieved by scanning both excitation and emission monochromators synchronously. The inverse of the corrected data is stored in the memory of the spectrofluorometer. The software program multiplies the inverse of the corrected data with the raw spectral data, so that the resultant corrected emission is in real-time.

The microprocessor-based instruments usually provide menu-driven operation, enhancing operating ease. Each set of programs for a particular application is labelled with a specific method number so that the operator can easily select the desired program. The operator can determine instrument conditions by means of interactive communication, with the instrument through the screen. The facilities available include real-time display and record of emission spectrum, excitation spectrum, concentration correction of spectra, data processing (such as smoothing, derivatives, peak selection, arithmetic manipulation), photomultiplier tube voltage control and self-diagnostic features.

A typical example of a microprocessor-based fluorescence spectrometer is the Perkin Elmer LS-5B. In this instrument, the excitation monochrometer covers the wavelength range 230–720 nm, while the emission monochromator covers the wavelength range 250–800 nm. A red sensitive photomultiplier tube is used for the 650–800 nm region. The instrument provides a choice of five scan speeds ranging from 30 to 480 nm/min. A pre-scan mode combines both the abcissa and ordinate controls, and is designed to search automatically for and display the maximum excitation and emission wavelengths of an unknown fluorescing or phosphorescing material. On completion of the search, the excitation and emission monochromators display the peak wavelengths. The spectra are recorded on a printer–plotter connected to the instrument. Through the use of an RS 232 C interface, the instrument can be connected to a computer for instrument control and external data manipulation.

Linking a microcomputer to a fluorescence spectrometer has several advantages. As we know, conventional single-beam fluorescence spectrometers produce excitation and emission spectra that are uncorrected, i.e. they reflect the wavelength dependence of the optical components of the instrument as well as the properties of the sample. Correction of the spectra requires the acquisition and storage of arrays of correction factors, which can be conveniently done by a microcomputer. Microcomputers are also useful for expediting the acquisition of derivative fluorescence spectra, fluorescence polarization values and difference spectra.

Figure 6.13 :: Typical spectrofluorometer: model cary eclipse (Courtesy M/s Varian Inc.)

Each fluorophore is characterized by two spectra, excitation and emission. A three-dimensional presentation of excitation wavelength, emission wavelength and intensity is required for its complete description. Such three-dimensional pictures can be conveniently obtained as isometric projections with a microcomputer. The advent of 16- and 32-bit computers ensures that the required high speed and large memory capacity are available in modern microcomputers.

Modern spectrofluorometers are generally PC-based. Figure 6.13 shows a typical instrument from M/s Varian Inc.

6.5.1 Perkin Elmer Fluorescence Spectrometer Model LS-3

The optical system of the model LS-3 Perkin-Elmer Fluorescence spectrometer is shown in Figure 6.14. The source is a pulsed xenon flash tube which provides a continuum of energy over the spectral range of interest. Energy from the source is selected by the ellipsoidal mirror (ME) and reflected by the toroidal mirror (MT) onto the entrance slit 2, from where it goes to the 1200 lines per mm grating, a spherical mirror (MS) and exit slit 1. The energy diffracted by the grating emerging from slit 1 is of narrow wavelength. The wavelength of the excitation beam is deter-

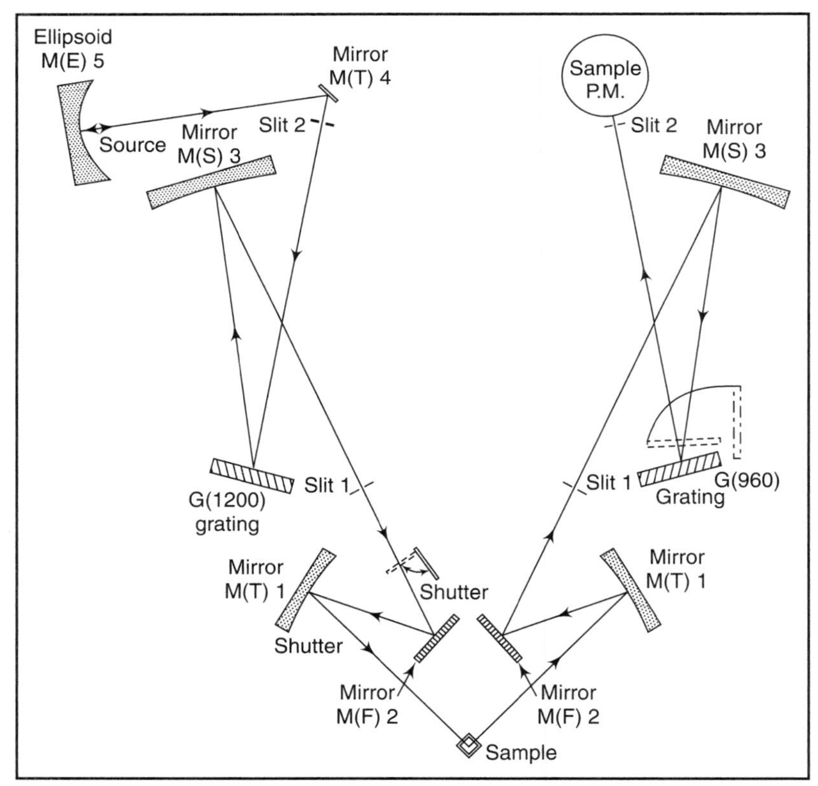

Figure 6.14 :: Optical diagram of Perklin Elmer Fluorescence Spectrometer Model LS-3

mined by the setting of the grating, the angle of which is controlled by means of a stepper motor. The energy emerging from the slit 1 is directed onto the sample through the toroidal and plane mirrors.

The energy emitted by the sample is focused by the toroidal and plane mirrors onto the entrance slit of the emission monochromator. The monochromator consists of the entrance slit 1, a spherical mirror, a 960 lines per mm grating and exit slit 2. A narrow wavelength band of energy diffracted by the grating emerges from slit 2. The wavelength of the light falling on the photomultiplier is determined by the setting of the grating, the angle of which is controlled by a stepper motor.

A microprocessor controls the stepper motors that move the excitation and emission monochromator gratings. The monochromators can be scanned over their ranges independently or synchronously. Also, they can be driven to selected points in their ranges. The normal range of the excitation monochromator is 230 to 720 nm, and that of emission monochromator as 250 to 800 nm. A red sensitive photomultiplier is necessary for measurements in the emission range of 650 to 800 nm.

The xenon source lamp is made of quartz and is highly susceptible to contamination. It is therefore recommended that the glass envelope of the source should not, at any time, be touched by hand. If the source becomes contaminated accidentally, it can be cleaned with ethanol or methanol by using a piece of soft non-fluffy material.

The replacement of the source lamp becomes necessary if it shows deterioration characteristics. The symptoms of source deterioration are a noticeable drop in intensity (caused by blackening of the tube), low sensitivity, audible misfiring of the trigger supply and consequently resulting in an erratic display.

6.6 Measurement of Phosphorescence

In the phosphorescence process, two transients are involved between the lowest excited singlet state and the ground state. This involves an inter-system crossing or transition from the singlet to the triplet state. A triplet state results when the spin of one electron changes, so that the spins are the same or unpaired. The first is a non-radiative transition, as the molecular energy level lowers to the vibrational level of a metastable triplet state (S to T_1). From the triplet state, the molecular energy level drops to the ground state; the radiation emitted during the transition T_1-G is called phosphorescence (Figure 6.1). The time involved in the transition is of the order of 10^{-3}s, which gives relatively long periods of phosphorescence emission. A characteristic feature of phosphorescence is an after glow, i.e. emission which continues after the exciting source is removed. Since the energy level of the triplet state lies below the excited singlet levels, the phosphorescence is emitted at wavelengths longer than those of fluorescence. Because of the relatively long life time of the triplet state, molecules in this state are much more susceptible to deactivation processes, and only when the substance is dissolved in a rigid medium can phosphorescence emission usually be observed.

Phosphorescence emission is characterized by its frequency, lifetime, quantum yield and vibrational pattern. These properties form the basis for qualitative analysis, whereas the correlation of intensity with concentration gives the bases for quantitative measurements.

6.6.1 Phosphorescence Spectrometer

Phosphorescence is measured in a similar manner as fluorescence, except for the use of a mechanical method used for distinguishing between them by their time–delay as shown in Figure 6.15. Light from the xenon lamp is passed through an excitation monochromator. It falls on the sample via a fixed slit system, and through a slotted cam type shutter, which surrounds the sample cell. The slotted cam is driven by a variable speed motor, having the sample cuvette at the axis of rotation. The slots are so arranged that the sample is first illuminated and then darkened. While it is dark, its phosphorescence is allowed to pass and go to the phosphorescence monochromator, placed at right angles to the excitation radiation. The mechanical device accomplishing the modulation of the radiation from the light source incident on the sample and simultaneously modulating the luminescence radiation from the sample, which is incident on the photodetector, is known as Phosphoriscope. The modulation is periodic and out-of-phase, so that no incident exciting or luminescent radiation reaches the photodetector during one phase, whereas only long decaying phosphorescence radiation reaches the photodetector during the other phase. Hence, the phosphoriscope allows measurement of phosphorescence in the presence of fluorescence and other scattered radiation. The phosphorescence spectrum is obtained by setting the excitation monochromator at the wavelength corresponding to a phosphorescence maximum and allowing be emission monochromator to sweep throughout its wavelength range.

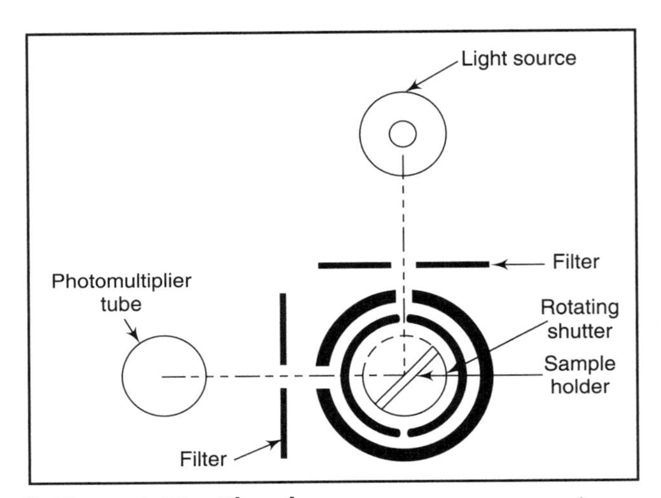

Figure 6.15 :: Phosphorescence measurement method

Phosphoriscope

Phosphorescence is not observed at room temperature, as the energy of the triplet state is readily lost by a collisional de-activation process involving the solvent. Therefore, phosphorescence is normally observed at reduced temperatures in solidified samples. The sample is generally placed in a small quartz tube, which is then placed in liquid nitrogen (77 K), and held in a quartz Dewar flask. The incident radiation passes through the unsilvered part of the Dewar flask, and phosphorescence is observed through the sample part of the flask at right angles to the incident beam. The sample cell is immersed directly into the coolant, which is usually liquid nitrogen. Several commercial fluorimeters are provided with phosphorescence attachments.

Until 1970, it was widely believed that solution phosphorescence could only be observed at cryogenic temperatures. More recently, several studies have demonstrated practicable ways of observing room temperature phosphorescence, and the method now holds a great promise in high performance liquid chromatography (HPLC). As in other photoluminescence methods, the main areas of application are likely to be in biochemical and environmental analysis.

Pulsed source-time resolved phosphorimetry, which has been commercialized by Perkin-Elmer in their Model LS-5 luminescence spectrometer, has the following advantages over the conventional mechanical chopper method:

- A pulsed source produces higher peak intensities than a continuously operated xenon lamp, resulting in greater phosphorescence emission intensity.
- Pulsed source phosphorimetry has the advantage of time resolution, compared with a mechanically modulated system, permitting the analysis of organic phosphors with short life-times in the range of 0.1–50 ms.

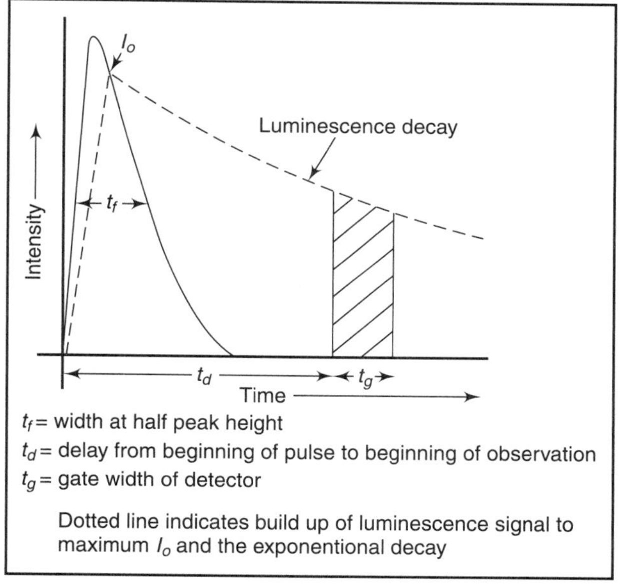

Figure 6.16 :: Schematic diagram of events occurring during the excitation of a sample with a pulsed xenon source in phosphorescence mode (after Williams, 1981)

Figure 6.16 illustrates the events occurring during the excitation of a sample with pulsed xenon source in the phosphorescence mode. The xenon lamp produces a burst of energy with a width at half-peak intensity of less than 10 μs. During this period, the phosphorescence intensity rises to the peak value (I_o) and then theoretically decays exponentially. The signals from the detector (photomultiplier) are gated and both the delay of the start of the gate after the start of the flash (t_d) and the duration of the gate (t_g) can be varied in multiples of 10 μs. During the cycle, a quantum-corrected reference photomultiplier is used to measure the flash intensity and the signals from the sample and reference photomultipliers are ratioed to compensate for any source instability, and to provide the corrected excitation spectra.

Chapter 7

Raman Spectrometer

Raman effect

The Raman effect was experimentally discovered in 1928 by C.V. Raman. It is a powerful analytical tool for the quantitative analysis of complex mixtures, for locating various functional groups or chemical bonds in molecules and for the elucidation of a molecular structure. The fields of application of Raman spectroscopy are quite diverse and include polymers, paints, semiconductors, corrosion, medicine, biology, ceramics, catalysis and photochemistry, etc. It has often been observed that Raman spectroscopy is most effective when used in conjunction with IR data. While this is true in some cases, it is also an undeniable fact that Raman spectroscopy is complimentary to IR and can be used very effectively even alone.

7.1 The Raman Effect

It was shown by Lord Rayleigh in 1871 that when a sample is irradiated with monochromatic light in the visible region, the majority of the light simply passes through the sample in the direction of the incident beam. However, a small amount (about 1 part in 10^5) is scattered by the sample in all directions, which can be observed by viewing the sample at right angles to the incident beam. The scattering of light at the same frequency as the incident radiation is called Rayleigh scattering. It can be shown that scattering from small particles is proportional to the inverse fourth-power of the wavelength. This beautifully explains the blue colour of the sky and the redness of the setting sun, wherein scattering is predominantly due to particles of molecular dimensions. The sky is illuminated by sunlight, which is scattered by the atmosphere; the blue light intensity predominates, because of the wavelength dependence of Rayleigh scattering.

Rayleigh scattering

It has been observed from spectroscopic investigations of light scattered from a sample illuminated with a monochromatic beam, that about 1 per cent of the total scattered intensity occurs at frequencies that are different from the incident frequency. This is called 'Raman scattering'. This phenomenon results from an interaction between the incident photons and the vibrational energy levels of the molecules. The interaction is not

Raman scattering

simply absorption, but rather a transfer of a part of the energy of the photon to the molecule or vice versa.

The equation for the energy interchange is:

$$h\upsilon_1 + E_1 = h\upsilon_2 + E_2$$

$$\Delta E = h\upsilon_1 - h\upsilon_2 = \pm\, h\Delta\upsilon,$$

where υ_1 and υ_2 are the frequencies of the incident photon and the scattered radiation respectively; E_1 and E_2 are the initial and final energy states of the scatterer and $\Delta\upsilon$ corresponds to energy transitions (rotational, vibrational or electronic), within the sample itself. The sample may be solid, liquid or gaseous. A plot of $\upsilon \pm \Delta\upsilon$ against photon intensity yields a spectrum of vibrational transition, which is analogous to, but not identical with, an IR spectrum. This is shown in Figure 7.1. The scattered lines, called Raman lines, are characteristics of the vibrational modes of the substance irradiated and represent a sort of fingerprint of that substance.

Raman lines

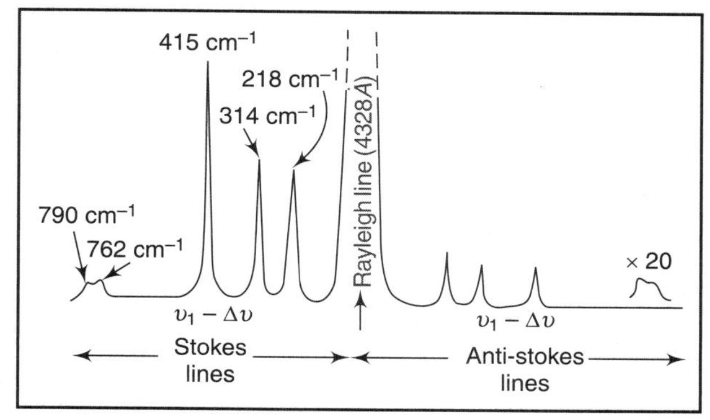

Figure 7.1 :: Raman spectrum of CCL_4. This is only a schematic showing the stokes and anti-stokes lines. The wave number values on the diagram are Δr, that is the Raman shift, and correspond to vibrations of the molecule

It may be noted that the shift in the frequency of the scattered Raman lines is proportional to the vibrational energy involved in the transition. The shift is, in fact, a measure of the separation of the two vibrational energy states of the molecules. It is independent of the frequency of the incident radiation.

Absorption spectroscopy such as UV, visible and IR, are resonance processes, in that they utilize photons which have energies equal to the differences in energy level spacings of the sample. In general, the Raman process is non-resonant, since the incident photon has much greater energy than the transition of interest. This is shown in Figure 7.2, in which the molecule is in virtual energy state. This is not a stationary state of the molecule, and

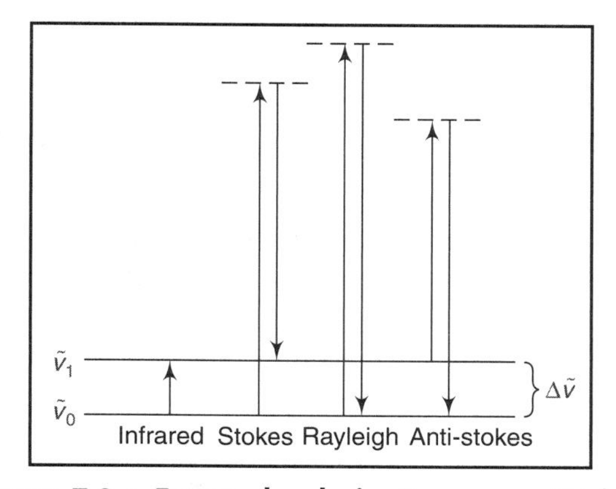

Figure 7.2 :: Energy levels in Raman scattering. The dashed lines show non-existent states at levels determined only by the energy of the incoming photons

one can regard the absorbed photon as recognising the unstable state and being re-emitted by the molecule which then returns to one of its stationary states. If the final and initial infrared states are identical, then this is Rayleigh scattering while if they are different, we have Raman scattering.

Stokes lines

The transition occurring below the Rayleigh or exciting line are termed Stokes lines ($\upsilon_1 + \Delta\upsilon$), while those above it are called anti-Stokes ($\upsilon_1 - \Delta\upsilon$). It is usual to scan the stokes side, because these bands are much more intense than the anti-Stokes lines. Figure 7.3 shows the principle of Raman spectroscopy.

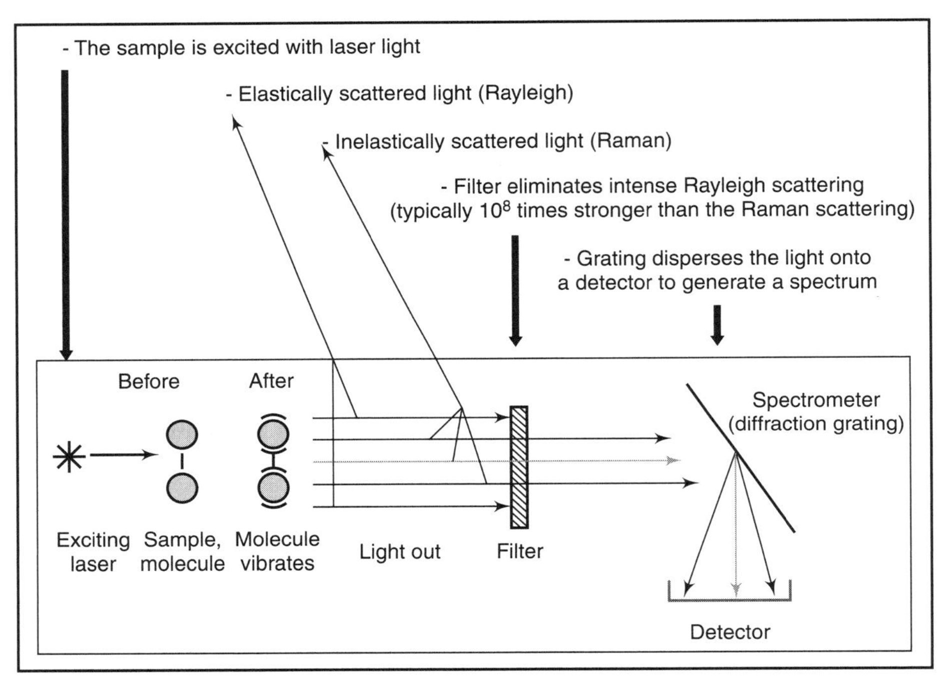

Figure 7.3 :: The principle of Raman spectroscopy (Windisch, 2003)

It may be observed that in Raman scattering, the absorption and emission of the photon are actually simultaneous events. The Raman effect differs from fluorescence in exactly this respect. In fluorescence, the incident photon is absorbed and the molecule attains an excited stationary state, which has a well-defined lifetime. Thus, the absorption and emission steps are actually sequential.

Raman spectroscopy has certain intrinsic advantages over IR spectroscopy. In Raman spectroscopy both the incident and scattered photons are typically in the visible region, while their difference ($\Delta\upsilon$) is in the infrared. This facilitates the use of excellent detectors, like photomultiplier tubes, which can have single photon counting capability and are more sensitive than the infrared counterparts.

7.1.1 Resonance-enhanced Raman Scattering

Raman spectroscopy is conventionally performed with green, red or near-infrared lasers. The wavelengths are below the first electronic transitions of most molecules, as assumed by the scattering theory. The situation changes if the wavelength of the exciting laser is within the electronic spectrum of a molecule. In that case, the intensity of some Raman-active vibrations increases by a factor of 10^2–10^4. This resonance enhancement or resonance Raman effect can be quite useful.

Resonance enhancement does not begin at a sharply defined wavelength. In fact, enhancement of 5X–10X is commonly observed if the exciting laser is even within a few hundred wave numbers below the electronic transition of a molecule. This pre-resonance enhancement is found to be experimentally useful.

7.1.2 Surface-enhanced Raman Scattering

The Raman scattering from a compound (or ion) adsorbed on or even within a few angstroms of a structured metal surface can be 10^3–10^6X greater than in solution. This surface-enhanced Raman scattering is strongest on silver, but is observable on gold and copper as well. At practical excitation wavelengths, enhancement on other metals is unimportant. Surface-enhanced Raman scattering (SERS) arises from two mechanisms.

The first is an enhanced electromagnetic field produced at the surface of the metal. When the wavelength of the incident light is close to the plasma wavelength of the metal, conduction electrons in the metal surface are excited into an extended surface electronic excited state called a surface 'plasmon resonance'. Molecules adsorbed or in close proximity to the surface experience an exceptionally large electromagnetic field. Vibrational modes normal to the surface are most strongly enhanced.

The second mode of enhancement is by the formation of a charge–transfer complex between the surface and analyte molecule. The electronic transitions of many charge transfer complexes are in the visible, so that resonance enhancement occurs.

The intensity of the surface plasmon resonance is dependent on many factors including the wavelength of the incident light and the morphology of the metal surface. The wavelength should match the plasma wavelength of the metal. This is about 382 nm for a 5 μm silver particle (Kerker, *et al.*, 1980), but can be as high as 600 nm for copper and gold (Morris, 1992), the other two metals which show SERS at wavelengths in the 350–1000 nm region. The best morphology for surface plasmon resonance excitation is a small (<100 nm) particle or an atomically rough surface.

7.2 Raman Spectrometer

Raman spectrum

The Raman spectrum can be observed by illuminating the sample with monochromatic light and observing the light scattered at right angles to the incident radiation. Raman intensities are approximately 0.01 per cent of the incident radiation and therefore, the Raman spectrometer must employ intense source of radiation, sensitive detector, high light gathering capability coupled with freedom from extraneous scattered incident light. The Raman spectrum can be recorded in the following two different ways:

- By focusing the spectrum from the prism or grating onto a photographic plate and measuring the line frequencies and intensities using external equipment after the plate is developed; and
- By focusing the spectrum produced by the monochromator on to a photomultiplier tube amplifying the detected signal and recording it. This method has the advantage that the response to the Raman line intensity is linear, which greatly simplifies quantitative analysis.

The earlier Raman instruments made use of a cylindrical sample cell (Raman tube) placed inside a helical mercury arc lamp, which radiates energy at 435.8 or 546.1 nm. Between the mercury lamp and the sample is a filter jacket (organic dye), which isolates monochromatic light of either 435.8 or 546.1 nm. This monochromatic radiation irradiates the sample tube from all sides, and the light which is scattered out of the end of the tube is introduced into the monochromator. The mercury arc source is quite inefficient, as a 2.5 kW arc lamp radiates about 50 W at 435.8 nm, of which only about 1 W is actually effective in exciting the sample. The mercury arc lamp also generates considerable heat, which must be dissipated by circulating the filter solution, that also serves as a coolant. Glass absorption filters are not suitable, as too much heat is generated.

Modern Raman instruments employ laser as an excitation source and use computers for data acquisition and control. The major components in a Raman system are:

- A source of monochromatic radiation,
- Sample compartment and associated optics,
- Spectrometer or monochrometer,
- Detection system, and
- Computer.

Figure 7.4 shows the arrangement of these components in a Raman system. The minimum system would consist of a laser, a double monochrometer and a photomultiplier tube detector with a recorder.

7.2.1 The Source

In modern Raman spectrometers, lasers are used as photon sources due to their highly monochromatic nature, and high beam fluxes. This is necessary as the Raman effect is weak, typically the Stokes lines are ~10^5 times weaker than the Rayleigh scattered component.

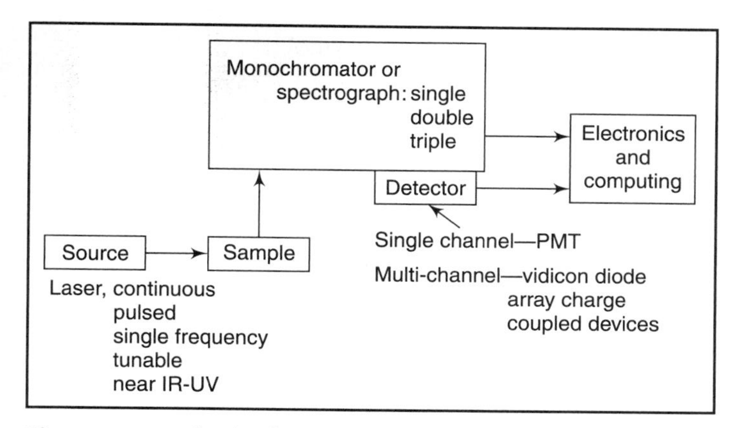

Figure 7.4 :: Block diagram of a laser-based Raman spectrometer. The diagram also shows some choices associated with the selection of a Raman system.

The helium–neon laser, which emits highly monochromatic light at 632.8 nm, is a commonly used excitation source in the modern Raman spectrometers. With this wavelength being in the red region of the spectrum, there is a loss of scattering intensity associated with the use of this longer wavelength and the photoelectric detectors are less sensitive. These problems can be overcome by using lasers with power outputs in the range of 5–30 mW. They produce as good intensity Raman spectra as the older arc sources. The longer wavelength, on the other hand, offers very significant advantages. Usually the Raman spectrum cannot be obtained, if the sample absorbs the exciting wavelength. At 632.8 nm, fewer samples have interfering absorption bands. Another difficulty in Raman spectrometry is that of the sample fluoresced when irradiated; the spectrum would be observed by the fluorescence spectrum. Since red radiation is frequently less effective in generating fluorescence, this problem is automatically minimized.

Howard (1986) explains that the argon–ion (visible) laser is the most commonly used source in Raman spectrometers. However, tunable dye lasers are increasingly being used and the present range of interest includes the visible and UV lasers. The type of laser required depends upon the sample and the Raman technique to be used. Specialized Raman methods may sometimes involve the simultaneous use of several lasers of different types.

7.2.2 Sample Chamber

In Raman spectrometers, samples may be examined as solids, liquids or solutions, or in the gas phase. However, a Raman spectrum is most easily obtained by using liquid samples. The narrow and readily collimated laser beam can be simply focused into a capillary tube containing the liquid. Water is an excellent solvent for Raman spectroscopy, because it has few interference bands and dissolves readily large numbers of inorganic materials.

For a study of the Raman spectrum of gases, the gas sample can be placed inside the laser cavity. In case of solid samples, the laser beam is focused into a capillary tube containing the powdered solid. Only a few milligrams of a powdered solid sample are adequate to give a good spectra.

Most Raman spectrometers for material characterization use a microscope to focus the laser beam to a small spot (<1-100 μm diameter). Light from the sample passes back through the microscope optics into the spectrometer.

7.2.3 The Spectrometer

It was explained that the intensity of Raman lines is much weaker than the exciting line. Thus, an excellent optical system would be needed to ensure stray light rejection. This is most important if it is required to measure close to the Rayleigh line. This is usually achieved by using two or three monochromators. The use of holographic gratings and multiple slits ensures proper levels of luminosity.

7.2.4 The Detector

The most commonly used detector is the photomultiplier tube, which provides excellent sensitivity, low noise and a large dynamic range. This is, however, a single-channel detector and requires point-by-point spectral data acquisition and consequently long scan times. Multi-channel detectors are presently assuming more popularity, with their obvious advantages for time resolved measurements. A multi-channel detector may be a one-dimensional diode-array, with 512 or 1024 pixels (each 2.5 mm high by 25 μm wide) or a two-dimensional type like vidicon or charge coupled device with arrays of 512 × 512 pixels. Multi-channel detectors offer certain advantages over photomultiplier tubes for many applications, as they permit the collection of large portions of a vibrational spectrum in seconds.

7.2.5 Computer

Computers incorporated in the modern instruments are essential for spectral manipulation (addition, substraction, self-deconvolution, etc.). They are indispensable when a multi-channel detector is being used and offer a major advantage with a single channel system. Good software and graphics facilities for Raman spectroscopy are constantly witnessing intensive development.

Raman spectroscopy is a rapidly developing analytical tool. Several new advanced techniques are being developed, with a view to enhance the utility. There has recently been substantial interest in using Fourier transform techniques, to collect Raman data. This technique would appear to have a bright future, particularly as it can be an add-on to IR spectrometer.

7.3 PC-based Raman Spectrometer

De Graff, *et al.* (2002) illustrate the details of a PC-based laser Raman spectrometer using CCD detection. The layout is shown in Figure 7.5. The beam from a 532-nm linearly polarized laser is passed directly into the sample without focusing. In order to observe polarization effects, a half-wave plate is placed between the laser and the sample to rotate the polarization of the beam. In order to measure depolarization ratios quantitatively, a polarization analyzer is placed between the sample and the fibre optic. Samples are held in a standard 1 cm glass fluorimeter cell and scattered light is collected at a 90° angle from the laser beam path. A notch filter is placed between the sample cell and the fibre optic coupling to reduce the Rayleigh scattered 532 nm excitation light. The notch filter provides a 2 nm bandpass width for the notch filters. This corresponds to ~ 70 cm^{-1} at the 532 nm Nd/YAG line; however, the wavelength of the notch is

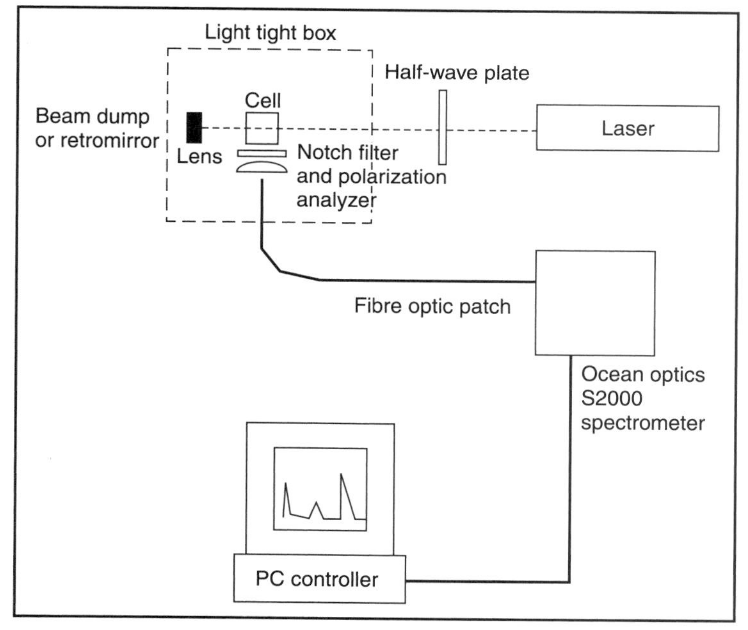

Figure 7.5 :: Modern Raman spectrometer (De Graff *et al.*, 2002)

rather angle-dependent. A collecting lens focuses the scattered light into the fibre optic patch. The fibre optic patch to the spectrometer has a 400 μm diameter. A box surrounds the sample, filter, lens, and fibre optic cable end in order to prevent room light from reaching the detector. The other end of the fibre optic cable is connected to the spectrometer which contains a 50 μm slit and a 600-groove/mm grating with a blaze wavelength of 400 nm. The multi-pin electronics cable from the spectrometer leads to an A/D converter card in a Pentium PC. The software is used to collect and process the data.

The spectrometer software allows integration times to be varied from 1 ms to 65 s. Because of the relative weakness of the Raman signal in comparison to pixel noise, a dark spectrum (with laser light blocked) must be subtracted from the overall signal. The software explicitly allows for dark spectrum subtraction.

Like infrared spectrometry, Raman spectrometry is a method of determining modes of molecular motion, especially the vibrations, and their use in analysis is based on the specificity of these vibrations. The methods are predominantly applicable to the qualitative and quantitative analysis of covalently bonded molecules rather than to ionic structures.

7.4 Infrared and Raman Microspectrometry

Both the Raman and the infrared spectrum yield a partial description of the internal, vibrational motion of the molecule in terms of the normal vibrations of the constituent atoms. Neither type of spectrum alone gives a complete description of the pattern of molecular vibration, and, an analysis of the difference between the Raman and the infrared spectrum, can help procure additional information about the molecular structure.

The coupling of optical microscopy to vibrational spectroscopy (infrared and Raman) enables the chemical characterization of samples or domains as small as 10 μm (IR) or 1 μm (Raman). These techniques are applied to analyze polymers, rubbers, papers, organic and inorganic materials. Results are obtained by spectral interpretation or by imaging the functional groups. This has opened up the possibility of Raman microstructural investigations on the μm scale. A schematic diagram of a typical system (ISA Jobin Yvon LABRAM micro-Raman) is shown in Figure 7.6.

The system uses two excitation lines 632.8 nm (He-Ne laser) and 784.8 (diode laser). It is completed with an XY motorized stage with spatial resolution of about 1 micron and with an auto focus system which allows one to obtain Raman mappings on a micrometric scale in three dimensions.

Although there are many different molecular spectroscopes, the most versatile ones for identification are infrared (IR) and Raman spectroscopies. The basic differences in the two methods are worth noting (Katon, *et al.*, 1986). Raman spectrometry is normally carried out in the visible region of the spectrum, and thus normal microscopes, with the usual glass optics are quite satisfactory for coupling to the spectrometer. Since the source for Raman spectrometry must be

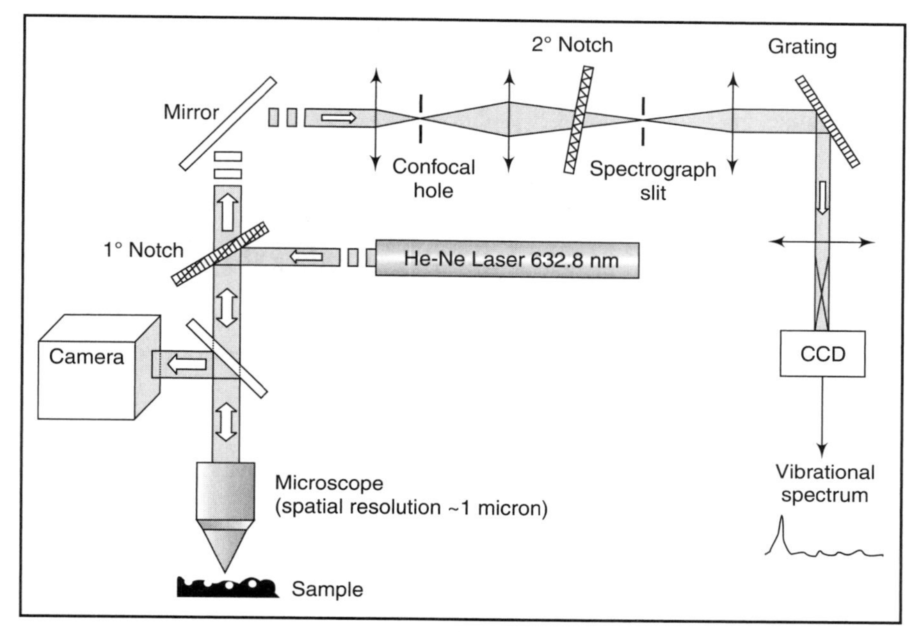

Figure 7.6 :: Micro-Raman apparatus

monochromatic, an argon ion laser is typically used. On the other hand, glass is not transparent in the IR region. Therefore, all IR microscopes must use reflecting optics and cassegrainian objectives or mirrors of the proper configuration, for focusing and collecting the incident and transmitted beams, respectively.

Chapter 8

Photoacoustic and Photothermal Spectrometers

8.1 Photoacoustic Spectroscopy

Photoacoustic spectroscopy (PAS) has generated considerable interest as a technique for the study of solid, semi-solid and liquid samples, especially when the samples are difficult to analyze by other methods. The technique allows one to undertake effective studies for samples, which are strongly or weakly absorbing, or where the scattering of incident radiation precludes the use of reflectance or transmission spectroscopy (Kreuzer, 1978).

The principle of photoacoustic spectroscopy is explained by Rosencwaig and Gersho (1976) and is illustrated in Figure 8.1. When a light absorbing substance is subjected to modulated incident radiation, a fraction of the radiation falling upon the sample is absorbed and results in

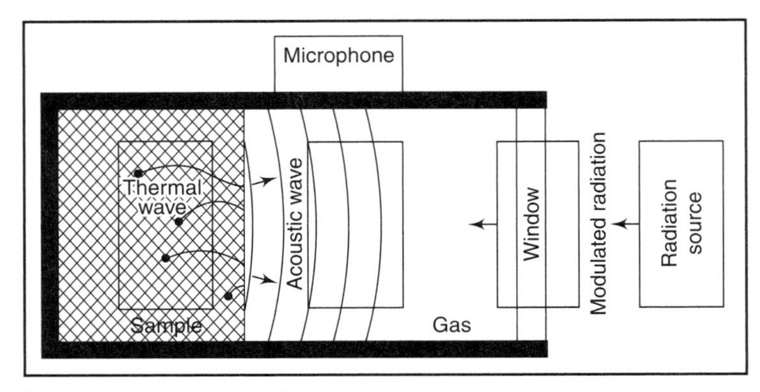

Figure 8.1 :: Principle of photoacoustic spectroscopy

excitation. When a molecule absorbs energy, it must lose an equivalent quantity of energy via photodecomposition, photoluminescence or thermal relaxation. The last process is usually predominant and therefore, non-radiative de-excitation processes, which normally occur, give rise to the generation of thermal energy within the sample.

If the incident radiation flux is periodically interrupted (modulated), the energy released by the sample will also be periodic, and if contained in a cell having a constant volume, it will give rise to acoustic pressure pulses. This is called a photoacoustic signal. Usually, a condenser microphone is used for the detection and measurement of the photoacoustic signal. In effect, the sample is enclosed with a microphone inside a sealed cell containing a gas and fitted with an acoustically isolated window for the illumination of the sample. The modulated variations in the temperature of the surface of the sample result in the generation of an acoustic wave in the gas. This wave propagates through the volume of the gas to the microphone, where the signal is produced.

Photoacoustic signal

It may be observed that the photoacoustic signal is a function of the following two processes occurring in the sample:

1. The absorption of electromagnetic radiation (specified by the absorption coefficient β), and
2. Thermal propagation in the sample (specified by the thermal diffusivity X).

It is possible to define a characteristic thermal diffusion length μ_s. The thermal wave produced in the sample is heavily damped and may be fully damped out within a distance $2\,\pi\mu_s$. Thus, only those thermal waves originating from a depth less than or equal to μ_s will give an appreciable contribution to the photoacoustic signal intensity.

8.1.1 System Components

The photoacoustic signal is normally of rather small magnitude and it is essential to maximize the signal-to-noise ratio. As the magnitude of the photoacoustic signal is proportional to the incident power of the radiation falling upon the sample, high intensity sources are generally employed in photoacoustic systems. Lasers operating in the near-UV, visible and mid-IR region of the spectrum have been used. However, in a majority of the applications, a relatively wide spectral range is required and continuum sources are essential. Usually, 300 W xenon arc lamps are used for studies in the UV-visible-near IR regions. The photoacoustic signal is produced when the incident radiation is modulated. Usually, amplitude modulation is employed, which is implemented either electronically or by a mechanical chopper. A grating monochromator is used to achieve wavelength dispersion of the incident radiation. Although condenser microphones are widely used as detectors in photoacoustic spectrometers, piezoelectric transducers are becoming more popular. (Wetsel, 1980)

The photoacoustic signal is a vector which has magnitude and phase components. Therefore, a lock-in amplifier is commonly used for signal recovery from the photoacoustic cell, which has the same frequency and a particular phase relationship to the modulation waveform.

The output power of the source in photoacoustic spectrometers is not constant with the variation in the wavelength. Therefore, any photoacoustic spectrum obtained will be observed as the product of the true photoacoustic spectrum, and the power spectrum of the source. The following methods are employed to obtain a corrected photoacoustic spectrum:

- Using a double-beam spectrometer with two matched photoacoustic cells, one normally containing finely produced carbon as a reference absorber. The ratio of sample and the reference signals provides the corrected spectrum.
- Using a double-beam system with a pyroelectric detector in the reference channel. The greater sensitivity of the pyroelectric detector results in only about 10–20 per cent of the source power being used in the reference channel. This has the advantage of giving a corrected photoacoustic spectrum in real-time, together with automatic correction for any fluctuation in source intensity.
- Having a single-beam system and producing digital storage for the uncorrected sample spectrum and a spectrum of carbon black. The corrected photoacoustic spectrum is obtained by calculation on a point-to-point basis. The technique, however, does not correct for source drift.

8.1.2 Typical Photoacoustic Spectrometers

Blank and Wakefield (1979) described a double-beam photoacoustic spectrometer for use in the ultraviolet, visible and near-infrared spectral regions. A functional diagram of the system is shown in Figure 8.2. The spectral range is 240 to 250 nm, and it is determined by the output of the lamp at both ends of the spectrum. The lamp used is a 300 W xenon arc lamp, with an integral parabolic reflector and a sapphire window. All transmitting optical components except filters are fabricated in UV grade quartz and UV grade sapphire, including lenses. The UV sapphire and quartz are stable and transmit well in the entire spectral region, except that UV quartz exhibits some absorbing bands in the near infrared.

Either the radiation beam can be modulated with an electromechanical chopper or the arc lamp can be current-modulated. Although the latter method facilitates higher modulating frequencies (from 100 to 10,000 Hz), desired for control of μ_s, most designs employ electromechanical modulation at frequencies from 10 to 1000 Hz, due to better signal-to-noise ratio capability. Hereafter, the radiation enters the Ebert monochromator, which carries four gratings mounted in a turret. The gratings have 2360, 1180, 590 and 295 grooves/mm and are blazed at 300, 500, 1200 and 2100 mm, respectively. Wavelength scanning is done by a stepper motor drive, and in a scan, the grating interchange is automatic. Near the exit slit is a wheel containing absorption and interference order sorting filters. The filters are automatically selected as wavelength is changed.

After emerging from the monochromator and filter, the beam is converged and divided into two parts by a beam divider. After division, each of the two beams is directed into its respective photoacoustic cell, reference cell and the sample cell. The pressure fluctuations in each of the cells

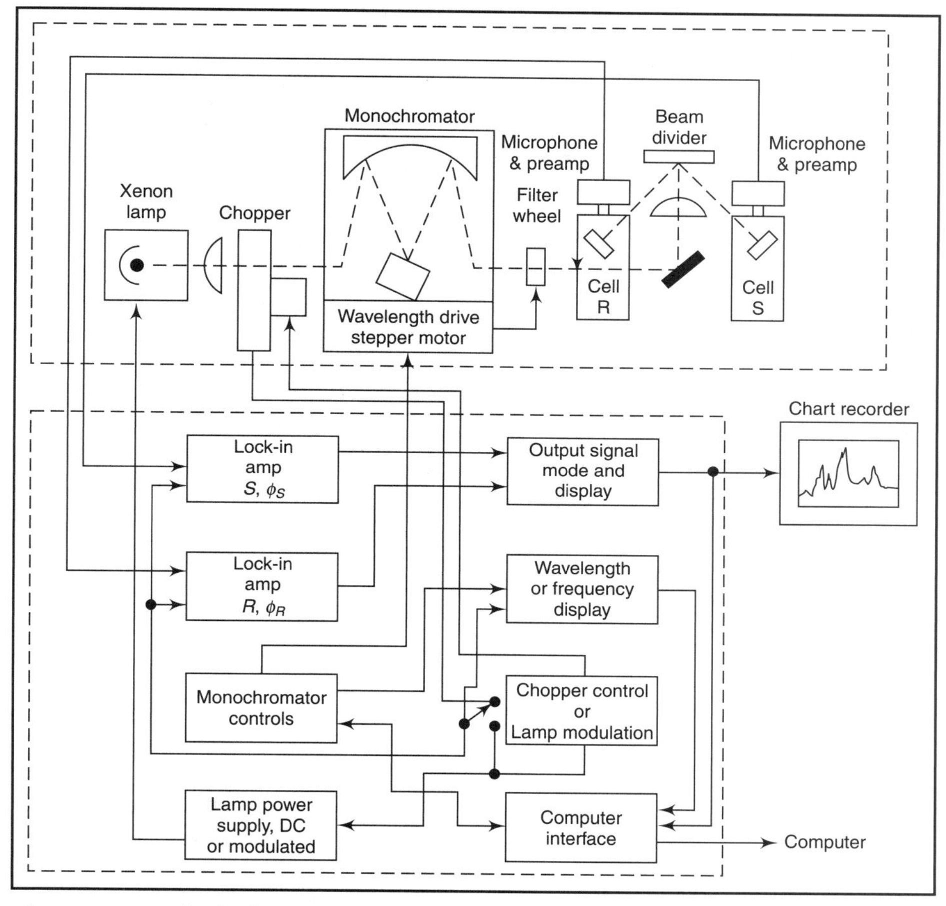

Figure 8.2 :: Block diagram of a double-beam photoacoustic spectrometer (redrawn after Blank and Wakefield, 1979)

are fed to the microphone and adjacent pre-amplifier signal conditioners, before it is sent to the lock-in amplifier. The lock-in amplifier can output the signal phase or magnitude at adjustable fixed phase. The phase is measured relative to that of the modulator. The signals from the sample and reference cells are called S and R, respectively. Although several modes of operation are possible, the most commonly used mode to date is S/R, wherein R is derived from a saturated reference material such as carbon black. The resulting signal from each cell consists of the sample spectrum, with a superimposed system power spectrum, due to the lamp output and optics. The reference spectrum from carbon black consists of the system power spectrum. The ratio S/R divides out the unwanted power spectrum leaving only the true sample spectrum. The chopping frequency is continuously variable and either wavelength or frequency can be displayed. Figure 8.3 shows a photoacoustic spectrometer from M/s McPherson Inc., USA.

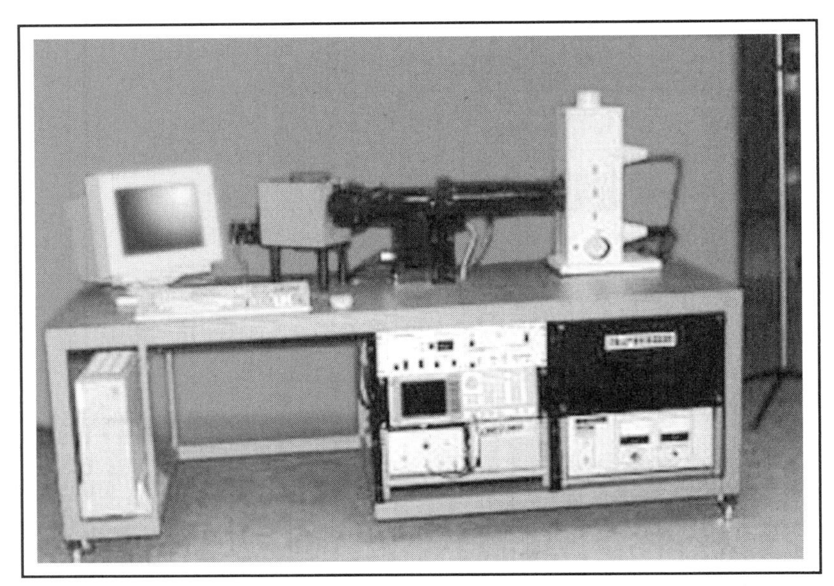

Figure 8.3 :: Photoacoustic spectrometer (Courtesy M/s McPherson Inc., USA)

Cox, *et al.* (1980) illustrated a double-beam photoacoustic spectrometer, which employed a linear scanner to alternately illuminate sample and reference materials contained in a cell, which employs a single microphone. Data acquisition requires only one pre-amplifier and one lock-in amplifier. The linear scanner is an analog device, for which the angle of deflection of the shaft is proportional to the power input. The basic function of the scanner is to move the monochromatic radiation, which is, in fact, an image of the source from sample to reference and back.

It has been found that there is a poor signal-to-noise ratio of the photoacoustic response, and therefore, repeated experiments and ensemble averaging of the results would be required. Kirkbright (1984a and b) reports advances in Fourier-transform infrared photoacoustic spectrometry and correlation techniques as a possible method of overcoming the limitations of the conventional photoacoustic spectroscopy.

A schematic view of the photoacoustic spectrometer (Arnott, *et al.*, 1999), which measures *in situ* light absorption by aerosol, is shown in Figure 8.4. The laser beam power is modulated at the acoustic resonance frequency of the photoacoustic spectrometer. Light absorbing components (gas or aerosol) convert laser beam power into an acoustic pressure wave through absorption-induced gas expansion. A microphone detects the acoustic signal, and hence a measure of light absorption is produced. The piezoelectric disk is used to determine the acoustic resonance frequency of the spectrometer and the resonator quality factor (gain) to calibrate the system. An acoustic notch filters block most of the air-flow pump noise and spurious sound produced by absorption of light on the windows from entering the spectrometer.

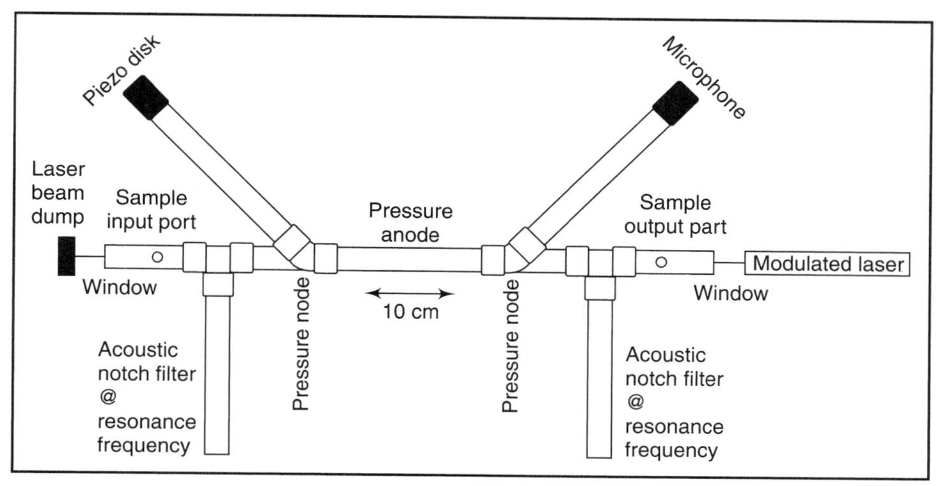

Figure 8.4 :: Schematic view of the photoacoustic spectrometer (redrawn after Arnott, *et al.*, 1999)

Light absorption in dimensions of inverse distance, is determined from the acoustic pressure measured with (calibrated) microphone and corrected for pre-amplifier gain, resonator quality factor Q; resonance frequency f_0; the Fourier component of laser beam power at f_0; and the resonator cross-sectional area.

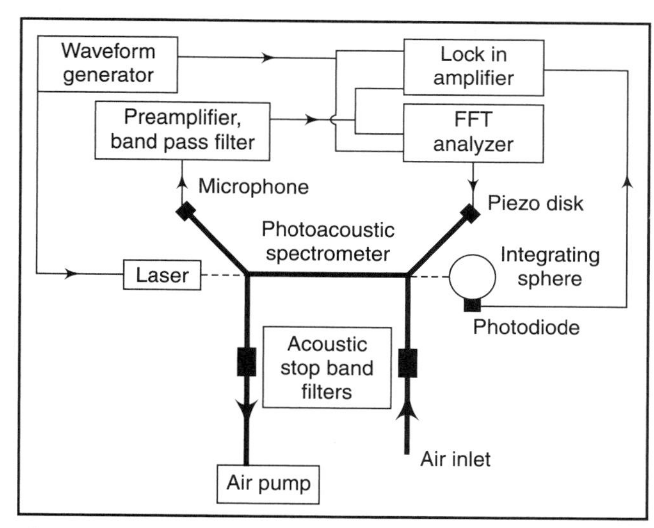

Figure 8.5 :: Block diagram of the photoacoustic spectrometer and detection electronics (adapted from Arnott, *et al.*, 1999)

A block diagram of the prototype photoacoustic spectrometer and detection electronics is shown in Figure 8.5. A pump was used to draw air from outside through the spectrometer with a volume flow rate of 3 L min^{-1}. Considering the volume of the spectrometer, this gives a flow rate time constant of about 8 sec. Acoustic notch filters were placed between the inlet and the spectrometer, and between the pump and the spectrometer, to reduce the ambient and pump generated sound spectral energy at and near the resonance frequency of the spectrometer. These are in addition to the filters on the spectrometer. The laser power is modulated at the spectrometer acoustic resonance frequency by using the square wave output of the waveform generator. The waveform generator provides timing for phase-sensitive detection of the microphone signal by the lock-in amplifier and the Fast Fourier Transform (FFT) analyzer. The microphone signal is amplified with a low noise pre-amplifier (typically by a factor of 100) and is bandpass-filtered (50 Hz band width centered at 500 Hz). Pre-amplification is necessary to provide adequate signal level for the FFT analyzer. The FFT analyzer determines the magnitude of the Fourier component of acoustic pressure at frequency f_0. The FFT analyzer is also used to calibrate the spectrometer by issuing a single cycle of a sine wave to the piezo disk and obtaining the ratio (transfer function) of the spectral resonator response to the issued spectrum. The computer receives the transfer function, and fits it to a standard resonance response function as a function of Q, f_0, and the peak of the transfer function. The lock-in amplifier is used to determine the phase of the microphone signal relative to the waveform generator timing signal, and also determines the photodiode response at f_0. The photodiode is calibrated with a laser power meter.

The entire measurement procedure is automated with the use of the computer. The pre-amplifier communicates to the computer via a serial port. All other instruments are equipped with IEEE 488 communication capability. National Instruments Lab View software is used to provide a convenient user interface. Each measurement of light absorption is preceded with a resonator calibration by using the FFT analyzer. The lock-in time constant was set at 10 seconds for light absorption measurements, and 128 averages lasting 4 seconds each are used with the FFT analyzer.

Muller, *et al.*, (2003) present an all solid state, photoacoustic spectrometer for highly sensitive mid-infrared trace gas detection. A complete spectral coverage between 3.1 and 3.9 μm is obtained by using a PPLN-based continuous-wave optical parametric oscillator pumped by a Nd: YAG laser at 1064 nm. A low threshold is achieved by resonating the pump, and spectral agility by employing a dual-cavity set-up. The output idler power is 2 × 100 mW. The frequency tuning qualities of the OPO (optical parametric oscillators) allow reliable scan over gas absorption structures.

The photoacoustic spectrometer (Figure 8.6) consists of the OPO, chopper photoacoustic cell (PAC), pyroelectric detector and the wavemeter, and is installed on a 120 cm × 75 bread-board. The amount of radiation absorbed by the molecules is measured by its conversion into heat. The 3 μm beam is modulated in amplitude at the resonance frequency of the photoacoustic cell, generating a standing acoustic wave. The signals from the microphone and the pyroelectric detector behind the cell are processed with two lock-in amplifiers. About 70 mW of idler power are available at the front-side of the PAC. For measurements, two different PACs are used. The small PAC (resonator length of 7 cm and diameter of 6 mm), which allows a fast gas exchange, has a Q factor of 17.4. The large PAC (16 cm resonator length and diameter of 14 mm) has a Q factor of 51.1. In order to avoid memory effects during gas exchange the inner surfaces of both cells are passivated against molecular adsorptions.

The power supply operates in the range from 0–50 mA and maximum output voltage of 12 kV. Two separate power supplies are used to obtain equal current on both arms of laser tube.

The photoacoustic cell consists of a polished brass resonator (l = 100 mm, ϕ = 15 mm) inside an aluminum housing. The first longitudinal mode frequency is around 1600 Hz. The cell has acoustic

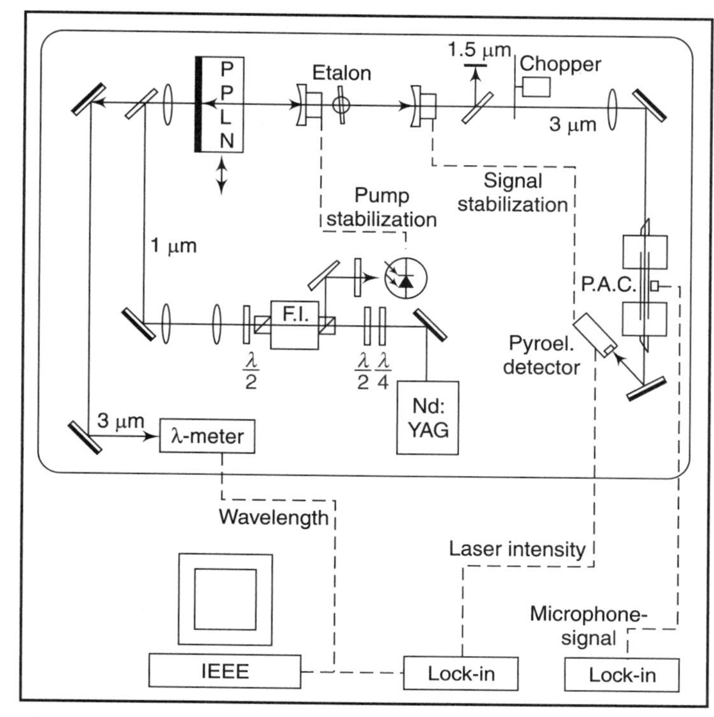

Figure 8.6 :: Schematic of the OPO-based photoacoustic spectrometer (Muller, *et al.*, 2003)

internal insulators ($L = N4$) close to its extremities that are closed with ZnSe Brewster windows. Two electret microphones (sensitivity 22 mV/Pa@1600 Hz) are glued in the centre of the resonator.

The best photoacoustic detection limits are usually achieved in instruments utilizing resonant photoacoustic cells. Resonant cells allow a build-up of the acoustic wave over many cycles of the resonant frequency. These resonances, which can be due to the radial, azimuthal, or longitudinal modes of the cell, are typically at a few kHz and the amount of build-up (the quality factor, or Q, of the resonator) typically exceeds the expected loss in photoacoustic signal that would be predicted in shifting to higher frequencies.

8.2 Photothermal Spectroscopy

Photothermal spectroscopy is a high sensitivity method used to measure the optical absorption characteristics of a sample. It is a more direct measure of optical absorption in terms of sample heating than optical transmission based spectroscopies. This has given rise to a new class of sensitive spectrophotometric techniques, in which the heat produced by non-radiative decay of

Photothermal absorption techniques

excited species acts to modify the optical properties of the sample. These photothermal absorption techniques can be categorized by their spatial temperature distribution and corresponding refractive index change produced by radiation absorption. Laser light, rather than other types of sources, is the usual energy source. In liquids or gases or weakly absorbing solids, the refractive index changes are probed in the absorbing medium itself. Opaque or highly scattering solids, such as powders are probed by the refractive index gradient generated in a coupling fluid in contact with the solid.

8.2.1 Basic Processes in Photothermal Spectroscopy

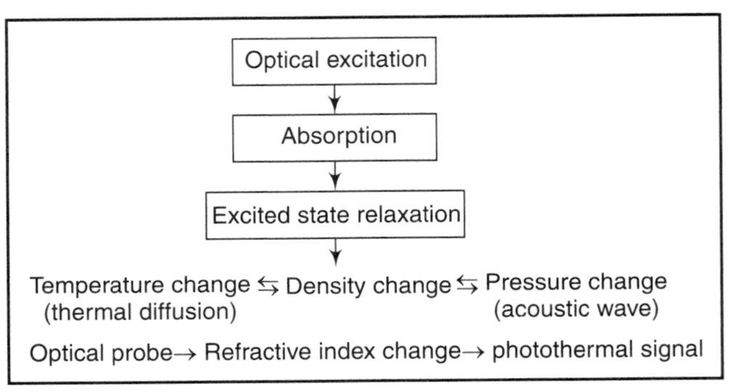

Figure 8.7 :: Basic processes in photothermal spectroscopy

The basic processes responsible for photothermal spectroscopy signal generation are shown in Figure 8.7. Optical radiation, usually from a laser, is used to excite a sample. The sample absorbs some of this radiation resulting in an increase in the internal energy. The increased internal energy results in a temperature change in the sample or the coupling fluid placed next to the sample. This temperature change results in a change in sample or coupling fluid density.

If the photothermal-induced temperature change occurs faster than the time required for the fluid to expand, then the rapid temperature change will result in a pressure change. The pressure perturbation will disperse in an acoustic wave. Once the pressure has relaxed to the equilibrium pressure, a density change proportional to the temperature will result.

In either case, there will be a change in temperature induced by the absorption of optical energy. This temperature change will, in turn, result in a density change in the sample. In combination, temperature and density changes affect other properties of the sample. Photothermal spectroscopy is based on a measurement of these properties. In particular, the sensitive photothermal methods are based on measurement of the refractive index change that occurs with changes in temperature and density of the sample.

Photothermal spectroscopy signals are directly related to light absorption whereas scattering and reflection losses do not produce photothermal signals. Therefore, photothermal spectroscopy more accurately measures optical absorption in scattering solutions, in solids and in interfaces.

The four widely used photothermal spectroscopic techniques are detailed below.

Thermal Lens: When a focused laser beam is used for excitation, the resulting temperature gradient produces a lens-like optical element, which can be measured by its effect on the divergence of the laser beam. This is shown in Figure 8.8.

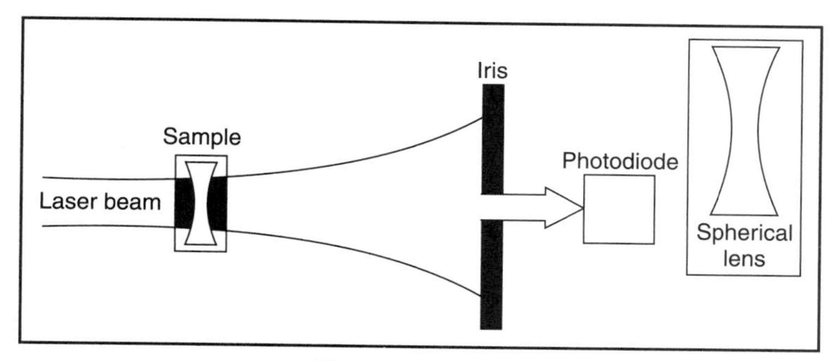

Figure 8.8 :: Photothermal lens effect

Transverse Photothermal Deflection: By creating a thermal gradient in a sample with optical excitation at an interface, the resulting thermal prism which can be detected by its deflection of a laser beam (Figure 8.9.)

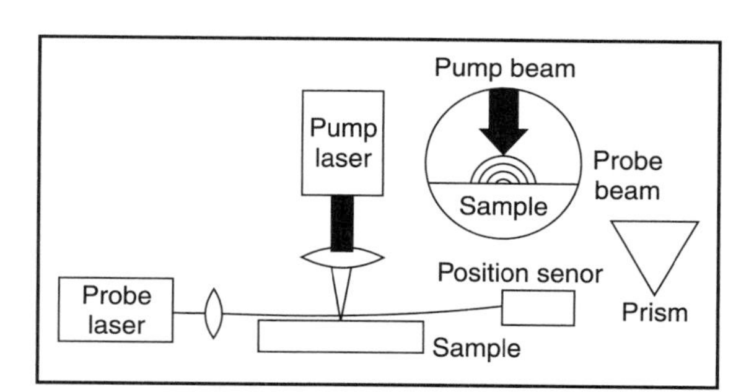

Figure 8.9 :: Transverse photothermal deflec-tion (prism effect)

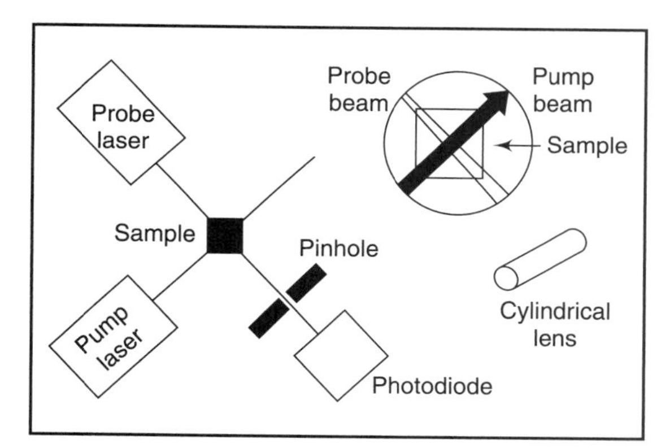

Figure 8.10 :: Photothermal refraction effect (cylindrical lens effect)

Photothermal Refraction: It involves probing the thermal lens in a sample at right angles to the laser beam that caused it, as shown in Figure 8.10, the thermal lens will appear as a cylindrical lens to the probe beam.

Photothermal Diffraction: A periodic temperature distribution can be generated by an excitation interference pattern and probed as a thermal transmission grating by diffraction of a laser beam into a detector. Thus, diffraction photothermal effects can be generated with two crossed laser beams and probed with a third (Figure 8.11).

It may be observed from the above that in each case, a different discrete optical element approximates the behaviour of the refractive index distribution as generated and measured. Since

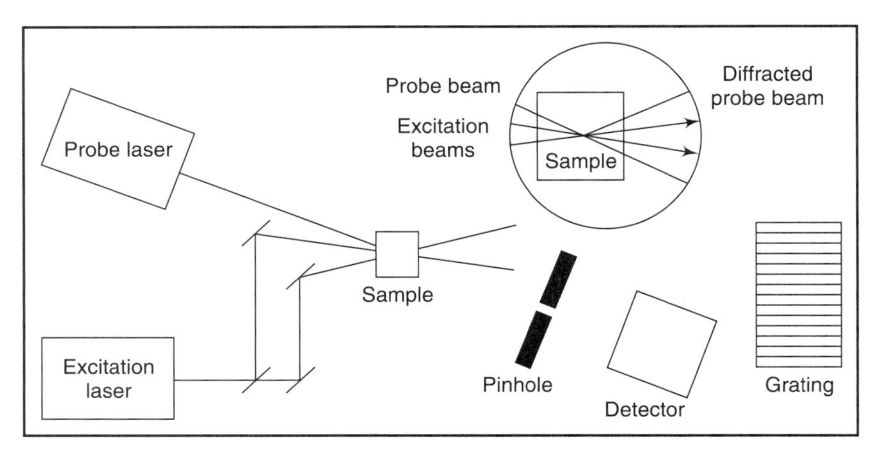

Figure 8.11 :: Photothermal diffraction effect (grating effect)

a laser beam can be de-focused or deflected by a refractive index distribution, no matter how it is generated, lasers are invariably used in photothermal instrumentation.

Unlike conventional transmission or reflection measurements, the sensitivity of photothermal techniques depends upon the power of the radiation used for excitation and the thermo-physical properties of the sample. Solvents which exhibit a large change in refractive index with temperature (dn/dt), are more suitable for study with this method, since a given increase in temperature produces a large change in optical path.

Photothermal effects are closely related to the photoacoustic effect. As we observed in the last section, in photoacoustic measurements, the heat generated by light absorption is detected as a pressure change, using a microphone or other pressure transducer in contact with the sample, whereas it is refractive index changes which are detected in photothermal spectroscopy. Both techniques are capable of measuring absorbances down to 10^{-9} units, although this level is rarely reached. By that standard, nominally transparent materials, which may have absorbances of 10^{-6} – 10^{-3} /cm, are strongly absorbing (Morris, 1986).

8.2.2 Photothermal Instrumentation

Laser light sources are usually used to perform photothermal spectroscopy due to their high spectral purity and power. For an excitation of a sample with a given absorption coefficient, the temperature change will be proportional to the optical power in the case of continuous excitation, or energy in the case of pulsed excitation. The photothermal spectroscopy signal is generally proportional to the temperature change. Thus the greater is the power or energy, the greater is the resulting signal. Lasers can deliver high powers or pulse energies over very narrow optical bandwidths thereby enhancing the photothermal signals. Another reason is the spatial coherence of the laser beam. The temperature change is not only proportional to the optical power or energy, but is also inversely proportional to the volume over which the light is absorbed since heat capacity scales with the amount of substance. The spatial coherence properties of laser sources allow the light to be focused to small volumes. The small volumes used in photothermal spectroscopy enhance signal magnitudes, allow photothermal spectroscopy to be used in small volume sample analysis, and allow for a microscopic analysis of heterogeneous materials.

The most popular technique in photothermal instrumentation is the transverse photothermal deflection system. A functional block diagram is shown in Figure 8.12. The heating beam is a low-power laser beam, which is modulated at 5–20 Hz. The modulation is achieved by a mechanical chopper. The typical power output of the argon laser (heating laser) is 20–100 mW. The heating beam is combined with a helium–neon probe beam. The power output of this probe beam is 0.5–2 mW. The combined beams are focused into the sample with a short (30–100 mm) focal length lens. A simple glass filter is used to isolate the probe beam beyond the sample. An optical fibre serves as the limiting aperture and also uses to relay the probe beam to a photodiode detector. This is followed by a lock-in amplifier for demodulation, before it is fed to a read-out system.

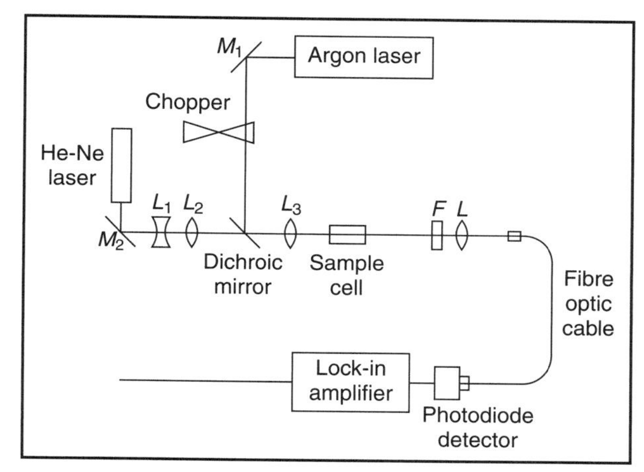

Figure 8.12 :: Thermal lens measurement system—basic building blocks

For quantitative analysis, the concentrations of samples can be obtained simply by calibration at a fixed wavelength with standards. If the wavelength of the pump beam is varied, the deflection of the probe beam is then a measure of the absorption spectrum of the sample investigated.

Several factors determine the signal-to-noise ratio in photothermal measurements. In general, the intensity of photothermal effect increases linearly with power density. Also, in the refractive-index distribution generated in any photothermal technique, there is a sample region with maximum refractive index change. Probing this region yields maximum sensitivity. However, in many analytical applications, photochemical reactions are potentially serious sources of error in transverse photothermal deflection measurements, or any other measurements of solids.

The key to the success of sensitive photothermal apparatuses lies in the measurement of a thermal change and not in the thermal state itself. Although apparatuses could directly or indirectly measure the thermodynamic parameters such as temperature, pressure, density, and energy state, the limiting absorption that could be measured would be imposed by thermodynamic fluctuations.

Sensitive photothermal spectroscopy methods circumvent direct measurements by measuring refractive index changes due to a non-equilibrium change in the energy of the sample. The change occurs in both space and time. Photothermal spectroscopy methods measure some effect that the spatially or temporally dependent refractive index change has on the propagation characteristics of light used to monitor the refractive index.

Photothermal effects are useful in characterizing the physical and chemical properties of a wide variety of solid, liquid and gaseous materials. Wider availability of compact and inexpensive lasers that are suitable for photothermal measurements, is likely to faster a rapid increase in the number of problems which can be readily solved by photothermal spectroscopy.

Chapter 9

Mass Spectrometer

9.1 Basic Mass Spectrometer

In a mass spectrometer, the sample to be analyzed is first bombarded with an electron beam to produce ionic fragments of the original molecule. These ions are then sorted out by accelerating them through electric and magnetic fields, according to their mass/charge ratio. A record of the numbers of different kinds of ions is called the mass spectrum. The uniqueness of the molecular fragmentation is the basis for identification of different molecules in a complex mixture, as no two molecules will be fragmented and ionized in exactly the same manner. Very small sample sizes, which may be of the order of a few tenths of a milligram, are generally sufficient as long as the material is able to exist in the gaseous state at the temperature and pressure existing in the ion source. Several other useful applications of mass spectrometry include the direct determination of the molecular weight, the placement of functional groups into certain areas of the molecules and their interconnection and investigation of reaction mechanism.

Mass spectrum

One of the main advantages of mass spectrometry over other spectroscopic techniques is its sensitivity. In some instruments, it is possible to obtain full mass spectra with 1 ng of material in about a second. Compounds now amenable to mass spectrometry vary from low molecular weight gas mixtures to high molecular weight natural products. Various ionization techniques are used to provide complementary structural information and molecular weight and elemental composition data are readily obtainable.

Mass spectrometry

A significant advantage of mass spectra is its suitability for data storage and library retrieval, since the positions (masses) of the peaks in the spectrum of a given compound are fixed. Another asset of mass spectrometry is its capability in handling complex mixtures, currently obtained via the gas chromatograph-mass spectrometer.

The procedure for analyzing a substance by mass spectrometry starts by converting the substance into a gaseous state by chemical means. The gas is introduced into the highly evacuated spectrometer tube, where it is ionised by means sof an electron beam. The positive ions thus formed are deflected and focused by means of suitable magnetic and electric fields. For a given accelerating voltage, only positive ions of a specific mass pass through a slit and reach the collecting plate. The ion currents thus produced are measured by using a sensitive electrometer tube. By varying the accelerating voltage, ions from other mass species may be collected, and the ion currents measured in such cases would be proportional to the amount of the given mass species present. A great advantage of mass spectrometry over other direct methods is that the impurities of masses different from the one being analyzed, do not interfere with the results.

Ion currents

9.2 Principle of Operation

Figure 9.1 shows the principle of operation of a mass spectrometer. The molecules in the gas sample (*A*) to be analyzed are bombarded with electrons to produce ions (*B*). These ions are accelerated in a high vacuum into a magnetic field (*C*), which deflects them into circular paths (*D*). Since the deflection for light ions is greater than that for heavy ions, the ion stream separates into beams of different molecular weight. A suitably placed slit (*F*) allows a beam of a selected molecular weight to pass through to a collection electrode (*E*). As the accelerating voltage on the ion source is gradually reduced, ion beams of successively greater mass pass through the slit. When these ions fall on the collector electrodes, they produce minute electric currents, which may be measured after suitable amplification. Their amplitude will indicate the number of ions in each beam. The proportion of molecules of different masses in the gas sample may thus be found and a complete analysis of the gas sample may be made; provided all the constituent gases have a different molecular weight. This is usually the case in respiratory gas analysis work.

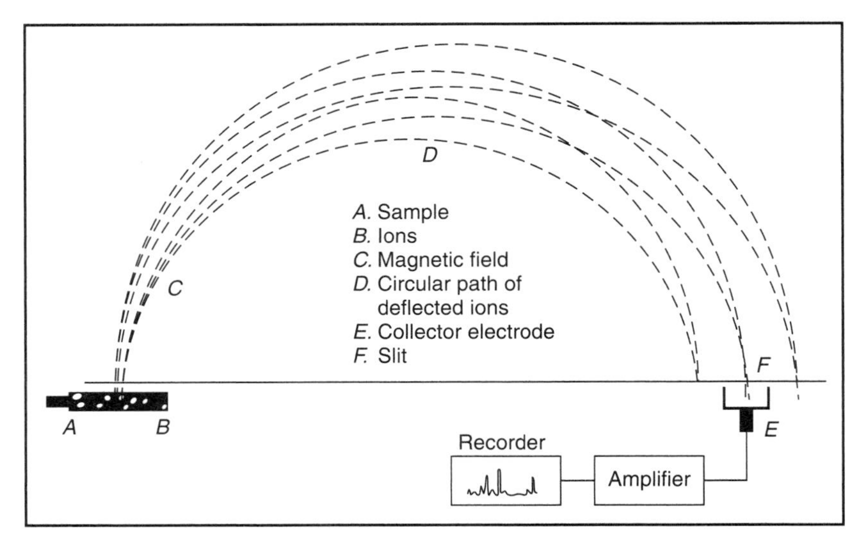

Figure 9.1 :: Principle of operation of a mass spectrometer

It is apparent that there is a close analogy between mass spectrometry and optical spectroscopy. This is shown in Figure 9.2.

Spectrometry and optical spectroscopy

In optical spectroscopy, the source is a visible light, which is composed of individual colours (different wavelengths of light) that are present at different intensities. A prism (or grating) is used to separate the light into its different wavelengths followed by a slit which selects the wavelength reaching the detector. The different wavelengths are then scanned across the detector slit and the light intensity is measured as a function of wavelength.

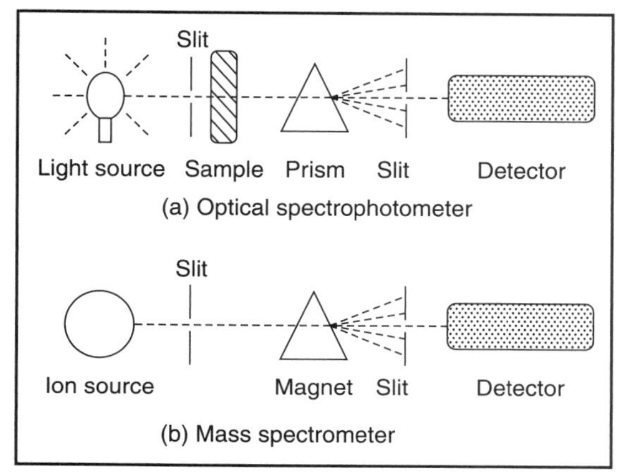

Figure 9.2 :: Analogy between optical spectroscopy and mass spectrometry

In scanning mass spectrometry, the starting point is a mixture of ions having different mass-to-charge ratios. A slit is used to select which mass-to-charge ratio reaches the detector. The different mass-to-charge ratios are then scanned across the detector slit and the ion current is recorded as a function of time (mass).

Mass spectrometers are used for all kinds of chemical analyses, ranging from environmental analysis (e.g. detection of poisons such as dioxin) to the analysis of petroleum products, trace metals and biological materials, including the products of genetic engineering.

Let us take water (H_2O) as an example to illustrate the principle of mass spectrometry. A water molecule consists of two hydrogens (H) and one oxygen (O) atoms. The total mass of a water molecule is the sum of the mass of two hydrogens (approximately 1 atomic mass unit per hydrogen) and one oxygen (approximately 16 atomic mass units per oxygen):

+H:	1 u (atomic mass units)
+H:	1 u
+O:	16 u
= H_2O:	18 u

Let's suppose that we put some water vapour into the mass spectrometer. A very small amount of water is all that is needed—the water is introduced into a vacuum chamber (the 'ion source') of the mass spectrometer. If we shoot a beam of electrons through the water vapour, some of the electrons will hit water molecules and knock off an electron. If we lose an electron (negatively charged) from the neutral water molecule, the water will be left with a net positive charge. In other words, we have produced charged particles, or 'ions' from the water:

$$H_2O + 1 \text{ (fast) electron} \rightarrow [H_2O]^+ + 2 \text{ electrons}$$

Some of the collisions between the water molecules and the electrons will be so hard that the water molecules will be broken into smaller pieces, or 'fragments'. For water, the only possible fragments will be $[OH]^+$, O^+ and H^+.

The mass spectrum of water is shown in Figure 9.3. It shows peaks that can be assigned to masses of 1, 16, 17, and 18.

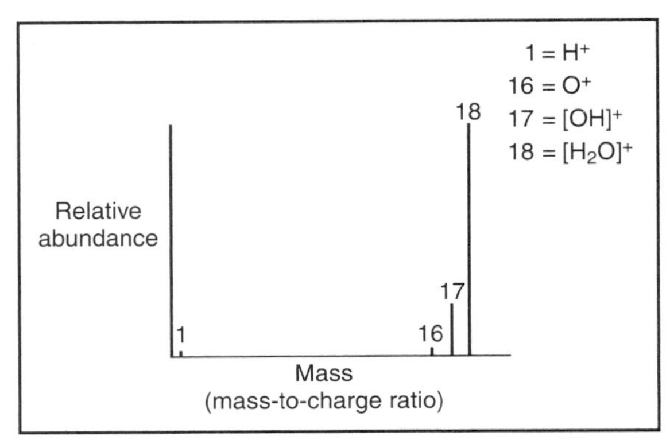

Figure 9.3 :: Mass spectrum of water

$1 = H^+$
$16 = O^+$
$17 = [OH]^+$
$18 = [H_2O]$

It is obvious that only certain combinations of elements can produce ions that have these masses. For example, the ammonium ion $[NH_4]^+$ also has an approximate mass of 18 atomic mass units, but there would be peaks at mass 14 and 15 in the mass spectrum of ammonia corresponding to a N^+ and $[NH]^+$ (nitrogen is atomic mass 14).

A trained mass spectrometrist can interpret the masses and relative abundances of the ions in a mass spectrum, and determine the structure and elemental composition of the molecule. Computer programs, such as those that search libraries of mass spectra for the best match can also be used to interpret a mass spectrum.

Mass spectra can provide other information as well. For example, a high-resolution mass spectrometer can determine the mass of an ion very precisely. If we knew that the mass of our hypothetical ion at mass 18 was actually mass 18.010, we could easily distinguish it from an ammonium ion, which would have an exact mass of 18.035. Given an accurate mass, and an estimated error tolerance, a computer can easily calculate the elemental composition of the molecule.

9.3 Types of Mass Spectrometers

Several methods of producing mass spectra have been devised. However, the principal difference between the various types of spectrometers lies in the means for separating the ions according to their mass-to-charge ratio. The important types of spectrometers are described below.

9.3.1 Magnetic Deflection Mass Spectrometer

Figure 9.4 shows the essential parts of a typical magnetic deflection mass spectrometer tube due to Nier (1940). The heated tungsten filament produces an electron beam, which passes between

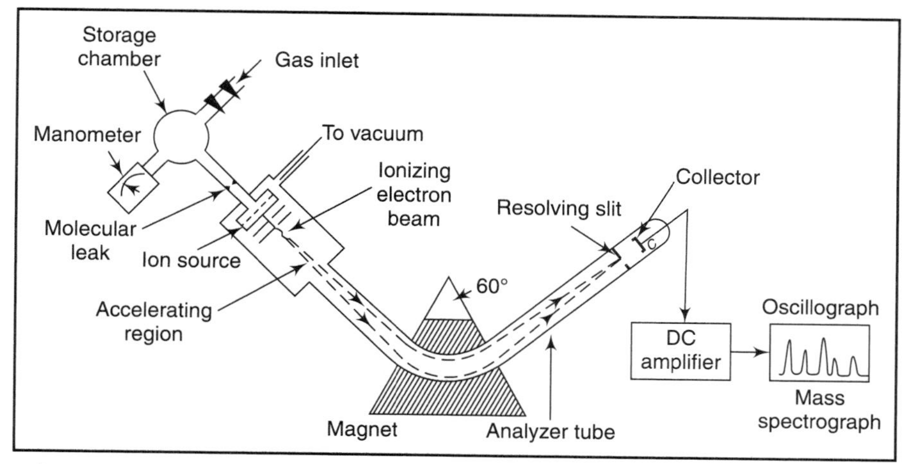

Figure 9.4 :: Schematic diagram of a Nier 60° sector mass spectrometer

plates A and B. A difference in electrical potential between A and B pulls ions out of the beam, so that they pass through the slit in B into the region between B and C. The potential difference between B and C is adjustable from 0 up to several thousand volts. The ion beam then enters the space between two trapezoid shaped magnet poles, where it is deflected through an angle of 60, 90, 120 or 180.

When ions of mass m and charge e pass through an accelerating electric field, they would attain a velocity V which can be expressed in terms of the accelerating voltage V as follows:

$$1/2\, mv^2 = eV, \qquad \text{(i)}$$

where $1/2\, mv^2$ is the kinetic energy of the ion as it leaves the electric field.

From Equation (i), v can be written as

$$v = [2\, eV/m]^{1/2} \qquad \text{(ii)}$$

If the ions next enter a magnetic field of constant intensity h, which is applied at right angles to their direction of motion, the ions would be diverted into circular orbits. From physics, it is known that when acceleration is applied perpendicular to the direction of ion motion, the objects velocity remains constant, but the object travels in a circular path. Therefore, the magnetic sector follows an arc. The radius and angle of the arc vary with different optical designs. Equating the centripetal and the centrifugal forces:

$$mv^2/r = hev \qquad \text{(iii)}$$

The radius of curvature of the trajectory is given by:

$$r = mv/eh \qquad \text{(iv)}$$

Substituting for v from Equation (ii), we get:

$$r = [2\,eV/m]^{1/2} \times m/eh$$

$$= [2\,eV/m \times m^2/e^2h^2]^{1/2} \qquad \text{(v)}$$

$$= [2V/h^2 \times m/e]^{1/2}$$

This equation shows that the radius of the orbit is a function of the mass/charge ratio of the particles.

In practice, all the quantities of Equation (v) are kept constant with the exception of m and V. By varying the accelerating voltage V, it is possible to cause an ion of any mass to follow the path which may coincide with the arc of the analyser tube in the magnetic field. Ions of different m/e ratio strike the tube at some point and would get grounded.

Under specified conditions, the ions which will be collected would follow the expression:

$$m/e = h^2r^2/2V \qquad \text{(vi)}$$

The relationship in Equation (vi) shows that for obtaining a mass spectrum, the accelerating voltage or the magnetic field strength can be varied. Usually, it is the magnetic field which is kept constant and the voltage is adjusted to bring to a focus specific m/e. The magnetic field must be very uniform over a large area. If electromagnets are employed, they would require large power supplies capable of providing several kilowatts of power and must be highly regulated. For a particular instrument, the angle of deflection (radius) is fixed for a given analyzer tube (180°, 90°, 60°). Therefore, Equation (vi) can be written as:

$$mV = \text{constant}$$

If this constant is, say 2000, then as V is set at 100 V, particles of mass 20 can reach the collector plate, while to direct particles of mass 25 to it, V must be set at 80 V.

Direction focusing

Direction focusing is achieved by deflecting the ion beam along a 180° trajectory through the magnetic field. A large magnet is required for an 180° mass spectrometer. When the ion source is in the uniform magnetic field, the gap between the poles must be large enough to contain the ion source. In order to overcome this difficulty, Nier (1940) employed the direction focusing properties of a 60° sector magnetic field. In this arrangement, the magnetic field does not envelop the ion source; a separate source magnet is required. With sector type instruments, a mass resolution of 200 to 600 mass units can be obtained.

The use of simultaneous direction and velocity focusing results in a high resolution mass spectrograph. This is done by placing an electrostatic analyzer between the ion accelerating slits and the magnetic field. After the usual acceleration of the beams in an electrostatic field, the ions are deflected through a tandem arrangement of an electrostatic analyzer and then a magnetic analyzer. The arrangement enables focusing of ions having the same mass/charge but different initial velocities and directions. Resolving power of 8500 with a 0.05 mm entry aperture have been achieved by utilizing the focusing properties of electric and magnetic sector fields in a parabolic configuration.

Commercial mass spectrometers appeared in the USA in the early 1940s. In these instruments, ions were separated according to the quotient mass/charge and the separated ion beams were recorded directly on a photographic plate. Later, the design of a mass spectrometer for precision mass measurements was developed. The instrument incorporated the arrangement, known as Nier-Johnson geometry, which involves a deflection of $\pi/2$ radians in a radial electrostatic field analyzer, followed by a magnetic deflection of $\pi/3$ radians. One ion beam at a time is brought to a focus on an exit slit and measured electrically. These types of instruments became commercially available in the early 1960s.

Nier-Johnson geometry

About the same time, commercial mass spectrographs using Mattauch-Herzog geometry also became readily available. In this arrangement, a deflection of $\pi/4\sqrt{2}$ radians in a radial electrostatic field is followed by magnetic deflection $\pi/2$ radians, and all ions of different mass can be simultaneously focused on a photographic plate. The spectrographs with photographic recording are used for analysis of solids and for the recording of organic mass spectra. An accuracy of 1 part in 10^9 has been obtained in precision mass measurements.

Mattauch-Herzog geometry

A magnetic sector alone will separate ions according to their mass-to-charge. However, the resolution will be limited by the fact that ions leaving the ion source do not all have exactly the same energy, and therefore do not have exactly the same velocity. In order to achieve better resolution, it is necessary to add an electric sector that focuses ions according to their kinetic energy. Like the magnetic sector, the electric sector applies a force perpendicular to the direction of ion motion, and therefore has the form of an arc. Figure 9.5 shows the schematic diagram of the JEOL M station double focusing mass spectrometer. This arrangement is called 'Reverse Geometry' magnetic sector mass spectrometer, which means that the magnetic sector precedes the electric sector.

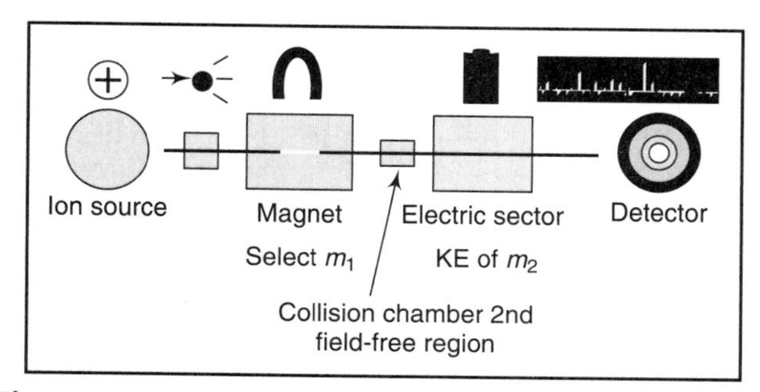

Figure 9.5 :: Reverse geometry mass spectrometers (hold magnetic fold strength constant and scan electric field

The electric sector is usually held constant at a value which passes only ions having the specific kinetic energy. Therefore, it is the magnetic fold strength 'B' which is most commonly varied. The magnetic fold is usually scanned exponentially or linearly to obtain the mass spectrum. An alternative is to hold 'B' constant and scan 'V'. The electric sector field is not subject to hysteresis, so the relationship between mass-to-charge ratios and accelerating voltage is a simple linear relationship.

9.3.2 The Time-of-Flight Mass Spectrometer

In a time-of-flight (ToF) mass spectrometer, ions of different mass/charge ratio are separated by the difference in time they take to travel over an identical path from the ion source to the collector. This requires the starting time (the time at which the ions leave the ion source) to be well-defined. Therefore, ions are either formed by a pulsed ionization method (which is usually matrix-assisted laser desorption ionization) or various kinds of rapid electric field switching are used as a 'gate' to release the ions from the ion source in a very short time.

Ppulsed mass spectrometer

In the pulsed mass spectrometer, ion packets of a few microseconds duration are emitted at intervals of a few milliseconds from a voltage source. The ions traverse an evacuated tube called the drift tube to reach the detector. The detector is sensitised for a brief instant to register their arrival. Since ions of different masses arrive at the detector at different times, the accurate measurement of the time between activating the source and sensitising the detector gives information concerning the mass of the ions. The signal from the ions reaching the detector is amplified and applied to the vertical deflection plates of an oscilloscope. The horizontal axis deflection of the oscilloscope commences as the ion packets start out. This produces a mass spectrum on the screen of the oscilloscope. The device, thus, gives a mass spectrum in a very short time. The essential parts of a time-of-flight instrument are shown in Figure 9.6. It consists of:

- An electron gun for the production of ions,
- A grid system for accelerating ions to uniform velocities in a pulsed mode,
- An evacuated tube, called the drift tube, and
- An ion detector and suitable electronic circuitry for translating the time-dependent arrival of ions of different velocities into a time base that is related to mass number.

If L is the length of the drift tube in centimeters and t is transit time in microseconds, for singly charged ions of mass m and constant energy Ve, then

$$t = L[m/2Ve]^{1/2} \qquad \text{(vii)}$$

$$= L[m/e \times 1/2\ V]^{1/2}$$

$$m/e = 2V/L^2 \times t^2$$

If the detector is sensitized for a period Δt at time t, the resolution $\Delta m/m$, for constant energy, is given by:

$$\text{Resolution} = \Delta m/m\ 2\Delta t/t$$

Equation (vii) shows that the time resolution will increase with increased drift tube length and will decrease with increasing accelerating voltage.

The current produced by the ions arriving at the collector may have a very short duration, which necessitates the use of a wide-band amplifier. A specially designed magnetic electron multiplier is used for this purpose. This multiplier uses a strip of semiconducting material for the multiplying surface instead of dynodes. A gain of 10^7 is attained with a dark current of only 3×10^{-21}A.

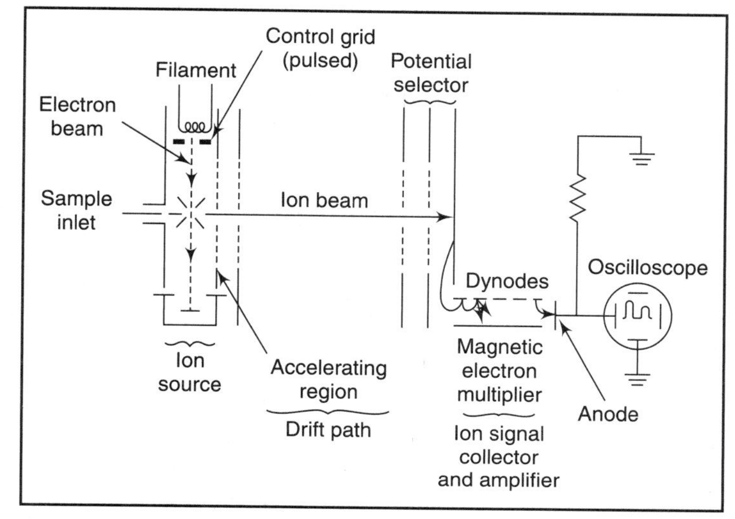

Figure 9.6 :: Schematic diagram of a time-of-flight mass spectrometer

The ions leaving the ion source of a time-of-flight mass spectrometer have neither exactly the same starting times nor exactly the same kinetic energies. Various time-of-flight mass spectrometer designs have been developed to compensate for these differences. One such design makes use of a device known as 'Reflectron'. The Reflectron is an ion optic device in which ions in a time-of-flight mass spectrometer pass through a 'mirror' (reflectron) and their flight is reversed.

Reflectron

The linear reflectron allows ions with greater kinetic energies to penetrate deeper into the reflectron than ions with smaller kinetic energies. The ions that penetrate deeper will take longer to return to the detector. If a packet of ions of a given mass-to-charge ratio contains ions with varying kinetic energies, then the reflectron will decrease the spread in the ion flight times, and therefore improve the resolution of the time-of-flight mass spectrometer. The arrangement is shown in Figure 9.7.

The main advantages of the time-of-flight (TOF) spectrometer include its speed and ability to record entire mass spectrum at one time. A conventional spectrometer detects only one peak at a time. Its accuracy depends on electronic circuits, rather than on extremely critical mechanical alignment and on the production of highly stable and uniform magnetic fields. The main disadvantage of the TOF spectrometers is their poor resolution due to display on an oscilloscope screen.

9.3.3 Radiofrequency Mass Spectrometer

The most popular radiofrequency mass spectrometers make use of the Bennett (1950) tube. The

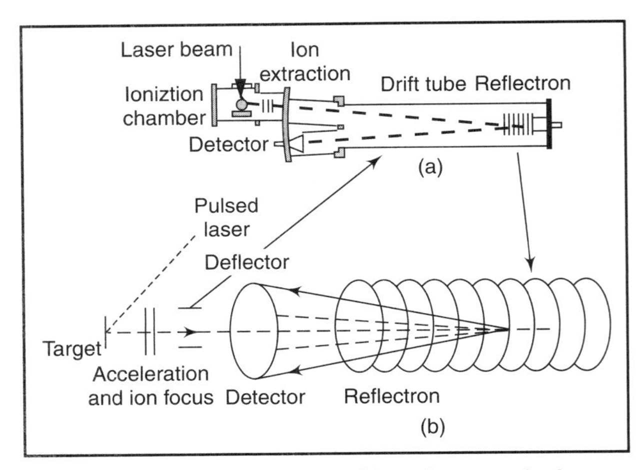

Figure 9.7 :: (a) Ablation of ions by use of a laser from a solid sample (b) Principle of working of a reflectron (adapted from *www.jeol.com/ms/docs/ms-analyzers.html*)

arrangement in the instrument is such that the charged particles emerging from the ion source are all accelerated to the same energy in an electrostatic field, and then they pass through a system of radio-frequency electrodes. The energy acquired by the ions in this process is a function of their specific mass/charge ratio. Maximum energy increment would be acquired by these ions, only if the ions start with the correct velocity at the optimum phase of the rf field. A potential energy selector is placed before the detector, which balances out the energy of the ion beam and the mass spectrum is recorded by detection of the ions with the highest energies as the frequency of the alternating rf voltage is varied.

The rf field is applied in one or more rf stages. Each stage is a series of three equally spaced parallel grids. An alternating rf voltage is applied to the central grid, with respect to the outer grids, which are kept at ground potential.

If v is the velocity attained by the ions in phase with the radiofrequency field, then

$$v = df,$$

where f is the frequency of the rf field in MHz and d is the spacing between adjacent grids in centimeters. The mass/charge ratio of the ion beam reaching the detector is given by:

$$m/e = 0.266V/d^2f^2 \text{ (in cgs units)}$$

The rf mass spectrometer does not require a magnet and is therefore comparatively lightweight and simple in construction. In the Bennett spectrometer, the rf voltage has a fixed frequency and is modulated at 10% at 1 kHz. The current received at the detector is amplified with an ac amplifier tuned to the modulation frequency. The dc ion-accelerating voltage is swept from 50 to 250 V, twice per second. The spectrum in the range $M = 10$ to 50 is reproduced twice per second.

Bennett spectrometer

It has been shown through investigations that the resolving power of a Bennett tube is primarily determined by the distances between the individual rf accelerating stages. A two-stage tube would be the most favourable, but the instrument tends to provide spurious lines. Separation of the principal line from the spurious lines with a retarding field would greatly affect the sensitivity of the spectrometer. The spurious lines may be suppressed in a three-stage tube, although resolving power is slightly reduced.

The resolution of the Bennett tube can be improved by using a square wave radiofrequency signal, in place of sinusoidal rf voltage. Several improvements have been incorporated over the original design and mass spectrometers for special applications have been built-up based on this principle.

9.3.4 Quadrupole Mass Spectrometer

Quadrupole type mass spectrometers are simple in construction and lightweight as compared to sector-type mass spectrometers, high speed electronic scanning and low cost. A quadrupole mass spectrometer consists of an ion source, a quadrupole mass filter and a lens system to focus the ions into the quadrupole filter. Figure 9.8 shows the principle of a quadrupole mass filter. The arrangement consisting of four cylindrical rod-shaped electrodes which provide a potential field distribution, periodic in time and symmetric with respect to the axis which will transmit a select mass group and cause ions of improper mass to be deflected away from the axis. This mass selection scheme uses a combination of dc potential plus a radio-frequency (rf) potential. By proper selection of potentials and frequency, an ion of the desired mass can be made to pass through the system, while unwanted masses will be collected on one of the electrodes.

Quadrupole mass filter

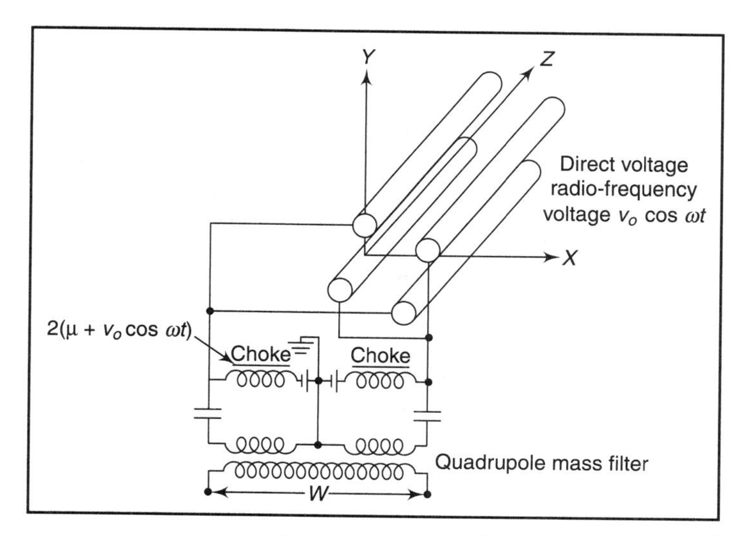

Figure 9.8 :: Quadrupole mass filter

In practice, opposite electrodes of the filters are connected together, and to one pair a potential $\phi(t) = u + v\cos(2\pi ft)$ is applied, where u is a dc voltage and v is the peak amplitude of the rf voltage at a frequency f. To the other pair of electrodes, the same potential but of the opposite sign is applied. Ions emitted out of the ion source and focused into the quadrupole filter are made to undergo transverse motion by the *rf* and dc field perpendicular to the Z-axis.

The motion of the ions in the X-Y plane can be described by the differential equations, and their solutions contain either exponential or oscillatory factors depending upon the value m/e of the ions. With proper selection of u and v, ions of a given m/e will have stable trajectories; these ions will oscillate about the Z-axis and ultimately emerge from the exit of the mass filter assembly. Ions with other values of m/e will have unstable solutions; they will move away from the Z-axis and ultimately strike the electrodes thus being removed.

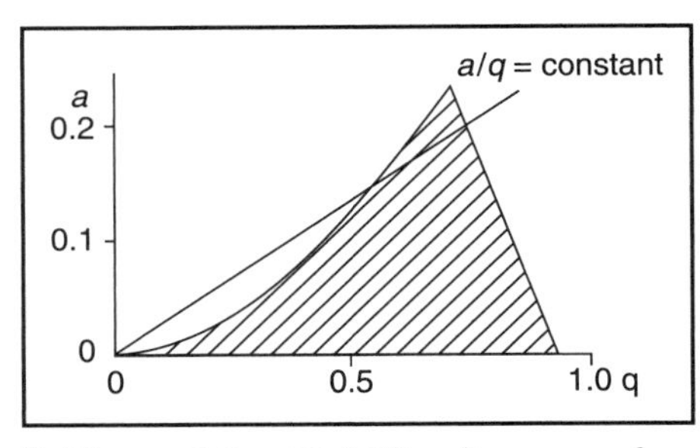

Figure 9.9 :: Stability diagram of quadrupole mass spectrometer

Conditions for stable trajectories are shown in Figure 9.9. Stable conditions are obtained if the values a and q, expressed by the following equations are within the hatched area.

$$a = 8\ eu/mr_o^2\,\omega^2$$

$$q = 4\ ev/mr_o^2\ \omega^2, \qquad (w = 2\ \pi f)$$

where e = electric charge
v = maximum value of rf voltage
m = mass number
r = radius of electric field
f = frequency
u = dc voltage.

When the frequency f and the ratio u/v are kept constant, the value a is linearly proportional to the value q, a/q is constant. Because the ratio a/q $(= 2u/v)$ is not mass-dependent, all ions of different masses in the stability diagram lie on a straight line through zero, with its gradient only depending on the ratio of u and v. Only ions of the q intervals inside the hatched region of the Figure are stable.

By increasing the ratio u/v, the stable interval, which corresponds to a stable mass interval is reduced, so that only ions of one mass can pass the quadrupole filter. In other words, the resolution of a quadrupole mass spectrometer is maximized when the straight line of a/q cuts the apex of hatched area. The lower the gradient, the lower is the resolution.

When the value of q is kept constant, the value of m/e is proportional to the value of v. Therefore:

$$m/e = k \cdot v/r_o^2 f^2,$$

where k is a constant.

By variation of the rf amplitude, the m/e varies. In short, in a quadrupole mass spectrometer, masses are scanned by changing the value v, with the u/v ratio kept constant.

Changing the slope of the scan line will change the resolution. Increasing the resolution decreases the number of ions that reach the detector. Good resolution also depends upon the quality of the machining for the quadrupole rods.

The performance of a mass filter depends upon the quality of the quadrupole rods, the stability of the applied voltage and the field characteristics at the two ends of the rods. Especially, it is essential that the quadrupole rod assembly is produced with super-high precision.

In cases where a differentiation is required between very similar substances, another technique, making use of quadrupole principle, is 'pyrolysis' mass spectrometry. In this arrangement, the sample to be analyzed, usually a solid or involatile liquid, is rapidly heated in a vacuum to a precise temperature. This causes thermal breakdown and subsequent gas phase reactions. The gas or pyrolysate is then analyzed directly in a mass spectrometer.

'pyrolysis' mass spectrometry

The block diagram of a pyrolysis mass spectrometer is shown in Figure 9.10 (Ottley, 1986). The sample is held on a metal substrate, usually a wire of foil, made of a magnetic alloy. This is heated by induction from an external coil operating in the 1 MHz range. This causes rapid heating to a specific temperature, known as Curie point at which the magnetic permeability drops abruptly. The gas produced during pyrolysis is then held in a small buffer volume before being formed into a molecular beam. A liquid nitrogen cooled shield surrounds the ion source and reduces source contamination.

Once the gas has entered the ion source of the mass spectrometer, electron impact ionization creates an ion beam which is then analyzed in a quadrupole mass analyzer. High speed pulse counting using an electron multiplier is employed to permit fast scanning over a mass range typically between 12–200 amu.

Spectra from several samples of the same substance are averaged to enable a library of known materials to be built up. Unknown samples may then be matched against the library and hence identified.

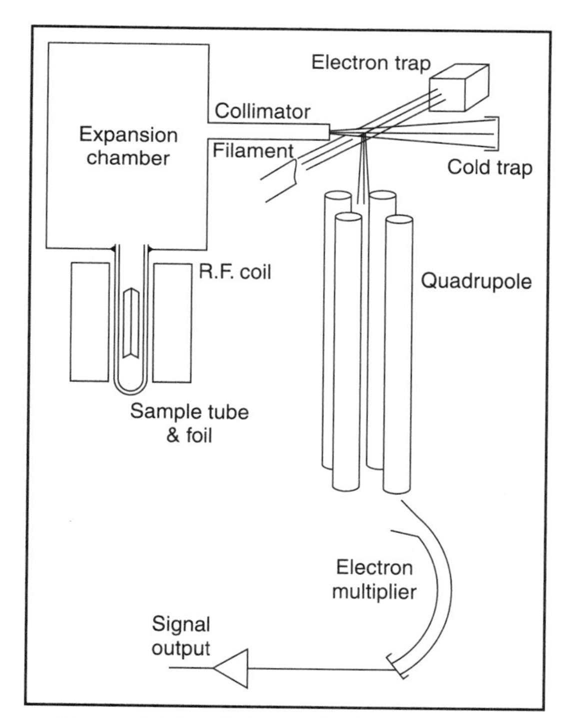

Figure 9.10 :: Schematic diagram of the pyrolysis spectrometer (Ottley, 1986)

9.4 Components of a Mass Spectrometer

The following five units are common to most mass spectrometer instruments:

- the inlet sample system,
- the ion source,
- the electrostatic accelerating system,
- the detector, amplifier and display system, and
- auxiliary equipment (pumping system).

9.4.1 The Inlet Sample System

Gaseous Samples: The introduction of gases merely involves the transfer of the sample from a gas bulb into the metering volume. The arrangement is a small glass manifold of known volume attached to a mercury manometer. The pressure range is generally from 30 to 50 mm of Hg.

The gas sample is introduced into the mass spectrometer ion source through a leak of some kind. Generally, the leak is a pin-hole in metal foil. Hogg (1969) explains the construction of a variable leak inlet system used in high resolution mass measurement

Liquid Samples: Liquid samples may be introduced by a hypodermic needle and injected through a silicon rubber dam, or by a break-off device which consists in touching a micropipette to a sintered glass disc under mercury. The low pressure in the reservoir draws in the liquid and vaporizes it instantly.

Solid Samples: Solid samples can be vaporized to gaseous ions by instantaneous discharges with a power up to 100 kW by using a radiofrequency (1 MHz) spark. Under these conditions, all constituents of the sample are converted to gaseous form at an equal rate without regard to their vapour pressure, thereby eliminating the possibility of preferential vaporization.

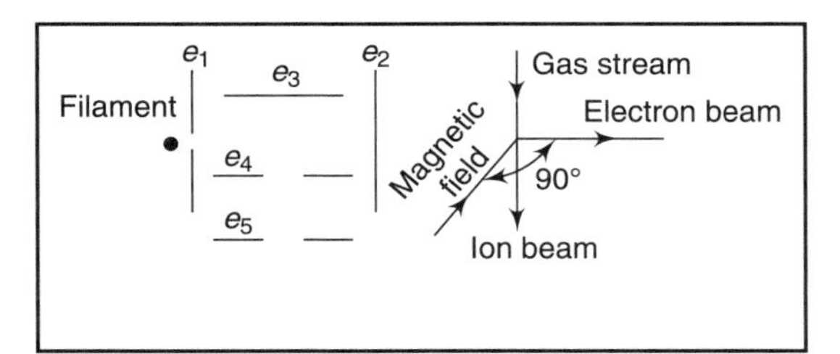

Figure 9.11 :: Schematic diagram of electron gun and ionization chamber

9.4.2 Ion Sources

Following the leak is the ionization chamber, which is maintained at a low pressure (10^{-4} to 10^{-7} mm Hg) and at a temperature of 200°C. The electron gun is located perpendicular to the incoming gas stream. Electrons are emitted from a filament (Figure 9.11) normally of carbonized tungsten, but for special purposes, tantalum or oxide coated filaments may be used. They are drawn off by a pair of positively charged slits, through which they pass into the body of the chamber. The potential present in the slits controls the electron emission and the energy of the electrons. An electric field applied between these slits accelerates the electrons, which on subsequent collisions with molecules of the passing gas stream, produce ionization and fragmentation. In order to obtain a mass spectrum, the electric field is kept between 50 and 70 V. The electron beam is usually collimated by a magnetic field, which is confined to the ionization region.

9.4.2.1 ICP as Ion Source

ICP (inductively coupled plasma) has been shown to be a potentially effective ion source (Houk, 1986). ICP hardware is designed to generate plasma, which is a gas in which atoms are present in an ionized state. The basic set-up of an ICP consists of three concentric tubes, most often made of silica. These tubes, termed outer loop, intermediate loop, and inner loop, collectively make up the ICP of the Torch. The torch is situated within a water-cooled coil of a radiofrequency (rf) generator. As flowing gases are introduced into the torch, the rf field is activated and the gas in the coil region is made electrically conductive. This sequence of events forms the plasma.

ICP Torch

The formation of the plasma is dependent upon an adequate magnetic field strength and the pattern of the gas streams follows a particular rotationally symmetrically pattern. The plasma is maintained by inductive heating of the flowing gases. The induction of a magnetic field generates high frequency annular electric current within the conductor. The conductor, in turn, is heated as the result of its ohmic resistance.

In order to prevent possible short-circuiting as well as meltdown, the plasma must be insulated from the rest of the instrument. Insulation is achieved by the concurrent flow of gasses through the system. Three gases flow through the system—the outer gas, intermediate gas, and inner or carrier gas. The outer gas is typically argon or nitrogen. The outer gas has been demonstrated to serve several purposes including maintaining the plasma, stabilizing the position of the plasma, and thermally isolating the plasma from the outer tube. Argon is commonly used for both the intermediate gas and inner or carrier gas. The purpose of the carrier gas is to convey the sample to the plasma. However, the ICP operates at atmospheric pressure, whereas mass spectrometer (MS) requires an operating pressure $\leq 10^{-5}$ torr for resolution and detection of ions. The solution acts as an interface between the two.

The conventional method of sample introduction for inductively coupled plasma mass spectroscopy is by aspiration, via a nebulizer, into a spray chamber. A small fraction of the resulting aerosol is swept by argon into the torch. Approximately 1 ml of sample is required per analytical run, about 99 per cent of which is wasted.

Recently, low cost, low uptake rate, high efficiency nebulizers have been employed to combat this problem. The high efficiency nebulizer operates more efficiently at 10–200 l/min. The detection limits and precision obtained with the high efficiency nebulizer are superior to conventional nebulizers.

Figure 9.12 shows the arrangement presently used for extracting ions from the ICP. The plasma flows around the tip of a water-cooled metal cone called the sampler. This cone has a circular orifice of 0.5-1.00 mm diameter drilled into its tip. Gas from the ICP enters this orifice from a cross-sectional area of diameter about three times the orifice diameter, so that a 1.5-3 mm wide

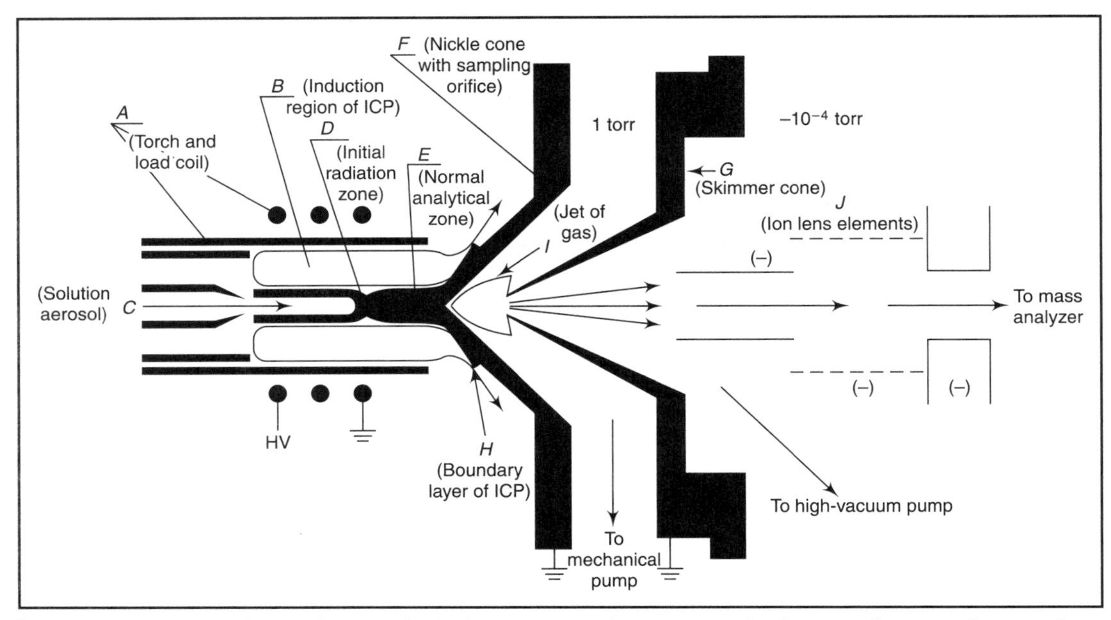

Figure 9.12 :: Inductively-coupled plasma as an ion source (redrawn after Houk, 1986)

section of the axial channel is sampled. Most of the gas flow is evacuated by a mechanical pump that maintains a pressure of the order of 1 torr.

The central orifice of a conical skimmer is located behind the sampler in such a way that the sampled beam is transmitted into a second vacuum chamber. The pressure here is low enough, and the mean free path long enough for ion lenses to collect, focus and transmit the ions to the mass analyser.

The formation of singly charged ions is very efficient in the ICP. Some 54 elements are expected to be ionized with an efficiency of 90 per cent or more. Even-metalloid or non-metallic elements such as P and As, for which ionic emission lines are either absent or not prominent, are ionized with reasonable efficiency. Therefore, a combination of the ICP with MS shows promise of extending the sensitivity and selectivity of MS to elemental analysis of solutions.

9.4.2.2 Glow Discharge Ion Source

In recent years, the glow discharge has been developed as a low-energy alternative ion source, which, coupled with a quadrupole mass filter, serves as a simple and inexpensive mass spectrometer. Basically, the glow discharge is a simple two electrode device, filled with a rare gas to about 0.1–10.0 torr. A few hundred volts applied across the electrodes cause breakdown of the gas and formation of the ions. Figure 9.13 shows the basic components and discharge regions. The sample to be analyzed serves as the cathode, whereas the anode material is not particularly critical (Harrison, *et al.*, 1986).

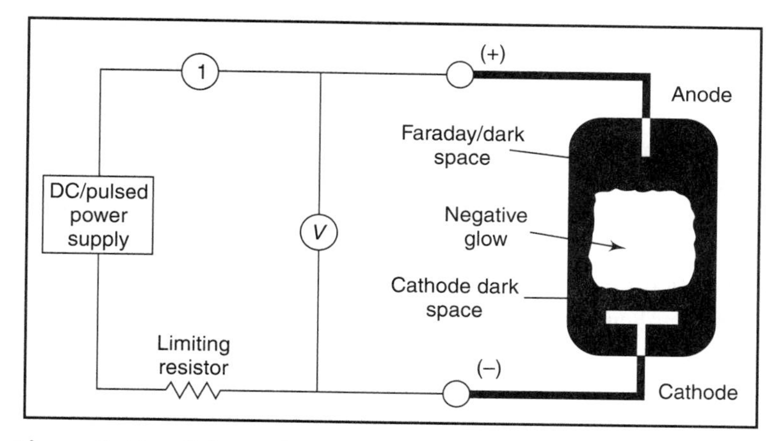

Figure 9.13 :: Schematic representation of a glow discharge ion source (redrawn after Harrison, *et al.*, 1986)

The sample is atomized into the discharge by a process known as sputtering. The glow discharge not only atomizes the solid sample, but also provides the means by which these atoms are ionized. Sputtered atoms diffuse into the negative glow, which contains energetic electrons, ions and metastable atoms. We are basically concerned with the ionization of the sputtered atoms, rather than the bulk ionization process, by which the discharge sustains itself.

The glow discharge is adaptable to many discharge types and ion source configurations. Sometimes, the sample type itself dictates a certain cathode geometry, whereas other source models arise from more fundamental considerations. Glow discharge ion sources have been operated in the dc, pulsed-dc and rf modes.

The glow discharge source presents a flowing gas load to the mass spectrometer, requiring differential pumping to interface the 1-torr source with the 10^{-6} to 10^{-7} torr analyzer chamber. Most of the reported work in glow discharge mass spectrometers has been carried out using quadrupole analyzer. This analyzer makes it possible to design instruments with low sampling voltages and high ion transmission, particularly at low mass ranges.

The glow discharge possibly ionizes only a small percentage of the sputtered sample atoms. Coupling the output of a tunable laser to the glow discharge plasma permits very selective ionization of a given element. The focused laser beam is capable of close to 100 per cent ionization efficiency. In this arrangement, the glow discharge functions primarily as an atomization source and the laser as the means of ionization. Harrison *et al.* (1986) have reviewed the developments in glow discharge mass spectrometry.

9.4.2.3 Other Ionization Methods

The various other ionization methods in use are described as follows.

Electron Impact: An electron impact source uses an electron beam, usually generated from a tungsten filament, to ionize gas-phase atoms or molecules. An electron from the beam knocks an electron off of analyte atoms or molecules to create ions.

Electro-spray Ionization: This source consists of a very fine needle and a series of skimmers. A sample solution is sprayed into the source chamber to form droplets. The droplets carry charge when they exit the capillary. As the solvent evaporates, the droplets disappear leaving highly charged analyte molecules. The method is particularly useful for large biological molecules that are difficult to vaporize or ionize.

Fast-atom Bombardment: In this method, a high energy beam of natural atoms, typically Xe or Ar, strikes a solid or low vapour-pressure liquid sample causing desorption and ionization. It is used for large biological molecules that are difficult to get into the gas phase. The sample is usually dispersed in a matrix such as glycerol. The method causes little fragmentation and usually gives a large molecular ion peak, making it useful for molecular weight determination.

The atomic beam is produced by accelerating ions from an ion source through a charge-exchange cell. The ions pick up an electron in collision with neutral atoms to form a beam of high-energy atoms.

Field Ionization: Molecules can lose an electron when placed in a very high electric field. High fields can be created in an ion source by applying a high voltage between a cathode and an anode called 'a field emitter'. A field emitter consists of a wire covered with microscopic carbon dendrites, which greatly amplify the effective field at the carbon points.

Laser Ionization: A laser pulse ablates material from the surface of a sample, and creates a microplasma that ionizes some of the sample constituents. The laser pulse accomplishes both vaporization and ionization of the sample.

Matrix-assisted Laser Desorption Ionization: This is a laser-based method of vaporizing and ionizing large biological molecules such as proteins or DNA fragments (Cotter, 1999). The biological molecules are dispersed in a solid matrix such as nicotinic acid or dihydroxybenzoic acid. A UV laser pulse ablates the matrix which carries some of the large molecules into the gas phase in an ionized form so they can be extracted into a mass spectrometer.

Thermal Ionization: Thermal ionization is used for elemental or refractory materials. The sample is deposited on a metal ribbon, such as Pt or Re, and an electric current heats the metal to a high temperature. The ribbon is often coated with graphite to provide a reducing effect.

9.4.3 Electrostatic Accelerating System

Positive ions, which are separated from electrons by a weak electric field, are accelerated in a strong electrostatic field between the first and second accelerating slits. Voltages of the order of 400–4000 volts accelerate the ions to their final velocities of up to 150,000 miles/s and they acquire a kinetic energy of a few thousand electron volts. Such a relatively high kinetic energy is imparted to the ions to produce an almost mono-energetic beam, when it finally emerges out of the final accelerating slit, which is approximately 0.076 mm in width. The electrostatic voltages are highly stabilized to an accuracy of better than 0.01 per cent.

9.4.4 Ion Detectors and Recording of Mass Spectrograph

Faraday cup: The ion beam passing through the exit slit of the analyzer tube is normally collected in a cylinder (Faraday cage), which is connected to the grid of an electrometer tube, whose output is in turn amplified. The use of an electrometer tube is necessitated because of an extremely low magnitude of the ion current (10^{-6} to 10^{-10} A). Vibrating electrometers have also been used in order to convert the dc current output into an ac signal. The amplified ion current is recorded as a function of the ratio m/e on an oscillograph or a pen-and-ink strip chart recorder. Since a Faraday cup can only be used in an analog mode, it is less sensitive than other detectors that are capable of operating in pulse-counting mode.

Channeltron: A channeltron is a horn-shaped continuous dynode structure that is coated on the inside with an electron-emissive material. An ion striking the channeltron creates secondary electrons that have an avalanche effect to create more secondary electrons and finally a current pulse.

Electron multiplier tube: Electron multiplier tubes are similar in design to photomultiplier tubes. They consist of a series of biased dynodes that eject secondary electrons when they are struck by an ion. They, therefore, multiply the ion current and can be used in analog or digital mode.

Micro-channel plate: A micro-channel plate consists of an array of glass capillaries (10–25 μm inner diameter) that are coated on the inside with an electron-emissive material. The capillaries are biased at a high voltage and like the channeltron, an ion that strikes the inside wall of one of the capillaries creates an avalanche of secondary electrons. This cascading effect creates a gain of 10^3 to 10^4 and produces a current pulse at the output. The schematic of a micro-channel plate is shown in Figure 9.14.

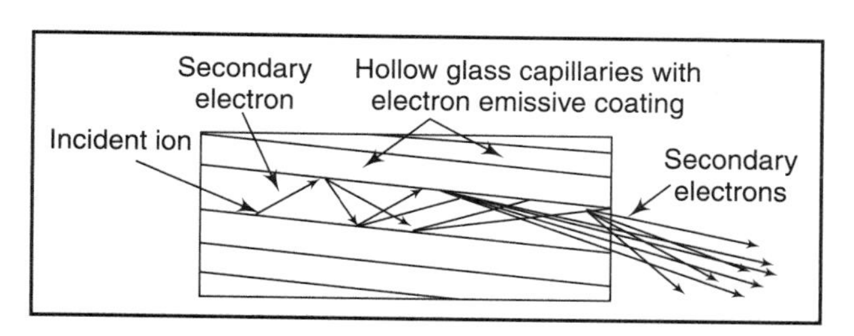

Figure 9.14 :: Schematic of a micro-channel plate

The recorder must have a provision for automatically recording peaks of widely varying amplitudes. This is achieved by using five separate recording channels with relative sensitivities of 1, 3, 10, 30 and 100, respectively. This arrangement should enable to record the height of any peak within better than 1 per cent accuracy, over a range of magnitude of 1 to nearly 1000.

9.4.5 Vacuum System

In order to prevent undue scattering by collision of ions with residual gas molecules, the mass spectrometer requires a good vacuum system. Generally, separate mercury or oil diffusion pumps are employed in the source and analyzing regions of the spectrometers.

Vacuum pump

The two important parameters of a vacuum pump are its lowest attainable pressure, and its pumping speed, typically listed as litres per minute (lpm) or cubic feet per minute (cfm). The lowest attainable pressure depends upon the design of the pump. The pumping speed of the different types of pumps depends on the physical size of the pump. Table 9.1 gives the types of vacuum pumps along with the lowest attainable pressure

Table: 9.1 :: Common Types of Vacuum Pumps

Pump	Lowest Attainable Pressure	Typical Use
Mechanical pump	10^{-3} – 10^{-4} torr	Rouging or backing pump
Diffusion pump	10^{-6} torr	Vacuum lines
Turbomolecular pump	10^{-9} torr	High-vacuum systems

Mechanical pumps consist of an inlet, and exhaust with a one-easy valve, and an off-center rotating piston in a cylindrical cavity. As the piston rotates, gas is pulled into the cavity, and forced out through the exhaust port. The rotating piston has spring-loaded vanes to create a seal with the cavity walls. This seal, and the exhaust port valve, are lubricated with a low-vapour-pressure oil. A two-stage mechanical pump consists of two pumping cavities in series to achieve a lower vacuum pressure. The accessories needed when mechanical pumps being used are a mist filter or vent to trap oil mist in the pump exhaust, and a trap to prevent oil vapour from backstreaming into the volume being evacuated.

A *diffusion pump* consists of a bath of boiling oil that streams through a jet-shaped volume. The oil entrains gas molecules and transports them in the direction of the oil flow. A mechanical pump can then pump away the exhaust from the diffusion pump.

A *turbomolecular pump* or just turbo pump contains a turbine that is spinning at a very high rate of revolution, typically tens of thousands revolutions per second. The turbine blades are spinning faster than the average speed of gas atoms or molecules, so that any gas-phase species that enter the turbo pump are physically forced out of the pump by the turbine blades. A mechanical pump is required to maintain a low pressure and pump away the exhaust from a turbo pump.

Many analytical instruments including mass spectrometry, X-ray photoelectron and Auger electron spectroscopies, and electron microscopy require a low pressure to operate. Special gadgets or instruments are required to measure such low level pressures.

By definition, pressure is the force per unit area, with units of pounds per square inch (PSI), or Newtons per square meter (N^*m^{-2}). In practice, pressure is usually reported in atmospheres (atm), which is the atmospheric pressure at sea level, or torr, which is 1.0 mm of mercury in an Hg manometer. One atm = 760 torr. Table 9.2 shows various types of pressure gauges and their range of measurement.

Table 9.2 :: Common Types of Pressure Gauges

Gauge	Pressure Range	Typical Use
Manometer	760–1 torr	Systems near atmospheric pressure
Thermocouple gauge	$1–10^{-3}$ torr	Monitoring mechanical pumps
Ionization gauge	$10^{-3} - 10^{-9}$ torr	High-vacuum systems

A *manometer* consists of a U-shaped tube that is closed and evacuated at one end, and filled with Hg or oil. One atmosphere of pressure at the open side of the tube pushes mercury in the tube to a height of 760 mm in the evacuated side of the tube. Lower pressure on the open side of the tube pushes the mercury to less than 760 mm, and provides a measure of the pressure.

Thermocouple gauges operate on the dependence of thermal-conductivity on gas pressure. In these gauges, a constant current is applied to a metal filament to heat the filament. The temperature of the filament depends upon the heat transfer to gas molecules, which depends upon the pressure. The temperature of the filament is measured by making a thermocouple junction with the filament. The pressure reading is based on a calibration, which depends upon the gas present in the vacuum system.

The most common type of *ionization gauge* is a thermionic, or hot-cathode gauge. It consists of an electrically heated filament and two electrodes. The filament (at ground voltage) emits electrons that are accelerated to the positively electrode. If the electrons collide with gas atoms or molecules they produce positive ions. Positive ions are collected at the negative electrode, creating an ion current which can be measured on the density and on the ionization cross-section of the gas-phase species. As with thermocouple gauges, a calibration must be used for different gases in the vacuum system.

Extreme cleanliness must be maintained on the surfaces in all regions of the evacuated system. Hands should not touch any interior surface, or any volatile lubricant used.

9.5 Inductively Coupled Plasma–Mass Spectrometer

Inductively coupled plasma mass spectroscopy (ICP–MS) was developed in the late 1980s to combine the easy sample introduction and quick analysis of ICP technology with the accurate and low detection limits of a mass spectrometer. The resulting instrument is capable of trace multi-element analysis, often at the part per trillion level (Wolrey and Kvech, 2005).

ICP technology uses the same principles as those used in atomic emission spectrometry. Samples are decomposed to neutral elements in high temperature argon plasma and analyzed on the basis of their mass to charge ratios. An ICP–MS consists of four main processes: (i) sample introduction and aerosol generation, (ii) ionization by an argon plasma source, (iii) mass discrimination, and (iv) the detection system. The schematic (Figure 9.15) illustrates this sequence of processes.

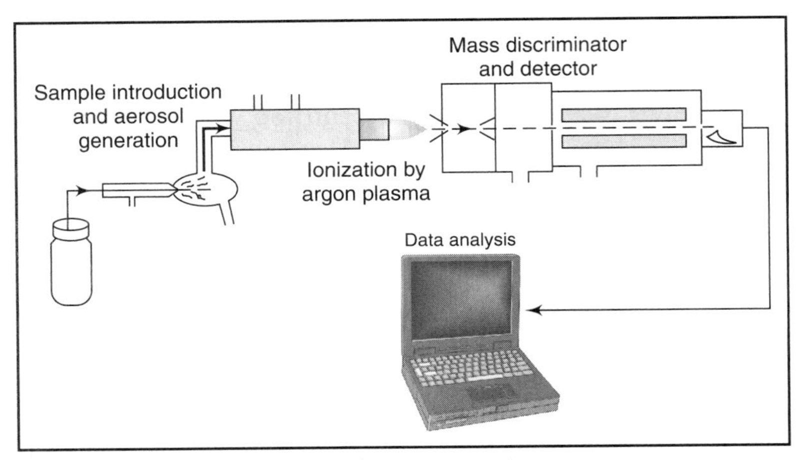

Figure 9.15 :: Schematic of ICP–MS main process

Unlike the atomic emission spectrometer, ICP–MS spectrometers can be used to analyze both solid as well as liquid samples. Solid samples are introduced into the ICP by way of a laser ablation system. Aqueous samples are introduced by way of a nebulizer which aspirates the sample with high velocity argon, forming a fine mist. The aerosol then passes into a spray chamber where larger droplets are removed via a drain (Jarvis, *et al.*, 1992). Typically, only 2 per cent of the original mist passes through the spray chamber (Olesik, 1996). This process helps to produce droplets that are small enough to be vaporized in the plasma torch.

After the sample passes through the nebulizer and is partially desolvated, the aerosol moves into the torch body where it is mixed with more argon gas. A coupling coil is used to transmit radio frequency to the heated argon gas, producing an argon plasma 'flame' located at the torch. The hot plasma removes any remaining solvent and causes sample atomization followed by ionization. In addition to being ionized, sample atoms are excited in the hot plasma. The next stage is an ICP torch (Figure 9.16). The aerosol moves into the bottom of the torch body. The ports on the right side of the torch body are where more argon is introduced to the flow. At the top are two high quality quartz tubes and an inner alumina injector tube.

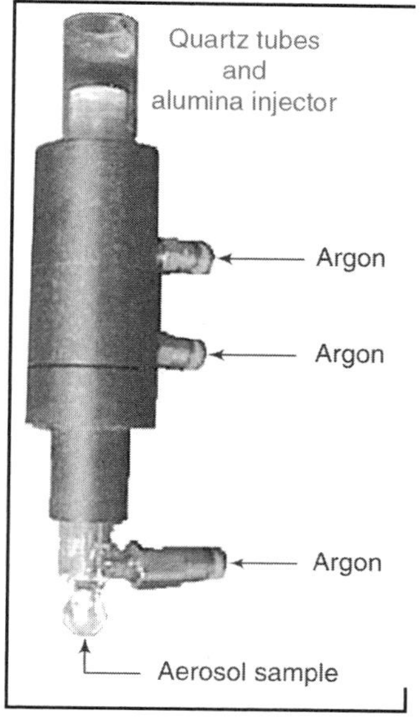

Figure 9.16 :: ICP torch

Atomization/ionization occurs at atmospheric pressure. Therefore, the interface between the ICP and MS components becomes crucial in creating a vacuum environment for the MS system. Ions flow through a small orifice of approximately 1 nm in diameter, into a pumped vacuum system. Here a supersonic jet forms and the sample ions are passed into the MS system at high speeds, expanding in the vacuum system. The entire mass spectrometer is kept in a vacuum so that the ions are free to move without collisions with air molecules. Since the ICP is maintained at atmospheric pressure, a pumping system is needed to continuously pull a vacuum inside the spectrometer. In order to most efficiently reduce the pressure, several pumps are typically used to gradually reduce pressure to 10^{-5} mbar before the ion stream reaches the quadrupole. If only one pump were used, its size would be excessive to reduce the pressure immediately upon entering the mass.

During the first stage of the mass spectrometer, ions are removed from the plasma by a pumped extraction system. An ion beam is produced and focused further into the actual unit. Different types of mass analyzers can be employed to separate isotopes based on their mass to charge ratio. Quadrupole analyzers are compact and easy to use but offer lower resolution when dealing with ions of the same mass to charge (m/e) ratio. A series of ion lenses, maintained at appropriate voltages, are used to direct the ions into quadrupole mass analyzer. The ions are transmitted through the quadrupole on the basis of their mass to charge ratios and then detected by an electron multiplier.

The use of a quadrupole mass analyzer gives better than unit mass resolution over a mass range up to $m/z = 300$. The inductively coupled plasma-mass spectrometry system is considered to be a sequential multi-element analyzer that has scan times less than 20 ms for one sweep. The signal intensity is a function of the number of analyte ions in the plasma and the mass-dependent transport through the mass spectrometer. Double focusing section analyzers offer better resolution but are larger and have higher capital cost.

The most common type of ion detector found in an ICP–MS system is the channeltron electron multiplier. This cone-or horn-shaped tube has a high voltage applied to it opposite in charge to that of the ions being detected. Ions leaving the quadrupole are attracted to the interior cone surface. When they strike the surface, secondary electrons are emitted which move farther into the tube thereby emitting additional secondary electrons. As the process continues, more electrons are formed, resulting in as many as 10^8 electrons at the other end of the tube after one ion strikes at the entrance of the cone. This results in an avalanche effect to create a current pulse.

The efficiency of the Inductively Coupled Plasma in producing singly-charged positive ions for most elements makes it an effective ionization source for mass spectrometry. Inductively coupled plasma–mass spectrometry is unique among the flame and plasma spectroscopy techniques in terms of its ability to discriminate between the mass of the various isotopes of an element where more than one stable isotope occurs. Isotope dilution, in which the change in isotope ratio for two selected isotopes of an element of interest is measured in a solution after the addition of a known quantity of a spike that contains enrichment of one of the isotopes, permits calculation of the concentration of the element. Isotope dilution is the most reliable method of accurate determination of elemental concentration.

Isotope dilution

ICP–MS has been used widely over the years, finding applications in a number of different fields including drinking water, wastewater, natural water systems/hydrogeology, geology and soil science, mining/metallurgy, food sciences and medicine (Qlesik, 1996).

The most important advantages of ICP–MS include its multi-element capability, high sensitivity, and the possibility to obtain isotopic information on the elements determined. The disadvantages inherent in the ICP–MS system include the isobaric interferences produced by polyatomic species arising from the plasma gas and the atmosphere. The isotopes of argon, oxygen, nitrogen, and hydrogen can combine with themselves or with other elements to produce isobaric interferences. ICP–MS is not useful in the detection of non-metals.

9.6 Trapped Ion Mass Analyzers

The ion–trap mass spectrometer uses three electrodes to trap ions in a small volume. The advantages of the ion–trap mass spectrometer include its compact size, and the ability to trap and accumulate ions to increase the signal-to-noise ratio of a measurement.

The mass analyzer consists of a ring electrode separating two hemispherical electrodes. A mass spectrum is obtained by changing the electrode voltages to eject the ions from the trap.

This class of instruments operate by storing ions in the trap and manipulating the ions by using dc and RF electric fields in a series of programmed timed events. This arrangement provides very high resolution and high sensitivity. There are two types of trappedion mass analyzers. They are:

- Static Traps: Ion Cyclotron Resonance mass spectrometer
- Dynamic Traps: Three-dimensional quadrupole ion trap mass spectrometer

9.7 Ion Cyclotron Resonance (ICR) Mass Spectrometery

It takes advantage of ion–cyclotron resonance to select and detect ions. Ions move in a circular path in a magnetic field. The cyclotron frequency of the ions circular motion is mass-dependent. By measuring the cyclotron frequency, the mass of the ion can be determined.

By equating the centripetal force mv^2/r and the Lorentz force evB experienced by an ion in a magnetic field, the working equation for ICR can be derived as follows:

$$mv^2/r = evB,$$

solving for the angular frequency (W), which is equal to v/r then

$$W = v/r = eB/m.$$

Ions of the same mass-to-charge ratio have the same cyclotron frequency, but they move independently and out of phase. If an excitation pulse is applied at the cyclotron frequency, the resonant ions will absorb energy and are brought into phase with the excitation pulse.

If several different masses are present, an excitation pulse that contains components at all the cyclotron frequencies must be applied. The ions re-emit the radiation, which is picked up by the receiver plates as image currents, amplified and digitized. The signal induced in the receiver plates depends upon the number of ions and their distance from the receiver plates.

The image currents induced in the receiver plates will contain frequency components from all the mass-to-charge ratios. The various frequencies and their relative abundances can be worked out mathematically by using a Fourier Transform which converts a time–domain signal (the image currents) into a frequency–domain spectrum (the mass spectrum).

Figure 9.17 shows a schematic representation of a cubic FTICR (Fourier Transform Ion Cyclotron Resonance) cell. The cell consists of three pairs of parallel plates. The functions of the excitation and receiver plates have been explained above. A small potential is applied to the trapping plates to keep the ions contained within the ICR cell because the magnetic field does not constrain the ion motion along with the direction of the applied magnetic field. Most FTICR mass spectrometers use superconducting magnets, which provide a relatively stable calibration over a long period of time.

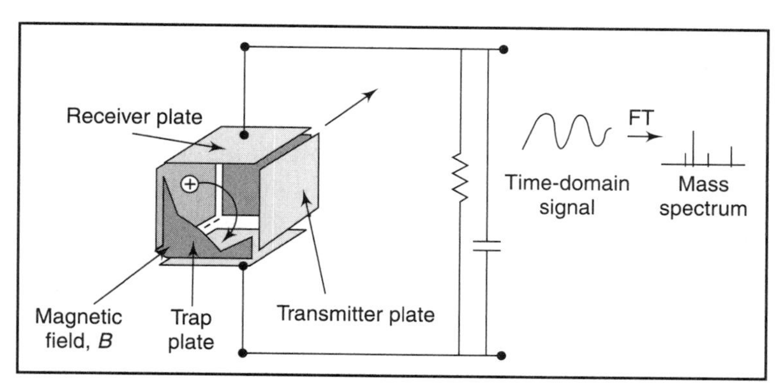

Figure 9.17 :: Schematic diagram of FT-MS (Tissue, 2000a)

9.8 Quadrupole Ion Trap Mass Spectrometer

In these instruments, ions are formed within the ion trap or injected into the trap from an external source. The ions are dynamically stored in a three-dimensional quadrupole ion storage device.

The RF and dc potentials can be scanned to eject successive mass-to-charge ratios from the trap into the detector (mass-selective ejection). The arrangement is shown in Figure 9.18.

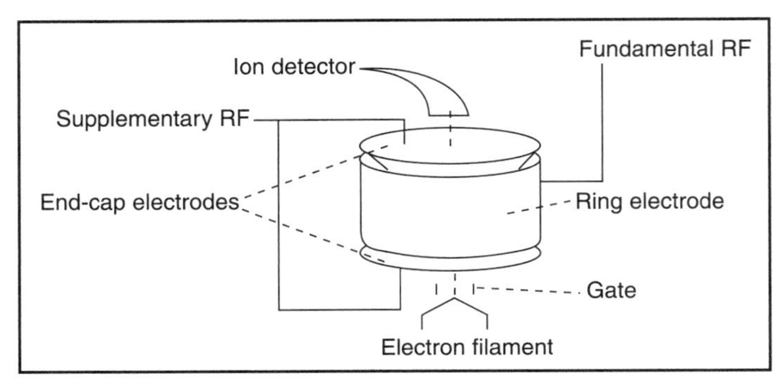

Figure 9.18 :: Ion trap mass spectrometer

9.9 Resolution in Mass Spectrometry

Several different definitions of resolution are used in mass spectrometry. Some of the commonly used definitions are delineated below.

The difference between each mass from the next integer is defined as a unit resolution: the capability of the equipment to distinguish mass, say 500 from mass 501. This definition is commonly used when describing resolution on quadrupole and ion trap mass spectrometers. Peak shapes in quadrupole mass spectrometers are usually 'flat-shaped' (Figure 9.19).

Resolution

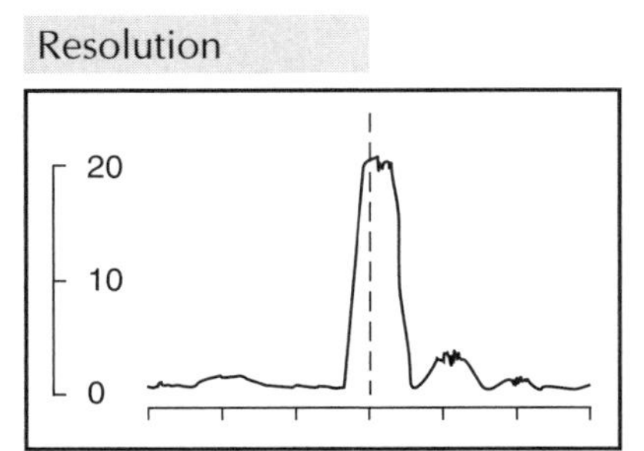

Figure 9.19 :: Flat-shaped peaks in ion trap mass spectrometer

In magnetic sector mass spectrometers, 'resolution' is defined as the mass divided by the difference in mass number between two distinguishable neighbouring lines of equal height in the mass spectrum. If two ions of mass M_1 and M_2 differing in mass by ΔM give adjacent peaks in the mass spectrum as shown in Figure 9.20 and the height of peak is H above the base line, and height of the valley h is less than or equal to 10 per cent of the peak H, i.e. $h \leq 10$, the resolution is then $M_1/\Delta M$. For example, if the peaks representing two masses 100.000 and 100.005, are separated by a 10 per cent valley, the resolution is 100.000/0.005, i.e. 20,000.

Time-of-flight mass spectrometers usually use the 50 per cent peak-height (h) definition. Similarly, Fourier Transform Ion Cyclotron Resonance (FTICR) mass spectrometers also usually use the 50 per cent valley definition for resolution because of the broadening near the base time of the peak shape.

Resolving power is inversely proportional to mass in FTICR. So, it is important to know the mass at which a given resolving power was obtained in order to determine what the resolution should be at another mass.

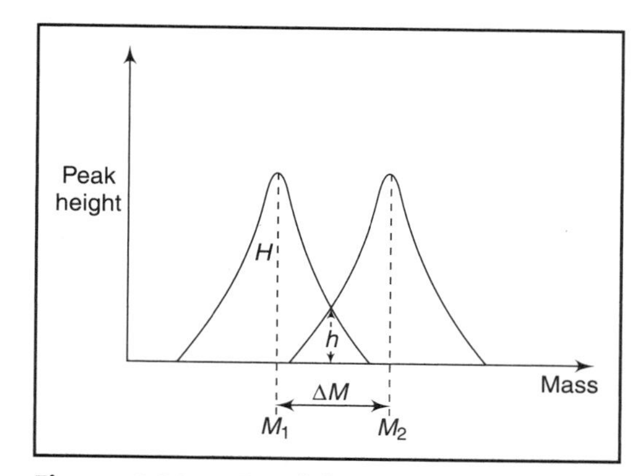

Figure 9.20 :: Resolving power calculation for mass spectrometer

9.10 Applications of Mass Spectrometry

In addition to general inorganic analysis and trace analysis in inorganic chemistry, rocket-borne mass spectrometers now analyze the upper atmosphere. Perhaps the most impressive experimental achievement in mass spectroscopy is the successful mass spectrometric analysis carried out on the surface of Mars. Process monitor mass spectrometers, with the idea of direct control of industrial process have been constructed.

An important requirement for studying respiratory physiology in routine clinical investigations is the continuous analysis of the gas flow in a patient's breathing cycle. A mass spectrometer would be capable of analyzing gases in the mass range 15 to 50, which covers the full range of respiratory gases from water to carbon dioxide. A mass spectrometer would require very little volume of gas for analysis, which may be only about 15 ml/mm. This is a negligible fraction of the patient's minute respiratory volume. The instrument provides a rapid response, with an overall accuracy of 2 to 3 per cent. For respiratory work, a mass spectrometer with an analyzer tube of 180° in a permanent magnetic field is quite suitable. This would provide a resolution of about 50. Figure 9.21 shows a typical mass spectrograph of exhaled air as seen on oscilloscope display. Mass spectrometers are widely used in biochemical analysis in medicine and other fields.

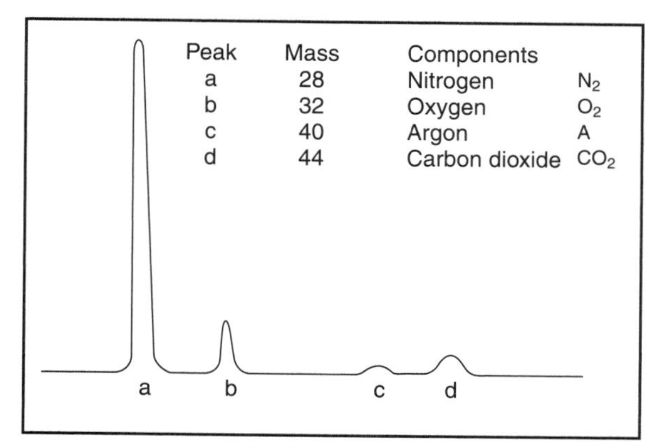

Figure 9.21 :: Typical mass spectrograph of expired air as seen on cathode ray oscilloscope

In the early 1950s, mass spectrometers were used for the quantitative analysis of mixtures of the lighter hydrocarbons containing molecules with up to six or seven carbon atoms. The mass spectra of the various components in a mixture were taken to be linearly additive, and the analysis was worked out by setting up a set of simultaneous linear equations and solving them, usually by matrix inversion, using desk calculators. Advances in digital computers have considerably transformed the situation concerning calculations of such equations. Mass spectrometry is presently used to make analyses of mixtures into different hydrocarbon types, even when many hydrocarbons are present with 20 to 30 carbon atoms in the molecule. Therefore, they are used in refineries for trace element investigations, analysis of lubricating oils and quantifying the substances in mixtures of organic compounds. In addition, they are used for detecting and measuring the concentrations of pollutants in air and water. Each mass analyzer has its own special considerations and applications and its own benefits and limitations. The choice of mass analyzer is usually based upon the application, cost and performance desired.

9.11 Gas Chromatograph–Mass Spectrometer (GC–MS)

Gas chromatography provides an excellent method for separation of the components of a mixture. However, the technique does not provide direct identification of the separated components. The only information presented for each component is its retention behaviour in comparison with that of the other constituents under a given set of GC operating conditions. The most usual and reliable method of identification is to isolate the compound by using spectral methods like IR, UV, NMR and mass spectrometry. Mass spectrometry is the most sensitive of the spectral methods and permits the direct introduction of a gas chromatograph's effluent stream into it. Moreover, both require samples in the vapour state. GC–MS can be thus recognized as an entity in itself rather than just a combination of the two.

Placed at the end of a chromatographic column in a manner similar to the other GC detectors, the mass detector is more complicated than, for instance, the FID because of the mass spectrometer's complex requirements for the process of creation, separation and detection of gas phase ions. A capillary column is most often used in the chromatograph because the entire MS process must be carried out at very low pressures (~10^{-5} torr) and in order to meet this requirement, a vacuum is maintained via constant pumping by using a vacuum pump. It is difficult for packed GC columns to be interfaced to an MS detector because they have carrier gas flow rates that cannot be as successfully pumped away by normal vacuum pumps; however, capillary columns' carrier flow is 25 or 30 times less and therefore easier to 'pump down'. GC–MS interfaces have been developed for packed column systems that allow for analyte molecules to be dynamically extracted from the carrier gas stream at the end of a packed column and are thereby selectively sucked into the MS for analysis.

One of the most obvious problems encountered in combining GC and MS is the considerable difference in their operating pressures. The pressure at the exit of the column in GC is atmospheric. The flow rate of the carrier gas is 15–50 ml/min, depending upon the type of column used. Mass spectrometers, on the other hand, accept ion source pressures, generally no higher than 10^{-4} torr (1 torr = 1 mm of Hg). Therefore, the total flow of the GC instruments cannot be introduced into any MS of commercial type. In the earlier days, some sort of splitting arrangement was used, which permitted the feeding of only a portion of the effluent into the mass spectrometer. More efficient linkages have been reported in literature, which, in principle, strip the sample from the carrier gas, discard the carriers and introduce the sample into the mass spectrometer. These are known as enrichment devices, or molecular separators, since the sample gets concentrated by passage of the effluent stream through the separator.

Molecular separator

Figure 9.22 shows the principle of a jet type molecular separator, which provides a pressure dip from approximately atmospheric to about 10^{-4} mm of Hg. Substances eluted from the GLC column are carried along with helium carrier gas into the system, which, in two stages, removes most of the helium (carrier gas), while the sample passes into the ion source chamber in a highly concentrated molecular beam. Each stage consists of an evacuated chamber connected to a pumping system. The column effluent enters the system through a very fine jet, which is aligned with a small exit orifice positioned at a short distance from the entrance jet. The column effluent would enter the separator as a very high speed stream of gas, with the entrance jet acting as a restrictor between the GC and the separator. The carrier gas with the low-molecular weight would diffuse at a higher rate than the higher molecular weight sample. Therefore, the carrier gas would diffuse away from the line of flow and would be pumped away. The sample molecules, along with the remaining carrier gas, then pass into the second stage, when the process repeats itself. This type of separator is used in most of the commercial GC–MS systems. Other methods for separating out the carrier from the sample include porous tubes, teflon tube and selective membranes.

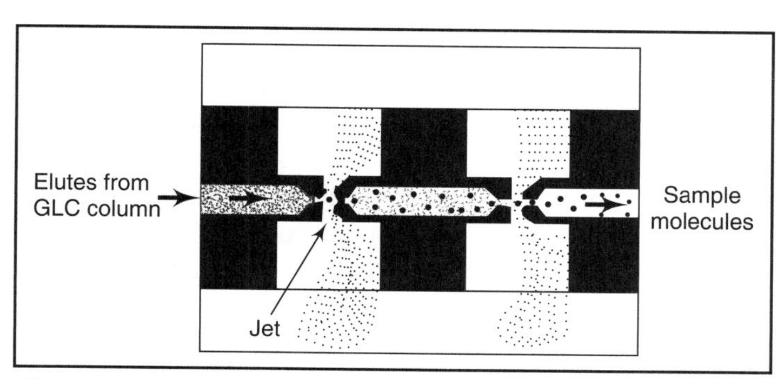

Figure 9.22 :: Principle of a jet type molecular separator

Figure 9.23 shows a simplified block diagram of a gas chromatograph–mass spectrometer, which is self-explanatory. The GC system incorporates a coiled column housed in a compact oven. The effluent stream of the GC enters the molecular separator. The separator is maintained at a controlled temperature to prevent sample condensation. Its output is connected to the MS system through a three-way valve system. The mass spectrometer is usually a single focusing instrument equipped with a 60° sector, 20 cm radius magnetic analyser. The system for measurement of ion intensity consists of an electrometer and a wide-band amplifier feeding a direct

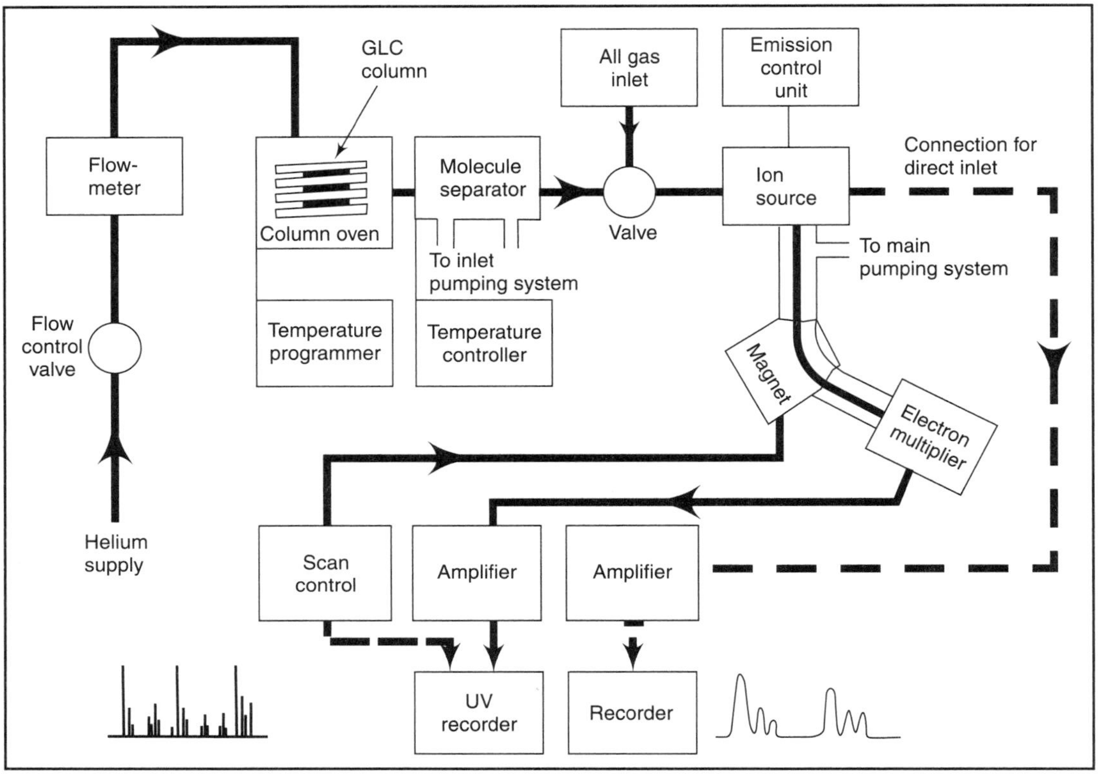

Figure 9.23 :: Block diagram of gas chromatograph-mass spectrometer

writing recorder. The high capacity vacuum system comprises of two isolated pumping assemblies, one for the analyzer tube and one for the inlet systems. A cold tray is provided for liquid N_2 or CO_2 but this is used only for mass spectral analysis of extremely small samples.

The optimization of the parameters controlling ion source sensitivity, mass spectral scan cycle time, and chromatographic elution profile as well, on-line computer systems to record, display in real-time and subsequently evaluate these sets of GC-MS data is mandatory, to ensure the full utilization of the potentiality of this combination.

9.12 Liquid Chromatograph–Mass Spectrometer

A potentially advantageous area is the development of combinations of liquid chromatographic systems with mass spectrometers. However, the levels of utilization from the point of view of the types of molecules which are amenable to HPLC separation versus interface and mass spectral characterization are still so significantly disparate, that these developments are just exploratory from an analytical point of view.

The problems associated with interfacing liquid chromatography with mass spectrometers in a practical way are considerable. The simplest method is the direct introduction of the liquid from the LC to the MS ion source region. However, it should be realized that the gas burden from the conventional LC flow rates (1 ml/min of water produces 1.2 l/mm of gas) creates nearly 20 times more gas than a cryo-pumped vacuum system can handle. Therefore, the introduction of total HPLC effluent into a MS is not feasible. Therefore, the interface should be in a position to split the total effluent, so that only 1–5 per cent is introduced in the MS. This, however, would be at the cost of detection limit and sensitivity.

Covey and *et al.*, (1986) review the various techniques used in interfacing LC with MS. The most common interface currently used is the heated nebulizer shown in Figure 9.24. The HPLC effluent passes through the central microbore throughput tube of the probe, while the nebulizer gas and make-up gas are introduced coaxially into the heated nebulization region. The combination of heat and gas flow dissolves the nebulized droplets, producing a dry vapour of solvent and analyte molecules.

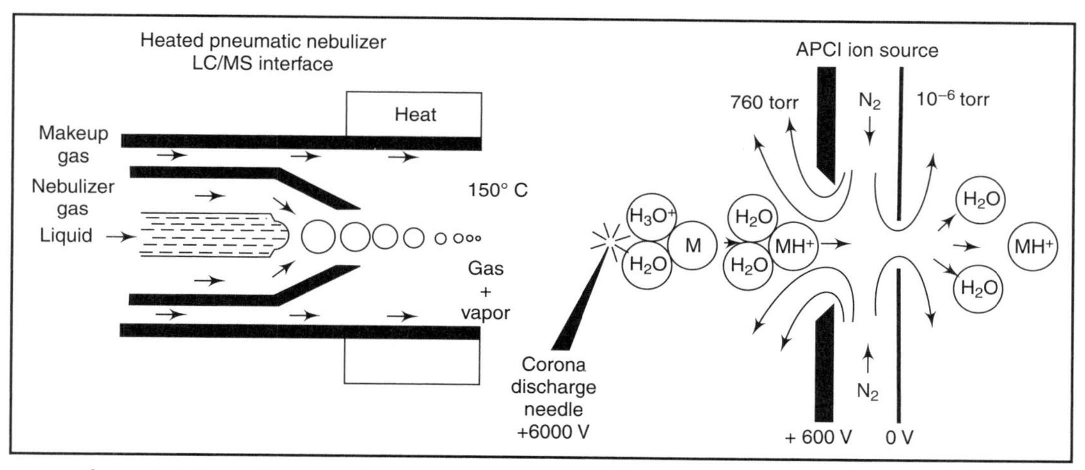

Figure 9.24 :: Chematic diagram of a heated pneumatic nebulizer LC/MS interface (after Covey, *et al.*, 1986)

The ionization of solvent molecules is produced by a corona discharge at the discharge needle. The solvent ions formed produce analyte ions by atmospheric pressure chemical ionization source. These ions are focused and de-clustered through a dry nitrogen region, and then through a 100 μm orifice into the high vacuum analyzer region of the MS, where they are mass analyzed. Typically, a nebulizer vapour temperature of 125–150°C is maintained, which is suitable for a variety of applications.

9.13 Tandem Mass Spectrometry (MS/MS)

Mass spectrometers are commonly combined with separation devices such as gas chromatographs (GC) and liquid chromatorgraphs (LC). The GC or LC separates the components in a mixture, and the components are introduced, one by one, into the mass spectrometer. MS/MS is an analogous technique where the first-stage separation device is another mass spectrometer.

Let us assume that we want to analyze a mixture of components. Each component of the mixture produces characteristic ionic species and has a unique molecular weight. The mass spectrum of the mixture contains peaks for each compound present in the mixture. However, the entire mass spectrum tells us the molecular weight, but we would really like to see fragment ions that provide structural information for the component of interest.

The simplest form of tandem mass spectrometry is shown in Figure 9.25. It combines two mass spectrometers. The first mass spectrometer is used to select a single (precursor) mass that is characteristic of a given analyte in a mixture. The mass-selected ions pass through a region where they are activated in some way that causes them to fall apart to produce fragment (product) ions. This is usually done by colliding the ions with a neutral gas in a process called collisional activation or collision-induced dissociation. The second mass spectrometer is used to separate the fragment ions according to mass. The resulting 'MS/MS' spectrum consists only of product ions from the selected precursor. The chemical background and other mixture components are absent.

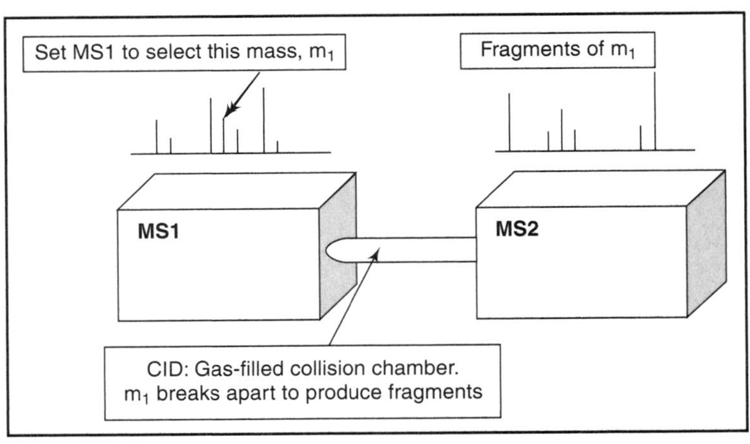

Figure 9.25 :: Tandem mass spectrometer (MS/MS)

Early work on MS/MS was done with reverse-geometry double focusing mass spectrometers. A reverse-geometry mass spectrometer is one in which the magnetic sector precedes the electric sector. A magnetic sector alone can be used as a mass spectrometer, with roughly unit resolution.

Therefore, the magnetic field strength can be adjusted to select a given precursor mass. The precursor mass is then activated by collisions in the second field-free region, just between the magnetic and electric sectors, and then the electric sector is used to analyze the kinetic energies of the product ions. This is referred to as 'mass-analyzed ion kinetic energy spectrometer' or MIKES.

In MIKES experiments, we measure product ion kinetic energies instead of product-ion mass-to-charge ratios. MIKES spectra provide a 'fingerprint' that can be used to identify a given analyte, and they are useful for ion chemistry studies.

Most magnetic sector analyses are done by setting the accelerating voltage and electric sector to a fixed accelerating voltage and electric sector. MS/MS can also be done by scanning the electric and magnetic sector together according to certain scan laws.

Consider what happens to ions during a product–ion linked scan where collisions occur in the first field-free region. Ions leaving the ion source are accelerated to a kinetic energy that depends only upon the accelerating potential and the number of charges on the ion:

$$T = eV = \frac{mv^2}{2}.$$

All ions with the same number of charges will have the same kinetic energy. Assuming that the ions have only a single charge, it is apparent from the above expression that ions with different masses must have different velocities if their kinetic energies are the same.

Let's assume that precursor ions with mass m_1 fall apart in the first field-free region to for product ions with mass m_2. Lets also assume that the velocity does not change when the ions fall apart. This is a safe assumption because we will only be observing ions that undergo grazing collisions, and any change in velocity will be small as compared to the total velocity of ions accelerated to, say, 10 kilovolt kinetic energies. The product ions will still have the same velocity as the precursor ions. If we can select ions according to their velocities, we can tell which product ions were formed from precursor ions with known velocities.

Recall that the magnetic sector separates ions according to their momentum (mv) while the electric sector selects ions according to their kinetic energy ($mv^2/2$). The ratio of B to E is related to the velocity:

$$\frac{B}{E} = \frac{mv}{mv^2/2} = \frac{2}{v}.$$

Thus, a B/E ratio can be chosen to select ions with a given velocity. If we scan B and E together, always keeping a constant B/E ratio, we will detect product ions from the precursor ion with the specified velocity. This means that a B/E scan is a product ion scan.

A tandem mass spectrometer for MS/MS studies consists of a quadrupole mass filter followed by a time-of-flight mass analyzer (Glish and Goeringer, 1984). This is shown in Figure 9.26. A collision cell is located in the region between the two analyzers. Low energy parent ions (≤ 100 eV) are mass analyzed by the quadrupole and injected into the collision cell and is accelerated by 1000 eV upon exiting the collision cell. These accelerated ions are then gated into the time-of-flight for mass analysis.

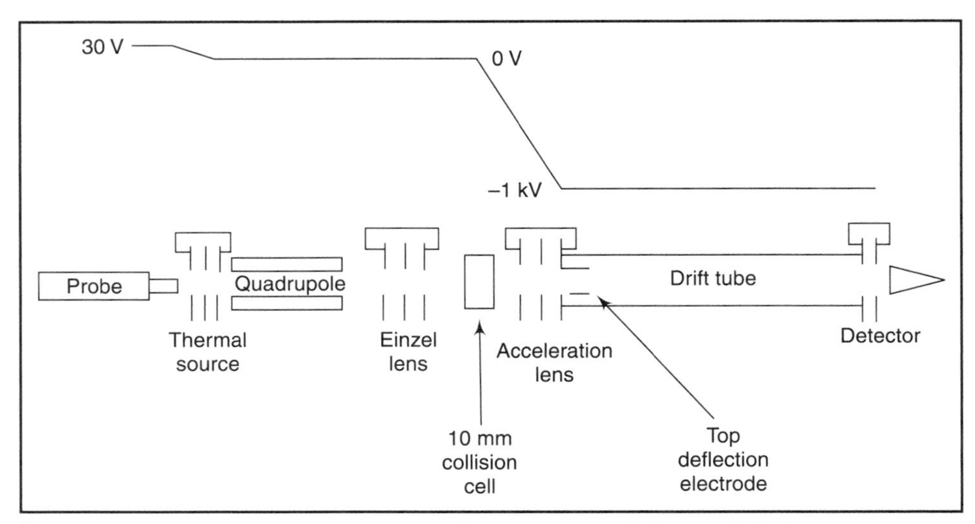

Figure 9.26 :: Schematic of the tandem quadrupole/time-of-flight instrument

Ionization or solvent molecules are produced by a corona discharge at the discharge needle. The solvent ions formed produce analyte ions by atmospheric pressure chemical ionization source. These ions are focused and de-clustered through a dry nitrogen region, and then through a 100 pm orifice into the high vacuum analyzer region of the MS, where they are mass analyzed. Typically, a nebulizer vapour temperature of 125–150°C is maintained, which is suitable for a variety of applications.

Chapter 10

Nuclear Magnetic Resonance Spectrometer

10.1 Nuclear Magnetic Resonance Spectroscopy

Nuclear magnetic resonance (NMR) spectroscopy is a non-destructive technique for mapping molecular structures and learning how molecules function and relate to each other. It is recognized as one of the most powerful techniques for chemical analysis. The importance of this technique is reflected in the efforts that have been made to extend its applicability to smaller and smaller sample sizes. The NMR spectrometer provides an accurate and non-destructive method of determining the structure of molecules in liquids and soluble chemical compounds.

The study of absorption of radiofrequency radiation by nuclei in a magnetic field is called nuclear magnetic resonance. For a particular nucleus, an NMR absorption spectrum may consist of one to several groups of absorption lines in the radiofrequency region of the electromagnetic spectrum. They indicate the chemical nature of the nucleus and the spatial positions of neighbouring nuclei (Grant, 1985).

Nuclear magnetic resonance

Nuclear magnetic resonance (NMR) spectroscopy uses radiofrequency radiation to induce transitions between different nuclear spin states of samples in a magnetic field. NMR spectroscopy can be used for quantitative measurements, but it is most useful for determining the structure of molecules along with IR spectroscopy and mass spectrometry. The utility of NMR spectroscopy for structural characterization arises because different atoms in a molecule experience slightly different magnetic fields and therefore, transitions at slightly different resonance frequencies in an NMR spectrum. Furthermore, splittings of the spectra lines arise due to interactions between different nuclei, which provide information about the proximity of different atoms in a molecule.

10.2 Principle of NMR

10.2.1 Nuclear Spin

Elementary particles such as electrons or a nucleus are known to behave as if they rotate about an axis and thus have the property of spin. The angular momentum associated with the spin of the particle would be an integral or a half-integral multiple of $h/2\pi$, where h is Planck's constant. The maximum spin component for a particular particle, is its spin quantum number I. Sub-atomic particles (electrons, protons and neutrons) can be imagined as spinning on their axes. In many atoms (such an C^{12}), these spins are paired against each other such that the nucleus of the atom has no overall spin. However, in some atoms (such as H^1 and C^{13}), the nucleus does possess an overall spin. The rules for determining the net spin of a nucleus are as follows:

(i) If the number of neutrons and the number of protons are even, the spin would be zero. Nuclei of this type do not give rise to an NMR signal, neither do they interfere with an NMR signal from other nuclei. Examples are C^{12} and 0^{16}.

(ii) Nuclei having either the number of protons or the number of neutrons as odd have half-integral spin. Examples are H^1, B^{11}, p^{31}, etc.

(iii) Nuclei which have both the number of neutrons and the number of protons as odd, would have integral spin. For example, H^2 and N^{14}.

10.2.2 Nuclear Energy Levels

Since a nucleus possesses a charge, its spin gives rise to a magnetic field that is analogous to the field produced when an electric current is passed through a coil of wire. The resulting magnetic dipole or nuclear magnetic moment μ is oriented along the axis of spin and has a value that is characteristic for each kind of particle.

When a spinning nucleus is placed in a strong uniform magnetic field (H) (Figure 10.1), the field exerts a torque upon the nuclear magnet. This would make the nucleus assume a definite orientation with respect to the external field. The torque is a vector with its direction at right angles to the plane of μ and H. This results in a rotation of the nuclear axis around the direction of the external field. This is called precessional motion.

Each orientation of the nucleus corresponds to a different energy level or state. The inter-relation between particle spin and magnetic moment leads to a set of observable magnetic quantum states given by:

$$m = I, I-1, I-2, \ldots, -(I-1), -1$$

The overall spin, I, is important. Quantum mechanics tell us that a nucleus of spin I will have $2I + 1$ possible orientations. A nucleus with spin 1/2 will have two possible orientations. In the absence of an external magnetic field, these orientations are of equal energy. If a magnetic field is applied, then the energy levels split. Each level is given a magnetic quantum number, m.

The spin number for both the electron and proton is 1/2. Thus each has two spin states, corresponding to $I = 1/2$ and $I = -1/2$. The spin number of a nucleus is related to the relative number of protons and neutrons it contains. Therefore, for heavier nuclei, the spin number range from zero to at least 9/2. When the nucleus is in a magnetic field, the initial populations of the energy levels are determined by thermodynamics, as described by the Boltzmann distribution. This is very important, and it means that the lower energy level will contain slightly more nuclei than the higher level. It is possible to excite these nuclei into the higher level with electromagnetic radiation. The frequency of radiation needed is determined by the difference in energy between the energy levels.

When brought into the influence of an external magnetic field, a particle that possesses a magnetic moment tends to become oriented such that its magnetic dipole and hence its spin axis are parallel to the field. The energy levels are a function of the magnitude of the nuclear magnetic moment and the strength of the applied magnetic field.

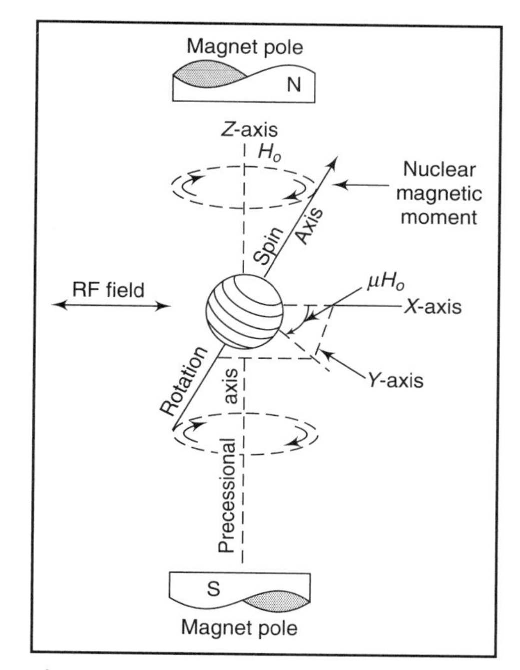

Figure 10.1 :: Spinning nucleus in a magnetic field

10.2.3 Resonance Conditions

When an alternating radiofrequency field, superimposed over the stationary magnetic field, rotates at exactly the frequency of an energy level, the nuclei will be provided enough energy to undergo a transition from a lower energy level to a higher energy level. In general, the energy difference between states is given by:

$$\Delta E = \mu\beta \cdot H_o/I,$$

where H_o is the strength of the external magnetic field in gauss, β is a constant called the nuclear magneton, 5.049×10^{-24} ergs – gauss^{-1}, μ is the magnetic moment of the particle expressed in units of nuclear magnetons. The value of μ for the proton is 2.797 nuclear magnetons.

The frequency υ of the radiation that will effect transitions between energy levels can be determined from Planck's equation:

$$\Delta E = h\upsilon = \mu\beta\, H_o/I$$

The frequency of the resonance absorption can be adjusted by varying the value of the applied magnetic field. Difficulties in construction of large magnets limit the field to approximately 23,000 gauss. In that case:

$$\upsilon = \mu/h \cdot \beta \cdot H_0/I \qquad h = 6.626 \times 10^{-27} \text{ erg.s}$$

$$= \frac{2.797 \times 5.05 \times 10^{-24} \times 23{,}000}{(6.6256 \times 10^{-27})(1/2)}$$

$$= 95 \times 10^6 \text{ Hz} = 95 \text{ MHz}$$

Therefore, the proton will precess 95 million times per second in a fixed field of 23,000 gauss. The frequency 95 MHz lies in the radiofrequency range of the electromagnetic spectrum.

With a field strength of 14092 gauss, the frequency would be 60 MHz. Similarly, for a fixed field strength of 10,000 gauss, the frequency is 40 MHz.

10.2.4 NMR Absorption Spectra

For understanding NMR absorption spectra, one has to first understand the behaviour of a charged particle in a magnetic field.

Imagine a nucleus of spin 1/2 in a magnetic field. This nucleus is in the lower energy level i.e. its magnetic moment does not oppose the applied field. The nucleus is spinning on its axis. In the presence of a magnetic field, this axis of rotation will precess around the magnetic field.

The frequency of precession is termed the Larmor frequency, which is identical to the transition frequency.

The potential energy of the precessing nucleus is given by:

$$E = -\mu B \cos \theta,$$

where θ is the angle between the direction of the applied field and the axis of nuclear rotation.

If energy is absorbed by the nucleus, then the angle of precession, θ, will change. For a nucleus of spin 1/2, absorption of radiation 'flips' the magnetic moment so that it opposes the applied field (the higher energy state).

It is important to realize that only a small proportion of 'target' nuclei are in the lower energy state and can absorb radiation. There is a possibility that by exciting these nuclei, the populations of the higher and lower energy levels will become equal. If this occurs, then there will be no further absorption of radiation. The spin system is saturated. The possibility of saturation means that we must be aware of the relaxation processes which return nuclei to the lower energy state.

NMR absorption spectra can be obtained either by changing the frequency of the radiofrequency oscillator or by changing the spacing of the energy levels by varying the magnetic field. Constructing a highly stable oscillator, whose frequency can be varied continuously is a difficult job. Also, there are no dispersing elements analogous to a prism and a grating for radiofrequency radiation. Therefore, it is more practical to hold the oscillator frequency constant and vary the magnetic field continuously. Since for a given nucleus, the frequency and field strength are directly proportional, the magnetic field (*H*) can be used equally well as abscissa for recording an NMR absorption spectrum.

10.2.5 Relaxation Process

When nuclei in the higher energy state return to the lower state, emission of radiation takes place. However, the emission is insignificant because the probability of re-emission of photons varies with the cube of the frequency. At radiofrequencies, re-emission is negligible. Therefore, non-radiative relaxation processes will give more useful information.
There are two major relaxation processes, which are:

- Spin–lattice (longitudinal) relaxation, and
- Spin–spin (transverse) relaxation.

10.2.5.1 Spin–Lattice Relaxation

Nuclei in an NMR experiment are held in a sample. The sample in which the nuclei are held is called the lattice. Nuclei in lattice are in vibrational and rotational motion, which creates a complex magnetic field. The magnetic field caused by the motion of nuclei within the lattice is called the 'lattice field'. This lattice field has many components. Some of these components will be equal in frequency and phase to the Larmor frequency of the nuclei of interest. These components of the lattice field can interact with nuclei in the higher energy state, and cause them to lose energy, thereby returning to the lower state. The energy that a nucleus loses increases the amount of vibration and rotation within the lattice resulting in a tiny rise in the temperature of the sample.

10.2.5.2 Spin–Spin Relaxation

Spin-spin relaxation describes the interaction between neighbouring nuclei with identical precessional frequencies but differing magnetic quantum states. In this situation, the nuclei can exchange quantum states; a nucleus in the lower energy level will be excited, while the excited nucleus relaxes to the lower energy state. There is no net change in the populations of the energy states, but the average lifetime of a nucleus in the excited state will decrease. This can result in line-broadening.

10.2.6 The Chemical Shift

This is the phenomenon that occurs in which a specification i.e. a carbon or hydrogen atom, in a given molecule resonates at a slightly different frequency based on its local chemical environment. In other words, the difference between the field necessary for resonance in the sample and in some arbitrarily chosen reference compound is called the 'chemical shift'. For protons, it is usual to refer spectra to tetramethyl silane (TMS_i) with extrapolation to infinite dilution in an inert solvent such as CCl_4. TMS_i gives a sharp resonance line at the high field end of the range of observed proton shifts and therefore, it does not obscure any other proton lines arising from the sample.

The chemical shift is expressed as:

$$\delta = \frac{H_{sample} - H_{TMSi}}{H_1} \times 10^6$$

where H_{sample} and H_{TMSi} are the positions of the absorption peaks for the sample and reference material, respectively in Hz and H_1 is the radiofrequency of the signal used. The chemical shift δ units' is expressed in parts per million.

NMR takes advantage of the phenomenon of the chemical shift so that it allows a chemist to obtain a picture of not only the number of particular atoms present in a molecule, but also an idea of how the atoms are joined together in the molecule.

10.3 Types of NMR Spectrometers

There are two NMR spectrometer designs, continuous-wave (cw), and pulsed or Fourier transform (FT–NMR).

10.3.1 Continuous-wave NMR Spectroscopy

A CW–NMR spectrometer consists of a control console, magnet, and two orthogonal coils of wire that serve as antennas for radiofrequency (rf) radiation. One coil is attached to an rf generator and serves as a transmitter. The other coil is the rf pick-up coil and is attached to the detection electronics.

Since the two coils are orthogonal, the pick-up coil cannot directly receive any radiation from the generator coil. When a nucleus absorbs rf radiation, it can become re-oriented due to its normal movement in solution and re-emit the rf radiation in a direction that can be received by the pick-up coil. This orthogonal coil arrangement greatly increases the sensitivity of NMR spectroscopy, similar to optical fluorescence.

Spectra are obtained by scanning the magnet and recording the pick-up coil signal on paper at the control console.

Continuous-wave NMR spectrometers have largely been replaced with pulsed FT-NMR instruments. However, due to the lower maintenance and operating cost of cw instruments, they are still commonly used for routine ^1H NMR spectroscopy at 60 MHz. Low resolution cw instruments require only water-cooled electromagnets instead of the liquid-He-cooled superconducting magnets found in higher field FT–NMR spectrometers.

10.3.2 Fourier Transform NMR Spectroscopy

Fourier transform NMR spectrometers use a pulse of radiofrequency (rf) radiation which causes nuclei in a magnetic field to flip into the higher energy alignment. Due to the Heisenberg uncertainty principle, the frequency width of the rf pulse (typically 1-10 μs) is wide enough to simultaneously excite nuclei in all local environments. All the nuclei will re-emit rf radiation at their respective resonance frequencies, creating an interference pattern in the resulting rf emission versus time, known as a free induction decay (FID). The frequencies are extracted from the FID by a Fourier transform of the time-based data.

Free induction decay

Since it allows NMR transitions to be observed simultaneously rather than serially, it has increased the sensitivity of NMR by more than a factor of 10 and has reduced the time required to obtain NMR data by a factor greater than 100.

An FT–NMR spectrometer consists of a control console, magnet and a coil of wire that serves as the antenna for transmitting and receiving the rf radiation. Only one coil is necessary because signal reception does not begin until after the end of the excitation pulse. Since the FID results from the emission due to nuclei in all environments, each pulse contains an interference pattern from which the complete spectrum can be obtained. Because of this multiplex advantage, repetitive signals can be summed and averaged to greatly improve the signal-to-noise ratio of the resulting FID.

10.4 Constructional Details of NMR Spectrometer

Figure 10.2 shows the block diagram of a nuclear magnetic resonance spectrometer. It is a complex system integrating several technologies into an analytically powerful, information rich system. The key parts of the system are:

- A magnet, which produces a magnetic field in the range 10,000 to 25,000 gauss,
- Radio-frequency transmitting system,
- The signal amplifier and detector,
- A display device, which may be a recorder or an oscilloscope,
- A non-magnetic sample holder, which holds the sample, and
- Computer workstation which stores and processes the NMR data using complex software to generate spectrum for the same.

The various sub-systems in an NMR spectrometer are shown in Figure 10.3.

10.4.1 Magnetic Field

The magnet used in these instruments may be a permanent magnet or an electromagnet. Alternately, the magnetic field may be produced from superconducting solenoids. An important

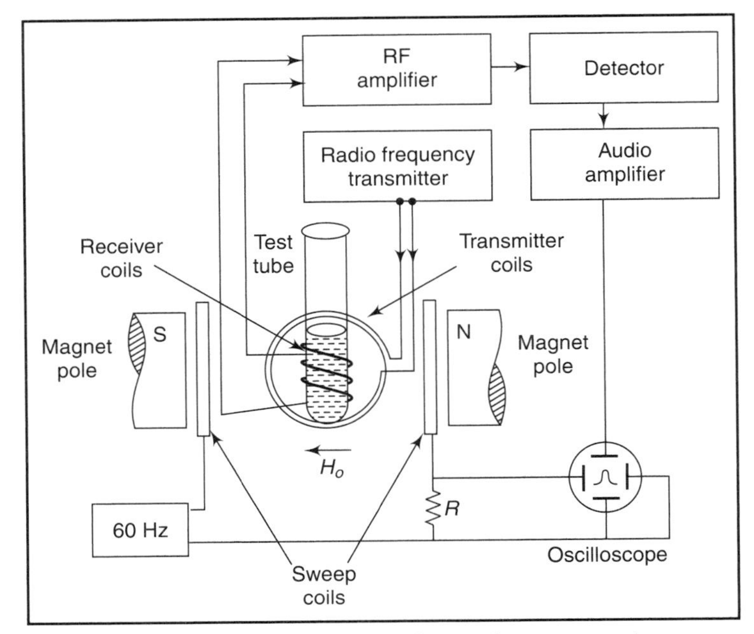

Figure 10.2 :: Block diagram of a nuclear magnetic resonance spectrometer

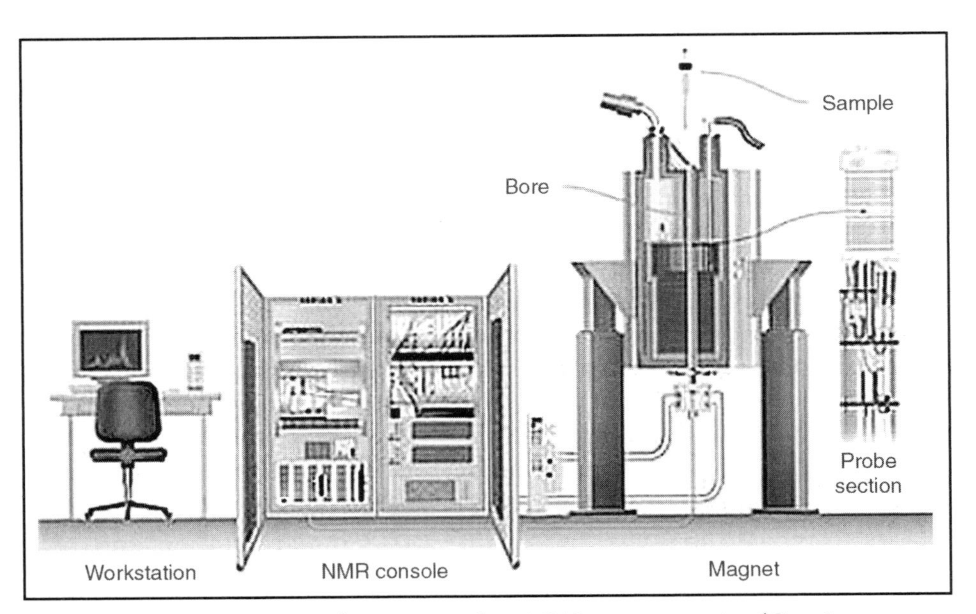

Figure 10.3 :: Various sub-systems of an NMR spectrometer (Courtesy M/s Varian, USA)

requirement of the magnet is that it should be stable and homogeneous. Stability of the magnetic field is achieved by continuously compensating for small rapid fluctuations in magnetic field with coils wound around the pole faces and by controlling the temperature of the magnet to minimize thermal fluctuations. Inhomogeneity is also compensated with small magnetic fields, produced by passing dc current through small electrical coils located on the faces of the magnet, and by spinning the sample to effectively average out field gradients in the direction perpendicular to the spin axis. This, however, produces spinning side bands, due to modulation of the magnetic field. They can be identified from the main signal by changing the spinning rate. The magnetic field must be highly homogeneous in the sample area and the order of homogeneity is 1 part in 10^8. Decidedly, a permanent magnet is less expensive, but it does not allow the observation of the resonance of different nuclei and of a given nucleus, at two different field strengths. In high resolution instruments, it is temperature compensated to ensure the field stability of the magnet. The magnet is placed in a thermostated oven and is surrounded by heavy thermal insulation. Special magnetic and rf shielding minimizes the effects of environmental perturbations, which can disrupt NMR operation.

It is not usually necessary for the magnetic field and the rf frequency to be stable to the degree mentioned above; it is sufficient if their ratio remains constant. The stability in field frequency can be conveniently achieved by continuously adjusting the magnetic field, or the frequency to maintain the resonance condition for a particular sample. A pair of coils located parallel to the magnet faces permit alteration of effective field by a few hundred milligauss without loss of field homogeneity. Generally, the field strength is changed linearly and automatically with time. For a 60 MHz instrument, the sweep generator periodically sweeps the magnetic field in the vicinity of 14,092 gauss, with a sweep of 1000 Hz and some integral fraction thereof.

For the study of nuclear magnetic resonance, there are great advantages in working at the highest possible applied magnetic field strength, especially for the resonances of nuclei other than hydrogen and fluorine. The expected signal-to-noise ratio increases rapidly, as the strength is increased, provided that the resonant frequency does not become so high as to make the problem of coupling radiofrequency power in and out of the sample very difficult. For higher field strengths, it is better to use a superconducting solenoid. Magnets of this kind also have the advantage of quite remarkable stability, when operated in the persistent mode. A typical superconducting solenoid is 26 cm long, with a 3 cm bore and an outside diameter of 8.36 cm. It is wound with Nb-Zr 25 per cent wire and produces a field of 5T (1T = 10,000 gauss) with a current of about 20 A. Permanent magnets can yield maximum magnetic fields, approximately 14,000 gauss, whereas electromagnets can produce fields of up to about 24,000 gauss. Superconducting solenoids have been used to give approximately 70,000 gauss. In case of electromagnets, the pole pieces are about 12 inches in diameter and are spaced about 1.75 inches apart.

Advances in magnet technology have resulted in the availability of commercial high resolution spectrometers with superconducting solenoids producing fields of 12 tesla (500 MHz ^{1}H resonance) with the probability of 14.4 T just around the corner. The principal advantages of these high fields are in the greater dispersion of complex ^1H spectra and the higher sensitivity of the basic NMR technique. However, the higher field strength magnets are of a comparatively high cost.

The strength of a superconducting magnet is typically specified in terms of the resonance frequency for the hydrogen atom expressed in megahertz (MHz). Depending on the magnet, the field strength can range from 200 MHz to 900 MHz. The 900 MHz NMR magnet is 10 times stronger than the most powerful magnetic resonance imaging (MRI) system used in hospitals and 200,000 times stronger than the earth's magnetic field.

10.4.2 The Radiofrequency Transmitter

The radiofrequency transmitter is a 60 MHz crystal-controlled oscillator. The rf signal is fed into a pair of coils mounted at right angles to the path of the field. The coil that transmits the radiofrequency field is made in two halves in order to allow insertion of the sample holder. The two halves are placed in the magnetic gap. For high resolution work, the transmitted frequency must be highly constant to about 1 part in 10^8. The oscillator is of low power, generally of less than 1 W. The basic oscillator (Figure 10.4) is usually crystal controlled, at a fundamental frequency of 15 MHz. It is followed by a buffer doubler, the frequency being doubled by tuning the variable inductance to the second (30 MHz). It is further connected to another buffer doubler, with the connector inductance, tuned to 60 MHz. A buffer amplifier T_4 is provided to avoid circuit loading on the tuned doubler. Precision resistors are used in the transmitter for their low noise characteristics.

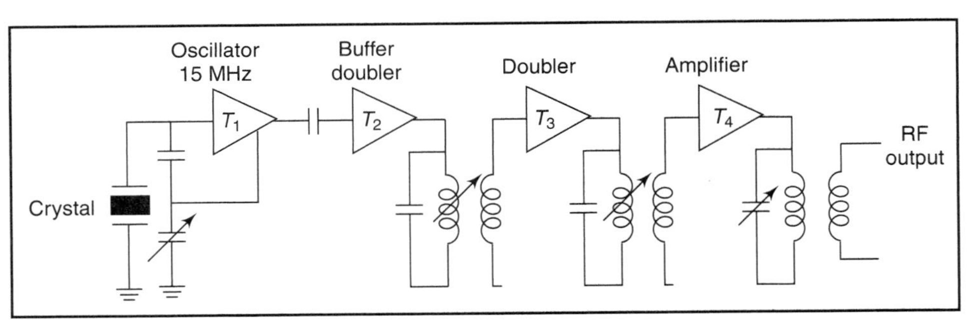

Figure 10.4 :: Transmitter circuit

10.4.3 The Signal Amplifier and Detector

The radiofrequency signal produced by the resonating nuclei is detected by means of a coil that surrounds the sample holder. This coil consists of a few turns of wire and is placed at right angles to the source coil and the stationary field to minimize pick-up from these fields. Even so, the coupling between transmitting and receiving coils cannot be completely eliminated and some leakage is always known to be present. The problem of this coupling is solved to some extent by using devices called paddles, which act as inductors mutually coupled to both receiver and transmitter.

The signal results from the absorption of energy from the receiver coil, when nuclear transitions are induced and the voltage across the receiver coil drops. This voltage change is quite small and must be amplified in a radiofrequency amplifier before it can be displayed.

Referring to Figure 10.5, after impedance matching in a transformer T_1, the input rf signal is amplified in a three-stage low noise cascade amplifier. The output of the last rf stage is applied to the AGC (automatic gain control) and spectrum detectors. The AGC voltage is fed back to the previous stage through an AGC amplifier, which is kept cut-off due to a delaying bias, until a pre-determined signal level is reached. This delayed AGC allows maximum gain on weak signals and prevents over-driving on strong signals.

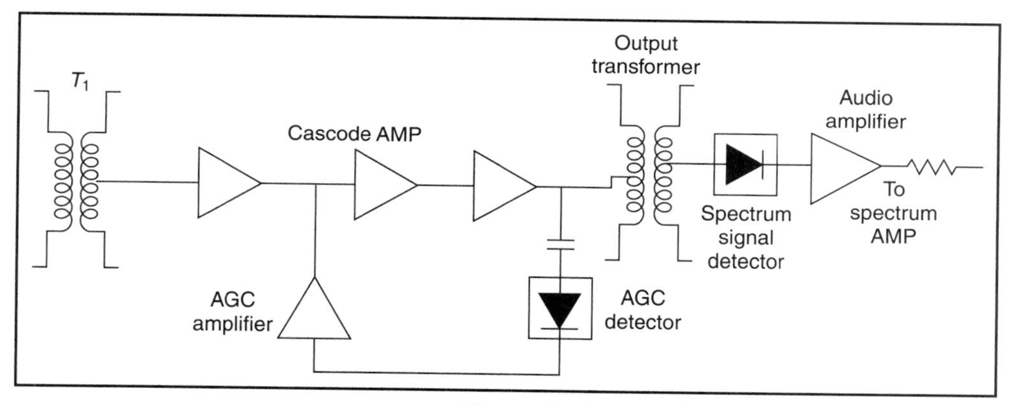

Figure 10.5 :: Receiver and detector circuit

10.4.4 The Display System

The detected signal is applied to the vertical plates of an oscilloscope to produce the NMR spectrum. The total amplification required is usually of the order of 10^5. The spectrum can also be recorded on a chart recorder. NMR spectrometers have built-in electronic integrators for measuring the relative areas under the peaks. The integrated spectrum is a step function, with the height of each step being directly proportional to the area under the peak corresponding to the step.

10.4.5 Recording Unit

The recorders with NMR spectrometers are usually the self-balancing null-type potentiometric recorders, having an FET chopper amplifier in the input stage. Phase-detected NMR signals are given to the amplifier. The chopper is driven at the line frequency and it produces a square wave output signal, which is phase compared to the input signal and applied to a source follower. The signal is amplified and applied to the drivers and power amplifiers. The power amplifier stage obtains filtered voltage from a rectifier circuit to establish a phase relationship with the chopper output signal and the power line frequency. The output of the drivers is zero when no NMR signals are present. A signal input causes a square wave at the chopper and depending upon the polarity, it results in a lead or lag voltage with reference to the power line frequency. The servometer then rotates in a direction, so as to cause the difference voltage between NMR and the reference signal to become zero. The horizontal movement of the pen carriage is controlled by synchronous sweep motor. A sweep potentiometer is turned by the gear train in synchronism with the horizontal movement of the pen carriage. Variable sweep width can be selected with a manual switch, by controlling the amplitude of the dc voltage applied to the probe sweep coils from the sweep potentiometer.

The chart size in NMR spectrometer is usually around 8×11 inches, with its abscissa calibrated in Hertz. The recorder has calibrated sweep width ranges and sweep speeds, which can be selected according to requirement.

10.4.6 Use of Computers with NMR Spectrometers

Computer-equipped NMR spectrometers have been used for a variety of tasks. Simple time-averaging for signal-to-noise enhancement has been extremely helpful in continuous wave NMR studies. It only requires relatively simple hardware. With Fourier transform NMR becoming more common, general purpose digital computers are required for data acquisition and data display. It is possible to place under computer control such functions as pulse timing, delay and acquisition timing, digitization rate, filter bandwidth, transmitter and decoupler offsets, receiver gain, noise bandwidth, plotter and pulse sequence. This leaves only a few essential manual adjustments, such as establishing the NMR lock for field frequency ratio stabilization and trimming up the field homogeneity controls. Varian CFT-20 Fourier Transform NMR Spectrometer provides automated unattended performance. This increases productivity and also improves signal-to-noise ratio.

In modern NMR spectrometers, the computer workstation and complex software directs the NMR experiment from start to finish. NMR signals are subjected to a complex digital signal processing algorithm, including the Fourier Transform, to convert the NMR information into a form that is easily interpreted by the end-user. The signals are displayed as a series of peaks or spectrum, on the workstations monitor (Hawkes, 1984).

Many of the peripheral devices once found only on larger mainframe computers are now standard features on modem NMR spectrometers. The data are usually stored on a disc and the processed data are plotted on a paper or displayed on a graphics display terminal.

10.4.7 The Sample Holder

The sample holder for NMR studies consists of a glass tube, generally of 0.5 mm outer diameter. Microtubes for smaller sample volumes are also available. The sample is invariably in the liquid form. If case studies are to be made on solid or gaseous samples, the solids may be studied above their melting point, and gaseous samples below their liquification point. Samples can be less than 1/1000 cubic inch of the gas, liquid or solid.

The sample holder is placed in a sample probe, which also contains the sweep source and detector coils. This ensures reproducible positioning of the sample with respect to these components. The sample probe is also provided with an air-driven turbine for rotating the sample tube along its longitudinal axis at several hundred rpm. This rotation averages out the effects of inhomogeneities in the field and provides better resolution.

Most high resolution spectrometers incorporate high sensitivity probes which are tunable over a significant frequency range (two or more octaves), thus simplifying the observation of resonances from the less common nuclei (other than ^{1}H and ^{13}C).

For continuous monitoring of a chemical process, it is necessary to use a special sample cell situated in the detector coil of the spectrometer, through which the solution is made to flow, continuously. It incorporates a flow system which uses the magnet of the spectrometer as the polarizing field. The sample cell is made from a soda glass tube of diameter 0.180 inch and is locked at the centre of the polarizing coil. The latter is made of nylon tube of inner diameter 0.20 inch.

The detector coil made from four turns of 40 SWG copper wire is wound directly on the outside of the cell.

In NMR studies, it is often necessary to determine the behaviour of the measured parameters of a sample as a function of temperature. Often the range from room temperature down to the boiling point of liquid nitrogen is appropriate for these studies. This necessitates that sample temperatures be kept stable for several hours, while allowing rapid changes to be made when passing a cooled gas over the sample or using a controlled heat leak from the sample to a liquid nitrogen reservoir.

In the Varian T60A NMR spectrometer, the sample tube, with spinner attached, is simply dropped into a chimney-like outlet and floats down on to a cushion of compressed air, eliminating the possibility of damage to inner components of the probe. Spinning speed is adjustable and the speed can be optically monitored and displayed on a directly calibrated meter.

NMR Probe: In practice, the antenna that provides the RF link between the sample and the instrument electronics is called the 'NMR' probe. Inserted in the magnet, the probe holds the sample at the centre of the magnetic field. It bombards the sample with RF energy and then receives the very weak RF responses from the sample, which it sends to the receiver.

M/s Varian have introduced Cold Probes that offer increase in sensitivity than that of standard probes. The increased sensitivity allows the user to either collect data over 10 times faster or to use 3–4 times more dilute samples. The probe can be converted between the tube and flow mode.

The internal electronics of the probe are chilled to approximately -250°C in the cold probes (*Custserv@varianinc.com*).

10.5 Varian T-60A NMR Spectrometer

The Varian T60A NMR spectrometer is designed for use with protons only over the limited chemical shift range of 2000 Hz below and 500 Hz above tetramethyl silane at 60 MHz. The crystal-controlled 60 MHz transmitter is coupled to the probe and sample receiver through the directional coupler. The sine wave from the 5 kHz oscillator is applied through the field modulator to the ac sweep coils, thereby modulating the H_0 field at a 5 kHz rate. The signal in the sample receiver coil would have a frequency of 60.005 MHz, which is both amplitude and phase modulated by the nuclear resonance being observed. After detection in the receiver, which removes the 60 MHz component, the signal is amplified in the spectrum amplifier and then compared in the phase detector, with a reference signal derived from the 5 kHz oscillator. Since an increase in rf power results in phase shift on NMR displays, the reference signal is made continuously variable in the phase shifter, which supplies a defined baseline for display. The resulting phase corrected absorption signal is applied to the dc amplifier, noise filter and recorder for spectrum display or to the integrator and recorder to obtain the corresponding integral.

An air turbine sample spinner is used to average the effects of field inhomogeneities, which may exist in the plane perpendicular to the tube axis, thereby improving the system resolution. The spinner speed tachometer samples the spinner speed and the rate is indicated on a front panel meter.

The 5 kHz signal which emerges from the receiver–detector can be amplified and phased, so that when applied to the modulation coils in the probe, a self-sustaining oscillation is established in the loop comprised of ac sweep coils, sample tube, receiver coil, receiver and field modulator. Such a closed feedback loop is called a nuclear sideband oscillator (NSBO) and the spectrometer is said to be locked to a nuclear resonance line in the sample. It is convenient to switch to NSBO operation when adjusting the magnetic field homogeneity for optimum resolution. After homogeneity has been optimised, the NSBO loop is opened and the spectrometer returned to normal operation.

10.6 Sensitivity Enhancement for Analytical NMR Spectroscopy

The sensitivity of an instrument is a measure of its ability to differentiate signal from the surrounding noise. It is usually measured as the ratio of peak signal amplitude to rms noise.

Sensitivity = S/N = Peak signal height/rms noise,

where rms noise = peak-to-peak noise/2.5 (i)

Equation (i) slows that the sensitivity may be improved either by increasing the signal amplitude or by reducing the observed noise. Both these are considered in the initial design of a NMR spectrometer. However, special operating techniques allow the NMR operator to optimise them for a particular sample and spectrometer system.

There are three operating techniques for sensitivity enhancement which, are: (i) optimization of sample volume, (ii) optimization of instrumental parameters, and (iii) time averaging. The first two enhancement techniques (optimization of sample volume and spectral parameters) can often enhance the sensitivity by a factor of 10 over the normal operating conditions. Further the use of signal averaging with a computer can achieve a further increase of ten within reasonable time limits. This can significantly increase the applicability of analytical NMR to the investigation of small samples to give the NMR operator the best possible use of spectrometer.

10.6.1 Optimization of Sample Volume

Better sensitivity can be achieved with good resolution by concentrating the sample inside a specially designed NMR microcell with a volume, which approximates the active volume of the receiver coil and with a shape, which eliminates the effects of the magnetic field distortion. The most common shapes used are a spherical cell with a 4 mm diameter and a 1 mm capillary cell centred along the axis of the coil. These cells can be positioned inside the standard 5 mm cylindrical tube. Preferably, a liquid of similar susceptibility is placed outside the cell to minimize movement of the microcell within the sample tube.

10.6.2 Optimization of Instrumental Parameters

It is obvious that noise may be reduced by proper choice of filter setting. Filtering operates on the basis that the frequency spectrum of the noise generally occupies a wider band than the signal spectrum. Therefore, suppression of frequencies, which contain no signal, increases the effective S/N ratio.

The resonant nuclei in the active region of the receiver coil absorb rf energy over a small range of applied field-to-frequency ratios because of the slight differences in their molecular magnetic environments. In a swept NMR experiment, either the applied magnetic field or frequency is varied to make the different types of protons resonate sequentially. The rate of change of the field or frequency may be adjusted by the operator for maximum sensitivity. Sensitivity can be enhanced by maximizing signal amplitude. The amplitude of the NMR signal caused by the absorption of rf energy at the resonance frequency, depends upon the power of the rf energy applied. It will, however, depend upon the sweep rate selected. Therefore, optimization of signal amplitude with rf power should only be considered for a given sweep rate.

10.6.3 Use of Signal Averaging Technique

The signal averaging technique is based on the principle that the signal, if added coherently, will increase linearly with the number of scans N, while the noise being random, will add as the square root of the number of the scans. Thus, the sensitivity increases by the square root of the number of scans ($N^{1/2}$). In order to have a long measuring time for increasing sensitivity, it is better to have several fast scans than a slow single scan.

Signal averaging basically involves:

- A system, which must provide to repetitively scan the spectral region of interest,
- Some storage device for storing the spectral information, and
- A system to coherently add the individual spectra.

These are usually accomplished by a small computer (multi-channel pulse height analyzer), suitably modified to provide an NMR sweep and coherent spectral addition. These special purpose computers are commercially available from manufacturers of NMR spectrometers.

10.6.4 Spin Decoupler

Nuclear magnetic double-resonance

A spin decoupler permits spin-coupled interactions between nuclei to be eliminated through the use of a technique called nuclear magnetic double-resonance. In this technique, specific regions of the spectrum are irradiated by a strong component of radiofrequency power or rf side-band power, causing the multiplet patterns from spin coupled resonances in other regions to coalesce into less complex patterns more easily assignable. In principle, double-irradiation helps to solve problem introduced by complicated proton couplings. The second radio-frequency field is adjusted to the resonant condition for the group, whose coupling is to be eliminated. Under these circumstances, the proton being split sees only one equivalent state and single peak results.

In addition to the usual H_1 rf field used to examine the nuclear magnetic resonance response to the sample, a second rf field H_2 is introduced simultaneously into another region of the proton spectrum. The spectrum is recorded by sweeping the main magnetic field H_o and the H_2 field, while the H_1 field is held constant. H_1 modulation of the H_0 field generates the side-bands for the observing field. For frequency sweep, a voltage proportional to the magnetic field is applied to a voltage-controlled oscillator, which supplies the decoupling frequency. The magnitude and frequency of the H_2 are controllable.

Figure 10.6 shows a block diagram of spin decoupler arrangement. The decoupling frequency to the field modulator is supplied by 5 kHz voltage controlled oscillator.

The oscillator is a Wien Bridge Oscillator, whose gain stability is provided by a thermistor in the feedback loop. This frequency is amplified and applied to the ac sweep coils to generate the H_2 or decoupling field. The decoupling power level is adjusted at the oscillator output. In frequency sweep operation, tracking of the oscillator with the sweep is accomplished through the frequency control circuit. As the dc component of the field is swept, a differential voltage is

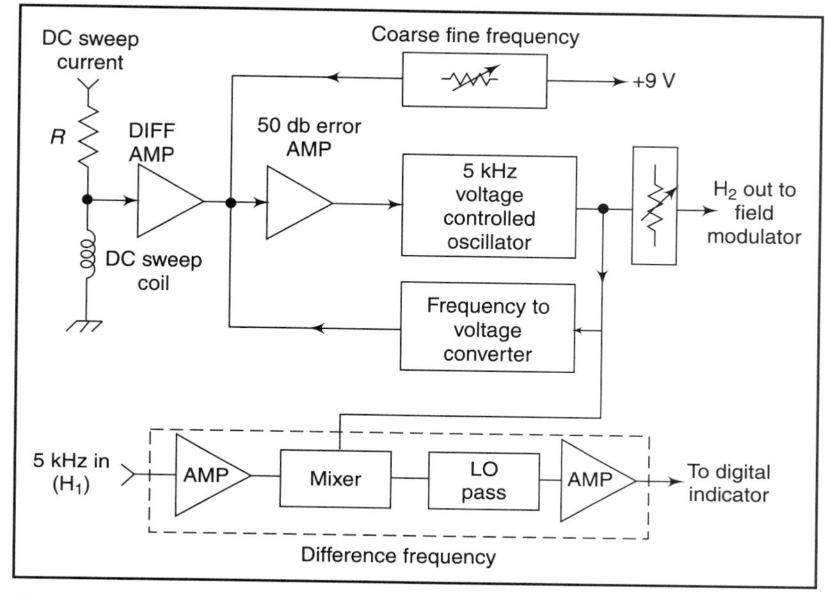

Figure 10.6 :: Block diagram of spin decoupler (Courtesy M/s Varian, USA)

developed across the resistor R placed in series with the sweep coils. The differential voltage is amplified in the differential amplifier and applied to the error amplifier. The error amplifier receives input from three different sources; the frequency control setting, the differential voltage from the dc sweep coils, and the feedback voltage from the frequency-to-voltage converter.

The frequency control sets the dc operating level that produces the desired frequency from the voltage-controlled oscillator. The dc sweep voltage from the differential amplifiers adds to the operating level and the oscillator follows the sweep. The frequency of the oscillator is sensed by the frequency-to-voltage converter, which applies an error signal to the error amplifier when the oscillator drifts or overshoots. Frequency-to-voltage converter provides a direct conversion of a sinusoidal signal to a dc signal. In this circuit, the input signal is squared by a Schmidt trigger circuit and differentiated to trigger a monostable multivibrator. The pulses from the monostable multivibrator are applied to a low-pass RC filter, which smoothens the pulses to a dc voltage.

A difference frequency circuit is included to provide a read-out frequency for the digital indicator. This circuit provides a means of mixing two audio signals and a filter to recover the low-frequency component. The read-out frequency is the difference frequency between the 5 kHz spectrometer control oscillator (H_1) and the spin decoupler oscillator frequency (H_2). The Spin Decoupler module temperature is maintained to ± 1°C by a temperature controller circuit. This circuit is a dc amplifier, which compares a dc reference voltage to the voltage change produced by a thermistor mounted in the Spin Decoupler oven. The output of the dc amplifier is applied to a power transistor, which controls the power supplied to the oven heaters.

10.7 Fourier Transform NMR Spectroscopy

In conventional NMR spectrometers for high resolution studies, the spectra are scanned by sweeping the frequency or the field through the region of NMR absorption. From the point of view of sensitivity, this is an inefficient mode of excitation, since only a very narrow band of frequencies is contributing a signal at any one time. Multi-channel excitation and detection would be expected to improve the sensitivity considerably. A spectrometer consisting of a large number (M) of transmitter frequencies, each matched by a suitable receiver channel can be built. However, this method proves to be highly uneconomical as M increases. The Fourier Transform accomplishes the same result in a much more satisfactory manner. In this method, a strong radiofrequency pulse excites the entire range of precision frequencies, and these frequencies can be detected in the receiver. Under certain conditions, the Fourier Transform of this precision signal is identical with the steady-state slow passage spectrum. The transformation process plays the role of a multi-channel receiver, eliminating the need for a cumbersome array of narrow-band filters and detectors.

Any complex waveform can be converted into a spectrum of frequencies by Fourier transformation. In NMR, the waveform in question is the superposition of a set of nuclear precision frequencies, with amplitude decaying due to relaxation and field inhomogeneity. Fourier transformation can be accomplished by using a spectrum analyzer. However, the spectrum would show only the absolute value of the XY component of magnetisation, with no account taken of the phase information, which often results in distortion of the lines. The transformation is carried out most satisfactorily on a digital computer. Usually dedicated computers are used for this purpose.

Sensitivity enhancement is the major attraction of the Fourier transform technique. Thus for proton NMR, it is possible to consider samples with an order of magnitude that is less concentrated than the previous limit, and this has great importance in the biochemical field. The XL-100 Pulsed-Fourier Transform NMR spectrometer from Varian is a high resolution 23.5 k gauss spectrometer, which can be operated in the frequency range from 6 to 100 MHz. This instrument is frequently designed around a fundamental clock. A 15.4 MHz crystal oscillator generates the deuterium resonance frequency at 23.5 k gauss. The deuterium resonance signal is then used to lock the magnetic field to the clock frequency. The choice of a 15 inch or 12 inch magnet system permits to accommodate a wide size range of sample tubes. This facilitates optimization of sensitivity versus the availability of samples. For Fourier transform, the spectrometer is provided with programmable rf pulse hardware and a digital computer programmed to control, acquire and transform the spectral data.

Chapter 11

Electron Spin Resonance Spectrometers

11.1 Electron Spin Resonance

Electron spin resonance involves detecting the detection of a physical phenomenon of absorption of electromagnetic radiation in the microwave region by paramagnetic species that are subjected to an external magnetic field. It is the study of magnetic dipoles of electronic origin by applying, usually, fixed microwave frequencies to a sample residing in a varying magnetic field. Also known as 'electron paramagnetic resonance' (EPR) spectroscopy, it is a valuable research and analytical tool in chemistry, physics, biology and medicine. It is used in the study of molecular structures, reaction kinetics, molecular motion, crystal structure, electron transport and relaxation properties.

The phenomenon of electron spin resonance is based on the fact that an electron possesses a spin, and associated with the spin, there is a magnetic moment, the value of which is called the Bohr magneton. The atoms, ions or molecules having an odd number of electrons exhibit characteristic magnetic properties, which arise from the spinning or orbiting action of the unpaired electrons about the nucleus. When a strong magnetic field is applied to the unpaired spins of an electron, the electrons will be split into two groups. In one group, the electron dipoles or magnetic moments of the electrons are aligned either parallel or anti-parallel to the direction of the external magnetic field. The electrons will precess about the axis of the magnetic field at a frequency that is proportional to both the applied magnetic field and the electron magnetic moment.

If a second, weaker radiofrequency alternating magnetic field, having the frequency of precession of the electron is applied at right angles to the fixed magnetic field, resonance occurs. At resonance, the absorption of energy from the rotating field causes the spin of the electrons to flip from the lower energy level to the higher level. The two levels are separated by:

$$E = h\upsilon = 2\ \mu H,$$

when h is Planck's constant and υ is the frequency. In comparison to NMR, the electron has a much smaller mass and larger magnetic moment than a proton. For a given magnetic field, the precession frequency is, therefore, much higher. For a free electron, the frequency of absorption is given by

$$\upsilon = 2\mu H/h$$
$$= (2\mu/h) \times H = (2.8026 \times 10^6) \times H$$

In a field of 3400 gauss, the precession frequency is approximately 9500 MHz.

In actual practice, the radiofrequency is in the microwave region and is held at a certain constant value; and the magnetic field strength is varied to obtain conditions wherein resonance occurs. The incident radiation is absorbed by the electrons at the lower energy level and they jump into a higher energy state.

In general, if n_1, the population of the ground state, exceeds n_2, the population of the excited state, a net absorption of microwave radiation takes place. The signal would be proportional to the population difference $(n_1 - n_2)$.

By the Boltzmann distribution law, the population ratio in the two states is given by:

$$n_1/n_2 = e^{-2\mu H/kt},$$

where k is the Boltzmarnn constant. The sensitivity of measurement is greatly enhanced by using a high magnetic field.

The expression tells us that ESR sensitivity (net absorption) increases with decreasing temperature and with increasing magnetic field strength. Since field is proportional to microwave frequency, in principle, sensitivity should be greater for K-band or Q-band or W-band spectrometers than for X-band. However, since the K-, Q-, or W-band waveguides are smaller, samples are also necessarily smaller, usually more than cancelling the advantage of a more favorable Boltzmann factor.

In most of the substances, chemical bonding produces paired electrons, as they would be either transferred from one atom to another to form an ionic bond, or electrons are shared between different atoms to form covalent bonds. The magnetic moments and spins of paired electrons point in opposing directions, with the result that there is no external spin paramagnetism. However, in a paramagnetic substance having an unpaired electron, resonance occurs at definite values of the applied magnetic field and incoming microwave radiation. The deviation from the standard behaviour of the unpaired electron due to the presence of magnetic fields in its surroundings, gives knowledge about the structure of the substance under study.

The ESR measurements are made by placing the sample under study in a resonant cavity positioned between the pole pieces of an electromagnet. The microwave frequency is set to a matched condition with the help of a tuning device. The magnetic field is varied to bring about resonant conditions and the microwave energy absorbed by the sample is plotted on a recorder. The spectrum is then analyzed to determine the behaviour and mechanisms associated with the unpaired electrons interaction with the external magnetic field and its environment.

For routine investigations with the ESR spectrometer, it is a general practice to employ an X-band frequency (8.5–10 GHz). This cannot be generated by using ordinary vacuum tubes, which is why special tubes capable of operating at microwave frequencies are employed.

Waveguides

Microwaves are normally handled using waveguides designed to transmit over a relatively narrow frequency range. Waveguides look like rectangular cross-section pipes with dimensions of the order of the wavelength to be transmitted. As a practical consideration, waveguides cannot be too big or too small—1cm is a bit small and 10 cm a bit large,—the most common choice, called X-band microwaves, has a wavelength (λ) in the range 3.0–3.3 cm (frequency 9–10 GHz). Although X-band is by for the most common, ESR spectrometers are available commercially in several frequency ranges as given in Table 11.1.

Table 11.1 :: Frequency Bands ESR Spectrometers

(*www.chm.bris.ac.uk/emr/phil/phil_1/p-1.html*)

Band	Frequency GHz	Wavelength λ, cm	B Magnetic Field Tesla
S	3.0	10.0	0.107
X	9.5	3.15	0.339
K	23	1.30	0.82
Q	35	0.86	1.25
W	95	0.315	3.3

Under ideal conditions, a commercial X-band spectrometer can detect the order of 10^{12} spins (10^{-12} moles) at room temperature. By ideal conditions, we mean a single line, of the order of 0.1 G wide; sensitivity goes down roughly as the reciprocal square of the line width. When the resonance is split into two or more hyper-fine lines, sensitivity goes down still further. Nonetheless, ESR is a remarkably sensitive technique, especially as compared with NMR.

The limitations of triode, pentode and similar tubes that arise at very high frequencies as a result of transit time effects, are avoided by employing types of tubes which make the use of transit time in achieving their normal operation. One such tube is the reflex klystron or reflex oscillator, as it is sometimes called. It requires only a single resonant cavity. Since it has an efficiency of only a few per cent, the reflex klystron is essentially a low-power device, typically being used to generate 10 to 500 mW. The reflex klystron is particularly satisfactory for use in the frequency range 1000 to 25000 MHz.

11.2 Basic ESR Spectrometer

Although many spectrometer designs have been produced over the years, the vast majority of laboratory instruments are based on the simplified block diagram shown in Figure 11.1. Microwaves

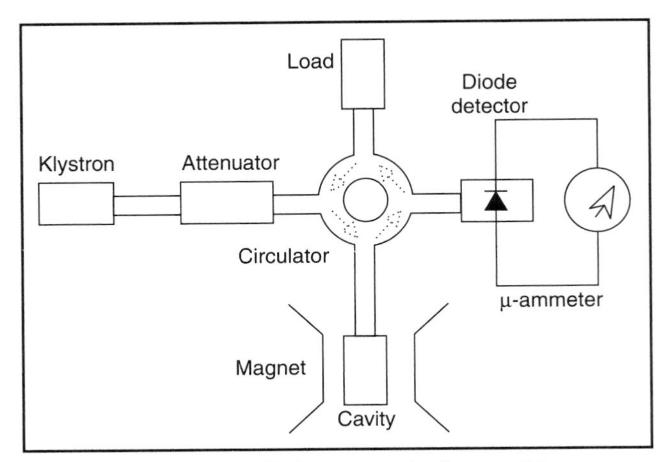

Figure 11.1 :: Block diagram of a basic ESR spectrometer

are generated by the klystron tube and the power level adjusted with the attenuator. The circulator routes the microwaves entering from the klystron towards the cavity where the sample is mounted.

Microwaves reflected back from the cavity are routed to the diode detector, and any power reflected from the diode is absorbed completely by the load. The diode thus produces a current that is proportional to the microwave power reflected from the cavity. Thus, in principle, the absorption of microwaves by the sample is detected by noting a decrease in current in the microammeter. In practice, such a dc measurement is too noisy to be useful.

The signal-to-noise ratio can be improved by introducing small amplitude field modulation. An oscillating magnetic field is superimposed on the dc field by means of small coils, usually built into the cavity walls. When the field is in the vicinity of a resonance line, it is swept back and forth through part of the line, leading to an ac component in the diode current. This ac component is amplified by using a frequency-selective amplifier, thus eliminating a great deal of noise. Thus the detected ac signal is proportional to the change in sample absorption.

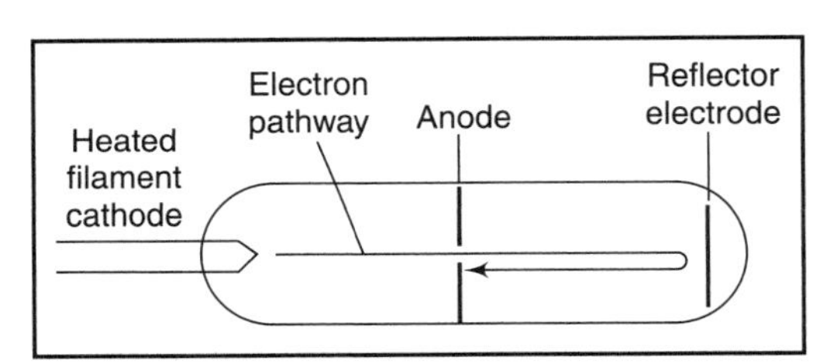

Figure 11.2 :: Schematic diagram of a microwave-generating klystron tube

Klystron tube

A schematic drawing of the microwave-generating Klystron tube is shown in Figure 11.2. There are three electrodes: a heated cathode from which electrons are emitted, an anode to collect the electrons, and a highly negative reflector electrode, which sends those electrons which pass through a hole in the anode back to the anode.

The motion of the charged electrons from the hole in the anode to the reflector and back to the anode generates an oscillating electric field and thus electromagnetic radiation. The transit time from the hole to the reflector and back again corresponds to the period of oscillation. Thus the microwave frequency can be tuned, over a small range by adjusting the physical distance between the anode and the reflector or by adjusting the reflector voltage. In practice, both methods are used: the metal tube is distorted mechanically to adjust the distance as a coarse frequency adjustment and the reflector voltage is adjusted as a fine control.

The sample is mounted in the microwave cavity as shown in Figure 11.3. The cavity is a rectangular metal box, exactly one wavelength in length. An X-band cavity has dimensions of about 1 × 2 × 3 cm. The electric and magnetic fields of the standing wave are also shown in the figure. Note that the sample is mounted in the electric field nodal plane, but at a maximum in the magnetic field.

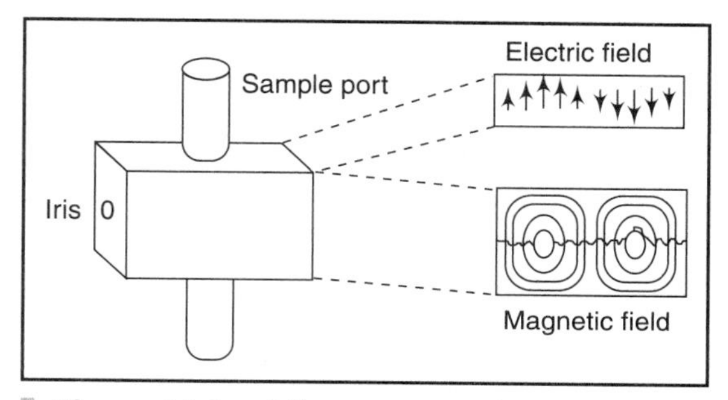

Figure 11.3 :: Microwave cavity

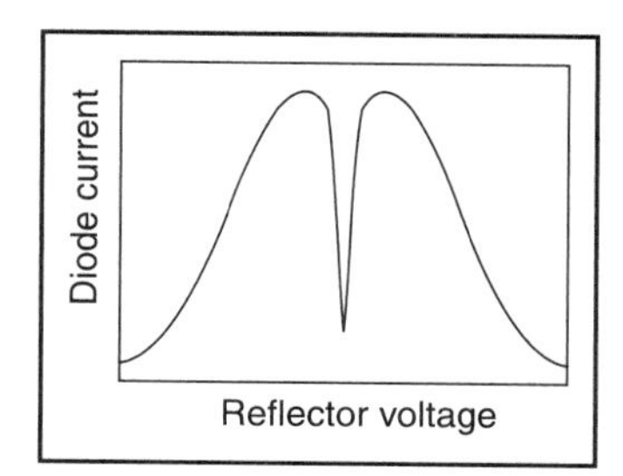

Figure 11.4 :: Klystron mode and cavity dip

Since the cavity length is not adjustable but must be exactly of one wavelength, the spectrometer must be tuned such that the klystron frequency is equal to the cavity resonant frequency. The tune-up procedure usually includes observing the klystron power mode, that is, the klystron reflector voltage is swept, and the diode current is plotted on an oscilloscope or other device. When the klystron frequency is close to the cavity resonant frequency, much less power is reflected from the cavity to the diode, resulting in a dip in the power mode as shown in Figure 11.4. The cavity dip is centred on the power mode using the coarse mechanical frequency adjustment with the reflector voltage used to fine tune the frequency.

11.3 Detailed Description of an ESR Spectrometer

Figure 11.5 shows the block diagram of an ESR spectrometer. The sample is irradiated by microwave energy from the klystron in the microwave bridge. A klystron normally operated at 9.5 GHz, generates a microwave field. The magnetic field at the sample is modulated at 100 kHz. The klystron output passes through an isolator, a power leveller and a directional coupler. The field is applied to the resonant cavity, which is connected to one arm of the microwave bridge. During the magnetic field scan, when field intensity reaches the value required to induce electron spin resonance in the sample, a change occurs in the amount of microwave energy absorbed by the sample. This causes a change in the microwave energy reflected from the cavity. The reflected microwave energy, which is modulated at the field modulation frequency, is directed to the

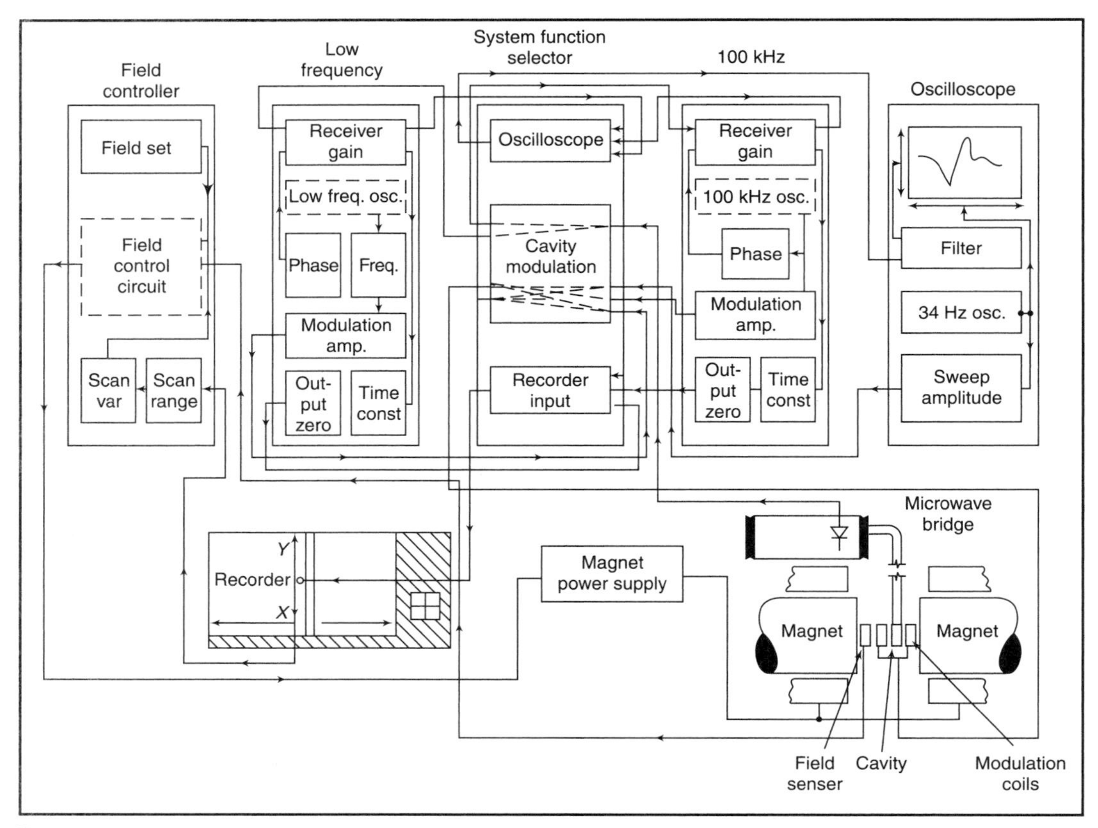

Figure 11.5 :: Block diagram of an ESR spectrometer (Courtesy M/S Varian, USA)

detector crystal. The detector is usually a silicon tungsten crystal rectifier. After detection at the crystal, the resulting 100 kHz, which contains the ESR information, is amplified in the pre-amplifier circuit and applied to the receiver section of the 100 kHz modulation unit. The amplified signal is phase-detected to obtain the spectrum, which appears as a deflection on the Y-axis of the recorder. The field scan potentiometer is linked mechanically to the recorder X-axis. The X-axis is calibrated in gauss. The ESR signal may also be applied to the oscilloscope for visual display. The field controller accurately controls the magnetic field to the desired set value.

The important operating parameters of a typical ESR spectrometer (Figure 11.6) are:

- Operating frequency — 8.8 to 9.6 GHz
- Sensitivity — 5×10^{10} H* Spins
- Modulation frequencies — 1000 kHz, 270 Hz, 35 Hz
- Modulation amplitude range — 5 mG to 40 G in steps
- rf power to cavity — 200 mW calibrated over full frequency range ±15 mW

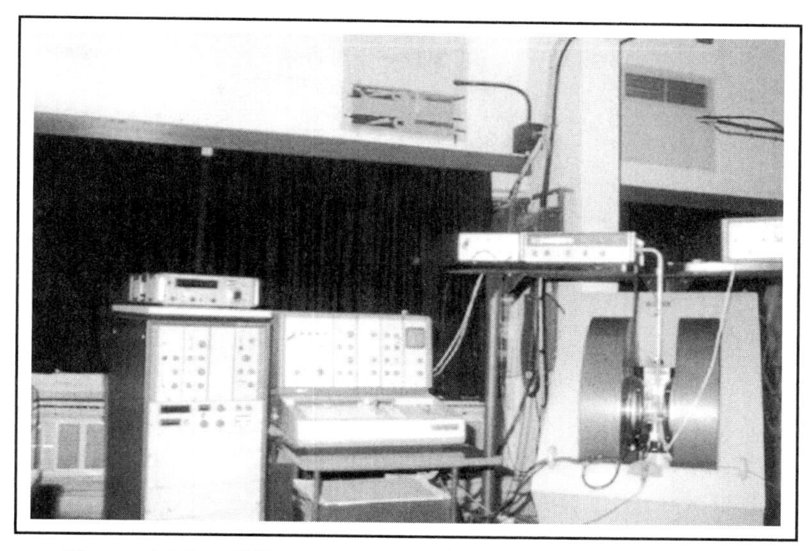

Figure 11.6 :: ESR spectrometer (Courtesy M/s Varian, USA)

• Field scanning rates	1/2, 1, 2, 4, 8, 16, min; 1/2, 1,2 4, 8, 16 h.
• Field scanning ranges	200 mG to 10 kG or 20 kG
• Receiver time constants	0.003 to 100 s
• Stability of field	With 1 ppm of set field or 3 mG, whichever is greater
• Oscilloscope sweep width	0.2 to 40 G (34 Hz sawtooth)
• Magnetic field homogeneity	6-inch magnets: within 50 mG at 3,400 G
	9-inch and larger magnets: 15 mG at 3,400 G

*H is defined as the signal linewidth in gauss at half-maximum absorption with one second integration time, the sample having negligible dielectric loss.

11.3.1 The Magnet and the Magnetic Field Controller

The magnet used in the ESR spectrometers is usually of the electromagnet type. It provides a homogeneous magnetic field, which can be varied from 200 mgauss to 20 kgauss, calibrated in steps. Stability of 1 part in 10^6 is satisfactory for adequate resolution of ESR spectra.

The magnetic field controller provides direct control and regulation of the magnetic field in the air gap. It is an ac carrier type servo system that accurately controls the magnetic field. Figure 11.7 shows the schematic diagram of the magnetic field controller in a spectrometer.

The driver amplifier supplies a 30 mA rms, 1230 Hz exciting current through the field set reference resistor (R), the Hall effect magnetic field sensor element and the primary of the scan voltage transformer (T_4). A field reference voltage is developed across R. Similarly, a field scan

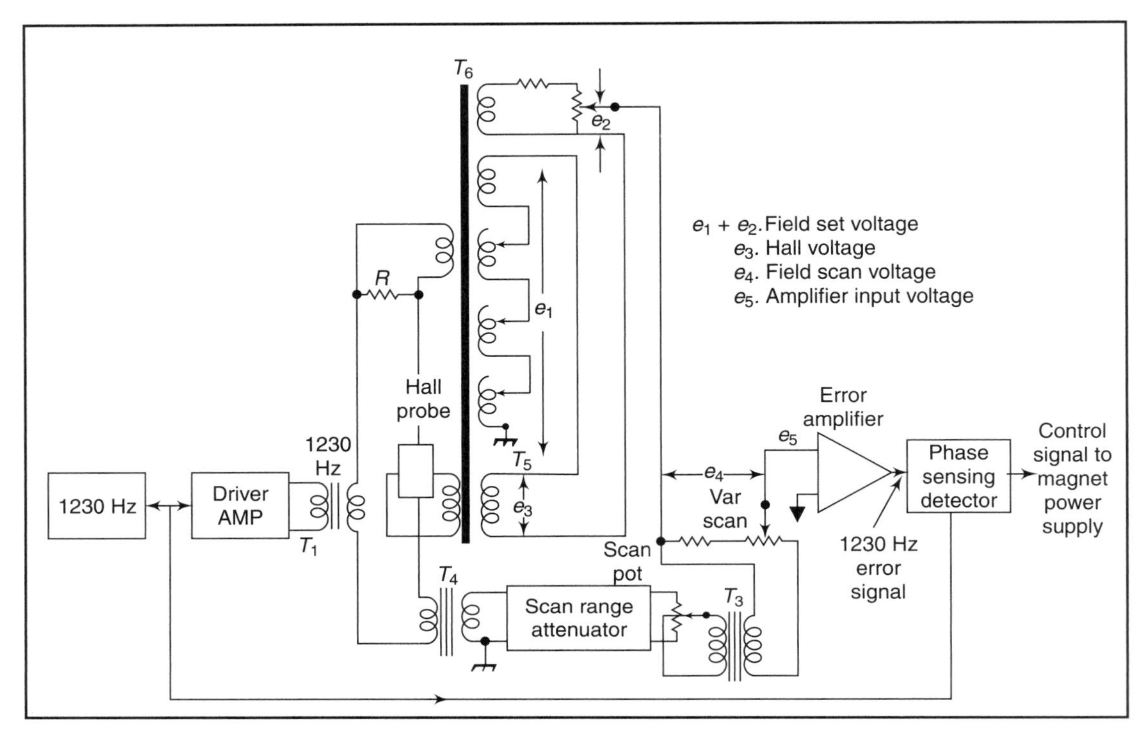

Figure 11.7 :: Schematic diagram of magnetic field controller

reference voltage is developed across the constant input impedance of scan range attenuator by the transformed reference current in the secondary winding of T_4.

The output of the summation circuit (i.e. input to the error amplifier) is the algebraic sum of the output of the Hall-effect field sensor (transformed by T_5) added to the field set and field scan voltage. The sum of these three voltages is amplified by the error amplifier and applied to the input of the phase-sensitive detector. When the output of the Hall-effect field sensor is less than the sum of the field set and field scan voltages, the output of the phase-sensitive detector is of a polarity which turns on the magnet power supply, to increase the magnet current and thus the magnetic field. In a similar manner, when the output of the Hall-effect field sensor exceeds the sum of the field set and field scan voltages, the current through the magnet winding is decreased and the magnetic field is decreased.

The field scan potentiometer is centre-tapped and the wiper is driven by the recorder's horizontal axis-drive mechanism. The voltage output of the scan potentiometer acts to oppose the field set voltage, when the recorder pen is in the left-half of its operating range, but adds to the field set voltage, when the recorder pen is moved into the right half of the recorder chart.

- The 1230 Hz oscillator is a stable Wein bridge oscillator, whose output level is internally regulated.
- The error amplifier is a high gain stable amplifier with a pass-band centred near 1230 Hz. It has a phase shifting potentiometer, which permits the phase to be adjusted, so that the signal component of the amplified error signal is exactly in-phase and the spurious quadrature component of the amplified error signal is exactly out-of-phase, so that the latter is completely rejected by the phase sensitive detector.
- The phase-sensitive detector converts the amplified 1230 Hz output signal of the summation circuit into dc control voltages to drive the magnet power supply.
- The Hall probe is maintained at a constant temperature. The temperature control is maintained by controlling current through a heater, sensed by a thermistor.
- The magnet power supply provides controlled dc current to the low impedance electromagnet. The magnet current is supplied by the main transformer and rectifiers.

The magnets used could be of 9, 12 or 15 inches. They are generally floor-mounted and require 2.5 to 22.5 kW power supplies to drive them. The magnets are equipped with ring-shim pole caps designed for maximum field homogeneity. A 3-inch clearance (air gap) provides ease of access. The pole caps are made compatible with tapered ring-shim tips, which reduce the air gap to 1.75 inch to achieve optimum field performance when operating at 35 GHz.

In order to improve magnet performance, the heat generated must be dissipated. To do so, coolant lines are run around the magnet, magnet power supply and microwave bridge. If the temperature of the untreated water is below the dew point, condensation may occur in the magnet power supply. In this event, the untreated water can be run through the magnet first and then through the power supply. Also, the coolant water flow should be correct for magnet system in use. This is ensured by checking the difference between inlet and outlet pressure if the difference is more than that specified for the magnet system, the cooling system may be clogged.

If the ambient room-air temperature or magnet cooling water temperature changes, the setting of the Hall probe temperature should be examined to ensure that the temperature control system is regulating, i.e. it is not saturated or cut off. This is particularly important in those installations wherein large annual variations in room or water temperatures take place. Noise and jitter can be induced into the field control system by transients on the power lines, that are coupled into the Hall probe circuit. As far as possible, the Hall probe cable should be isolated from power lines connecting the console and the magnet power supply.

It is often difficult to detect or troubleshoot instabilities of the magnetic field, as normally no convenient external field measuring device is available to monitor the magnetic field with sufficient stability or accuracy. However, the manufacturers of instruments suggest several test techniques that may be employed with easily obtainable auxiliary test equipment. If the magnet current rises beyond normal, the Hall sensor probe may be improperly phased. This is corrected by removing power from the system and reversing power leads to magnet coils.

11.3.2 Microwave Bridge

The microwave bridge detects the ESR signal reflected from the sample cavity. The bridge contains microwave circuitry, a klystron power supply, a pre-amplifier and an automatic frequency control circuit (Figure 11.8).

- The klystron generates microwave energy at 9.5 GHz, which is used to irradiate the sample. The klystron output is applied to an isolator which allows signal flow only in one direction. The klystron is water-cooled through its mounting flange. 35 GHz systems are also available.
- The output of the isolator is given to a four port circulator, wherein the microwave power is directed to the sample cavity. The circulator then directs the reflected power from the sample cavity back to the microwave detector. The crystal dc bias current is indicated by the detector level meter.
- When all the power is absorbed and none is reflected, the detector receives no signal. At resonance, the sample absorbs microwave energy and the unbalances the impedance of the cavity. The absorption in the sample is detected as a dip in the output of the crystal detector. The crystal output containing the ESR signal and the 70 kHz AFC (automatic frequency control) signal is amplified by the pre-amplifier. It is then coupled to the receiver and to the AFC amplifier.
- The automatic frequency control circuit contains a 70 kHz crystal-controlled oscillator, which generates an AFC carrier and the AFC phase detector reference voltage. The AFC carrier is

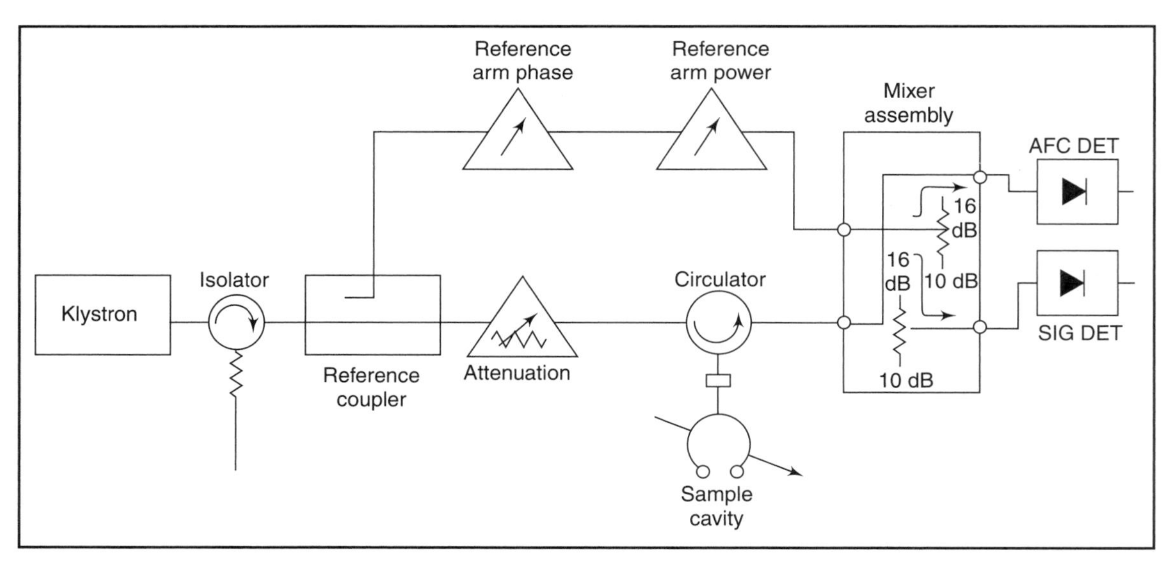

Figure 11.8 :: Block diagram of the microwave bridge

superimposed on the klystron reflector voltage, which results in a 70 kHz frequency modulation of the klystron output. The frequency modulated rf (radio-frequency) is applied to the sample cavity, resulting in an fm (frequency modulation) to am (amplitude modulation) conversion. The resultant am modulated microwave power is detected and amplified in a preamplifier, followed by a 70 kHz tuned amplifier. The output of this amplifier is phase-detected, amplified, filtered and applied to the klystron reflector tracking network. Another parallel output is given to an integrator, which quickly returns the phase detector dc output to zero. This arrangement permits a very accurate lock, over a range of at least ±15V reflector voltage correction.

Noise elimination is an important factor in enhancing the sensitivity of plotting the ESR spectrum. Most of the noise is eliminated in the phase-sensitive detector, because only that pad of the noise which is at the same frequency and in-phase with the reference signal, is allowed to pass through it. The noise is further eliminated by using a long-time constant filter, which averages out noise at frequencies greater than the reciprocal of its time constant. A better method of noise elimination is through continuous averaging by a computer, which employs a multichannel pulse-height analyser. This technique of using a computer for average transients efficiently removes both low and high frequency noise.

In order to achieve advantages of higher frequency operation such as increased sensitivity for small samples, minimization of second order shifts, increased resolution for powder samples with different g-values, the klystron frequency is set at 35 GHz instead of 9.5 GHz. The higher frequency also permits the observation of transitions that require higher energy.

The klystron requires +650V for the klystron beam and up to −400 V for the klystron reflector. The supply voltages are obtained from the 20 V dc inputs in inverter circuits, working at a frequency of 35 kHz. This frequency is obtained by dividing the 70 kHz AFC modulation frequency by 2. The klystron body or tuning shaft should not be touched when the bridge is in tune or in the operate position because the klystron body is at +650 V when operating. It is insulated from ground by an insulating gasket placed between the klystron flange and the water cooling flange.

11.3.3 Modulation Unit

The 100 kHz modulation unit acts as a transmitter and receiver. The transmitter provides the power to drive the cavity modulation coils. The receiver processes the ESR signal from the bridge pre-amplifier and converts it into a dc voltage for application to the recorder Y-axis. Figure 11.9 shows the block diagram of the 100 kHz modulation unit.

- The 100 kHz oscillator is crystal-controlled. Its output is attenuated by the modulation amplitude control and is applied to the output modulation amplifier. The gain of the amplifier is controllable. The 100 kHz modulation signal is transmitted to the cavity modulation coils.

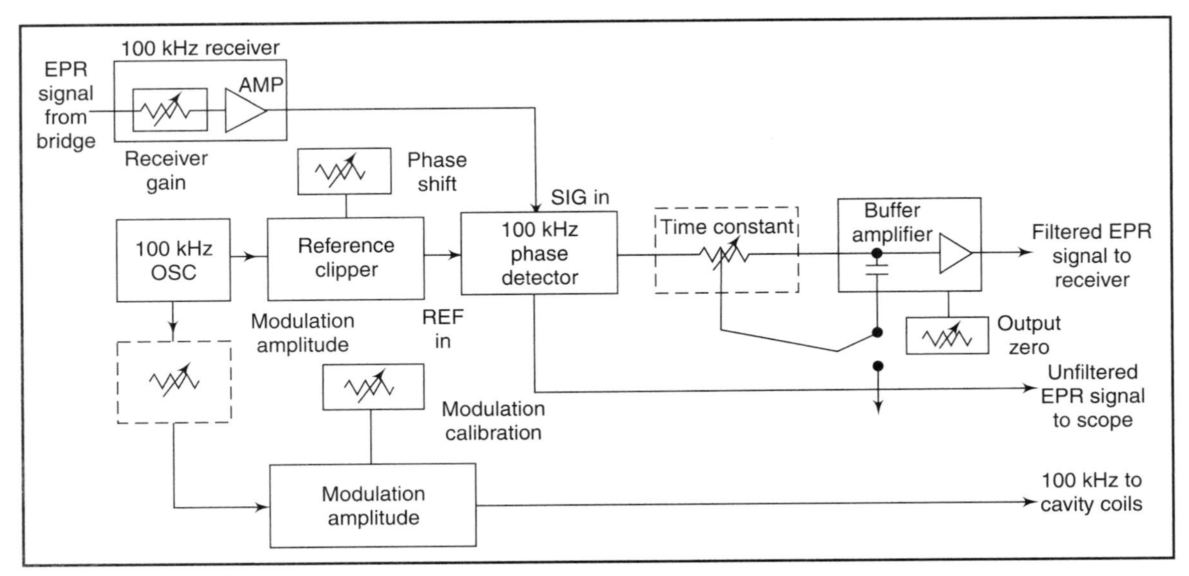

Figure 11.9 :: Block diagram of 100 kHz modulation unit

- The 100 kHz ESR signal from the detector crystal in the bridge is filtered and amplified in the 100 kHz receiver. The receiver gain control can be set in 1 dB steps.
- The amplified 100 kHz ESR signal is applied to the input of the phase detector. The amplitude of the dc phase detector output is proportional to the amplitude of the 100 kHz signal input. A low-pass filter removes any residual 100 kHz or harmonics from the phase-detected signal. Additional noise filtering is provided by an *RC* network, which provides a variable time constant. This is followed by a buffer amplifier, which provides low output impedance stage connection to the recorder.

Modulation may also be carried out at low frequencies like 10 kHz, 1 kHz, 270 Hz or 35 Hz. The low frequency module shown on the block diagram serves to accomplish this function.

11.3.4 System Function Selector

The system function selector provides the selection of signal inputs to the recorder and oscilloscope and cavity modulation to the cavity. When the oscilloscope monitor and the recorder input switches are in the 100 kHz or low frequency position, the phase-detected ESR signal passes from the receiver to the oscilloscope and recorder respectively. In the external position, the recorder input may be supplied from an external source and the oscilloscope may be fed an external signal.

The cavity modulation switch allows the operator to apply either 100 kHz or a low frequency modulation to the cavity coils. If two cavities are used, a low frequency can be applied to one cavity and 100 kHz to the other cavity, or the two modulation signals may be added and applied simultaneously to either cavity. The 100 kHz modulation and the 34 kHz sawtooth sweep are combined in the modulation summation circuit. Two decade capacitors series resonate the cavity modulation coils at 100 kHz for the most efficient energy transfer to the cavity. The ESR signal can be observed on the oscilloscope with 100 kHz or 10 kHz modulation, but not with any of the other lower frequencies.

11.3.5 Recorder

The recorder is used to produce a hard copy of the ESR spectrum, as a function of time on a 30 × 40 cm flat chart. By selecting the synchronized frequency to the stepper motor, the recording time can be selected from a few seconds to several hours.

Horizontal travel (X-axis) of the pen is achieved by a pulsed stepper motor and belt-driven capstan, which moves the pen carriage horizontally. The vertical (Y-axis) travel of the pen is controlled through a second system of pulleys and wire cable from a servo motor. Y-axis movement results when the servo loop, of which motor is a part, is unbalanced by a signal applied to the unbalanced by a signal applied to the recorder amplifier. The recorder amplifier is turned on by the servo motor to balance the servo loop.

11.3.6 Oscilloscope

The oscilloscope permits direct visual observation of rapidly changing or decaying signals as well as the optimization of instrument parameters. The oscilloscope supplies 34 Hz sawtooth modulation to the sample cavity modulation coils and displays the resulting ESR spectrum. The amplitude of the field modulation is kept adjustable and may be selected by sweep width control. Noise on the displayed spectrum may be reduced by the addition of an input filter.

The oscilloscope is also used to display the klystron mode. It is accomplished by applying the same 34 Hz sawtooth modulation to the klystron reflector and displaying the output of the microwave detector on the oscilloscope.

11.3.7 Sample Cavities

Signal intensity from a particular sample can vary greatly, depending upon the microwave cavity configuration. This is important when signal-to-noise poses a serious problem. The two main effects arise from: (i) the difference between cylindrical (TE_{011}) and rectangular cavities (TE_{012}), (ii) the effect of quartz such as Dewar inserts, on the microwave field distribution. The use of a cylindrical cavity results in a net improvement over the rectangular cavity. Both types of cavities are, however, transverse electric that the electric field

Transverse electric modes

lines are confined to the plane that is perpendicular to the longitudinal axis. No such restriction exists for the magnetic field lines. The sample cavity should be so designed that it can be held in the volume of maximum homogeneity of the magnetic field. The cavities are available for a wide variety of ESR signals, ranging from solids to liquids. They are designed to be compatible with other accessories like aqueous solution sample cell, flow mixing chamber, electrolytic cell, liquid nitrogen Dewar and the variable temperature accessory.

ESR studies involving comparative measurements are made by using a dual cavity. This cavity does this by allowing a reference sample of known magnetic resonance characteristics to be exposed to microwave energy simultaneously with and in the same cavity as the sample being studied.

A rotating cavity is designed to facilitate ESR studies of crystal anisotropy. The cylindrical rotating cavity has exterior modulation coils, which are free to turn with the rotating electromagnet and thus provide a modulation field, which is always parallel to the dc magnetic field. The sample tube remains fixed in the cavity, which keeps the Q constant and eliminates the need for re-tuning after rotation.

11.3.8 Sample Cells

Aqueous solution sample cells are specially designed to carry out ESR studies in lossy or aqueous solutions. Further, as optical irradiation experiments are often performed with these cells, the quartz is selected to give a maximum optical transmission in the UV-visible region of the spectrum. Special types of flow cells, tissue cells and electrolytic sample cells are also used when required. Generally, the standard sample tube is made of high purity quartz and measures 3 mm ID and 4 mm OD. Glass is not used because it contains traces of FE^{3+}. For maximum sensitivity, the tube may be filled to a height of 2.5 cm for a rectangular cavity and 5 cm for a cylindrical cavity.

The single crystal method of ESR analysis requires a large number of measurements with precise orientations of the sample. Such measurements are often difficult, because of the small size of microwave cavities. Some goniometers for ESR studies have been described in literature, but either they only permit 360° rotation about a single axis, or the angular variation of the signal/noise ratio (modulation coils not aligned with H_o), prevents the spectrum from being recorded for some orientations.

All sample tubes should be cleaned before being inserted into the cavity. Only standard ESR sample tubes should be used to prevent breakage and to minimize contamination signals. Sources of contamination such as cigarette ashes or smoke should be kept away from the immediate vicinity of the cavity, whether the cavity is installed in the air gap or not. Loose magnetic materials must be kept away from the gap of the magnet to prevent damage to the cavity, when it is in the air gap. Watches should be kept away from the magnet gap.

Sample materials or solvents (water-acetone) with high dielectric loss require flat cells. For materials with low dielectric, a regular 3 mm sample tube is used.

Chapter 12

Electron and Ion Spectrometers

12.1 Surface Spectroscopic Techniques

Electron and ion spectroscopic techniques find applications for surface* analysis. They can provide chemical information, which the classical methods like microscopy, reflectivity and adsorption isotherms cannot. With electron and ion spectroscopy, one can obtain elemental analyses, information about oxidation states and organic functional groups; quantitative analyses either as elemental ratios or oxidation state ratios and distributions of materials either across the surface or toward the bulk from the surface inward, (Hercules and Hercules, 1984a).

The surface spectroscopic techniques can best be explained by Figure 12.1. When a beam is incident on the surface, it penetrates to some depth within the surface layer. A second beam exits from the surface, which can be analysed by a spectrometer. The beams shown on the diagram may be photons, electrons or ions. Obviously, by varying the nature of the beams in and out of the surface, a large number of surface analytical techniques can be generated. Table 12.1 summarizes the various methods which emanate from the electron and ion spectroscopy. In this chapter, only the following techniques are covered:

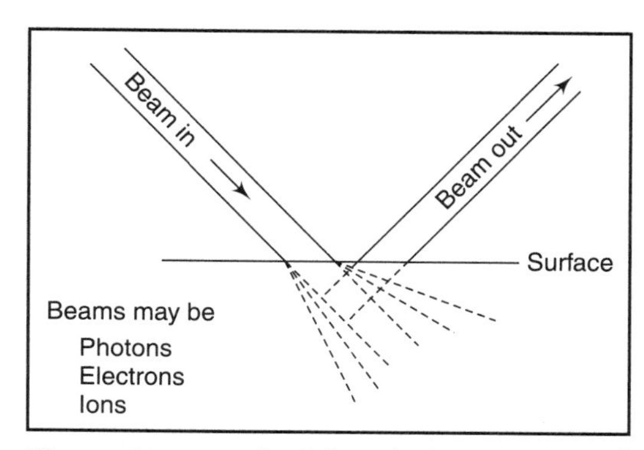

Figure 12.1 :: Principle of electron and ion-beam spectroscopy

*A surface is normally considered as the boundary layer of one phase at its interface with another. The surface is usually more than one atomic layer deep and is a region of non-uniform atomic potentials. Some workers also refer to the outermost layer of atoms as a surface, the transition layer as the selvedge, and the remainder as the hulk.

Table 12.1 :: Techniques developed by a Combination of Electron and Ion Beams

Beam in	Beam out		
	X-rays	Electrons	Ions
X-rays	X-ray fluorescence X-ray diffraction	• Photoelectrons • Auger electrons • Electron spectroscopy for chemical analysis (ESCA) • X-ray photoelectron spectroscopy (XPS) • Induced electron emission (IEE)	Ionization
Electrons	X-ray emission	• Secondary emission • Auger electrons • Auger electron spectroscopy (AES), • Electron impact spectroscopy (EIS) • Low energy electron diffraction (LEED)	Electron-induced ion desorption (EIID)
Ions		Ion-induced Auger electrons	• Secondary ion mass spectrometry (SIMS) • Ion scattering spectroscopy (ISS)

(a) Electron spectroscopy
   (i) Electron spectroscopy for chemical analysis (ESCA)
   (ii) Auger spectroscopy (AES)

(b) Ion spectroscopy
   (i) Secondary-ion mass spectrometry (SIMS)
   (ii) Ion scattering spectroscopy (ISS)

12.2 Electron Spectroscopy

12.2.1 Electron Spectroscopy for Chemical Analysis (ESCA)

Electron Spectroscopy for Chemical Analysis (ESCA), also known as X-ray Photoelectron Spectroscopy (XPS), is an effective technique for detecting the elements and their bonding states on the surface of solids. The methods use soft X-rays to eject electrons from inner-shell orbitals. The kinetic energy of these photoelectron energies are dependent upon the chemical environment of the atom, it makes XPS useful to identify the oxide state and ligands of an atom. On the other hand, however, the binding energy may shift (up to several eV) reflecting the chemical state of the atom, it is also possible to get information on the valence number and oxidation number, and the type of functional group present. Moreover, the ion etching technique provides the depth profiling from the surface.

Electron spectroscopy is based on the ionization phenomenon brought about by either a photon (X-ray) or an electron (Hercules and Hercules, 1984a). An X-ray photon ionizes an atom, thereby producing an ejected free electron. The kinetic energy of the ejected photoelectron is dependent upon the energy of the impinging photon expressed as

$$E_K = h\upsilon - E_b - \Phi,$$

where E_K is the kinetic energy of the photon ejected, $h\upsilon$ is the X-ray energy, E_b is the binding energy of the parent atom relative to the ejected electron and Φ is the work function, typically < 2 eV and often ignored.

The binding energy is specific for a given electron in a given element and can help in serve the identification of that element. Measuring the E_K and calculating the E_b yields the 'fingerprint' of the parent atom. This forms the basis of ESCA—the electron spectroscopy for chemical analysis.

ESCA gives sufficient chemical information up to a depth of about 5-20 Å in metals, 15-40 Å in oxide, and 40-100 Å in polymers. Thus the sensitivity is sufficient enough to identify and detect a fraction of a monolayer. It can identify elements in the periodic table above helium and adjacent elements are clearly distinguished.

In case of electron ionization of an atom, both an excited ion and a second electron are produced. However, because of electron–electron interactions, discrete electron energies are not observed. Therefore, ESCA is not observed when using electron ionization.

In ESCA, the goal is to catch the electrons in order to find out the atom from which they are coming. The information we are interested in is the so-called binding energy which they had before leaving the atom. All what we do is counting these electrons versus their binding energy and we obtain a spectrum looking like the one shown in Figure 12.2.

In this spectrum, two main peaks at 284.6 and 532.5 energy counts are observed. The unit used for counting energy is electronvolt-abbreviated as *eV*. Accordingly, peaks are designated as 284.6 eV and 532.5 eV.

Each energy matches a specific atom type, e.g. 284.6 eV matches carbon and 532.5 matches oxygen. From this, we can conclude that this specimen contains carbon and oxygen. Each peak area is proportional to the number of atoms being present in the studied element. By calculating the respective contribution of each area we obtain the specimen chemical composition, for example: 25 per cent oxygen and 75 per cent carbon. In other words, among 100 atoms present at the surface of the material, 75 are carbon atoms. By studying the energy of this carbon peak, it is possible to find out if the surface of this material corresponds to C–O or C=O chemical form.

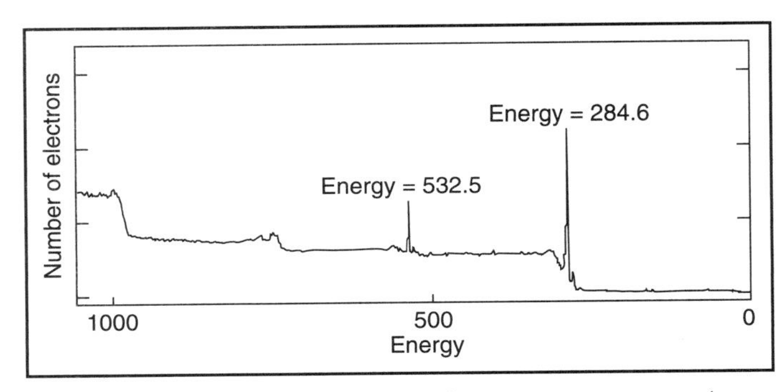

Figure 12.2 :: Electron count plot versus energy (energy spectrum) (adapted from *www.lasurface.com*)

Figure 12.3 shows a typical equipment for electron spectroscopy for chemical analysis.

Figure 12.3 :: ESCA-Electron spectroscopy for chemical analyzers equipment

12.2.2 Auger Electron Spectroscopy (AES)

Auger Electron Spectroscopy (Auger spectroscopy or AES) was developed in the late 1960s, and derives its name from the effect first observed by Pierre Auger, a French Physicist. It is a surface technique that utilizes the emission of low energy electrons in the Auger process and is one of the most commonly employed surface analytical techniques for determining the composition of the surface layers of a sample. Auger spectroscopy facilitates determination of the chemical composition of a surface. This characterization can be achieved up to a depth of 1 nm. The smallest surface that can be characterized is a few nm wide for the best instruments.

Auger electron spectroscopy can be used for surface chemical analysis in a way that is very similar to XPS. Since core levels are involved, the energy of the Auger is also very characteristic for the various elements. AES is, however, rather limited when it comes to very high resolution studies. On the other hand, a qualitative chemical analysis of the surface is still very often performed using AES.

Auger transitions

The Auger nomenclature follows the old X-ray notations. The Auger transitions are labelled ABC for the initial state hole (A) and the two final state holes (B) and (C), with the letter denoting the shell. A KLL Auger transition would be a transition starting from a hole in the 1*s* levels which would be filled up from the 2 p level. A 2 p electron would also be emitted. The more complicated nature of the upper levels causes a multiplet splitting in the Auger spectra.

An important point in AES is that how the initial core hole is created does not make any difference. In most practical cases, this is achieved by bombarding the sample with electrons of 2-3 keV kinetic energy. The Auger electrons are detected with the electron analyzers. For a quantitative analysis, the energy of the exciting electrons does, however, come into play because of the energy-and element-dependent ionization.

Auger electron

It may be noted that photo-ionization can produce either an ESCA electron or an Auger electron, while electron ionization can produce only Auger electrons. Further, the kinetic energy of the Auger electron does not depend upon photon energy, whereas that of the ESCA photoelectron does. It may be noted that the Auger phenomenon characterizes a de-excitation process leading to the loss of one electron. This electron is called an Auger electron.

Figure 12.4 illustrates the relationship between ESCA, X-ray and Auger processes for photo-ionization of a 1S electron. In the ESCA process, the photon ejects a 1 S electron from the atom. In case of X-ray, an electron drops from the 2 p orbit to fill the 1 S hole and a photon is emitted, resulting in $K\alpha$ X-ray emission. In the Auger process, a 2S electron drops to fill the 1 S hole, simultaneously expelling a 2 p electron.

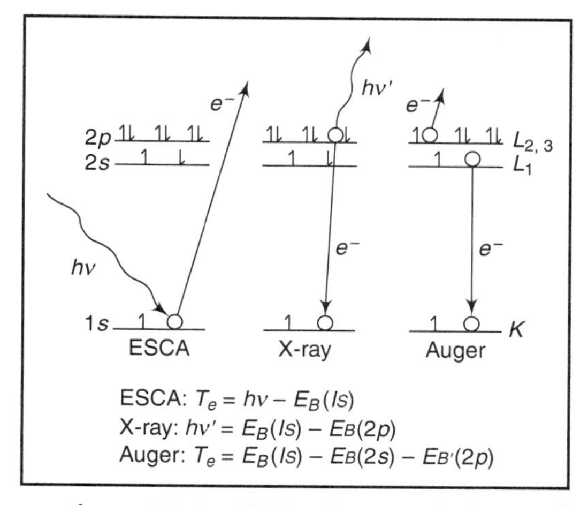

Figure 12.4 :: ESCA, X-ray and Auger processes

It is obvious from this diagram that Auger and X-ray emission are competitive processes. In general, for low energy (1000 eV) processes, the Auger effect is predominant, while for higher energy (10,000 eV) processes, X-ray emission will be dominant.

Auger electron spectroscopy involves the irradiation of the surface to be analyzed with a beam of electrons of energy in the 1–2 KeV range. Beam currents are typically 5-50 μA in a beam of diameter 0.5 mm. Since an electron with primary energy 1 to 2 keV can penetrate only a few atomic layers, AES is basically a surface chemical analysis technique providing chemical information from 4 to 10Å depth. The technique is very sensitive and generally allows detection and identification of less than 0.1 per cent of a monolayer of atoms. Table 12.2 gives the analytical characteristics of ESCA and AES.

Table 12.2 :: Analytical Characteristics of ESCA and AES (After Hercules and Hercules, 1984b)

Parameter	ESCA	Auger (AES)
Energy Range		
Kinetic energy	100–1500 eV	50–2500 eV
Escape depth	20A	20A
Peak locations	± 0.1 eV	±1 eV
Elemental sensitivity		
Elements Z>2		Z>2
Specificity	very good	good
Sensitivity variations	50 X	50 X
Quantitative Analyzer		
Absolute	± 30%	± 30%
Relative	±5%	±5%
Detection limit	0.1% monolayer	0.05% monolayer
Other aspects		
Vacuum	10^{-5}–10^{-10} torr	10^{-8}–10^{-19} torr
X-y resolution	none	0.5 μ
Speed	Slow, typical run is 30 min	Fast, takes minutes
Sample destruction	None in 95% of the sample	Frequent, bad for organics

The basic technique of Auger Electron Spectroscopy (AES), which is primarily a surface-sensitive technique used for elemental analysis of surfaces, has also been adapted for use in:

- *Auger Depth Profiling:* Providing quantitative compositional information as a function of depth below the surface. Depth profiles are obtained by employing a controlled sputtering process which enables elemental concentration to be plotted as a function of depth. 'Sputtering' is a process in which an ion gun is used to remove a few angstroms of the topmost surface of a sample. Sputtering and analysis are alternated until the desired depth is reached.
- *Scanning Auger Microscopy (SAM):* Providing spatially-resolved compositional information on heterogeneous samples. Therefore, the Auger multiprobe is capable of producing

elemental composition spectra, surface images, selective elemental line scans and maps, and depth profiles.

AES is a technique that is often described as being more sensitive than XPS. This difference in sensitivity between the two is primarily due to the difference in electron kinetic energies. For example, in the case of carbon, in XPS the kinetic energy is close to 1000 eV as opposed to 250 eV in AES. Since the mean free path changes with the electron kinetic energy, the depth of analysis will be smaller in AES than in XPS in the case of a carbon containing specimen. As a matter of fact, the surface contamination (primarily carbon and oxygen) is a much more sensitive factor in AES. An ion etching is sometimes necessary to study surfaces with this spectroscopy.

12.3 Instrumentation for Electron Spectroscopy

Figure 12.5 shows a block diagram of an electron spectrometer for either ESCA or Auger measurements. It contains the following components:

(a) A source of radiation with which to excite the sample,
(b) An electron energy analyzer,
(c) An electron detector
(d) A read-out system, and
(e) A high vacuum system.

In addition, the entire system must he shielded from the earth's magnetic field.

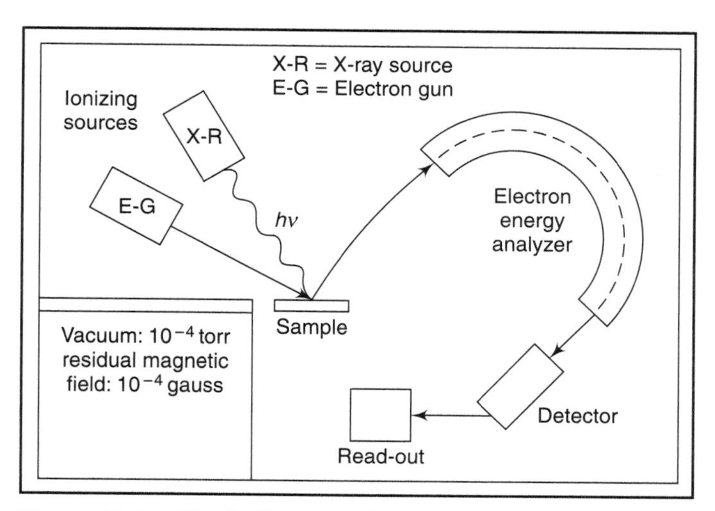

Figure 12.5 :: Block diagram of an electron spectrometer

12.3.1 Radiation Sources

12.3.1.1 X-ray Sources

Traditionally, ESCA instruments have used X-rays as radiation sources and Auger spectrometers as electron guns. It is desirable that the radiation be homogeneous in energy.

X-ray source

The basic X-ray source includes a heated filament and a large target anode. Electrons from the filament are accelerated toward the anode to produce radiation consisting of a continuum of bremsstrahlung radiation with characteristic X-ray lines superimposed upon it. The anode is held at a high positive potential, while the filament is held near ground potential. The positive potential on the anode ensures that scattered electrons do not enter the sample chamber but are retracted to the anode. The anode materials most commonly used, which give nearly monochromatic radiation sources, are magnesium or aluminium. These elements give high intensity and narrow wavelength bands of the $K\alpha$ lines. The most intense lines are called $K\alpha_1$ and $K\alpha_2$, according the old X-ray nomenclature (Figure 12.6). Often the doublet is viewed as one line and is called $K\alpha_{12}$. It has an energy of 1253.6 eV and 1486.6 eV, for Al and Mg, respectively (Hofmann, 2003).

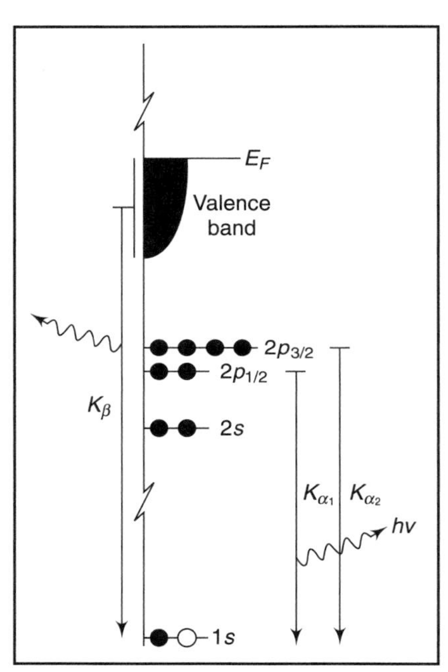

Figure 12.6 :: Nomenclature in the X-ray decay in Al and Mg

The $K\alpha_{12}$ is the most intense line in the X-ray spectrum. The $K\alpha_1$ and $K\alpha_2$ lines do also have a certain width themselves which, together with their separation, determines the ultimate resolution achievable with an X-ray source. It is determined by the lifetime of the core hole. The total width for the Al and Mg $K\alpha_{12}$ line is of the order of 1 eV.

As illustrated for a detailed chemical analysis, it is highly desirable to have a higher energy resolution. In order to achieve this, the X-ray source can be equipped with a monochromator. This will increase the energy resolution and at the same time remove the 'satellite' lines which, due to photo-emission, are induced by, for example, the $K\alpha_{34}$ line.

A thin X-ray transmitting window separates the excitation region from the specimen and prevents the entry of scattered electrons from the X-ray source into the sample chamber. The most commonly used window material is high purity aluminium or beryllium foil, which removes the K_β line and much of the background.

Another means of removing the background produced by bremsstrahlung radiation and satellite peaks produced by the less intense characteristic X-ray transitions is through the use of an

X-ray monochromator

X-ray monochromator. The most popularly used X-ray monochromator exploits Rowland's Theorem, which states that if a diffracting crystal with a radius of curvature R is tangent to a circle of diameter R, and if an X-ray source and the sample replaced on this circle, so that they both make angle θ relative to the crystal surface, then the radiation striking the crystal surface, which satisfies the Bragg diffraction law, will be reflected and brought to a focus on the sample, and all the other radiation in the X-ray spectrum will not be reflected.

The Bragg diffraction law is:

$$n \lambda = 2 d \sin \theta,$$

where n is any integer, λ is the X-ray wavelength and d is the spacing of the atomic planes within the crystal.

Figure 12.7 shows the arrangement of the monochromator. There are actually three crystals in the system, which help to collect more X-ray intensity than is possible with a single crystal. Each crystal has a separate Rowland circle in a different plane from the others. These three planes intersect in the line passing through the anode and the sample. The reflection of the Al $K\alpha$ radiation from the crystal planes takes place at a Bragg angle θ of 78.5°. A Bragg angle close to 90° has the advantages of minimizing the geometrical aberrations and maximizing the Bragg diffraction intensity and the dispersion of the monochromator.

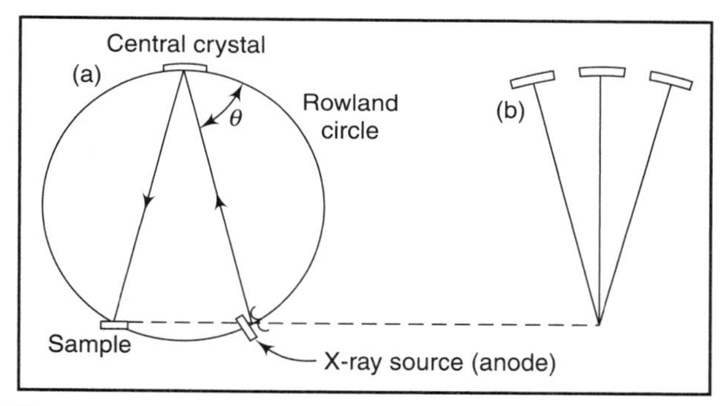

Figure 12.7 :: X-ray monochromator. It has three spherically bent crystals with three Rowland circles that intersect at the anode and sample

An electron beam for AES is produced in an electron gun, which basically consists of an emitting surface called cathode, an accelerating electrode (the anode) and one or several focusing electrodes, which control the characteristics of the electron beam. An electron beam can be used directly or with a monochromator. Generally the electrons from a heated cathode are fairly homogeneous, though there may be a small spread due to a range of kinetic energies of the emitted electrons. The beam can be made more homogeneous by employing an energy filter, of which there are several types.

Electron energy filters

Figure 12.8 shows the typical arrangements used as electron energy filters. If the electrons are subject to a retarding field between two grids [Figure 12.8(a)], only those electrons which have sufficient energy to overcome the field will go to the right. This arrangement essentially acts as a high pass filter, in that there is a lower but not an upper limit to the energy of the emerging beam. If the electron beam enters the space at an angle of 45° into the space between two parallel plane conductors [Figure 12.8(b)], only those electrons within a specified narrow energy band will emerge through the exit slit. The energy band will depend upon the distance between the plates and the potential applied.

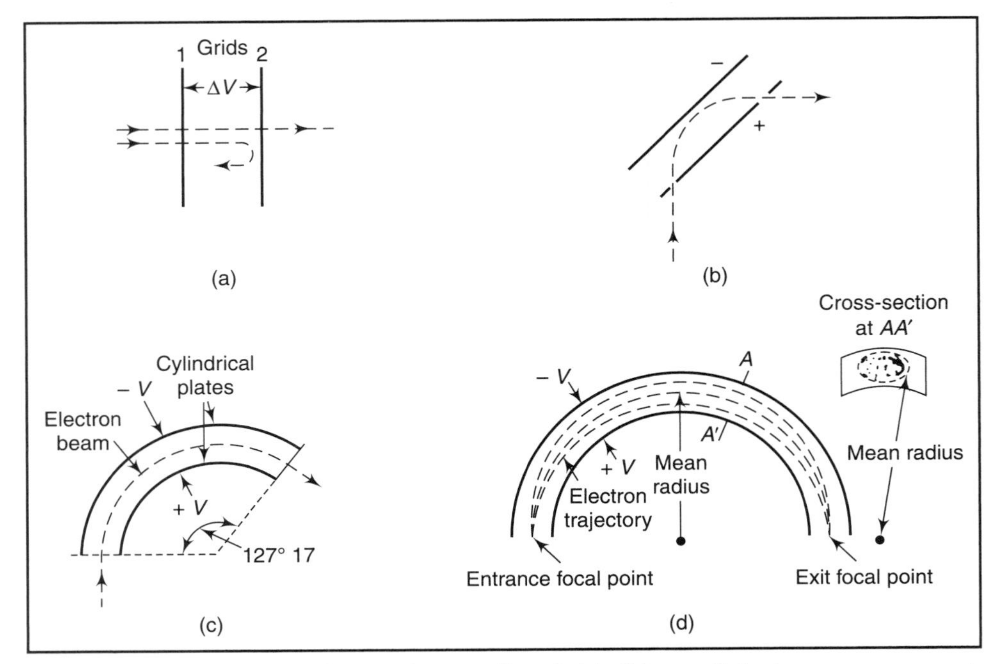

Figure 12.8 :: Energy filters: (a) retarding field, (b) parallel plate arrangement, (c) cylindrical electrode arrangement, (d) spherical electrode arrangement

The use of cylindrical electrodes (Figure 12.8(c)) instead of plane electrodes results in good energy discrimination. Even those electrons which enter the filler at a slightly divergent angle will be focused on the exit slit. For a double focusing effect, the angle $\pi/\sqrt{2}$ radians (127°17′) is required. Equally good focusing properties can be obtained by employing 180° spherical rather than 127° cylindrical segments as electrodes [Figure 12.8(d)].

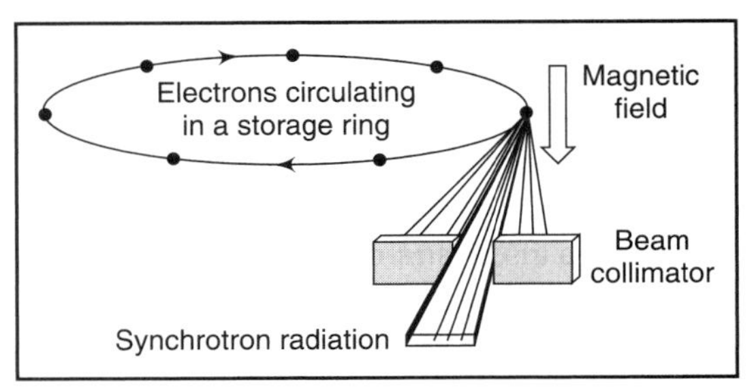

Figure 12.9 :: Synchrotron radiation emission mechanism

12.3.1.2 Synchrotron Radiation

Alternatively, one can use synchrotron radiation as an X-ray source. This radiation is caused by accelerating charged particles (mostly electrons), typically by forcing them to go around the corners of a storage ring (Figure 12.9).

Synchrotron radiation (SR) is emitted when charged particles moving with relativistic speeds are forced to follow curved trajectories in magnetic fields. In general, three kinds of magnets are used to make the necessary magnetic fields: bending magnets, wigglers and undulators.

For bending magnets, a sample dipole structure is used to constrain the electrons in a curved path. The radiation emitted is extremely intense and extends over a broad wavelength range from the infrared through the visible and ultraviolet, and into the soft and hard X-ray regions of the electromagnetic spectrum.

Wiggler magnets

High-field wiggler magnets are often used as sources in order to increase the flux at shorter wavelengths. A wiggler can be considered as a sequence of binding magnets of alternating polarities which gives a $2N$ enhancement in the flux, where N is the number of poles. The properties of wiggler SR are very similar to that of dipole radiation with a reduction in the critical wavelength as a consequence of the higher field. For superconducting wiggler magnets, a value of 6 Tesla, as opposed to around 1.2 Tesla for conventional dipoles, would be typical.

Undulators

Undulators, consisting of periodic magnetic arrays, cause small electron deflections comparable in magnitude to the natural emission angle of the SR. The radiation emitted at the various poles interferes coherently resulting in the emission of a pencil-shaped beam peaked in narrow energy bands at the harmonics of the fundamental energy. For N poles, the beam's opening angle is decreased by $N^{1/2}$ and thus the intensity per solid increases as N^2.

Synchrotron radiation has several advantages over conventional sources—the resolution can be very high, the radiation is polarized and most important, the photon energy can be changed. This allows the peaks in an X-ray spectrum be shifted to exactly the desired kinetic energy. In surface science, this is the energy where the mean free path of the electrons is shortest. The obvious disadvantage of synchrotron radiation is that you have to build a storage ring to get it.

12.3.2 Energy Analyzers

The function of the energy analyzers is to measure the number of photoelectrons as a function of their energy. This is done by using an electrostatic or magnetic analyzer. Magnetic deflection analyzers are effective, but less convenient to design and use than electrostatic types. The most widely used analyzer in commercial instrumentation is double-pass cylindrical mirror (CMA) and 180° spherical sector analyzers (SSA).

Cylindrical mirror analyzer

Figure 12.10 shows a schematic diagram of the double-pass cylindrical mirror analyzer. Basically, the CMA consists of two conical cylinders with angular entrance and exit apertures cut in the inner cylinder. A negative potential is applied to the outer cylinder. The potential applied between the inner and outer cylinders produces a cylindrical retarding potential. From the theory of this analyzer, it is found that optimum focus will be obtained with the angle between the electron beam and the axis of symmetry within a few degrees of 42° 20'. In the CMA, electrons, which leave the sample positioned at the focal point of the analyzer, pass through an annular slit, then pass into the radial field between the cylinders to be focused back to the axis by the negative potential. The electrons pass into the second cylinder mirror analyzer to be focused onto an electron multiplier.

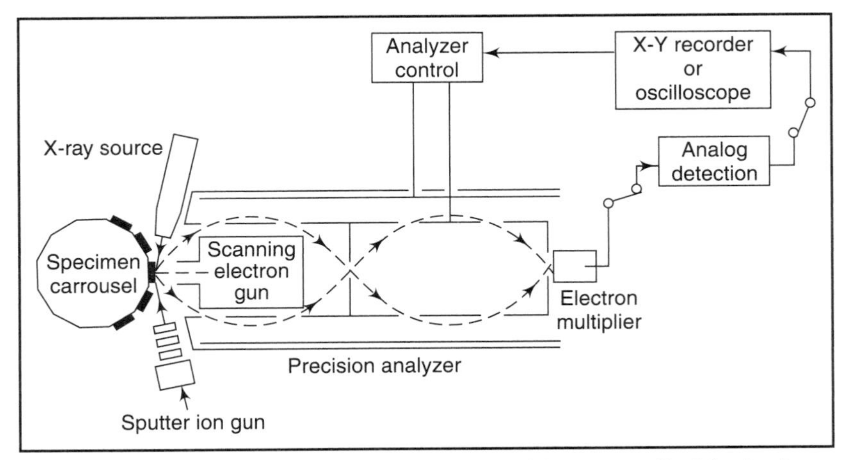

Figure 12.10 :: The arrangement of a double-pass cylindrical mirror analyzer used for X-ray photoelectron spectroscopy

Sspherical sector analyzer

In the spherical sector analyzer (Figure 12.11), electron energy is analyzed by passing the electrons between two hemispherical domes that have a potential difference between them. Electrons of the desired energy follow a circular orbit between the domes and reach the detector, while electrons having higher or lower energies strike one of the domes and are not detected. The number of electrons striking the detector for a given potential difference is counted and plotted as a function of energy.

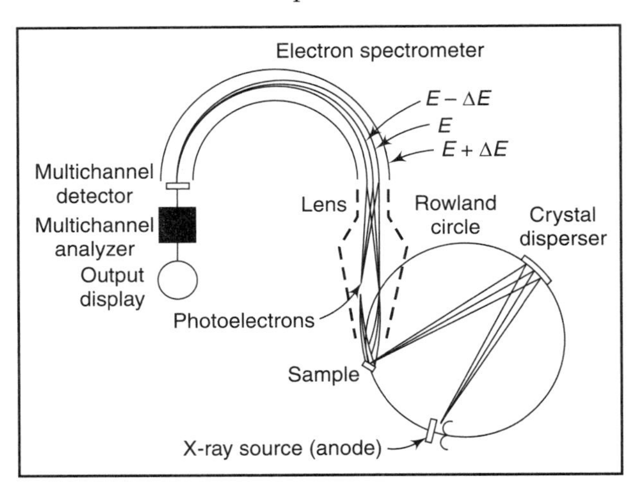

Figure 12.11 :: Schematic diagram showing 180° hemispherical analyzer with electrostatic lens and dispersion compensated electron optics

The energy of the X-ray photon incident on the sample varies across the sample by about one electron volt (which is the width of the X-ray line). This results in variation of the kinetic energy of the photoelectrons from a given energy level by one electron volt. This inherent variation is cancelled by using an electron lens. The electron lens has four elements or three gaps, corresponding to three degrees of freedom in the electron optics. The first is used to retard the photoelectrons of interest to 115 eV kinetic energy, before they enter the electron spectrometer. The second forms a focused electron image of the target at the spectrometer entrance. The last degree of freedom is used to image, so the dispersion of the electron spectrometer at 115 eV kinetic energy precisely cancels the dispersion of the X-ray monochromator across the sample. This cancellation, called dispersion compensation, removes the inherent line width limitation on instrument resolution.

The use of a retarding field formed by the lens improves the resolving power of the lens. If electrons leaving the sample with kinetic energy E_k are retarded to an energy E_R for transmission through the analyzer, then the resolving power of the complete system will be improved by a factor equal to the retarding ratio $E_K/E_{R.}$ Retardation also gives rise to improvement in sensitivity at a fixed energy resolution, since a larger slit may be used at the lower transmission energies.

The fringe fields at the entrance and exit to the electrodes of both the CMA and SSA need to be suitably terminated. Fringing fields at the ends of CMA are usually terminated by a series of rings coupled by dividing resistors or by resistively coupled ceramic disks. In case of a hemispherical analyzer, the fringing fields are terminated by concentric wires or shaped electrodes biased at the electron pass energy.

12.3.3 Electron Detectors

Electron multipliers

The most commonly employed detectors in ESCA and AES instruments are electron multipliers. They are similar to photomultipliers, but accept electrons directly instead of from a photocathode. One type of such a multiplier is the channel electron multiplier described by Evans (1972) and shown in Figure 12.12. It consists of a lead-doped glass tube with a secondary semiconducting coating possessing a high secondary electron yield. A voltage of 2 to 3 kV is applied between the ends of the multiplier to produce a gain of $10^6 - 10^8$ due to the cascade of collisions, as electrons travel down the inside of the tube. The device accepts electrons at one end and emits more electrons at the other, thereby acting as a current amplifier. The output of the multiplier is a series of pulses that are fed into a pulse amplifier–discriminator, then into a digital-to-analog converter, and stored in a multi-channel analyzer or a computer.

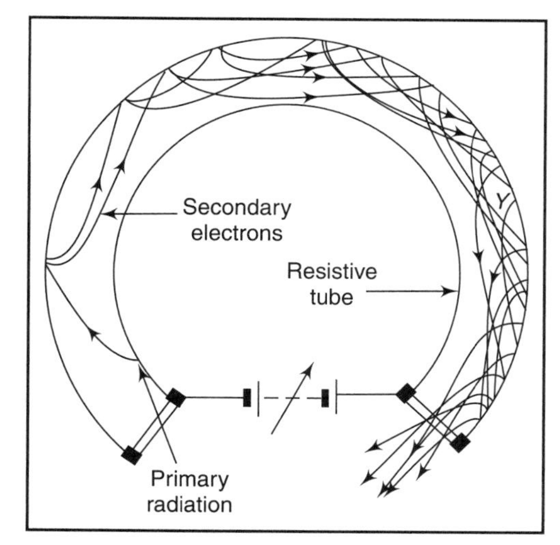

Figure 12.12 :: Channel electron multiplies

The spectrometers in which electron energy can be related to the position in the exit plane of the analyzer are equipped with multi-array detectors. One such system is shown in Figure 12.13. The conventional exit slit at the output of the spectrometer is replaced with a 1.1 inch-square imaging electron multiplier. Whenever a single photoelectron strikes the input of this active surface, a corresponding pulse of 10^8 electrons exits from

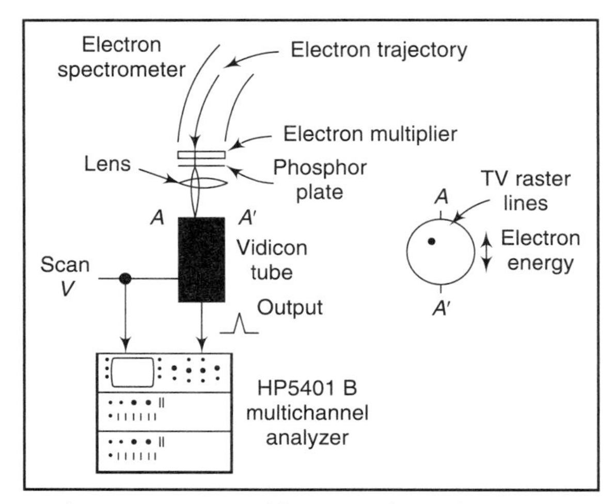

Figure 12.13 :: Wide-area detections system. An incoming electron strikes a multi-channel electron multiplier, resulting in a flash of light from a phosphor plate. The flash is imaged onto a vidicon tube. As the vidicon scans, each flash produces a pulse that is counted by the multi-channel analyzer and classified according to the incoming electron. The result is the ESCA spectrum, a plot of number of photoelectrons versus energy.

the same position on the output surface and strikes a phosphor plate. The resulting flash of light is imaged onto and stored within the target of a vidicon television tube. The event remains stored until it is erased by the conventional TV raster scan of the vidicon electron beam, which converts the event into an electrical pulse at the vidicon output. These pulses are transferred to a multichannel analyzer (1024 channels), which sorts out the pulses according to the energies of the detected photoelectrons.

12.3.4 Read-out System

Counts are accumulated in the various channels for an appropriate time and the resulting spectrum (a plot of counts per channel versus channel energy level) drawn by an X-Y plotter can be displayed. The information can also be put into a computer for analysis or further processing. Figure 12.14 shows a typical XPS spectrum, in which the data are presented as intensity or count rate versus either kinetic energy or binding energy. Since the analyzer determines electronic kinetic energy, therefore, a display of intensity versus binding energy requires correction as per Equation (1). Since the atomic structure of each element in the Periodic Table is distinct from all the others, measurement of the electron binding energies enables identification of the presence of elements on the sample surface.

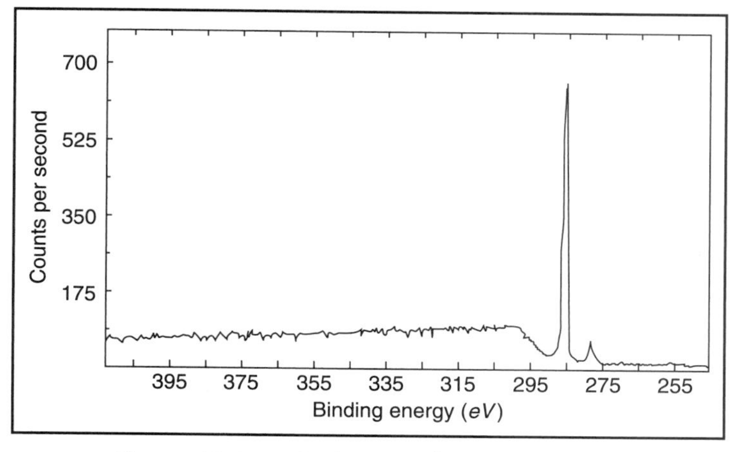

Figure 12.14 :: Carbon IS photoelectron spectrum

12.3.5 Vacuum Systems

Electron spectrometers must operate under a vacuum of 10^{-6} torr or lower, and 10^{-10} torr is ideal. At pressures higher than 10^{-6} torr, the electrons would be scattered en route from the sample to the detector. Pressures of these orders can be achieved by a variety of techniques. However, the most common system is a getter-ion pump complemented by a sublimation pump, with a cryogenic shroud and sorption forepump. The vacuum system is of stainless steel construction, with crushed metal gaskets.

12.3.6 Magnetic Shielding

Electron spectrometers are high resolution instruments and therefore, it is necessary to reduce the effect of stray magnetic fields within the volume of the analyzer. This is required, since the path of the electron is disturbed by stray magnetic fields, including that of the earth. Magnetic shielding can be provided in several ways. Ferromagnetic shielding is usually preferred in commercial instruments, as it is simple and less sensitive to magnetic field variations. An alternative method is to use Helmholtz coils, which are adjusted to produce a field exactly equal and opposite to the stray field present at any time. The system is made automatic and includes a magnetometer probe placed on the vicinity of the spectrometer and connected electrically to adjust the current in the Helmholtz coils as required to maintain a constant field.

12.3.7 Sample Handling

Surface techniques usually require the sample to be analyzed to be placed in a special environment, which, in the case of ESCA, should be a good vacuum. This is because electrons escaping from a solid at atmospheric conditions would be able to travel only a negligible distance before being stopped by air molecules, which would render analysis of their energy impossible. Moreover, if the sample is composed of a reactive material (for example, aluminium), the pressure on the sample chamber must be very low, to prevent the surface being oxidized or otherwise contaminated. Figure 12.15 shows a technique for inserting a sample originally at atmospheric pressure in a vacuum chamber. Three samples are attached to a rod, which slides through four seals. As the samples pass the second of these seals, they are evacuated to about 10^{-2} torr in a few seconds. They are then transported into the sample preparation chamber, where the pressure is approximately 10^{-7} torr. Here they can be cleaned by a beam of ions to remove layers of oxidation, or be otherwise prepared if desired.

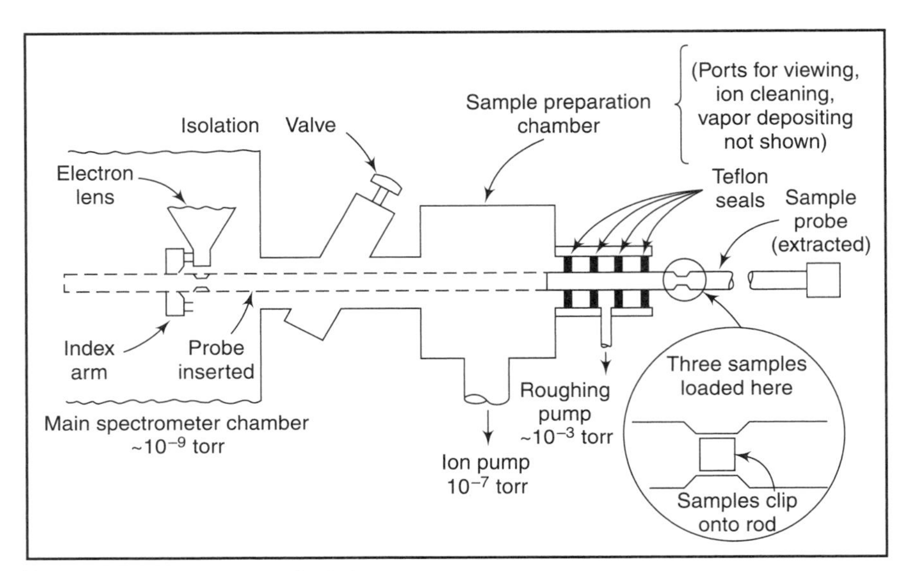

Figure 12.15 :: Sample inlet system

The rod is then inserted further into the main spectrometer chamber, and one of the three samples is placed in the proper position for analysis. The rod automatically positions the sample in the X-ray beam, so the emitted electrons can enter the electron lens.

The sample can be analyzed at any desired temperature between -150 and 300°C by using a special sample rod. A controller is supplied with a variable temperature probe, that automatically regulates the sample temperature at the desired value. Another special probe that permits samples to be vapour-deposited on a substrate is also provided with some instruments. Any material volatile enough to be vaporized below 1700°C can be evaporated. This probe is useful for extremely reactive samples. Samples can be cleaned to remove layers of oxide or other surface impurities with an ion gun attached to the sample preparation chamber.

Charging effect

A source of difficulty in ESCA and AES is the sample charging effect. The X-rays incident on the sample cause electrons to be emitted, so the sample surface acquires a net positive charge. If the sample is electrically conductive and in contact with the metal parts of the spectrometer, the positive charge cannot accumulate. But if the sample is an insulator, the positive charge will quickly build up and this charging effect could be very serious in some situations.

The charging effect can be turned off by supplying a flood of electrons having a uniform low energy to the sample. The surface potential can thus be clamped to a potential determined by the energy of these electrons. This capability actually results in additional useful information about the sample.

Improvements in computer technology and X-ray optics have made XPS the technique of choice for surface chemical characterization, achieving spatial resolution of the order of 3–5 micrometers, good enough for chemical imaging.

12.4 Ion Spectroscopy

Ion spectroscopy is another powerful technique applied to surface characterization. The two techniques usually adopted are ion scattering spectroscopy (ISS) and secondary ion mass spectrometry (SIMS). The principle of ion spectroscopy is shown in Figure 12.16. When primary ion, usually an inert gas ion, having a kinetic energy of 0.3–5 keV is incident on a surface, one of the following phenomena can occur (Hercules and Hercules, 1984b):

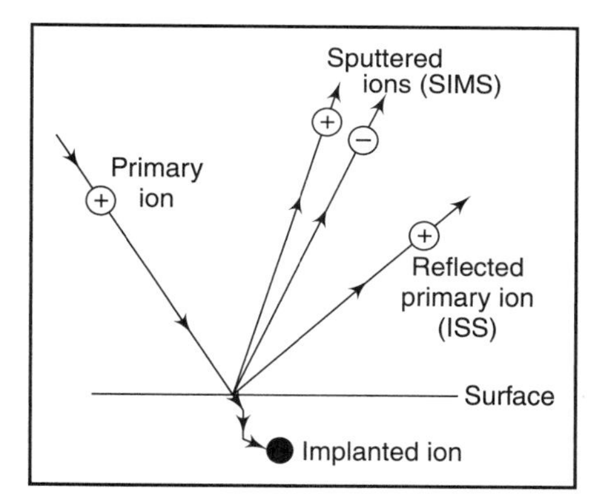

Figure 12.16 :: Fundamental processes important for SIMS and ISS

(i) The primary ion can be elastically scattered by a surface atom, resulting in a reflected primary ion. The kinetic energy of the reflected primary ion will depend upon the mass of the surface atom involved in the scattering process. It is the reflected primary ion which is measured in ISS.

(ii) The primary ion can penetrate the surface and become embedded in the solid. The penetration of the ion in the lattice results in considerable disruption by transfer of momentum to lattice atoms or molecules. This is called sputtering. For approximately 1 keV ions, this process takes place only several atomic layers deep, resulting in expelling atomic and molecular fragments. These fragments can be either neutral atoms or ions, which could be both positive and negative ions. The ions are referred to as secondary ions, and thus the term secondary ion mass spectrometry (SIMS).

Table 12.3 gives the analytical characteristics of ISS and SIMS.

Table 12.3 :: Analytical Characteristics of ISS and SIMS (after Hercules and Hercules, 1984c)

Parameter	ISS	SIMS
Range	Energy range: 1 keV	Spectral range 0–500 amu
Analysis depth	Samples top atomic layer	40 Å (dynamic), monolayer (Static)
Elemental sensitivity,		
Elements	Li to U	all
Specificity	variable	good
Sensitivity variations	30 X	10^5
Quantitative analysis		
Absolute	±30%	not possible
Relative	± 10%	±50%
Detection limit	10^{-3} % monolayer	10^{-4} % monolayer
Other aspects		
Matrix effects	some	Severe
Vacuum	10^{-5} torr of scattering gas	10^{-5} torr of ionizing gas
Depth profiling capability	yes, slow	yes, rapid (dynamic SIMS)
X-Y resolution	poor, 100 μ	1 μ with ion microprobe
Sample destruction	some sputtering	yes, sputtering of surface

12.4.1 Instrumentation for Ion Spectroscopy

A block diagram of a typical ISS/SIMS instrument is illustrated in Figure 12.17. Ions are formed by bombarding gas atoms with electrons. The positive ions are accelerated and focused on the sample at an angle of 45°. Ions are scattered in all directions. However, only those electrons which are in a selected small solid angle are received in the 127° electrostatic analyzer. The detector can be a channel electron multiplier or a solid state (Si) device. In order to obtain ISS spectra, the backscattered primary ions are sampled by the cylindrical mirror analyzer and their kinetic energies are measured.

The fundamental equation for single-event ion-scattering is:

$$E_1 = E_0 \left[\frac{\cos\theta \pm \sqrt{\alpha^2 - \sin^2\theta}}{1+\alpha} \right]^2, \qquad \text{(ii)}$$

where E_0 = energy of the incident ion,

E_1 = *energy* after collision with a surface atom,

$\alpha = M_2/M_1$, where M_1 = the mass of the incident ion

and M_2 = mass of the target surface.

Equation (ii) is a complex function of the angle (θ) between the initial direction of the ion and the direction the ion takes after the scattering process. For a scattering angle of 90°.

$$E_1 = E_0 \left[\frac{M_2 - M_1}{M_2 + M_1} \right]$$

This is valid only if $M_1 < M_2$. The energy E_1 is most sensitive to small differences in M_2 if M_1 is only slightly smaller than M_2. So, it is advantageous to use ions from a variety of gases. The noble gases, helium and argon are most frequently selected to avoid side effects.

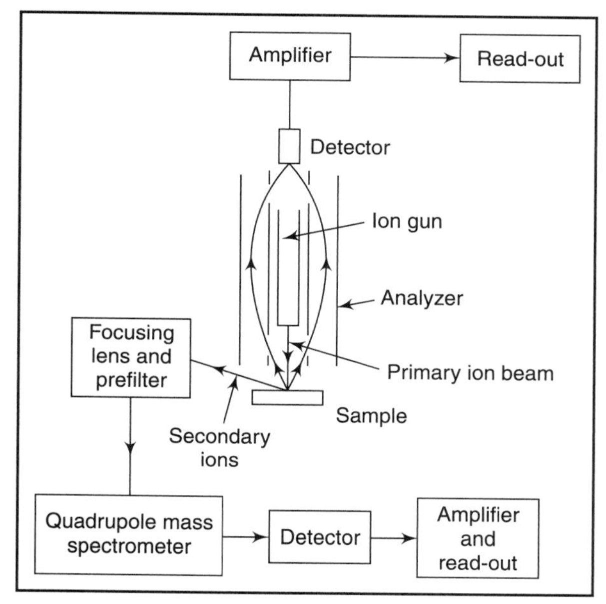

Figure 12.17 :: Block diagram of ISS/IMS instrument

ISS is sensitive to every element heavier than helium, since the lightest isotope used as a primary ion is ^{3}He and the scattering element must be heavier than the scattering gas. The specificity (ability to separate two particular elements) will vary depending upon the scattering gas used. The detection limit of ISS is probably of the order of 10^{-3} per cent of a monolayer, i.e. less sensitive than SIMS, but more sensitive than either Auger or ESCA. ISS can be used effectively for depth profiling, particularly in the range 0-100Å.

For recording of the SIMS spectrum, the secondary ions are collected and their masses are determined in a small dedicated mass spectrometer. The secondary ions emitted from the sample are first focused through a lens and a pre-filter. The pre-filter is effectively a discriminator used to select ions having a particular kinetic energy range. The ions are then passed on to a quadrupole mass spectrometer. The ions coming out of the mass spectrometer are detected by an electron multiplier, amplified and given to a read-out system. As with ESCA and Auger spectrometers, ISS/SIMS instrument must operate in high vacuum.

The quadrupole mass spectrometer used in SIMS instruments operates in the range 500–1000 amu with a resolution of 1 amu. SIMS shows good specificity, though there is some overlap between peaks of different elements. Because of severe matrix effects, sensitivity variations between the most sensitive and least sensitive elements are about 10^5, therefore, performing absolute analyses with SIMS generally is not possible. Also, the relative standard deviations even for calibrated systems are poorer than for the other surface analysis techniques.

The analysis depth of SIMS varies depending upon whether one is operating in the dynamic or static mode. In the dynamic mode, the primary ion current density is approximately 10^{-6} A/cm^2 and scrambling occurs in a layer at least about 40Å at best. Thus, dynamic SIMS has a depth resolution of 40 Å at best (Hercules and Hercules, 1984b). When operating in the static mode (10^{-9}A/cm^2), each sputtering event comes from a virgin surface, and thus one can assume that the SIMS spectrum is characteristic of the top few atomic layers.

12.5 Scanning Tunnelling Microscopy

Scanning tunnelling microscopy (STM) can image surfaces of conducting materials with atomic scale resolution. It uses an atomically sharp metal tip that is brought very close to the surface. When the tip and sample are connected with a voltage source, a small tunnelling current flows between the tip and sample surface. This current can be measured, whose magnitude depends on the distance between the tip and the surface. As the tip is moved laterally across the surface, a feedback mechanism moves the tip up and down to maintain a constant tunnelling current. Rastering the tip across the surface, therefore, produces a topographic map of the surface. (Brian, 2004a).

12.6 Atomic Force Microscopy

Atomic force microscopy is similar to scanning tunneling microscopy (STM) in that it can image surfaces at atomic scale resolution (Brian, 2004b). The difference between AFM and STM is that AFM does not require the sample to be an electrically conducting material. Like STM, it uses an atomically sharp tip that is brought very close to the surface. The tip will feel a chemical attraction or repulsion and will move up or down on its supporting cantilever. The key to the sensitivity of AFM is in monitoring the movement of the tip. A common means of monitoring the tip movement is to use a laser beam that is reflected or diffracted by the tip or cantilever. The up or down movement of the tip is then detected by changes in the laser beam position. As in STM, rastering the tip across the surface produces a topographic map of the surface with atomic resolution.

Chapter 13

Radiochemical Instruments

13.1 Fundamentals of Radiochemical Methods

The use of radioactive isotopes has led to the development of radiochemical methods for analyses, and has facilitated the examination of phenomena, the measurement of which was formerly complicated or even impossible. These operations are based on the fact that radioisotopes (isotopes of elements with unstable atomic nuclei) emit radiation which can be detected by suitably located detectors. The proportion of radioactive atoms in the volume of material perceived by the detector can thus be determined by the measurement of the intensity of such radiation. Radiochemical methods offer the advantage of elimination of chemical preparation that usually precedes the measurement. These methods are both sensitive and specific, and are often characterized by good accuracy.

Radioisotopes

In one method, a pure but radioactive form of the substance to be determined is mixed with the sample in a known amount. After equilibrium, a fraction of the component of interest is isolated and the analysis is then based on the activity of this isolated fraction. Alternatively, activity is induced in one or more elements of the sample by irradiation with suitable particles. The measurement of this activity gives information about the element of interest.

Before the advent of the cyclotron and more recently the chain-reacting pile, most of the work with radioisotopes was done with naturally occurring radioactive elements. However, it is now possible to obtain artificially produced radioisotopes of most of the elements. Also, it is possible to obtain these in large quantities and with extremely high activity.

13.1.1 Time Decay of Radioactive Isotopes

Each radioactive isotope is characterized not only by the type and energy of radiations emitted, but also by the characteristic lifetime of the isotope. This

Lifetime of the isotope

is most conveniently designated by the half-life or half-period of the isotope. The half-period of a radioactive isotope is the time required for half the initial stock of atoms to decay. Thus, after one-half period has elapsed, the total activity of any single radioactive isotope will have fallen to half its initial value; after two periods, the activity will be one-quarter its initial value and so on. After 6.6 half-periods, the activity will be 1 per cent of the initial activity.

Decay is a random process which follows an exponential curve. The half-life of a radioactive isotope is given by:

$$t^{1/2} = 0.693/\lambda,$$

where λ is the decay constant for a particular radioisotope. In practice, the disintegration rates are determined by counting the number of disintegrations over a certain time t_m and finding the ratio of the number of disintegrations to the time t_m.

13.1.2 Units of Radioactivity

The unit of radioactivity is curie. This was originally defined to represent the disintegration rate of one gram of radium, but is now used as the standard unit of measurement for the activity of any substance, regardless of whether the emission is alpha or beta particles, or X or gamma radiation. When used in this way, the curie is defined as an activity of 3.7×10^{10} disintegration. The curie represents a very high activity. Therefore, smaller units such as millicurie or microcurie are generally used.

Energy: The basic unit used to describe the energy of a radiation particle or photon is the electron volt, eV. An electron volt is equal to the amount of energy gained by an electron passing through a potential difference of one volt. The energy of the radiation emitted is a characteristic of the radionuclide.

13.1.3 Types and Properties of Particles Emitted in Radioactive Decay

The theory of atomic structure proves that some elements are naturally unstable and exhibit natural radioactivity. On the other hand, elements can be made radioactive by bombarding them with high energy charged particles or neutrons, which are produced by either a cyclotron or a nuclear reactor, respectively. This process will alter the ratio of photons to neutrons in the atoms, thus creating a new unstable nucleus which could undergo radioactive decay. The extra neutron disintegrates and in the process, releases energy in the form of gamma radiation.

Radioactive emissions take place in three different forms, which are:

Alpha emissions: Alpha particles are composed of two protons and two neutrons. They are least penetrating and can be stopped or absorbed by air. They are most harmful to the human tissue.

Beta emissions: These are positively or negatively charged and are high speed particles originating in the nucleus. They are not as harmful to tissue as alpha particles, because they are less ionizing, but are much more harmful than gamma rays.

Gamma emissions: Emissions like X-rays constitute electromagnetic radiation that travels at the speed of light. They differ from X-rays only in their origin. X-rays originate in the orbital electrons of an atom, whereas gamma rays originate in the nucleus. They are due to an unstable nucleus. X-rays and gamma rays are also called 'photons' or packets of energy. As they have no mass, they have the greatest penetrating capability. Gamma rays are of primary interest in radiochemical methods of analysis.

Different types of particles are distinguished by their penetration power, with particles with the greatest mass and charge being the least penetrating. The energies of alpha and beta particles and gamma radiations are expressed in terms of the electron volt. One election volt is the energy that an electron would acquire, if it were accelerated through a potential difference of one volt. Radioactive emissions have energies of the order of thousands or millions of electron volts. Alpha emission is characteristic of the heavier radioactive elements such as thorium, uranium, etc. The energy of alpha particles is generally high and lies in the range 2 to 10 MeV (million electron volt). Their penetrating power is low and are completely stopped by foils, solid materials like aluminium. Due to larger ionizing power of alpha particles, they can be distinguished from beta and gamma radiations on the basis of pulse amplitude they produce on a detector.

Beta emission consists of a very energetic electron or positron (beta particles that carry a unit positive charge). Their penetration power is substantially greater than alpha particles and have energy range 0–3 MeV.

Gamma rays are high energy photons with high penetrating and low ionizing power.

13.1.4 Interaction of Radiations with Matter

Beta particles interact primarily with the electrons in the material through which they pass. The absorption depends mainly upon the number of electrons in their path. The molecules of the matter may be dissociated, excited or ionized. However, it is the ionization which is of primary importance in the detection of beta particles.

Ionization

Alpha particles have a relatively large mass and higher charge, and the specific ionization produced by them is much larger than for beta particles.

Upon interaction with matter, gamma rays ionize energy through three modes: the photoelectric effect transfers all the energy of the gamma ray to an electron in the inner orbit of an atom of the absorber. This involves the ejection of a single electron from the target atom. This effect predominates low gamma energies with target atoms having a high atomic number.

The Compton effect occurs when a gamma ray and an electron make an elastic collision. The gamma energy is shared with the electron and another gamma ray of lower energy is produced, which travels in a different direction. The Compton effect is responsible for the absorption of relatively energetic gamma rays. When a high energy gamma ray is annihilated following interaction with the nucleus of a heavy atom, pair production of a positron and an electron results. Pair production

Compton effect

Pair production

becomes predominant at the higher gamma ray energies and in absorbers with a high atomic number. The number of ion pairs per centimetre of travel is called 'specific ionization'.

13.2 Radiation Detectors

Several methods are available for the detection and measurement of radiation from radionuclides. The choice of a particular method depends upon the nature of the radiation and the energy of the particle involved.

If the radiation falls on a photographic plate, it would cause darkening when developed after exposure. The photographic method is useful for measuring the total exposure of workers to radiation, who are provided with film badges. Better methods are available for an exact measurement of the activity.

13.2.1 Ionization Chamber

The fact that the interaction of radioactivity with matter gives rise to ionization makes it possible to detect and measure the radiation. When an atom is ionized, it forms an ion pair. If the electrons are attracted towards a positively charged electrode and the positive ions to a negatively charged electrode, a current would flow in an external circuit. The magnitude of the current would be proportional to the amount of radioactivity present between the electrodes. This is the principle of the ionizing chamber.

An ionization chamber consists of a chamber which is filled with gas and is provided with two electrodes. A material having a very high insulation resistance such as polytetrafluoroethylene is used as the insulation between the inner and outer electrodes of the ion chamber. A potential difference of a few hundred volts is applied between the two electrodes. The radioactive source is placed inside or very near to the chamber. The charged particles moving through the gas undergo inelastic collisions to form ion pairs. The voltage placed across the electrodes is sufficiently high to collect all the ion pairs. The chamber current will then be proportional to the amount of radioactivity in the sample. lonization chambers are operated either in the counting mode, in which they respond separately to each ionizing current, or in an integrating mode involving collection of ionization current over a relatively long period.

Figure 13.1 shows an arrangement for measuring the ionizing current. The current is usually of the order of 10^{-10}A or less. It is measured by using a very high input impedance voltmeter, which has a MOSFET in the input stage. The current is indicated on a moving coil type ammeter. Alternatively, the null method can also be used. In this method, the change in voltage produced across a capacitor by the ionizing current is counter-balanced by an equal and opposite voltage supplied from a potentiometer. A potentiometric recorder of the self-balancing type can be used to record the signal.

The magnitude of the voltage signal produced can be estimated from the fact that the charge associated with the 100,000 ion pairs produced by a single alpha particle traversing approximately 1 cm in air would be around 3×10^{-14} coulomb. If this average charge is made to pass through a resistance of 3×10^{10} Ω in 1s, a difference of potential of approximately 1 mV would develop across the high resistance. This voltage is a function of the rate of ionization in the chamber.

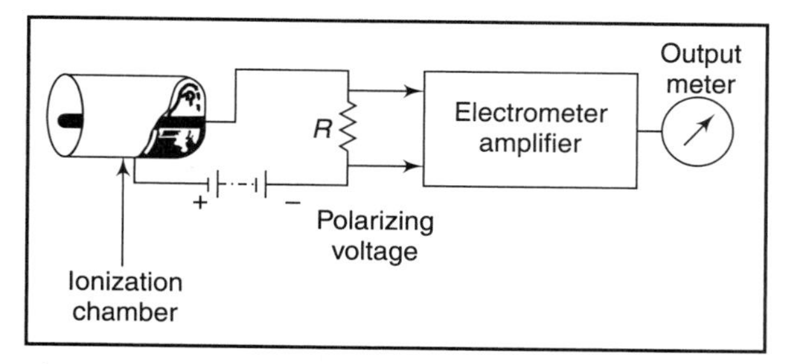

Figure 13.1 :: Schematic diagram for measuring ionizing current using DC ionizing chamber

Liquid samples are usually counted by putting them in ampoules and placing the ampoules inside the chamber. Gaseous compounds containing radioactive sources may be introduced directly into the chamber. Portable ionization chambers are also used to monitor personnel radiation doses.

13.2.2 Geiger–Muller Counter

GM tube

The Geiger counter is commonly called the GM tube. This tube consists of a metal cylinder (Figure 13.2), which acts as a cathode and is about 1 to 2 cm in diameter. It has an axial insulated wire working as an anode and is capable of being maintained at a high positive potential of the order of 800 to 2500 V. This assembly is placed in a tubular glass envelope containing a gas or mixture of gases, which is easily ionizable. The envelope is gas-tight and is

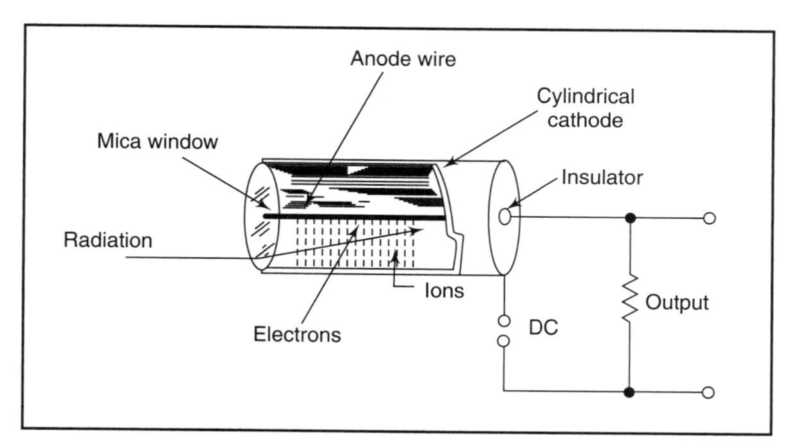

Figure 13.2 :: Construction of a Geiger-Muller tube

typically filled to a pressure of 80 mm of argon gas and 20 mm of alcohol. Alcohol, butane or bromine acts as a quenching gas and argon as the ionizing gas. The tube contains a window of thin mica or other suitable material, which permits effective passage of beta and gamma radiation, but not of alpha radiation.

As the gas is ionized in the counting tube, migration of ions takes place towards the appropriate electrodes under the voltage gradient. They soon acquire sufficient velocity, cause further ionization and give rise to an avalanche of electrons traveling towards the central anode. As a result, ion multiplication spreads to a complete sheath around the anode, and the same pulse size is observed for each primary ionizing event. The process, in fact, produces a continuous discharge, which fills the whole active volume of the counter in less than a microsecond. Each discharge builds up to a constant pulse of 1–10 V. This pulse amplitude is sufficient to operate a ratemeter or scaler without using any amplifier.

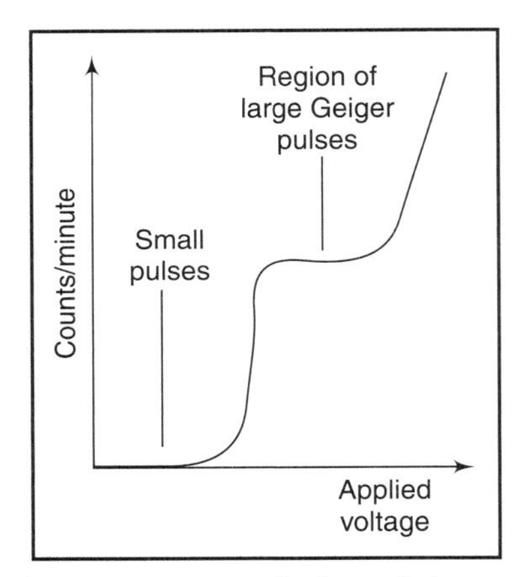

Figure 13.3 :: Variation of the count rate with voltage using Geiger–Muller counter

Figure 13.3 shows the variation of the count rate recorded by a typical Geiger counter, when the polarizing voltage is altered. The tube works in the voltage range exhibited by the plateau. Below the starting voltage, no counts are recorded. Between the starting voltage and beginning of the plateau, the voltage is too low to produce constant pulse size. Also, beyond the plateau, the count rate increases, because of breakdown and spurious discharges through the tube. The plateau is observed between 800 to 1400 V for commercial tubes. The slope of the plateau is generally expressed in terms of a percentage of the count rate per volt. It should not be more than 0.1 per cent per volt for a counter in a satisfactory condition. In order to minimize the counts due to background, the counter is normally placed inside a lead shield.

Since the positive ions produced by ionization, are much heavier, they have much lower mobilities than electrons. Therefore, they move only a very short distance towards the cathode in the time interval required for the electrons to cross the space between electrodes. This travel time may be several hundred microseconds and can vary from counter to counter. During this period, the positive ions form a sheath around the anode wire, which effectively lowers the potential gradient to a point wherein the counter becomes insensitive to the entry of fresh ionizing particles. This is called dead time of the counter. The dead time of the self-quenched GM counter is usually reduced by rapidly dropping the counter voltage below the starting potential. The use of the quenching circuit is preferred because it offers two advantages: reduction of the counter dead time, and prolongation of the lifetime of the counter by reducing the number of organic molecules dissociated in each discharge.

Dead time

In a GM tube containing ethyl alcohol as a quenching agent, some of the vapours are dissociated each time a count is recorded, so that the counter has a limited life. The life of these tubes is limited to perhaps 10^8 to 10^{10} counts. The tubes containing halogens have much longer life, because halogens simply dissociate during ionization and re-combine afterwards. The counting tubes containing halogens (bromine) can be used at low temperatures. The halogens, however, may be consumed by reaction with electrodes and other metallic parts of the tube. Geiger counters are utilized more often than any other counter. They have also been made in the miniature form with dimensions of less than one inch and a diameter of 2 mm.

13.2.3 Proportional Counter

The proportional counter is an ionization chamber that is operated at voltages beyond the ordinary ionization chamber region, but below that of the Geiger region. These counters are called proportional counter, because the output pulse from the chamber stalls to increase with the increase in the electric field strength at the central electrode, but is still proportional to the initial ionization. In these counters, the pulse produced is amplified by a factor of 1,000 or more. The design of the counter and the value of the applied voltages are so chosen that a high voltage gradient exists close to the anode. The radius of cathode is about 1 cm and that of the anode 0.001 cm, with a polarizing voltage of the order of 1000 V. The output pulse is generally of a few millivolts and therefore, requires amplification before the signal can be given to a scaler for counting.

The radioactive source may be placed inside the counter or outside it. In the former case, it avoids window absorption. In the continuous gas flow type counters, an argon–methane mixture flows at atmospheric pressure from compressed gas tank at a rate of 200 ml/min. Counter-life is, therefore, unlimited as the filling gas gets constantly replenished.

Figure 13.4 shows the schematic diagram of the counting equipment used with a flow type proportional counter. The pre-amplifier is a voltage follower, which provides high input

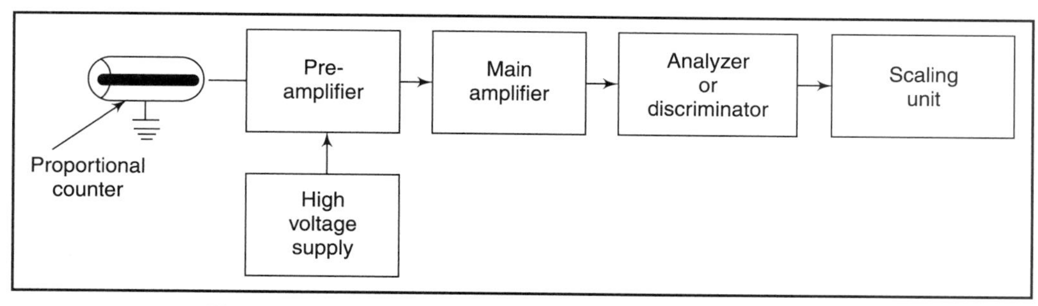

Figure 13.4 :: Counting equipment used with a proportional counter

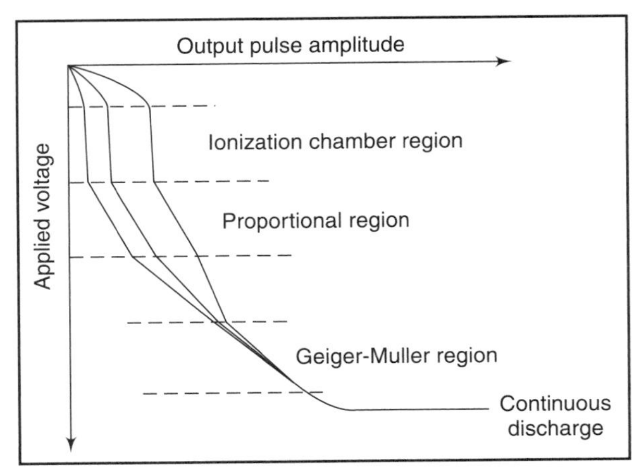

Figure 13.5 :: Range of operation of different types of counts

impedance. This is followed by a low noise linear amplifier having a very stable gain in the range of 500 to 1000. The amplifier has to be of the non-overloading type, since large pulses from a gamma ray background may overload a conventional amplifier for an appreciable time, causing the counts to be missed. In order to avoid attenuation and distortion of pulses due to capacitance of the long connecting cables, the pre-amplifier should be placed very near the detector. Figure 13.5 shows the range of operations of different types of counters.

In the proportional counter, different particles would yield output pulses of amplitude proportional to the isotope energy. By employing a pulse–height analyzer, which counts a pulse only if its amplitude falls within certain specific limits, a proportional counter can be made to respond to beta rays or X-ray frequencies, etc. In single-channel pulse height analyzers, provision is made for lower and upper energy discrimination, so that only pulses having amplitudes between the levels are passed. The voltage between the discriminating settings is called window, gate or width.

Scaling unit

The scaling unit counts down the pulses from the analyzer, so that they are digitally displayed. A decade system of counting is employed, which displays units, tens, hundreds, thousands and ten thousands of counts. Most of the scalers incorporate a counter/timer, which displays the time taken to record a definite number of counts, or the number of counts which occur within a definite time interval. Scalers can often be replaced by rate meters, which continuously indicate or record the mean value of the rate, at which pulses from the analyzers are applied to it.

13.2.4 Scintillation Counter

Scintillation is the process of turning radioactive energy (e.g. the energy associated with a beta particle) into light by using a 'scintillator'. A scintillator is thus a substance which produces minute flashes of light in the visible or near ultraviolet range, when it absorbs ionizing radiation. In such cases, the number of fluorescent photons is proportional to the energy of the radioactive particle. The flashes occur due to the re-combination and de-excitation of ions and excited atoms produced along the path of the radiation. The light flashes are of very short duration and are detected by using a photomultiplier tube, which produces a pulse for each particle. A scintillator along with the photomultiplier tube is known as a scintillation counter.

Scintillator

There are two types of scintillators. These are:

- Liquid scintillators, and
- Solid scintillators.

Liquid Scintillators: These are used for low energy β^- emitters such as ^{14}C, S^{35} and He^3. The radioactive material is mixed into a scintillation fluid referred to as 'cocktail'. Liquid scintillation fluids are organic molecules dissolved in an organic solvent. The mixture is placed between two photomultipliers which record the light produced. Figure 13.6 shows the simple arrangement of a liquid scintillation counter. A count is recorded only when a pulse of light is simultaneously detected by both PMTs, thus acting as a coincidence counter. This helps to reduce the background noise.

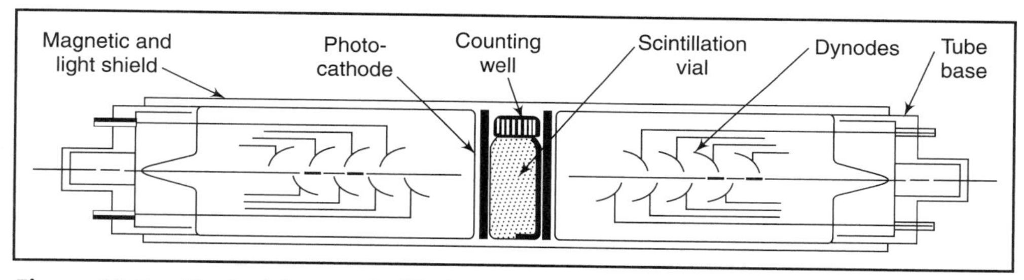

Figure 13.6 :: Typical beta scintillation counter showing only the counting well and photomultiplier tube detectors

Solid Scintillators: These are used for high energy β^- emitters such as Hg^{203}, Fe^{59}, Zn^{65} and Cd^{109}. The most commonly used solid scintillator is NaI (Sodium Iodide). Here, the radioactive sample is placed in a well (Figure 13.7) cut out of a crystal whereas the photomultiplier tube is mounted on one face of the crystal. Each β^- particle produces several thousand photons of light (wavelength: 400 nm) as it passes through the crystal, which are counted.

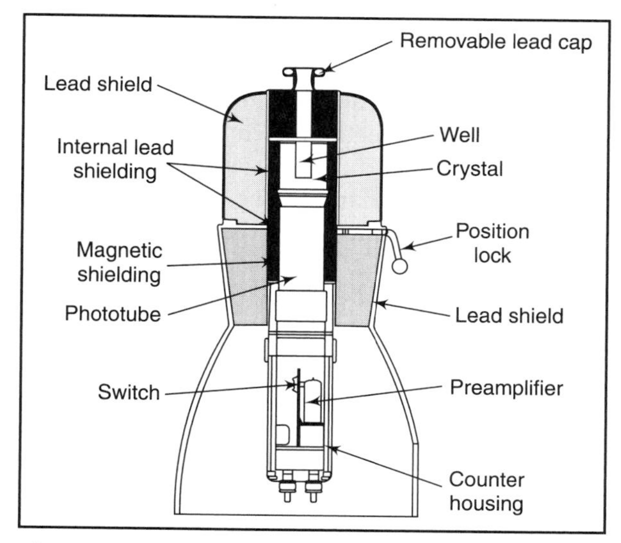

Figure 13.7 :: A well–type scintillation counter (Courtesy Texas Nuclear Division, Ramsey Engg. Co., Austin)

13.2.5 Gamma Counters

Gamma radiations cannot be detected directly in a scintillating material, because gamma rays possess no charge or mass. The gamma ray energy must be converted into

Scintillating material

kinetic energy of electrons present in the scintillating material. Thus the conversion power of the scintillating material will be proportional to the number of electrons (electron density) available for interaction with the gamma rays. Because of its high electron density, high atomic number and high scintillating yield, the scintillating material which is generally used as a gamma ray detector is a crystal of sodium iodide activated with about 0.5 per cent of thallium iodide. For counting beta particles, scintillator crystals of anthracene are employed. Being hygroscopic in nature, the crystal is usually mounted in a sealed aluminium container having a glass window on the side, which is in contact with the face of the photomultiplier.

For *in vitro* counting, the geometry of the scintillation crystal plays an important role. When it is convenient, the sample is placed in a well within the crystal and the crystal is coated on all sides with reflecting material, except for the side which is bonded to the face of the photomultiplier tube. In order to reduce the background counts, the crystal–photomultiplier tube assembly is mounted inside a cylindrical lead shield having a lead lid.

Gamma counting spectrometers

The instruments used for counting gamma particles are called gamma counting spectrometers. They may include an oscilloscopic display which is called energy scope. The energy scope provides a visual indication of the isotope spectrum. The Beckman Biogamma counting system provides a means of selecting the counting window by adjusting the variable discriminators and aids in selecting the proper high voltage. The vertical gain can be set in two positions to determine the display height of the scope. The LOW position is used when measuring high-energy isotopes such as Fe^{59} and Na^{22} and the high position is used with low-energy isotopes such as I^{125} and Co^{57}. The activity indicator indicates the amount of activity in terms of flashes of light. If the light glows constantly, a highly active source is in the counting chamber. If the light flashes, a less active source is in the chamber. Two variable discriminators permit adjustment from 0 to 1000 divisions to cover any part of an isotope spectrum. They are in fact two ten-turn potentiometers and are marked as upper and lower. The time selector switch is used to select one-of-ten time intervals (in minutes) that determine the length of time that each sample is counted. Count selector switch selects minimum number of counts that must be accumulated in the first 0.1 min. If the minimum is not reached, counting is terminated and the system moves to the next sample. Gamma counting systems generally include automatic sample changers which may hold 20 vial trays with 10 vials each. The vials are moulded from polypropylene to reduce gamma absorption in the vial walls. A printer can also be included to have a sample print-out of the counts made.

Samples containing weak beta emitters, such as H^3 and C^{14}, can be counted more efficiently by mixing the sample with a liquid scintillator, so that the scintillator is in intimate contact with the short range beta rays. Counting is carried out with a one inch diameter photomultiplier. The compound containing the radioactive source is dissolved in toluene or xylene, to which is added a primary scintillator and a secondary scintillator to increase the pulse height by acting as a wavelength shifter. The photomultiplier tube is dipped directly into the scintillator solvent and the counts are made.

Normally five adjustments are made in a scintillation counter to ensure proper counting of the ionizing particles. They are: high voltage setting, pulse-height-analyzer threshold voltage, analyzer channel width, amplifier gain and time constant of the rate meter.

Figure 13.8 shows the block diagram of an automatic gamma counting system MR 1032 from M/s Roche Medical Electronics. The maximum capacity of the sample changer in this system is 86 racks, each holding 12 sample tubes. The bidirectional sample changer has facilities for advance and reverse travel and up-down movement for single measurements. The sample changing cycle takes approximately 7 s.

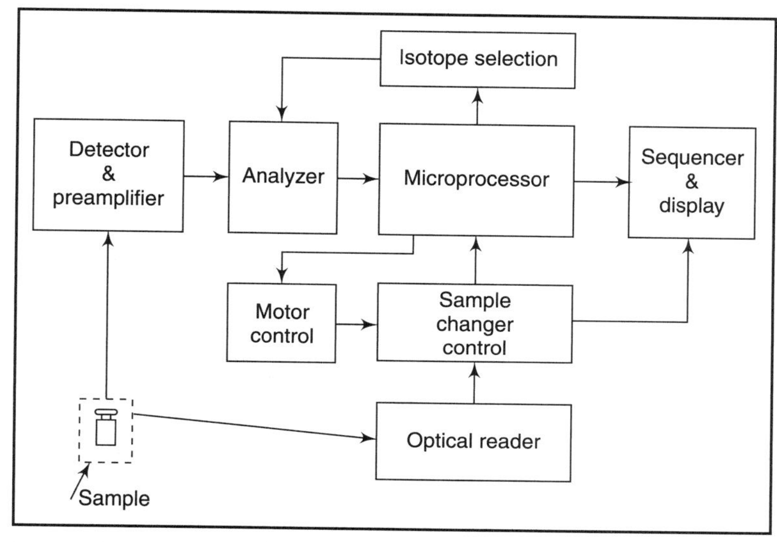

Figure 13.8 :: Block diagram of the automatic gamma counting system (Courtesy M/s Roche Medical Electronics, USA)

Sample and rack identification are sensed by on optical reader. Thus, each sample is identified and the rack number can be matched to a laboratory, or to a specific experiment. The detector is a well-type sodium iodide-crystal activated with a trace of thallium and coupled to a photomultiplier tube. The crystal is hermetically sealed in a thin can of aluminium and optically coupled to the sensitive face of an end-on PMT. Crystal and tube are covered with lead shielding for low and constant background.

The low energy models incorporate a pulse height analyzer with four pre-set windows, which discriminate between the following low-energy isotopes: ^{125}I ^{57}Co, ^{75}Se or ^{131}I. The high energy models incorporate a 7-isotope pulse–height analyzer. The capacity of the scalar is 1,000,000 counts. There is one scaler associated with each pulse-height analyzer. An increase in the number of counts accumulated will lead to a reduction in the percentage of statistical error and in increase in measuring accuracy. The display is a 7-digits one, which shows the counts per minute, elapsed time of the count in progress or identification number and the number of the sample being counted.

The system is directed and controlled by a microcomputer. This parallel 8-bit modular microcomputer has up to 65 k bytes memory capacity in PROM and RAM. The microcomputer incorporates a central processor module, a RAM module, a PROM module and an input-output module, which is designed for interface with a keyboard. The results are printed out in numbered A-4 formats.

13.2.6 Semiconductor Detectors

A great deal of development work has been done on semiconductor radiation detectors. These detectors can be made very small and robust. Silicon and germanium crystals have been employed mainly for counting alpha and beta particles. They function in a manner similar to that of the gas ionization chamber. On absorption of radiation in the crystal, electrons and positive holes are formed, which move towards opposite electrodes under the influence of applied potential. The resulting current is proportional to the energy of the ionizing radiation.

13.3 Liquid Scintillation Counters

Figure 13.9 shows a block diagram of a typical liquid scintillation counter which illustrates the manner in which the emitted radiation interacts with the cocktail (a mixture of a solvent and solute), leading to a count being recorded by the system.

Beta particles are emitted in a radioactive decay. In order to ensure the efficient transfer of energy between the beta particle and the solution, the solution is a solvent for the solvent material. The kinetic energy of the beta particles is dissipated and gets absorbed by solvent molecules thereby making them excited. The energy of the excited solvent is emitted as UV light and the solvent molecule returns to the general state, and thereby transfers energy to the solute.

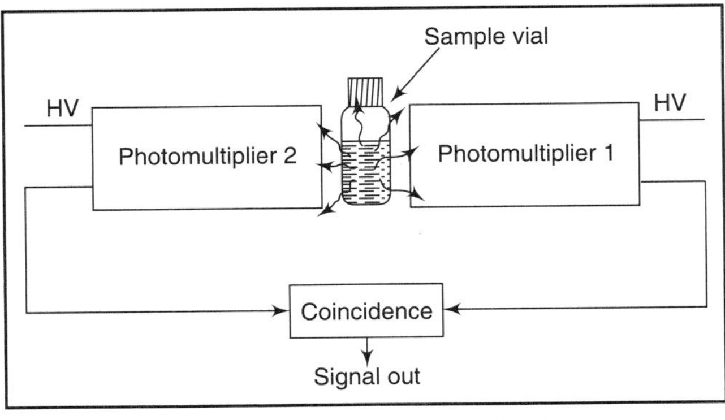

Figure 13.9 :: Block diagram of a scintillation counter

The solute is a fluor. As the excited orbital electrons of the solute molecule return to the ground state, a radiation results, in this case a photon of UV light. The UV light is absorbed by the fluor molecules, which emit blue light flashes upon return to the ground state.

Nuclear decay events produce approximately 10 photons per keV of energy. The energy is dissipated in a period of time of the order of 5 nanoseconds. The total number of photons from the excited fluor molecules constitutes the scintillation. The intensity of the light is proportional to the beta particles' initial energy.

Blue light flashes fall on the photocathode of the photomultiplier tube (PMT), which get converted into electrical pulses. The counter normally has two photomultiplier tubes. The signal from each PMT is fed into a circuit which produces an output only if the two signals occur together, i.e. within the resolving time of the circuit, which is generally 20 nanoseconds. By summing the amplitude of the pulses from each PMT, an output is obtained which is proportional to the total intensity of the scintillation.

The amplitude of the PMT pulse depends upon the location of the event within the vial. The pulse is large if it is closer to the PMT than if the event is more remote.

The analog electrical pulses are converted into a digital value. The digital value represents the beta particle energy. The pulses are then passed to an analyzer, where it is compared to the digital values for each of the channels of the counter. Each channel is the address of a memory slot in a multichannel analyzer which consists of many storage channels or storage slots covering the energy range from 0–200 keV. Instruments generally employ automatic data reduction including averaging of repeat sample counts, low count rejection and result normalization.

Modern liquid scintillation counters, such as the Beckman Model LS3801 (Figure 13.10) are programmable instruments. Parameters which can be programmed include the count time, sample repeat and count channel as well as radioisotope, which are to be observed. The instruments count samples in liquid form, either in the standard 20 ml glass or plastic vial or in a 6 ml miniature vial. Some instruments, like the Beckman Model 6500 can be programmed to accurately count dry samples, for which the instrument has to be specially designed.

Figure 13.10 :: Scintillation counter Model LS 3801 (Courtesy M/s Beckman)

The ability to sort the pulses detected according to their amplitude allows discrimination between emissions of different energies. For example, as shown in Figure 13.11, three commonly used isotopes H^3, C^{14}, and P^{32}, have beta energies of 18.3, 156 and 1,710, keV respectively. By setting windows judiciously, it is possible to specify which energy range has to be measured.

The number of pulses in each channel is displayed on a CRT or printed as a hard copy. In this manner, the sample is analyzed and the spectrum can be plotted

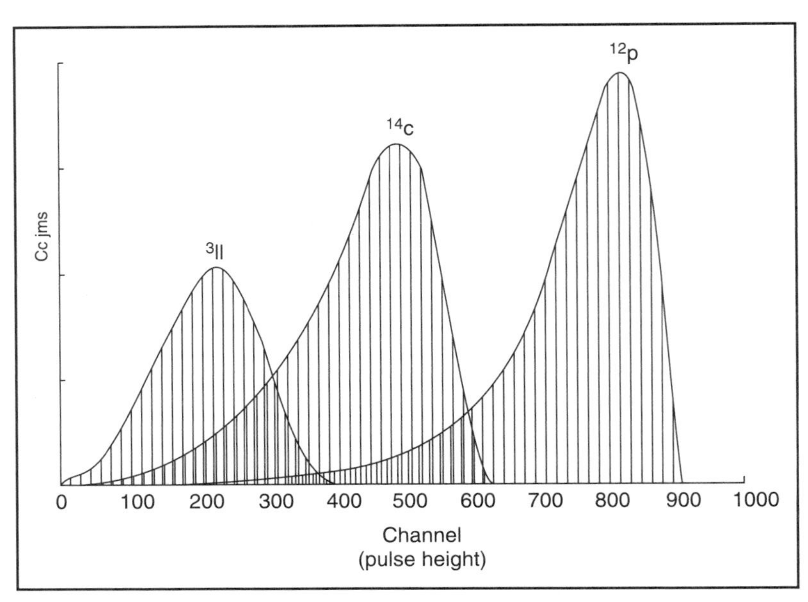

Figure 13.11 :: Pulse–height spectra of three radioisotopes from a liquid scintillation counter

to provide information about the energy of radiation or the amount of radioactive material dissolved in the cocktail.

Anything added to a counting vial (colour, solvents, filters, swabs) can reduce the efficiency of the scintillation process. This reduction in counting efficiency is called '*quenching*'. The three major forms of quench are:

Quenching

- *Chemical quench*: Some chemicals will affect the transfer of energy between the solvent and the fluors resulting in a reduction in the amount of light and a subsequent reduced counting efficiency.
- *Colour quench*: Solid and liquid scintillators emit light in the blue region of the spectrum. Red, green and yellow colours in the counting vial may absorb the light, resulting in reduced efficiency.
- *Self-absorption:* This occurs when radiation emitted by an isotope remains undetected due to absorption of the radiation by the sample itself, e.g. in precipitates, cells.

In practice, all samples are quenched to some degree. The counts per minute observed (cpm) may differ substantially from the true radioactive decay rate, disintegrations per minute, (dpm), depending upon the efficiency of the counting process. The counting efficiency is by definition:

$$\text{Counting Efficiency} = \frac{\text{cpm}}{\text{dpm}}$$

$$\text{Therefore, dpm} = \frac{\text{cpm}}{\text{Counting efficiency}}$$

In order to determine the counting efficiency, the amount of quenching has to be known. Several methods are used to characterize and quantify the quenching for a particular sample. A common method uses the so-called 'H-number' (*proprietary technique by Beckman Instruments) which is assigned by an on-line analysis of the Compton electron spectrum generated in the sample by shining briefly an external standard source on this sample.

A quench calibration curve can be constructed by plotting the counting efficiency versus the H-number, using a set of samples of known constant activity but containing varying amounts of quenching agent. In modern instruments, the quench curve can be stored in the machine's electronic memory such that the quenching correction is made automatically to directly provide the output as dpm.

Sometimes, spurious, non-reproducible counts will be observed, the cause of which may be static electricity. The handling of plastic scintillation vials with surgical gloves can build up a charge on vial. It is advisable not to wear gloves when loading the scintillation counter. Modern instruments incorporate electrostatic controllers to minimize the effect of static electricity.

It is known that photomultipliers used in liquid scintillation counters are a source of instability. The sources of these instabilities are: (i) variation of gain with temperature, (ii) variation of gain with tube current, (iii) variation of gain with sudden large changes in tube current, and (iv) effects due to aging. Stabilization systems based on the use of radioisotopes are common in scintillation counters used in nuclear spectroscopy. However, such systems are not practicable in a liquid scintillation counter, intended for counting low activity beta radiation sources. Several workers have reported stabilization systems which are not based on radioisotopes, and which, therefore, are potentially useful for liquid scintillation counters. These include systems based on the use of gas discharge tubes and those based on the use of modulated and other light sources. Light emitting diodes have also been employed for the stabilization of photomultipliers. This is done by the use of LEDs for stabilization of gain in photomultiplier, by means of adjustment of high voltage or pre-amplifier amplification.

13.4 Pulse–Height Analyzer

In radioactivity measurements, the individual particles are detected as single electrical impulses in the detectors. Also, various types of detectors can be set up to operate in a region in which the particular particle produces an electrical impulse having a height proportional to the energy of the particle. The measurement of pulse height is thus a useful tool for energy determination. In order to sort out the pulses of different amplitudes and to count them, electronic circuits are employed. The instrument which accomplishes this is called *pulse height analyzer*. These analyzers are either single or multiple-channel instruments.

Figure 13.12 shows a block diagram of a single-channel pulse height analyzer. The output pulses from the photomultiplier are amplified in a high input impedance low-noise pre-amplifier. Amplified pulses are fed into a linear amplifier of sufficient gain to produce output pulses in the amplitude range of 0–100 V. These pulses are then given to two discriminator circuits. A discriminator is nothing but a Schmitt trigger circuit, which can be set to reject any signal below a certain voltage. This is required for excluding scattered radiation and amplifier noise. The upper discriminator circuit rejects all but signal 3, and the lower discriminator circuit rejects signal 1 only and transmits signals 2 and 3. The two discriminator circuits give out pulses of constant amplitude. The pulses having amplitudes between the two triggering levels are counted. This difference in two levels is called the window width, the channel width or the acceptance slit and is analogous to monochromators in optical spectrometry.

Discriminator

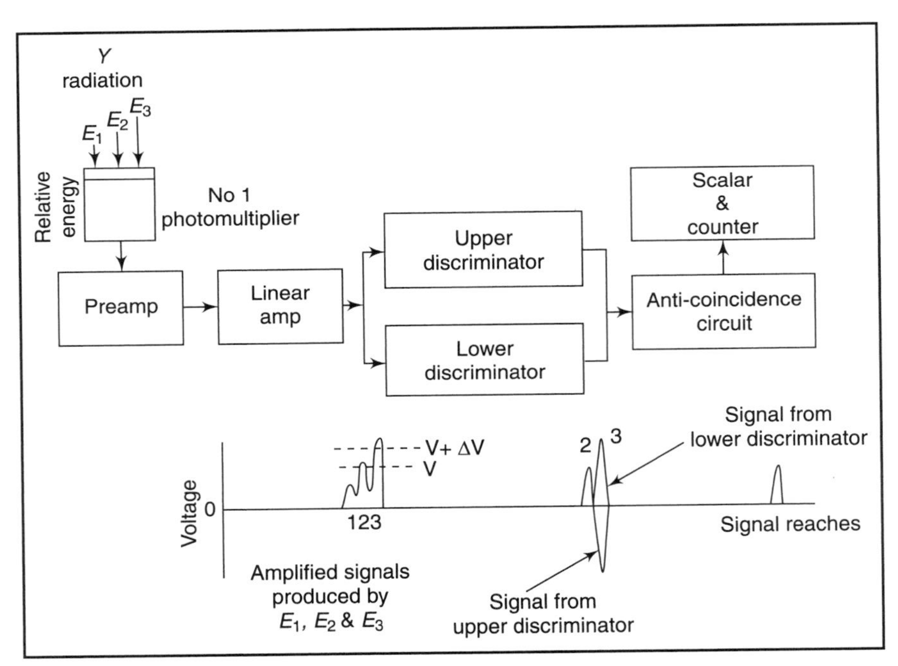

Figure 13.12 :: Block diagram of a pulse–height analyzer

Schmitt trigger circuits are followed by an anti-coincidence circuit. This circuit gives an output pulse when there is an impulse in only one of the input channels. It cancels all the pulses which trigger both the Schmitt triggers. This is accomplished by arranging the upper discriminator circuit such that its output signal is reversed in polarity and thus cancels out signal 3 in the anti-coincidence circuit. As a consequence, the only signal reaching the counter is the one lying in the window of the pulse height analyzer. The window can be manually or automatically adjusted to cover the entire voltage range with a width of 5–10 V. Scaler and counter follow the anti-coincidence circuit.

Multi-channel pulse height analyzers are often used to measure a spectrum of nuclear energies and may contain several separate channels, each of which acts as a single-channel instrument for a different voltage span or window width. The Schmitt trigger discriminators are adjusted to be triggered by the pulses of successively longer amplitude. This arrangement permits the simultaneous counting and recording of an entire spectrum.

A parallel array of discriminators is generally used, provided the number of channels is ten or less. If the number of channels is more than ten, the problems of stability of discrimination voltages and adequate differential non-linearity arise.

13.5 Gamma Spectrometry

From an analytical standpoint, a great majority of radioactive nuclei release gamma rays in their headlong tumble to a more stable existence, that is to say when they undergo radioactive decay. It is only a small minority that is not detectable by gamma spectrometry. Some of these are ^{3}H and ^{14}C and, of course, the measurement of their beta decay by liquid scintillation is very well established.

Figure 13.13 shows the decay scheme showing the origin of gamma radiation. In case of P^{32}, the energies of the beta particles emitted range up to a maximum of 1710 keV. In case Co^{60}, (Figure 13.13 b), beta decay occurs again, but not to the ground state of Ni^{60} but to an intermediate energy state. It is the dc-excitation of the state, which, in this case, occurs through an intermediate energy level, which is the source of gamma radiation. It may be observed that the gamma emission is not a decay process in the same sense as beta decay and alpha decay; but is one of the ways in which the excitation energy that the nucleus usually has, after one of those processes has occurred, is dissipated (Hemingway, 1986).

The essence of gamma spectrometry is the fact that the energies and abundances of gamma rays emitted are specific to a radionuclide. For example, in the figure shown, the energies of the two gamma rays released are Y_2 = 1173.2 keV (2505.7–1332.5) and Y_2 = 1332.5 keV (1332.5–0). If a sample emits two gamma rays of these energies, then you can be confident that the sample contains ^{60}Co, as these are unique to the decay of this nuclide. In general, every gamma-emitting radionuclide would emit its own characteristic pattern of mono-energetic gammas, but there are often dozens of different energies, each energy with its own particular abundance.

Semiconductor detector

Gamma spectrometry is based on the use of semiconductor detectors, i.e. germanium and silicon. The germanium detector is the most common, which, in the earlier days, required lithium compensation for the residual charge carriers due to impurities. These were called jelly [Ge (Li)] or lithium-drifted germanium. In these detectors, Li is incorporated into the semiconductor lattice by annealing the semiconductor with Li at a high temperature (~ 500°). A voltage of approximately 1000 V is placed across the semiconductor material with two electrodes and the electron cascade produced by a photoelectron is detected as an electrical pulse

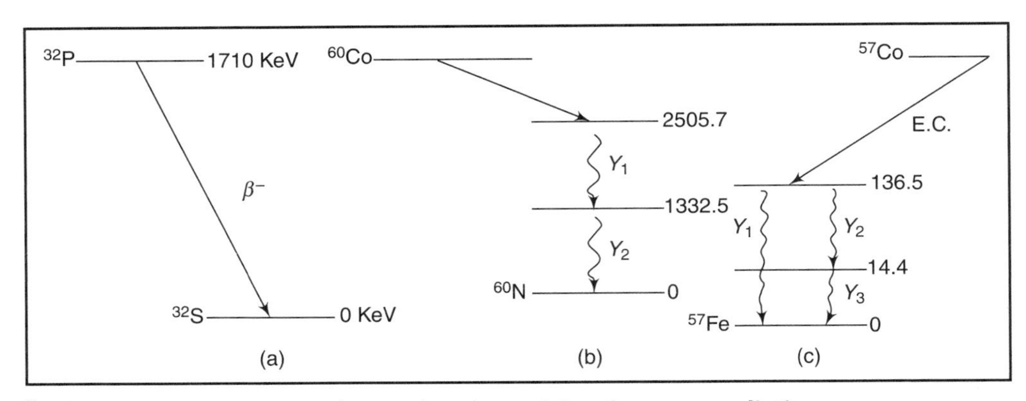

Figure 13.13 :: Decay scheme showing origin of gamma radiation

at the anode. But in high purity germanium, lithium compensation is not required and the detectors are called intrinsic or high purity.

In addition to being more robust than gas-filled or scintillator detectors, these semiconductor detectors also provide a much higher resolution. However, they need cooling to decrease the dark noise of the detector and current-to-voltage pre-amplifier.

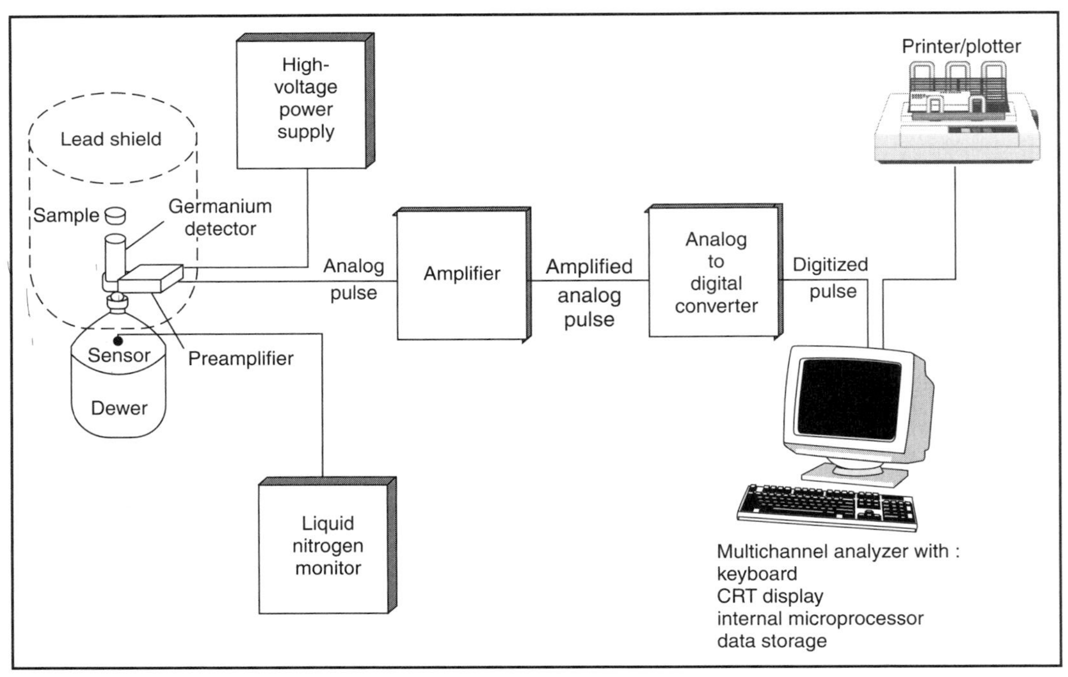

Figure 13.14 :: General arrangement of gamma spectrometry system equipped with a high purity germanium detector (Courtesy M/s Nuclear Data, IL)

Both types of detectors are used at liquid nitrogen temperature (77 K), where the thermally induced electronic noise is reasonably low. Most manufacturers have developed portable detectors, which can use a small Dewar flask of 0.5 to 2l volume of liquid nitrogen and have a start holding time of 24 h. While the Ge (Li) detectors must be maintained at 77 K all the lime, the intrinsic detectors can be allowed to warm up to room temperature without causing damage, and can thus be transported easily without depending on guaranteed supply of liquid nitrogen.

Figure 13.14 shows a block diagram of a gamma spectrometer. When gamma radiation is absorbed in the sensitive region of the detector, the energy deposited promotes electrons across the band gap, with the number of electron holė pairs thus produced being proportional to the absorbed energy. This charge is collected by a potential of a few thousand volts supplied by a high voltage power supply. A pre-amplifier covers the charge collected to a pulse, which is passed to the main amplifier. The pre-amplifier is necessarily mounted close to the detector. The main amplifier amplifies the pulse and suitably shapes it to be passed on to the multi-channel analyzer. The multi-channel analyzer sorts the pulses by height to give a pulse–height spectrum, which is equivalent to an energy spectrum.

Most modern gamma spectrometers use personal computers (PCs) to control and operate the multi-channel analyzer (MCA). The PC displays the gamma spectrum and the disk units to store the spectra. Packages containing MCA hardware and spectrum analysis software that are contained on a plug-in card into a PC are available. The information obtained from analysis would be the energy of peak maximum, which identifies the nuclide and the area of the peak, which, in turns when divided by the counting time to give a counting rate, quantifies the nuclide.

Chapter 14

X-Ray Spectrometers

X-rays are the short wavelength, high energy electromagnetic radiation associated with electronic transitions in atoms. The generation of X-rays and the interactions of X-rays with matter have numerous consequences of chemical importance.

X-radiation provides powerful tools for analytical purposes, which have advantages over other techniques. For example, in elemental analysis, it is, in principle, if not necessarily in practice, a non-destructive technique. Thus, a material may be analyzed for its elemental constituents without the need for conversion (as for example, in atomic absorption spectroscopy) to some particular chemical form. Several distinct ways in which X-radiation can be employed for this purpose are based on their following characteristics:

- *Absorption* of X-rays gives information about the absorbing material just as in other regions of the spectrum.
- *Fluorescence* emission of X-rays enables one to identify and measure heavy elements in the presence of each other and in any matrix.
- *Diffraction* of X-rays facilitates analysis of crystalline materials with a high degree of specificity and accuracy.

14.1 X-Ray Spectrum

X-ray tube

The most commonly used device for obtaining X-radiation is an X-ray tube. The conventional tube consists of a cathode (electron emitter) and an anode (target). Electrons from the cathode are accelerated by a high voltage (say 40 KV) applied to the target, which is usually a heavy metal like molybdenum. The electrons, upon impact on the anode, produce a broad range X-ray spectrum. It is due to the deceleration of the impinging electrons by successive collisions with the atoms of the target material. It has been observed that the emitted quanta are generally of a larger wavelength than the short wavelength cut-off. The intensity of the continuum rises to a broad maximum and falls off gradually with increasing wavelength. This cut-off wavelength is given (in nanometres) by:

$$\lambda_{\min} = \frac{hc}{Ve} = \frac{1240}{V},$$

where, h = Planck's constant

c = velocity of electromagnetic radiation

e = electronic charge

V = X-ray tube voltage in volts.

When the potential across the tube is increased, a stage is reached wherein the energy is sufficient to eject an electron from one of the inner shells of the atoms constituting the target material. The place of the ejected electron is filled up promptly by an electron from an outer shell, releasing a photon of X-radiation, whose wavelength is dependent upon the energy levels and hence characteristic of the material in the anode, or a specimen applied on the target. When originating in an X-ray tube, these lines will be superimposed on the continuum. As the inner electrons are not related to the state of chemical combination of the atoms (except for lighter elements), it follows that the X-ray properties of the elements are independent of chemical combination or physical state to a close degree of approximation.

X-ray wavelengths are normally specified as kX units, where 1 kX unit = 0.100202 nm. The commonly used wavelength unit in X-ray studies is the angstrom (Å). However, in SI units, the X-ray wavelengths are expressed in nanometres (nm) or picometers (pm). For analytical purposes, the useful X-ray wavelength range is 70 to 200 pm.

14.2 Instrumentation for X-Ray Spectrometry

Figure 14.1 shows the components of the instrumentation associated with X-ray methods. The important parts are:

- X-ray generating equipment,
- Collimator,
- Monochromators, and
- Detectors.

14.2.1 X-Ray Generating Equipment

Just as in the case of spectroscopy based on visible and UV radiation, it is possible to have both time and continuum sources for X-ray spectroscopy. An increasingly important source of high intensity, continuum X-radiation is the synchrotron. However, a synchrotron is an extremely expensive X-ray source and consequently, it is not employed in routine elemental analysis based

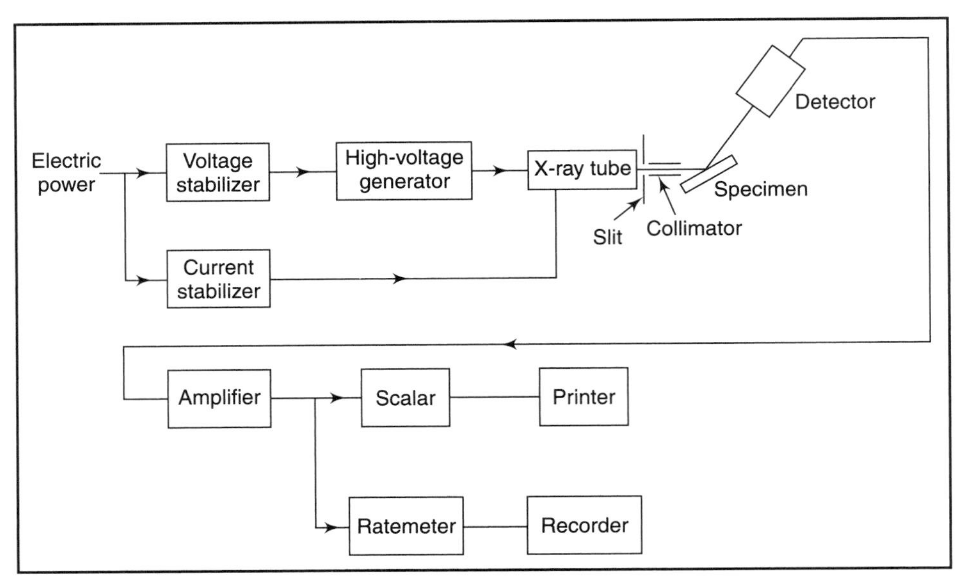

Figure 14.1 :: Basic components of instrumentation for X-ray spectroscopy

on X-ray spectrometry. For that, X-ray generators used in analytical instruments are basically the same as the system first discovered to emit X-rays by Röntgen in 1996.

Thus, X-rays are generated in specially designed high-vacuum sealed-off tubes, which have heavy metals such as copper or molybdenum as target. Targets made of tungsten, iron, chromium, nickel are also sometimes used for special purposes. Due to the bombardment of electrons under the influence of the high voltage, the target becomes hot, which can be either rotated or cooled with circulating water. The X-ray beam passes out of the tube through a thin window of beryllium.

The target is usually ground to a slight angle, so that it gives the desired shape of the focal spot from the side. For fluorescence work, the focus is of larger size (about 5 × 10 mm) and is viewed at an angle of about 20°. However, for diffraction studies, wherein sharper definition is required, the focal spot is a narrow ribbon and the source appears to be very small when viewed from the end. X-ray tubes are normally operated at 50 to 60 kV. Tubes operating at higher voltages, in the range of 100 kV, result in enhanced sensitivity, as the intensities of all lines increase with an increase in voltage. High voltages are generated in specially constructed transformers and employing full wave rectification.

14.2.2 Collimators

Collimation of an X-ray beam is achieved by using a series of closely spaced, parallel metal plates, or by a bundle of tubes, 0.5 mm or less in diameter. Increased resolution is obtained by reducing the separation between the metal plates of the collimator, but it will be at the cost of intensity. Collimators may be used: (i) between the specimen and the analyzer crystal, and (ii) between the analyzer crystal and the detector.

14.2.3 Monochromators

An X-ray tube generates not only lines but also continuum radiation. This is because the deceleration of high energy electrons within the target leads to the emission of Bremsstrahlung of all energies up to a maximum determined by the voltage applied to the tube. True X-ray emission from the target atoms, of course, does not occur until the critical inner shell ionization energy is achieved. Once this is exceeded, the intensity of line emission increases much more rapidly with increasing tube voltage than does the intensity of the background Bremsstrahlung or white radiation. Only the sharp features are determined by the nature of the target and, hence, they are termed characteristic lines.

An X-ray tube, however, can be used as a near monochromatic source in a rather simple way, which depends upon the special features of X-ray absorption by matter. Quantitatively, X-ray absorption at a given wavelength by a pure element is described by a law equivalent to the Lambert-Beer Law of UV and visible spectrophotometry, i.e.

$$I_t = I_o e^{-\mu m \rho x},$$

where

I_t = incident intensity of X-ray
I_o = transmitted X-ray intensity
μ_m = mass absorption coefficient (characteristic of the absorber)
ρ = density, and
x = thickness of the absorber.

Mass absorption coefficients are found to show a complex wavelength dependence showing sharp discontinuities, known as absorption edges, which correspond to the particular, allowed energy states of the absorbing atom.

It is, however, not so simple to construct a monochromator which can be varied as desired in the X-ray region, though narrow bands of wavelengths of various discrete points are simpler to obtain. Narrow bands can be achieved by different methods, which are discussed below.

(i) Using filters which consist of an element or its compound, which has a critical absorption edge (sharp discontinuities) at just the right wavelength to isolate a characteristic line from a source target. It is a common practice to insert a thin foil in the primary X-ray beam, to remove K_β-line from the spectrum, while transmitting the K_β-line with only a small loss of intensity. For example, a 35 kV molybdenum emission spectrum results in *Mo*-K_α (70.9 pm) and *Mo*-K_β (63.2 pm). Therefore, if a material could be found with a critical edge between these two wavelengths, it would absorb the β-line, while passing the α-line. Such a filter can be made of zirconium, which has its K-absorption edge at 68.9 pm. Figure 14.2 shows the

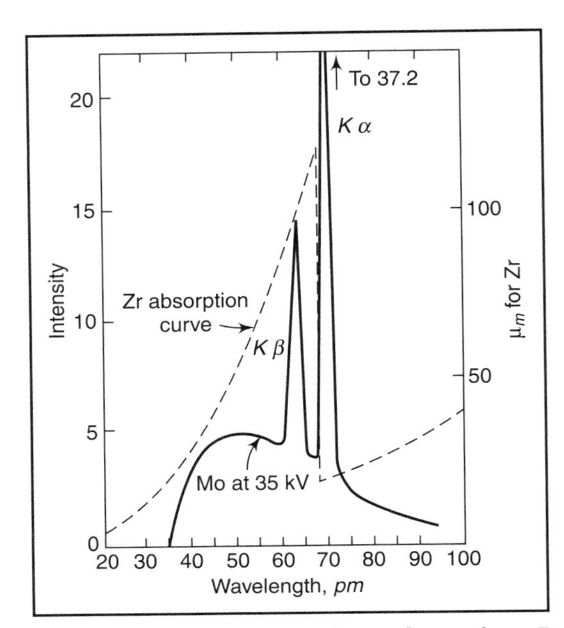

Figure 14.2 :: The zirconium absorption curve and the molybdenum emission spectrum

zirconium absorption curve along with the molybdenum emission spectrum. Characteristic wavelengths and filters for common targets of X-ray tubes are given in detail in Liebhafsky (1972).

(ii) Using a monochromator, in which an analyzing crystal of known spacing acts as a diffraction grating. This system could produce a narrow band of wavelengths at any point in the spectrum, if a continuous source of X-rays were available. The monochromator gives much better signal-to-noise ratio as compared to filters.

Since X-rays are fundamentally electromagnetic waves, the equation for diffraction by a grating which is $n\lambda = d \sin \theta$ is also applicable to X-rays. As the X-ray wavelengths are smaller by a factor of 1000 or more, the grating space (d) should be smaller by about the same factor to obtain reasonable values of θ. This requirement is difficult to meet by a grating, as it is impractical to rule it so finely. Fortunately, such orders of spacing exist between adjacent planes of atoms in crystals. The crystals suited for X-ray gratings are lithium fluoride, sodium chloride, topaz, calcium fluoride, gypsum, etc.

The monochromator arrangement consists in reflecting the X-ray beam from a plane crystal and selecting the wavelength by varying the angle. Referring to Figure 14.3, since the waves

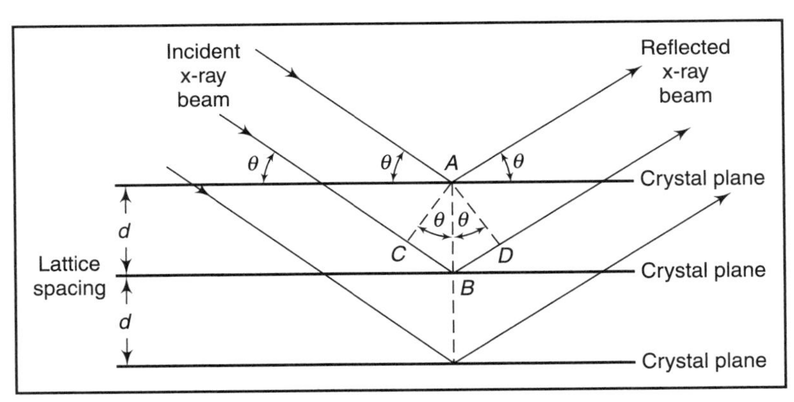

Figure 14.3 :: Diffraction of X-rays from a set of crystal planes

reflected at successive crystal planes must pass twice across the space between planes, the diffraction equation becomes:

$$n\lambda = 2d \sin \theta,$$

where d is the distance between adjacent planes in the crystal, and angle between the direction of the incident beam and that of the diffracted beam is 2θ. Obviously, the range of wavelengths obtainable from various analyzing crystals are governed by the d-spacing of the crystal planes and by the geometric limits to which the goniometer (on which the crystal is mounted) can be rotated.

(iii) Using radioactive sources which emit radiations in the X-ray region, in which case an X-ray tube with associated high voltage power supply, etc. is not required. These sources are not preferred, because they cannot be turned off and a radiation hazard is always present.

14.2.4 X-Ray Detectors

The earliest method of detecting X-rays was the use of a photographic film, which is no longer used for quantitative purposes, except for certain diffraction studies. Modern equipment use detectors, which are based on the properties of X-rays to produce flashes of light (scintillation) in certain materials or to cause ionization in others. Details of their construction and working are given in Chapter 13. However, it may be noted that the spectral sensitivity of each detector varies with the wavelength of the X-radiation. This is illustrated in Figure 14.4, which shows the relative spectral response of the three most important detectors. Each detector gives a somewhat broadened peak for a given monochromatic wavelength as shown in Figure 14.5. The abscissa scale is given

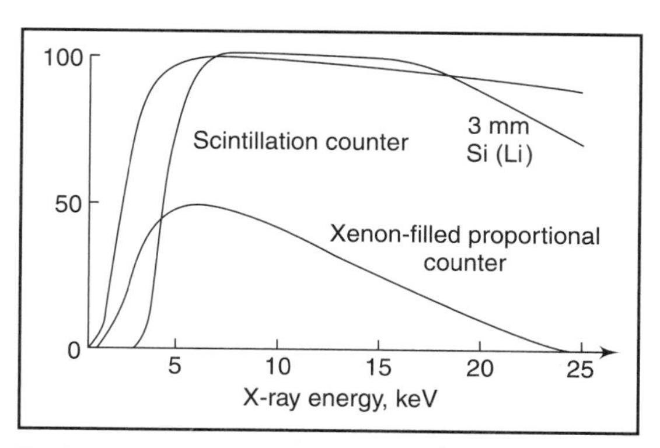

Figure 14.4 :: Spectral response of solid state [Si (Li)], scintillation and gas-filled proportional counters

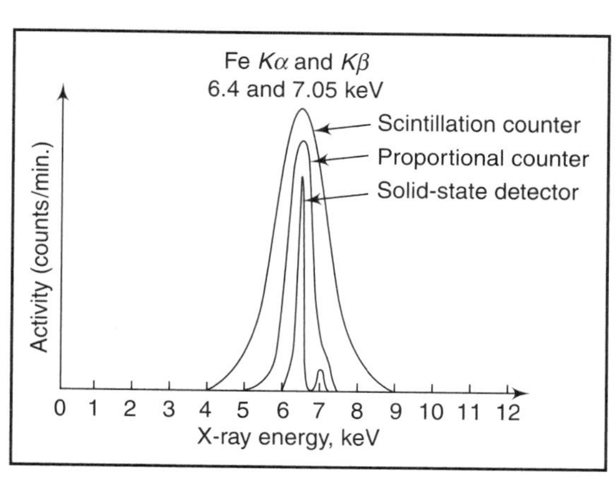

Figure 14.5 :: Comparison of resolution between scintillation, proportional and Si (Li) detectors

in energy units (reciprocal of wavelength). It may be appreciated that the Fe-K_α and Fe-K_β lines at 6.40 and 7.05 keV (193.7 and 175.7 pm), respectively, are reproduced only by Si(Li) detector and not by other detectors.

Solid state detectors

Solid state detectors are now the most widely used detectors. They employ silicon or germanium, which is sensitized to ionizing radiation, such as X-radiation, by the addition of lithium. This diffusion of lithium into the crystalline material is called drifting. When an X-ray photon enters the crystal, it dislodges electrons from the lattice, causing vacancies or holes. When a potential is applied across the detector, a potential gradient is set up. When an X-ray photon hits the detector, electron hole pairs are formed, which are collected at the electrodes. A pulse which is proportional to the energy of the X-ray is generated when the electron reaches the electrodes. The number of pulses indicates the intensity of the radiation.

The electrical output from the solid state detectors is much smaller than the corresponding signal from a scintillation detector, or a gaseous ionization detector. Therefore, high-gain, low-noise, amplification is required when using these detectors. In order to diminish noise pick-up, the pre-amplifier is generally located close to the detector in the same housing. Using a pulse–height discriminator, it is possible to discriminate electronically unwanted signals of different wavelengths.

Gas proportional counters and sodium–iodide scintillation counters suffer from poorer energy resolution. They are therefore, unable to differentiate between adjacent metals on the periodic table, which makes them unsuitable in such applications. Solid state detectors prove useful in such a situation.

14.3 X-Ray Diffractometers

Diffraction

Diffraction is a wave property of electromagnetic radiation that causes the radiation to bend as it passes by an edge or through an aperture. Diffraction effects an increase as the physical dimension of the aperture approaches the wavelength of the radiation. Diffraction of radiation results in interference that produces dark and bright rings, lines or spots, depending upon the geometry of the object causing the diffraction.

These interference effects are useful for determining dimensions in solid materials, and therefore, crystal structures. Since the distance between atoms or ions is of the order of 10^{-10}m (1Å), diffraction methods require radiation in the X-ray region of the electromagnetic spectrum.

The diffraction of X-rays is of great analytical significance, as it is applied to the study of the crystalline material producing the diffraction. No two chemical substances would form crystals, in which the spacing of planes is identical in all analogous directions. Thus, every crystalline substance would scatter the X-rays in its own unique diffraction pattern, giving a fingerprint or its atomic and molecular structure. X-ray diffraction is, therefore, adaptable to quantitative applications, because the intensities of the diffraction peaks of a given compound in a mixture are proportional to the fraction of the material in the mixture. This technique also furnishes a rapid, accurate method for the identification of the crystalline phases present in a material. In fact, sometimes it is the only method available, for determining which of the possible polymorphic forms of a substance are present, e.g. carbon in graphite or in diamond.

Diffraction and Bragg's Law

Diffraction occurs as waves interact with a regular structure whose repeat distance is about the same as the wavelength. The phenomenon is common in the natural world, and occurs across a broad range of scales. For example, light can be diffracted by a grating having scribed lines spaced on the order of a few thousand angstroms, about the wavelength of light (*www.geosci.ipfw.edu*).

X-rays have wavelengths of the order of a few angstroms, the same as that of typical inter-atomic distances in crystalline solids. This means that X-rays can be diffracted from minerals which, by definition, are crystalline and have regularly repeating atomic structures.

When certain geometric requirements are met, X-rays scattered from a crystalline solid can constructively interfere with each other, thereby producing a diffracted beam. In 1912, W.L. Bragg recognized a predictable relationship among the following factors:

- The distance between similar atomic planes in a mineral (the inter-atomic spacing), which is the d-spacing measured in angstroms.
- The angle of diffraction theta measured in degrees. For practical reasons, the diffractometer measures an angle twice that of the theta angle. Not surprisingly, the measured angle is '2-theta'.
- The wavelength of the incident X-radiation, symbolized by the Greek letter lambda, which is equal to 1.54 angstroms.

These factors are combined in Braggs' Law:

$$n\lambda = 2d \sin\theta,$$

where

n = an integer -1,2,3,..., etc. (n = 1 for our calculations)
λ (lambda) = wavelength in angstroms (1.54Å for copper)
d (d-spacing) = inter-atomic spacing in angstroms.

X-ray diffractometers

X-ray diffractometers are basically analogous to an optical grating spectrometer, with the difference that lenses and mirrors are not used with X-rays. Therefore, they appear to be quite different from their optical counterparts. The X-ray tubes, in some equipment, have several ports in different horizontal directions (when the tube is so mounted that the anode is vertical), permitting a number of experiments to be carried out simultaneously.

Debye-Scherrer Powder Camera

The diffracted beam of X-rays may be detected by one of the detectors described previously or photographically. A typical example of the use of a photographic method is in the Debye-Scherrer Powder Camera, which is illustrated in Figure 14.6. The sample in the powdered form, applied on any non-crystalline supporting

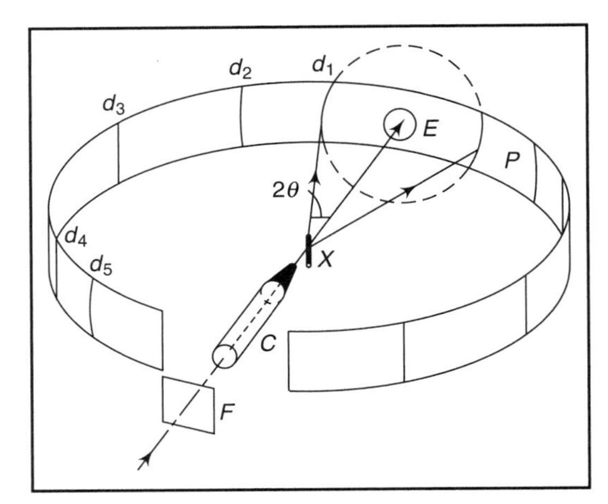

Figure 14.6 :: Debye–Scherrer X-ray powder camera—geometrical features

material (such as paper), is placed in the path of the X-rays. The particles of the sample are randomly placed, so it is quite possible that these are oriented in every possible direction relative to the beam of X-rays. This will result in diffracted rays corresponding to all sets of planes on the crystals.

A strip of X-ray film is mounted in a circular position around the sample. Radiations are made to fall on the sample x, but before that, it passes through filter F and collimator C. The undeviated central beam passes out through a hole E cut in the film strip P. Diffracted beams fall on the film at various points like d_1, d_2, d_3, etc. When the film is developed, it will show a series of lines on both sides of the central spot (produced by the undeviated beam). Obviously, the distance on the film from the central spot to any line will be a measure of the diffraction angle θ. So, if the wavelength and order n are known, the Bragg equation can be used to calculate the spacing d.

Cameras are usually constructed so that the film diameter has one of three values, 57.3, 114.6 and 143.2 mm. Cameras of a larger diameter make it more convenient to measure the separation of lines, provided the lines are sharp. The sharpness of the lines is greatly determined by the quality of the collimating slits and size of the sample. The slit should be able to produce a fine beam, and the sample size should be small to give a point source of the diffracted beam. These will obviously need long exposures and a compromise must be arrived at among all these factors.

The choice of radiation wavelength is an important factor in X-ray diffractometry. Radiation having a wavelength just short of the absorption edge of an element contained in the sample of interest, should not be used, because then the element absorbs the radiation strongly. The absorbed energy is emitted as fluorescent radiation in all directions, which increases the background and makes it difficult to see the diffraction maxima. It is because of this reason that multi-window tubes with anodes of different materials such as tungsten, copper, molybdenum or silver are used in some commercial instruments.

Single crystal diffractometers are quite complex instruments, as they have to provide a wide angular range for orienting and aligning the crystal under study and for providing a sufficiently large traverse of the detector on a spherical surface. A typical example of a modern diffractometer is that of the Norelco goniometer, whose geometrical features are shown in Figure 14.7. The X-ray tube provides a line source 0.06 by 10 mm in size with high intensity. The angular aperture (approximately 1°) of the beam is defined by a single divergence slit. Flat specimens up to 10 by 20 mm can be usually handled. The receiving slit defines the width of the reflected beam, which finally falls on the detector. Two sets of parallel slit assemblies (metal foils equally spaced) limit the divergence of the beam in any

Norelco goniometer

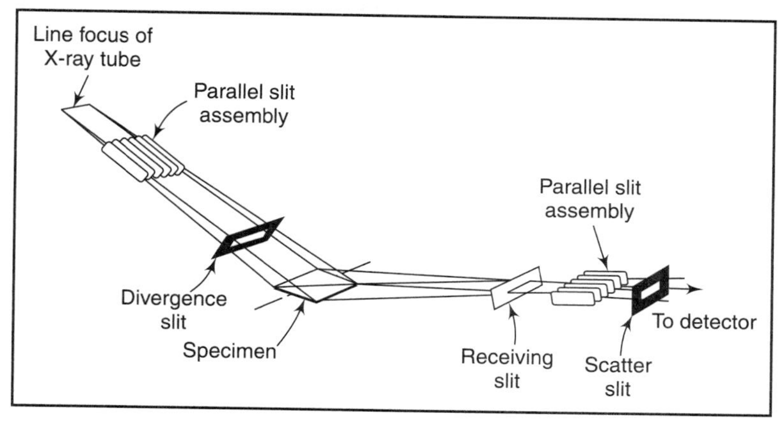

Figure 14.7 :: Geometrical features of Norelco goniometer

plane parallel to the line source. For producing an automatic record, the signal from the detector is amplified and fed into a pen recorder. The recording paper and the detector-bearing arm are turned by synchronous motors. The recorded graph gives a plot of the intensity of the diffracted beam versus angle of diffraction, usually denoted by 2θ.

X-ray powder diffraction has been found to be a convenient method for the identification of any compound that can be obtained in essentially pure crystalline form. The power of a diffracted beam is dependent upon the quantity of the corresponding crystalline material in the sample. It is, therefore, possible to obtain a quantitative determination of the relative amounts of various constituents of a mixture of solids.

14.4 X-Ray Absorption Meter

Like other regions of the electromagnetic spectrum, X-rays are absorbed by matter and the degree of absorption is determined by the nature and amount of the absorbing material. The fundamental equation for absorbance of a monochromatic X-ray beam follows Beer's law, which may be expressed as:

$$p = P_0 e^{-(\mu/p)p \cdot x},$$

where P = radiant power after passage through an absorbing sample, × cm in length.
P_o = initial power of the radiation
ρ = density of the material
μ = linear absorption coefficient.

The term μ/p is called the mass absorption coefficient, generally expressed as μ_m. The quantity depends upon the wavelength and the atomic properties of the absorbing substance as follows:

$$\mu_m = \frac{C \cdot Z^4 \cdot \lambda^4 \cdot n}{A},$$

where N is the Avogadro's number, A is the atomic weight of the absorbing material, Z the atomic number, λ the wavelength, n the exponent between 2.5 and 3, and C a constant over a range between characteristic absorption edges. The mass absorption coefficient μ_m is independent of the physical or chemical state of the specimen. In a compound or mixture, it is an additive function of the mass absorption coefficients of the constituent elements, i.e.:

$$\mu_m T = \mu_{m1} W_1 + \mu_{m2} W_2 + \cdots$$

where μ_{m1} is the mass absorption coefficient of element and W_1 is the weight fraction and so on, for the other elements present.

The variation of the mass absorption coefficient with wavelength follows an exponential law. Thus if the logarithms are plotted, a straight line should result with the slope equal to the exponent of λ. A typical absorption curve of argon is shown in Figure 14.8. This graph shows a discontinuity at λ = 387.1 pm, which is known as the k-critical absorption wavelength of argon. Radiation of greater wavelength has inadequate energy to eject k-electrons of argon. Hence, it is not absorbed so greatly, as is radiation of greatly shorter wavelength.

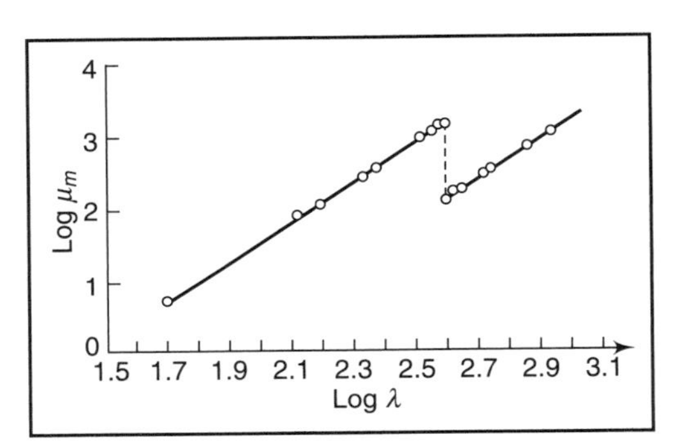

Figure 14.8 :: X-ray absorption spectra of Argon

X-ray absorption is of the most value, where the element to be determined is the only heavy component in a material of low atomic weight. Examples of such analysis are lead in gasoline and chlorine in organic compounds. It may be noted that X-ray absorption spectrophotometers that provide a continuously variable wavelength of X-radiation are not commercially available. Instead, multi-channel instruments are available. For industrial control purposes, absorption apparatus is usually designed specifically for each installation.

An atom absorbs an X-ray when the photon energy is sufficient to eject a photoelectron. Below this threshold energy, there is no absorption. Photons with energies greater than the threshold energy to produce a photoelectron are absorbed because the excess energy is conserved by transferring it to the kinetic energy of the photoelectron. However, the probability of the absorption occurring decreases as the photon energy increases above the threshold.

The use of X-ray absorption as an analytical method is not very uncommon because other techniques such as X-ray fluorescence are more sensitive. The absorption of X-rays by a certain element is often used in analytical instrumentation as a filter to block some X-ray wavelengths. For example, the absorption edge of Zr will block K_β and most of the continuum radiation of X-rays from an Mo target as illustrated in Figure 14.3.

14.5 X-Ray Fluorescence Spectrometry

The absorption of electromagnetic radiation by matter quite commonly results in the emission of radiation of the same or longer wavelengths (lower energies) and this phenomenon is referred to as 'luminescence'. If the time over which luminescence occurs, is very short it is termed 'fluorescence', and if longer, 'phosphorescence'. Thus, the prompt emission of an X-ray by an atom ionized by a higher energy X-ray is a form of fluorescence. Since such fluorescent emissions are characteristic of a particular element and virtually independent of its state of chemical combination, the measurement of the intensity of fluorescent X-rays can provide a simple, non-destructive means of quantitative analysis. The virtue of fluorescence as a means of X-ray generation is that, unlike electron bombardment, it is not associated with an X-ray continuum as a background to the characteristic lines. Hence, intensity measurement is inherently more sensitive.

X-ray fluorescence

X-ray fluorescence (XRF) spectroscopy is an extremely powerful tool for qualitative and quantitative determination of heavy elements in the presence of each other and in any matrix. Characteristic X-ray spectra are excited when the sample is irradiated with a beam of primary X-rays of greater energy than the characteristic X-radiation. XRF is based on the principle that the energy of the emitted X-rays depends upon the atomic number of the atom (*Z*) and their intensity depends upon the concentration of the atom in the sample.

Figure 14.9 shows the atomic processes involved in XRF. When a photon or charged particle of sufficient energy interacts with an atom, the atom may be excited releasing a specific electron out of an inner, *K* or *L* shell. In this situation, an outer shell electron can fall into the vacated inner shell, releasing energy as an X-ray. By measuring the photon energy of this fluorescent X-rays, the atom can be identified.

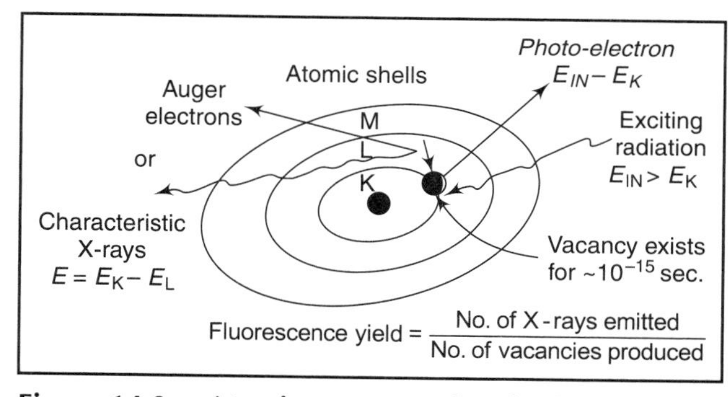

Figure 14.9 :: Atomic processes involved in X-ray fluorescence

The energy of the emitted radiation depends upon the atomic energy level separation and on the atomic number. In 1913, Mosley showed the relation between atomic number Z and the reciprocal of the wavelength λ for each spectral line belonging to a particular series of emission lines for each element in the periodic table. This is given as:

$$C/\lambda = a(Z - \sigma)^2,$$

where, C = Speed of light

a = constant of proportionality, and

σ = a constant whose value depends on the electronic transition series.

The sensitivity and specificity of the XRF depends upon the following fundamental physical factors:

- The probability that the incident radiation will produce the desired excitation.
- The probability that the resulting readjustment of the atom will produce fluorescence X-ray emission.

The relationship between the excitation intensity and the intensity of fluorescence depends upon several factors, such as: (a) the spectrum of the incident radiation, (b) angle of radiance, (c) molecular weight and matrix of the analyte, and (d) absorption path length.

The intensity of fluorescent X-rays are smaller by a factor of approximately one thousand than the X-ray beam obtained by direct excitation with a beam of electrons. Therefore, the fluorescent X-ray method would need high-intensity (tungsten-target) X-ray tubes, very sensitive detectors and suitable X-ray optics. The sensitivity of the analysis will depend upon the peak-to-background ratio of the spectral lines. Due to the relative simplicity of the X-ray spectra, the problem of spectral interference is not severe.

14.5.1 X-Ray Fluorescent Spectrometer

X-ray fluorescence analyzers consist of:

- Excitation source—which bombards the sample with sufficient energy to induce fluorescent X-radiation,
- Sample holder,
- X-ray spectrometer (energy analyzer), and
- Detection system.

Excitation Sources

Any source of radiation capable of producing vacancies in the inner shells of an atom can be used as an excitation source for X-ray fluorescence. It is, however, necessary to use a mono-energetic source in order to reduce unwanted background due to scatter occurring over a broad range of wavelengths obscuring the fluorescence of interest. The commonly used sources for obtaining X-ray photon sources are X-ray tubes. Special devices are used to make the X-ray beam from the conventional X-ray tube more nearly monochromatic. This can be done through the use of a (i) transmission-anode X-ray tube, (ii) secondary fluorescent target, or (iii) filters.

Transmission—anode X-ray tubes

Transmission-anode X-ray tubes operate on the principle that any metal is a good transmission filter for its own X-rays. The anode material chosen is such that it produces characteristic radiation of the desired energy. The accelerating potential selected is such that the characteristic to bremsstrahlung X-radiation is maximum. The thickness of the anode is such that it will filter out the lower energy bremsstrahlung. In secondary fluorescence, an X-ray beam from a primary X-ray tube is used to excite secondary fluorescence from a target, whose material is such that it produces the most appropriate exciting radiation. Filters make use of a thin metal foil to isolate a more nearly mono-energetic excitation beam. For example, by using a nickel filter which has an absorption edge at 1.488Å, the K_β of copper (1.392Å) can be filtered leaving the $K\alpha$ at 1.541 Å.

Spectrometer

The two main types of spectral analyzers used in XRF are: (i) energy dispersive, and (ii) wavelength dispersive type.

Energy dispersive system

The energy dispersive system is shown in Figure 14.10. It consists of an excitation source, a sample and a semiconductor detector. The fluorescent X-radiation resulting from irradiation of the sample reaches a detector, which produces an electrical pulse proportional to the energy of the X-rays. The energy level indicates that the element involved, and the number of pulses counted at each energy level over the entire counting time is related to the concentration of the element.

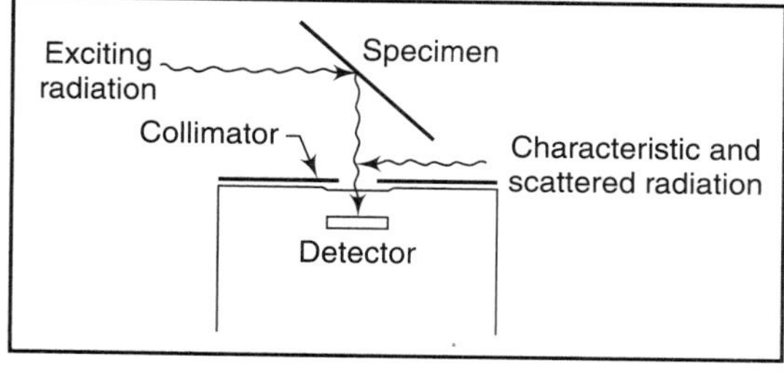

Figure 14.10 :: Energy dispersive system

In the wavelength-dispersive systems, X-rays emitted by the sample are diffracted by a crystal to an angle according to the Bragg equation: $\sin\theta = n\lambda/2d$.

Wavelength-dispersive systems

The detector receives the diffracted wavelength of interest and counts the pulses over the period of excitation. The range of wavelengths involved is scanned. Wavelengths at which the intensities peak indicate the types of atoms involved and the areas under the peaks are related to the concentration.

The technique derives its name from the fact that the analysis of the X-ray beam by diffraction is similar to spectrum analysis carried out with a diffraction grating. In this case, a crystal is used as a diffracting element. Essentially, the crystal possesses regular three dimensional lattice arrays of atoms and acts as an X-ray grating from which photons can be coherently scattered, so that they are in phase at certain angles. They reinforce each other producing a diffraction pattern.

A wide variety of wavelength dispersive systems are available. The two commonly used arrangements are shown in Figure 14.11 and Figure 14.12.

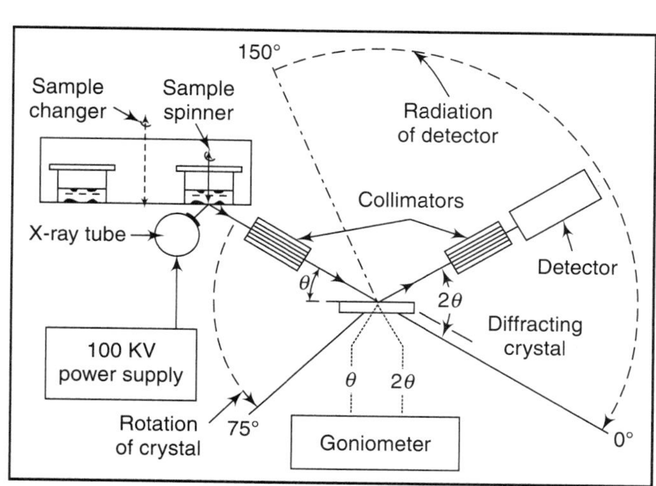

Figure 14.11 :: Plane crystal wavelength dispersive X-ray fluorescence spectrometer

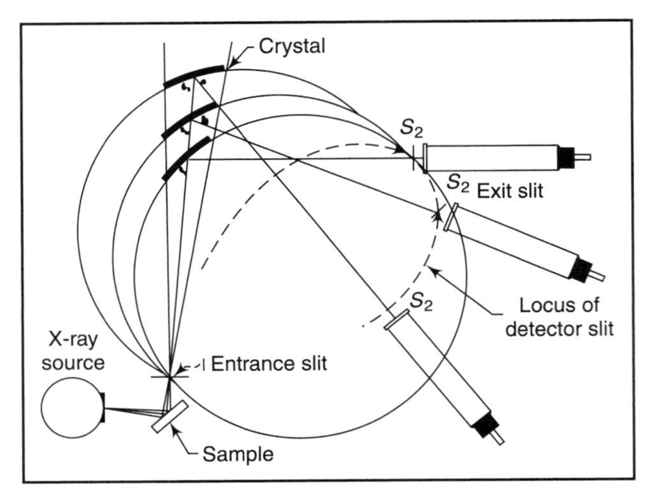

Figure 14.12 :: Geometry of curved crystal wavelength dispersive spectrometer

Basically, the two systems differ only in the shape of the wavelength dispersing crystal, which may be flat, curved or of any other geometries. Details on choice of crystals and the relative advantage of flat, curved and other types of crystals are explained by Liebhafsky (1972).

In the flat crystal arrangement (Figure 14.11), the primary and secondary slits and the analyzer crystal are placed on the focal circle, so that Bragg's law will always be satisfied, as the goniometer is rotated. The detector is rotated by an angle that is twice the angular change in the crystal setting. The crystal is a flat plate 2.5 cm in width and 7.5 cm in length. The specimen is held in an aluminium cylinder, though plastic material is used for examination of acid or alkaline solutions. Due to absorption of long wavelength X-rays by air and window materials, some intensity losses of X-rays take place, which can be reduced by evacuating the goniometer chamber; or the air in the radiation - from the sample surface to the detector window—may be replaced by helium. In some cases, vacuum spectrometers are also used.

The curved crystal arrangement (Figure 14.13) is more suitable for the analysis of small specimens. In this technique, collimators are not required, but increase in intensity is obtained by focusing the fluorescence lines. The crystal is bent to the diameter of the focusing circle and its inner surface is ground to the radius of the focusing circle. The radiation of one wavelength diverging from the entrance slit will be diffracted for a particular setting of the crystal and converge to a line image at a symmetric point on the focusing circle. The angular velocity of the detector is twice that of the crystal and the two of them move along the periphery of the circle. The X-ray spectral lines are dispersed and detected just as in the flat-crystal arrangement.

Detectors

A variety of detectors are used for the detection of X-radiation with the most common being the ionization chamber, gas proportional counters, scintillation detectors and semiconductor devices. For wavelength dispersive analysis, the resolution is determined by the complex mechanism, which aligns the detector and therefore, it is possible to use scintillation and gas proportional counters, which have relatively poor resolution, but very high count rates as detectors. For energy dispersive analysis, it is necessary to use a detector which has good resolution and low noise and for this, the semi-conductor detector is most suitable. In silicon semiconductor devices [Si(Li)], lithium drifted to compensate for electrically active impurity centres are now the most widely employed detectors for energy dispersive analysis. They are stable, show a high count-rate capability, have a small size and low background. They give adequate resolution (150 eV at

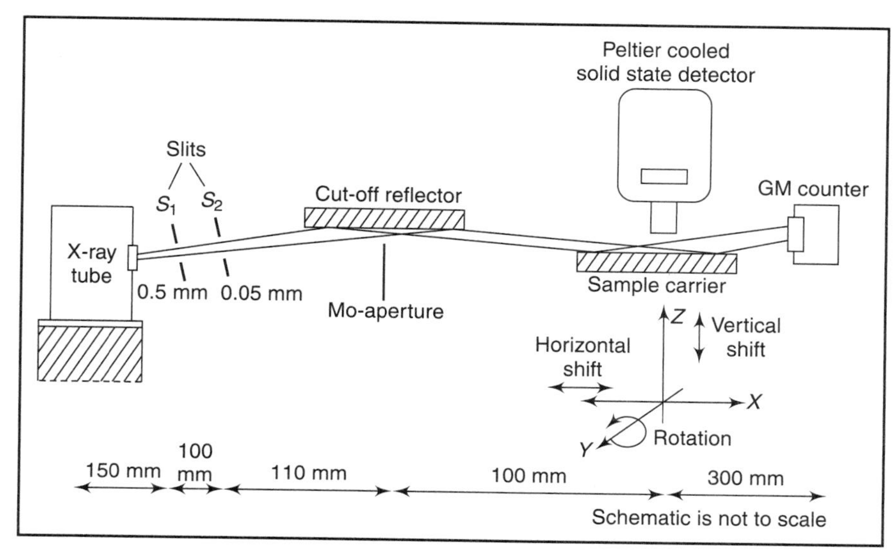

Figure 14.13 :: Schematic of the total reflection X-ray fluorescence (TXRF) set-up (adapted from Tiwari, *et al.*, 2002)

3.3 keV), which allows the separation of element pairs, as low in the periodic table as carbon/nitrogen. A resolution of 3 eV at 3.3 keV or better is reported with wavelength dispersion systems, which eliminates most line interferences and is able to distinguish between various metallic electronic states. The disadvantage of such detection is that they require low temperatures (say 77 K) for efficient operation and hence, are somewhat inconvenient to maintain. Similarly, germanium detectors, though they provide potentially better energy resolution, have the disadvantage that they must be operated at low temperatures and suffer from high leakage.

For qualitative analysis, the angle θ between the surface of the crystal and the incident fluorescence beam is gradually increased. At certain well-defined angles, the appropriate fluorescence lines are reflected. Additional factors for identification of a particular element may be obtained from relative peak heights of fluorescence lines, the critical excitation potential and pulse–height analysis. For quantitative analysis, the intensity of a characteristic line of the element is measured. The goniometer is set at the 2θ angle of the peak and counts are collected for a fixed period of time. Alternatively, the time is measured for the period required to collect a specified number of counts. The goniometer is then set to the portion of the spectrum, where only a background count is obtained. The net line intensity (peak minus background) in counts per second is then related to the concentration of the element with the help of a calibration curve.

14.5.2 Total Reflection X-Ray Fluorescence Spectrometer

Total reflectance X-ray fluorescence (TX RF) spectroscopy is a modern technique used for the determination of ultra-trace amount of elements in various kinds of samples. Here, the specimen is excited by the primary X-ray beam at a grazing angle less than the critical angle at which total external reflection occurs and primary excitation radiation is incident on a plane, polished surface serves either as a sample support or is itself the object to be examined. The design of a TXRF is illustrated by Tiwari, *et al.*, (2002).

A schematic representation of TXRF spectrometer is shown in Figure 14.13. It essentially comprises an X-ray generator, a slit–collimator arrangement, a monochromator stage, a sample reflector stage (i.e. sample carrier) and an X-ray detection system. The primary beam emitted by an X-ray tube from its line-focus is collimated by the slit-collimator arrangement, which generates an incident X-ray beam. A polished silicon crystal of dimensions 30 × 34 mm, and 0.3 mm thickness, is mounted on the first reflector stage to act as a cut-off reflector. The grazing angle on the cut-off reflector is set to such a value that it cuts off the higher energy part of primary radiation, thereby making the primary X-ray beam quasi-monochromatic.

The suppression of high energy bremsstrahlung radiation improves the signal-to-background ratio. The second reflector stage serves as a sample carrier through which the fluorescence measurement of a sample material can be performed under external total reflection condition. A float glass is mounted as the sample carrier on this stage in the upward looking direction. A 2 mm thick Mo aperture is placed near the cut-off reflector stage to shield the sample-reflector from the direct primary X-ray beam.

The detection system for energy dispersive measurement consists of a peltier cooled solid state detector, a spectroscopy amplifier and a multi-channel pulse–height analyzer installed in a personal computer. The solid state detector has an energy resolution of 240 eV at 5.9 KeV and operates without liquid nitrogen. The detector is placed very close and normal to the sample substrate, so that a large solid angle is intercepted which thereby maximizes the counting efficiency. To monitor the intensity of the primary beam, a GM counter is placed in the specular reflection direction.

Slit-collimator assembly

The slit-collimator assembly comprises a pair of precisely aligned Mo slits of dimensions 10 × 0.5 mm (S_1) and 10 × 0.05 mm (S_2), respectively, which are separated by 100 mm. This slit-collimator assembly makes primary beam dimensions 10 mm wide and 0.05 mm thick with a beam divergence of ~0.01°, which is small enough to obtain total external reflection. The vertical motion (Z-motion) as well as tilting motion (ø-motion) for whole slit-collimator assembly is provided by means of two micrometers. The angular tilt motion can tilt X-ray beam around X-axis in a controlled manner by using sine-bar arrangement.

The two reflector stages have separate mechanical arrangements for rotation (θ-motion) as well as for the vertical shift (Z-motion). Two micrometers of least count 1μm are used for each reflector stage—one for vertical movement and the other for incident angle adjustment using sine-bar arrangement.

All mechanical assemblies, including the slit-collimator arrangement, the first reflector stage and the second reflector stage, are mounted on a rigid stainless steel base plate. The TXRF module is connected directly to the window of an X-ray generator. The spectrometer has provision for the mounting of various types of optical elements such as cut-off-reflector, multi-layer filter or natural crystal at the first reflector stage. Therefore, by choosing appropriate optical element, the energy bandwidth of primary radiation can be tailored to suit the excitation of a given sample.

The most critical aspect of XRF analysis is sample preparation. The major problem is associated with matrix effects, i.e. absorption of both the primary and emitted radiation by the elements present in addition to those of interest. Unless matrix absorption can be made negligible or constant, apparent emission intensities do not vary linearly or reproducibly with concentration.

Matrix effects

A number of different procedures may be used to correct the matrix effects. For example, in computer-controlled iterative correction, initial raw intensity data may be used to estimate and elemental composition which can then be used to calculate absorption corrections. The cycle is then repeated until consistent results are obtained. The procedure is complex, but where an unknown matrix cannot be avoided, it is the only reliable technique.

Although XRF is, in principle, a non-destructive analysis, the appropriate preparation of a sample generally necessitates its destruction. The sensitivity of the method is highly dependent upon matrix effects, though under optimum conditions, concentrations as low as 1 ppm may be detected. For maximum emission intensity, as large an area as possible (~10 cm^2) is irradiated, with the source being ~3 cm from the sample. New developments such as the synchrotron X-ray microscope do, however, allow levels as low as 1 ppm in a volume as small as 10^{-9} cm^3 to be determined.

The X-ray fluorescence technique is inherently very precise and rivals the accuracy of wet chemical methods. It is attractive for elements which lack reliable wet chemical methods, such as tantalum and rare earths and simultaneous analysis of several elements is possible with automated equipment.

14.6 Electron Probe Microanalyzer

The electron microprobe is basically a combination of an electron microscope and an XRF spectrometer, with a visual microscope attached. Its use offers primarily the only method for detailed elemental analysis of materials which are heterogeneous at the (visual) microscopic level. By bombarding the sample with high energy electrons in a vacuum, the sample surface may be mapped by detection of scattered electrons (electron microscopy) and its elemental constitution determined by detection of emitted X-rays.

Electron probe micro-analysis is a technique used for non-destructive elemental analysis by utilizing a finely focused electron beam to excite X-rays in a solid sample as small as 1 μm in diameter. Figure 14.14 shows a schematic diagram of a typical apparatus, in which three types of optics, are employed: (i) electron optics, (ii) light optics, and (iii) X-ray optics.

Electron optics

The electron optics consists of an electron gun followed by two electromagnet lenses, which form the electron beam probe. This section of the instrument is very similar to an electron microscope. The specimen is mounted inside the vacuum column of the instrument and under the beam as the target.

The electrons are accelerated by a voltage between 5 and 50 kV maintained stable to at least 1 part in 10^4 and the whole system operates under a high vacuum (< 10^{-10} torr). Electromagnetic lenses focus the electron beam to a diameter between 0.1 and 4 mm, but as at the energies

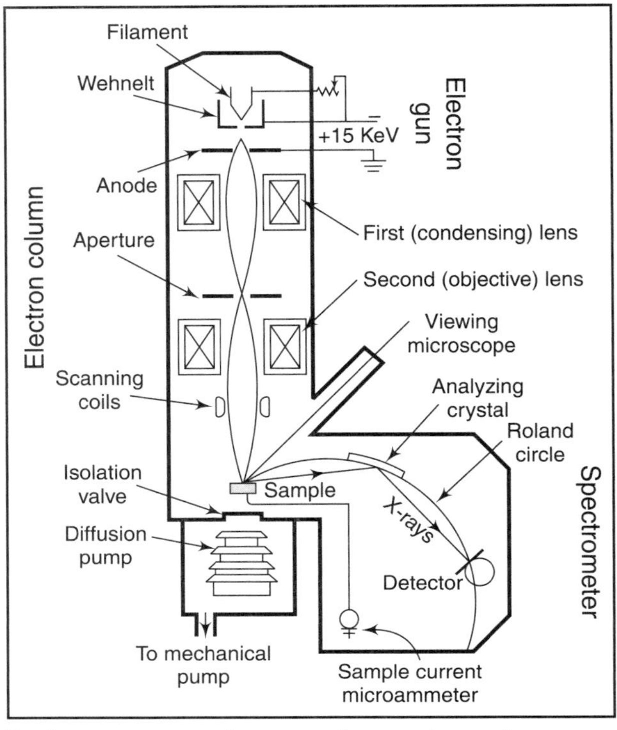

Figure 14.14 :: Electron microprobe analyzer

required to cause X-ray generation, the electrons spread laterally and longitudinally in the sample by approximately 1 mm, thus the effective resolution limits is 1 mm. The microprobe has the same limits in terms of elements conveniently analyzed as an ordinary XRF spectrometer, but its sensitivity is poorer (100-500 ppm) because the characteristic X-ray emissions must be detected against a background of white radiation. This sensitivity difference is offset, however, by the fact that the microprobe can be used on inhomogeneous material and focused to a very small area, so that as little as 10^{-15} g of material in 10 m^3 can be detected. In conventional XRF analysis, the detection limit is ~10^{-8} g.

The electron bombardment excites characteristic X-rays, which are analyzed in an X-ray spectrometer, that is of the curved crystal type. The optical microscope helps the operator to precisely locate the desired spot on the sample. A point-by-point microanalysis is obtained by moving the sample across the beam. This method allows the analysis of extremely small objects with the limit of detectability (in a 1μ size region) about 10^{-14} g. It is particularly useful for phase studies in metallurgy and ceramics for following the process of diffusion in the fabrication of semiconductors and corrosion studies, where excitation is restricted to thin surface layers, as the beam penetrates to a depth of only one or two microns into the specimen.

Matrix corrections constitute a major problem in quantitative microprobe analysis and demand computer-controlled iterative application of numerous empirically determined correction factors.

While the power of the electron microbe in analysis is readily appreciated, its deficiencies in micro-analysis should also be noted. These deficiencies stem from the large irradiation area, the inherent limitations of XRF (to the heavier elements), lack of sensitivity, inability to deal with thin films, and inability to provide information on isotopic composition of an element.

Electron probe micro-analyzers are made by several commercial firms. However, an alternative method using lasers for vaporising materials from small focused areas and analyzing the vapours by mass spectrometry, optical emission or absorption photometry, is gaining popularity.

Chapter 15

Automated Chemical Analysis Systems

15.1 Why Automate?

The automation of analysis techniques for either discrete samples for laboratory analysis or online sampling for process monitoring is an important development in recent decades. It is established that there are a number of possible benefits of an automated technique for a particular procedure, which may include greater accuracy and reproducibility of results due to the facts that samples are analyzed in a closed system free from contamination and all are subject to the same constant analytic process. In addition, there are reduced operator errors that often occur in manual analysis or the analysis of results. In addition, automation uses smaller sample and reagent volumes, which reduces cost.

Automated technique

Automation has enabled procedures to obtain far greater and more in-depth data analysis, which has opened up possibilities for completely new procedures. The user has been relieved of the tedious routine procedures, normally associated with manual procedures. Similarly, in industrial processes, greater quality control can be obtained through continuous accurate monitoring.

15.1.1 Benefits of Automation in Chemical Analysis

The following are the benefits of automation in chemical analysis:

- Faster analyses upto 120 samples per hour,
- Upto 300 samples can be analyzed in a batch,
- Automatic data recording and preparation,
- Being a closed system, automation reduces contamination, for example, from atmospheric gases,
- Greater accuracy and reproducibility of results as all samples are subject to the same processes,

- Smaller sample and reagent volumes, which reduces cost,
- Automatic range changing (for over-range samples), drift control and automatic sample preparation, and
- Out of hours operation with automatic start-up and shut-down procedures.

15.1.2 Types of Automatic Analysis Techniques

Basically, there are two types of automatic analysis techniques. These are described below.

15.1.2.1 Segmented Flow Analysis (SFA)

Fig. 15.1 shows the segmented flow analysis technique. The technique involves mixing a liquid sample with reagents pumped in a continuously flowing stream and segmented with bubbles. The sample and reagents are mixed by passing through glass coils (chemistry module) and also through a temperature controlled heating coil, if required, to speed up colour development before detection using a colorimeter, or fluorescence or other detector. A dialyzer with a cellophane dialysis membrane is used to remove high molecular weight components contained in samples which may interfere with chemical reactions.

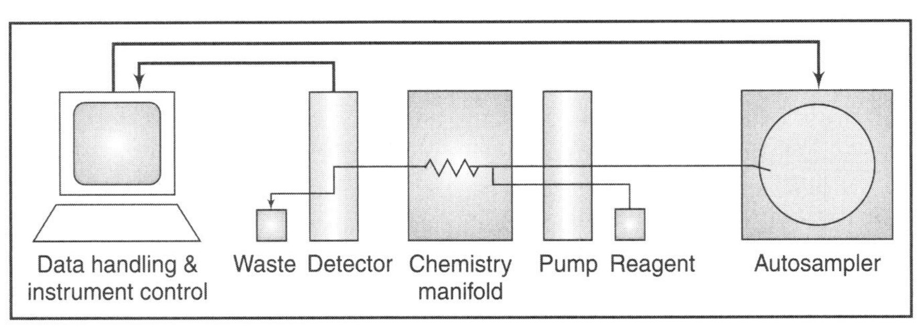

Figure 15.1 :: A typical segmented flow analysis system

SFA is one of the most reliable and widely used methods for analysis in routine and research analytical laboratories. The technique enables upto 16 determinations simultaneously at speeds of upto 120 samples per hour.

15.1.2.2 Flow Injection Analysis (FIA)

The principle of flow injection analysis technique is shown in Fig. 15.2. This involves the injection of a liquid sample into a moving non-segmented continuous carrier stream of a suitable liquid, which is usually the colour reagent. A high precision pump is used to pump samples and re-agents. The injected sample forms a zone which is then passed though mixing coils and carried to

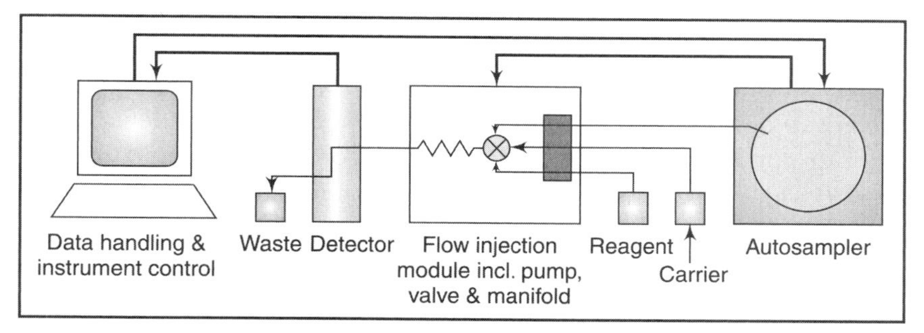

Figure 15.2 :: A typical flow injection system

a flowcell in the detector, usually a colorimeter, flame photometer or flourimeter for the measurement of optical density.

The technique offers advantages such as precision sample introduction, small sample volumes (typical 10–50 ml) low reagent consumption and fast sample throughput.

15.1.3 Basic Automatic Analysis System

Irrespective of whether the application is for discrete laboratory analysis or for on-line process analysis, there are the following system components:

- Sampling,
- Pump and Chemistry module,
- Detector, and
- Data recording and analysis.

15.2 Automated Biochemical Analysis System

The chemical analysis of blood and other body fluids is one of the earliest forms of diagnostic criteria leading to the investigation of diseases. In the early years of clinical biochemistry, most of the required analyses were performed on urine as it can be made available easily and in large quantities. However, with the development of semi-micro and micro-analytical techniques, analysis can be carried out with minute quantities of samples. Therefore, the analysis of the blood serum or plasma is becoming more frequent. A great increase in the number of determinations coupled with heavy workload in many laboratories and the development of optical and electronic techniques have stimulated attempts to introduce a high degree of automation in the clinical biochemistry laboratories. A majority of the blood analysis can now be performed by an automated system.

Developments in new concepts and more advanced techniques in analytical methodology have resulted in estimations of blood constituents as a group, whose metabolic roles are related and which collectively provide more meaningful information than individual analyses. For instance, the group of important anions and cations of blood plasma (electrolytes) like sodium, potassium, chloride and bicarbonate, which along with serum urea, form a related set of tests, that are useful to perform on patients with electrolyte disturbances. Another group consists of the several analyses—protein, bilirubin, alkaline phosphatase and SGOT, which together assess liver function. The effect of this trend is in the replacement of single isolated analysis by groups of analyses, all of which are carried out routinely on each sample with highly reproducible and accurate results.

Automatic analysis equipments

With this object in view, automatic analysis equipments have been designed and put to use. Automated analysis systems are available in multi-channel versions and a full description of the detailed working instructions and details of techniques for individual substances are given in literature supplied by the manufacturers to the purchaser of their equipment.

The major benefit of automation in the clinical laboratory is to get rid of the tasks that are repetitive and monotonous for a human operator, which may lead to improper attention thereby causing errors in analysis. However, it may be remembered that improvement in reproducibility does not necessarily enhance the accuracy of the results, because accuracy is basically influenced by the analytical methods used.

Any automated biochemical analysis system usually has the following steps:

- Specimen identification,
- Specimen transport and delivery,
- Removal of protein and other interfering components,
- Reagent handling and delivery,
- Chemical reaction phase,
- Measurement procedure,
- Signal processing, data handling, and
- Report preparation.

These steps are usually carried out sequentially under the control of microprocessors, which also facilitate to collation of results, besides performing many statistical and quality control tasks.

Automated systems are usually considered to be more reliable than normal methods, due to individual variations that may appear in handling various specimens. However, many possibilities exist for mismatching specimens and final results. The risks begin right from the time the sample is collected from the patient and are compounded with every step a specimen undergoes in the analysis chain. The risk is particularly great when accessioning, labelling, re-labelling and preparation of load lists are done by hand transcription. Any incorrect action can cause test results to be attributed to the wrong patient. Even though several codes have been developed to identify the specimen with the right patient, it needs utmost care in those parts, where human action is involved.

15.2.1 The System Details

The automated system is usually a continuous flow system, in which individual operations are performed on the flowing stream as it moves through the system. The end-product passes through the colorimeter, where a balance ratio system is applied to measure concentrations of various constituents of interest. The final results are recorded on a strip chart recorder along with a calibration curve, so that the concentration of the unknowns can be calculated. The output may also be connected to a digital printer along with the graphic record.

The automated system consists of a group of modular instruments (Figure 15.3) interconnected by a manifold system and electrical systems. The various sub-systems are:

- Sampling unit,
- Proportioning pump,
- Manifold,
- Dialyzer,
- Heating bath or constant temperature bath,
- Colorimeter/flame photometer/fluorometer, and
- Recorder.

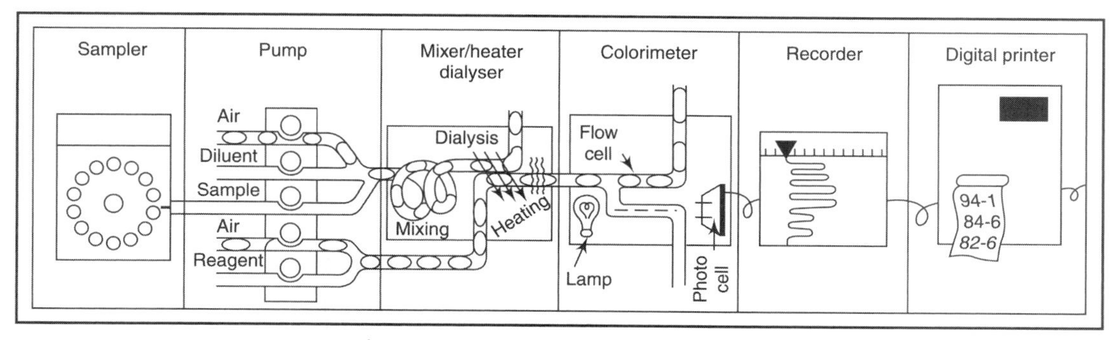

Figure 15.3 :: Schematic diagram of a continuous flow system

The sample to be analyzed is introduced into a stream of diluting liquid flowing in a narrow bore of flexible plastic tube. The stages of the analytical reaction are completed by the successive combination of the other flowing stream of liquids with the sample stream, by means of suitably shaped glass functions. Bubbles of air are injected into each stream, so that the liquid in the tubes is segmented into short lengths separated by air bubbles. This segmentation reduces the tendency for a stationary liquid film to form on the inner walls of the tubes and decreases interaction between a sample and the one which follows it. The diluted samples and reagents are pumped through a number of modules in which the reaction takes place, giving a corresponding sequence of coloured solutions, which then pass into a flow-through colorimeter. The corresponding extinctions are plotted on a graphic recorder, in the order of their arrival into the colorimeter cell. The

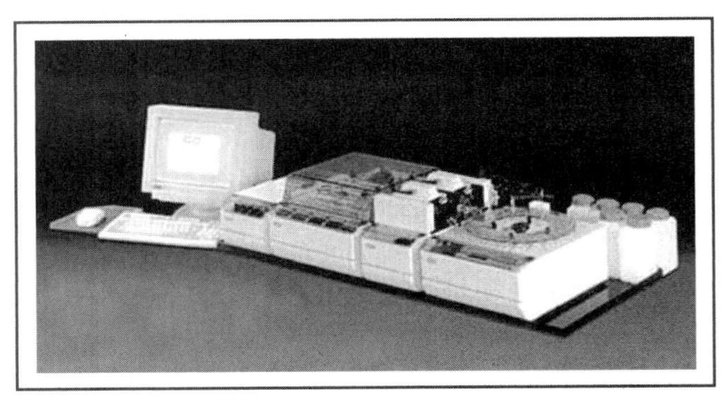

Figure 15.4 :: Multi-channel segmented flow analyzer (Courtesy M/s Burkard Scientific, UK)

air bubbles are removed before the liquid enters the colorimetric cell or flame emission.

A typical multi-channel segmented flow analyzer is the Model SFA2000 from M/s Burkard Analytical, UK. This is shown in Figure 15.4. This is a modular system and can be expanded with ease. The equipment is software controlled and has multi-tasking offering real-time analysis where primary data acquisition, graphics, peak detection, calibration and calculation take place, so results are produced immediately upon a peak being detected. Details of the individual units are described below.

15.2.1.1 Sampling Unit

The sampling unit enables an operator to introduce unmeasured samples and standards into the auto analysis system. The unit in its earlier form consisted of a circular turntable (Figure 15.5) carrying around its rim 40 disposable polystyrene cups of 2 ml capacity. The sample plate

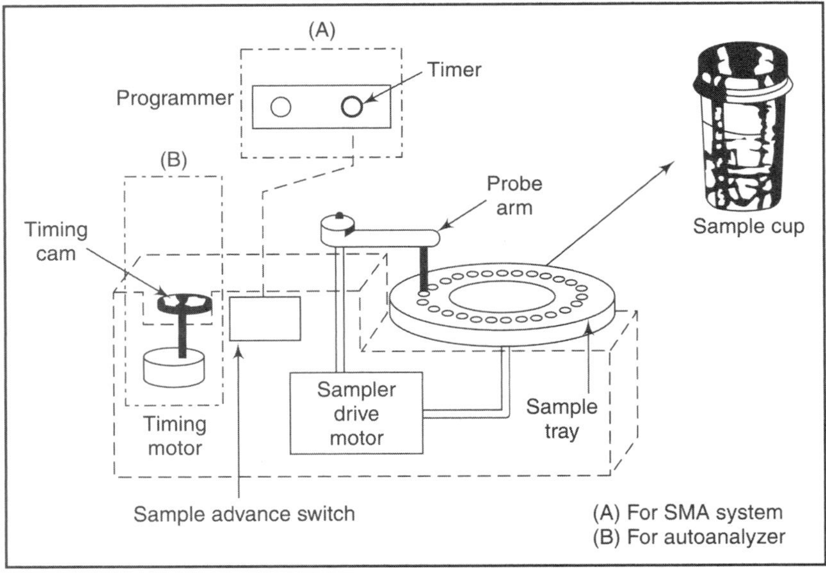

Figure 15.5 :: Sampler controls (Courtesy M/s Technicon Corp., USA)

carrying these cups rotates at a pre-determined speed. The movement of the turntable is synchronized with the movements of a sampling crook. The hinged tubular crook is fitted at a carrier of the base. The crook carries a thin flexible polythene tube, which can dip into a cup and allow the contents—water, standard or test solution, to be aspirated. At regular intervals, the crook is raised so that the end of the sample tube is lifted clear of the cup.

Sampling crook

Between each sampling, the crook enters a receptacle of water or other suitable wash fluid, to reduce cross-contamination of one sample with another. The ratio of sampling time to wash time is normally 2:1. The plate then rotates a distance sufficient to allow the tube, when it next moves down, to dip into the next cup. One complete rotation of the plate thus presents 40 samples. As the sample plate completes a cycle, a switch is operated, which stops the rotating action of the plate and the sampling action of the sample probe. The sampling rate can be adjusted to 20, 40 or 60 per hour. Accordingly, the above ratio, the time during which a sample is being drawn in, will be two minutes, one minute or 40s respectively. The volume of liquid taken up in most cases ranges from about 0.2 to 1.0 ml. This depends upon the rate at which the plate is run and the diameter of the pump tube.

The earlier version of the sampling system has been replaced by a more versatile form of the sampler, wherein during the time the sample tube is out of the specimen, the crook quickly comes down into water, and thus successive samples are separated by a column of water instead of air. This provides a better separation between them. With this sampler, the sample size may range from 0.1 to 8.5 ml. It utilizes cups of sizes 0.5, 2, 3 and 10 ml. The sample plate is kept covered to prevent evaporation, which may sometimes lead to errors up to 5 per cent. Sampling and washing periods are controlled by a programming cam. The sample speed and sample wash cycles are selected by the markings on the cam, such as 40 and 2:1. This implies that the speed is 40 per hour at a sample wash ratio of 2:1.

Sample mixer

The samplers are fitted with a sample mixer, which enables the sample to be mixed before and during aspiration. This is important in cases such as when whole blood is used for analysis. The mixer prevents separation of the plasma from the cells. The sample base plate is rotated by means of a motor drive gear. When this gear drives the second drive gear for the sample probe assembly, it also rotates a gear chain of internal gears to the Geneva cam. The Geneva cam is an index type cam that may be designed for any acceleration, deceleration or dwell period. During the complete cycle of the sample wash probe, the cam makes one complete turn. As the probe moves back into the wash reservoir, the second point enters the next hole in the drive assembly plate. This rotates the drive assembly, sampler housing and sample plate at a controlled speed. The sample speed is controlled by the sampling rate cam on the programmer. A new sample is aspirated every two minutes and the sample wash ratio of 95 per cent sample and 5 per cent wash is maintained.

Mechanical cams were used in the earlier modules of analyzers to initiate and control sample aspiration and wash cycles. Modern systems use electronic timers to do the same function. These timers provide greater flexibility in control of sample-to-wash ratios, which, in turn, allows flexibility in setting up parameters for analyses.

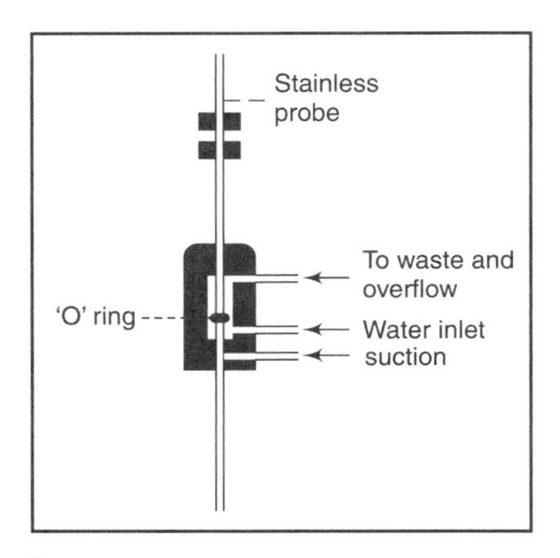

Figure 15.6 :: Sampling mechanism of the automatic sampler

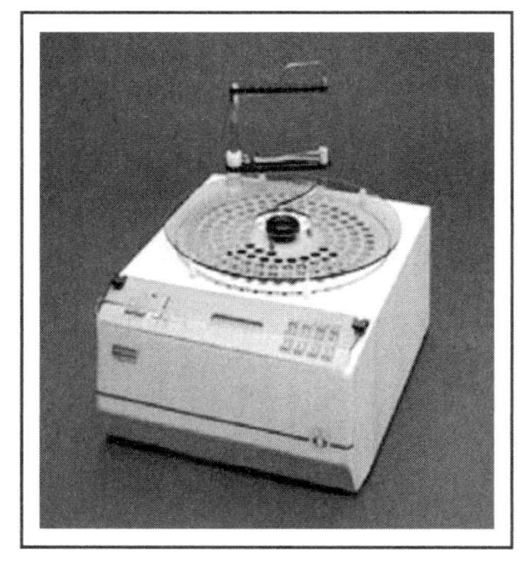

Figure 15.7 :: Electronic autosampler (Courtesy M/s Burkard Scientific, UK)

Automatic sampler

The automatic sampler (Figure 15.6) employs a different probe washing action between samples. Whereas most automatic samplers operate by dipping the probe alternately into the sample cup and then into a wash pot, the sampler has a probe which simply moves up and down.

The probe washing device consists of a washing chamber through which the probe moves vertically. When the probe is in the sampling position, a 0 ring at the lower end of the internal wash reservoir seals the outlet. Water is pumped through the reservoir by the peristaltic pump used in the analytical system. At this stage, the water flows upwards through the reservoir and then through the outlet at the top of the device. After sampling, the probe moves to the top of the reservoir and the 0 ring seals the outlet. The probe then aspirates water and the excess water runs down the probe exit hole, where it is sucked away through the annular channel which surrounds the channel. There is a sub-atmospheric pressure at the probe exit and entry hole, which causes any liquid adhering to the probe to be sucked away as it passes through the chamber. The sample and wash times can be adjusted from 0 to 120s.

M/s Burkard, UK, have introduced an auto-sampler, which incorporates a spiral table configuration that provides 140 places for unknowns, standards, drift controls and blanks. This is shown in Figure 15.7. The sampler is microprocessor-based and all electronic functions are achieved via the instrument keyboard or computer. Sequence programming and sampling is user controllable which include setting of various parameters including sample and wash times, numbers of samples set, and 'start' and 'end' cup/tube positions. A close-fitting tray cover protects the samples from aerial contamination and reduces evaporation from the sample cups. This is useful feature for long sample runs.

The samplers are usually designed to complement the automatic dilutor and standards preparation unit, which automatically creates a range of standards from a single working standard as well as creating all drift and blank controls.

15.2.1.2 The Proportioning Pump

The function of the proportioning pump is to continuously and simultaneously push fluids, air and gases through the analytical chain. In fact, it is the heart of the automatic analysis system. Here, all the sample and reagent streams in any particular analysis are driven by a single peristaltic pump, which consists of two parallel stainless steel roller chains with finely spaced roller thwarts.

A series of flexible plastic tubes, one from the sampler, the others from reagent bottles or simply drawing in air, are placed lengthwise along the platen spring-loaded platform. The roller head assembly is driven by a constant speed gear motor. When the rollers are pressed down and the motor switched on, they compress the tubes containing the liquid streams (sample, standard and reagents) against the platen. As the rollers advance across the platen, they drive the liquid before them.

The roller head rotates at a constant speed. The different flow rates required in the several streams (0.15 to 4 ml/min) are achieved by selecting tubes of appropriate internal diameter, but of constant wall thickness. Since the proportions of the various reagents are fixed by the tube sizes, no measurements are needed.

Proportioning pumps are available either for single-speed or for two-speed operation. The single-speed pump has the capacitor synchronous gear head utilizing 10 rpm output shaft at 50Hz. The two-speed pump has a non-synchronous 45 rpm motor. The slow speed in this pump is used for ordinary working during a run and a much quicker one for filling the system with reagents before a run and for rapid washing to clear out reagents after the run. It is also utilized for rapid cleaning of the heating bath, or of the complete system, when fibrin (an insoluble protein) problems are evident and are disturbing the run. High speed is not used for analysis. Heavy-duty pump is also available, which enables 23 pump tubes to be utilised simultaneously.

The plastic tubes are tautly held between two plastic blocks with locating holes, which fit on to pegs at each end of the platen. Before beginning a run, the tubes are stretched. With use and time, the tubes lose elasticity and pumping efficiency is reduced. Therefore, each block has three sets of holes, so that the tubes can be increasingly stretched and the tension thus maintained. The tubes are replaced at the first sign of aging. In tact, they should be replaced at regular intervals to forestall failure. When not in use, one of the blocks is removed, so that the tubes are not kept in tension.

Actually the sample or reaction stream is separated by air bubbles into a large number of distinct segments. The air bubbles completely fill the lumen of the tubing conducting the flow, thereby maintaining the integrity of each individual aliquot. In addition, the pressure of the air bubble against the inner wall of the tubing wipes the surface free of droplets which might contaminate the samples which follow. The proportioning include an air bar device (Figure 15.8), which adds air bubbles to the flowing streams in a precise and timed sequence. The air bar is actually a pinch valve connected to the pump rollers that occludes or opens the air pump tubes at

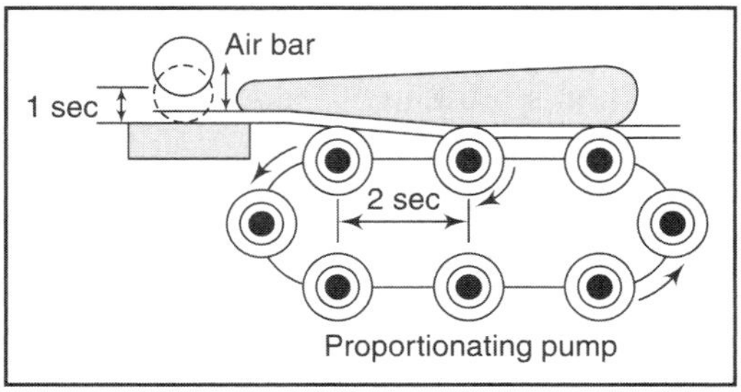

Figure 15.8 :: Principle of air segmentation in the continuous flow system

timed interval. Everytime a roller leaves the pump platen—and this occurs every two seconds—the air bar rises and lets a measured quantity of air through. The release of air into the system is carefully controlled, thereby ensuring exactly reproducible proportioning by the peristalitic pump.

The continuous flow analyzers make use of liquid reagents. Large volumes of reagents are stored in the systems and their quantity is adequate for the operation of the analyzer for several hours or days. Some automated systems use reagents in a dry tablet form. When required, the tablet is dispensed into a one test reaction vessel and dissolved. The sample is then added for the reaction to take place. This is basically a unit-dose concept, which offers several advantages like less storage space and operator time, long stability of reagents and lesser wastage.

15.2.1.3 Manifolds

A variety of chemistry manifold types and combinations are generally offered in the automatic analysis systems. A manifold mainly consists of a platter, pump tubes, coils, transmission tubing, fittings and connections. A separate manifold is required for each determination and the change can be effected within a few minutes. The pump tubings and the connected coils are placed on a manifold platter, which keeps them in proper order for each test. The pump tubings are specially made, they are of pre-measured length and are meant to introduce all constituents of an analysis into the system. The physical and chemical properties of the tubing are extremely important in the correct functioning of the pump. It must not be so flexible as to expand beyond its normal internal dimensions on release of pressure, which may lead to variation in the flow, thereby affecting reproducibility and accuracy of the system. The tubes should be chemically inert for the constituents which are expected to flow through the tube. The constant and correct tension also provides the continual delivery of a constant volume. The inside diameter of the pump tubing determines the flow rate per minute.

Several other tubes are required to introduce reagents and to transport the specimen from one module to another. There are five types of such tubings. They are of varying sizes and are to be selected according to the requirements. These are: standard transmission tubing (Tygon), solvaflex tubing, acidflex tubing, polyethylene tubing and glass tubing.

Two types of coils are employed in the system-mixing coils and delay coils. Coils are glass spirals (Figure 15.9) of critical dimensions, in which the mixing liquids are inverted several times, so that complete mixing can result.

Mixing coils

Mixing coils are used to mix the sample and/or reagents. As the mixture rotates through a coil, the air bubble along with the rise and fall motion produces a completely homogeneous mixture. The mixing coils are placed in a horizontal position to permit proper mixing. Delay coils are employed when a specimen must be delayed for completion of a chemical reaction before reaching the colorimeter. These coils are selected in length according to the requirements. The standard delay coil is 40 ft long, 1.6 mm I.D., and has a volume of approximately 28 ml. The time delay can be calculated by dividing the volume of coil by the flow rate of specimen plus bubbles.

Delay coils

With twelve tests to be recorded on each sample and a sampling rate of 60 samples per hour, it follows that 5 s are allowed to record each steady state plateau. The reaction streams in the twelve channels and up to four blank channels must, therefore, be phased to arrive at the colorimeter in waves 5 s apart. For example, if the cholesterol stream arrives at X time, calcium must arrive at X + 5 s, total protein at X + 10 s, albumin at X + 15 s etc. In order to ensure proper sequencing for presentation of the results, a number of devices have been provided to make this adjustment an extremely simple operation. Phasing coils are used to permit the channels to enter the colorimeter in the proper sequence.

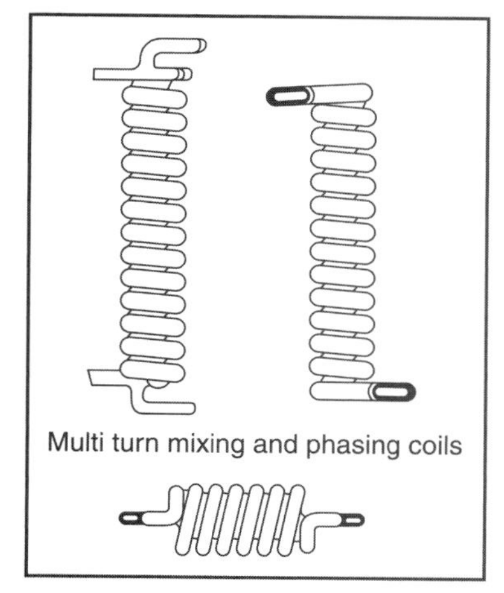

Figure 15.9 :: Typical phasing and mixing coils

Phasing coils

In the earlier models, sleeving was employed for interconnection of manifold tubes and cyclohexamone was applied to seal the joints. Polyethylene nipples are now used for these connections. Various types of fittings (Figure 15.10) are employed to join stream of liquids to split a stream or to introduce air segmentation to the stream.

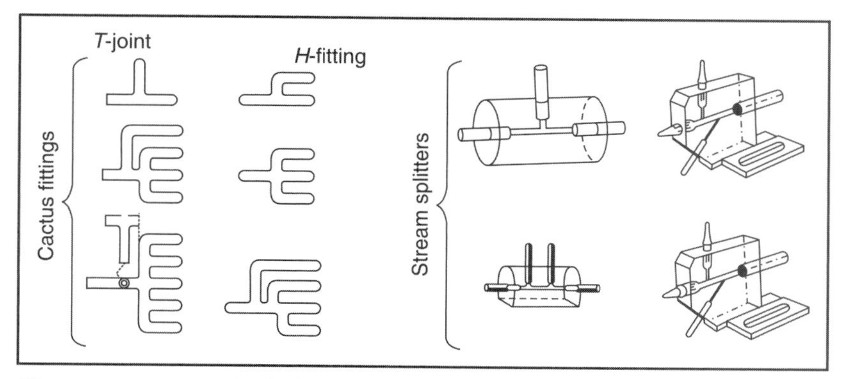

Figure 15.10 :: Typical glass fittings in an automated analysis system

Mixing with other reagents begins on leaving the pump. The first reagent with which the test specimen is mixed is usually a simple diluent. The reagent lines are segmented by introducing air through one or more additional tubes into the manifold. This produces a series of bubbles at regular intervals in the liquid stream. This is designated as bubble pattern. A uniform bubble pattern is essential for accurate analysis. While it is not necessary that every bubble be absolutely identical in length, a firm consistent flowing segmented stream is required.

Bubble pattern

15.2.1.4 Dialyzer

In analytical chemistry, it is often necessary to remove protein cells to obtain an interference-free analysis. This is accomplished by dialysis. The dialyzer module (Figure 15.11) consists of a pair of perspex plates, the mating surfaces of which are mirror grooved in a continuous channel, which goes in towards the centre on itself and returns to the outside. A semi-permeable cellophane membrane is placed between the two plates and the assembly is clamped together, similar to the kidney dialyser. The continuous groove channel thus gets divided into two halves and the dialysis occurs across the membrane. A solution containing substance to be analysed passes along one-half, usually the upper one, of the channel, while the solvent that is receptive to the substance to be removed enters the other half. The substance to be separated from the sample diluent stream, will diffuse through the semipermeable membrane by osmotic pressure into the recipient stream and the non-diffusable particles will be left behind.

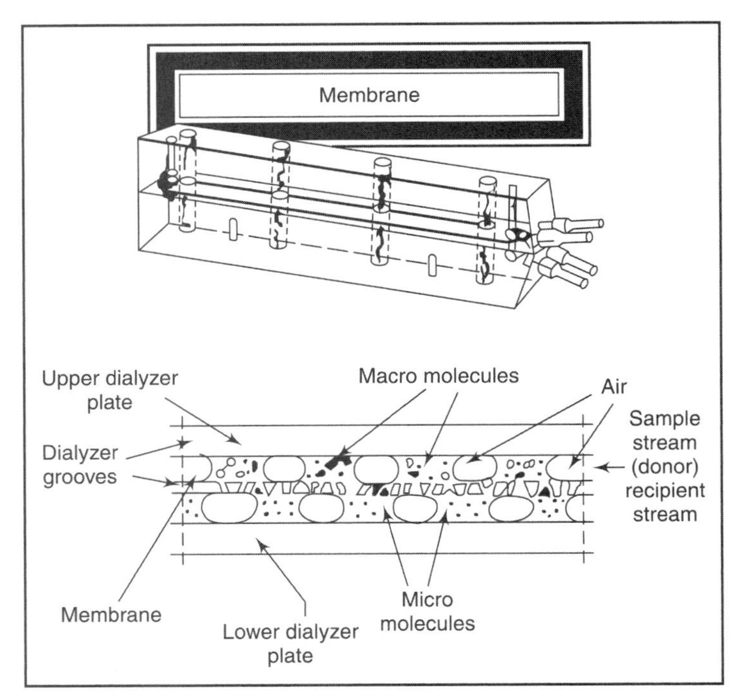

Figure 15.11 :: Simplified diagram of the dialysis process

Cellophane membrane

The cellophane membrane usually used in the dialyzer has a pore size of 40-60 Å. The rate of dialysis is dependent upon temperature, area, and concentration gradient. For this reason, the dialyzer unit is usually immersed in a water bath maintained at a constant temperature (37 ± 0.1°C). The temperature is kept constant with a thermostatically controlled heater and a motorised stirrer. Both streams pass through pre-heating coils, before entering the dialyzer unit. The channel path is 2.2 m long, which provides a large surface presentation to the dialyzing membrane. The plates of the dialyzer must be a matched set. If the plates are not matched set, the channels may be slightly off, causing leakage, poor bubble patterns and loss of dialyzing area, which would ultimately result in loss of sensitivity.

The quantity of solute that passes through the membrane in the dialyzer is determined by the concentration gradient across the membrane, the duration of contact of the two solutions, the area of contact, the temperature and by the thickness and porosity of the membrane. Other factors which affect the rate of transfer are the size and shape of the molecules, their electrical charge and the composition of the fluids across the membrane.

A decrease in the flow rate of the liquid streams increases sensitivity in continuous flow systems, since more concentrated samples and thinner membranes can be used. Modern dialyzers, therefore, have shallower and shorter grooves, resulting in reduced sample interaction and carryovers. Membranes have been found to age with use and time due to protein deposition on their surface and therefore, need periodical replacement. In the recent systems, the computer informs the operator to investigate the need for membrane replacement.

15.2.1.5 Heating Bath

On leaving the dialyzer, the stream may be combined by one or more additional reagents. It is then passed to a heating bath. This module is not used in all the tests performed by the analyzer. The heating bath is a double-walled insulated vessel, in which a glass heating coil or helix is immersed in mineral oil. A thermostatically controlled immersion heater maintains a constant temperature within ± 0.1°C. Inside the bath, the stream passes along a helical glass coil about 40 ft. long and 1.6 mm I.D., immersed in oil, which is constantly stirred. The heating bath may have fixed temperature, as 95° or 37°C or an adjustable value. Passage through the heating coil takes about five minutes, but it would obviously vary with the rate at which the liquid is moving, which, in turn, depends on the diameter of the tubes in the manifold.

Some systems have dry heating bath which operate at temperatures upto 165°C controlled to ± 0.1°C. The dry bath heater uses a standard glass coil, approximately 6 ml volume, with pre-set digital temperature control adjustable from 37°C. The dry heating bath avoids the use of bulky oil baths and lengthy glass connections.

15.2.1.6 Measurement Techniques

Although automated analyzers mostly use absorption spectrophotometry as the major measurement technique, several other alternative photometric approaches have been utilized in recent years. These are reflectance photometry, fluorometry, nephelometry and fluorescence polarization. The use of ion-selective electrodes and other electrochemical measurement techniques is also becoming popular.

In the absorption photometers used in automated systems, the radiant energy sources employed include tungsten, quartz, halogen, deuterium, mercury and xenon lamps and lasers. The spectrum covered is usually from 300 to 700 nm. Spectral isolation is generally achieved with interference filters in most automated systems. These filters have a peak transmittance of 30-80 per cent and bandwidths of 5-15 nm. The filters are usually mounted on a rotating wheel and the required filter is brought in position under the command of a microprocessor. Some automated systems also make use of monochromators, which obviously provide greater flexibility for the development and addition of new assays. The most popular detector used in the automated systems is the photomultiplier although some of the recent systems also employ photodiodes.

Proper alignment of flow cells or cuvettes with the light path is as important in automated systems as in manual methods. Stray energy and internal reflections are required to be kept as low as possible, to approximately less than 0.2 per cent. This is usually done by careful design of the wavelength isolation filters, or monochromator, or by the use of dual filters to increase rejection of stray light

Colorimeter

The colorimeters used in automated systems continuously monitor the amount of light transmitted through the sample. They employ flow-through cuvettes. In the earlier designs of flow cells, the arrangement was such that as the incoming stream entered the cell, the air bubbles escaped upwards through an open vent, so that a continuous stream of liquid could fill the cell before going to waste. The flow cell size varied from 6 to 15 mm. The later designs of flow cells are all of tubular construction. This requires a much smaller volume of fluid so that a smaller volume of sample can be used. Being completely closed, a colorimeter does not require separate cleaning. Before the stream enters the flow cell, it is pumped to a de-bubbler, where the air bubbles are removed. The stream is then pulled through the flow cell under the action of another pump.

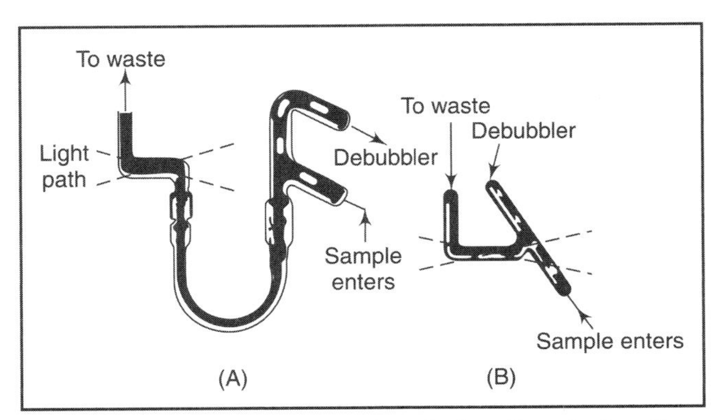

Figure 15.12 :: Constructional details of the flow cells (Courtesy M/s Technicon Corp., USA)

Figure 15.12 shows the construction of the two types of flow cells used in the auto-analyzer colorimeter. They are designed to exhibit optimum wash (clean out) characteristics. A debubbler is coupled with the flow cell, so that the entrapped air bubble may leak out. The cell shown in Figure 15.12 (B) is of one piece construction. It has smaller volume, while it maintains the same optical path length.

Multi-channel colorimeter

Some automatic continuous flow chemical analysis systems incorporate a multi-channel colorimeter. The colorimeter employs a fibre-optics system, using a single high-intensity quartz halogen lamp, which is coupled to a highly stabilised power supply. This lamp passes its energy through lights guides (Figure 15.13), which are connected to up to five independent colorimeter modules. Each colorimeter has two detectors one for the sample cell and the other for the reference beam. An insulating wall is placed between the light source and the colorimeter modules, so as to maintain good temperature and electrical stability. The colorimeter contains a glass continuous flow sample cell with integral debubbler.

Light guides from the main light source plug into one side of the cell block and the photo-detectors are placed on the other side of the block. The filters are inserted into a slot between the light guide and flow cell. The detectors are, closely matched and their output is connected to a highly stabilized bridge circuit. The output fed to the recorder is logarithmic, hence for solutions which obey Beer's Law, the recorder output is linear in concentration over the normal range of optical densities observed in continuous flow automatic analysis system.

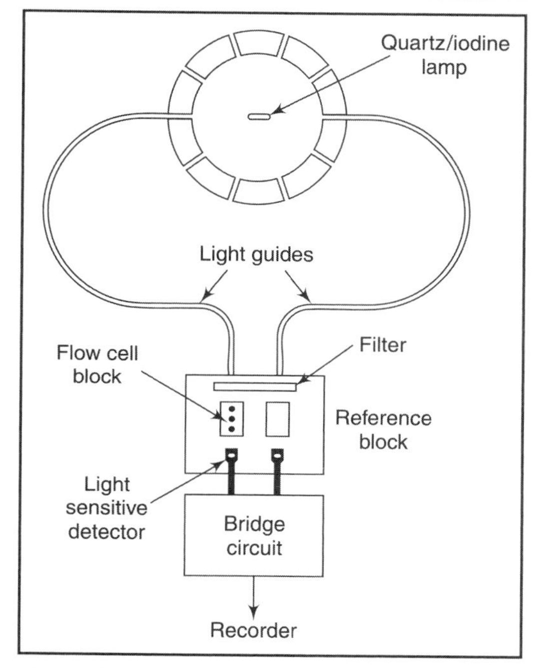

Figure 15.13 :: Colorimeter for automated analysis (Courtesy Chemlab, UK)

Flame Photometer

The measurement of sodium and potassium for clinical applications is carried out by a flame photometer. This comprises an oxygen propane burner, into which samples for analysis are pumped and atomized. The light output from the flame passes through and falls on three photocells. One filter transmits the wavelength of sodium flame, one the potassium wavelength, and the third the light emitted by lithium. An acid lithium nitrate solution is used as sample diluent and the output of the lithium sensitive photocell is used as a reference. The ratios of the outputs of sodium and potassium detectors to this reference signal is recorded on a strip chart recorder. This minimizes any minor signal changes caused by small variations in gas pressure, air or oxygen pressure, dialysis rate and sample flow rate. This is because these variations would affect the signal output of both the sample and reference photocells to the same degree.

The instruments are generally filled with automatic flame failure detection for user safety and automatic dilutors. Serum samples need to be diluted 200:1 or 100:1 prior to presentation to the flame photometer. Additional filters are available for the measurement of lithium (Li), barium (Ba) and calcium (Ca).

Fluorimeter

Fluorimetric analysis permits measurements to be made at concentrations as low as 0.01 part per billion. The fluorimeters used for automated work, like colorimeters, have flow-through system. The continuous flow cuvette is made of Pyrex glass, which transmits light from the visible region to approximately 340 nm. For the ultraviolet region below 320 nm, quartz cuvettes are used. The fluorimeter used are generally self-balancing, double-beam type instruments.

15.2.1.7 Signal Processing and Data Handling

The availability of low-cost microprocessors has had a major impact on the signal processing and data handling of analytical procedures in automated systems. Specific algorithms have made real-time acquisition and processing of data possible so that the output is immediately useful and meaningful. Transformation of complex, non-linear standard responses into linear calibration curves have allowed automation of procedures, such as reflectance spectrometry.

Specifically, microprocessors are now being used in automated methods for the following functions:

- Complete control of the electromechanical operation of the analyzer in relation to transfer of solutions, selection and placement of proper filters and continuous monitoring of operation. This ensures that all functions are performed uniformly, repeatably and in correct sequence.
- Acquisition assessment, processing and storing of operational data from the analyzer.
- Providing effective communication between the analyzer and the operator through alpha-numeric display on the CRT. Some systems even monitor the equipment function and give out message describing the site and type of problem in a malfunctioning equipment.
- Facility to communicate to mainframe computers through RS-232 interface for integration of instrument with laboratory information.
- Facility to communicate over the telephone lines, using a modem, with the manufacturer's central service department, thereby enhancing ability of the on-site operator to service and repair the analyzer.

Modern automatic analysis systems make use of a PC as a multi-tasking data processor and system controller. However, such multi-tasking is achieved with the use of a microprocessor-based analogue interface card which fits neatly into a standard extension slot inside the computer. Primary data acquisition, graphics, peak detection, calibration and calculation all place in the background, so results are produced immediately upon a peak being detected. These systems have comprehensive software support which includes method development, special functions, interfacing requirements and system control of power values, pump, colorimeters, etc. Peaks are captured to have similar dimensions (height and width), independent of the channel or analysis speed, so that most peak-picking parameters do not need to be changed from their default values. Upto 16 channels of independent data can be processed simultaneously.

15.2.1.8 Recorders

All twelve tests for each sample are reported in directly readable concentration units, on a single strip of pre-calibrated chart paper. Since the normal ranges for each parameter are also printed, the physician does not have to remember the normal values. Thus, each abnormality stands out clearly. The most common type of recorder used with automated systems is the dc voltage null-balance potentiometric recorder.

Significant fluctuations in the flow pattern may result in irregularities of the base line on a recorder. A good bubble pattern with air bubbles of equal size and equal spacing between them is one of the important factors in producing a smooth line with good peaks. The tubes and reagents must be clean. Irregularities are also produced due to a drift in the electronic circuit. Pulse suppressors are used to smoothen out fluctuations due to pump action and are specially required when micro-sample tubing is used.

After obtaining a good base line, the test run can be begun with a series of standards followed by test samples. It is usual to introduce standards at the end of the run and at intervals during the course of runs to ensure whether any change has occurred.

Since concentration of a substance in a sample is related logarithmically to the percent transmission, when this is plotted on a graph, the curve between them will not be linear. Therefore, a linearizing recorder with a logarithmic slide wire is employed, which plots chart readings and yields a straight line, if Beer's law is strictly applicable. Alternatively, sufficient standards are run to give an accurate standard curve and the test results are then read from the same.

Actual measurements are made only after the analytical curve (Figure 15.14) reaches its steady state plateau (equilibrium condition in the system at which there are no changes in concentration with time). At this steady-state plateau, all effects of possible sample interaction have been eliminated, and the recorded signal gives a true reflection of the concentration of the constituent being measured. Herein lies the importance of segmenting each of the sample and reagent streams with air bubbles. In effect, the air bubbles act as barriers to divide each sample and reagent stream into a large number of discrete liquid segments. Equally important, the air bubbles continually scrub the walls of the tubing. This sequential wiping of the walls diminishes the possibility of contamination in succeeding segments of the same sample. Thus, should there be any interaction between two samples, it can easily be seen that the effects of this interaction will occur only in the first few segments of the second sample. In the middle segments, the air bubbles immediately preceding will have effectively cleansed the system and prevented further interaction. It is these middle and final segments, free from interaction, which are recorded as the steady-state plateau and appear as flat lines on the graph.

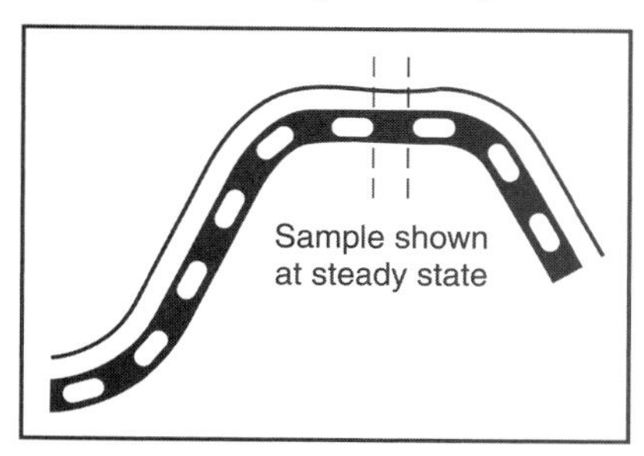

Figure 15.14 :: Typical curve showing the steady state conditions when measurements are made in a continuous flow system

15.3 Advanced Versions of Multiple Analysis System

The advanced version based on the auto-analyzer concept can perform more than 20 simultaneous analyses on one specimen. Since as many as 150 specimens can be loaded on the system, they can carry out more than 2500 test results per hour.

In these systems, after the specimen is sampled, diluted with distilled water and segmented with air bubbles, it is passed through a series of analytical cartridges. Each cartridge is a single-channel analyzer and includes a peristaltic pump, pump tubes, mixing coils, injection blocks and flow cells. Each cartridge draws the diluted sample from the central riser and reagents from the corresponding containers, the amount depending upon the internal diameter of the tubes. Capacities of reagent containers range from 20 to 1000 ml and reagent consumption may vary from 30 to 60 ml per channel per hour. Pump tubes usually need replacement after about 200 h of operation. If required, an on-line dialyzer for protein separation or an on-line heating bath for colour development are included in the analysis chain.

Spectrophotometry is used for quantitation of most of the analytes, except for estimation of sodium and potassium, which are measured by in-line ion-selective electrodes. Methods for glucose and uric acid employ specific enzymes immobilized in nylon coils and indicator reagents for spectrophotometry. Optical fibres carry light from one of two light sources, UV and visible, to each flow cell and from each flow cell to a single photomultiplier tube detector. The wavelength for each flow cell is selected from a rotating wheel through the computer. The transmittance signal from the photomultiplier is converted to absorbance by a logarithmic amplifier and the analogue signal is digitized in an analog-to-digital converter. Digital data are processed for peak detection, which is checked and tested for abnormal waveforms against method specific curve characteristics stored in memory. Abnormal curve shapes due to insufficient sample, poor mixing or any other irregularity in flow, etc., are detected. The control system informs the operator about the problem with a message on the video display, and the remedial measures are suggested.

The analysis is initiated by the operator through the keyboard of the control console. Start up procedures such as automatic delivery of water and reagents are directed from the console. Calibration cycles, frequency of control specimens and intervals for re-adjustment of reagent baselines are all controlled through a keyboard. During the run, the operator can call for display of analytical curves of 16 channels at one time on the display.

The field of automated biochemical analysis systems is changing rapidly. Several manufacturers are producing systems, which are based on different automation concepts like continuous-flow parallel, batch, discrete and random access analysis. The systems are undergoing continuous improvements, particularly in reduction of the number of manipulative steps in analytical systems. This reduction has become possible through the use of non-liquid reagents in tablets or in thin-films or fibre matrices, bound enzymes, ion-selective electrodes coupled to reagent active membranes.

15.4 Flow Injection Analysis (FIA) Technique

The use of air bubbles to separate samples in the SFA technique is an effective method of avoiding sample carryover. An alternative automatic method, which could be operated by the reproducible injection of small samples into a continuously flowing, unsegmented stream of carrier or regent, is described by Miller (1983). The scheme is shown in Figure 15.15. At first sight, it appears that the samples will rapidly spread out in the flowing stream and there will be unacceptable carryover. It has been observed, both theoretically as well as practically, that though dispersion of the sample does occur, it is under the control of the laboratory technologist, being regulated by such factors as the carrier stream flow rate, the length and diameter of the flow tubing and so on. This controlled dispersion is the distinctive feature of the FIA technique.

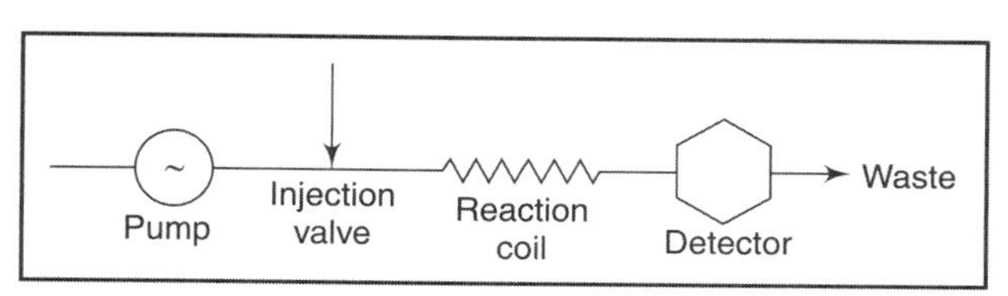

Figure 15.15 :: Single channel flow injection analysis technique

The major differences between FIA and SFA are that there is normally only one sample at a time between the injection point and the detector in an FIA system and that by making use of a precision of timing that is impossible in an air segmented stream. Also, FIA can measure partial responses, i.e. from incompletely formed response peaks in the detector. This means that very high sampling rates can be attained with the FIA technique.

The equipment needed for a flow injection analysis system is comparatively very simple and cheap. In most FIA systems, the carrier stream is impelled by a simple peristaltic pump. Alternatively, in place of the pump, a constant head device may be used. Flow tubing generally has an internal diameter of 0.5–0.8 mm. Flow rates and tube lengths will vary according to the required dispersion. Depending upon the reaction requirements, the flow manifolds are immersed in a thermostat bath.

As in the sequential flow techniques, specialized modules can also be incorporated into FIA manifolds. These include dialysis blocks, solvent extraction modules and several types of packed bed reactor.

One of the major advantages of the FIA technique is the possibility of using a wide variety of detector systems. Almost all the common atomic and molecular spectroscopy techniques as well as electrochemical detectors have been employed.

Dispersion

The type of applications in FIA system depend upon the dispersion (D) in the system, which is defined as C_0/C, where C_0 is the uniform concentration of a sample slug injected into the flowing carrier stream and C is the concentration of the sample by the time it reaches the detector. The dispersion can be measured as H_0/H, where H is the peak height from the detector in the FIA system and H_0 is the peak height observed if the sample is directly injected into the detector.

In practice, three ranges of D can be considered:

- *Low Dispersion* (0<3): These experiments are of value in studying the intrinsic properties of a sample, e.g. pH, colour, turbidity, etc.
- *Medium Dispersion:* (0 = 3 to 10): The medium dispersion mode accounts for the maximum application areas. These experiments involve substantial mixing of the sample with the carrier stream, it normally affects detection sensitivity.
- *High Dispersion* (*D*>10): This procedure finds fewer applications except for providing the means of obtaining calibration curves. It is of particular value when deliberate dilution of a sample is necessary.

The FIA technique allows analytical procedures to be carried out in a few seconds, with high precision and minute consumption of reagents. It appears to be one of the most potential and fascinating analytical procedures developed in recent years.

The rate of use of FIA is fast increasing and research papers and reviews are now being published with astonishing frequency.

15.5 Lab-on-Chip Technology for Automated Analysis

Microchips or microfluidic devices for chemical and biochemical analyses have been developed with the progress in microfabrication techniques. This new field of chemistry is known by the name of micro-total analysis system (μ-TAS), labs-on-a-chip or integrated chemistry lab (Manz and Becker, 1998). The concept of these microchip-based systems employs the integration of various chemical operations involved in conventional analytical processes done in a laboratory, such as mixing, reaction and separation, into a miniaturized flow system.

A liquid micro-space has several characteristic features that are different from the bulk scale: for example, short diffusion distances, high interface-to-volume ratio and small heat capacity. These characteristics in micro-space are the key to controlling chemical unit operations, such as mixing, reaction, extraction and separation, and constructing the integrated chemical systems. In addition, the transport time takes from several hours to one day when the diffusion distance is 1 cm, since the diffusion coefficient of typical molecular ions is of the order of 10^{-5} cm^2/s. In contrast with that case, it takes only several tens of seconds when the diffusion distance is 100 μm. These kinds of scale merits become remarkable below a scale of about 250 μm.

Focusing primarily on high applicability, Sato, *et al.*, (2003) have developed pressure-driven continuous flow chemical processing (CFCP) on microchips. In order to realize CFCP, a combination of micro unit operation (MUO) and multi-phase laminar flow was conceived. The components of MUO are shown in Figure 15.16. By combining these MUO components, various kinds of chemical processing can be integrated into the microchips. The authors have demonstrated the integration of fundamental MUOs, such as mixing and reaction, two and three-phase formations, solvent extraction, solid phase extraction, heating, and cell culture. Moreover, the formation of a stable multi-phase laminar flow network in micro-channels has also been reported.

microchip

The microchip is generally made from a glass plate, a silicon wafer, ploydimethy-lsiloxane (PDMS), polymethy-lmethacrylate (PMMA), other polymers, or their combination. Because of the chemical and physical stability and optical transparency for detection, a glass microchip is preferred for various applications.

Glass microchips are fabricated by using a photolithographic wet etching method (Hibara, *et al.*, 2001).

The description of an automated microchip-based platform for fabrication of microchip channel networks, which combines the electro-osmotic pumping and capillary electrophoresis for fluid transport and separation is available on. The system enables the on-chip integration of the key elements in analytical processing: injection, mixing, separation, detection and waste elimination within about three minutes. Normally these steps when done manually would require between 30-60 min.

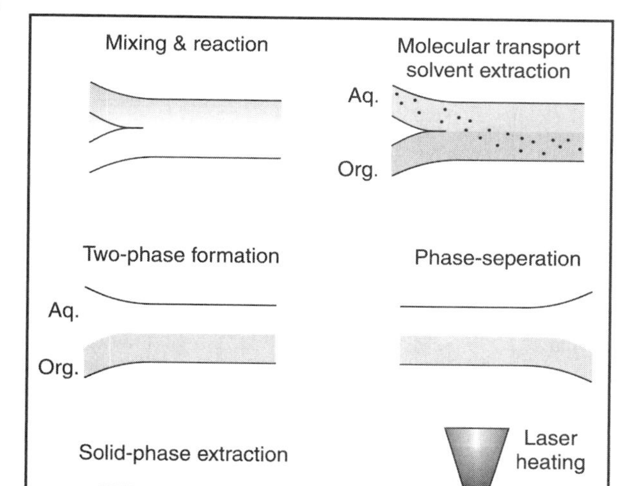

Figure 15.16 :: Sample introduction technique in a microchip (Hibara *et al.*, 2001)

Development work has been carried out for complete on-chip chemical processing for immuno-assays of the protein ovalbumin. The ability of the micro-machined glass plated to function as automated chemical processors rests on the integration of three peripheral subsystems with the chip: high voltage power supplies and relays for on-chip fluid control, fluidic interface for transporting liquids to and from the chip, and epiluminescent optical assembly for detection of products by laser-induced fluorescence.

The optics comprised a solid state laser which has an output of 8 mW at 635 nm and a circular Gaussian beam profile. The microscope assembly contains an aspheric lens with negative elements. Emission is collected at 670 nm using a series of bandpass and long pass optical filters and a photomultiplier.

For sample introduction (Figure 15.17) a low flow-resistance interface has been constructed on-chip. A large sample introduction channel (SIC) was fabricated 1 mm wide and 0.3 mm deep. The remainder of the on-chip channels were about 50 μm wide and 15 μm deep. Thus, the larger difference in channel dimensions provides a large difference in resistance to flow, so that sample can be pumped through the low resistance sample channel without the risk of contamination of the narrower chemical processor portion of the channel network. Injection of sample can be controlled precisely by electro-osmotic pumping. A fluidic interface assembly has been designed to facilitate connection of the chip to an external fluidic source.

Microchip

The microchip assembly is installed into the interface carriage by using a lever cam clamp. Sample analyses are delivered to the microchip network by a peristaltic mini-pump through the wide sample introduction channel. Programmable high voltage power supplies and switchable relays control the flow of solution and the fluidic processing. The experimental work described was carried out on the automated microchip platform. The single-channel design provides complete on-chip chemical processing of direct immuno-assays.

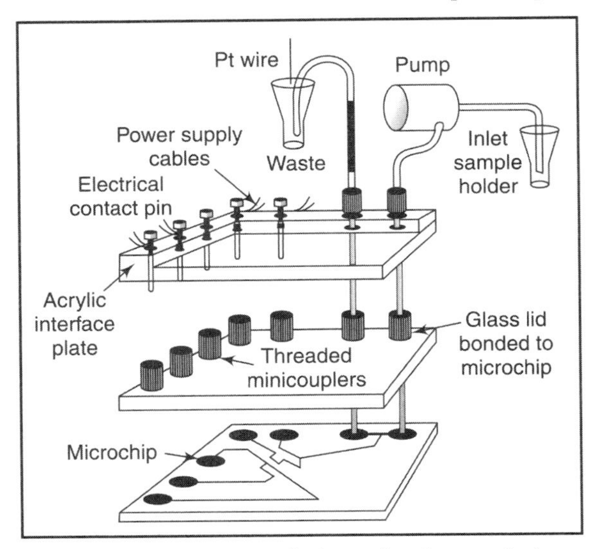

Figure 15.17 :: Sample introduction technique in a micro-chip (*http://www.suffield.drdc-rddc.gc.ca.*)

The chip-on-lab technology has reached a stage of commercial production. The Agilent 2100 Bioanalyzer, which is an automated analysis system, is powered by major advances in micro-fabrication technology with the development of a fully automated laboratory on a chip. The chip is designed to integrate sample preparation, fluid handling and bio-chemical analysis. The lab-on-a-chip technology uses semiconductor-like microfabrication techniques to translate experimental and analytical protocols into chip architectures consisting of interconnected fluid reservoirs and pathways. The movement of fluids, or molecules within fluids, is effected by electrodes that create electrokinetic forces capable of driving fluids through selected pathways.

Automated regulation of voltage between these electrodes controls the speed and direction of fluid movement. In this way, it is possible to create the functional equivalent of valves and pumps capable of performing manipulations such as dispensing, mixing, incubation, reaction, sample partition and detection.

Figure 15.18 illustrates the process on the lab-chip integrating:

- sample handling,
- separation and detection, and
- data analysis within a single, compact system.

The system is PC-based and primarily developed for DNA, RNA and protein analysis. The software quantitates each DNA fragment automatically against internal standards. This ensures increased accuracy and reproducibility. Similarly, for automated detection of ribosomal RNA contamination, the software integrates the messenger RNA smear and calculates the percentage of ribosomal RNA contamination.

Many merits and uses of microchip systems with pressure-driven flow have been reported. By utilizing several characteristics of microspace, i.e., a large specific interface, a short molecular diffusion distance, and so on, the performances of several analysis systems were greatly

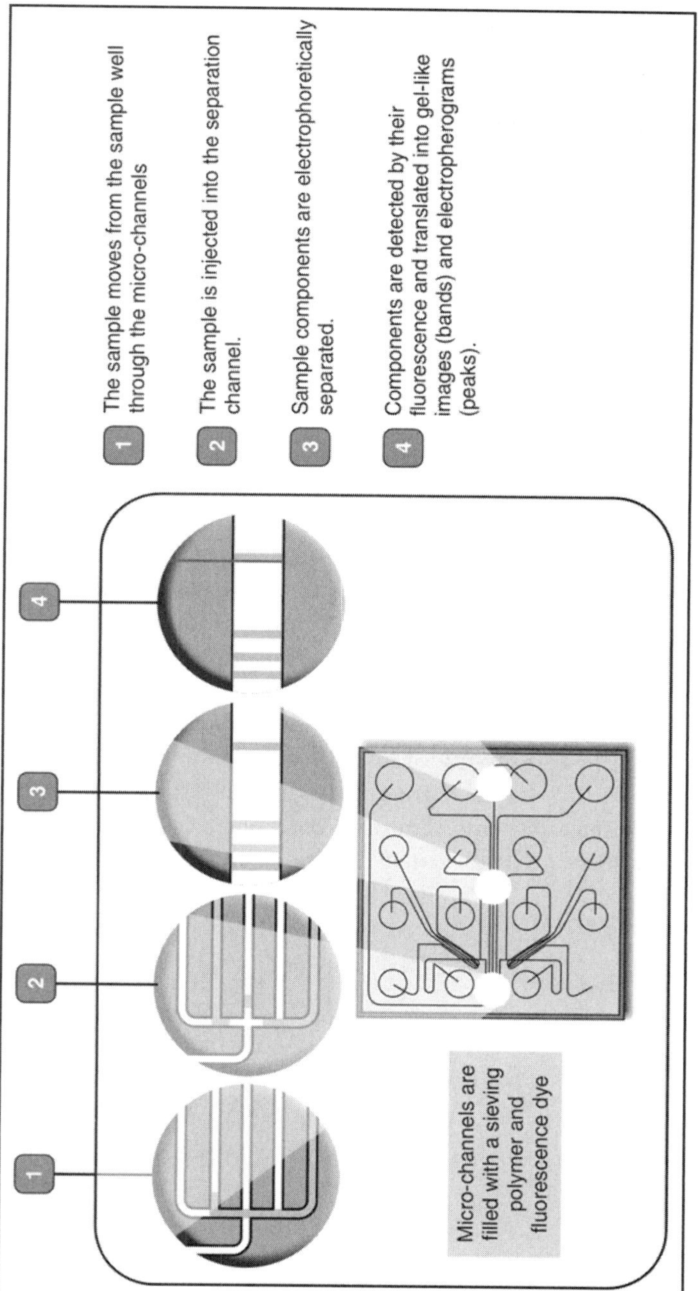

Figure 15.18 :: Working principle of Agilent 2100 Bioanalyzer—Lab-on-chip technology

improved by microchip integration. With a lab-on-chip, the following advantages have been established:

- Miniaturized fluid pathways give shortened run times.
- Strong electrokinetic driving forces improve analyte resolution.
- Microfabricated chips yield better reproducibility than conventional technologies.
- Versatile chip design enables flexible experiment design on one system.
- Micro-scale format minimizes sample use.

In the near future, these microchip-based systems will be widely spread and highly beneficial to our daily life by further efforts of analytical chemists and engineers.

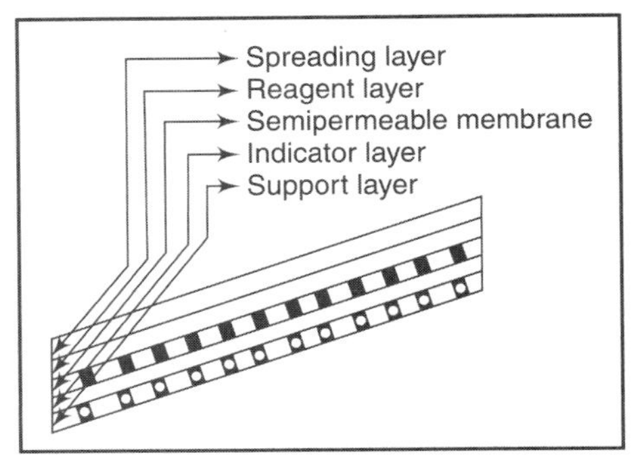

Figure 15.19 :: Exploded view of the multilayer dry film reagent used in Kodak Ektachem analyzers

For measurement, the Ektachem (Eastman Kodak Co) analyzer uses a multi-layer dry-film reagent approach for colour development (Stockwell, 1985). The system uses a 16 mm square slide, in which reagents are dispersed in emulsions which are achieved by diffusion of the sample fluid into the layers. Figure 15.19 shows an exploded view of a typical slide, the type of which is now available for different tests. Slides incorporating miniature ion-selective electrodes for measuring sodium, potassium, carbon dioxide and chloride are available.

The dry reagent systems eliminate dispensing and mixing of liquid reagents. Nevertheless, the systems still require a mechanism to maintain a stable temperature and to provide accurate positioning of the reaction unit for optical measurements. Thermal control in the Ektachem is provided through an intimate contact of the slides, with a heated carousel that supports the slide and rotates it to the measurement position.

Chapter 16

Gas Chromatographs

16.1 Chromatography

A sample that requires analysis is often a mixture of many components in a complex matrix. For samples containing unknown compounds, the components must be separated from each other so that each individual component can be identified by using standard analytical methods. The separation properties of the components in a mixture are constant under constant conditions, and therefore once determined they can be used to identify and quantify each of the components. Such procedures are typical in chromatographic separations.

Chromatography

Chromatography is a physical method of separation of the components of a mixture by distribution between two phases, of which one is a stationary bed of a large surface area and the other a fluid phase that percolates through or along the stationary phase. Chromatography was first reported by a Russian botanist, Tswett, who separated leaf extract into coloured bands by using a column of inulin and a solvent of ligroin. The technique remained largely ignored until the 1930s, when chromatographic separations of carotenes and xanthylls further demonstrated the possibilities of the technique and accelerated its development. The high separating power of gas chromatography has made possible the analysis of samples, which were hitherto regarded as difficult or impossible. This is evident from the wide range of applications found in routine and research work in medical and industrial fields. However, the applicability of the technique is limited to those substances which may be vaporized without decomposition or which may be thermally decomposed in a reproducible manner. Several techniques are used in chromatography as shown in Figure 16.1.

The process of chromatographic separation involves the transport of a sample of the mixture through a column. For this purpose, the mixture may be in the liquid or gaseous state. The stationary phase may be a solid adsorbent or a liquid partitioning agent. The mobile phase is usually a gas or a liquid and it transforms the constituents of the mixture through the column.

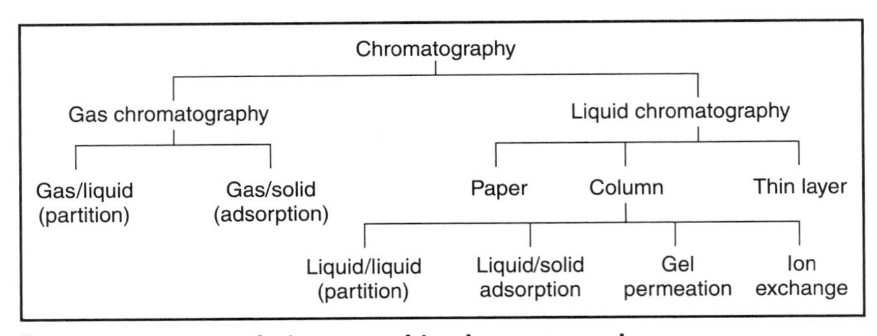

Figure 16.1 :: Techniques used in chromatography

During such transport, the material in the column (stationary phase) exercises selective retardation on the various components of the sample. This retardation may be due to adsorption, solubility, chemical bonding, polarity or molecular filtration of the sample. Therefore, the components of the mixture tend to move through at different effective rates, and thereby tend to segregate into separate zones or bands. In general, all chromatographic procedures isolate, detect and characterize these bands at some point, usually the column exit. Upon emerging from the column, the gaseous phase immediately enters a detector attached to the column. At this place, the individual components register a series of signals, which appear as successive peaks above a baseline on the recorded curve, called a chromatogram. A typical chromatogram is shown in Figure 16.2. The area under the peak gives a quantitative indication of the particular component and the time delay between injection and emergence of the peak serves to identify it.

Chromatogram

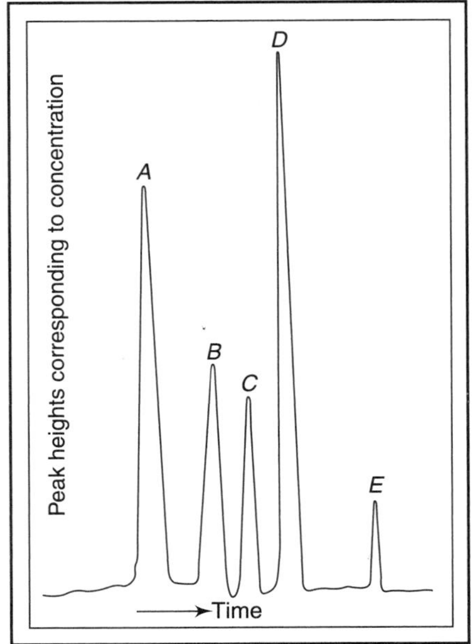

Figure 16.2 :: Typical gas chromatogram

The nature of the mobile phase is frequently used to label some of the major forms of chromatography, such as those designated by the acronyms GC (Gas Chromatography), LC (Liquid Chromatography) and SFC (Supercritical Fluid Chromatography).

Other classifications are based on the way the stationary phase is supported. For example, when the stationary phase is spread as a thin, adherent layer on a flat plate of an inert material such as glass or aluminium, the technique is referred to as planar chromatography, the best known variant being Thin Layer Chromatography or TLC. TLC can be used for quantitative analysis and preparative work but is most commonly applied to qualitative analysis of complex mixtures and assessment of purity. In most forms of chromatography that are important in quantitative analysis, the sstationary phase is supported in a long, tubular column usually made of glass on stainless steel. When the acronyms GC, LC and SFC are used, it is usually understood that particular forms of column chromatography are being referred to. Thus, the standard chromatograph is a column into which a mixture, transported by some fluid, is passed and from which, at some later times, the spatially separated components of the mixture emerge.

16.2 Basic Definitions

Retention Time (t_R)

The retention time is the total time that a compound spends in both the mobile phase and stationary phase. In other words, the time between sample injection and an analyte peak reaching a detector at the end of the column is termed the retention (t_R). Each analyte in a sample will have a different retention time. Retention time is generally expressed in minutes.

The time taken for the mobile phase to pass through the column is called t_M. This is shown in Figure 16.3.

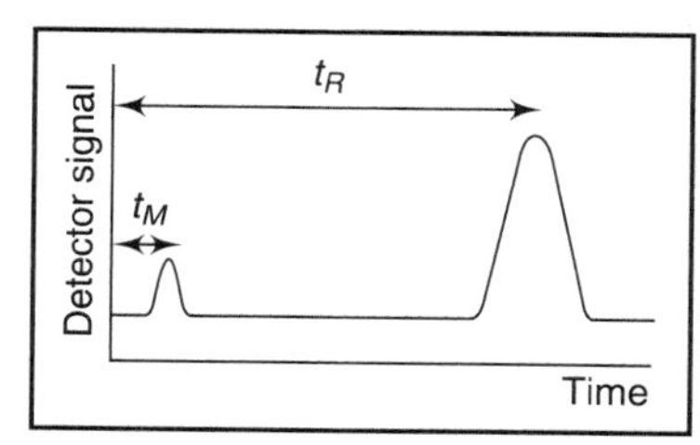

Figure 16.3 :: Illustration of 'Retention Time'

Dead Time (t_m)

The dead time is the time that a non-retained compound spends in the mobile phase, which is also the amount of time the non-retained compound spends in the column. Dead time is generally expressed in minutes.

Adjusted Retention Time (t_R')

The adjusted retention time is the time that a compound spends in the stationary phase. The adjusted retention time is the difference between the dead time and the retention time for a compound.

$$T_r' = t_r - t_m$$

Capacity Factor (or Partition Ratio) (k')

The capacity factor is the ratio of the mass of the compound in the stationary phase relative to the mass of the compound in the mobile phase. The capacity factor is a unitless measure of the column's retention of a compound.

$$K = \frac{t_r - t_m}{tm}$$

Phase Ratio (b)

The phase ratio relates the column diameter and film thickness of the stationary phase. The phase ratio is unitless and constant for a particular column and represents the volume ratio β.

$$\beta = r/2 \cdot d_f$$

Distribution Constant (K_D)

The distribution constant is a ratio of the concentration of a compound in the stationary phase relative to the concentration of the compound in the mobile phase. The distribution constant is constant for a certain compound, stationary phase and column temperature.

$$K_D = \frac{\text{concentration of compound in stationary phase}}{\text{concentration of compound in mobile phase}} = k \cdot \beta$$

Selectivity (or Separation Factor) (alpha)

The selectivity is a ratio of the capacity factors of two peaks. The selectivity is always equal to or greater than one. If the selectivity equals one, the two compounds cannot be separated. The higher the selectivity, the more will be the separation between two compounds or peaks.

$$\alpha = KA/KB$$

Linear Velocity (u)

The linear velocity is the speed at which the carrier gas or mobile phase travels through the column. The linear velocity is generally expressed in centimeters per second.

$$u = L/t_m$$

Efficiency

The efficiency is related to the number of compounds that can be separated by the column. The efficiency is expressed as the number of theoretical plates (N, unitless) or as the height equivalent to a theoretical plate (HETP, generally in millimeters). The efficiency increases as the height equivalent to a theoretical plate decreases, and thus more compounds can be separated by the column. The efficiency increases as the number of theoretical plates increases, and thus the column's ability to separate two closely eluting peaks increases.

$$N = 5.545.\ [t_r/W_h]^2$$

$$H = L/N$$

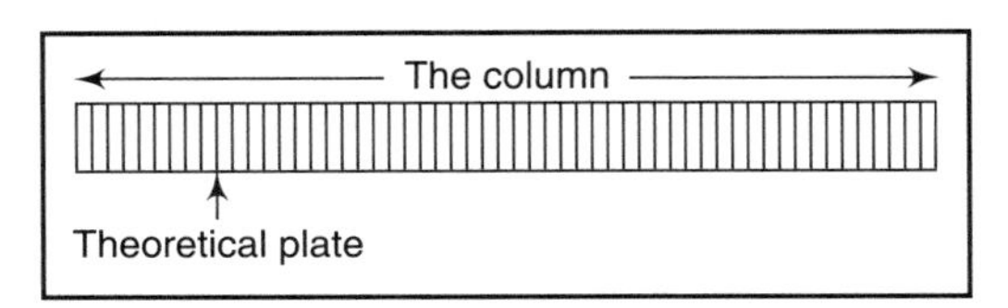

Figure 16.4 :: Theoretical plate model of chromatography

The theoretical plates concept (Figure 16.4) supposes that the chromatographic column contains a large number of separate layers. Separate equilibrations of the sample between the stationary and mobile phases occur in these plates. The analyte moves down the column by the transfer of the equilibrated mobile phase from one plate to the next. It may be noted that plates do not really exist, but the concept helps us understand the processes at work in the column. They serve as a way of measuring column efficiency.

The nomenclature of chromatography normally used in practice is given by Ettre (1979).

16.3 Gas Chromatography

Gas chromatography (GC) is an analytical technique used for separating compounds based primarily on their volatilities. Gas chromatography provides both qualitative and quantitative information for individual compounds present in a sample. Compounds move through a GC column as gases, either because the compounds are normally gases or they can be heated and vaporized into a gaseous state. The compounds are partitioned between a stationary phase, which can be either solid or liquid, and a mobile phase (gas). The differential partitioning into the stationary phase allows the compounds to be separated in time and space.

The particular advantage of using a gas as a mobile phase is that high flow rates are possible even with long columns. Another important practical advantage is that several methods of detecting components in a flowing gas stream are available. A further advantage is that it is relatively easy to find a chemically innocuous gas which acts as no more than a carrier of various vapours, though this can also be regarded as a disadvantage in that the mobile phase cannot be used to modify discrimination between materials.

In its simplest form, gas chromatography can be regarded as a multiple distillation process, with separations determined by relative volatilities, i.e. differences in the liquid vapour equilibrium of various materials. Temperature variation is the obvious way of controlling volatility and this is basically why temperature control is an important aspect of the operation of gas chromatograph.

Volatility is, however, not the only factor of importance, since specific interactions with the stationary phase, which is usually a liquid adsorbed on a solid but which can be a simple solid, do occur. The separation of simple gases such as O_2, N_2, CH_4, CO and CO_2, for example, can be achieved on a molecular sieve column, wherein the size of pores within the stationary phase determines whether or not a molecule enters and diffuses within the solid, so being retarded relative to the carrier gas.

One of the recent advances in gas chromatography, in fact, has been a change to capillary glass columns rather than the sometimes chemically reactive stainless steel columns that were originally predominant. This use of glass capillaries can be regarded as an offshoot of the development of fibre optic technology. Another recent development has been the use of open tubular rather than packed columns. In these, the stationary phase is present as a very thin film on the walls of only the column rather than being adsorbed on a finely divided solid which fills the column void. The open column allows high flow rates and the thin film allows very efficient mass transfer, so that extremely small values of H result.

16.4 Basic Parts of a Gas Chromatograph

The basic parts of a gas chromatograph are shown in Figure 16.5. It consists of the following parts:

- Carrier gas supply along with pressure regulator and flow monitor,
- Sample injection system,
- Chromatographic column,
- Thermal compartment or thermostat,
- The detection system, and
- Recorder.

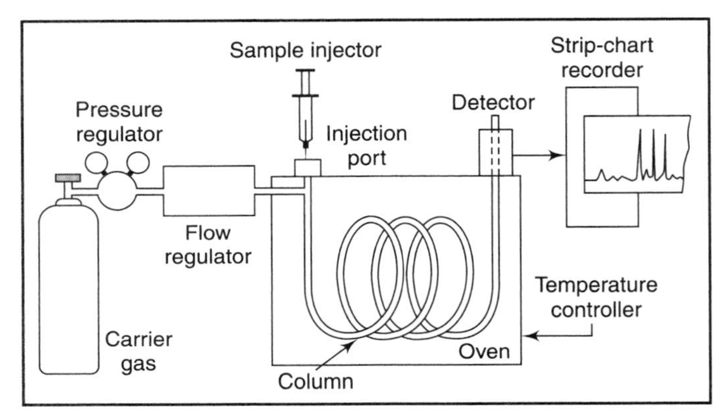

Figure 16.5 :: Block diagram of a gas chromatograph

The carrier gas, normally N_2, Ar or He, is usually available in a compressed form in a cylinder fitted with a suitable pressure regulator. The gas is conducted from the cylinder through a flow regulator, to a sample injection port maintained at a certain temperature T_1, which is such that it ensures rapid vaporization, but not thermal degradation of the solute. Gas and liquid samples are almost always injected by syringe through a self-sealing silicon rubber diaphragm in the injection port. The solute vapour mixes almost instantaneously with the flowing carrier gas and is swept into the chromatographic column, which is the heart of the chromatograph. It is there that the different solutes in the vaporized sample are separated from each other, by virtue of their different interaction with the column packing. The column is maintained at another temperature T_2. This temperature determines the time for the passage of the solutes and to some extent, the resolution and efficiency obtained with a particular column. At the end of the column, the solutes emerging individually enter the detector, which produces an electrical signal corresponding to the quantity of solute leaving the column. The detector signal is supplied to a recorder and a plot of the time–signal amplitude called chromatogram, is obtained. This record is used to determine the identity of the components in the mixture and their respective concentrations. The various parts of a gas chromatographic system are described below:

16.4.1 Carrier Gas Supply or the Mobile Phase

In a gas chromatograph, the mobile phase is formed by the continuous supply of a carrier gas. This supply is taken from commercially available cylinders, in which they are stored at pressures up to 2500 lb/sq. in. They pass through the column at low rates of flow (20–50 ml/min), at pressures not much greater than the atmospheric pressure. The carrier gas supply system comprises a needle valve, a flow meter, a pressure gauge and a few feet of metal capillary restrictors.

Types of Gases

Several gases like hydrogen, helium, nitrogen, argon and carbon dioxide have been tried as carriers. The carrier gas affects both the column as well as the detector performance. The carrier gas which is best for a particular detector, may not always be the best for the required separation. The choice of the gas is mostly determined by the type of the detector and the ready availability of the gas. For example, helium and hydrogen are preferred when thermal conductivity detection is employed, since their thermal conductivities are much higher than those of the compounds to be separated. Similarly, argon is used with argon ionization gauge detectors. Carbon dioxide is used with integral detection systems involving the removal of carrier gas by absorption in an alkali solution. On the basis of separation power, nitrogen, argon and carbon dioxide are slightly better than the lighter gases, as the latter have a tendency to enhance the axial diffusion of the solutes, a factor that could seriously affect the efficiency of the column. Nitrogen is particularly used where separating power is more important than high detector response.

Purity of Gases

The presence of contaminants in the carrier gas may affect column performance and detector response, particularly when ionization detectors are used. Carrier gases may be purified by the inclusion of a trap containing a molecular sieve of 5 Å. This is usually adequate for the removal of hydrocarbons and water vapour. Ultra pure N_2 for use with a flame ionization detector can be generated by commercially available apparatus. A low oxygen content of the carrier gas is essential for obtaining repeatable retention times and peak widths because a small amount of oxygen is enough to change the liquid phase and affect the overall partition ratio. In programmed temperature gas chromatography, a high purity carrier gas is even more important than in isothermal operation. The column is saturated at every temperature with the carrier gas impurities. During the heating up cycle, the saturation level changes and the impurities are given off to a certain extent. These impurities, which are liberated, are eluted and will temporarily increase the zero signal and noise, resulting in a broad peak.

The Gas Flow Rate

The rate of the gas flow to be used in a particular analysis depends upon column diameter, among other factors. The flow is generally in the range of 10-400 ml/min—very low and very high flow rates may affect the efficiency adversely. Flow rate should be controlled within 1 per cent, in order to reduce analytical errors. It should also be constant in order to give reproducible retention times. The flow rate of the carrier gas also affects the detector signals, because fluctuations would produce variable heat removal from the Katharometer filaments and hence cause variable filament temperature response and sensitivity.

The flow of the carrier may be maintained constant by inserting a capillary before the column, so that a pressure drop which is much larger than the pressure drop in the column, is created. When this capillary is kept at a constant temperature, the flow in the whole system will be mainly determined by this capillary and not by the varying flow resistance of the column. Alternatively, the gas is passed through a short piece of capillary tubing, resulting in a small pressure drop. A regular pressure regulator connected across the capillary keeps the pressure drop constant, resulting in a constant flow. With a view to speed up analysis, flow programming in which the flow through the column continuously increases, can be adopted. This would have the same effect, as in temperature programming of producing closely spaced peaks at the end of the chromatogram. For columns of different sizes, the volumetric gas flow rate should be varied in proportion to the square of the diameters, so as to maintain the average linear rate at approximately the same value.

16.4.2 Sample Injection System and the Size of the Sample

The purpose of the sample injection system is to introduce a reproducible quantity of the sample to be analyzed into the carrier gas stream. The transfer of the sample should be made rapidly to ensure, that the sample occupies the smallest column volume and thus prevents excessive peak broadening, which affects the overall resolution of the system. Samples can be introduced in their gaseous, liquid or solid states and many methods have been suggested for the purpose. However, the choice of the method of sample injection depends upon the pressure in the column at the point of introduction, the type of detector to be used and the source of the sample.

Injector The injector is basically a hollow, heated, glass-lined cylinder wherein the sample is introduced into the GC. The temperature of the injector is controlled so that all components in the sample will be vaporized. The glass liner is about 100 mm long and has a 4 mm internal diameter.

Theoretically, for optimum separation efficiency, the sample must be introduced into the column as rapidly and in as concentrated a band as possible. This is called Plug insertion that chromatographers consider ideal. The purpose of the sample injection system is to insert, volatilize and move the resulting gaseous sample into the column. The design of this system is a critical gas chromatograph performance factor, because a column capable of efficient component resolution may appear ineffective, due only to the inadequacy of the sample insertion system.

Theoretical considerations enjoin that the efficiency of separation in the GC column improves as the size of the sample is reduced. With a normal analytical column of 44 mm (internal diameter), a liquid sample of 2-20 μl and a gas sample of 0.5–5 ml at atmospheric pressure is generally satisfactory. An exception to this general practice is seen in the case of trace analysis, when liquid samples up to 200 μl are necessary to obtain a sufficient response. Wybrow (1985) describes a microcomputer-based injection system for use in the chromatographic analysis of gases.

Liquid Samples

The usual method is to inject liquid samples with a micro-syringe, through a silicon rubber septum. Syringes of various capacities are commercially available and are generally employed for injection of samples between 0.1 and 10 μl. A sample is injected into the hot zone of the column, so that the liquid gets rapidly transferred into the gaseous phase.

A typical arrangement for injecting liquid samples is shown in Figure 16.6. The metal block containing a capillary is heated by a controlled resistance heater. Here, the sample is vaporized and carried into the column by the carrier gas. Care should be taken to insert, inject and remove the needle quickly.

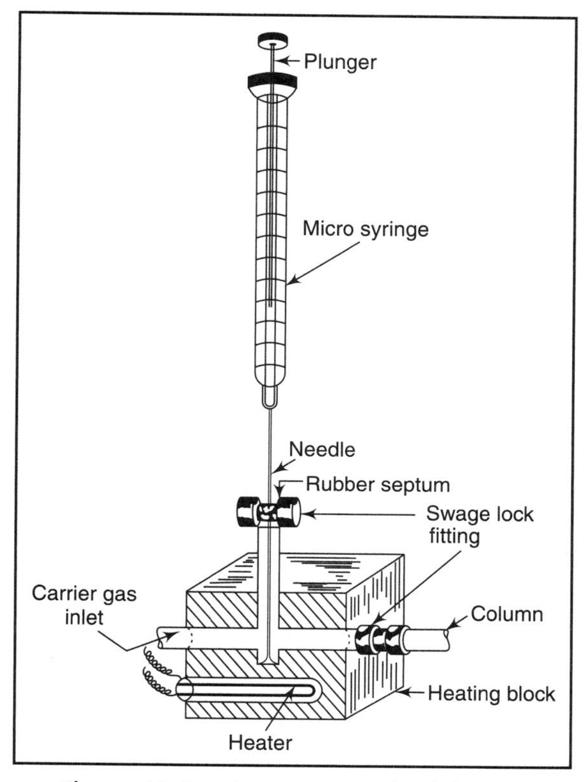

Figure 16.6 :: Arrangement for injecting liquid samples

Gas Samples

Gas samples are injected by a gas-tight syringe suitable for delivering 0.1–10 ml of sample. They are usually difficult to handle and often cause inaccuracies.

The other method of injecting gaseous samples is the bypass system, also called a stream splitter. This system has been found to be most valuable and gas sampling valves using this technique are used extensively. The principle of the system is to fill a loop of known volume with the sample. By operating a valve, the loop is placed directly in the carrier gas line. The valves were earlier made of glass, but they have now been replaced by polytetrafluoroethylene designs.

Stream splitter

Figure 16.7 shows a schematic of this type of arrangement. Basically, it is an arrangement of three stopcocks, between two of which there is a standard volume, in which the gas sample is enclosed. Gas from the bypass capillary loop is introduced into the column by a rotating or sliding valve, so that the loop is connected with the stream of the carrier gas.

Split/splitless injector

Figure 16.8 shows the schematic of the most popular split/splitless injector suitable for capillary columns. It has the same requirements as the packed column injectors: carrier gas inlet, a septum, septum purge, injector insert, heater block and column connection, but the improvement is another set of gas lines out of the injector—another path that the vaporized sample can take. This is called the split line. Thus, the carrier gas enters the chamber and can leave by three routes, when the injector is in split mode. The sample vaporizes to form a mixture of carrier gas, vaporized solvent and vaporized solutes. A portion of this mixture passes onto the column, but most of it exits through the split outlet. The septum purge outlet prevents septum bleed components from entering the column. In order to prevent column overloading, the amount of sample reaching the column is reduced and very narrow initial bandwidths can be obtained. For maximum sensitivity, the injector can be used in the splitless mode, then all the injected sample will reach the column.

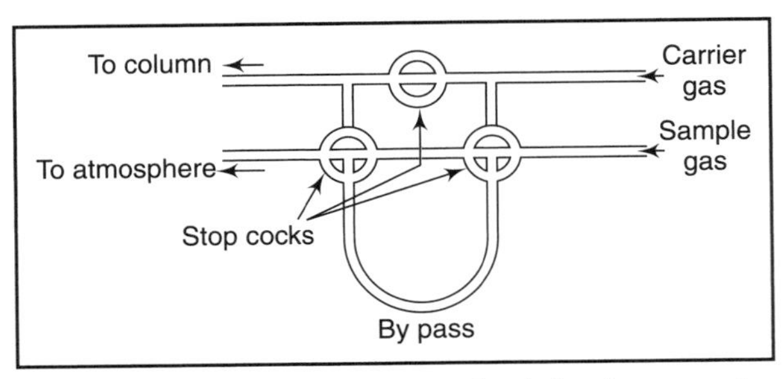

Figure 16.7 :: Bypass system for injecting samples

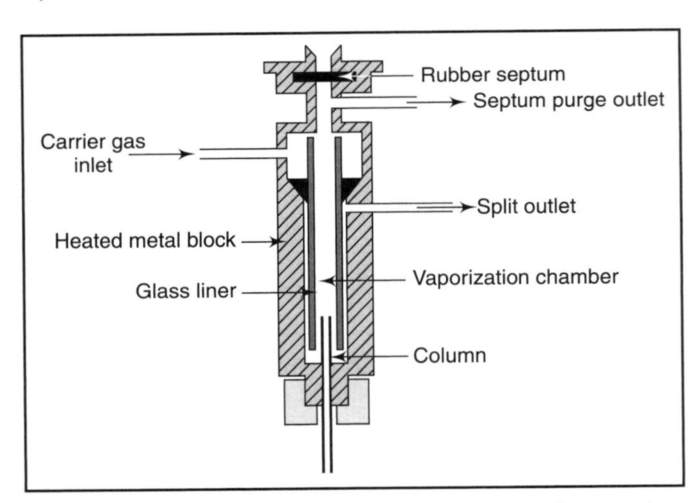

Figure 16.8 :: The split/splitless injector (redrawn after Chasteen, 2000)

Solid Samples

Solid samples may also be injected by using solid injection syringes, where the sample is deposited on the end of the plunger and withdrawn inside the needle. After piercing the injection septum, the plunger is extended to place the solid in the hot zone of the column, where it is vaporised. Alternatively, the solid is deposited in a glass tube or gauze from a solution. After the evaporation of the solvent, the sample holder is dropped into the column, thus making the injection. Another method is to dissolve the solids in volatile liquids or temporarily liquify them by exposure to infrared heat.

16.4.2.1 Pyrolysis

Pyrolysis offers a technique for the injection of certain types of materials, which are low or non-volatile, but which may be thermally decomposed in an inert atmosphere to offer a qualitatively and quantitatively reproducible mixture of volatile fragments. The pyrolysis products are transferred to a chromatographic column and separated in the usual manner. Pyrolysis has been accepted as a valuable technique for sample injection in rubber, plastic, polymers and adhesive industries.

16.4.3 Chromatographic Column

The column is the heart of a gas chromatograph, wherein the fundamental process of separation takes place. Its action is based on the fact that when a sample of gas or vapour is introduced into a column, it spreads by molecular diffusion to yield a concentration profile. As the sample moves through the column, additional spread takes place, but the band maintains its general shape, which is detected and recorded as the familiar chromatographic peak. The degree of peak broadening with respect to time and column length is an indication of column efficiency. Column performance is usually measured by the number of theoretical plates, which may be determined from the dimensions of peaks.

A theoretical plate is defined as a layer at right angles to the column of such a thickness, that the solute at its mean concentration in the stationary phase in this layer, is in equilibrium with its vapour in the mobile phase leaving the layer. Laboratory columns of 20–100 plates are widely in use and normally have a height equivalent to a theoretical plate (HETP) of about 1 cm. Longer practical columns of up to 1,00,000 plates have been reported in literature.

16.4.3.1 Types of Columns

There are two types of columns which are commonly used. They are: (i) packed columns, and (ii) capillary columns or open tubular columns.

Packed Column: The packed column is a tube packed with a suitable material, which performs the separations. Columns may be made from any suitable tubing. Glass, stainless steel or copper are the materials most frequently used for making columns. For moderate temperatures, polyvinyl chloride tubing is satisfactory. The internal diameter of the column is usually between 4 and 8 mm. The length of the column may be between 1 and 50 m. However, for most of the applications, a length of about 2 m is adequate. Columns longer than 3 m are difficult to pack uniformly.

Therefore, very long columns are best constructed by coupling short (less than 3 m) sections, to obtain any desired length. Standard swagelok unions are adequate for the purpose. Sample sizes for packed analytical columns vary between 0.1 and 10 μ. By increasing the column diameter, greater sample sizes, of the order of 100 μl to 3 ml can be used for preparative purposes. For this type of chromatography, the column has an internal diameter from 6 to 25 mm and the packing material specifications are also changed to cope with increased loading.

For convenience, the column is made in the form of a U or helix or it can even be straight. Straight and U-shaped columns can be repacked more easily. A helical column is normally difficult to fill. For this reason, copper tubing column is preferred, because it can be filled while straight and then bent into the helical shape afterwards. A definite advantage of a coiled or helical column is that they are more compact for a given length of column and are therefore easier to heat to an even temperature, The helix type of column is usually 50–250 mm in diameter and 2 m in length. As long as the column is packed sufficiently tightly before coiling, there is no significant difference in performance as compared to a straight tube. Columns are quite difficult to pack after they are coiled. Instruments designed for glass columns have tall ovens to accommodate U shapes, that can be easily packed for maximum performance. Columns with diameter larger than 12.5 mm are made in straight sections connected by small diameter tubing.

Capillary Column: Capillary columns are open tube columns of tubing that are approximately 0.25 mm in diameter. Their lengths may run anywhere from 30 to 300 m. Very high efficiencies have been achieved with capillary columns, since the cross-diffusion of sample molecules is minimized by the narrow diameter. Capillary columns contain no packing and the stationary phase is coated directly on the inside of the tubing. Capillary columns cannot handle samples more than 0.1 μl. Larger samples are handled by the use of inlet splitters. With capillary columns, better separations can be achieved at lower temperatures and in a shorter time.

Capillary columns are of two types: *wall-coated open tubular* (WCOT) or *support-coated open tubular* (SCOT) types. Wall-coated columns consist of a capillary tube whose walls are coated with a liquid stationary phase. In support-coated columns, the inner wall of the capillary is lined with a thin layer of support material such as diatomaceous earth, onto which the stationary phase has been adsorbed. SCOT columns are generally less efficient than WCOT columns. Both types of capillary columns are more efficient than packed columns.

Figure 16.9 shows a new type of WCOT column called the fused silica open tubular column. These columns have much thinner walls than the glass capillary columns, and are given strength by the polyimide coating. These columns are flexible and can be wound into coils. They have the advantages of physical strength, flexibility and low reactivity.

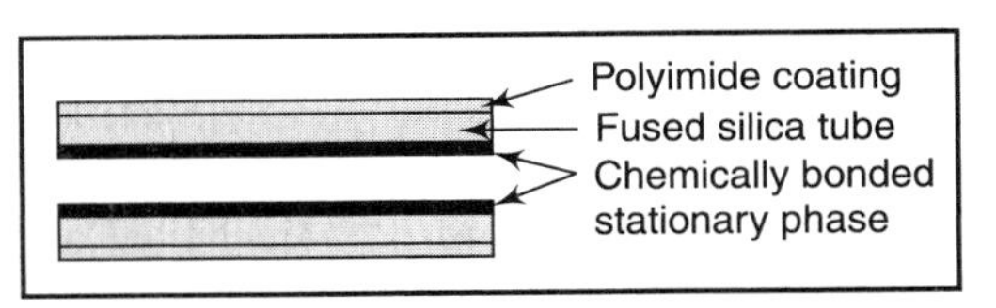

Figure 16.9 :: Cross-section of a fused silica open tubular column

The great analytical strength of capillary gas chromatography lies in its high resolution. Capillary columns have more theoretical plates (a measure of column resolving power or efficiency) per meter as compared to packed columns. The average capillary column (30 meters long) has approximately 100,000 theoretical plates while the average packed column (3 meters) has only 2500 plates. They have less resistance to flow and can consequently be longer than packed columns. Also, capillary columns require a smaller amount of sample than packed columns. While the average sample mass of each component in a mixture that is separable by packed column GC can be in the microgram range (10^{-6} grams) per injection, capillary columns routinely only handle 50 nanograms (10^{-9} grams) of a particular component or less.

However, some limitations are associated with the capillary columns. Since they have smaller diameters (0.05 to 0.53 mm) than those of packed columns (2 to 4 mm), they require relatively specialized injectors and flow controllers.

Capillary column gas chromatography is rapidly developing to meet many of the analytical needs such as environmental sections, analysis at trace levels for clinical and biomedical diagnostics, the separation of natural products and other complex mixtures.

16.4.3.2 Support Material

A support material is used in partition chromatography to provide a thin liquid film, with as large an interface as possible between the gas and liquid phases, so as to facilitate partition between them. A primary requirement of solid support material is that it should not possess adsorptive properties towards sample components. Besides being inert, the support should have the structure and surface characteristics to hold the liquid phase uniformly over a large area.

A number of supports have been tried, including glass balls, sodium chloride and pumice powder. However, only diatomaceous earth has been found to be satisfactory. It has been in use under the trade name Celite, with an average particle diameter of 40 μ. Its trade names also include GC-22 Super Support, Sil-O-Cell, C-22, Fire brick and Chromosorb P, W or G. Chromosorb is prepared by calcining diatomaceous earth obtained from a marine deposit in Lompoc, California. Celite is the most widely used support material.

Glass beads are also available in a variety of mesh ranges. Because of their low surface area and light liquid phase loading required, they will elute high boiling point compounds at a lower temperature than is normally required. Teflon 6 is another material, which is one of the most inert and non-adsorptive. It is best suited for the analysis of highly reactive materials, which are difficult to chromatograph on other supports. However, they offer lower efficiency and a temperature limit of 25°C. Particle sizes may range from 10 to 100 mesh.

Peak tailing

The most commonly encountered problem due to support participation with the column is that of peak tailing, which appears as a long tailing edge of a peak due to adsorption. In extreme cases, if the peaks tail badly, separation is impaired and the determinations based on peak area measurements become difficult to make and thus give unreliable results. In these situations, samples may be totally adsorbed. It is possible to minimise tailing effects by modification of the support. The Celite surface, which is prone to cause sample adsorption problems, resulting in peak tailing, may be covered in hydroxyl groups. The adsorption effects can also be minimized if the Celite is treated with a silating agent, such as hexamethyldisilazine or dimethyldichlorosilane.

The column must be so loaded that it has an even packing and the gas flow does not vary either across the column or irregularly along its length. Experience is needed to achieve an even packing of the column to obtain a high efficiency.

16.4.3.2 The Stationary Phase

The separation of the sample into its components is achieved by a partition process involving the stationary phase and the moving carrier gas phase. The stationary phase is either liquid or solid, Therefore, there are two possible methods with the gas as the mobile phase. These are:

(a) Gas–Liquid Chromatography, and
(b) Gas-Solid Chromatography.

Gas–Liquid Chromatography: Here, the stationary phase is liquid, which is distributed on a solid support material. The stationary phase must be involatile at all temperatures, at which the column will be operated for the analysis and should be coated as a thin even film on to the support. It is chosen for the selective retention characteristics of components in the sample that it will be used to separate. In general, highly polar stationary phases are employed to selectively retard polar compounds. On the other hand, non-polar stationary phases offer little selectivity and components tend to be eluted due to differences in boiling points of the sample components.

At temperatures above 150°C, special difficulties arise, as normal solvents become highly volatile or even unstable. In such situations, substances like silicon polymers may be used, especially for temperatures above 250°C.

An important requirement of the stationary phase liquid is a certain compatibility with the components of the sample under analysis. Generally it is found that a polar substance is most satisfactorily analyzed on a polar stationary phase. Similarly, a non-polar compound will give the best results on a non-polar phase. For example, for separations of polar components like alcohols, amines, etc., it is preferable to choose a polar liquid like polyethylene glycol. Nevertheless, departure from the above rule of similarity is quite often necessary.

Normally, for analytical packed columns, 1 to 10 per cent w/w of stationary phase on the support is employed. The analysis time is approximately proportional to the quantity of stationary phase in the column. Higher rates induce diffusion phenomena that would impair the separation. On the other hand, at low ratios, the inert support might manifest considerable residual absorptivity to cause tailing of elution peaks.

For applications in the biomedical field, for the analysis of sugars, bile acids and steroids, a low percentage of stationary phase may be used to produce a very fast column, which will pass high boiling point samples at moderate temperatures to avoid thermal decomposition. The choice of stationary phase is extremely important for the successful analysis of each sample mixture. Sometimes, in practice, brief experiments are necessary with a number of trial columns, in order to make a suitable section. It is here that the experience of the operator would prove useful.

Gas–Solid Chromatography: The stationary phase in this type is a solid material with surface active properties. The separating principle is based on the variation in the extent to which the constituents of a mixture are adsorbed on the adsorbent packed in the column. Therefore, the separation is obtained because of the different adsorption affinities which the column packing has towards the sample components. This type of chromatography is used in the analysis of inorganic gases and low molecular weight hydrocarbon gases. Among the most commonly used adsorbents in gas–solid chromatography are silica gel (SiO_2), alumina (Al_2O_2), charcoal and molecular sieves (sodium or calcium aluminium silicates).

16.4.4 Thermal Compartment

16.4.4.1 Isothermal Operation

The column is not normally operated at room temperature, because it would then be suitable only for the analysis of gases or extremely volatile liquids. Therefore, it must be heated in some form of thermostat. Moreover, it is desirable to keep the column at a precisely constant temperature. This is essential because the quantitative response of the detector is often affected by column temperature. For this purpose, the column is housed in an oven, whose temperature is controlled to an accuracy of 0.1°C. Various other methods have been tried, namely vapour jackets, electrically heated air baths, liquid baths or metal blocks. Usually, an air bath chamber surrounds the column and air is circulated by a blower through the thermal compartment. The temperature of the oven may be controlled accurately by using a proportional temperature controller with a platinum resistance thermometer as a sensing element. The oven is thermally insulated so that heat loss to the atmosphere is minimized. However, this factor is balanced against the thermal capacity of the insulating material, which, if too high, would affect the rate of cooling of the oven.

Normally, the temperature is so chosen that it gives a satisfactory time for analysis. Approximately, a temperature in the vicinity of the average boiling point of the components in the sample will be convenient, so as to effect an elution period of 10 to 30 min.

Temperature controller

Figure 16.10 shows a schematic diagram of an oven temperature controller. The temperature sensing is done by the platinum resistance R_1, which is placed in the oven. The temperature setting is done by adjusting the potentiometer VR_1, calibrated in terms of temperature. This control is provided on the front panel of the instrument. When a setting is made, the bridge gets unbalanced and the amplifier, the synchronous rectifier and the UJT oscillator are actuated to open the gate of SCR. Thus, the current is supplied to the oven heater and the oven temperature begins to rise. As the oven temperature approaches the pre-set value, the sensor resistance becomes higher. The bridge approaches nearer the balanced state and the heater current decreases. When the oven temperature reaches the pre-set value, the heater current would not flow and the bridge would be balanced. The oven temperature is thus kept constant. A thermal fuse placed in the circuit prevents the oven from overheating.

16.4.4.2 Temperature Programming

When the column temperature is kept constant, it is difficult to analyze samples having components of a wide boiling range. This difficulty can be overcome by using programmed heating of the

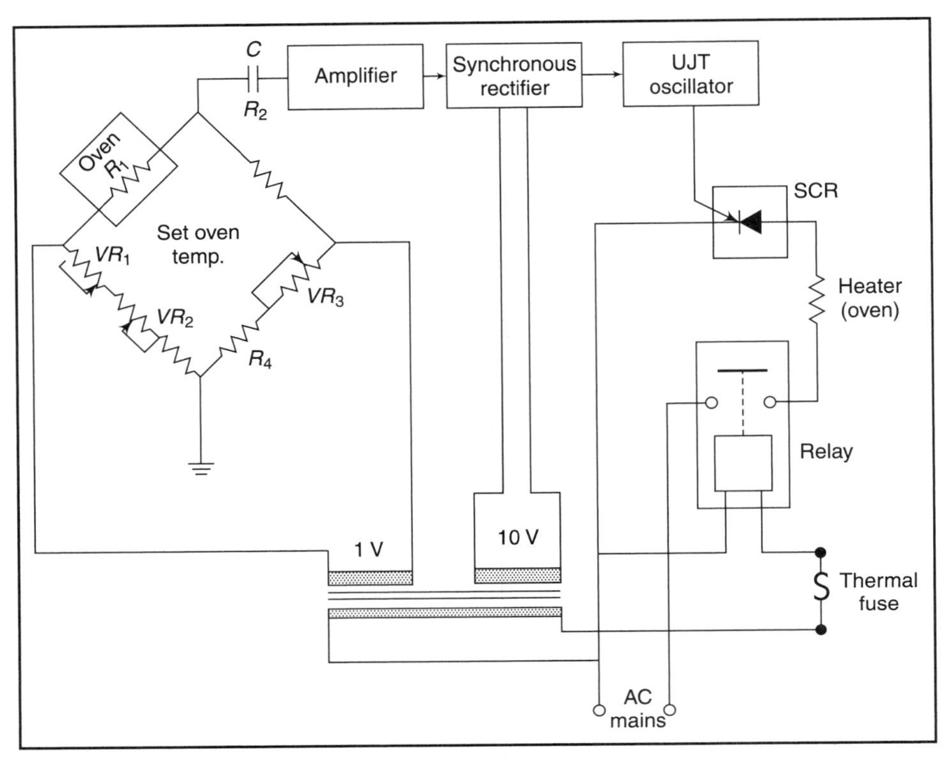

Figure 16.10 :: Temperature control circuit for ovens

column, so that its temperature is not kept constant but is subjected to an exactly controlled temperature rise, while a separation is in progress. The technique combines in it the advantage of a low temperature for better separation of low-boiling point components with that of high temperature for more rapid elution of high-boiling point components, thereby shortening the time of analysis and sharpening the resultant chromatographic peaks. The temperature increase may be programmed to be carried out either continuously or in steps, or abruptly to a pre-determined higher level between two peaks. Programs are available, which give linear and non-linear temperature programming of ovens. The temperature can be raised at various rates. Generally, linear rates of temperature programming in the range of 1 to 20°C per min are used. The rates 5-7°C per min are most typical. Some applications require non-linear temperature programming in an exponential or ballistic manner.

For temperature programming, the program according to which the temperature is to be varied is taped or ink recorded on special Mylar format sheets. Curved rates of temperature rise, linear sections and isothermal operation can be plotted as desired. The recorded sheet is fixed on a rotating drum and the program line is followed by an optical scanner. The scanner is linked to a servo system, which continuously controls the wattage supplied to a proportional heating system. At the completion of a run, column temperature must be dropped from about 300°C to less than 100°C in a few minutes, so that the column may become ready for the next run. Baudean, *et al.* (1977) explains the use of microprocessors for temperature programming.

16.4.4.3 Use of Two Columns

When the programmed temperature technique is being used, the behaviour of the column itself is influenced with the change in temperature. There is an increasing tendency for the stationary phase to bleed from the column, as the temperature rises and it is reflected as a baseline drift, with a chromatogram superimposed on it. When working at high sensitivity, it is possible that this drift in base line may severely limit the use of temperature programming. This problem can be partially offset by using two matched columns and operating two detectors in a differential mode. One column is called the sample column and the other the reference column, to which no sample is added. The signals from the detectors are combined to balance the bleed effects and give a straight baseline. However, a careful setting up procedure is necessary to balance the two columns.

The following are important considerations, which are kept in view when designing column ovens:

- The oven must have minimum thermal gradients so that the temperature is uniform over the whole column.
- The oven must have a fast rate of heating. For this, it must be constructed from low-mass materials. This requirement is particularly significant for changing the column temperature rapidly as in operations involving temperature programming.
- Temperature controlling facility up to 400°C is necessary.
- Power consumption should be kept low. For this, heat loss by all means must be minimized.
- The door of the oven should be large enough to facilitate installation and removal of column and its accessories.

16.4.5 Detection Systems

The detector is placed at the exit of the column. It is employed to detect and provide a quantitative measurement of the various constituents of the sample, as they emerge from the column in combination with the carrier gas. The detector, in fact, acts as a transducer and converts the changes in some physical property to changes in an electrical signal, which can be conveniently recorded.

The choice of a particular type of detector is governed by the following factors:

- The detector should have a high sensitivity, sufficient enough to provide an adequate signal for all components with a small sample. It should also permit the use of lower column temperatures.

- It is desirable to be able to measure components from the fractional ppm to almost 100 per cent in one sample. The response of the detector should be linear over the whole range.
- A small internal volume ensures that the resolution of components, which are separated by the column, is not lost and that the shape of peaks is not distorted by the detector.
- Detector temperature should be such that an appreciable amount of the eluted vapours does not condense in it.
- The detector should be insensitive to changes in the rate of flow of the carrier gas.
- The detector should give good reproducibility of the base line.

Several detection systems are used in gas chromatography. Quite often, the fields of application of these detectors overlap to a certain extent, but one of the detectors will usually have characteristics that make it most suitable for a particular analysis.

Signals from various types of gas chromatographic detectors can be conveniently amplified by employing operational amplifiers. With flame ionization detectors, where the signal levels are of the order of 10^{-11} A and even lower, it is necessary to utilize electrometric (high input impedance) input operational amplifiers in the input stage. These amplifiers have input bias currents of the order of 10^{-13} to 10^{-14}A. A wide range of signal amplitudes can be handled by a logarithmic electrometric amplifier. Linearity of the over-all response is then restored by an exponential converter following the logarithmic amplifier.

16.4.5.1 The Katharometer or Thermal Conductivity Detector

The thermal conductivity detector is a simple and most widely used type of detector. It is based on the principle that all gases have the ability to conduct heat, but in varying degrees. This difference in heat conduction can be used to quantitatively determine the composition of a mixture of gases. By definition, the thermal conductivity of a gas is the quantity of heat (in calories) transferred in unit time (seconds) in a gas between two surfaces 1 cm^2 in area, and 1 cm apart, when the temperature difference between the surfaces is 1°C.

Thermal conductivity

In its simplest form, the detector may consist of a hollow tube with an electrically heated coil mounted axially in its centre. When only the carrier gas flows over it, a thermal balance can be attained at a certain temperature. However, when a gas or vapour differing in thermal conductivity from the carrier gas flows past the heated coil, the temperature of the coil gets altered and a proportionate change in the electrical resistance of the wire takes place. Such changes in resistance arising from the components of the sample are used for detection and estimation of the unknown sample components.

In actual practice, the detector consists of two temperature sensing elements arranged in a Wheatstone Bridge circuit, one in the reference and the other in the measuring arm. The heat-sensitive elements are either thermistors or resistance wires, like platinum or tungsten. Figure 16.11 shows a typical circuit arrangement for measuring the changes in the resistance produced in the Katharometer cell elements. Resistances R_1 and R_2 are the Katharometer wires while resistances R_3 and R_4 are the ratio arms of the bridge. Resistances R_7 and R_8 are used for

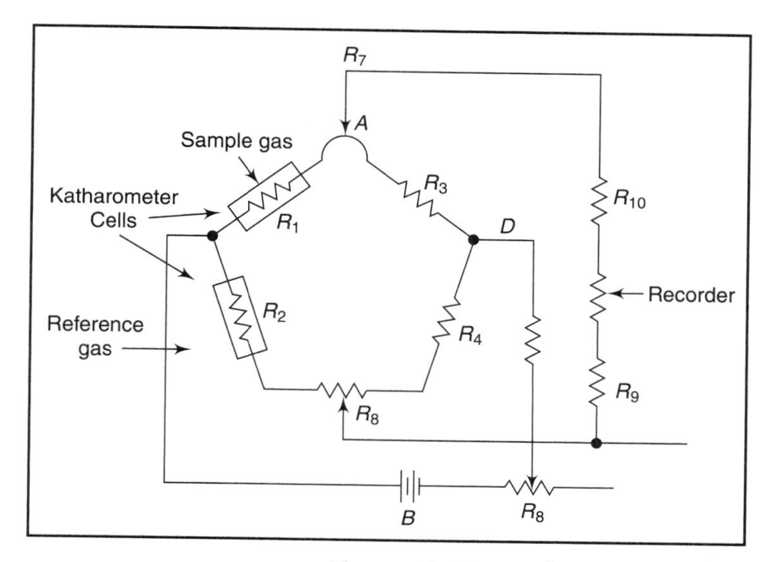

Figure 16.11 :: Katharometer cell

making the base line adjustment and are made of manganin wire. The output of the bridge is fed to the recorder through an attenuator, so that if signal is greater than the span rating of the recorder, full-scale reading may be adjusted.

For the balanced bridge conditions, when the carrier gas flows through the two cells, no current would be flowing between A and C and:

$$R_1/R_2 = R_3/R_4$$

However, when the resistance R_1 changes due to the components of the sample gas, it causes an unbalance current to flow from A to C. The magnitude of the current serves to detect and measure the magnitude of the gas component vapour passing over the measuring cell. If the Wheatstone bridge is excited with an ac current, it can be made many times more sensitive, because the ac signal can be conveniently amplified before it is given to the recorder.

The sensitivity of a thermal conductivity detector depends upon the nature of the carrier gas. When helium gas is used, 10^{-7}g of inorganic gases can be detected. A Katharometer is fairly satisfactory for a wide variety of analytical applications. The technique is non-destructive and therefore, the components of the sample can be further trapped for other forms of analysis.

16.4.5.2 Flame Ionisation Detector (FID)

The flame ionization detector is by far the most widely used detector in gas chromatography. It responds with high sensitivity to almost all organic compounds. Its linear dynamic range is approximately 10^7, which is much wider than that obtained from other detectors.

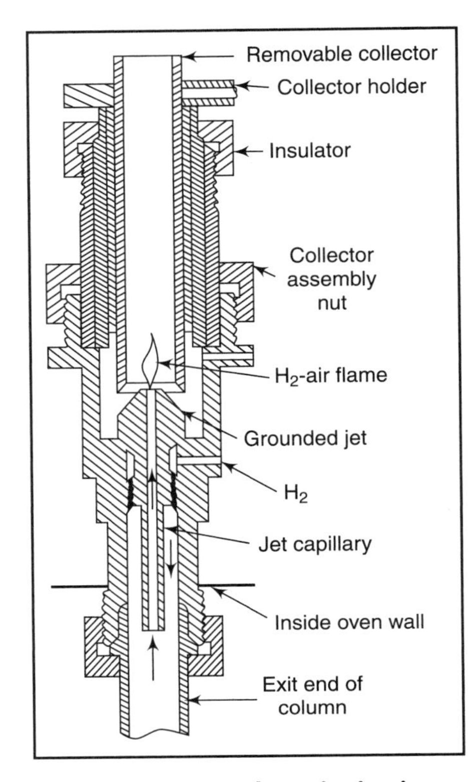

Figure 16.12 :: Flame ionization detector

In this detector, the effluent from the column is led into an oxy-hydrogen flame (Figure 16.12). An electrical potential is applied across two electrodes placed in a stainless steel housing. The hydrogen flame burns at the tip of a capillary, which also functions as the cathode and is insulated from the body by a ceramic seal. The collector electrode consists of a loop of platinum and is located at about 6 mm above the burner tip.

The current across the electrodes remains constant, when only the inert carrier gas passes the flame. However, when the vapour of a compound emerging from the column passes the flame, the vapour molecules are broken into ions by the hot flame. These ions result in the ionization current and there would be a consequent change in the current flowing across the electrodes. The magnitude of the variation in current would be directly proportional to the number of ions or electrons formed in the flame gases, which, in turn, would be proportional to the carbon content of the organic molecules in the vapour.

The flame ionization detector has a high output impedance similar to that obtained with glass electrodes, when making pH measurements. Commercial pH meters can, therefore, be easily adopted for use with this detector. A vibrating reed electrometer is often used in the input stage of the amplifier to attain sensitivities up to 5×10^{-13}A. By placing a set of high resistors across the flame and changing their resistances, enables the sensitivity to be varied. The sensitivity is high, because of the inherently low noise level of this detector.

Figure 16.13 shows a current amplifier with which a flame ionization detector can be used at all practicable sensitivity levels. The amplifier makes use of metal-oxide-silicon transistors in the input stage. The amplifier gives a high degree of stability. The amplifier has a sensitivity of 100 mV pA^{-1} and a noise output equivalent to 2×10^{-15} A rms at the input, thermal drift is typically 1 mV/°C. The input stage of the amplifier consists of a matched pair of MOSFETs, which is followed by a matched pair of p-n-p transistors. The output stage is an emitter follower, while the first stage is a long-tailed pair.

Current amplifier

There are certain limitations in the use of the flame ionization detector. These are:

- The FID does not respond to inert gases and inorganic compounds.
- The emerging components get destroyed in the flame.
- The response to sample weight has to be separately determined for each component.

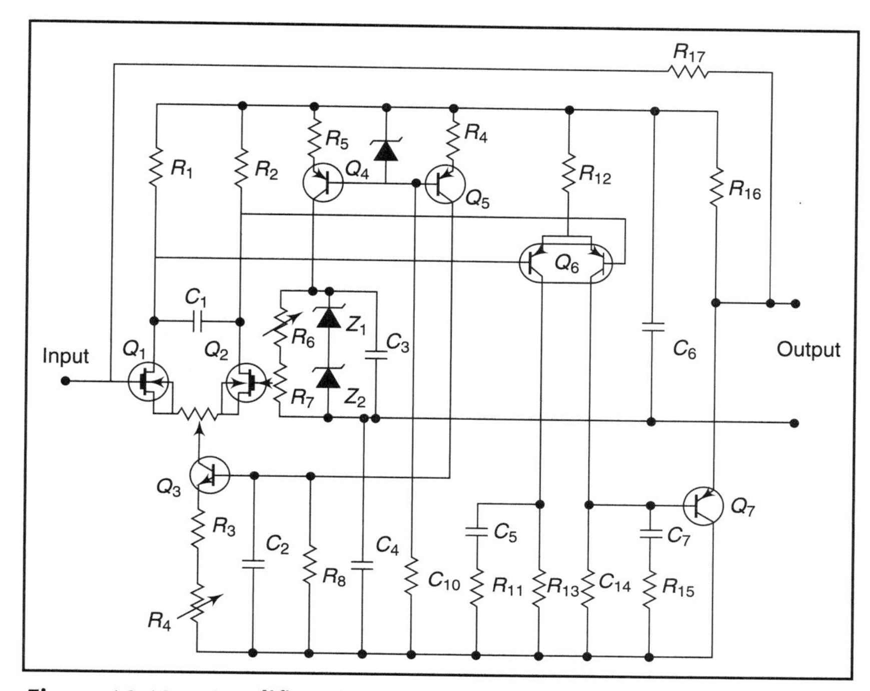

Figure 16.13 :: Amplifier circuit used with flame ionization detector

In a flame ionization detector, it is not only the sample components which are ionized in the hydrogen flame, but the liquid phase escaping from the column also participates in producing ions. The ionic current due to the liquid phase is recorded on the chromatogram as a background signal. In case of the programmed temperature analysis, the background signal changes with time, because the amount of liquid phase escaping from the column increases gradually in response to the raised column temperature. This causes great inconvenience to the programmed temperature analysis of trace substances. In order to eliminate this difficulty, a hydrogen Differential Flame Ionization Detector has been devised. It consists of two FIDs based on the same working principle. The schematic diagram of the DFID is shown in Figure 16.14.

Differential Flame Ionization Detector

Two columns are used in this arrangement. If the same packed column as the sample side one is employed on the reference side, the carrier gas is fed through the reference side column at the same speed as on the sample side, and if the reference side column is mounted very near the sample side one, under these conditions, the amount of liquid phase flowing from the reference side column can be considered to be the same as that of the liquid phase flowing from the sample side. On applying the voltages to the electrodes, the signals produced are shown in Figure 16.15 indicating that the sample side and reference side signals offset each other, producing the base line as straight. However, a certain degree of base line drift cannot be avoided even with DFID, when the sensitivity of FID or the amount of liquid phase differs between the sample side and the reference side.

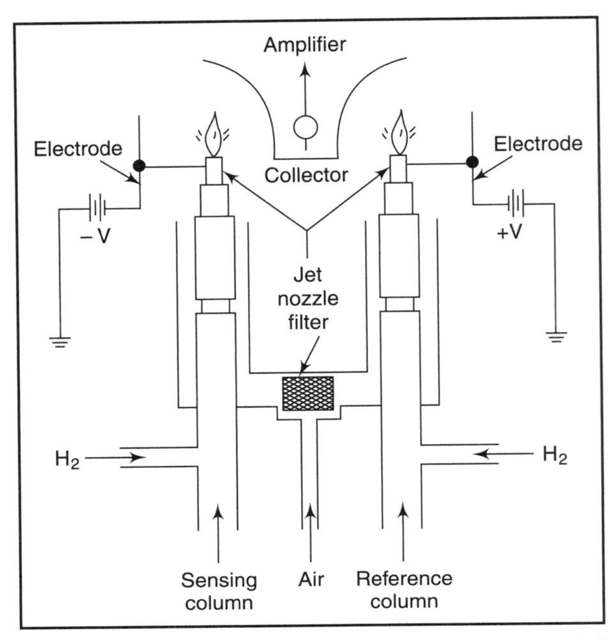

Figure 16.14 :: Working principle of differential flame ionization detector

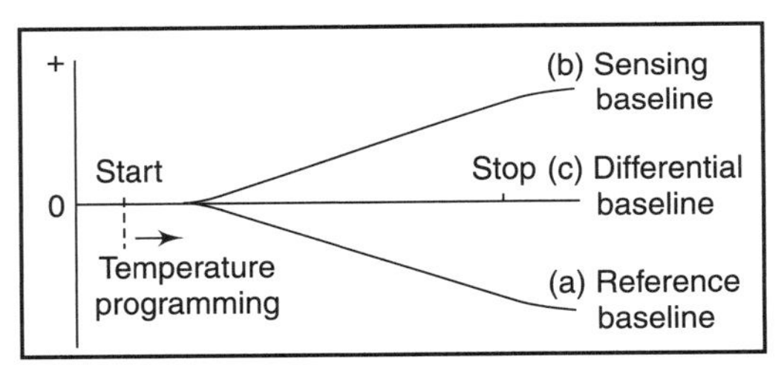

Figure 16.15 :: Ideal case of base line compensation in dual FID arrangement

16.4.5.3 Flame Photometric Detector

A flame photometric detector is primarily used for the determination of sulphur or phosphorus containing compounds. This device uses the chemiluminescent reactions of these compounds in a hydrogen/air flame as a source of analytical information that is relatively specific for substances containing these two kinds of atoms.

Figure 16.16 shows the basic components of gas chromatographic flame photometric detector (Lecture Notes by Dr. T.G. Chasteen, *http://www.shsu.edu/~chemistry/FPD/FPD.html*), which are:

- A combustion chamber to house the flame,
- Gas lines for hydrogen (fuel) and air (oxidant),
- An exhaust chimney to remove combustion products, and
- Thermal (bandpass) filter to isolate only the visible and UV radiation emitted by the flame. Without this, the infrared radiation emitted by the flame's combustion reactions would heat up the PMT and increase its background signal.

The λ_{max} for emission of excited S_2 is approximately 394 nm while it is 510–526 nm for phosphorus compounds in the flame. In order to selectively detect one or the other family of compounds as it elutes from the gas chromatic column, an interference filter is used between the flame and the photomultiplier tube to isolate the appropriate emission band. The disadvantage of this detection system is that the filter must be exchanged between chromatographic runs if the other family of compounds is to be detected.

Pulsed Flame Photometric Detector

An improvement over the flame photometric detector is the Pulsed Flame Photometric Detector (PFPD), which is capable of analyzing many other elements, besides sulphur and phosphorus. Figure 16.17 shows a schematic

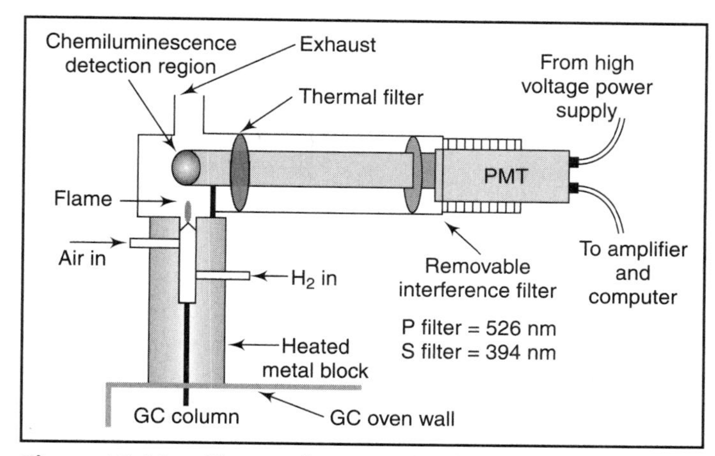

Figure 16.16 :: Flame photometric detector (Courtesy Dr. Sam, Houston State University)

of the PFPD. In this arrangement, two different combustible gas flows enter the bottom of the combustion chamber through narrow gas lines, as against one fuel line in the normal FPD. The second incoming gas flow's job in the PFPD is to help fill up the outer volume of the combustion chamber, while the analyte and the primary combustion gas flow into that chamber (*http://www.shsu.edu/~chemistry/PFPD/pfpd.html*).

The detector contains an ignition wire, which continuously stays red hot. When the gases flowing into the combustion chamber, including the analytes exiting the GC column, reach a flammable mixture, they are ignited by the ignition wire and the flame propagates back down the combustor. The flame front uses up all the quickest burning flammable material in the combustion chamber in less than 10 milliseconds and the flame goes out. It is after this short flame pulse that the slower burning analytes are excited and emit the light that is characteristic of their elements. After about 300 milliseconds, the flame pulses again as new flammable material fills the combustion chamber from the inlet tubes and GC column and that combination once again constitutes a flammable mixture. In this way, about 3 flame pulses are recorded per second. By

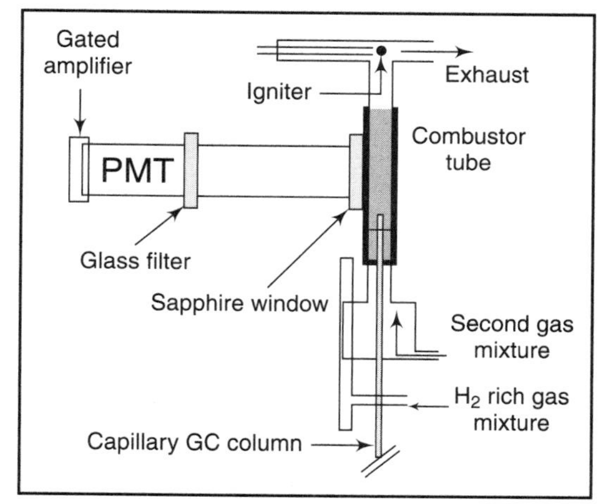

Figure 16.17 :: Pulsed flame photometric detection (after Chasteen, 2000)

using a gated amplifier, controlled by a computer, the part of each pulse to be amplified and recorded can be chosen.

The analytical discrimination gives the PFPD the ability to selectively and sensitively detect some analytes co-eluting in the presence of others and the ability to produce element-specific chromatograms.

16.4.5.4 Photo-ionization Detector

The photo-ionization detector is mainly used for the selective determination of aromatic hydrocarbons or organo-heteroatom species. The device uses ultraviolet light as a means of ionizing an analyte exiting from a GC column. The ions produced by this process are collected by electrodes. The current generated is therefore a measure of the analyte concentration.

If the amount of ionization is reproducible for a given compound, pressure, and light source, then the current collected at the PID's reaction cell electrodes is reproducibly proportional to the amount of that compound entering the cell. The reason for using PID for analysis of compounds like hydrocarbons or heteroatom is that they have ionization potentials that are within reach of commercially available UV lamps. The available lamp energies range from 8.3 to 11.7 ev, that is, λ_{max} ranging from 150 nm to 106 nm. Although most PIDs have only one lamp, lamps in the PID are exchanged depending upon the compound selectivity required in the analysis. (Dr. T.G. Chasteen, *http://elchem.kaist.ac.kr/vt/chem-ed/sep/gc/detector/pid.htm*).

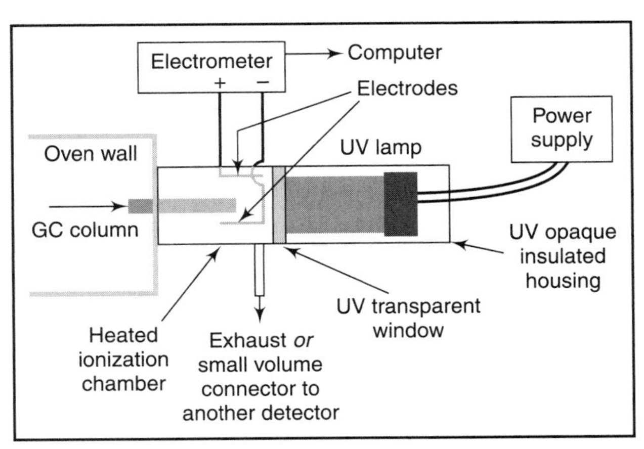

Figure 16.18 :: Photo-ionization detector (after Chasteen, 2000)

Figure 16.18 shows the schematic of a gas chromatographic photoionization detector. The major advantage of this detector is that only a small fraction of the analyte molecules are actually ionized in the PID chamber, thus this can be considered as a non-destructive GC detector. Therefore, the exhaust port of the PID can be connected to another detector in series with the PID. In this way, data from two different detectors can be taken simultaneously, and selective detection of PID responsive compounds augmented by response from, say, a flame ionization detector or an electron capture detector.

16.4.5.5 Electron Capture Detector (ECD)

This detector works on the principle that the ionization current set up by certain radioactive sources like Ni^{63} or H^3 gets reduced when an electron capturing compound is introduced into the cell. In effect, the ECD measures the loss of signal due to the re-combination phenomenon rather than measuring a positively produced electrical current.

Ettre (1978) explains the construction of an electron capture detector. The detector consists of two electrodes (Figure 16.19) across which a potential difference of 10 to 100 V can be applied. A radiation source of β-rays (tritium) is mounted on a tantalum wire saturated with the radioactive isotope of hydrogen, so that the emitted β-rays encounter the effluent from the GC column.

As the carrier gas (nitrogen) flows through the detector, β-particles from the tritium source ionize the nitrogen molecules and form slow electrons. These slow electrons migrate to the anode under a fixed voltage. When these electrons are collected at the collector electrode, they produce a steady current, which provides a base line on the recorder.

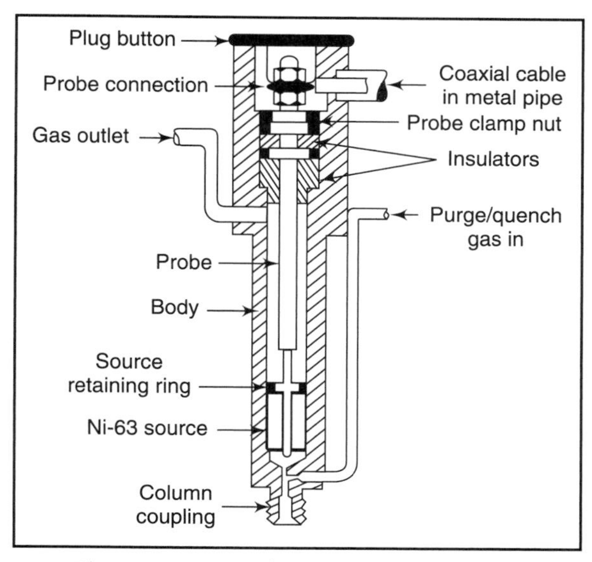

Figure 16.19 :: Electron capture detector

The organic compounds containing halogen, nitrogen and phosphorus, have the property of capturing electrons, resulting in a variation in the number of electrons reaching the collector electrode, thereby producing proportionate signals in the detection device.

Detector discrimination can be regulated through the potential applied to the collector electrode. In fact, the response of weakly capturing compounds can even be abolished since the response for different classes ceases at well defined applied potentials.

This detector has a sensitivity of 3×1^{-14}A. However, the linear range is limited to less than 10^3. Nitrogen and hydrogen are the best carrier gases with this type of detector. Hydrogen should be used with caution, lest there be an explosion.

16.4.5.6 Argon Ionization Detector

For the argon ionization detector, argon is used as the carrier gas. The detector contains two electrodes placed parallel to each other and a potential difference is applied across them. With the carrier gas emerging out of the GC column, no current passes across the electrodes under normal conditions, as the gas is a non-conductor. A radioactive source (tritium) is placed in the approach region to the electrodes, so that the rays emitted by it excite the argon atoms and electrons are produced by this bombardment action. These electrons are accelerated under the influence of a potential of about 1000 V and upon collisions with other argon atoms, raise them to the metastable state. Such metastable argon atoms collide with organic molecules of the sample emerging from the GC column, resulting in these molecules becoming ionized and consequently conducting. This results in the flow of proportionate current across the electrodes and produces signals, the magnitude of which would depend on the quantity of organic samples passing through the detector.

The argon detector responds to most of the organic and inorganic compounds although it is inert to water vapour, oxygen, methane, carbon dioxide and oxygen. The sensitivity of the detector is 0.08 μg/ml and the linear dynamic range is 10^5. The detector is suitable for measurement of organic molecules present over a wide range of concentrations.

16.4.5.7 Cross-Section Ionization Detector

The cross-section ionization detector is one of the most useful detectors for gas chromatography, which has proved to be a very reliable method for the separation of gases in mine air. It is precise, reliable, robust, insensitive to changes in the carrier gas flow rate and characterized by a response, which is linearly dependent upon the concentrations of the components under investigation, over a wide range of change in concentration. The vapour concentration for any given molecule can be calculated from the known properties of its constituent atoms.

The ionization current in the chamber is small, when it is filled with light gas such as hydrogen, but increases with the addition of any other gas. This increase in current is because of the increased total ionization cross-sections of the gas mixture inside the chamber. The cross-section ionization of a gas is a quantity determined by the size of a gas molecule and the number of electrons in the atoms forming the molecule. Therefore, the action of the cross-section ionization detector is based on differences in the cross-section affecting the ionization of the analyzed components.

In this type of detector, a radiation source (Sr^{90}) mounted in the approach region of the two electrodes separated by the carrier gas, generates ion-pair from the organic molecules passing with the carrier gas. The application of a potential of 300–1000 V ensures collection of the electrons. A variation in the current flowing across the electrodes leads to proportionate signals and the magnitude of the signal depends on the concentration of the component emerging from the GC column.

In this detector, hydrogen or helium is usually the carrier gas. The detection is non-destructive, though its sensitivity is low (about 10^{-7}g/s). Response to any substance can be calculated from the values of the atomic cross-sections of its constituent atoms.

16.4.5.8 Atomic Emission Detector (AED)

The atomic emission detector has the ability to simultaneously determine the atomic emissions of many of the elements in analytes that elute from a GC capillary column. As eluants come off the capillary column, they are fed into a microwave powered plasma cavity where the compounds are destroyed and their atoms are excited by the energy of the plasma. The light that is emitted by the excited particles is separated into individual lines via a photodiode array. The computer then sorts out the individual emission lines and can produce chromatograms made up of peaks from eluants that contain only a specific element.

The atomic emission detector basically includes the following components, which are shown in Figure 16.20:

- An interface for the incoming capillary GC column to the microwave-induced plasma chamber,

- The microwave chamber,
- A cooling system for the microwave chamber, which is required because much of the energy focused into the cavity is converted into heat,
- A diffraction grating and associated optics to focus and then disperse the spectral atomic lines, and
- A position adjustable photodiode array interfaced to a computer.

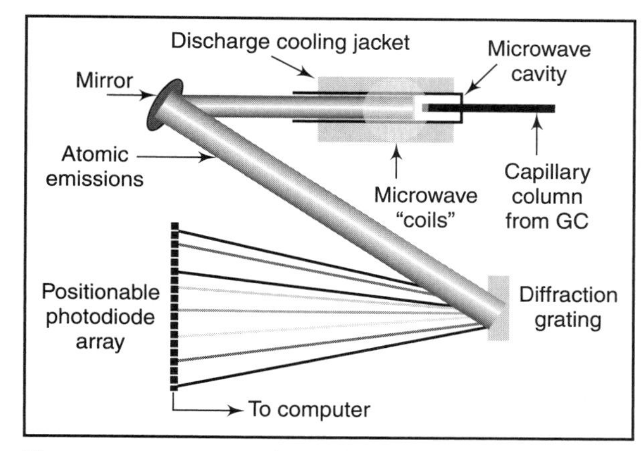

Figure 16.20 :: Atomic emission detector (after Chasteen, 2000)

This detector is based on the atomic emissions instead of measuring simple gas phase ions created in a flame as with the flame ionization detector or the change in background current because of the electronegative element capture of thermal electrons as with the electron capture detector (Courtesy Dr. T.G. Chasteen, *http://www.shsu.edu/~chemistry/AED/AED.html*). It therefore has much wider applicability.

16.4.5.9 Chemiluminescence Spectroscopy-based Detectors

Chemiluminescence uses quantitative measurements of the optical emission from excited chemical species to determine analyte concentration. However, chemiluminescence is usually emission from energized molecules instead of simply excited atoms. The bands of light determined by this technique emanate from molecular emissions and are, therefore, broader and more complex than bands originating from atomic spectra. Furthermore, chemiluminescence can take place in either the solution or gas phase, whereas AES is almost strictly as gas phase phenomenon.

Liquid phase chemiluminescence

Although liquid phase chemiluminescence plays a significant role in laboratories using this analytical technique, often in conjunction with liquid chromatography, the instrumental components are somewhat simpler in gas phase chemiluminescence reactions. These detectors are also often used as detectors for gas chromatography. A schematic of the components necessary for a gas phase chemiluminescence detector interfaced to a capillary gas chromatograph is shown in Figure 16.21.

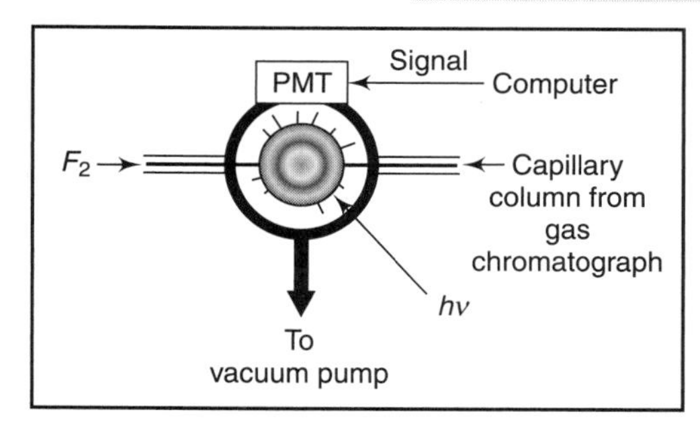

Figure 16.21 :: Chemiluminescence detector

Like fluorescence spectroscopy, the strength of chemiluminescence lies in the detection of electromagnetic radiation produced in a system with very low background. In addition, because the energy necessary to excite the analytes to higher electronic, vibrational, and rotational states (from which they can decay be emission) does not come from an external light source like a laser or lamp, the problem of excitation source scattering is completely avoided. A major limitation in the detection limits achievable by chemiluminescence involves the dark current of the photomultiplier (PMT) necessary to detect the analyte light emissions.

Gas phase chemiluminescence

In gas phase chemiluminescence, light emission is produced by the reaction of an analyte and a strongly oxidizing reagent gas. The reaction occurs on a time scale such that the production of light is essentially instantaneous; therefore, most analytical systems simply mix analytes and the reagent in a small volume chamber directly in front of a PMT. If the analytes are eluting from a gas chromatographic column, then the end of the column is often fed directly into the reaction chamber itself. Since as much of the energy released by the reaction should be used to excite as many of the analyte molecules as possible, loss of energy via gas phase collisions is undesirable, and therefore a final consideration is that the gas pressure in the reaction chamber be maintained at a low pressure (~ 1 torr) by a vacuum pump in order to minimize the effects of collisional de-activation.

16.4.5.10 Nitrogen–Phosphorus Detector (NPD)

The design of a nitrogen–phosphorus detector (NPD) is similar to a flame-ionization detector (FID). The major difference is that the hydrogen/air flame of the FID is replaced by a heated rubidium silicate bead in the NPD. The effluent from the GC column passes through the hot bead. The hot rubidium salt emits ions when nitrogen and phosphorus containing compounds pass over it. The ions are collected on a collector above the heated bead to produce a current, similar to the FID.

16.4.5.11 Other Types of Detectors

A number of new detectors have been introduced recently. There has been a resurgence in the use of electrochemical detectors for gas chromatography. The majority of these include electrolytic conductivity detectors, micro-coulometry and ion-selective electrodes. Some workers have also used radiochemical radio-ionization detectors for gas chromatography.

16.4.5.12 Calibration of the Detector

Before the analysis of the unknown sample is carried out, it is important to calibrate the detector and calculate response factors for components to be determined. This is done by preparing mixtures of known composition by accurate weighing or mixing and them analyzing them under the same conditions, which would be used for the unknown samples. From these results, calibration curves can be drawn and the response factors determined, which are applied to the samples to be analysed.

16.4.6 Recording Instruments

The data recorder plots the signal from the detector over time. This plot is called a chromatogram.

The retention time, which is when the component elutes from the GC system, is qualitatively indicative of the type of compound. The data recorder also has an integrator component to calculate the area under the peaks or the height of the peak. The area or height is indicative of the amount of each component.

Chromatogram recording is usually done on the self-balancing type potentiometric single-pen graphic recorders. The span of these recorders may be 0.5 or 1 mV. These recorders require a low impedance input and therefore, impedance converters are used with the high impedance detectors. The chart widths are 25 to 30 cm and response time is about 1 s. For multi-speed operation, a gear box for changing the speed is necessary.

All ionization detectors generate some background signal (with the carrier gas only) ranging from 10^{-8} to 10^{-11} A. The maximum signal in the presence of vapour is in the range of 10^{-6} to 10^{-8} A. Therefore, the apparatus for the measuring current must respond to all current in the range of 10^{-6} to 10^{-13} A. It should also have the means of offsetting the background current of the detector in use. The response over a large current range is achieved by putting a series of high stability resistors across the input of the potentiometric recorder.

16.4.6.1 Qualitative Analysis

The elution of a component from the chromatographic column appears as a peak on the graphic recorder. Under specified column conditions, a component has a characteristic retention time (the time a component is retained in the column) or retention volume (the volume of carrier gas passed during the retention time). This forms the basis of qualitative analysis.

16.4.6.2 Quantitative Analysis

Quantitative analysis by gas chromatography depends upon the measurement of areas under the component peaks. The area contained by a peak is proportional to the quantity of the component present in the sample. Every peak is measured and the areas calculated. These are summed and the components are expressed as a percentage. The following conditions are necessary for this:

- The flow rate must be constant, so that the time abcissa may be converted to volume of carrier gas.
- The output of the detector system must be linear with concentration.

Retention Measurement

The time taken for a given species to pass through the column is referred to as the retention time, T_R, for that species. Most desirably, retention times are short while separations are efficient, i.e. the resolution is high. Achieving these characteristics for complex mixtures of similar materials is a challenging task, and the extraordinary utility of chromatography lies in its successful application to such tasks.

The identity of an unknown component in a mixture, when analyzed with GLC, can be established by knowing its time of elution or retention volume. It is known that under given operating conditions of column temperature and flow, the time of elution or retention volume for a component is constant. In practice, the retention times of known and unknown materials are compared and a tentative identification of the component is established. However, when more than one component has the same retention time, it is possible that wrong results may be reduced.

Relative Retention Measurement

In this technique, relative retention data are made use of instead of the absolute values of the retention parameters. This method is more reliable, does not depend upon analysis conditions to be set very precisely, and is suitable for day-to-day comparisons. Unknown components are compared with the standard to give a relative retention volume. In this way, an operator may build up a library of retention data for future use.

16.5 Methods of Measurement of Peak Areas

Gas and liquid chromatography provide fast and convenient means for analyzing the chemical components of complicated mixtures. However, identifying and quantifying the raw chromatographic information obtained on a conventional strip chart recorder requires a major effort. A typical chromatogram would comprise a series of peaks separated in the time domain, each peak corresponding to a chemical component detected. The time of occurrence of each peak corresponds to the travel time through the column and can be used to identify the corresponding chemical component. The area enclosed by the peak corresponds to the concentration of that chemical. The area of the peak is measured by counting squares, using a planimeter or cutting out the peaks and weighing the paper. Obviously, a lot of effort goes into reducing the data. For accurate determination of the concentration of the sample, the peaks should be completely resolved. This is slightly difficult, as one would have to depend upon extrapolation. The peak areas may be determined by various methods. These are detailed below.

Width at 50 per cent height × height: In this method (Figure 16.22), the peak height (h) is measured and multiplied by the width at half-height (w).

$$\text{Area} = w \cdot h$$

Triangulation: This method consists in drawing tangents to each side of the peak and to extend them so as to form a triangle with the base line. The area of this triangle is approximately proportional to the area under the peak.

$$\text{Area of the triangle} = 1/2 \cdot w \cdot h,$$

where w = *peak* width
and h = *peak* height.

Weighing of the chart paper: In this method, the peak area is determined by cutting out the peak from the chart and weighing the paper. This method assumes a uniform weight distribution in paper, so that, the area is proportional to weight.

Plain paper is preferable to a line chart, because the latter varies in weight. Some check about the uniformity of the paper must be made. This is done by the following:

- *Use of Planineter:* This method is tedious and requires practice for obtaining satisfactory results.
- *Electronic Integration:* Electronic computing circuits may be used for integrating the area under the peaks.

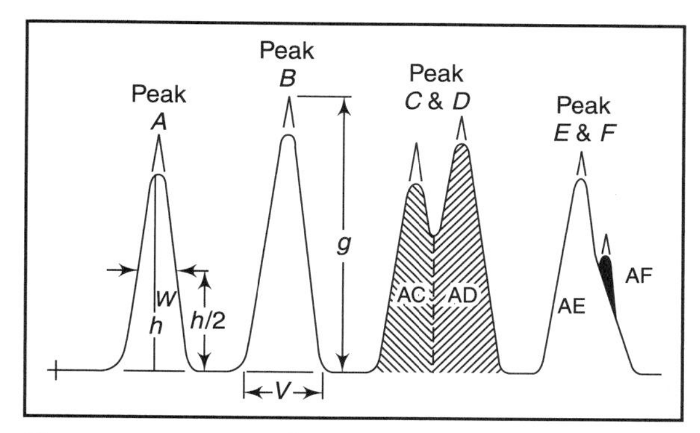

Figure 16.22 :: Peak area measurement by width at 50% height × height method

The area under the chromatographic peaks can be measured by analog or digital techniques. In the analog technique, the detector current is amplified in an operational amplifier and then integrated. It gives a curve with waves, the wave heights corresponding to the peak areas. In the digital method, a circuit shown in Figure 16.23 can be employed.

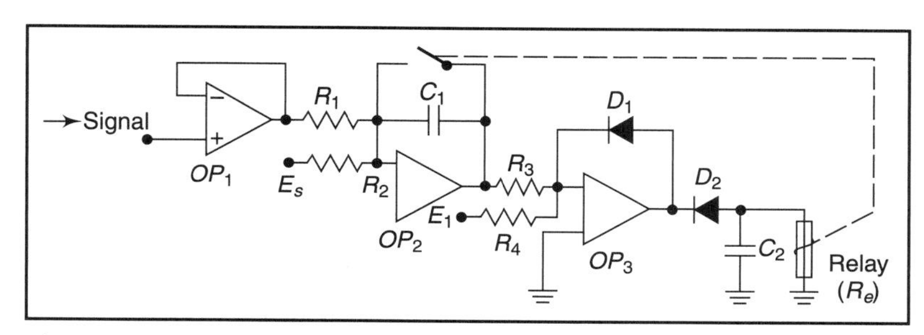

Figure 16.23 :: Circuit diagram for measurement of area under the chromatographic peak

The output from the pre-amplifier after the detector is brought to resistance R_1 of the integrator. To the E_1 point is brought a voltage used to eliminate zero shift. As soon as the absolute values of the two voltages differ, charging of the integrating capacitor begins. The output of OP_2 would be a voltage increasing linearly with time, the slope of which will correspond to the difference between the signal voltage and the E_S. As soon as the integrator output voltage reaches or slightly exceeds the voltage value applied to input E_1, the comparator (OP_3) output jumps to the saturation voltage and the relay (R_e) is closed. The integrating capacitor is thus discharged. This process can be repeated and the circuit can be used for the conversion of an analog signal into a digital signal. As each relay closure is recorded by a counter, the number of pulses registered in the counter gives a measure of the area under the peak.

Chromatographic peaks distorted by tailing can be modified by summing the original signal and its first derivative. The development of electronic integrators speeded up data reduction by automatically computing the areas under the peaks and print the areas and retention time for each peak. Basically, these integrators are voltage-to-frequency converters that monitor the output of the chromatograph detector and drive a counter activated by, rather complex, peak-recognition logic.

Integrators

The next development was to derive final results with the aid of a microprocessor working directly from an analog-to-digital converter. This processor provided means of adding automatic calibration, so that the integrator could identify the peak belonging to the calibrating signal and then scale results. It also incorporates means for reducing the effects of detector noise, and for selecting the optimum slope sensitivity automatically; so it can be sensitive to small peaks while ignoring noise peaks.

The data processor includes an analog-to-digital converter, which uses the integrating digital voltmeter circuit to measure the average amplitude of the chromatograph detector output five times per second. To smooth noisy chromatograms, a running average of consecutive samples is calculated by a weighted averaging method. This smoothens the high frequency noise without distorting true peaks.

Integrators universally select peaks on the basis of the slope. If the slope threshold is set too low, noise on the baseline can trigger integration. If it is set too high, integration starts high on the peak and a significant part of the peak is lost. The digital processor can be set to start integration at the slightest hint of a peak, but it discards the count if the peak presence is not confined. Thus, even with the slope threshold set high, the total peak area is integrated. This is done as follows: the microprocessor measures the slope by continuously comparing each new averaged value to the previous value. If the difference is positive and exceeds a certain minimum for several successive samples, the processor judges that a peak is being detected. If the sample-to-sample comparison indicates that the slope reverses before the threshold criteria is reached, then it is assumed that a noise peak had been encountered and the total count is discarded. Once the processor has made the decision that a peak is being detected, counts continue to be totalled, until the sample-to-sample difference indicates that the detector output has returned to the base line. The processor memory is capable of holding counts obtained from 54 peaks in any one chromatograph run. The processor compares the beginning and end of each peak to detect the base line drift. It then adjusts the readings to account for drift, if present.

A difficult problem for designing integrators lies in finding the true areas of peaks that overlap or merge on the chromatogram. Two cases of merged peaks are shown in Figure 16.24. The processor judges when the sample-to-sample comparison indicates that the slope of the chromatogram changes slope before it reaches the base line. The accumulated count is stored up to that point and starts a new count if the trace returns to the baseline on the next down slope, the two counts obtained are stored, as the area counts for the two peaks. This is called the dropline method of merged peak separation.

When a small peak rides on the tail of a larger peak, the method known as the tangent skim is preferred. Here, the end of the second peak is determined by continuously calculating the slope of a line drawn from the start of the second peak to the latest sample and by comparing the line's slope to the slope of the chromatograph curve. When the two slopes coincide, the end of the peak is indicated. The processor then calculates the area.

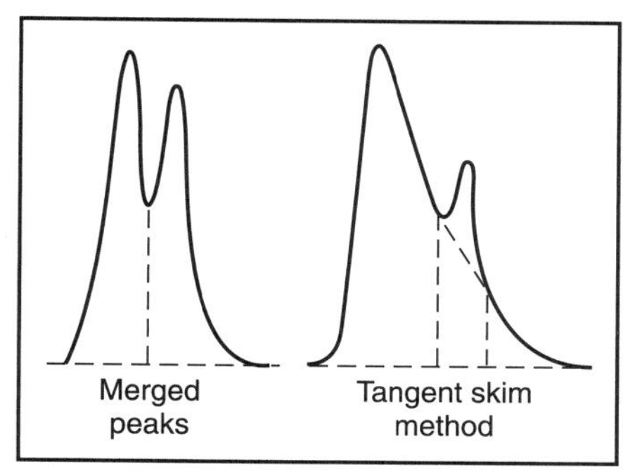

Figure 16.24 :: Areas of merged peaks (left) are separated by a line dropped from the valley to the base line. The area of a small peak riding on a tailing peak (right) is computed by using the tangent line as a base line.

The parameters are entered in the processor through a keyboard. Some parameters may even be entered through slide switches. In fact, through the keyboard, all aspects of the analysis are controlled, including the column oven temperature program, the temperature of other heated zones, the integration parameters, the calibration and the type of computation. Operation of backflush valves, a change in recorder speed, a change in detector and other parameters can be programmed to occur at specific times following the start of a run.

Modern gas chromatographs are microprocessor-/PC-based, which are controlled through a simple keyboard and high resolution video display units. They provide real-time screen graphics display of the chromatogram and integral data handling. Such instruments offer flexibility of application through the use of a range of easily interchangeable components. For injection, packed, split/splitless, on-column capillary or programmable temperature vaporizer systems are available, which can be extended by a range of manual and automated gas and liquid-sampling valves.

High speed gas chromatic systems have been developed (Sacks, *et al.*, 1998), which combine technologies for dramatically reducing separation times by using electronically controlled inlet systems, temperature and pressure programming with closed-loop control and electronically controlled column selectivity.

Chapter 17

Liquid Chromatographs

17.1 Liquid Chromatography

Liquid chromatography (LC) is an analytical chromatographic technique which is useful for separating ions or molecules that are dissolved in a solvent. If the sample solution is in contact with a second solid or liquid phase, the different solutes will interact with the other phase to differing degrees due to differences in adsorption, ion-exchange, partitioning, or size. These differences allow the mixture components to be separated from each other by using these differences to determine the transit time of the solutes through a column.

The early separations by chromatographic technique were performed using a bed of solid, powder absorbent, such as alumina or charcoal through which the sample was passed in a stream of solvent. Since these techniques used liquid as the percolating agent, liquid chromatography can be considered as the oldest of all chromatographic processes. Later developments included liquid/liquid partition chromatography, paper chromatography and the ion-exchange chromatography. Gel-permeation chromatography and thin-layer chromatography were developed in the 1950s.

Simple liquid chromatography consists of a column with a fritted bottom that holds a stationary phase in equilibrium with a solvent. Typical stationary phases (and their interactions with the solutes) are: solids (adsorption), ionic groups on a resin (ion-exchange), liquids on an inert solid support (partitioning), and porous inert particles (size exclusion). The mixture to be separated is loaded onto the top of the column followed by more solvent. The different components in the sample mixture pass through the column at different rates due to differences in their portioning behavior between the mobile liquid phase and the stationary phase. The compounds are separated by collecting aliquots of the column effluent as a function of time.

Until a few years ago, liquid chromatography was not quite commonly used, because of the non-availability of high sensitivity detection systems. With the introduction of such detectors, the analytical potential of liquid chromatography is greatly enhanced and sophisticated liquid chromatographs are now commercially available.

Conventional LC is most commonly used in preparative scale work to purify and isolate some components of a mixture. It is also used in ultra trace separations where small disposable columns are used once and then discarded. Analytical separations of solutions for detection or quantification typically use more sophisticated high-performance liquid chromatography instruments. These instruments use a pump to force the mobile phase through and provide higher resolution and faster analysis time.

High pressure liquid chromatography (HPLC) is similar to gas chromatography in that the chemical components of a mixture are separated as the mixture is forced through a column, packed with fine particles. In gas chromatography, the substance is carried through the column in vaporized form by an inert gas, whereas in HPLC it is carried through in liquid form by a solvent. Since the substance need not be vaporized, HPLC can be used on a broad range of substances that are not analyzable by gas chromatography.

High pressure liquid chromatography presents some unique problems such as the need for maintenance of the solvent flow accurately for obtaining repeatable results. Problems in controlling solvent flow arise because the solvents differ in viscosity, compressibility and other characteristics. In addition to this, the volume of the solvent mixture is not necessarily equal to the sum of the volumes of the individual solvents. Also, as the column must be tightly packed with small, uniform particles to obtain adequate separation of the component substances, high pressure of the order of 3000 psi or more is needed to force the substance through the column in a reasonably short time.

Liquid chromatography has been performed in a column and on an open bed (paper chromatography and thin-layer chromatography), whereas HPLC has been performed almost totally in columns. However, thin-film chromatography was introduced recently, as a high speed method for thin-layer chromatography.

17.2 Types of Liquid Chromatography

Liquid chromatography can be classified as under:

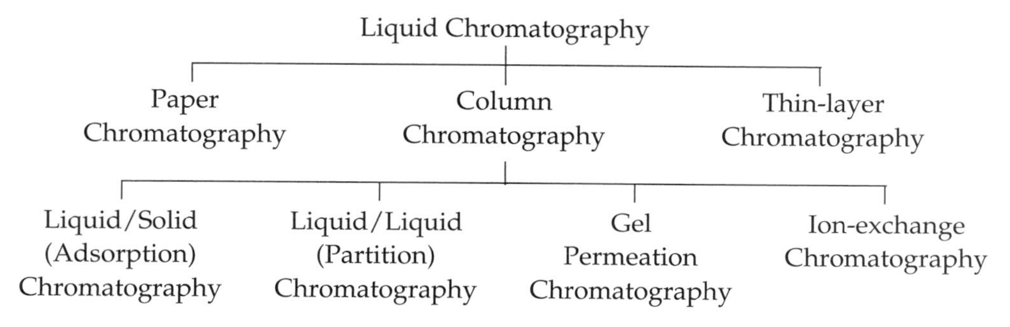

17.2.1 Column Chromatography

17.2.1.1 Adsorption Chromatography (Liquid/Solid)

In adsorption chromatography, a solid adsorbent, usually in powder form, is the stationary phase, through which a mobile liquid phase carrying the mixture to be analyzed is allowed to percolate. Adsorption chromatography is carried out in columns with the adsorbent supported by a plug of glass or cotton wool, or by a sintered glass filter.

The stationary phase in adsorption chromatography consists of silica or alumina particles. Analytes are separated due to their varying degree of adsorption on to the solid surfaces. The main advantage of adsorption chromatography is in separating isomers, which can have very different physisorption characteristics due to steric effects in the molecules.

17.2.1.2 Partition Chromatography (Liquid/Liquid)

In this technique, the mobile liquid phase is made to percolate through a column containing the stationary liquid phase, which is deposited on a solid surface as a thin film. The solid supports usually used are silica gel, porous glass and cellulose.

17.2.1.3 Gel Permeation Chromatography

Gel permeation chromatography is a recently developed separation technique in liquid chromatography. The separation is based on molecular size and shape. The gel permeation column is packed with a stationary phase in the form of a gel which contains pores of a specific size. As the sample is carried through the column bed by the carrier liquid, the sample molecules penetrate the pores in the packing gel, depending upon the size and shape of the molecules. Large molecules do not penetrate the gel and are consequently quickly eluted. Elution takes place in inverse order of their degree of gel permeation and consequently of decreasing molecular size.

17.2.1.4 Ion-exchange Chromatography

Ion-exchange chromatography involves the exchange of ions between a solution, and as solid insoluble material in contact with the solution. Many naturally occurring solid materials have the ability to exchange ions. Also, many artificial ion-exchange materials have been developed. The ion-exchange process is reversible and this fact is made use of in ion-exchange chromatography. When a sample is introduced at the top of an ion-exchange column (Figure 17.1), the ions exchange rapidly with the ions in the resin. If a mobile phase is used, the sample ions are displaced into the solution again and then re-exchange on to the resin. This process continues until the sample ions emerge from the end of the column. If the various sample ions are held on to the resin to different extents, then the time taken for them to pass through the column will be different and a separation will be achieved.

This type of chromatography depends upon molecular ion-exchange instead of liquid surface adsorption. The separation of acidic and basic organic substances from mixtures is achieved by using synthetic resins, which have highly selective action for certain substances, particularly amino-acids and allied compounds. Essentially, the process involves interchange of anions and cations between the components of certain resins.

Ions in a solution can be detected by measuring the conductivity of the solution. In ion chromatography, the mobile phase contains ions that create a background conductivity, making it difficult to measure the conductivity due only to the analyte ions as they exit the column. This problem can be greatly reduced by selectively removing the mobile phase ions after the analytical column and before the detector.

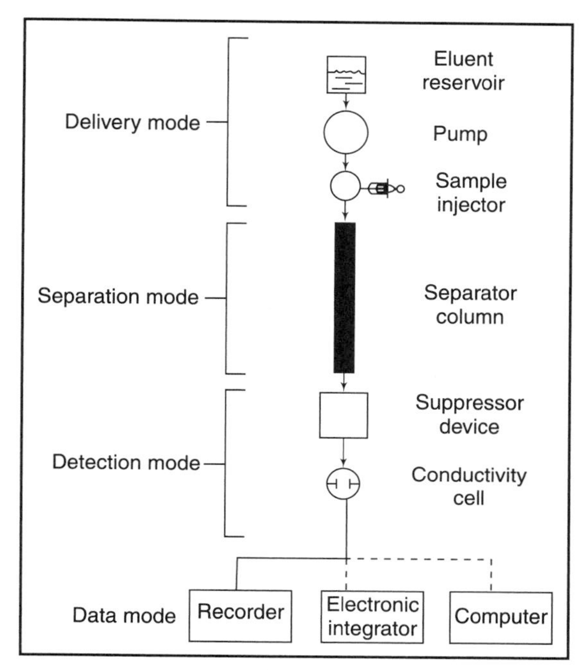

Figure 17.1 :: Typical ion chromatography configuration

In the early stages of development of ion chromatography, dedicated ion chromatographs were considered necessary for the application of the technique. Advances in recent years, particularly in the field of detection systems, have however, eliminated the need for this dedicated approach, and enabled ion chromatography to be carried out using conventional HPLC instrumentation (Browne, 1986). For identifying and quantifying various ionic species in solution, ion chromatography is a competitor to well-established electrochemical methods such as ion-selective electrodes and polarography (George and Adam, 1984).

17.2.2 Thin Layer Chromatography

In thin layer chromatography (TLC), the stationary adsorbents are applied to a planar glass or plastic surface and the solvent flows over them. All of the basic types of action, like adsorption, partition, ion-exchange, gel-filtration can be used on TLC plates, while solvents are applied in a chamber similar to that used in paper chromatography.

This technique is useful for separating organic compounds. Because of its simplicity and rapidity, TLC is often used to monitor the progress of organic reactions and to check the purity of products.

Thin layer chromatography consists of a stationary phase immobilized on a glass or plastic plate, and an organic solvent. The sample, either liquid or dissolved in a volatile solvent, is deposited as a spot on the stationary phase. The constituents of a sample can be identified by simultaneously running standards with the unknown. The bottom edge of the plate is placed in a solvent reservoir, and the solvent moves up the plate by capillary action. When the solvent front reaches the other edge of the stationary phase, the plate is removed from the solvent reservoir. The separated spots are visualized with ultraviolet light. The different components in the mixture move up the plate at different rates due to differences in their portioning behavior between the mobile liquid phase and the stationary phase.

TLC can be automated by using a forced solvent flow, running the plate in a vacuum-capable chamber to dry the plate, and recording the finished chromatogram by absorption or fluorescence spectroscopy with a light source. The ability to program the solvent delivery makes it convenient to do multiple developments in which the solvent flow for a short period of time, the TLC plate is dried, and the process is repeated. This method refocuses the spots to achieve higher resolution that in a single run. (Poole and Poole, 1994).

17.2.3 Paper Partition Chromatography

Paper partition chromatography is a simplified version of column chromatography, which makes use of strips or hollow cylinders of filter paper to hold both the solid and liquid phases. Here, drops of the solutions containing unknown mixtures are applied to a number of parallel strips, a few inches from the end of each test paper, and allowed to dry. The strips of paper are placed in a chromatography chamber with a saturating and equilibrating vapour and hung from a solvent reservoir, so that the downward movement of the solvent can be timed and the relative partition of the different substances measured.

17.3 High Pressure Liquid Chromatograph (HPLC)

An HPLC chromatograph consists of the following parts as shown in Figure 17.2:

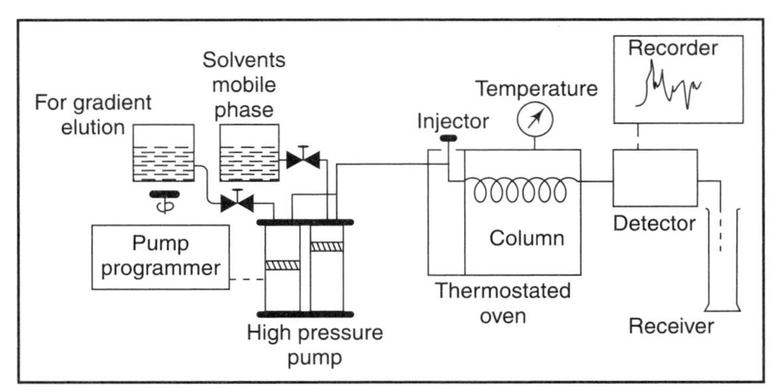

Figure 17.2 :: Block diagram of a high pressure liquid chromatograph

- A high pressure pump system to force the liquid mobile phase through the column,
- Gradient elution or solvent programmer,
- The sample injection system,
- The column,
- The detection system including display or recording devices, and
- The computer for data processing and storage.

Figure 17.3 shows the physical layout of various components of an HPLC system.

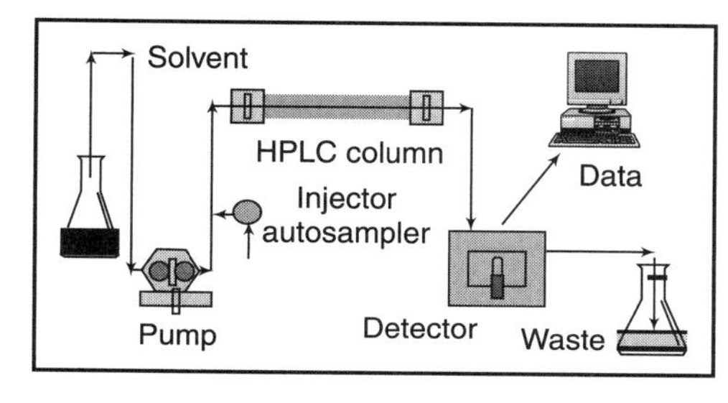

Figure 17.3 :: Physical layout of various components of a HPLC (Courtesy, Waters Corp.)

As in other chromatographic techniques, the sample is introduced into the column with the help of a sample injection system. Various components of the sample are fractionated during their passage through the column. The detection system senses these components as they elute from the column and produces a signal proportional to the amount of solutes passing through the detection system. The detector determines what separation has taken place and provides data permitting the qualitative and quantitative evaluation of the results. This can be accomplished by simply recording the response of the detector in the form of a chromatogram and/or with the help of a data handling equipment. The individual sample components separated in the column can also be collected.

17.3.1 High Pressure Pump System

Liquid chromatographs of the early type made use of wide diameter columns packed with coarse mesh packing material. They required very little pressure to obtain an adequate flow rate of the mobile phase liquid. Modern instruments, which employ smaller diameter columns filled with fine mesh particles, have necessitated the use of high pressure solvent delivery systems.

Solvent delivery

The most commonly used methods for solvent delivery are gravity feed systems. The gravity feed systems, though simple, are not able to deliver solvent at high pressure. They are therefore, not used with narrow bore columns packed with fine mesh particles, which need high inlet pressures to yield the required flow rate. Various types of pump systems incorporating piston pumps, peristaltic pumps, diaphragm pumps and syringe pumps, etc. are therefore used in HPLC systems. Several types of pumps are available for use with HPLC analysis, including constant flow pumps (reciprocating piston pumps, syringe type pumps), and constant pressure pumps.

17.3.1.1 Constant Flow Pumps

Reciprocating Piston Pumps: They consist of a small motor-driven piston which moves rapidly back and forth in a hydraulic chamber that may vary from 35-400 μl in volume. On the back stroke, the separation column valve is closed, and the piston pulls in solvent from the mobile phase reservoir. On the forward stroke, the pump pushes solvent out to the column from the reservoir. A wide range of flow rates can be attained by altering the piston stroke volume during each cycle, or by altering the stroke frequency.

Figure 17.4 shows the schematic of a single piston type reciprocating pump. Here, a rotating eccentric cam forces the piston to expel liquid through a one-way valve, called the check valve. The pumping rate is usually adjusted by controlling the distance the piston retracts, thus limiting the amount of liquid pushed out by each stroke. The purpose of the check valve is to ensure that the liquid moves only in one direction. These pumps obviously deliver a series of pulses of the mobile phase, which may disturb the detector. It is thus necessary that these pulses may be eliminated, for which several methods have been developed. Dual and triple head pumps consist of identical piston chamber units which operate at 180 or 120 degrees out of phase. This type of pump system is significantly smoother because one pump is filling while the other is in the delivery cycle.

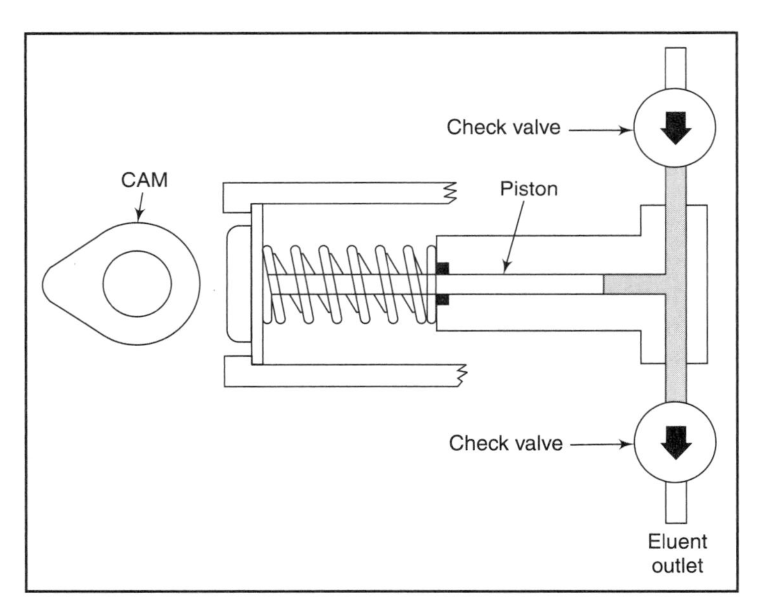

Figure 17.4 :: Principle of single-piston reciprocating pump

The schematic diagram of a dual head reciprocating pump which provides automatic pulse damping is shown in Figure 17.5. In this arrangement, both the pump chambers are driven by the

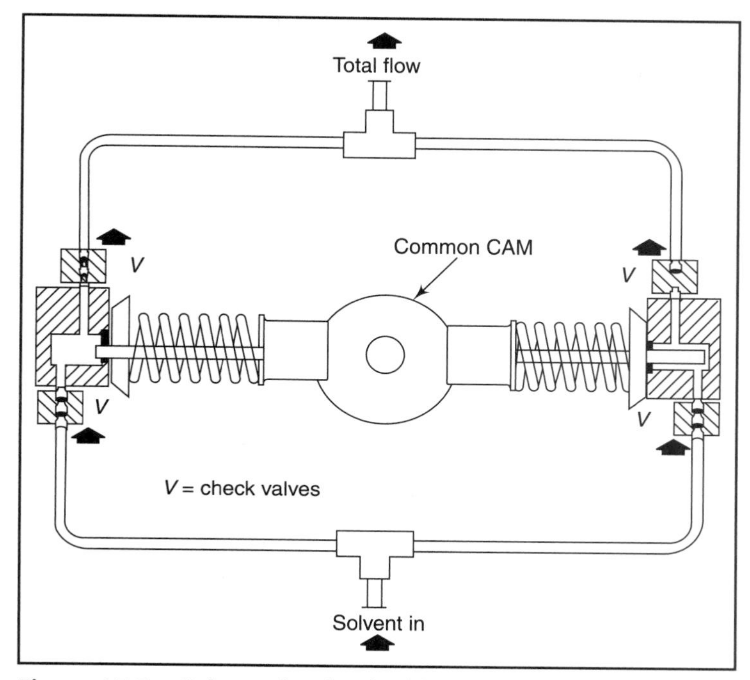

Figure 17.5 :: Schematic of a dual-lead reciprocating pump

same motor through a common cam, gears or hydraulic linkage. This common drive allows one piston to pump while the other is refilling, as a result, the two flow-profiles overlap each other, significantly reducing the pulsation, downstream of the pump. Usually, a computer-designed camshaft is used to achieve maximum overlap of pump strokes, resulting in virtually undetectable pulsation or ripple.

Syringe Type Pumps: A most commonly encountered piston type pump is the syringe pump. In these pumps, a constant and reproducible flow can be obtained by using a gear mechanism. Spring loaded teflon seals are used in the plungers to minimize leakage around the pistons at high pressure. They are most suitable for small bore columns because this pump delivers only a finite volume of mobile phase before it has to be refilled. These pumps have a volume between 250 to 500 ml. The pump operates by a motorized lead screw that delivers mobile phase to the column at a constant rate. The rate of solvent delivery is controlled by changing the voltage on the motor.

Figure 17.6 shows the schematic of a typical syringe pump. The cylinder holds the mobile phase which is expelled by a piston. The piston is advanced by a motor connected through worm gears, to produce smooth pulseless flow. In syringe pumps, pressure capability is generally quite high and maintenance is infrequent. However, they have limited reservoir capacity and a slight change of flow rate when extremely high pressure compresses the solvent.

17.3.1.2 Constant Pressure Pumps

In these pumps, the mobile phase is driven through the column with the use of pressure from a gas cylinder. A low pressure gas source is needed to generate high liquid pressures. The valving arrangement allows the rapid refill of the solvent chamber whose capacity is about 70 ml. This provides continuous mobile phase flow rates.

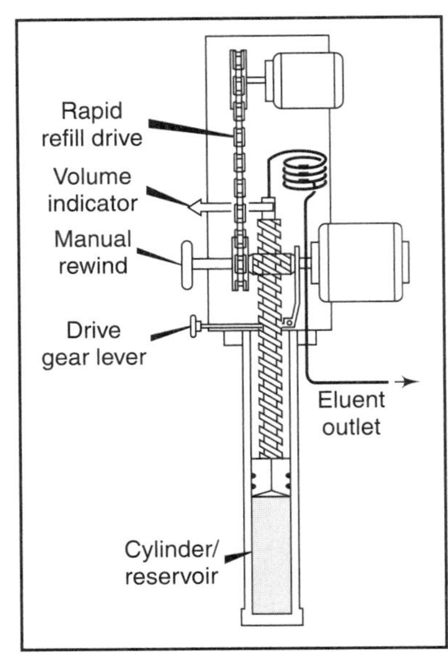

Figure 17.6 :: Schematic of a positive displacement syringe pump (after Yost, *et al.*, 1980)

High pressures of several hundred atmospheres are required in high resolution high speed liquid chromatography. The Varian Aerograph Model 8500 Liquid Chromatograph employs pump system, which gives pressure up to 600 atmospheres. Generally used pumps are the piston type, which provide very high solvent pressures. The flow rate can be set to the desired rate by adjusting the pump stroke length and the motor speed. The pressure is observed as a dependent variable. The flow of solvent from a piston pump is usually in the form of a series of pulses. This type of ripple in the flow is likely to affect column resolution and detector stability. In order to smoothen out the ripple, a long nylon tube of about 1.5 mm diameter may be used between the pump and the chromatographic column. A ripple can also be reduced by using bellows, restrictors or multi-piston pumps, wherein the action of the individual pistons is arranged at regular intervals of a complete stroke cycle.

> The metric unit of pressure measurement is the Pascal, abbreviated as Pa. One pascal is equal to one Newton per square meter (N/m^2) and, in previously used units, to 1×10^{-5} bar.
>
> Most scientists use the Kilopascal (KPa = 10^3 Pa) or mega Pascal (MPa = 10^6 Pa). Since 1 bar equals 0.9869 atm, 14.5038 psi and 1.0197 Kg/cm^2, one mega Pascal equals 10 bar, 9.869 atm, 145.038 psi and 10.197 kg/cm^2.

A flow control system for high pressure liquid chromatography, which maintains constant flow, irrespective of differing solvent viscosities and compressibilities, utilizes a hydraulic capacitor (Figure 17.7), which smoothens the high pressure pump pulsations, normally encountered in piston-operated pumps. It consists of a rigid vessel filled with fluid of a known compressibility. A small fraction of the space is separated from the compressible fluid by an impermeable membrane. The solvent mixture passes through this separated space.

A restriction in the solvent flow path, which could be the chromatographic column itself, is the hydraulic analog of a resistor, so that the unit can function analogously to an RC filter. With a sufficiently large fluid volume (large *C*), adequate smoothening of the pump pulsations can be obtained with a relatively low value of *F* on the output side. The flow measurement and control system used by HP in their model 1010B HPLC is shown in Figure 17.8.

A pressure transducer is installed in the hydraulic capacitor. Measurement of the pressure is synchronized with the pump, so that measurement is made only during the discharge phase. The time integral of the ac component of the transducer output is proportional to changes in pressure and is thus proportional to flow. The ac component is fed to a voltage-to-frequency converter. A counter totals the output of v to f converter, effectively integrating the transducer output. At the end of each 12 s integration period, the counter's contents are compared digitally to the set point. Any error is then used to adjust the pump stroke setting to bring the flow rate to the set point value. The averaged dc output of the pressure transducer represents absolute pressure, and is used as a constant to compensate for the influence of pressure on the compressibility constant of capacitor *C*. The effects of temperature on fluid compressibility are circumvented by using a temperature control system to maintain the capacitor at a constant temperature.

The flow rates in the range of 0.10-9.99 ml/min are controlled by this system with repeatability of ± 1 per cent. The flow rates of the two pumps are controlled indepen-

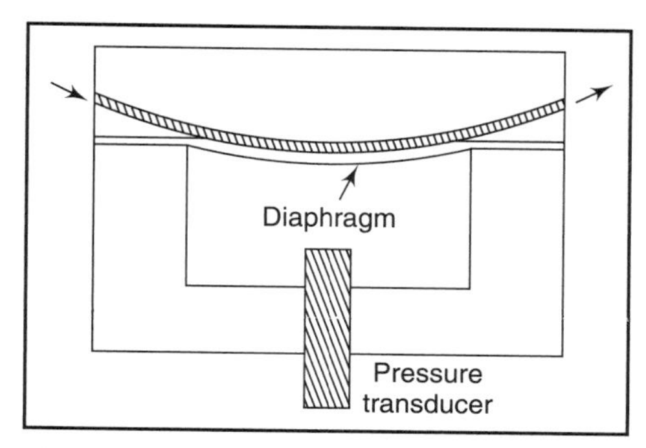

Figure 17.7 :: Hydraulic capacitance for smoothening of flow pulsations (after Schrenker, 1975)

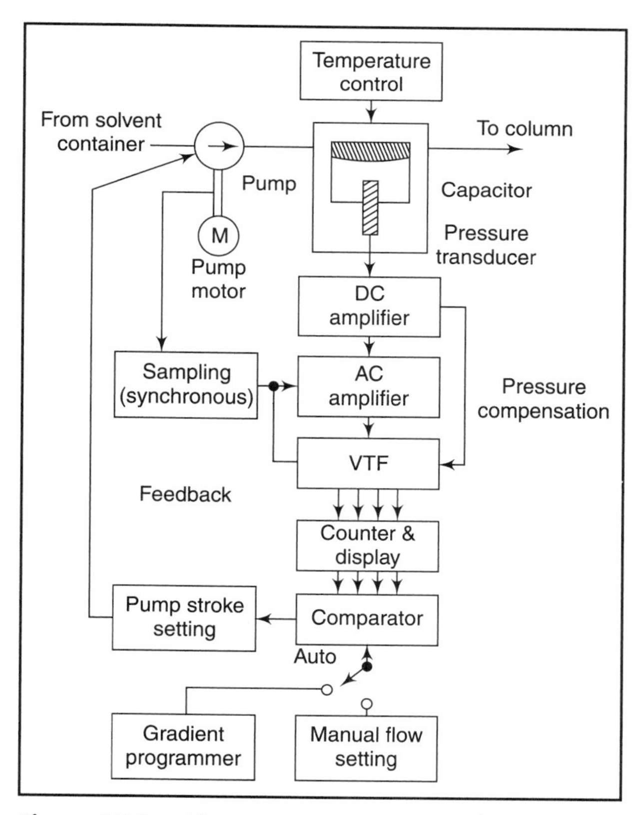

Figure 17.8 :: Flow measurement and control system (after Schrenker, 1975)

dently and then outputs are mixed in a low volume mixing chamber, immediately upstream of the sample injection port.

The use of microprocessors in LC pumps has facilitated a combination of features of uniform flow, accurate solvent proportioning and low system volume. The series 410 microprocessor-based LC pump from Perkin Elmer, whose details are shown in Figure 17.9, is a positive volume proportioning pump, which ensures accurate and repeatable solvent delivery and composition, with no adjustments required by the chromatographer. The pump actually measures the volume of solvent delivered by the pump. This information is monitored by the microprocessor, which automatically makes any necessary correction to compensate for volume reductions due to solvent mixing. This results in the solvent blend being essentially the same as the one that the experienced chromatographer would mix. The pump provides four solvent gradient capabilities including step, linear, convex and concave gradients to cover the full range of LC applications, with the flow rate ranging from 0.01 to 10.0 ml/min for microbore, analytical high speed and semi-preparatine. The multi-tasking software enables ease of operation by allowing modification of both current and inactive methods, while the pump is running.

Positive volume proportioning pump

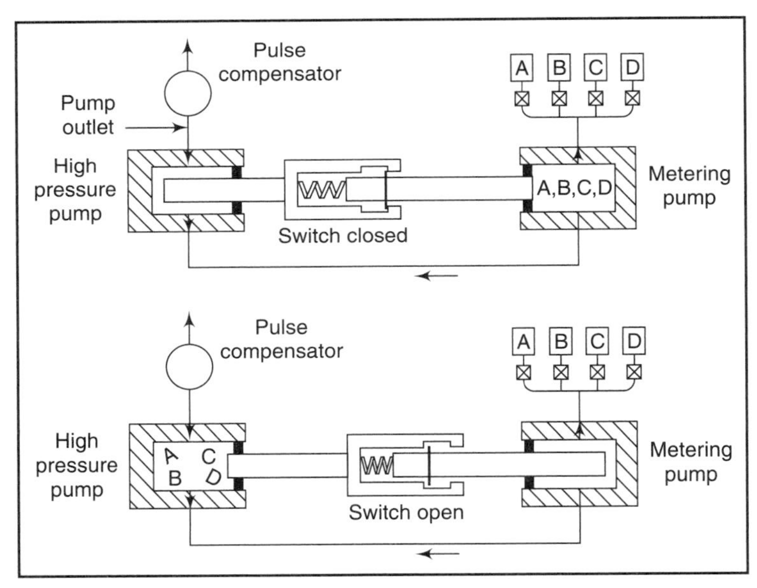

Figure 17.9 :: Principle of positive volume proportioning pump (Courtesy M/s Perkin Elmer, USA)

17.3.1.3 Gradient Elution or Solvent Programming

In liquid chromatography, a single substance may be used as a mobile phase during an analysis of the mixture of two or more substances to properly adjust the characteristics of the phase. Also, one may maintain a constant mobile phase composition during analysis or change it. The first mode is called the Isocratic operation, while the second is called the gradient elution:

Figure 17.10 represents the functional schematic of the system providing the mixed mobile phase to the column.

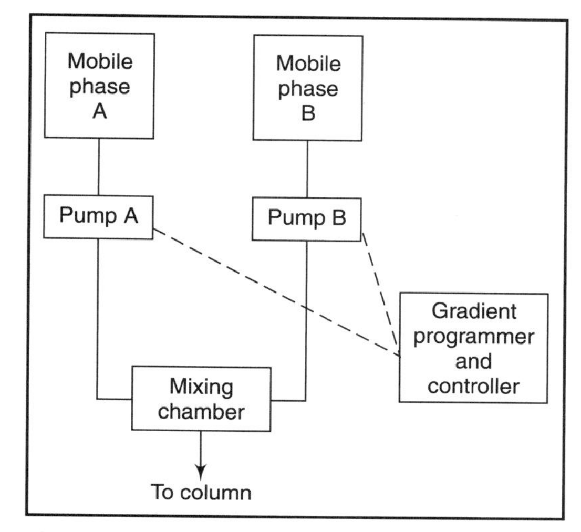

Figure 17.10 :: Functional schematic of a system for gradient elution

Gradient elution is often required to resolve complex mixtures, especially those containing components with significantly different chromatographic behaviour. A solvent programmer helps to control the composition of the mobile phase according to a pre-determined program as the analysis proceeds. Solvent programming is generally carried out by continuously adding a more polar solvent to the mobile phase feed reservoir, thereby increasing the polarity of the eluant as a function of time. The technique involves the use of separate pumps, feeding different solvents or solvent mixtures concurrently into the column and programming the output of each pump.

Figure 17.11 shows the arrangement usually employed for solvent programming. The supply of the solvent from the pump is given to a T-connection through solenoid valves. One solvent is used as a feed to the pump and the other is introduced into the bellows assembly. By properly programming the time intervals between the opening and closing of the solenoid valves, the

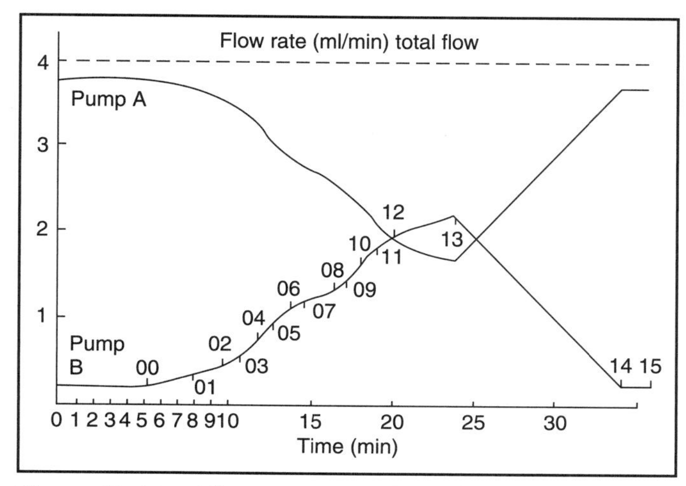

Figure 17.11 :: Flow programming of two pumps to give a constant total flow output

desired gradient to the column can be obtained. Standard linear, convex and concave gradients are usually sufficient. However, some manufacturers also provide facilities for non-linear gradients. With the help of this system, it becomes possible to generate gradients of virtually all complexities that one may need to resolve difficult mixtures. The rate of change of solvent composition is selected with the help of suitable programme. The solvent strength can be increased, decreased or held constant. It is held constant at any point or points in the program simply by omitting a rate pin in a given step. This ability to hold constant solvent strength at any point in the program is highly useful for maximizing resolution in difficult parts of an analysis, for example, in separating isometric compounds.

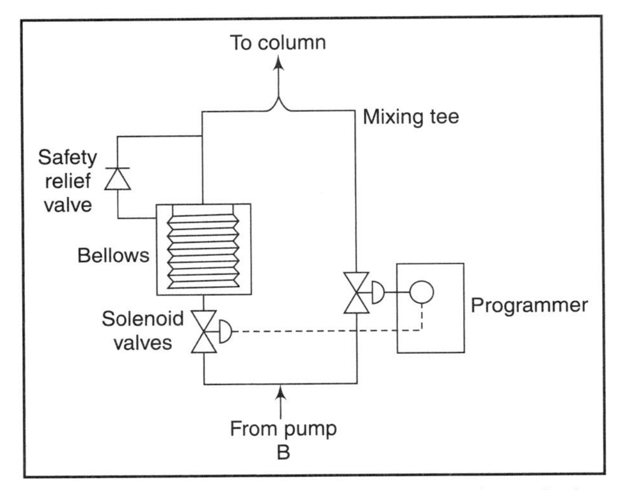

Figure 17.12 :: Arrangement employed for solvent programming

In the recent instruments, a gradient program is entered by way of a calculator-like keyboard on the gradient programmer. The program is entered as a series of linear segments (Figure 17.12) that approximate the desired program curve. Each segment is specified by these program entries:

- Flow rate at the beginning of the segment,
- Flow rate at the end of the segment, and
- Time duration of the segment.

This information is stored in the digital memory. Since the flow rate programming for each pump is independent of the other, the following three program modes are possible:

- Change the mixing ratio from the two pumps, while maintaining constant column flow by programming equal but inverse flow rate changes for the two pumps (gradient programming).
- Change the column flow rate while maintaining a constant mixing ratio by programming equal percentage changes of flow in A and B (flow programming).
- Combine (i) and (ii), resulting in flow program superimposed on a gradient proram.

17.3.2 Sample Injection System

There could be several methods for the introduction of the sample on the top of a liquid chromatographic column. One method is to disconnect the solvent supply, to add the sample in solution and reconnect the solvent supply to the column. This mechanism is simple, but the method is tedious to operate. More recently developed methods fall into two categories, namely, the syringe injection method and the injection valve method. Both these methods enable the sample to be introduced directly into the column packing, without interrupting the solvent flow.

17.3.2.1 Syringe Injection Method

Injection port

Samples are injected into the HPLC via an injection port. The injection port of an HPLC commonly consists of an injection valve and the sample loop. The sample is typically dissolved in the mobile phase before injection into the sample loop. The sample is then drawn into a syringe and injected into the loop via the injection valve. A rotation of the valve rotor closes the valve and opens the loop in order to inject the sample into the stream of the mobile phase. Loop volumes can range between 10 μl to over 500 μl. In modern HPLC systems, the sample injection is typically automated.

Stopped flow injection

Stopped flow injection is a method whereby the pump is turned off allowing the injection port to attain atmospheric pressure. The syringe containing the sample is then injected into the valve in the usual manner, and the pump is turned on. For syringe type and reciprocation pumps, flow in the column can be brought to zero and rapidly resumed by diverting the mobile phase by means of a three-way valve placed in front of the injector. The method can be used up to very high pressure (Willard, *et al.*, 1988).

The syringe injection method basically involves the insertion of the syringe needle through a rubber septum at the top of the column (Figure 17.13). This method, however, cannot be used for the injection of large sample volumes into high pressure solvent systems. At pressures greater than about three atmospheres, the pressure has to be reduced by turning off the solvent supply before the injection of the sample can be carried out.

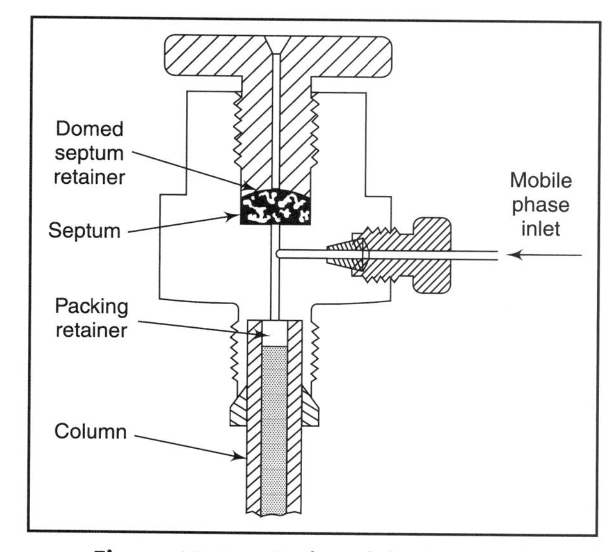

Figure 17.13 :: Syringe injection method

17.3.2.2 Injection Valve Method

In this method, the injection valve containing sample loops is connected in the solvent supply pipe work at the top of the column. The sample loop can be introduced into the solvent stream when desired, without turning off the solvent flow. After sufficient flushing of the loops with solvent has taken place, the sample is completely carried to the column. The loop can then be removed from the solvent stream for refilling with the next sample. Injection valves can be used for sample introduction into very high solvent pressure systems. By changing the volume of the sample loop, the sample size can be easily varied. Also, this method can be conveniently automated for automatic injection of the samples.

17.3.3 The Column

The column is, by far, the most important part of any chromatographic system, since the ultimate performance of the chromatograph is determined largely by the column. Most of the early analysis work, which was carried out by liquid chromatography, made use of large columns with an internal diameter of 1 cm or more. With the development of highly sensitive detection systems, it has become possible to analyze minute quantities of sample and to reduce the column diameter. The reduction of sample size and the column diameter result in an improvement in separation efficiency.

Band broadening

There is another factor which necessitates the use of small diameter columns. A large contribution to the band broadening in the chromatographic peaks is known to be made by the large scale unevenness of flow, which becomes worse as the column diameter increases. The effect of uneven flow may be reduced by decreasing the diameter of the column. The columns in current use are generally in the 0.1–2.0 cm internal diameter range. However, practical problems occur when the column diameter becomes less than 0.05 cm, because very small particle sizes and very high pressure drops will have to be used. As regards the column length, they may be 1–4 m long, but most of the applications can be performed on columns of less than 100 cm length.

Various materials have been used for the construction of columns. Glass columns are usually preferred on account of their inert nature and the facility of being able to observe the packing visually. High performance liquid chromatography (HPLC) columns are stainless steel tubes, typically of 10–30 cm in length and 3–5 mm inner diameter. Short, fast analytical columns and guard columns, which are placed before an analytical column to trap junk and extend the life time of the analytical column, are 3–10 cm long.

Theoretical considerations have revealed that much smaller particles should be used for column packing in liquid chromatography than in gas chromatography. Generally, the particle size ranges from 20 to 50 μ. Occasionally, even smaller particle sizes are used. For example, in the Varian 8500 LC system, the columns are packed with particles of diameter less than 10 μ. Columns packed with small sized particles are more efficient, since the solute mass transfer takes place at a very rapid rate. Small packing particles used in liquid chromatography, however, present great problems in getting homogeneous packing of the columns. Also, the method of column packing is dependent on the type, regularity and the particle size of the packing used.

In most of the separations, liquid chromatographic columns are operated under ambient temperature conditions. However, some researchers have shown that improvements in column efficiency can be achieved in certain cases by working at elevated temperatures. Columns are, therefore placed inside ovens capable of operation up to 25°C. The temperature of these ovens is controllable to a high degree of constancy.

17.3.4 Detection Systems

The detector for an HPLC is the component that emits a response due to eluting sample compound and subsequently produces a peak on the chromatogram. It is positioned immediately posterior to the stationary phase in order to detect the compounds as they elute from the column (Frel, *et al.*, 1985).

High sensitivity detection systems are necessary for achieving optimum column separating performance, by the use of small sample volumes. The absence of a versatile and economical detector has been one of the main reasons for the limited development of liquid chromatography in the past. The earlier methods of detection used in liquid chromatography were of a non-continuous nature. However, continuously monitoring detection systems are more convenient in operation, and therefore, all modern liquid chromatographic detectors incorporate this feature.

Several detection systems have been developed, which are mostly dependent upon the measurement of a physical property of the column elute. These physical properties could be changes in ultraviolet absorption, infrared absorption, heat of adsorption, refractive index or electrical conductivity.

In principle, any of these measurement techniques can be used, but in practice there are several constraints when a detector is interfaced to LC (Yeung and Synovec, 1986). These constraints are:

- Good sensitivity is essential to deal with the low concentrations of typical analytes.
- The volume of the detector must be small to avoid additional band broadening due to extra column effects, particularly in the recently introduced techniques of micro column LC and super-critical fluid chromatography. Volumes below 1 μl for 1 mm internal diameter (packed micro-bore) columns and below 1 nl for open micro-tubular columns of 10 μm internal diameter are required.
- The detector must be able to function in the presence of a large background signal and be able to null out this signal and to maintain it at a stable level to reduce noise.
- The response time of the detector must be compatible with the chromatographic event.
- Detector selectivity is more important in LC as compared to GC, as chances of overlapping in LC are much higher. A selective detector can effectively resolve components without physical separation.

Many types of detectors can be used with HPLC. Some of the more common detectors include: Refractive Index (RI), Ultraviolet (UV), Fluorescent, Radiochemical, and Electrochemical, Near-Infrared (Near-IR), Mass Spectroscopy (MS), Nuclear Magnetic Resonance (NMR), and Light Scattering (LS).

All present-day detectors are on-stream systems, continuously monitoring specific characteristics of the column effluent in a flow through cell. Modern LC systems employ fairly low flow rates and provide high resolution in a short time. The detectors utilize low volume cells, which are efficiently flushed to prevent band broadening and tailing. A measure of the efficiency of these cells is the so-called 'instrument bandwidth.' The instrument bandwidth, expressed in microlitres, is the minimum peak volume that the detector produces, no matter how small the sample volume introduced into the detector.

Instrument bandwidth

17.3.4.1 UV–Visible Spectrophotometric Detectors

With this type of detector system, it is possible to detect and analyze compounds that absorb at any wavelength in the UV-visible range from about 200 nm to 800 nm with a bandwidth of 5 nm. Almost any spectrophotometer suitable in this range can be modified to work as a detector. The cell in this ease would be of the flow-through type, with a path length of 1 cm and a cell volume as low as 8-20 µl. Cells may be made of quartz, Teflon and KELF. They are usable up to a pressure of 500 psi. Stainless steel cuvette assemblies are used for higher pressures. When using a recording spectrophotometer, it is possible to stop the flow and scan the spectrum of individual peaks in the chromatogram. Just like conventional spectrophotometers, a choice of light sources is usually available, depending upon the wavelength desired. Deuterium lamp for the wavelength range of 200 to 400 nm and Tungsten lamp for 350 to 800 nm can be selected when required. These instruments are calibrated in absorbance units over these ranges: 0.01, 0.02, 0.04, 0.1,0.2, 0.4 ,1.0 and 2.0 absorbance units. Three transmission ranges may also be provided to measure 100 per cent , 10 per cent and 1 per cent full-scale on any suitable recorder having a full-scale sensitivity of 10 mV. For resolving small events in peaks, the instruments are provided with a ten-turn fine adjustment to achieve zero suppression over the full range of two absorption units or a 120 per cent transmission range.

Spectroflow monitors

Spectrophotometric instruments used in liquid chromatography are often called spectroflow monitors. Figure 17.14 shows the block diagram of spectroflow monitor Model SF 770 of Schoeffel Inst. Corp., USA, used for UV-visible liquid chromatography. The system makes use of a double-beam principle and a chopping system to measure the transmission through the sample and reference cells alternately. A reflection mode attachment is also available with this instrument. Conventional log function generating amplifiers are employed to get direct readings of optical density. This instrument gives a Noise figure which is better than 5×10^{-4} O.D. (at $\lambda = 280$ nm) and stability better than 5×10^{-4} O.D. per hour (at $\lambda = 280$ nm).

Detection systems are also available for the UV range alone (Figure 17.15). Most of these systems operate at fixed wavelengths of 254 or 280 nm. These wavelengths offer excellent sensitivity for many compounds, yet permit the use of a wide range of solvents without interference, even with gradient elution. The UV detector employed by Varian Aerograph Model 8500 LC has a special thermal isolation design, which gives a low noise level of only $\pm 5 \times 10^{-5}$ absorbance units equivalent to low nanogram sensitivity. The output of the detector is linear in absorbance and thus linear in concentration for solutes obeying Beer's Law. The wide linear dynamic range (10^4) enables one to measure both trace and major components in the same chromatogram.

The usefulness of spectral information has led to new commercial instrumentation based on diode arrays, rapid scan spectrometers and even Fourier transform (visible) interferometers. Commercial accessories for Fourier transform infrared (FT-IR) detection following LC are also available, either as thin flow cells or as off-line collection devices based on solvent removal. This is because, unlike GC, the LC eluent absorbs strongly in the IR region to obscure analyte signals. The most widely used detectors currently used in HPLC are the ultraviolet, visible and fluorescence spectrometers. Although normally limited to the detection of only a single wavelength element at a time, these single-channel devices have proved to be sensitive and robust detectors. For multiple component analysis, new techniques such as rapid scanning detectors based on the linear photodiode array (LDA), the silicon vidicon tube or the charge coupled detector presents an alternative technology for rapid wavelength detection in HPLC (Fell, 1985). Currently, there are a number of commercially produced HPLC detectors based on the LDA, usually in combination with a microcomputer. The design uses a 256-element diode array and an 8 μl flow cell to acquire spectral data during elution.

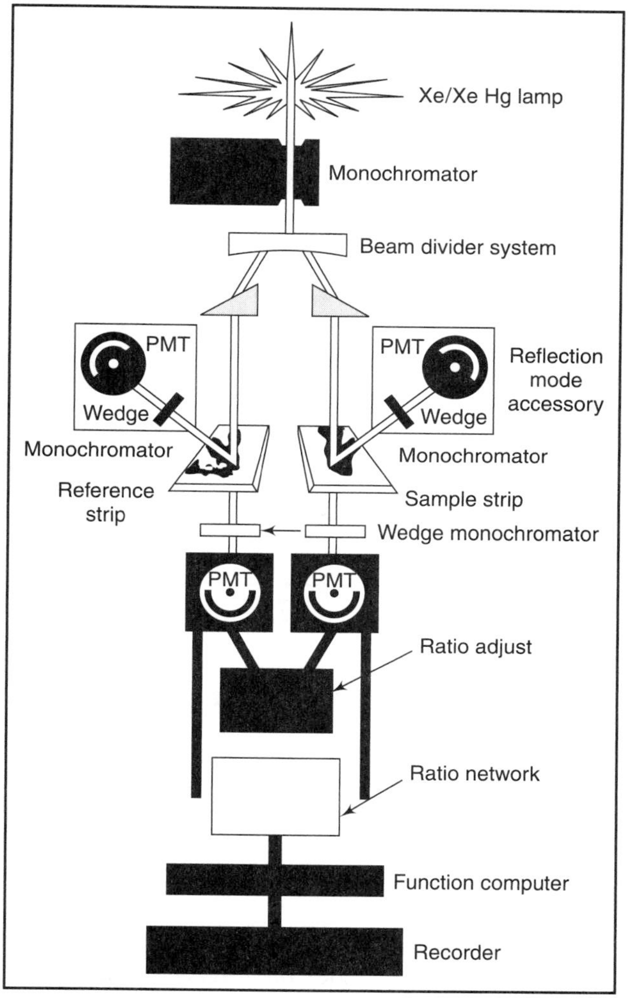

Figure 17.14 :: Optical diagram of spectroflow model SF 770 (Courtesy Schoeffel Instruments Corp., USA)

Microcomputers provide the necessary memory and speed of operation to handle in real-time the data generated by diode arrays, thereby providing formidable facilities as shown in Figure 17.16 (Miller, 1984). These include comparison of the experimental spectrum with a library of spectra for rapid sample identification, subtraction of one spectrum from another, and the calculation of first, second and higher derivative spectra. The graphic display facility enables to have spectrochromatograms, normally presented as isometric projections, which show in three dimensions, axes representing time (which can be related directly to chromatographic retention parameters), spectral intensity and

Spectrochromatograms

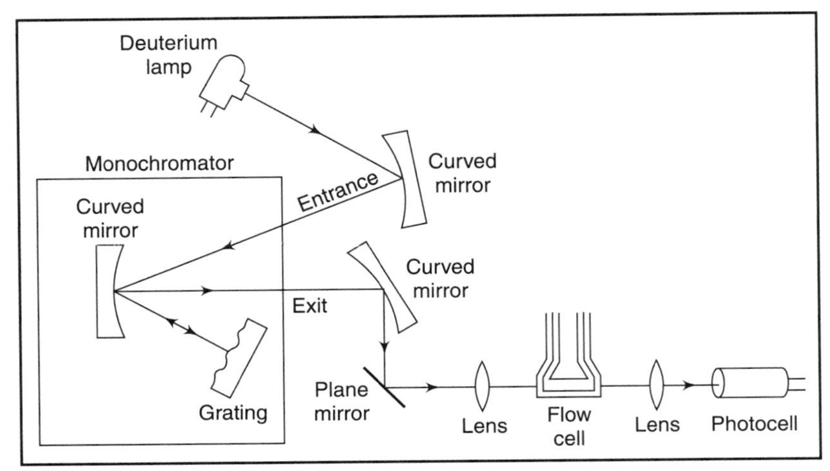

Figure 17.15 :: Optical schematic of a typical UV liquid chromatography detector

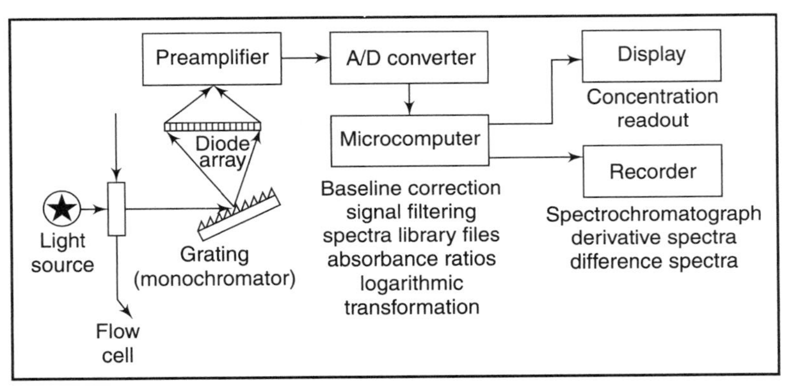

Figure 17.16 :: Facilities offered by a diode array spectrometer with a microcomputer (after Miller, 1984)

wavelength. Spectrochromatograms are thus able to present simultaneous images of the chromatographic separation and the spectral properties of the separated fraction. Moreover, the computer will calculate sections through the total image to generate chromatograms recorded at specific wavelengths, and spectra recorded at a particular time.

17.3.4.2 Fluorescence Detector

Fluorescence measurements of minute quantities, as encountered in liquid chromatography, differ greatly in technique and behaviour, if compared to standard flow through absorption monitors or spectrophotometers. In a fluorimeter, the presence of a fluorescence-emitting substance is measured.

The emitting substance is only present occasionally in liquid chromatography and its emission of energy is detected by a highly sensitive photomultiplier.

Figure 17.17 shows the optical schematic of a typical fluorescence detector for liquid chromatography. Commercially available detectors differ in the manner in which the wavelengths are controlled. Less expensive instruments utilize filters, medium-priced units offer monochromator control of at least one function, usually excitation wavelength, and full capability research grade instruments provide monochromator control of both excitation and emission wavelengths. A spectrofluorimeter specifically designed for LC applications offers continuously selectable monochromatic excitation energy over the entire UV–visible spectrum, utilizing a highly stabilized deuterium or tungsten–halogen lamp. The monochromator makes use of a grating system. Its continuously variable wavelength extends analytical application far beyond the limited areas normally offered by line spectra of Hg lamps. Excitation energy from the monochromator enters the cuvette and emission from a 5 μl cavity is collected and directed towards the photomultiplier. A set of easily interchangeable filters is provided to select the emission spectra of interest. The filter set contains filters of wavelengths 370, 389, 418, 470, 550 and 580 nm. Transmittance is greater than 0.9 in these areas compared to less than 10^{-5} below the cut-off point and because of this high efficiency, virtually no emission energy gets absorbed before reaching the photomultiplier.

Spectrofluorimeter

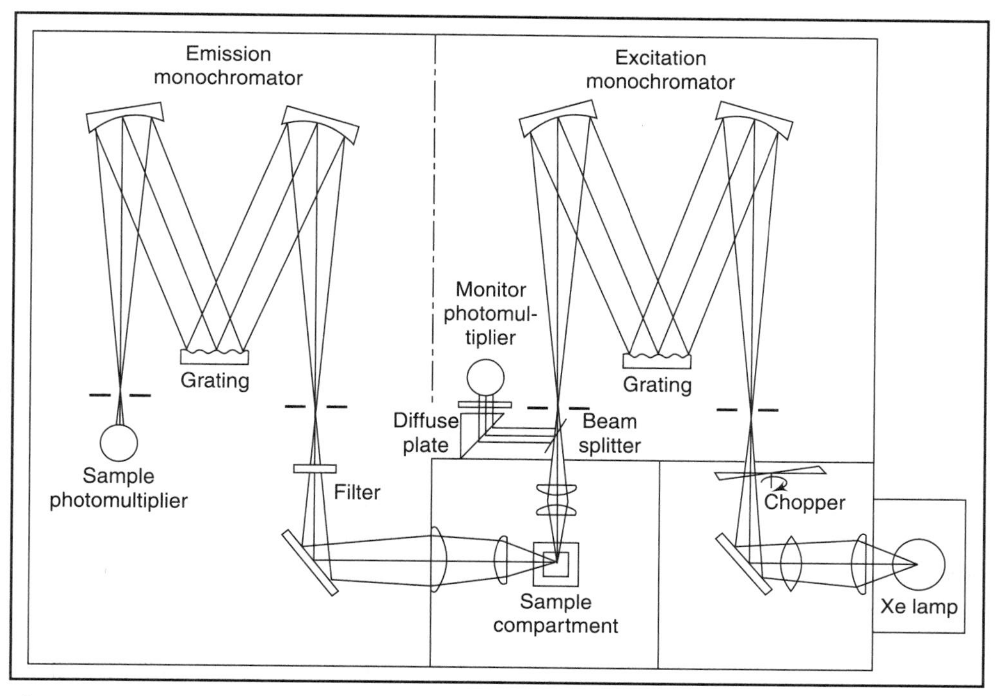

Figure 17.17 :: Optical schematic of a typical fluorescence detector for liquid chromatography

In normal fluorimeters, which utilize standard cuvette, volumes of several cc's are illuminated by larger excitation light beams of at least 1 cm^2. The emitting material acting as secondary light source to be detected, originates from a much larger area or volume than what is available in the microlitres type cuvette of fluorimeters used in high pressure liquid chromatography. Since emission occurs in all directions from an excited sample surrounding this sample with an efficient optical collector is a must. M/s Schoeffel have introduced a 2π Steradian cuvette (Figure 17.18) which has provided a solution to this problem. This cuvette is of stainless steel and is of the flow-through type.

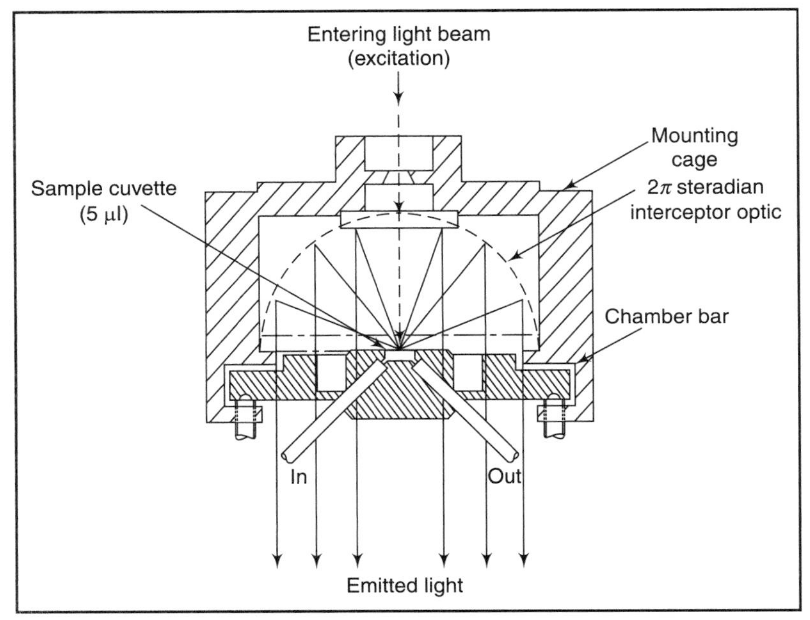

Figure 17.18 :: Steradian cuvette used in fluorescence detector

Fluorescence detection is more selective than absorption detection, because all species that absorb light do not necessarily fluoresce. However, fluorescence detection is limited by the presence of background light, which includes various types of light scattering, luminescence from the flow cell walls and emission from impurities in the solvent. Although fluorescence intensity generally increases with excitation intensity, all the above-mentioned factors increase with excitation intensity to produce no net gain. The situation is worse when lasers are used as excitation sources because laser intensities are inherently much less stable than conventional light sources.

The use of laser-excited fluorescence is getting rapidly established, but the developments are really attributable to new designs of flow cells and optics that reduce stray light. Even though fluorescence intensity increases with path length, high power lasers can be used to provide sufficient signal levels over short path lengths, enabling one to achieve small detector volumes.

The advantage of laser fluorometry is thus its mass detectability, which can be 100 times lower than that obtained from conventional excitation sources. Moreover, better monochromaticity of the lasers can be advantageously used to reject stray light. Yeung and Sepanak (1980) have detailed the designs of flow cells for laser fluorometric detection in HPLC.

17.3.4.3 Refractive Index (RI) Detectors

The common refractive index (Rl) detectors are based on refraction, reflection or interference of light beams (Yeung and Synovec, 1986). Refractive Index (RI) detectors measure the ability of sample molecules to bend or refract light. This property for each molecule or compound is called its refractive index. For most RI detectors, light proceeds through a bi-modular, flow cell to a photodetector. One channel of the flow cell directs the mobile phase passing through the column while the other directs only the mobile phase. Detection occurs when the light is deflected due to samples eluting from the column. This is read as a disparity between the two channels. Refraction type RI detectors depend upon Snell's law at the interface between the cell in wall (glass) and the flowing liquid to deflect a light beam. Changes in RI are monitored at the far field by a position sensor, or as an intensity change on a small area photodiode. Since the effect is produced at the interface, very small volumes are possible if the proper cells are made. Reflection type RI detectors are based on Fresnel's laws at the interface between the cell (glass) and the liquid, which has a smaller RI. Sensitivity will increase as the incidence angle approaches the critical angle. Interference type refractometers provide the best concentration limits of detection in commercial instruments. This is because a longer path length can be used to compensate for lower concentrations.

Refractive index

Figure 17.19 shows the optical schematic of refractive (Fresnel) type detector. Here, the light beam is focused on and reflected from both the liquid- prism cell interfaces and a polished back plate (which forms the near surface of both sample and reference cells) onto the detecting photocell. As the intrusion of sample into the one cell causes light to be refracted at a different angle, this light changes in intensity rather than position, with the unbalance once again being detected as a change in electrical energy. This difference between sample-cell signal and reference-cell signal is given to a recorder. The major advantages of this type of detector include its higher potential sensitivity, ability to operate at extremely low flow rates, with very low volume flow cells, easy cell accessibility and low cost. Its disadvantage is the need for changing prism to accommodate either high or low RI solvents (Yost, *et al.*, 1980).

In RI detectors, the limit of detection is limited by fluctuations in the RI of the LC effluent. Solvent delivery systems must be pulse-free to avoid pressure effects. Also, thermostated cells must be used to avoid temperature effects.

The use of lasers has helped reduce detector volumes and has provided added selectivity and sensitivity. The development of small columns to minimize contributions from the solvent and to allow cost-effective use of exotic mobile and stationary phases have also resulted in improvements in detection systems. Therefore, the development of separation processes has, in turn, benefited from the new generation of detectors.

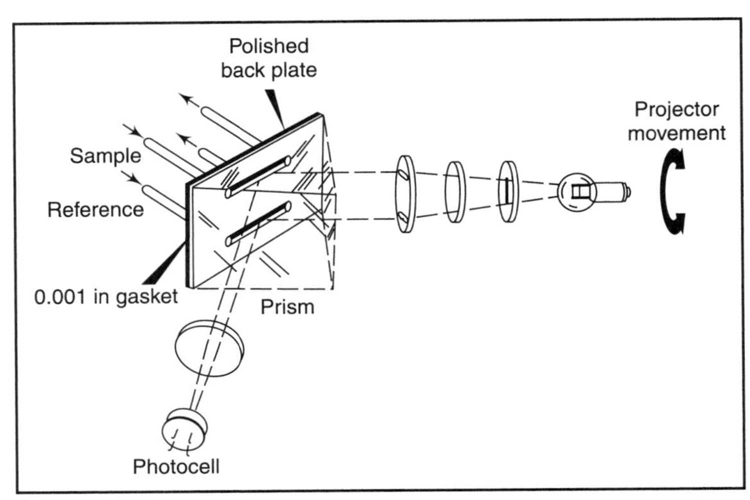

Figure 17.19 :: Optical schematic of the fresnel-type refractive index detector

Compounds without significant UV or visible absorbance can be monitored with a refractive index detector. This detector may be a dual-beam refractometer of the Fresnel type. These instruments measure the intensity of reflectance, which in turn, is inversely proportional to the refractive index. Careful control of the temperature is necessary in these instruments. Small cell volumes of the order of 5 μl are possible.

Differential refractometers are based on the measurement of the difference in refractive index between the solution and of the pure solvent. This is usually carried out by measuring the bending of monochromatic light beam as it passes through sample cuvette. Temperature control to within 0.01°C is necessary for measuring absolute refractive indices to within 10^{-5} units. Cell volumes of the order of 75 μl and higher are necessary. The sensitivity of these refractometers alters with variations in flow. The range of refractive index covered is 1.30 to 1.60.

17.3.4.4 Adsorption Detector

This type of detector depends upon the measurement of the evolution of the heat of adsorption and the heat uptake at desorption as the solutes in the effluent stream come into contact with an adsorbent. The measurement of heat is carried out by a thermistor placed in the stream. The detector operates differentially by having another thermistor in the stream and temperature changes as low as 10^{-4} °C can be detected by this system. One detector cavity is packed with an adsorbent, such as alumina, porous glass beads or silica. The other cell, called the reference cell is packed with inactive glass beads. The thermistors placed in these detectors form a part of the Wheatstone bridge, which gives out electrical signal suitable for recording purposes. These detectors are flow-sensitive and therefore, centre discs are used to decrease this effect. The dual detection system cancels out baseline shifts due to changes in flow rate.

17.3.4.5 Electrical Conductivity Detector

The electrical conductivity detector is perhaps one of the most important and promising detectors for the future of liquid chromatography. Areas in which this mode of monitoring are particularly attractive are aqueous and non-aqueous gel filtration, ion exchange chromatography and many applications of liquid–liquid chromatography. The conductivity cells used in liquid chromatographs have three electrodes with an internal diameter of under 2 mm and a length of 10 cm. The electronics comprises a stabilized and isolated AC bridge and phase detector system. The unit is sensitive to a change of approximately 1 part in 10,000 in conductivity and a measuring range of 10 to 100,000 μ mhos/cm.

Figure 17.20 shows a typical arrangement of a conductivity detector for LC applications. It is a three-electrode design. The dimensions of the cell are 10 cm length and 2 mm ID. The cell constant of the cell is 10 cm^{-1}. The total effluent stream passes through the flow-through cell, though the stream can be split to pass only the required amount through it.

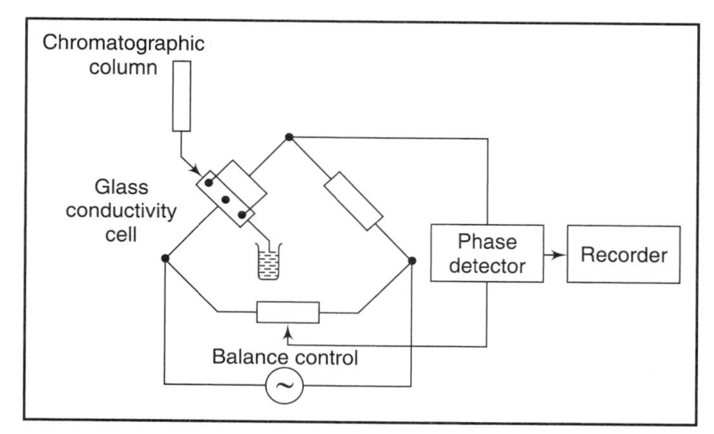

Figure 17.20 :: Conductivity detector

17.3.4.6 Thermal Detectors

Thermal detectors are also known as micro-adsorption detectors. The principle of operation of these detectors depends upon temperature changes taking place due to the heat of adsorption on an active solid source. These detectors are now manufactured by a number of companies. Generally, a portion of the adsorbent column packing is contained in a small chamber at the column outlet. A second chamber filled with an inert material such as glass micro-beads is used to achieve a reference signal. This assembly is carefully thermostated. Both chambers contain matched thermistors, which constitute the measuring arms of a Wheatstone bridge. As the eluted sample is adsorbed on the solid, a local temperature change takes place, which initiates a signal.

These detectors can be used in applications involving liquid–solid, liquid–liquid, ion exchange and gel permeation chromatography. They are non-destructive. However, they are subject to other thermal effects such as thermal conductivity and heat capacity of the solvent. They, therefore, require accurate calibration before use.

17.3.4.7 Mass Detector

A mass detector for HPLC based on the scattering of light is described by Turner (1986). It is found that the commonly used UV absorption and refractive index detectors are unresponsive to certain classes of compounds and are also not useable with certain solvent/solute combinations which again restricts their use. Figure 12.21(a) shows a schematic of the mass detector, with its cross-section shown in Figure 17.21(b). In this detector, the solvent stream enters the detector through a small bore hypodermic tube at the top. This is fed into the atomizer assembly where a venturi jet operated by inert gas or air converts the solution into a uniform dispersion of droplets which passes, as a continuous sample into the evaporation.

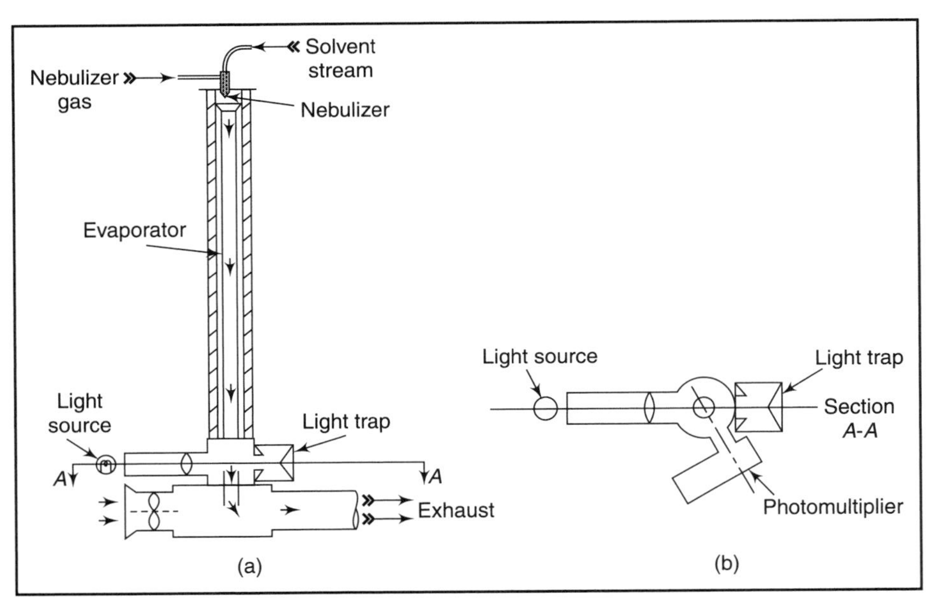

Figure 17.21 :: (a) Schematic mass detector (b) Cross-section at A–A (after Turner, 1986)

After nebulization, the atomized solvent spray passes rapidly down the evaporator. This produces a slightly negative pressure at the base of the evaporator column and ensures that a mixture of gas and solvent present in the evaporator is drawn down and out through the exhaust vent.

Light from a lamp is collimated and passed through the instrument at right angles to the direction of air flow. A light trap is located opposite the source of light to eliminate internal reflection. The amount of light scattered to the photomultiplier does not cause a detectable base line drift on the recorder. If, however, a solute is present, a particle cloud passes through the light path. Light refracted by this particle cloud is detected by a photomultiplier located at an angle of 120° to the incident light beam. The signal from the photomultiplier is amplified and displayed on a recorder. The output is proportional to the concentration.

17.3.4.8 Radiochemical Detection System

This system involves the use of radio-labelled material, usually tritium (^{3}H) or carbon-14 (^{14}C). It operates by the detection of fluorescence associated with beta particle ionization and is most popular in metabolic research.

17.3.4.9 Laser-based Detectors

Detectors based on the use of lasers offer several beneficial properties in chromatographic studies. They offer improved sensitivity and selectivity over other detectors. Also, lasers are capable of better resolution than most liquid chromatographic needs, and have improved signal-to-noise ratio. Some of the types of laser-based detectors are:

- Laser-induced Absorption,
- Laser-induced photoacoustic detector,
- Laser-induced fluorescence,
- Laser-based refractive index detector,
- Laser light scattering detector, and
- Laser-induced Raman scattering detector.

High energies of lasers can, however, cause thermal distortions, and sensitivity can decrease due to scattering at the optical sections of the system. They are not widely used in routine chromatographic practice.

17.3.4.10 Dual Detector Systems

In many cases, the use of a single detection might give ambiguous or even incorrect results. This is particularly true for complex, natural samples. For this reason, the simultaneous use of two detectors is often preferred. The following are the advantages of using dual detection system:

- By combining a RI and an UV detector, one can achieve an almost universal detecting capability.
- When a selective detector is combined with a general purpose detector, one can obtain the general fingerprint-type chromatogram of the sample and at the same time, identify certain peaks.
- One can combine two UV detectors operated at two different wavelengths, e.g. a fixed wavelength and a variable wavelength detector.

In dual detector systems, the instrument may be fitted with a second injector and column permitting the use of either one column with two detectors in series or two columns each with a separate detector as separate LC systems. Naturally, dual pumps are needed for this purpose.

17.3.5 Programmers and Read-Outs

Liquid chromatographs are finding on-line applications for monitoring chemical processes and providing information for the desired adjustments. Thus, a liquid chromatograph must operate automatically. Unless it is computer-controlled, it requires a programming unit. The programmer controls the analysis time, injects the sample, selects and measures the peaks, and presents the results for display or control. In addition, the programmer controls all column switching functions, including valves used for back flushing and washing. The programmer is usually placed in the control room along with the strip-chart recorder. The computerized system also includes a keyboard and CRT terminal. A dedicated computer can preferably replace a programmer, which can normally control several L.C. instruments.

The computer or programmer calibrates the read-out for each component of interest. Furthermore, the programming unit controls all functions related to calibration, including re-zeroing the base line at selected times.

17.4 Amino-Acid Analyzers

Amino-acids

Amino-acids are the components of all proteins and as such are essential for the growth and well-being of living organisms. Some are metabolized from the diet by hydrolysis resulting from enzyme activity, while others are synthesized within the organism. Therefore, the presence, distribution or absence of any amino-acid in living tissues signifies normality or abnormality in the organism. Over 40 of these complex molecules are known to exist naturally, and many more have been synthesized.

Amino-acids are distinguishable from each other only by one or two atoms in their structure. However, they may be found bonded together in manifold diversity, in the organization of proteins. Although the concentration of any free amino-acid can be accurately measured colorimetrically, all except two must be measured at the same wavelength. Therefore, they cannot be distinguished from each other, unless they have first been separated in a predictable sequence. The micro-technique of paper electrophoresis, when applied in this application, lacks resolution and reproducibility. Gas chromatography, though potentially fast and simple in operation, loses accuracy during the conversion of amino-acids to the gaseous phase and cannot be used for samples containing proteins or polypeptides.

The most widely accepted earlier technique was ion exchange chromatography on sulphonated polystyrene resin, followed by colorimetric analysis of the separated fractions. The method gave high accuracy, but the inconvenience of handling many individual samples and the long and laborious process of cleaning of glassware limited the deployment of the technique in routine analysis. Subsequently, an apparatus was developed in which ion exchange resin chromatography was followed by the automatic introduction of colour reagent into the separated fractions, as they emerged from the column. So, a continuous flow of solution, coloured at intervals according to the concentration of each fraction as it occurred, could be passed through a colorimeter to produce a graphical record on a multi-point recorder.

Resin column chromatography

Resin column chromatography depends upon the characteristic strength of the ionic bond formed with the resin by each acid, as it is applied to the top of the column. This bond breaks again successively, causing a characteristic delay, as the acid is eluted through the column by a buffer of known pH and ionic strength. Thus, the column effluent consists of a series of discrete amino-acids, separated by buffer in which no amino-acids are present. Ninhydrin colour reagent is metered into the column effluent. The colour reaction develops during a known period, at a controlled temperature. The flow energizes a multi-point recorder. The record is a continuous chromatogram, interpreted quantitatively against the chromatogram of a standard amino-acid mixture.

Proline and hydroxyproline are measured at 440 nm and all other amino-acids at 570 nm. Therefore, two photometric systems are employed. One photometer serves the high precision section, measuring at both wavelengths mentioned, through a path length of typically 2 mm. A secondary 570 nm channel has a path length of only 0.7 mm to accommodate more highly-concentrated fractions. Two other photometers measure at the same wavelengths through 20 mm. They are used to obtain the same sensitivity from the high-speed section, as the high precision section provides, from samples which are ten times more concentrated. Alternatively, they may be used to improve ten-fold the detection limit of the high precision section.

During measurement, two columns are used, of which the short one is for the slow moving basic amino-acids, which elute more quickly. At the end of a separation, the basic amino-acids remain in the long column and must be removed before the next hydroxide, followed by automatic changeover to 3.25 pH buffer to restore the equilibrium of the column. Apart from the amino-acids, which are commonly known to be the constituents of proteins, the substances that can be analyzed on analyzers based on the ion-exchange chromatography on resin are amines, amino compounds, peptides and any other chemical compounds which are ninhydrin positive.

17.4.1 Automatic Amino-acid Analyzer

For the separation, identification and quantization of amino-acids and related compounds, automatic instrumentation has been introduced. These instruments permit complete protein or peptide hydrolysate analyses with reduced sample amounts and increased sensitivity of 0.05 μmole or less.

Figure 17.22 shows the schematic diagram of the components and the flow system of a typical amino-acid analyzer. The system may employ seven chromatographic ion exchange columns, having a precision bore 0.9 cm inner diameter. At the bottom of each of these tubes is a porous Teflon disc, upon which the resin is supported. The disc can be easily removed for cleaning or replacement, and facilitates draining of the resin from the column.

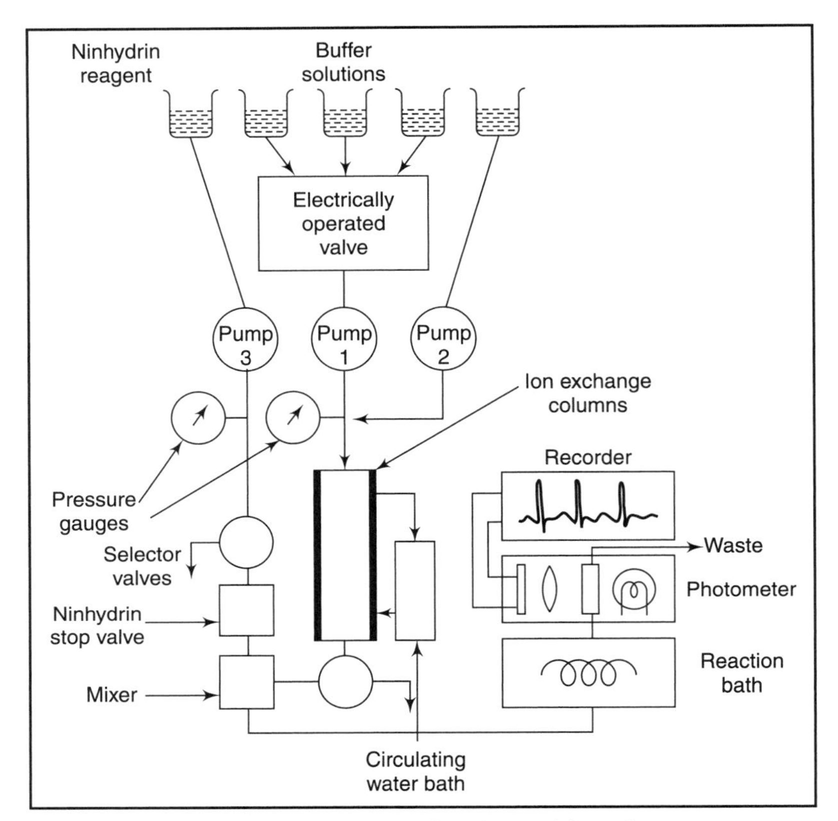

Figure 17.22 :: Block diagram of amino-acid analyzer

The sample containing 0.05 to 2.0 μmoles of each amino compound is introduced at the top of an ion exchange column. Buffer is supplied to the column from a reservoir by an accurate non-corrosive metering pump at a constant rate of 80.0 ml per hour.

Flow pumps are the most important of all the functional elements in the liquid chromatographic system. Unless pumps that are capable of producing a constant liquid flow are provided, any attempt to perform a quantitative or qualitative substance is impossible. Highest mechanical perfection is essential in order to attenuate all adverse effects due to changes in the kind of liquid pumped, its viscosity and back pressure, etc.

The pump used is a double-action, piston-driven one, in which each stroke performs both the intake and discharge functions simultaneously, and in which a valve is switched in synchronization with the piston movement, so as to achieve a constant liquid flow. The flow rate is varied by the replacement of a set of gears. Before the buffers enter the pumps, trapped air is eliminated with a bubble-trap type de-aerator to ensure constant volume delivery by the pump, and to prevent release of air in the resin bed of the ion-exchange column. All columns are enclosed in thermostated water jacket.

The purpose of the circulating bath is to keep the column temperature at a constant level. For this purpose, the water inside the bath, which is controlled to a fixed temperature, is circulated around the column by a circulating pump. The bath is equipped with a 400 W heater, which switches ON and OFF under the control and action of a temperature-sensing electronic circuit. Another heater of 500 W is provided to raise the temperature faster than usual.

The circulating water is pumped at a flow rate of about 2 lit./min. The bath is equipped with a cooling coil, which is used in case water temperature lower than the room temperature is required. The complete analysis proceeds unattended after the sample is applied and initial adjustments are made.

As the buffer and sample are pumped through a column, the amino-compounds in the mixture separate according to the difference in the affinity that they have for the adsorbent. The resolved amino-compounds emerge from the bottom of the column and flow through the capillary tubing to a column selector manifold. Here the column effluent is mixed with a ninhydrin reagent supplied by a pump at a constant rate (typically 40.0 ml/h). The mixture then flows through a reaction coil of Teflon capillary tubing contained in a reaction bath maintained at a temperature of 100°C. The reaction coil is volumetrically calibrated so that the time required for a given portion of solution to pass through the reaction coil is optimum for complete reaction between the amino-acids and ninhydrin reagent to produce colours. The stream then passes through a three unit photometer, where the absorbance values at wavelengths 440 nm and 570 nm are measured for each reaction product. Two of the three photometer units have different absorption path lengths at 570 nm, enabling a wide range of colour concentrations to be measured.

Alternatively, a single photometer is equipped with three interference filters, which are mounted, 120° apart, on a revolving wheel. They are switched in cycles by synchronizing signals generated by the recorder. A special wheel is used to turn the filter revolver by 120°, as a servomotor which drives the cam makes one complete turn. The time required for the motor to make one complete turn is about 0.5 s. The absorbance flow cell has a standard thickness of 2 mm. This is a continuous flow type cell and the liquid flows upwards from the bottom to the top.

A tungsten lamp having a rated capacity of 10 V and 3A is used as a source and is operated to emit a constant light output by using a current regulating device. The photoelectric element is a selenium cell, whose output goes to the recorder after due amplification. The absorbance values are measured by sensitive photodetectors, which transmit signals to the strip chart recorder. The different peaks correspond to different amplitudes. The base line is established at zero absorbance on the recorder chart, when a blank buffer ninhydrin solution is passed through the photometer. The recorder used is a self-balancing type potentiometric recorder, having a chart width of 12 inch, which graphically records the electrical outputs of the photometer for each wavelength. The recording sequence associated with the wavelength is generally printed out in a different colour. In addition, the recorder has a marker pen, which automatically marks the chart whenever a new set is advanced to the filling position. The entire analytical process is automated with the aid of a programming device and a complete array of electrical controls, which automatically safeguard the operation day and night.

Chapter 18

Thermo-analytical Instruments

18.1 Thermo-analytical Methods

A number of analytical methods have been developed, in which some property of the system is measured as a function of the temperature. In fact, the data are obtained as continuously recorded curves, which may be termed as thermal spectra. These spectra characterize a system; single or multi-component, in terms of the temperature dependence of its thermo-dynamic properties and physico-chemical reaction kinetics. Some of the commonly used thermo-analytical methods and the property used in them are as follows:

Thermal spectra

Method	**Property Used**
Thermogravimetric analysis (TGA)	Change in weight
Derivative thermogravimetric (DTG) analysis	Rate of change of weight
Differential thermal analysis (DTA)	Heat evolved or absorbed
Differential Scanning Calorimetric (DSC) Analysis	Heat evolved or absorbed
Thermometric titration	Change of temeprature

This chapter gives details of the most important thermo-analytical techniques, namely TGA, DTA and DSC.

18.2 Thermogravimetric Analysis (TGA)

Thermogravimetric analysis involves the measurement of the weight of a sample under investigation as the temperature is increased at a pre-determined rate. The sample may either lose weight to the atmosphere or gain weight by reaction with the atmosphere. There are two techniques in TGA. These are:

- Dynamic thermogravimetry, in which the sample is subjected to continuous temperature changes, usually linear with time, and
- Static thermogravimetry, wherein the sample is maintained at a constant temperature for a period of time during which any changes in weight are recorded.

The thermogravimetric analysis record is generally in the form of an integral curve, with absolute weight (W) as the Y-axis and time (t) or temperature (T) as the X-axis.

Figure 18.1 shows the typical thermogravimetric curves. The weight of a precipitate of silver chromate, placed in a crucible is shown. The initial drop in weight with temperature is due to loss of excess wash water. Just above 92°C, the weight becomes constant and continues to be so till about 812°C. A further increase in temperature to 945°C results in the loss of weight due to release of oxygen according to the reaction:

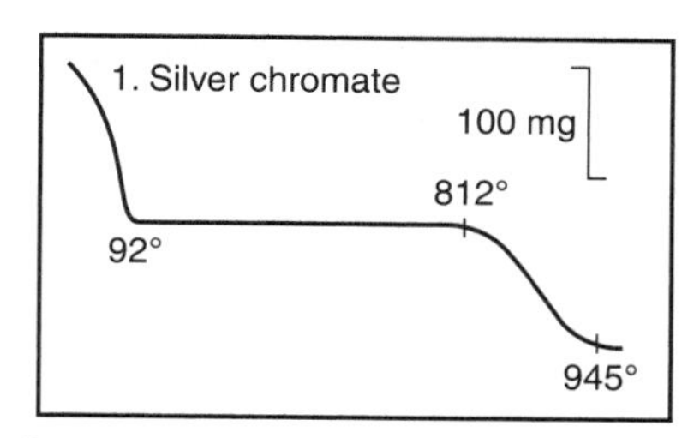

Figure 18.1 :: A typical thermogravimetric curve

$$2Ag_2CrO_4 \rightarrow 2O_2 + 2Ag + Ag_2Cr_2O_4$$

The residue is a mixture of silver and silver chromite.

The thermogravimetric analysis technique usually finds application in testing materials that are actual or potential analytical standards, and, of course, in the direct application of the technique to analytical determinations, particularly to evaluate kinetic parameters of weight changes in reactions.

It is important to note that the shape of the thermogravimetric curve is influenced by several factors, the most important among them being the following:

- *Heating Rate:* At any given temperature, the extent of decomposition is greater at a slow rate of heating than for a similar sample at a fast rate of heating. When successive reactions are expected to take place, the rate of heating will determine whether or not these reactions will be separated.
- *Sample:* Equilibrium between the sample and the product gas and with the furnace temperature will depend upon the bulk of the material in the sample cup. Better results are observed if a small and finely divided sample is used.
- *Atmosphere:* The composition of the atmosphere immediately surrounding the reacting particles influences the thermogravimetric curve.

While all these factors influence this analytical technique, the temperatures of salient features on the curves are somewhat different, as observed on different instruments, or on the same instrument at different rates of temperature scanning, or with different size samples, etc.

18.2.1 Instrumentation

The apparatus for thermogravimetric analysis comprises:

- Precision balance,

- Heating arrangement—a furnace programmed for a linear rise of temperature with time,
- Temperature measurement system, and
- Recorder.

Balance: If very large samples are to be avoided, precision and sensitive balances must be employed to detect the sample weight change. Balances may be either the null-point or deflection type instruments. In null-type instruments, any deviation of the balance beam is detected by a suitable sensing element. The beam is subsequently returned to the original position by the application of a restoring force, which would be proportional to the change in weight. This restoring force is recorded either directly or through a transducer. Deflection type instruments transfer the deviations in beam position into a record of the weight change. These are based on conventional analytical balance, a helical spring, a cantilever beam, a strain gauge or a torsion balance.

Several innovative arrangements have been employed to follow weight change, continuously and precisely. However, most modern systems make use of a torque motor. This is basically a galvanometer with the sample attached to the needle. The weight of the sample is proportional to the current required to restore the needle to some null position. Null is detected by a pair of photocells behind a vane in a light beam.

Thermobalances

Thermobalances used in TGA have a provision for the recording of weight. Those of the non-recording type are intended mainly for the determination of superficial moisture in bulk materials, wherein the sample is heated by an infrared lamp while on the pan of the balance, specially designed to minimize error produced by air currents.

Heating arrangement is provided by a furnace, which should maintain either a linear heating by program (10-600°C/h) or a fixed temperature. Control is usually achieved by using a thermocouple or resistance thermometer placed as close to the furnace winding as possible. The heating material is selected depending upon the temperature required. For example, nichrome winding gives a maximum temperature of around 1100°C, platinum-rhodium up to 1450°C, and a graphite tube furnace is used for higher temperatures. At higher temperatures, an inert gas blanket is usually employed.

Temperature measurement of the sample is perhaps one of the most ticklish problems in TGA. For example, during the reaction of a sample with the balance atmosphere, the temperature of the surface of the sample is important. The temperature of the gas around the surface of the sample is not important to the reaction kinetics. On the other hand, the temperature of the interior of the sample is important during auto-decomposition. Therefore, it is essential to arrive at a compromise for temperature measurement problems with the rest of the balance system.

Thermocouples are usually employed as transducers for temperature measurement. They can be small enough to be located within the sample. In some cases, use has also been made of bolometers and optical pyrometers for temperature measurement. The Perkin Elmer thermobalance makes use of the changes in furnace winding resistance to measure furnace temperature.

The recording system should be able to record both temperature and weight continuously and to make a periodic record of the time.

18.3 Differential Thermal Analysis (DTA)

Differential thermal analysis is the measurement of the difference in temperature between a sample and a reference as heat is applied to the system. It is a fingerprinting technique which provides information on the chemical reactions, phase transformations and structural changes that occur in a sample during a heat-up or a cool-down cycle. The DTA measures the differences in the energies released or absorbed, and the changes in heat capacity of materials as a function of temperature. All materials behave in certain, predictable ways when exposed to certain temperatures, so that the resulting DTA curve is an indication of the materials and phases present in the sample. For example, the DTA is used to indicate the relative magnitude of reactions and phase transitions of ceramic materials or batches that can be destructive so that safe drying and firing schedules can be determined. The DTA identifies the temperature regions and the magnitude of critical events during a drying or firing process.

DTA basically involves heating or cooling a sample and a reference material in close proximity at a linear heating rate, monitoring and recording the temperature of the furnace, and the difference in temperature between the sample and reference. The differential thermogram consisting of a record of the difference in sample and reference temperature (ΔT) plotted as a function of time (t), sample temperature (T_s), reference temperature (T_r) and furnace temperature (T_f) provide valuable information on phase transitions and chemical reactions. The DTA technique is especially suited to studies of structural changes within a solid at elevated temperatures, where few other methods are available.

Differential thermogram

An endothermic process will cause the thermocouple junction in the sample to lag behind the junction in the reference, and hence develop a voltage, whereas an exothermic process will generate a voltage of opposite sign. Conventionally, the exotherms are plotted upwards, whereas the endotherms are plotted downwards. Figure 18.2 shows a typical DTA curve for the decomposition of calcium oxalate monohydrate in air and in carbon dioxide.

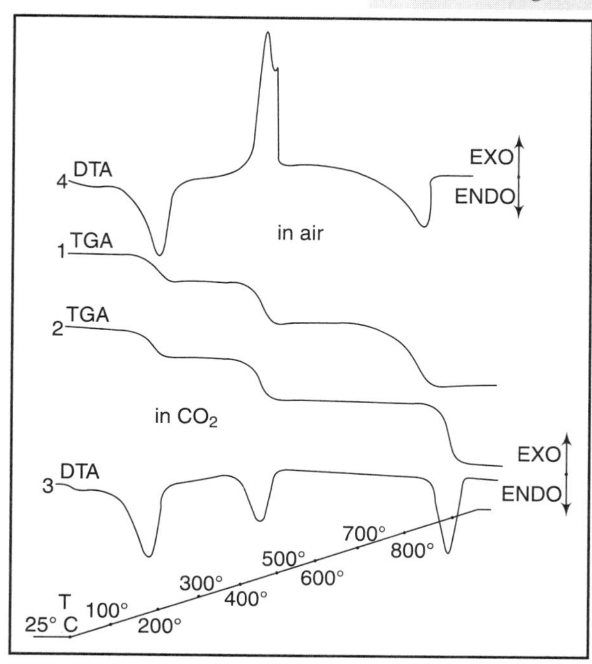

Figure 18.2 :: Simultaneous DTA-TGA oxalate monohydrate in air and in carbon dioxide

18.3.1 Instrumentation

A modern thermal analyzer instrument (Figure 18.3) consists of a furnace for heating (or cooling) the sample at a controlled rate and a selective transducer to monitor changes in the substance. The transducer can be a thermocouple to measure temperature changes (heat flow), a balance to monitor weight changes, or a linear variable differential transducer (LVDT) to detect changes in dimensions. The transducer generates a voltage signal which is amplified, stored on a disk along with a direct temperature response from the sample and recorded on a printer or plotter.

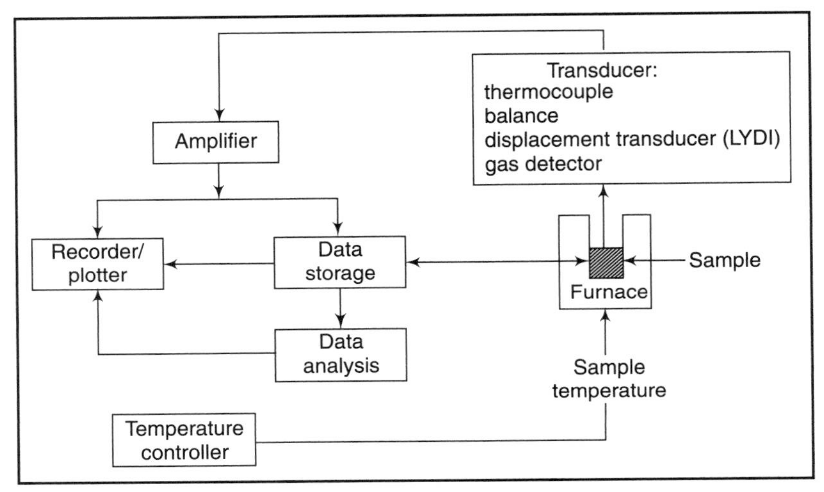

Figure 18.3 :: Schematic of a modern thermal analyzer instrument (Gill, 1984)

DTA apparatus is available from several companies; their products differ with regard to such parameters as sample, size, temperature range, selectable scanning rates and precision, etc. However, the following are the five basic components in all of them:

- Sample cell with temperature detector,
- Furnace assembly,
- Temperature programmer,
- Amplifier and recorder, and
- Atmosphere control.

Figure 18.4 shows a schematic diagram of a typical DTA apparatus.

The sample cell is just a disposable tube, about 2 mm in diameter which can contain 0.1–10 mg of sample. A similar tube carries the reference material such as alumina or quartz sand. The two tubes are inserted into a metallic or ceramic block. Very thin thermocouples are inserted in the sample and reference tube. In the low temperature range (–100 to 600°C), the thermocouples used

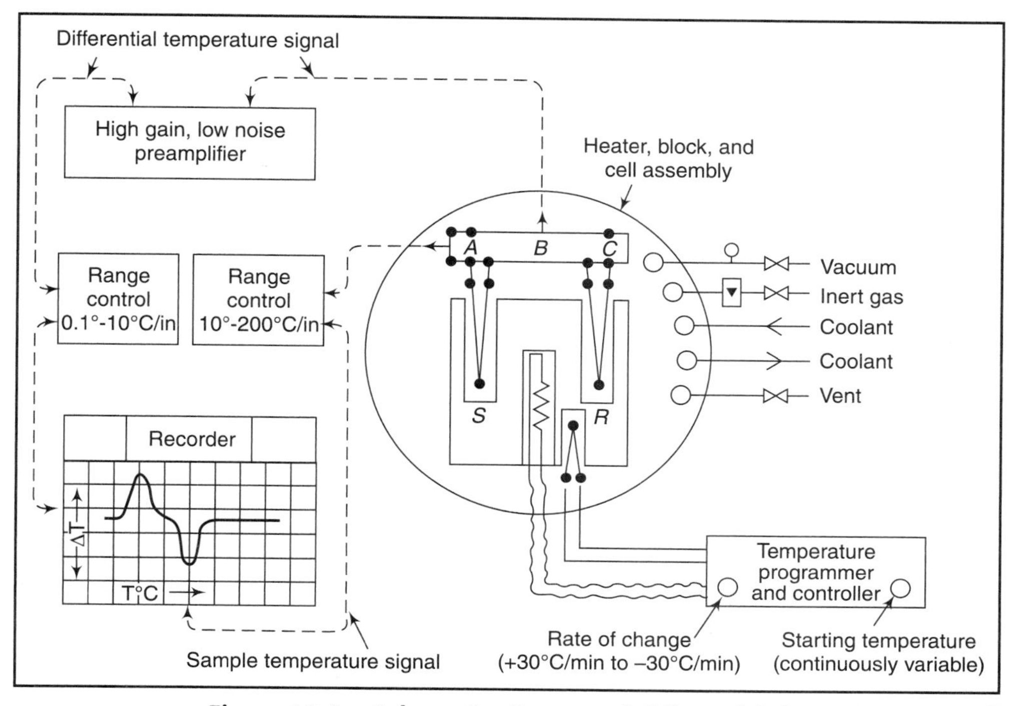

Figure 18.4 :: Schematic diagram of differential thermal analyzer

are copper–constantan, iron–constantan and chromel–alumel. For higher temperatures, Platinum/platinum—10 per cent rhodium thermocouples are necessary. The use of other detectors such as thermistors and platinum-resistance thermometers have also been made in some commercial instruments.

The furnace used is generally of the high-temperature, pressure-vacuum electric type. Sample temperatures up to 500°C are usual, while the maximum temperatures used are around 1000°C. Relatively, small sample volumes minimize thermal gradients and make evacuation easy.

Temperature programmer

A temperature programmer must provide a smooth and oscillation-free heating or cooling rate, which should be reproducible to ±0.1°C or better. Heating or cooling rates usually provided on the instruments are 0–30°C/min. The simplest temperature programmer is a variable transformer, in which various programmers consisting of motors attached to the variac shaft are incorporated. However, modern DTA instruments incorporate solid state electronic temperature controllers. In these controllers, the signal from the thermocouple in the furnace is compared electronically against a reference potential, which can be programmed to correspond to a variety of heating modes and heating rates.

Amplifiers are used for suitable amplification of the dc signal obtained from the thermocouple. The amplifier is necessarily a high gain, low noise and high input impedance circuit. The amplification before the signal is given to the recorder is about 10,000. The recorder should be dual-pen strip chart recorder, though some systems use two recorders. Usually ΔT (change in temperature) is recorded as Y-axis whereas T (furnace temperature) is the X-axis. The recorder sensitivity is usually 1 to 0.1 mV/inch and a time constant of 1 s or less.

For some types of samples, the atmosphere must be controlled either to suppress an undesirable reaction such as oxidation, or to learn the nature of a reaction, as for example, by varying the pressure of a gaseous reaction period. Four types of systems are in current use for control of the atmosphere: (i) inert gas purge for the protection of the unencapsulated sample, (ii) inert gas covering within the encapsulation, (iii) inert or reactive gas for thermodynamic and reaction kinetic studies, and (iv) flowing inert or reactive gas for study of reaction kinetics.

The cooling system is usually separated from the temperature programmer in most of the commercial instruments. The most satisfactory arrangement for cooling is with a gas as the heat exchange medium.

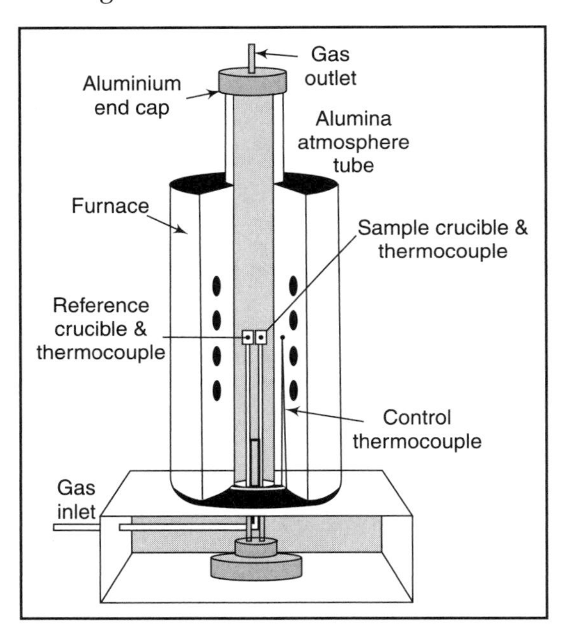

Figure 18.5 :: Differential Thermal Analysis System (Courtesy M/s Orton, USA)

Figure 18.5 shows the physical layout of a DTA apparatus from M/s Orton, USA. It uses a pair of ceramic cups that are supported by a pair of vertical thermocouples (differential thermocouple) positioned on the module base. After both cups are placed on top of the differential thermocouple stalks, the furnace is lowered over the cups, and the furnace is heated and cooled according to the programmed thermal cycle. The differential thermocouple output, i.e. DTA signal in micro-volts, is displayed on a PC monitor, and stored on the PC hard drive as a function of time and temperature for post-testing analysis.

Temperature control is effected with a user-programmable PID controller which is used to control the thermal cycle of the furnace. The system includes a Windows 98/2000 compatible data acquisition/analysis software and an analog to digital interface card for the personal computer system. The software displays the test progress on the monitor, stores the data and enables the user to perform the standard DTA analyses on the data after the test is completed.

18.4 Differential Scanning Calorimetry

DTA is capable of giving good qualitative data about the temperatures and indications of transitions, but it is difficult to obtain quantitative data, i.e. the heat of transition if the purity is known, or the amount of constituent in a sample if the heat of transition is known. This could be due to certain unknown and uncontrollable factors, such as the specific heat and thermal conductivity of the sample before and after the transition. Several other parameters such as the rate of heating, placement of thermocouple and other instrumental parameters also create problems in carrying out quantitative analysis. These difficulties have been solved by using another closely related technique to DTA, which is called differential scanning calorimetry (DSC).

Differential scanning calorimetry is a thermal analysis technique, which is used to measure the temperature and heat flows associated with transitions in materials as a function of time and temperature. Such measurements provide qualitative and quantitative information about the physical and chemical changes that involve endothermic and exothermic processes or changes in heat capacity on DSC technique usually cover the range from 60° to 1600°C with variable atmospheres.

In this technique, the sample and reference are mounted on two separate small heaters. The temperatures of the two cells are monitored by platinum-resistance thermometers, the two heater windings are supplied with current, so that both windings heat or cool at the same rate, which would be up to 80°C/min. The power difference required for two heaters is measured (ΔP) and recorded as a function of the program temperature. This record shows the physical or chemical transitions in the sample and the power difference is equal to the thermal energy absorbed or released during the transition.

DSC differs fundamentally from DTA in that the sample and reference are both maintained at the temperature pre-determined by the program even during a thermal event in the sample. The amount of energy which has to be supplied to or withdrawn from the sample to maintain a zero temperature differential between the sample and the reference, is the experimental parameter displayed as the ordinate of the thermal analysis curve. The sample and reference are placed in identical environments, metal pans on individual bases each of which contain a platinum resistance thermometer (or thermocouple) and a heater (Figure 18.6). The temperatures of the two

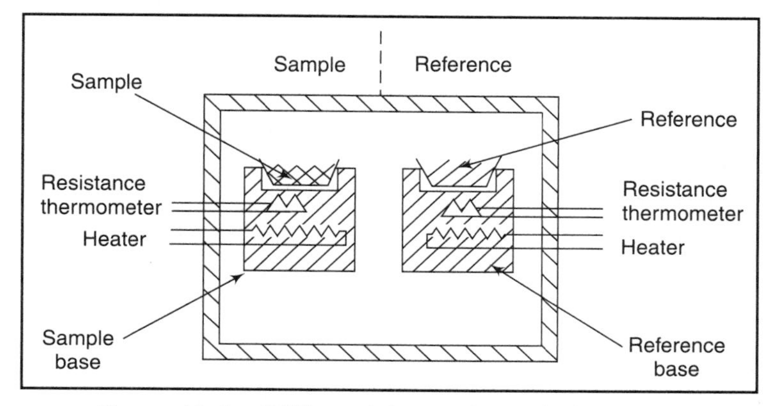

Figure 18.6 :: Differential scanning calorimetry set-up

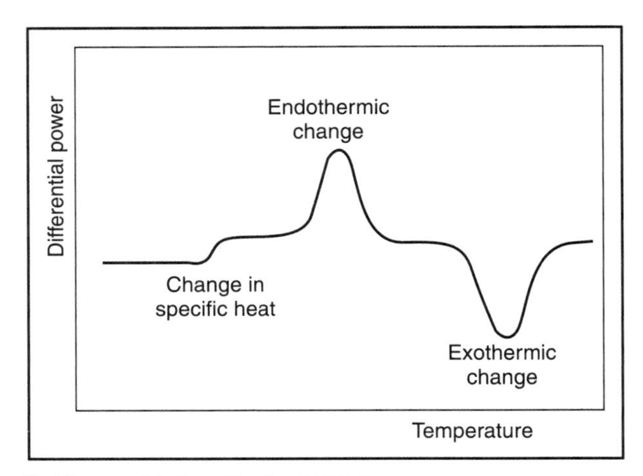

Figure 18.7 :: Typical DSC curve

thermometers are compared, and the electrical power supplied to each heater are adjusted so that the temperatures of both the sample and the reference remain equal to the programmed temperature, i.e. any temperature difference which would result from a thermal event in the sample is 'nulled'. The ordinate signal, the rate of energy absorption by the sample (e.g. millicalories/sec.), is proportional to the specific heat of the sample since the specific heat at any temperature determines the amount of thermal energy necessary to change the sample temperature by a given amount. Any transition accompanied by a change in specific heat produces a discontinuity in the power signal, and exothermic or endothermic enthalpy changes give peaks whose areas are proportional to the total enthalpy change (Figure 18.7).

In other words, in DSC, the measuring principle is to compare the rate of heat flow to the sample and to an inert material, which are heated or cooled at the same rate. Changes in the sample that is associated with absorption or evolution of heat cause a change in the differential heat flow, which is then recorded as a peak. The area under the peak is directly proportional to the enthalpic change and its direction indicates whether the thermal event is endothermic or exothermic. Donovan (1984) discusses the use of scanning calorimetry for biological studies.

18.5 Simultaneous Thermal Analysis/Mass Spectrometer

Simultaneous thermal analysis (STA) techniques comprise both differential thermal analysis (DTA) and thermogravimetry (TG). The equipment operates in a manner similar to the differential scanning calorimeter (DSC). Two sample crucibles are heated or cooled at a precisely controlled rate in a controlled environment. One crucible contains a standard of known thermal response; the unknown is placed in the second crucible. Differences in the thermal behaviour of the two materials caused by differences in specific heat, occurrence of an exothermic or endothermic reaction, or a phase change result in a temperature difference between the two crucibles.

Temperature differences are measured with a Pt. vs. Pt-10Rh differential thermocouple that enables properties of the unknown to be determined relative to that of the standard. Simultaneously, any change in mass of the specimen during a heating cycle can be measured with a microbalance as a function of temperature. A mass spectrometer (MS) is attached to the STA instrument for evolved gas analysis. During the thermal cycle, a capillary leak allows samples of the gaseous environment in the immediate vicinity of the specimen to be drawn into a quadruple mass spectrometer. Evolved gases can thus be identified by the mass-charge ratios of molecules, fragments, or atoms in the gas. Figure 18.8 shows the schematic diagram of STA/MS system.

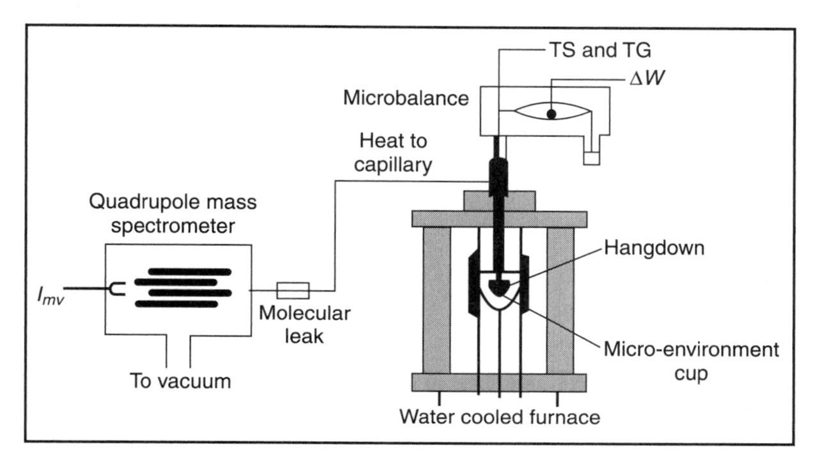

Figure 18.8 :: Schematic diagram of simultaneous thermal and mass analyzer (*Courtesy: www.ms.ornl.gov/htmlhome/tpuc/sta.html*)

- The arrangement can be used to make DTA and TG measurements from 25 to 1500°C with mass change determinations to +/– 0.01 mg.
- Simultaneous quadruple mass spectrometric (MS) facilitates analysis of evolved gases with sensitivity in ppm level and mass range to 600 amu.

STA can be used to follow the course of chemical reactions, thermal decompositions or phase changes as a function of temperature. The sensitive balance associated with the TG capability of the system allows the mass change of a specimen to be measured as a function of temperature. Simultaneous DTA, TG and mass spectrometer measurements provide information about the cause of the mass changes.

Chapter 19

Electrophoresis Apparatus and Densitometers

19.1 Electrophoresis

Electrophoresis is an old established method of analytical chemistry. It is based on the principle that the individual components of the colloidal solution migrate in a liquid at different speeds when subjected to an electric field. Separations are possible because particles of similar geometry but different charge, and particles of like charge but different geometry migrate at different rates towards an oppositely charged electrode. Therefore, when the current is passed for a certain time through such a solution, various components present in the solution would move through different distances in their effort to migrate towards the electrodes. Therefore, a substance which may be a mixture is thus separated into its components along the migration distance, according to a definite law. Measurement of the concentration along this migration distance would therefore, provide the quantitative result of the analysis.

Accordingly, electrophoresis is a separation technique that is based on the mobility of ions in an electric field. Positively charged ions migrate towards a negative electrode and negatively charged ions migrate toward a positive electrode. For safety reasons, one electrode is usually at ground and the other is biased positively or negatively. Ions have different migration rates depending upon their total charge, size and shape, and can therefore be separated.

Historically, some of the earliest reports described characteristic electrophoretic mobilities of bio-colloids such as proteins or enzymes. However, the technique received little note until 1937, when Tiselius published a paper introducing the moving boundary concept. The moving boundary method utilizes the migration of particles in free solution and observation of the various molecular boundaries through sensitive refractometric techniques. With this, the value of electrophoresis in obtaining distinct and measurable fractions of a variety of substances got well-established, particularly in clinical laboratories.

Moving boundary concept

Basically, the electrophoresis technique separates the molecules based on the size and charge under the influence of an electric field. If E is the strength of the electrical field, Z is the charge on the molecule and F is the frictional force on the molecule, then V the velocity of migration is given by

$$V = \frac{EZ}{F}$$

The frictional force can be defined as

$$F = 6\,\pi\eta r,$$

where η is the viscosity of the medium and 'r' is the stokes radius of the molecule. Therefore,

$$V = \frac{EZ}{6\,\pi\eta r}$$

This implies that the electrophoretic mobility is proportional to the charge on the molecule and inversely proportional to the radius of the molecule, i.e. larger the radius translates to lower electrophoretic mobility.

Normally, with the moving boundary method, only two components of a mixture, one with the highest mobility and the other with the lowest mobility, can be separated in pure form. If it is desired to recover components other than those of the highest and lowest mobilities in pure form, multiple separations have to be carried out.

19.2 Electrophoresis Techniques

Various electrophoresis techniques have become popular for determining the relative amount of components in certain body fluids. The commonly used techniques are:

- *Paper electrophoresis*, in which the separation of different particles occurs on a piece of filter paper that is saturated with an electrolyte.
- *Cellulose acetate electrophoresis*, which makes use of cellulose acetate filter, that offers advantages such as sharp and quicker separations and colourless background. This technique has almost replaced the filter paper techniques.
- *Gel electrophoresis*, which involves the use of agar gels for immuno-techniques, while starch and polyacrylamide gels exhibit the peculiar property of acting as molecular sieves, as opposed to other media, which generally act as inert supports for buffers. This results in better resolving power in the gel electrophoresis technique.
- *Micro-immuno electrophoresis*, which is a combination of two techniques: electrophoretic and immuno-diffusion. It has been developed to overcome the difficulty experienced in separating and identifying highly complex protein mixtures by other chemical or physical means.
- *Thin layer electrophoresis*, which employs thin layer materials such as silica, gel and aluminas as electrophoresis media. This technique offers a number of advantages in many types of separations such as amines, amino-acids, phenols, etc.

- *Cooled platen electrophoresis*, which is designed for those applications, wherein it is essential to keep temperature as low as possible, to avoid denaturation of the sample and drying out of the support medium. The technique is particularly useful in the separation of enzymes, which can be severely affected by higher temperature. Constructed in anodized aluminum, it incorporates an effective water circulation system.

19.3 Paper Electrophoresis

The moving boundary method has a disadvantage in that complete separation of the components of a mixture is never achieved. There is always a dead volume of the original solution at the bottom of the U-tube, and only the fastest and slowest moving components may be obtained in one operation. This limitation is due to the very manner in which boundaries are stabilized, namely by gravity. The density of the solution must increase downward at every boundary, or else there will be some convection tending to destroy the boundaries.

If electrophoresis is carried out in columns packed with powders, gels, filter paper or on a piece of paper, the undesirable effects of gravitational convection may be eliminated, as the supporting medium itself would stabilize the various boundaries. A complete separation of the various components leads to the formation of separate and distinct zones on the supporting medium. Here all that is required is to apply a potential difference to the ends of a strip of paper or a column of gel impregnated with some suitable buffer solution. If a drop of the solution to be analyzed is put on it, the differently charged particles will move with different velocities. The migration process can be followed and the velocities measured by suitable means.

Supporting medium

Of all the media presently used for supporting electrophoresis, the most practicable and widely used is the filter paper. The basic principle of the method using paper as the supporting medium is the same as that of the moving boundary method. Under the action of the electric field, the charged molecules migrate through the capillaries of the paper, just as they might through an unbounded solvent. However, the advantage of the method is that it is possible to obtain a complete separation into zones of different migration and not as boundary separations of overlapping zones in the liquid phase. The separation zones are located by applying various active reagents.

Paper electrophoresis has developed into a generally applied and valuable routine clinical method. Its main advantage is the small amount of substance required, which may be 1 mg of albumen per analysis and on the smaller outlay on apparatus and technology to carry out a larger number of electrophoretic analyses simultaneously.

During the past few years, several new approaches have been tried in the field of electrophoresis. Among these are the use of hydrolyzed starch and polyacrylamide as molecular sieving type gels, which yield numerous fractions from serum and other heterogeneous proteins. A great deal of interest has also been evinced in the matter of supports, such as cellulose acetate, but they are also subject to many of the same deficiencies as the filter paper methods.

19.3.1 Methods of Zone Localization

The procedure for localizing the zones, after an electrophoretic separation on paper has been achieved, plays an extremely important role in the detection and quantitative estimation of the components. The simplest and the most convenient method of detection and estimation is a colour reaction, and therefore the main problem then remains to find a suitable reagent. A large number of reagents for colour reactions have been proposed. For example, amino-acids are detected by the use of ninhydrin, either in alcohol or acetone solution. For visualizing the amino-acids, the paper at the end of electrophoresis is first dried and dipped into the reagent. The coloration appears after drying and develops slowly. The colour may then be preserved by fixing the paper in copper nitrate in alcohol solution, which has been acidified by the addition of nitric acid. The most commonly used reagents for proteins are dyes. Some of the commonly used dyes are bromopheno blue, naphthalene blue-black and azocarmine G.

19.3.2 Quantitative Considerations

After staining or dyeing, the next step is to measure the quantity of the dye bound. This may be done by elution or by photometric scanning of the dyed strip.

In the elution method

In the elution method, the paper is cut into thin sections, say 5 mm strips, and the substances are eluted from them. The technique is accurate, but the difficulty lies in deciding the number of sections that the strip should be cut into obtain good results. After the sections are cut, they are eluted for about 15 to 30 mm with 0.01 m caustic soda in dilute alcohol solution. The amount of dye present in each section is then measured in a spectrophotometer, in terms of optical density.

Direct photometer recording of transmitted light through the strip is based on the Beer-Lambart Law. This law applies only to a clear, transparent, homogeneous and isotropic medium. In practice, these conditions are not generally fulfilled by coloured filter paper and therefore, the measurements are confined to a limited range of the calibration curve, which approximates closely to a straight line.

Since optical density is a function of the wavelength of the incident light, measurement is approximately made at the absorption maximum of the dye stuff by using colour filters. The recording can be more conveniently made with reflected incident light. In that case, the strips do not need to be transparent. This method is based on the relation between the colour intensity and the intensity of the light reflected from the illuminated paper. Measurement by reflected incident light is very easy and produces curves free from distortion.

After the optical density is measured photometrically, the readings are plotted against the distance of the paper section from the origin. The curve obtained expresses the variation in colour density along the paper strip and hence, the variation in protein concentration.

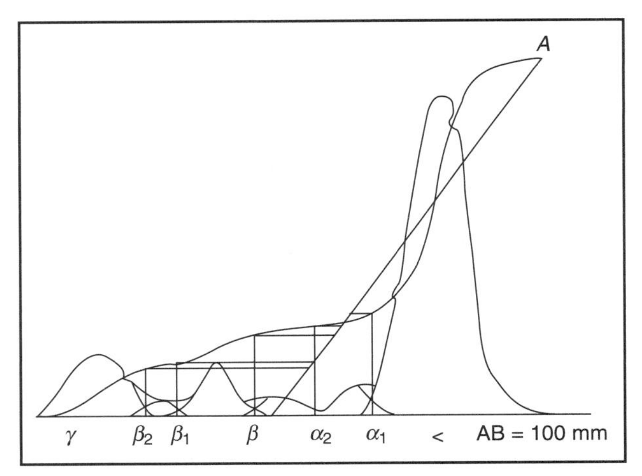

Figure 19.1 :: Extrapolation of the optical density curve according to the Gaussian distribution technique

19.3.3 Evaluation of the Curves

The area under the peaks of an experimental extinction curve obtained may be easily measured on a planimeter, or the curve is extrapolated according to the Gaussian distribution curves. Vertical lines up to the integral curve are drawn (Figure 19.1) through the points of intersection of the curves and from these lines, parallel to the zero lines are drawn. A ruler of length 100 mm should be placed with one end at the highest point (*A*) of the integral curve and its other end on the zero line (*B*). Now it is possible to read from the ruler the percentage proportions of the individual components of the mixture, as differences between the points of intersection with the parallel lines.

An error is always present, when we try to determine the concentration of the components from the area under each of the peaks. This is because all the proteins may not bind the dye in the same proportion. For exact determinations of protein concentration, it is necessary to multiply the photometric value for a particular protein by an affinity factor of the dye used. This factor is usually obtained experimentally, by measurements of the colour density on highly purified protein solutions of known concentration.

Affinity factor

19.4 Electrophoresis Apparatus

An electrophoresis apparatus consists of a high-voltage supply, electrodes, buffer, and a support for the buffer such as filter paper, cellulose acetate strips, polyacrylamide gel [Figure 19.2(a)], or a capillary tube. Open capillary tubes are used for many types of samples and the other supports are usually used for biological samples such as protein mixtures or DNA fragments. After a separation is completed, the support is stained to visualize the separated components. A complete electrophoresis apparatus [Figure 19.2(b)] comprises:

- Electrophoresis cabinet,
- Power supply, and
- Densitometer or scanner.

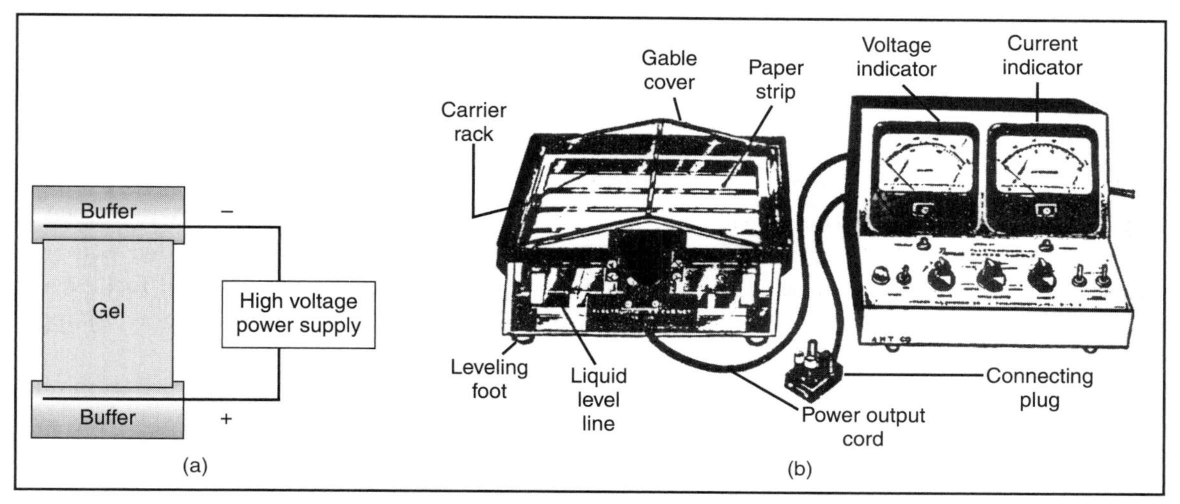

Figure 19.2 :: (a) Gel electrophoresis set-up (b) Paper electrophoresis apparatus (Courtesy, M/s Arther H. Thomas, USA)

19.4.1 Electrophoresis Cabinet

The electrophoresis cabinet consists of a methacrylate plastic cabinet and gable cover, and a carrier rack of phenolic plastic. The gable cover prevents condensation droplets from falling on paper strips. The carrier supplied for use with paper or cellulose strips up to a total width of 18 cm is suitable for supporting paper strips. End surfaces of the carrier are roughened to grip wet paper.

Some electrophoresis baths are designed to take up to six strips, with each being 34 cm long and 5 cm wide. The strips are supported in pairs on three removable plastic bridges, which are provided with a number of supports to carry the paper and to prevent undue sagging. At each end of the bath, there are two compartments to contain the buffer solution. The ends of the paper strips dip into the solution in the outer of these two compartments, whilst three platinum electrodes are mounted in the inner compartments. The electrodes are wired to a lead and polarized plug for connecting to the power unit.

The tank should not leak electrically or mechanically. In order to check electrical leakage, the buffer solution is poured into the tank without placing any filter strip. The electrodes are connected to a power supply at 500 V and a microammeter is connected in series with the supply. The current should not exceed 1 μA in a good cell. A tank with a higher leakage should be discarded or cemented at its inter-surface joint.

Isoelectric focusing

Resolution can be greatly improved by using isoelectric focusing. In this technique, the support gel maintains a pH gradient. As a protein migrates down the gel, it reaches a pH that is equal to its isoelectric point. At this pH, the protein is neutral and no longer migrates, i.e., it is focused into a sharp band on the gel.

19.4.2 Regulated Power Supply

Regulated power supply allows one to select either constant voltage (0–250 V, 0–500 V or 0–1000 V) or constant current (0–20 mA, 0–100 mA or 500 mA range). The supplies are provided with two meters, one for indicating current and the other for voltage. The polarity reversal facility and overload protection are provided in the power supplies. The ripple content in the power supply should be less than 0.1 per cent.

Power supplies designed specifically for gel electrophoresis offer extensive programming capabilities, including pre-set and customer preferred settings. They provide constant voltage, constant current or power setting facilities. Most of them are fully protected offering open circuit detection, short-circuit protection, thermal shutdown, over-voltage, current wattage detect, etc.

19.4.2.1 Constant Voltage Power Supply

A constant voltage power supply maintains a constant voltage across the load, irrespective of the current drawn by the load. Therefore, it should have output impedance as close to zero as possible.

Figure 19.3 shows a voltage regulated power supply. It basically consists of a conventional rectifier and filter circuit to convert ac from the mains to dc with low ripple content. The rectified dc is fed through a series regulator to the output terminals of the load. A reference circuit applies a voltage to one terminal of a comparison amplifier, equal to the desired output voltage. If the output voltage does not equal the reference voltage, the input voltage to the comparison amplifier

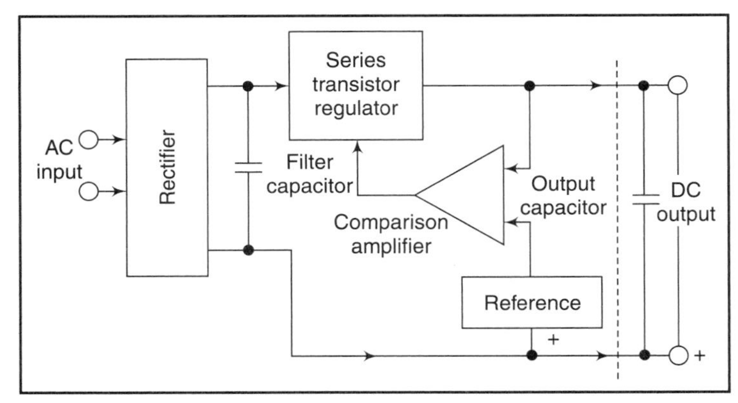

Figure 19.3 :: Block diagram of a constant-voltage power supply

is not zero and the amplified output of the comparison amplifier changes the conduction of the series regulator. This results in a change of current through the load resistor until the load voltage is equal to the desired output value.

19.4.2.2 Constant Current Power Supplies

An ideal constant current source is the one for which the current remains constant, regardless of the value of the output voltage demanded by the load. Since it is possible that the load resistance connected to a constant current power supply may vary, the ideal constant current source must have an infinite source impedance at all frequencies.

Figure 19.4 shows the block diagram of a constant-current regulated power supply. This supply resembles, in many respects, the block diagram of a constant voltage regulated power supply. However, instead of comparing the reference voltage with the output voltage, the comparison amplifier of a constant current power supply, compares the reference voltage with an IR drop caused by the output current flowing through a fixed resistor. The action of the feedback loop and consequently the conductance of the series regulating element is adjusted, so as to maintain the IR drop across the series monitoring resistor constant and equal to the reference voltage, and thereby holding the output current to some constant value as desired.

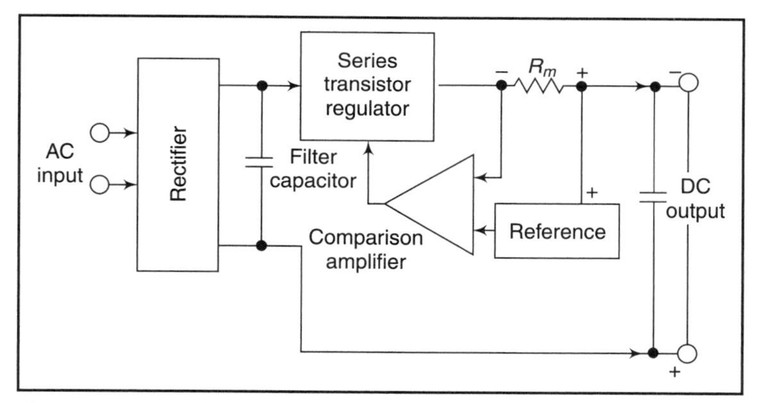

Figure 19.4 :: Block diagram of a constant-current power supply

It is desirable for many reasons to limit the maximum instantaneous current, which the series transistors can pass to some pre-determined value. The primary reason for doing so is to protect the series transistors themselves from damage due to excessive heating, and to be able to charge a large load capacitor without damaging the power supply. Consequently, a protection circuit is provided to many power supplies to limit the maximum output current under any load condition. This is achieved by having an automatic crossover between the two modes of operation, namely constant voltage and constant current.

19.4.2.3 Constant Voltage, Constant Current (CV/CC), Regulated Power Supply

Figure 19.5 shows the block diagram of CV/CC power supply. The diodes are connected in the circuit such that when the supply is in constant voltage operation, the upper diode is forward biased or shorted, while the lower diode is reverse biased or open. Conversely, when the supply is in constant current operation, the upper diode is reverse biased and the lower diode is forward biased. Thus, the series regulator is only called upon to respond to either the constant voltage comparison amplifier or the constant current amplifier. The effectiveness of one amplifier is not diluted by the shunt presence of the other. In this circuit, the same output terminals are used for both constant voltage and constant current operation. The supply delivers constant voltage with a continuously variable current limit, or of constant current operation with a continuously variable voltage limits.

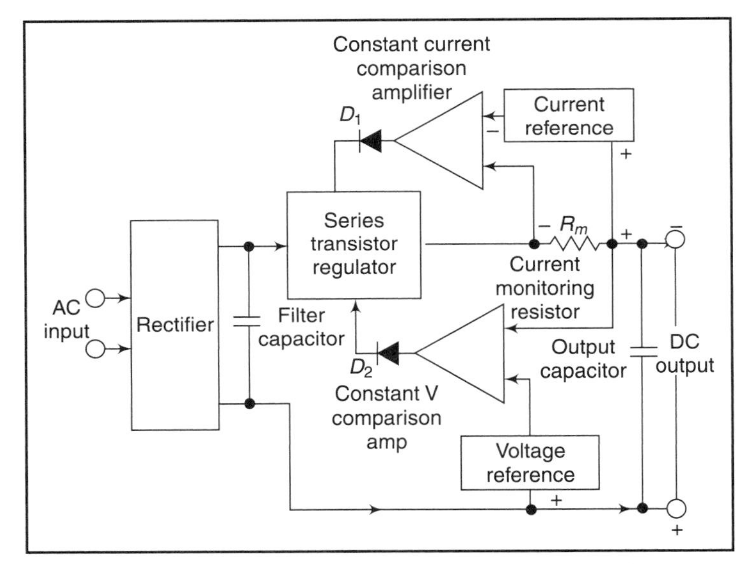

Figure 19.5 :: Block diagram of a constant-voltage/constant-current power supply

Whether the supply is in constant voltage or constant current operation at any instant, depends upon the relationship between the dc load resistance and the critical value of the load resistance, defined as the ratio of the front panel voltage control setting, to the front panel current control setting. If the load resistance is greater than the critical load resistance, the supply will be in constant voltage operation and if the load resistance is less than this critical load resistance, the supply will be in constant current operation.

19.4.2.4 High Voltage Electrophoresis

High voltage electrophoresis is a rapid and highly effective technique for the separation of a wide range of compounds, especially those with low molecular weights. It is invaluable for work with many biological materials, of which amines, amino-acids, indoles, peptides, sugars, organic acids, dansyl derivatives, are only a few.

The very high voltages necessary to take advantage of the technique give rise to problems like heating, which are overcome by a water cooling system or some other special technique. The high voltages normally employed are 10,000 V at 100 mA or 5000 V at 200 mA.

19.4.3 Densitometer

The quantitative estimates obtained by elution of the dye from 5 mm strips and measurement of the optical densities of the resulting solutions in a suitable absorptiometer offer a lengthy and laborious method and give only approximate results. The method has been simplified and made more accurate by the direct measurement of the optical density on the dyed paper. In this method, the stained electrophoretic strip is made transluscent by prolonged immersion in a 1:1 mixture of paraffin oil and a bromonaphthalene and then placed between two parallel glass plates and moved, before the slit of a photometer and the reading of the optical densities of the sections of strips are measured. While taking the readings, it is advisable to use a light filter in doing the colorimetry, since the wavelength of light used has a rather large effect on the readings obtained.

Scanners

A number of scanners are commercially available for making direct photometric recordings. All of them work more or less on the same principle. Figure 19.6 shows the typical arrangement. The stained strip is placed between two glass plates and made to move in front of a 1 mm slit. Light from a constant source after passing through the slit falls on the transluscent strip mounted on a slide. A photoelectric cell is placed in line with the light source and the slit, but on the other side of the paper having a suitable optical filter in front of it.

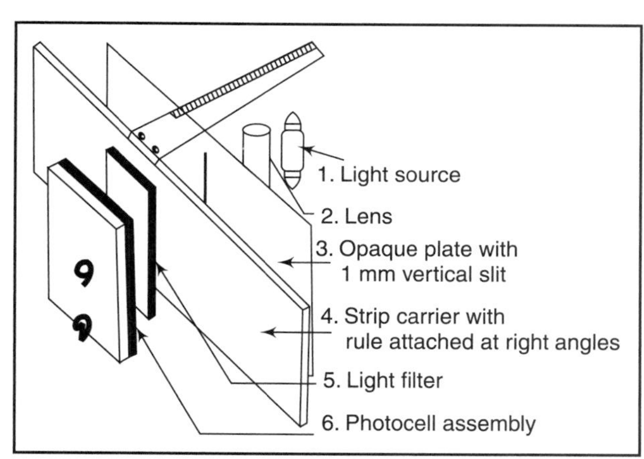

Figure 19.6 :: Arrangement of various components in a single-beam densitometer

The photocell produces a small but measurable electrical current in direct proportion to the intensity of light falling on it, which can be directly read from a microammeter.

At right angles to the paper holding slide is attached a transparent rule, suitably graduated so that readings from the microammeter may be plotted at required intervals. The operator only has to move the transparent rule and therefore, the slide, and at each movement, plot along the rule the microammeter reading, which records the intensity of dye at that point on the strip. By joining the points so obtained, a clear picture of the pattern of fractionation is immediately obtained.

When light from the source lamp passes through the blank portion of the oily filter paper, the meter is set to read zero corresponding to zero OD (optical density) and 100 per cent transmission. When the paper moves before the light path, the portion which is richest in protein concentration will offer maximum opacity to the light path and thus the photocell will receive minimum light corresponding to which the recorder will trace a large area curve on OD scale.

A serious problem with densitometers using a single photocell is their instability against line fluctuations. With the changes in the input voltage, the source lamp voltage will change, and affect the amount of light falling on the photocell. Hence, the interpretations of the concentrations from the graph will not be accurate. The stabilization of the source lamp can be achieved either by using a saturation core transformer or by incorporating electronic regular circuit. However, the preferred method is to have a double-beam densitometer having facilities for automatic recording. Figure 19.7 shows the arrangement of different components of such an instrument. One of the photocells acts as a reference photocell, which receives light directly from the source lamp through a variable slit. The other photocell, called a measuring photocell, receives light through a fixed slit after passing through the stained electrophoretic paper strip. The photocells are so connected electrically that the net output from them is the difference between the photovoltage of the two cells. A dc amplifier is used to amplify this difference to a level sufficient to drive a servo-recorder.

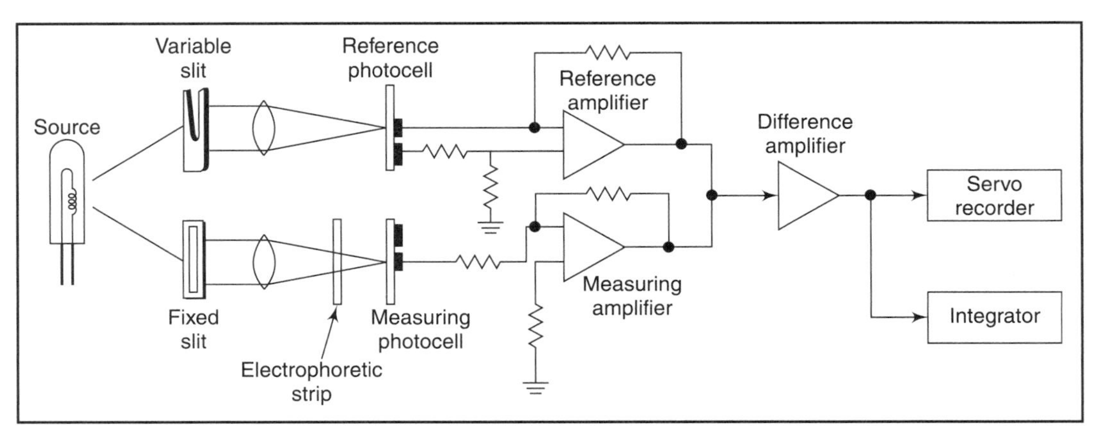

Figure 19.7 :: Optical arrangement of a double-beam densitometer

Initially, the zero optical density is set on the recorder or meter by allowing the light to pass through the transluscent portion of the stained paper placed in front of the measuring photocell.

Then, as the stained paper is made to move across the light source, the recorder will trace curves on the graph paper according to the concentrations of the protein samples. Any fluctuations in the lamp intensity are received simultaneously by both the photocells and the effect is automatically nullified. Some instruments include attachments like electronic integrators for measuring peak height, in which the results are available directly in concentration.

Densitometers are available to work in either transmittance or reflectance mode (Greenhalgh, 1983). However, identical samples will give different results when scanned in two modes. Again, the ability of the densitometer to work in fluorescence mode is a desirable optional attachment. It may be noted that the ability of a densitometer to resolve two lines a set distance apart is entirely dependent upon the quality of the optical system. The most consistent results are obtained by using a slit width approximating to the width of the bonds on the electrophoretic support material. The light source for most densitometers is a tungsten filament lamp and some densitometers depend upon this lamp down to the near UV.

19.5 Spectrodensitometers

The introduction of agar gel and subsequently of polyacrilamide gel electrophoresis for the fractionation of ribonucleic acids and the need for a direct ultraviolet densitometry of the gels has stimulated the development of electrophoretogram scanning techniques. A double-beam recording spectodensitometer is generally used for measuring the optical density of dry agar electrophoretograms.

A double-beam ratio type scanning spectrodensitometer incorporates a high intensity continuous xenon light source and a low stray-light quartz prism monochromator. Thin layers as well as gels and paper strips of all types, and preparative forms can be rapidly and very accurately scanned for quantitative determinations.

The instrument also facilitates liquid samples in static or flow cells for a large number of investigations in the field of column chromatography, spectodensitometry and kinetic and denaturation research. Figure 19.8 shows the optical system of the instrument. A 150 W xenon high pressure short arc lamp or 200 W xenon–mercury lamp (1) is fitted in a special air-cooled housing, which is equipped with high transmission suprasil optics (2), and a front surface spherical reflector (3) for greater efficiency operation. The reflector is adjustable, so that the primary light image and reflected image can be aligned to be coincident.

A light tube attached to the lamp housing focuses the light beam onto the entrance slit of the monochromator. A slot located just past the focusing adjustment is provided for the insertion of an interference or other type of filter, and a special split ring, which serves to cover this slot while not in use. The light tube also contains the beam diverging optics, all necessary beam aperture and balance adjustments and a deflector mirror to divert the beams down to the exit post (8), adjustment which includes a calibrated slit width control (9). The details of the optical components of the monochromator are given in (17). The light beam emanating from the source strikes

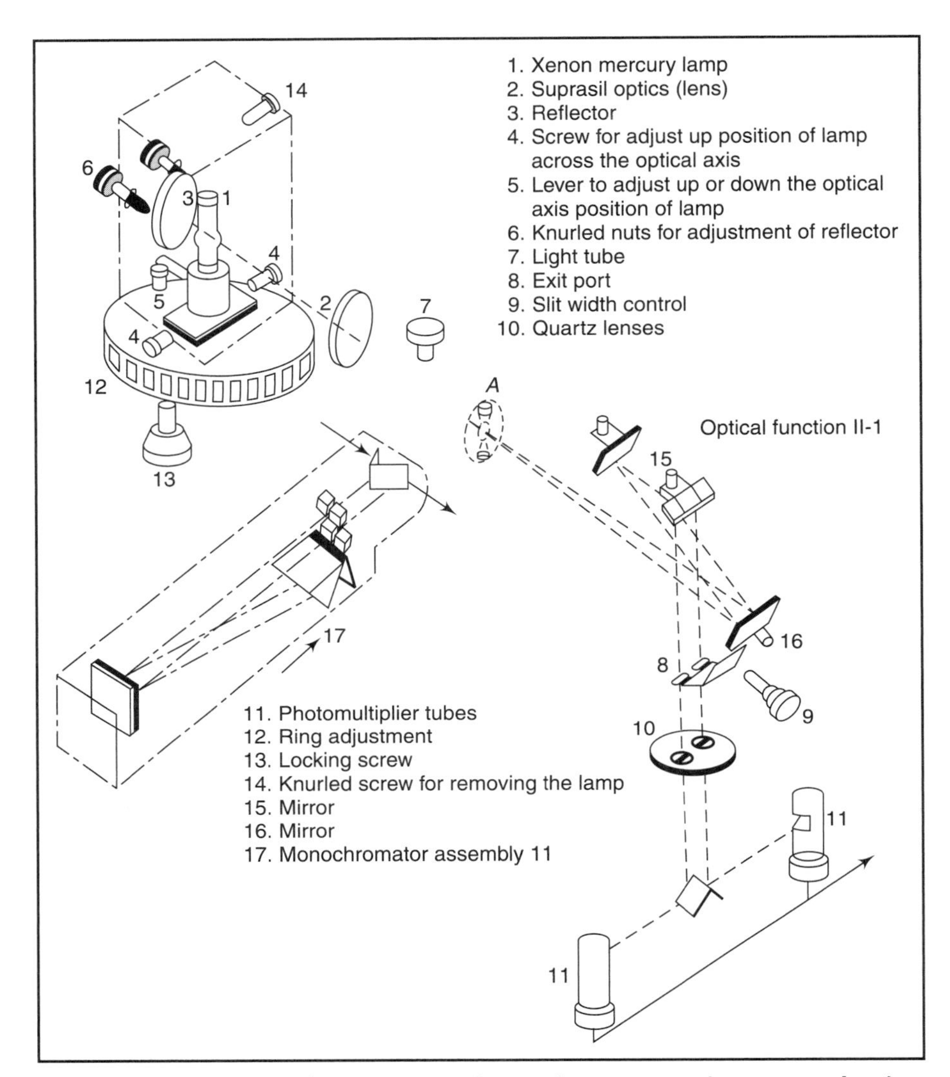

Figure 19.8 :: Optical arrangement in a ratio type scanning spectrodensitometer (Courtesy, M/s Schoeffel Instrument Corp., USA)

the TLC plate or other media. The media scatters the light, making it as a pair of secondary light sources. These light beams then pass through protective quartz lenses (10) and are maintained separately by a baffle between the lenses which is also in close proximity to the plate. These lenses collect the light and focus it onto the individual photomultiplier tubes (11). The lenses are located in such a way that only the incident beams or the scattered light resulting from the incident beams are directed onto the photomultiplier tubes. The photomultipliers are equipped with individual dynode chain divider circuits and are mounted in a sturdy cast aluminium housing inside the main instrument. They are arranged, so that both angular and lateral adjustments can be made at the factory.

The photomultiplier tubes are supplied with a highly regulated, low ripple power supply. The gain control is adjusted by varying the amount of high voltage being applied to both photomultiplier tubes. The balance control changes the voltage on the signal photomultiplier, with respect to the voltage applied to the reference photomultiplier. The balance control does not affect the reference signal. The amplifiers comprise of high gain amplifiers, which act as current-to-voltage converters. They are adjusted to have equal output from the sample channel and reference channel, with a zero balance setting on the controls. This means the ratio would be 1, i.e. 100 per cent *T* or zero optical density. Changes in the sample channel will affect this ratio and can, therefore, be measured. The ratio system compensates for any changes in the light due both to the lamp power supply and characteristics of the lamp itself. In case optical density is to be obtained, the log function of the ratio of the two input signals coming from the reference and sample photomultiplier is obtained by using a log amplifier. This gives a linear output for recording and integration.

The monochromator used is calibrated in nm 200 to 700 in 10 nm increments, linearly. It incorporates high transmission quartz optics in a modified Littrow mount. The wavelength cam drive can linearize non-linear dispersion over the spectral range. The monochromator can be fitted with a motor drive assembly for spectral scanning. The scanning is done at a speed of 200 nm/min. The wavelength drive should have zero backlash design. A slip clutch permits manual override of the motor drive, permitting any selected wavelength to be set manually. The drive will automatically take over upon release of the manual wavelength control.

19.6 Microprocessor-based Densitometer

The inclusion of microprocessors in densitometers enables one to perform certain operations while carrying out two-dimensional density scanning of electrophoretic strips. It enables one to obtain a graphical representation of the density profile, with trough detection marks, peak positions, peak heights, peak integrals, relative percentages and total integral. Chopra (1984) illustrates the use of a microprocessor-based scanning densitometer, whose block diagram is shown in Figure 19.9.

The apparatus essentially consists of a large flat-bed specimen table which can accommodate a wide variety of samples, besides providing straight-forward sample acquisition. The specimen table is driven in both X and Y axes by incremented stepper motors under microprocessor control, with an incremental step size of 50 µm. Automatic sample acquisition is accomplished by software-based microprocessor control. The densitometer accepts all standard media including paper, acetate, agarose and polyacrylamide gel strips. The scan pattern is fully programmable for efficient processing of all media. Linear scans of up to 200 mm can be carried out for single and multi-

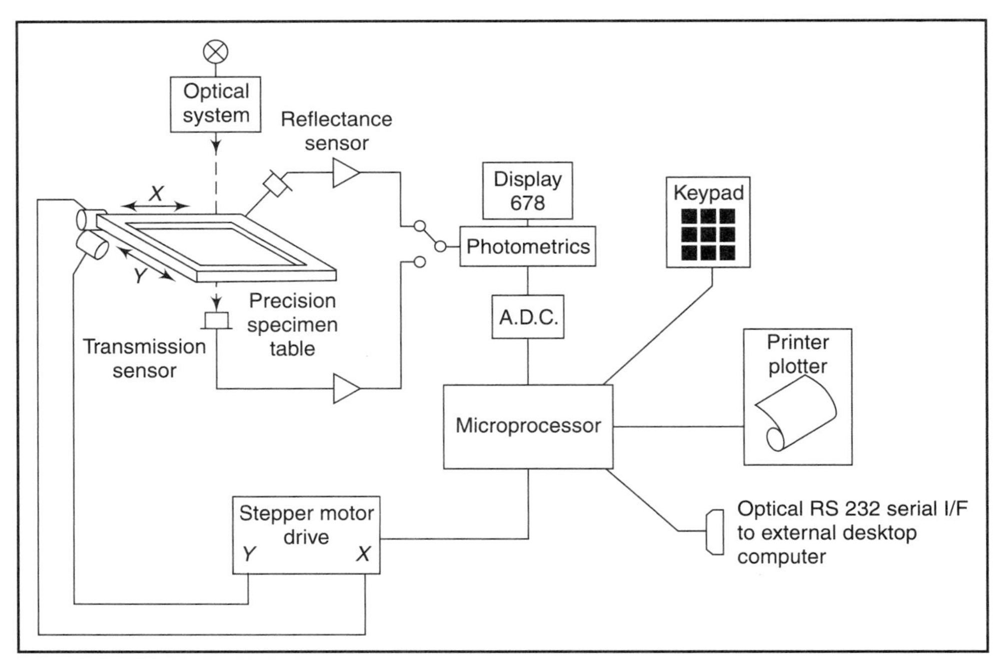

Figure 19.9 :: Microprocessor-based scanning densitometer (after Chopra, 1984)

band samples. For analysis of TLC and HPTLC and multiple-spot traces, raster scans with programmable facilities for raster size reading interval and line interval up to 200 mm. Different detection modes like visible and UV transmission and reflection are employed. In fluorescence and quench fluorescence are used. The scanning aperture is selected by means of interchangeable aperture holders and the scanning wavelength is selected by means of interchargeable filters. The filters may be located for either pre-sample or post-sample filtering. The measurement range is 0–3 AU in the transmission mode and 0–2 AU in the reflectance mode. A background adjustment is provided in the equipment.

The scanning mode and scan parameters can be set up by the user via a push-button control panel, which is interfaced to the microprocessor control and data analysis. The system can be interfaced to an external computer via an RS-232C interface. After scanning, the printer/plotter provides fully annotated graphical hard copy results of each scan. Migration in any electrolyte is influenced by several factors, such as (i) size and shape of the particle, (ii) concentration, (iii) pH, (iv) temperature and viscosity of the electrolyte, (v) intensity and distribution of the electric field, and above all, (vi) the technique used. The energy in the field tends to produce temperature differences in electrolyte which, in turn, create mild convection currents. In electrophoresis experiments, these currents will disturb the orderly movement of the particles unless they are controlled.

19.7 Capillary Electrophoresis

One of the major drawbacks of gel electrophoresis is the speed of analysis. Speed can only be improved by increasing the electric current of the system. However, with an increase in current, a large amount of heat would be generated and an efficient cooling system would be required.

The development of capillary electrophoresis (CE) has solved the heating problem. Silica fused capillaries ranging from 0.150 to 0.375 mm in outer diameter efficiently dissipate the heat that is produced. Increasing the electric fields produces very efficient separations and reduces separation times.

In a CE separation, a very small amount of sample (0.1 to 10 nl) is required. The sample solution is injected at one end and an electric field of 100 to 700 volts/centimeter is applied across the capillary. Proteins in the solution migrate through the capillary due to the applied electric field (electrophoresis). Differing electrophoretic mobility drives each of the components into discrete bands. Quantitative detectors such as fluorescence and absorbance detectors can be used to identify and quantify the proteins in the solution.

Figure 19.10 shows the basic instrumental set of a capillary electrophoresis apparatus. It consists of a high voltage power supply (0 to 30 KV), a fused silica (S_1 O_2) capillary, two buffer reservoirs, two electrodes, and an on-column detector. Sample injection is done by temporarily replacing one of the buffer reservoirs with a sample vial. A specific amount of sample is introduced by controlling either the injection voltage or the injection pressure.

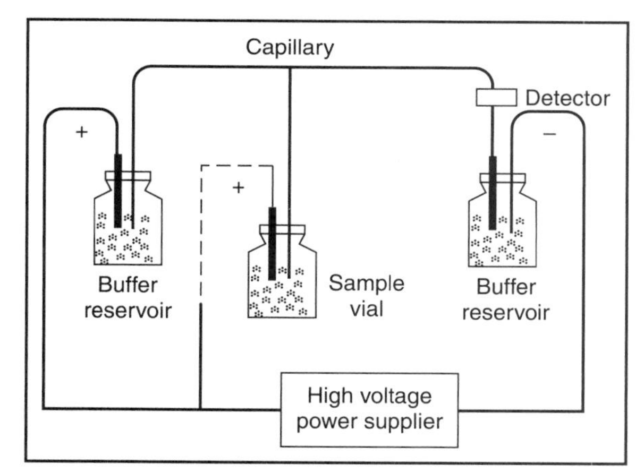

Figure 19.10 :: Instrumental set-up of a capillary electrophoresis apparatus (Xu, 1996)

Capillaries are typically of 50 µm inner diameter and 0.5 to 1 m in length. Capillary electrophoresis uses an electromotive force rather than a pump, to drive the mobile phase through the capillary. Due to electro-osmotic flow, all sample components migrate towards the negative electrode. A small volume of sample (10 nl) is injected at the positive end of the capillary and the separated components are detected near the negative end of the capillary. Capillary electrophoresis detection is similar to detectors in HPLC, and include absorbance, fluorescence, electrochemical, conductivity measurements and mass spectrometry. Campana, *et al.* (1997) point out that when combined with capillary electrophoresis, chemiluminescence offers excellent analytical sensitivity and selectivity.

Capillary electrophoresis is an attractive technique for DNA analysis because the narrow-bore, gel-filled capillaries provide high-speed, high resolution separations, increased separation efficiency, and automated gel and sample loading (Khetarpal and Mathies, 1999). The throughput of the system can be increased many times by using capillary array electrophoresis for running multiple capillaries in parallel. The instruments based on this design, which run capillaries, are being used in genome centres and pharmaceutical companies for DNA sequencing and analysis. The authors describe details of detection instrumentation used for this purpose which are primarily based on scanning or imaging technology.

The capillary can also be filled with a gel, which eliminates the electro-osmotic flow. Separation is accomplished as in conventional gel electrophoresis but the capillary allows higher resolution, greater sensitivity and on-line detection.

The surface of the silicate glass capillary contains negatively-charged functional groups that attract positively-charged counter-ions. The positively-charged ions migrate towards the negative electrode and carry solvent molecules in the same direction. This overall solvent movement is called electro-osmotic flow. During a separation, uncharged molecules move at the same velocity as the electro-osmotic flow (with very little separation). Positively-charged ions move faster while negatively-charged ions move slower.

Electro-osmotic flow

19.8 Micro-Electrophoresis

The electrophoresis cell used in normal applications requires a relatively large volume of solution, about 20 ml, which might mean as much as 100 to 200 mg of material, if accurate measurements are desired. Such large material requirements are not easily arrangeable in some biochemical investigations. This requirement has led to the development of methods for micro-electrophoresis, which have the advantage of reducing the volume and concentration of the solution required.

The use of sensitive optical methods can also aid in achieving a reduction in the volume of solution required. One of the methods used for this purpose is the Jamin interferometric system. The method is based on the formation of interference fringes, which are produced by the interference of two beams, one of which has passed through some homogeneous reference medium and the other through the cell under investigation. In the Jamin interferometric system, an image of the cell is obtained, which is crossed by fringes. The variation in spacing of the hinges represents the variation in concentration along the cell. The fringes will be crowded at a boundary and in effect, each component of a mixture will make itself seen by a corresponding group of fringes. A simple counting of the fringes gives an approximation of relative concentrations.

Jamin interferometric system

A slight variation in the measurement technique using an interferometer is incorporated in the micro-electrophoresis apparatus shown in Figure 19.11. In this method, the light from the source (*A*) is split by the first mirror (*B*) into two coherent light beams, passes through the measuring and reference channels of the cell (*C*). The two light beams are in the rotatable mirror (*D*).

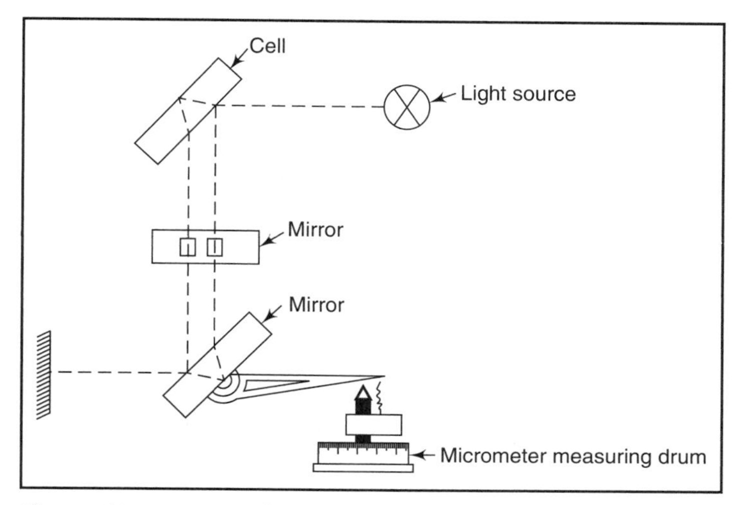

Figure 19.11 :: Interferometric method for measurement of density on electrophoretic strips

Whenever the two coherent beams encounter variations in optical density in the cell because of the migration of the proteins into the buffer solution in the measuring channel and the presence of buffer in the reference channel, they pass through optically unequal paths. This difference in optical path length can be compensated for by rotating the mirror (*D*). Correct compensation can be checked by means of a spectrum and an index line in the eye piece. The mirror is rotated by hand through a micrometer screw, whose movement can be read off and recorded. The rotation of the micrometer screw corresponds directly to a difference in the concentration of the solutions in the two channels. The channels are measured out at intervals of 0.1 mm from the top downwards, along the protein migration distance.

Micro-electrophoresis apparatus can be used to determine the electrophoretic mobility of a wide range of suspended particles. The apparatus uses conventional quartz–iodine illumination unit (which can make particles visible down to about 0.2 μm). For lowering the limit of particle visibility, a laser illuminator (3 mW He-Ne laser) is preferred, which can make particles as small as 0.09 μm visible. The apparatus uses a binocular microscope or closed circuit television system to aid the operator (Figure 19.12).

19.9 Steps in the Electrophoresis Procedure Requiring Quality Control

With appropriate knowledge of sample handling and with a useable control specimen, the following points should be borne in mind to reduce error and improve reproducibility:

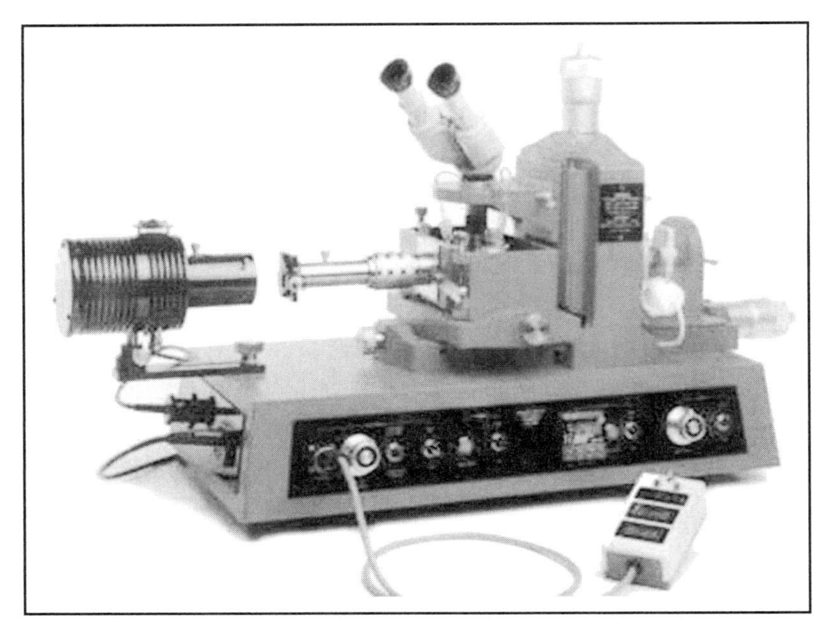

Figure 19.12 :: Microelectrophoretic apparatus (Courtesy, M/s Rank Brothers Ltd., UK)

- Sample application is the most critical portion of the electrophoresis procedure. Overloading of the samples should be avoided. This can result in an inability of the densitometer to read the most intense peaks due to the optical density of the dye protein complex, being in a region of non-linearity with respect to concentration. This is readily checked by making 5 per cent, 10 per cent and 25 per cent dilutions of the sample and carrying out electrophoresis on undiluted and diluted samples.
- Likewise, it is important to assess sample applicator repeatability is likewise important to assess. This is especially true in those laboratories where several applicators are used interchangeably.
- Buffer pH and composition should be carefully checked. Ideally, conductivity measurements should be made on each new batch of buffer made. Variations in buffer conductivity as large as ± 0.10 per cent can generally be accepted. Buffer pH should be maintained within ± 0.10 of the given value for buffer.
- Power supplies should be checked regularly to determine if the actual output of the electrophoresis cell is the same as the reading on the supply meter. It is important to monitor and record initial and final current levels for each electrophoresis run. Buffer and media changes, open or short circuits and unstable power supplies can be evaluated by comparing current levels between runs. Applied voltages should be controlled to within ± 5 per cent of the recommended voltage conditions.

- Electrophoresis running times must be controlled to within ± 1 min of the recommended time for a given procedure. The combination of buffer voltage, running time and medium interact to determine the degree of sharpness of the separated fractions.
- Invariably, it is not possible to control the properties of the electrophoretic medium. Therefore, within the lot and lot-to-lot, variations in performance must be assessed.
- Post-electrophoresis conditions such as staining, rinsing and clearing times should be maintained within ± 30 s. These steps directly control the degree of transparency, which, in turn, is reflected in the uniformity of densitometer scanning and the resultant values.

Chapter 20

Electrochemical Instruments

20.1 Electrochemical Methods for Analysis

Electrochemical methods are characterized by high sensitivity, selectivity and accuracy. The analytical sensitivity attainable even exceeds the 10^{-10} molar level, and analyses at the sub-nanogram range are possible. Electrochemical methods have been extensively developed and each basic electrical parameter, namely current, resistance and voltage, has been utilized alone or in combination with each other for analytical purposes. The simplest and the most commonly used electrochemical technique involves passing of the current through a solution for exhaustive electrolysis. The solution gets electrolyzed until the entire product is deposited on a mercury electrode or platinum gauze. The electrode is weighed to complete the analysis. Besides direct analysis, electrical measurements are excellent indicators in all areas of titrimetry. Reagents are added volumetrically by automatic or manual means, to an end point that is conveniently detected electrometrically. Sometimes, even the reagent can be coulometrically generated within the sample to obtain a complete electrochemical system. Electrochemical sensors are based on EMF (potentiometric), resistive (conductometric) or current (amperometric) principles, each having its own use in a given situation.

Electrochemical technique

When using electrochemical techniques for chemical analysis, the potential at an electrode is a function of the concentration of some analyte, the current that flows during the measurement, a time function, and any mass transport rate that might exist at the electrode and the area of the electrode.

20.2 Electrochemical Cell

An electrochemical cell (Figure 20.1) consists of two electrodes of the same metal, or different metals 1 and 2 immersed in the same or in different electrolytes. Each compartment is a half-cell.

Electrodes 1 and 2 are connected to an electrical instrument P, which can exchange electrical energy with the cell. The electrical energy exchanged between the cell and instrument P is liberated or consumed by reactions involving transfer of charge at the electrodes. Electrons are consumed at the cathode, where reduction occurs, and are supplied by the anode 1, where oxidation occurs. Free electrons generally do not exist in electrolytes and electricity is transported through the cell by migration of ions. Positive ions migrate toward the negative electrode, and negative ions toward the positive electrode.

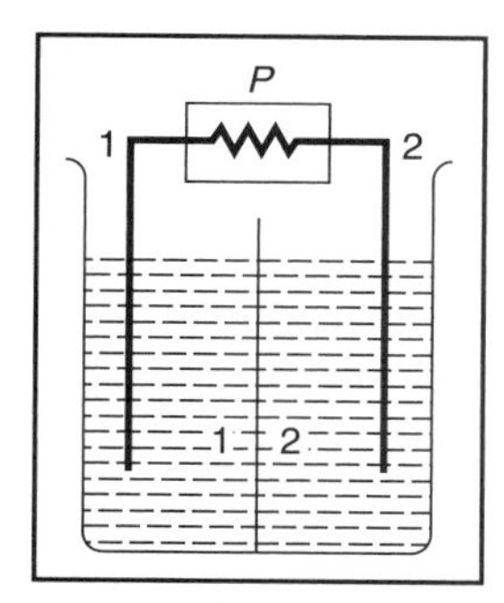

Figure 20.1 :: Schematic diagram of an electrochemical cell

Electrochemical cells

There are two types of electrochemical cells: galvanic (voltaic) and electrolytic. A galvanic cell consists of two electrodes and one or more solutions (two half-cells). In these cells, a chemical reaction involving an oxidation at one electrode and a reduction at the other electrode occurs. An electrolytic cell is one in which the electrical energy is supplied from an external source, the cell through which the current is forced to flow.

Half-wave potential

In an electrochemical cell, the potential at which the current is half its limiting value, is known as the half-wave potential ($E_{1/2}$), which is one of the most useful properties of an electroactive substance. When fast electron transfer rates are involved, $E_{1/2}$ is equal to E^0 of the Nernst equation. For slow rates, more voltage than predicted by the Nernst equation is required to reach the $E_{1/2}$ point.

Figure 20.2 shows the components of a titration cell. This cell comprises of an indicator electrode and a reference electrode. The reference electrode is usually the commercial calomel electrode. Continuous stirring of the solution is done with a magnetic stirrer. It is usual to carry out many oxidation-reduction titrations in an air-free medium. Nitrogen is then bubbled through the solution before and during titration.

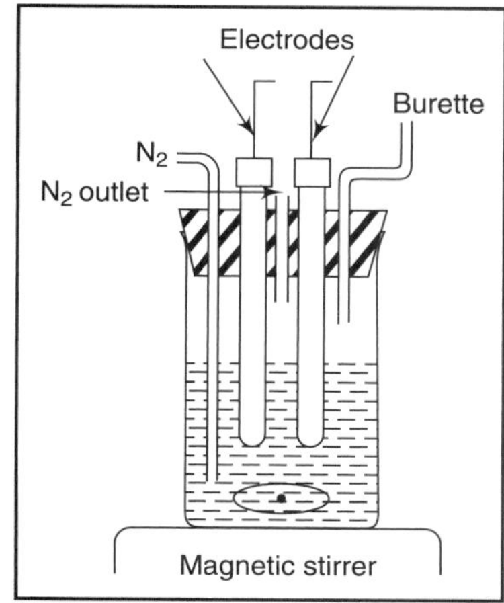

Figure 20.2 :: Titration cell

20.2.1 Types of Electrodes

A wide variety of electrodes are used for different electrochemical techniques. Basically, they can be classified as working, auxiliary and reference electrodes.

20.2.1.1 Working Electrode

The working electrode is the controlled carbon electrode at which the desired reaction occurs. Platinum disc, paste and dropping mercury electrode are examples of working electrodes. The primary reaction takes place at the surface and the change in current or potential resulting from this reaction is measured.

Solid surface electrodes and mercury electrodes normally cover different voltage ranges. In general, metal electrodes cover the positive voltage range (+1.0 to 0.5V versus SCE) and mercury the negative voltage range (0.0 to –2.0V versus SCE). Carbon electrodes cover a much wider range, where SCE is the saturated calomel electrode.

Mercury is used in various forms. The dropping mercury electrode (DME) is the most common. The mercury pool offers an about ten-fold increase in sensitivity over the DME, but requires the use of a large pool (2 cm diameter) to minimize changes in curvature and area with a change in the applied potential due to inter-facial tension. The mercury plated electrode provides a very high resolution and operation over extended negative potentials. Nickel, silver and platinum are most often used as the plating surface. The solid electrode is usually in the form of a wire, disc or wire mesh. Both platinum wire and disc are often rotated to obtain increased current signal-to-background ratio, obtained by the enhanced mass transport or convection of the ions.

More work has been done with platinum than with any other solid electrode. However, a platinum surface is subject to formation of oxide films on its surface by either chemical or electrolytic oxidation. A gold surface is less susceptible than platinum, to attack by some chemical oxidizing agents. Gold is probably a better electrode material than platinum for general utility.

The carbon paste electrode is one of the most practical solid electrodes and for routine applications, the carbon paste surface is superior to gold or platinum. Vitreous carbon is highly resistant to chemical attack and is relatively insensitive to changes in pH. It has however, higher residual current than the carbon paste electrode.

20.2.1.2 Reference Electrode

This electrode provides a stable potential which may be taken as reference. Most literature states the potential measurements in relation to a saturated calomel electrode (SCE), which is a convenient electrode to use. Another common reference electrode is the silver-silver chloride electrode. Reference to hydrogen electrode, the difference in potential between a calomel (–0.244V) and silver-silver chloride (–0.200V) electrode is 44 mV.

20.2.1.3 Auxiliary Electrode

The electrode required for completing the electrolysis cell is called the auxiliary electrode. It performs the reverse reaction of that, which takes place at the working electrode. Usually, a platinum disc may be used as an auxiliary electrode. A frit separates the auxiliary compartment solution and sample compartment containing the working electrode. This frit is normally sufficient to prevent contamination in quiet solution techniques. However, an agar plug can be used for stirred solutions, such as are required in coulometry.

20.3 Potentiostats

Potentiostats are amplifiers used to control a voltage between two electrodes, a working electrode and a reference electrode, to a constant value. Reference electrodes are electrodes which maintain a constant voltage referred to the potential of the hydrogen electrode. A silver wire, covered with a silver chloride layer, dipping in a chloride solution, is a simple reference electrode. However, as soon as a current passes this electrode, it is polarized. This means that its potential varies with current, and so it is not possible to maintain a constant potential difference between a working electrode and reference. Therefore, another electrode, called the 'Auxiliary electrode' is introduced into the system.

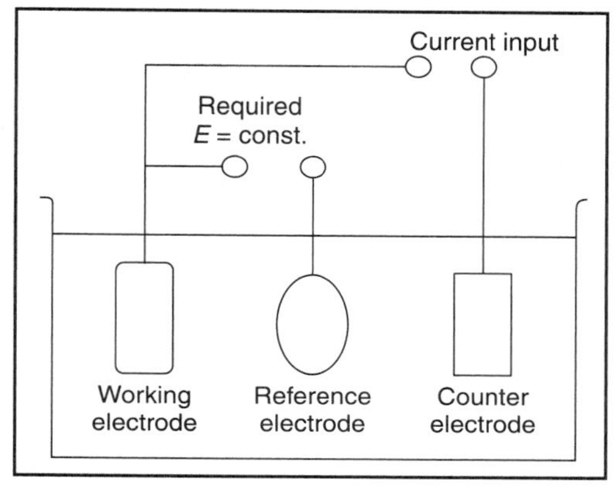

Figure 20.3 :: Principle of a potential-controllable electrochemical cell

Figure 20.3 shows the principle of a potential-controllable electrochemical cell, called the Potentiostat. The potentiostat has two functions, which are:

- To measure the potential difference between working electrode and reference electrode without polarizing the reference electrode, and
- To compare the potential difference to a preset voltage and force a current through the auxiliary electrode towards the working electrode in order to counteract the difference between preset voltage and existing working electrode potential.

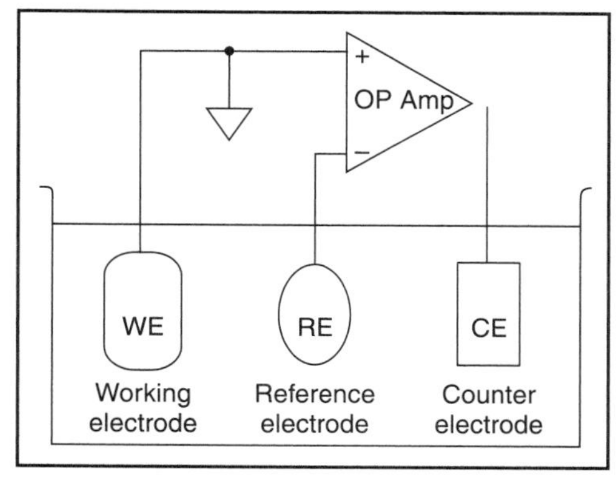

Figure 20.4 :: Using an amplifier as potentiostat (adapted from Potentiostat: Bank Electronic, GmbH)

This can be realized by the arrangement shown in Figure 20.4, which uses an operational amplifier. Here, the difference voltage between working electrode and reference electrode is amplified and inverted by the operational amplifier. A matching current is fed to the auxiliary electrode. The control circuit is closed by the cell, where the current passes through the electrolyte from the auxiliary electrode to the working electrode. This polarizes the working electrode exactly so that the difference between the reference electrode input and the working electrode input is set to zero. By doing so, the potential of the working electrode can be kept exactly on the potential of the reference electrode. Sophisticated circuits are used to get a good potentiostat. There is a variety of potentionstat models for a variety of application fields.

20.4 Types of Electrochemical Methods

There are five main types of electrochemical methods. These are:

- *Potentiometry*: This measures the electrical potential developed by an electrode in an electrolyte solution at zero current flow. The concentration of ions in the solution is calculated by using the Nernst equation.
- *Voltametry*: This determines the concentration of ions in solutions from current flow as a function of voltage when polarization (depletion of concentration caused by electrolysis) of ion occurs around the electrode. When dropping mercury electrode is used, the method is called 'polarography'.
- *Coulometry*: This measures the charge (Q) over a fixed time and controlled voltage, i.e. it involves electrolysis of a solution and the use of Faraday's Law relating the quantity of electrical charge to the amount of chemical change. Faraday's Law states that it takes 9.65×10^4 coulombs of electrical charge to cause electrolysis of 1 mole of a univalent electrolyte species.
- *Amperometry*: This measures the concentration of an electroactive substance by applying a fixed voltage across the electrodes and then measuring the current passing through the cell.
- *Conductometry*: This measures the conductance for controlled concentration.

20.5 Potentiometers

Potentiometry involves the measurement of the difference in potential between an indicating electrode and a reference electrode immersed in a solution of the ions to be determined. The potential E (half-cell potential) of any electrode is given by the generalized form of the Nernst equation:

$$E = E^{\circ} + RT/nF \log a_{ox}/a_{red},$$

where E° = reduction potential of the half-cell under standard conditions

E = the potential of the half-cell

R = a constant, 8.314 J/deg.

T = absolute temperature

F = Faraday Number (96,494 C)

n = number of electrons transferred in the electrode

a_{ox} and a_{red} are the activities of the oxidized and reduced forms, respectively, of the electrode action. Substituting concentrations for activities and various other constants and assuming the temperature to be 25°C, the Nernst equation becomes:

$$E = E^o + 0.0591/n \log (ox)/(red)$$

where $[ox]$ is the concentration of the oxidized form of the ion and $[red]$ is the concentration of the reduced form of the ion.

From this equation, it is obvious that the measurement of the potential developed by a half-cell serves as a measure of the concentration of the components in a solution. In practice, the half-cell to be measured is connected to a standard or reference half-cell to form a complete cell. The method of potentiometry is used to measure the voltage of the complete cell and the potential of the complete cell is calculated from the relationship:

$$E \text{ (half-cell)} - E \text{ (standard half-cell)} = EMF \text{ (observed)}$$

The EMF of the standard half-cell being known, the unknown potential can be determined. A potentiometer is used to measure the voltages instead of voltmeter, as it draws negligible current and hence does not produce any depolarization during the measurement.

Although potentiometry is useful in several analytical areas, its most common application lies in the determination of pH. Potentiometric measurements are also applied to the detection of the end-point of titrations. Another area where potentiometry is being increasingly applied, is that of the use of ions-selective electrodes for the direct measurement of cations and anions.

20.5.1 Principle of a Potentiometer

Although the voltage of an electrochemical cell can be measured by connecting a voltmeter to the cell, it has the disadvantage that it is not very precise and draws a relatively large current from the cell. With the flow of current through the cell, a reaction may occur at the electrodes and alter the composition of the solution. The problem can be solved, to a certain extent, by using FET input amplifiers with a very high input impedance, so that they draw very little current from the cell. They too may lack precision due to their rather short scale length. Very accurate measurements can be made by the potentiometric method. This method is a comparison technique, in which the unknown cell emf to be measured is compared with a known emf source.

The principle of a simple potentiometer is shown in Figure 1.4. The standard voltage across the slide wire is provided by using a standard cell, which is usually a Weston cadmium cell. The positive terminal in this cell is a mercury electrode, whereas the negative terminal is a cadmium amalgam with an excess of cadmium to maintain saturation at all temperatures. Voltage recording is common in potentiometric methods and instrumental analysis in general. Self-balancing null type poteniometric recorders are used for this purpose.

When potentiometry is used as an end-point detection system, various techniques can be employed. These are detailed as follows.

20.5.2 Zero Current Potentiometry

No net electrochemical reaction occurs at the indicating electrode and the measurement is independent of mass transport. In this method, the course of a titration is followed by measuring the change in potential of an indicator electrode at zero current.

20.5.3 Constant Current Potentiometry

In this method, a small but constant current flows through the indicator electrode system during the titration. The potential change during the titration is monitored as usual. In a stirred solution, the current is sufficiently small, so that a stable potential is achieved due to diffusion of the electroactive species present.

20.5.4 Null-point Potentiometry

In this method, the potential difference between the two electrodes is brought to zero by having same concentration of the investigated substance in both the half-cells. This is done on adding titrant to the unknown sample, in order to bring the concentration of the species of interest to the same concentration of the same species in the other half of the cell system. Identical indicator electrodes are used in both sides of this two-chamber cell.

20.5.5 Cyclic Chronopotentiometry

In chronopotentiometry, a constant current is passed through an electrolytic cell. The potential of the working electrode is monitored. The variation of this potential with time takes the form of a wave, having a characteristic transition time (Figure 20.5). Information about the reactions occurring at the electrode can be obtained from the shape and duration of this wave.

Cyclic chronopotentiometry

Cyclic chronopotentiometry is a useful extension of the technique, in which current is reversed at the end of each transition time, so that a sequence of forward and reverse transitions is produced. The current reversal is made, when the electrode potential reaches either of two preset values V_1 and V_2, so that potential oscillates between these values. The current is reversed at the appropriate electrode potentials using a relay.

If the potential drop across the cell is small, a series resistor can be used to control the current adequately, using a stable high voltage supply. If the cell resistance is high (e.g. with non-aqueous electrolytes), a high value of series resistor and a high voltage would be needed. In order to avoid this, an electronic current stabilizer (galvanostat) is generally used. The potential of the working electrode is measured against a suitable reference electrode. Ideally, negligible current should be drawn from this type of electrode. The emf between the reference and the working electrodes is

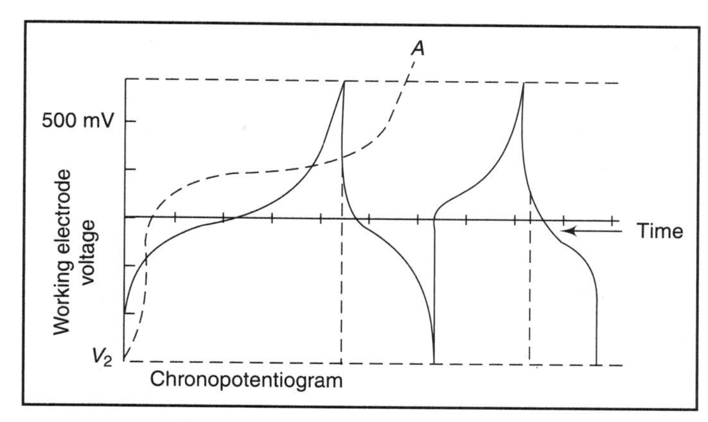

Figure 20.5 :: Typical chronopotentiogram

measured using a high input impedance, fast response amplifier. The design of such an amplifier is simplified by holding the working electrode at earth potential.

Figure 20.6 shows the circuit arrangement used for cyclic chronopotentiometry. A stabilized voltage supply is connected to the working electrode through a resistance *R*. Amplifier 1 amplifies the voltage difference between the working electrode and earth and the output is applied to the

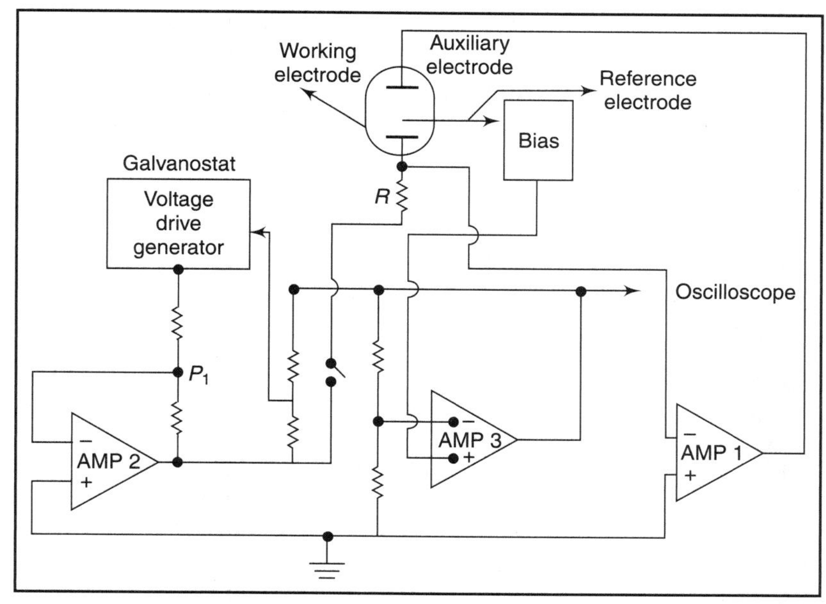

Figure 20.6 :: Circuit arrangement for making cyclic chronopotentiometric measurements (after English, 1976)

auxiliary electrode, thus maintaining the working electrode at earth potential. Since the amplifier draws negligible current in the input circuit, all the current passing through the resistor also passes through the cell. The cell emf is supplied by the amplifier. The cell current is therefore determined by the supply voltage and the series resistor. The supply to the resistor is supplied by amplifier 2, which is an inverter following the output of the galvanostat voltage drive generator. The cell current is reversed at the desired switching potentials (V_1 and V_2) by reversing the potential, which controls the amplifier 2. The switching potentials are controlled by reference potential amplifier 3. The voltage swing is determined by the ratio at P_2 and it can be extended over a wide range by adjusting P_3. With a 1 μF capacitor connected between the auxiliary electrode and the working electrode terminals and proportion of the auxiliary electrode voltage fed back to the reference electrode terminal, a steady triangular wave of good form can be obtained.

20.6 Conductivity Meters

The conductivity of an electrolyte is a measure of the ability of the solution to carry electric current. The current through the solution takes place through the movement of electrically charged particles called ions. When a potential difference is applied to two electrodes immersed in the solution, ions are almost instantaneously accelerated towards the electrodes. Since the conductance of a solution of electrolyte is related to the concentration of electrolyte, analytical applications of conductance are possible.

Like a metallic conductor, electrolyte solutions obey Ohm's law. The reciprocal of the resistance R of the electrolytic solution ($1/R$) is called the conductance. It is expressed in reciprocal ohms or mhos. The resistance of a solution depends upon the length l, area a and the intrinsic properties of the solution. It can be expressed as:

Conductance

$$R = \rho l/a,$$

where ρ is known as specific resistance. Since conductance is the reciprocal of resistance:

$$1/R = 1/\rho(a/l) = K(a/l)$$

The constant K is called the specific conductance. It is expressed in $ohm^{-1}\,cm^{-1}$.

The specific conductance of an electrolyte is a function of concentration. As the solution is diluted, the specific conductance will decrease. This is because fewer ions are present to carry the electric current in each cubic centimeter of solution. The ability of individual ions to conduct is usually expressed by a function called the 'equivalent conductance', which is the conductance of a hypothetical solution containing one gram equivalent of an electrolyte per cubic centimeter of solution. The equivalent conductance (A) is connected with the specific conductance and concentration (in gm equivalent per 1000 cm^3) as follows:

Specific conductance

Equivalent conductance

$$A = 1000\,K/C, \qquad K = AC/1000$$

The conductance of an electrolyte between two electrodes, when expressed in terms of equivalent conductance and concentration is given by:

$$1/R = K(a/l)$$
$$= A\,C\,a/1000\,l$$

The equivalent conductance of a salt is the sum of the equivalent ionic conductances of its ions:

$$\lambda = \lambda_+ \lambda_-$$

The total conductance of a solution at infite dilution is:

$$A_\infty = \Sigma\lambda_+ + \Sigma\lambda,$$

which shows that the migration of ions is theoretically independent of each other and the ions are not affected by other ions in the solution. This is, however, not strictly true and there are very slight differences in ionic conductance of an ion in the presence of various other ions.

20.6.1 Measurement of Conductance

20.6.1.1 Null Method

Conductivity is usually determined by measuring the resistance of a column of solution. This is done by using a Wheatstone bridge, in which the conductivity cell forms one arm of the bridge (Figure 20.7). In order to avoid changes in ionic concentrations due to net chemical reactions at the electrodes, alternating current rather than direct current devices are employed. The choice of frequency is not critical and may range from 50 and 10,000 Hz. However, most commonly employed frequency is 1000 Hz. This ac source may be a low-voltage tapping on the 50 Hz transformer or transistor oscillator.

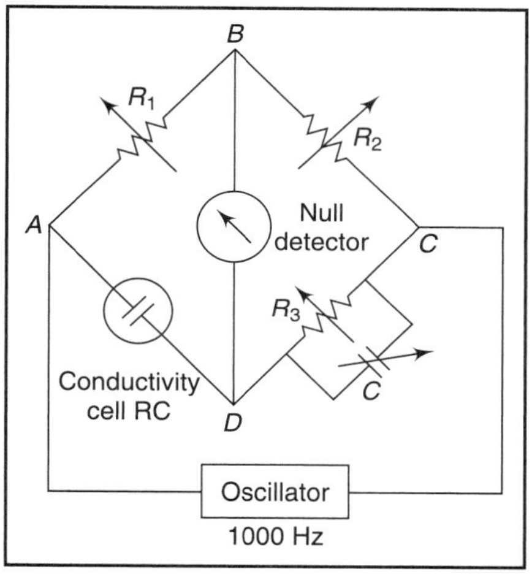

Figure 20.7 :: Wheatstone bridge circuit for measurement of conductivity

When the bridge is balanced, and it is assumed that the conductivity cell behaves as a pure resistance, then the voltage between B and D is equal to zero.

$$R_c = \frac{R_1}{R_2} R_3$$

By adjustment of the ratio R_1/R_2, a wide range of resistances can be measured. However, whenever possible, this ratio is kept unity. This condition is the most favourable to precise measurements. The arms AB and BC represented by resistance R_1 and R_2 are usually in the form of a single calibrated slide wire resistor, with a sliding contact connected to the null detector.

As the conductivity cell contains electrodes separated by a dielectric, an appreciable cell capacitance is invariably present. This capacitance is balanced out by providing a variable capacitor, in parallel with resistance R_3. It is so adjusted that the detector gives a sharply defined balance point.

20.6.1.2 Direct Reading Method

In practice, direct reading instruments are preferred over the null-balance type instruments. In these instruments, the necessity of converting resistance readings into conductance readings is eliminated. The unbalanced bridge current is amplified in an electronic amplifier and displayed on a calibrated panel meter.

Kinetic studies of reactions in solution can also be advantageously made by conductance measurements. Direct reading instruments are preferred for this purpose, which is shown in Figure 20.8. The excitation for the conductivity cell is a square wave, generated by a simple astable multivibrator constructed around an operational amplifier. Its frequency is adjusted to 10 kHz. The conductance amplifier is a wide-band high-gain operational amplifier, in which difference feedback resistors can be selected to obtain different conductance ranges between 0.2 and $10^{-7}\ \Omega^{-1}$. This is followed by a full-wave rectifier for rectifying the output of the conductance amplifier. The output stage of the circuit is a scaling amplifier. The dc signal from the rectifier is, in volts, 0.5×10^n times the conductance of the solution, where n is a scale factor, introduced by the conductance amplifier feedback resistors. This is multiplied by 2 to obtain an output signal, that is a simple factor of 10 of the conductance. The output of the amplifier is fed to the oscilloscope, via a logarithmic amplifier. The arrangement gives a straight line, with a slope equal to the rate constant of the reaction with reasonably good linearity.

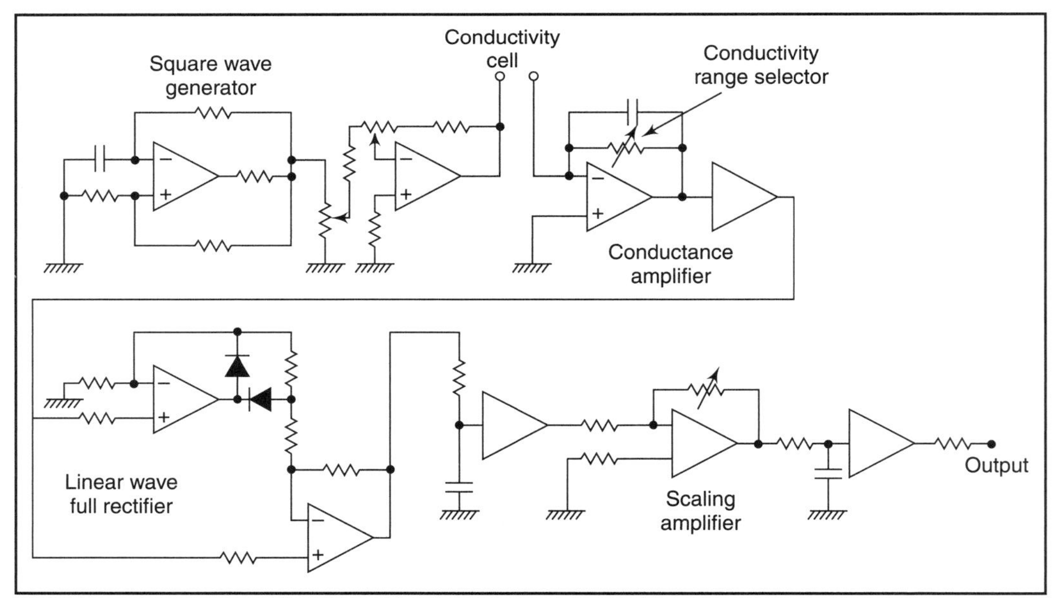

Figure 20.8 :: Direct reading type conductivity meter (after Knipe, *et al.*, 1974)

20.6.2 Conductivity Cells

A conductivity cell comprises two electrodes, which may be two parallel sheets of platinum fixed in position by sealing the connecting tubes into the sides of the measuring cell. In order to reduce the polarization effects which produce a large cell capacitance, the effective area of the electrode is greatly increased by coating the electrode with platinum black. This deposit can be obtained by immersing the electrodes in a solution containing 7.5×10^{-2} M chloroplatinic acid and about 8×10^{-4} M lead acetate and applying a direct current, reversing the direction every half-minute. When a black deposit is obtained, the electrodes are washed in water, and the occluded gases are removed by electrolyzing in dilute H_2SO_4 for about half an hour with current reversal every minute. Electrodes are washed again and stored in distilled water overnight before using. Also, the electrodes should be stored in distilled water when not in use. This prevents the platinum black from drying out.

Most of the conductivity cells are of such a design that the solution completely surrounds the electrodes. In such cases, the conductance of the cell is given by:

$$1/R = \rho[a/l], \quad \rho = 1/R\ (l/a)$$

The term l/a is called the 'cell constant' and may be denoted by θ.

$$\rho = \theta/R$$

The effective value of θ for a cell is not simply related to the cell geometry. However, it has a constant value for electrolytes, measured in that particular cell. The cell constant can be determined by measuring R for a solution of known specific conductance.

Solutions of potassium chloride of known concentrations are invariably employed for this purpose. The specific conductance of these solutions is determined once and for all with a cell, in which the cross-sectional area is uniform and known with accuracy.

Conductivity cells are available in different types, sizes and shapes. The simplest is the dip type, which is immersed in the liquid to be tested. The solution may be an open container and may have volumes in the range of 5ml. Pipette cells permit measurements of conductivity with small volumes of solution, which may be as small as 0.01 ml. Epoxy cells are employed for high temperature use.

For the majority of applications the cells used are made from a specially developed high density carbon that has the same desirable quality as platinised platinum of eliminating electrochemical errors, but without the need for frequent re-platinisation and recalibration. The annular carbon electrodes are fitted within the tubular bore of the cell. The cell body is moulded from an epoxy resin. The cells are now available for screw-in, flow-line and dip-type installations.

Two-terminal conductivity cells are commonly used. These cells are quite satisfactory in many applications, but with dirty solutions, fatty acids or other sticky deposits, fouling takes place. This modifies the surface area and thus results in change of the cell constant, resulting in incorrect readings. This problem has been largely overcome by the four-terminal conductivity cell. Here, a four-electrode cell is used, two outer electrodes being for current and the two inner ones as voltage electrodes.

20.6.3 Temperature Compensation in Conductivity Measurements

The conductivity of electrolytic solutions varies with temperature. This is because the ionic mobilities are temperature-dependent. The temperature coefficient is usually of the order of 1.5 to 2 per cent/°C at room temperature. Control of temperature is thus very essential in precision work. This is usually done by introducing into the bridge circuit a resistive element, which will change with temperature at the same rate as the solution under test. The temperature compensating resistor may be a rheostat calibrated in temperature, which can be manually adjusted. Automatic temperature compensation can be provided by using thermistor and resistance combination in contact with the solution, which would automatically offset the effect of changes of temperature of the solution under test.

An important accessory to a precision type conductivity meter is a thermostatic bath capable of providing very high long-term temperature stability. A proportional controller is employed in preference over the conventional on-off methods. With this method, a temperature stability of 0.02°C may be achieved. Thermostatic baths are very essential for continuous measurement of electrical conductivity in liquid streams, especially where very small changes in salt concentrations take place.

20.6.4 Conductivity Measurements Using High Frequency Methods

High frequency method of measuring the conductivity of solutions offers the advantage of placing the electrodes outside the solution container and out of direct contact with it. This eliminates the possibility and danger of electrolysis or electrode polarization.

The method consists in placing the container with the sample to be analysed between the plates of a capacitor, which forms a part of the high frequency generator circuit, functioning at a frequency of a few megacycles per second. Since the capacitor is a part of the oscillator circuit, any changes in the composition of the solution will result in the changes in the plate and grid currents and voltages due to change in the conductance and capacitance of the cell. The frequency of a parallel resonance circuit is given by:

$$f = \frac{1}{2\pi} \sqrt{LC}$$

The sample cell is usually placed in parallel with a calibrated capacitor. In order to achieve the resonant frequency, the exact amount of capacitance which is added by the sample is removed by adjustment. This gives a measure of the conductivity.

Beat frequency method

The beat frequency method can also be used to measure the output frequency of two oscillator circuits. One of these circuits (Figure 20.9) contains the sample cell as a part of the oscillator capacitance and the other is of fixed frequency. The output of the two when given to a mixer unit, would be the difference of the two frequencies (f-f_0), which would be directly proportional to the changes in high frequency capacitance of the cell. The difference in frequency is measured directly with a beat indicator.

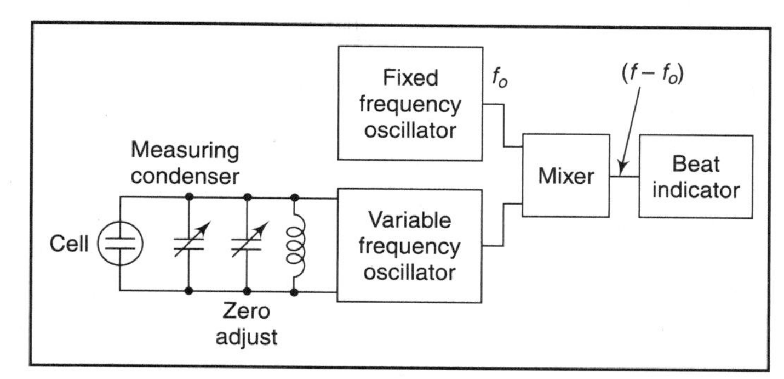

Figure 20.9 :: Beat frequency method for measuring conductivity

A typical sample cell for high frequency conductivity measurement may be composed of two metallic plates sealed on to the wall of a rectangular container. When the solution is put in the container, the metal plates act as a condenser with solution and glass as the dielectric. The equivalent circuit of the cell is as shown in Figure 20.10. C_g represents the capacitance of the glass walls of the cell, C_s is the capacitance of the sample and R_p is the resistance in parallel with C_s. The resistive component is very high and offers negligible contribution. Capacitive effect is the major factor in high frequency measurements, whereas resistive balance is more important in low frequency measurements.

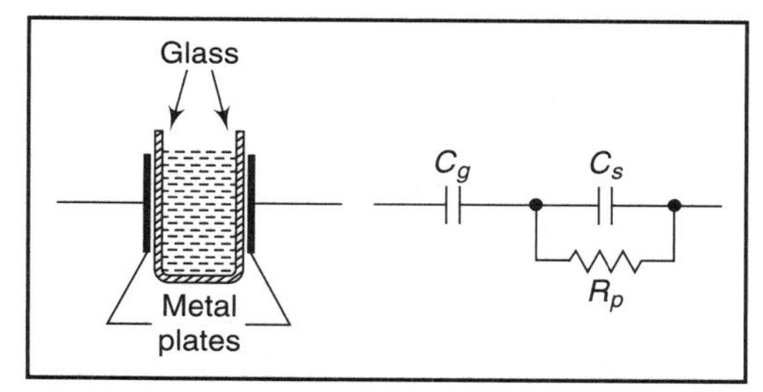

Figure 20.10 :: Equivalent circuit of the conductivity cell when used in high frequency method

20.7 Voltametry

Voltametry refers to the measurement of current that results from the application of potential to drive an electrochemical reaction. Unlike potentiometry measurements, which employ only two electrodes, voltametric measurements make use of a three electrode electrochemical cell. The use of three electrodes (working electrode, reference electrode and auxiliary electrode) along with the potentiostat instrument allow accurate application of potential functions and measurement of the resultant current (Hacke, 1985).

There are different volumetric techniques which are distinguished from each other primarily by the applied potential function to drive the reaction and by the material used as the working electrode. Examples are polarography, and its various forms and linear sweep voltametry.

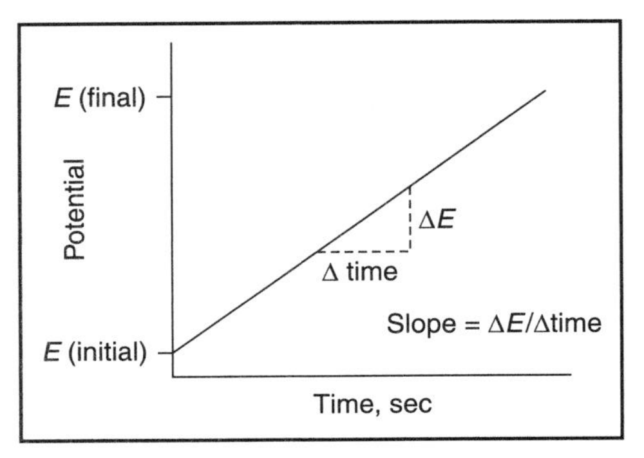

Figure 20.11 :: Principle of linear sweep voltametry (Anderson, 1996a)

Linear sweep voltametry is a general term applied to any voltametric method in which the potential applied to the working electrode is varied linearly in time. These methods include polarography, cyclic voltametry and rotating disc voltametry.

Figure 20.11 shows the type of waveform used in linear sweep voltametry. In this figure, the slope of the ramp is called scan rate and is represented as volts per unit time. The value of the scan rate may be varied from as low as a few mV/sec to several thousand volts per sec.

20.8 Polarographs

When a voltage is applied to a pair of inert electrodes, placed in a solution, a specific relationship exists between current and voltage, which depends upon the electroactive species present in solution. Current-potential curves can be plotted, which prove useful for chemical analysis. These curves can be plotted by varying the voltage applied to a cell and measuring the current flowing through it. Usually, it is assumed that the ohmic drop in the cell is negligible and the potential of one electrode is independent of current. This electrode is said to be unpolarised and the other electrode as polarised.

Polarography

Polarography is the name given to the technique, in which a dropping mercury electrode (DME) is used as an indicator electrode. The basic procedure deals with the measurement of current as a function of electrode potential. Recording instruments are available for directly having a plot of these characteristics.

In polarography, the electric potential or voltage is varied in a regular manner between two sets of electrodes (working and reference) while the current is monitored. The shape of a polarogram depends on the method of analysis selected, the type of working electrode used and the shape of the potential waveform that is applied. Figure 20.12 shows different methods of polarography in which various types of potential ramps are applied to a mercury electrode and the shapes of the resulting waveforms are compared.

20.8.1 Basic Polarographic Instrument

The essential instrumental requirements of a polarograph are few and simple. The arrangement must have:

- The means of applying a variable but known voltage ranging from 0 to ± 3V to the cell, and
- A method for measuring the resultant current which is usually in the range of 0.1–100 μa.

A block diagram of a polarographic set-up is shown in Figure 20.13.

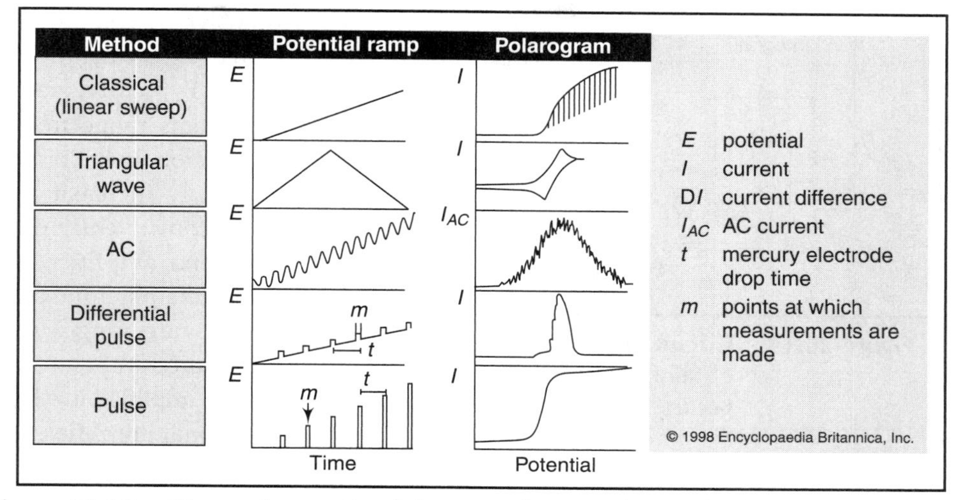

Figure 20.12 :: The various potential ramps that can be applied to a mercury indicator electrode during selected forms of polarography, along with their typical polarograms (Encyclopaedia Britannica Inc., 1998) (*http://chem.ch.huji.ac.il/~eugniik/polarography.htm*)

The electrolysis cell is shown in the diagram as *G*. The beaker contains the test solution. *K* represents a dropping mercury electrode consisting of a narrow capillary from which mercury emerges at a rate of 20–30 drops per minute. *A* is a non-polarizable electrode, e.g. a mercury pool at the bottom of the beaker. Instead of this, it is common to use a calomel half-cell.

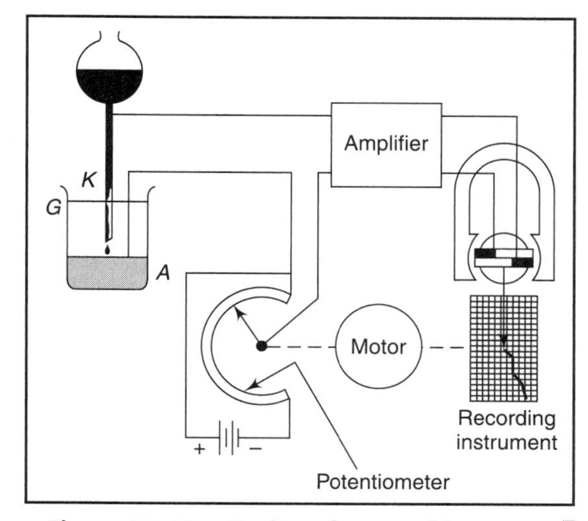

Figure 20.13 :: Basic polarographic set-up

Polarographic method

The polarographic method is based on recording of the variations in the current flowing through the electrolysis cell, as the potential between the electrodes is gradually increased from one value to another, say from 0 to 2 V. This process is achieved through a motor, which simultaneously drives the potentiometer and feeds the chart, on which the electrolysis current is recorded as a function of the electrolysis voltage.

A large variety of polarographic instruments are commercially available. The simpler instruments require manual control and point-by-point plotting of current-voltage curves. They are less

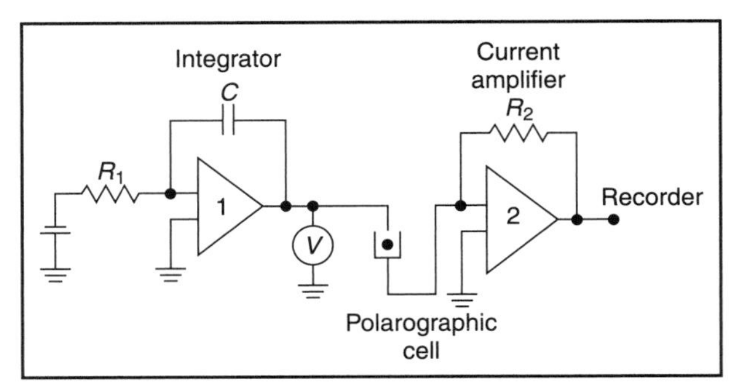

Figure 20.14 :: Circuit arrangement for obtaining linearly increasing polarizing voltage

expensive as compared to those featuring graphic recording of the current-voltage traces. Typically, sensitivities of these instruments range from 0.025 to 0.003 μa per division on calibrated scale.

Linearly increasing polarizing voltage can be conveniently obtained by using an operational amplifier as an integrator. This enables the potential scan-rate to be varied over a wide range. Also, small currents encountered in polarography are advantageously amplified in a high input impedance operational amplifier, without loading the system with an undesirable voltage drop, when connected with grounded positive input and negative feedback. The input bias current of the operational amplifier used as an integrator must be at the most of the order of 10^{-11} A, if integration times of the order of minutes are to be achieved. Figure 20.14 shows a simple two-electrode polarographic system, with amplifier 1 connected as an integrator, to the output of which the dropping mercury electrode is connected. The counter electrode is connected to the input of amplifier 2. The integrator output is monitored on voltmeter *V*. The output of current amplifier is connected to a recorder.

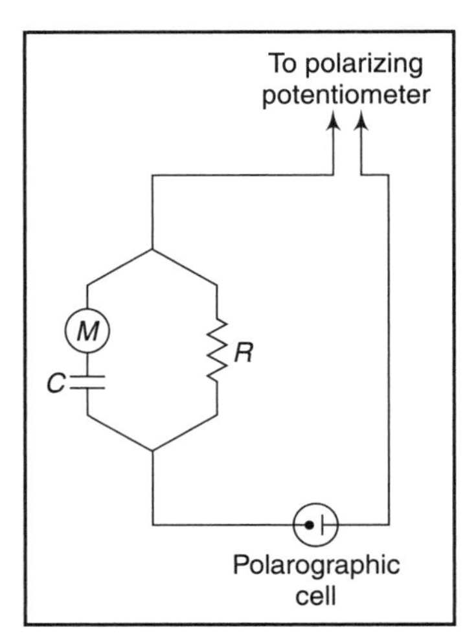

Figure 20.15 :: Damping arrangement in recording polarograms

The polarographic current pulsates between zero and a maximum value during the growth and fall of the mercury drops. Rapid oscillations are not followed by the moving coil meters. The electromagnetic currents are generated, which cause overswing of the needle. This effect is controlled by connecting a suitable resistor across the meter, which is adjusted to provide slight over-damping. In order to smoothen out excessive fluctuations, which may still be present, a capacitor-resistance circuit is usually connected in parallel with the microammeter.

Commercial polarographs are often provided with the facility of plotting derivate polarograms. If the gradient *di/dv* is plotted against applied potential, the peaks become sharp. In recording instruments, the derivative of the current can be recorded by means of a slight circuit modification (Figure 20.15). A capacitor is put in series with the microammeter and a resistor is connected in parallel across them. Several modifications have been suggested over this basic circuit to improve resolution.

It is obvious that the maximum sensitivity cannot be achieved for measuring the diffusion current of a very small wave, which follows a large one, unless the first wave can be cancelled out by passage of a counter-current through the microammeter, equal and opposite to the first diffusion current. Counter currents are applied from a compensating potentiometer connected directly across the microammeter. Zero setting controls are supplied on some instruments, which may be used to obtain a balance at any desired voltage.

Differential current compensation is another technique for applying counter currents. In this method, two polarographic cells with identical electrodes and solutions are connected in series with their like poles connected, the opposing currents would balance each other giving zero current. However, it is difficult to obtain and maintain perfect sunchronisation of drop fall in two DME's. This is achieved preferably by a suitable smoothing circuit, to clamp out the oscillating current.

20.8.2 Dropping Mercury Electrode

The dropping mercury electrode functions exactly as its name implies. It consists of a length of marine barometer tubing with a fine capillary and a head of mercury above it. Mercury, usually under force of gravity, is forced through a section of very fine glass capillary. A mercury drop starts, grows and finally falls off as another drop starts. The measured current will naturally tend to follow this process of increasing steadily, dropping sharply and finally increasing again. The mercury head is so adjusted that it gives a drop time of 2–5 s. The head is generally kept between 40 and 80 cm. The internal diameter of the capillary tube is of the order of 0.03–0.05 mm and the length of the capillary is about 8 cm. A platinum wire is immersed in the mercury reservoir and the dropping mercury electrode is coupled with an unpolarised electrode. This electrode is useful over the range from 0.4 to –2.8 V; referred to the normal hydrogen electrode. Above 0.4 V, mercury dissolves and gives an anodic wave. At potentials more negative than –1.5 V, the electrolytes begin to discharge.

The dropping mercury electrode is a truly elegant electrode and has the following advantages over other types of electrodes:

- Hydrogen has a much larger over-potential than mercury. It renders possible the deposition of substances difficult to reduce, as for example, the alkali ions.
- It provides nearly ideal conditions for obtaining a diffusion-controlled limiting current, which is reproducible.
- It provides a continuously refreshed surface, which is conducive to a high degree of reproducibility for the current measurements. The constant renewal of the electrode surface eliminates passivity or poisoning effects.

The dropping mercury electrode, however, cannot be used for dilutions of less than 10^{-5} M, due to the presence of a relatively large charging current. Nevertheless, high sensitivity derivative instruments may be 200 times more sensitive, as they compensate for the effect of charging current.

20.8.3 Reference Electrode

A mercury pool at the bottom of the polarographic cell acts as a reference electrode. It has a large area and therefore, the current is generally very small. The concentration over-potential at this electrode is negligible and its potential may be regarded as constant. Though convenient, the mercury pool never possesses a definite known potential. Therefore, the reference electrode is usually a saturated calomel electrode (SCE). It is almost an universal practice in polarography to express half-wave potentials with reference to this electrode.

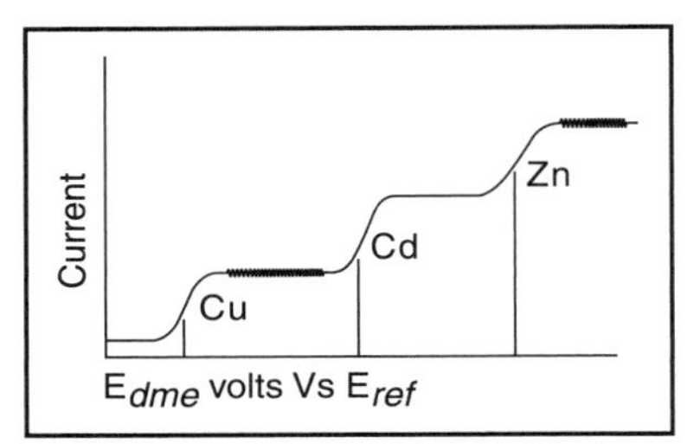

Figure 20.16 :: Typical polarogram

20.8.4 Typical Polarographs

Figure 20.16 shows a polarogram of a solution containing copper, cadmium and zinc. Initially, the electrolysis current is practically zero. The current then starts rising rapidly with the voltage. The increase is brought about, as an increasing number of Cu^{2+} ions are discharged at the dropping mercury electrode. The current soon reaches a constant value, which is governed by the velocity of diffusion at which the Cu^{2+} ions are transferred to the cathode. This difference would reach the maximum value when the Cu^{2+} ions are discharged as fast as they appear at the mercury cathode. This means that the height of the copper wave is proportional to the Cu^{2+} connection. As the potential is increased further, the Cd^{2+} ions will begin to discharge at the cathode, which would result in a wave corresponding to the Cd^{2+} concentration. The Zn^{2+} is recorded at a still higher potential. This figure shows that the height of the wave corresponds to the concentration of the ion concerned, whereas the potential at which the wave is produced, is characteristic of the ion.

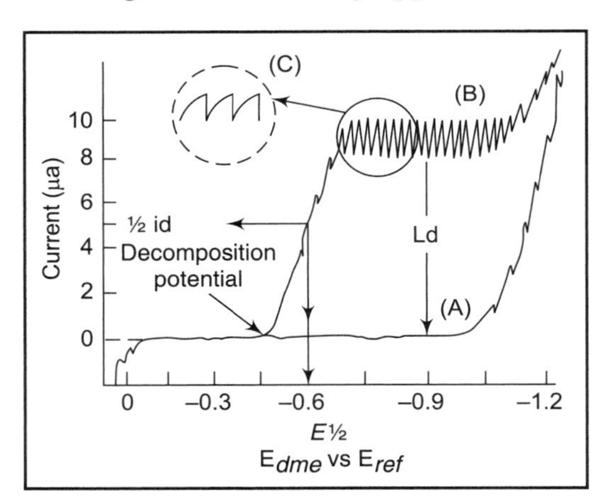

Figure 20.17 :: Typical polarogram showing growth and fall of the mercury drop and corresponding change in the current

The spiked oscillations (Figure 20.17) seen on the recorded polarograms are due to the growth and fall of the mercury drops.

As shown in the inset, the current increase as the drop grows, and then drops sharply when the mercury drop detaches itself and falls.

20.8.5 Quantitative Aspects of Polarography

The quantitative aspects of polarography are based on the equation given by Ilkovic and are as follows:

$$(I_d)_{max} = 708(n)(m^{2/3})(t^{1/6})(D^{1/2})C \text{ at } 25°C,$$

where $(I_d)_{max}$ is the maximum diffusion current.

The rate of diffusion of the substance being reduced or oxidized at the electrode controls the current given by the above equation, which is therefore called the diffusion current.

Diffusion current

n = Number of electrons in the electron transfer reaction
m = Average mercury flow rate in mg/s through the capillary
t = Drop life (s)
D = Diffusion coefficient of the electroactive species
C = Concentration of the analyte in the solution (mM)

The linear dependence between diffusion current and concentration forms the basis for quantitative polarography. Ilkovic's equation also shows quantitatively the influence of other factors, which also directly affect the diffusion current.

When a fast recording system is used having less than 1 s full scale response, the peaks of the oscillations equal $(I_d)_{max}$. When the recording system is damped, the average diffusion current $(I_d)_{ave}$ can be determined from the average of the recorded oscillations. The relationship between average and maximum diffusion currents is given by:

$$(I_d)_{max} = 7/6\ (I_d)_{ave}$$

While the electroactive material in polarography is quantitatively related to the limiting diffusion current, it is quantitatively characterized by the half-wave potential (Figure 20.18).

This is the potential at the point of inflection of the current-voltage curve, one-half the distance between the residual current and the final limiting current plateau. Thus, an $E_{1/2}$ value from the polarogram of an unknown sample can be indicative of possible substances.

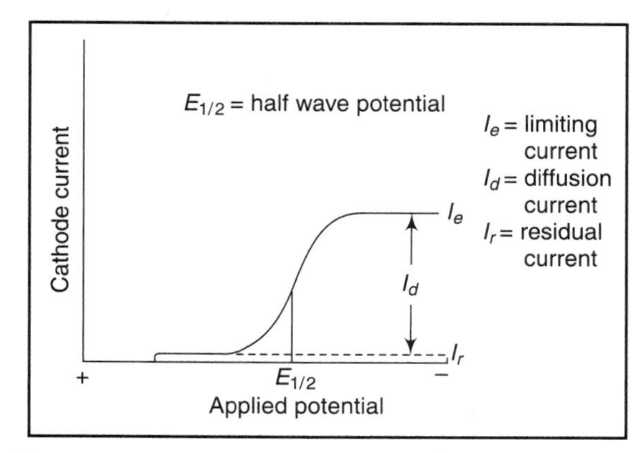

Figure 20.18 :: Current potential curve in a polarogram

Quantitative polarography has been most successfully applied in the 10^{-2} to 10^{-5} formal region. At still lower concentrations, the residual current is often as large as the diffusion current. The selectivity depends greatly upon the solvent used. When analyzing a multi-component system, species with $E_{1/2}$ values differing by 150 mV or greater can easily be resolved. Usually, the first species to be electrolyzed can be determined with the greatest reliability.

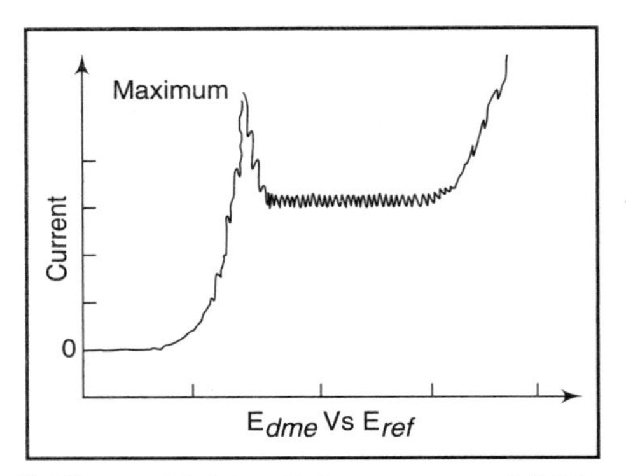

Figure 20.19 :: Polarogram exhibiting maxima

Polarograms often exhibit current maxima as shown in Figure 20.19. This is due to the streaming of solutions past the mercury surface at certain potentials. These maxima can often obscure the wave of interest. They can be often eliminated by adding a small amount of surface-active material, such as gelatin.

20.8.6 Types of Polarographs

Polarography has advantages over non-electrochemical methods in terms of speed, selectivity and sensitivity. An earlier method of achieving these was by the single-sweep cathode ray polarograph, in which the scan of a quickly increasing dc potential applied across the polarographic cell, in the last quarter of the life of every mercury drop, led to sensitivity improvement by avoiding the serrations caused by scanning over the lives of several drops. The curves do not resemble a conventional polarogram, but possess peaks of characteristic shape with the value of the current at the summit.

ac polarography

Another technique known as ac polarography was later developed. In this, a small ac voltage is superimposed upon the potential scan of conventional polarography. The superimposed voltage may be sinusoidal. Here, the disadvantage is that, the small polarographic waves are obscured by the capacity current, with consequent sensitivity restrictions. The undesirable effect of capacity current on the sensitivity of ac polarography is eliminated by applying a square wave (pulse) voltage, in place of the sinusoidal voltage, in addition to the linearly increasing voltage, which is effectively applied to each drop. The amplitude of the ac component of the cell current is measured shortly before each sudden change in the applied voltage.

Square-wave polarography

The method of square-wave polarography gave derivative type polarograms, with peaks proportional to concentration down to a possible detection limit of approximately 10^{-8} mol dm^{-3} for the reactive species. However, the instrumentation is complex and expensive. This technique was improved by the introduction of derivate pulse polarography, in which one pulse of rectangular voltage, usually having a duration of 1/25 s and amplitude 30 m V, is applied to the slowly and linearly increasing potential, after a definite time in the life of each mercury drop and the current is measured during the second-half of the pulse, the recorded curves having the shape of peaks. Pulse synchronization permits very short time intervals between sample points in drop lifetimes; fast scan rates are also possible.

Pulse polarography

In the integral varient of pulse polarography, the dropping mercury electrode is polarized with a pulse of rectangular voltage having linearly increasing amplitude and the current measured during the second-half of the pulse. The polarograms obtained in this way are identical in shape with those obtained in classical polarography, but since the time between the moment of application of pulse to the moment at which the current is measured, is much shorter than the drop-time, the limiting value of the measured current is much higher than that observed in classical polarography.

Pulse polarography can be carried out in two modes; normal pulse polarography and differential pulse polarography. Normal pulse polarography involves the imposition of square-wave voltage impulses or increasing magnitude upon a constant dc voltage. In normal pulse polarography, the contribution of background capacitance to the current is minimized by eliminating the continuously varying potential ramp, and replacing it with a series of potential steps of short direction. This is shown in Figure 20.20. Here, each potential step begins at the same value and the amplitude of each subsequent step increases in small increments. This method offers excellent discrimination against the background capacitive current. The pulse polarography method increases the analytical sensitivity by 1-3 orders of magnitude. In the differential pulse mode, fixed magnitude pulses superimposed on a dc voltage ramp are applied to the working electrode; the currents are measured prior to pulse application and just before termination of the voltage pulse. For a given pulse, the output is recorded as the difference between the two current flows. Figure 20.21 shows the voltage waveform applied in differential pulse polarography. This method has even better ability to discriminate against capacitive current because it measures a difference current.

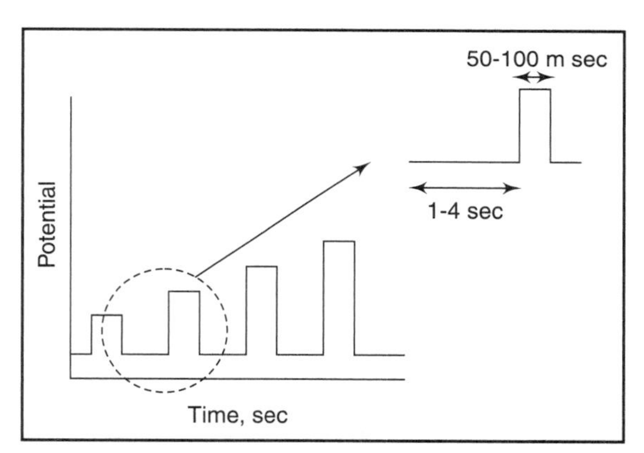

Figure 20.20 :: Principle of normal pulse polarography (Anderson, 1996b)

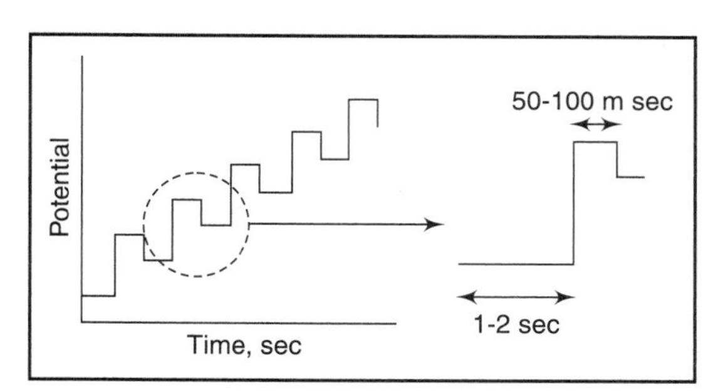

Figure 20.21 :: Principle of differential pulse polarography (Anderson, 1997)

Figure 20.22 shows the various types of pulses involved in pulse polarography. The cell current in pulse polarography would consist of:

- A background current due to reductions occurring at the applied dc voltage.
- A current due to the charging and discharging of the double-layer capacity around the mercury drop when the pulse is applied. If the cell impedance is low enough, this falls to zero within 20 ms.
- A Faradaic current associated with the change of reaction rate, as the pulse is applied.

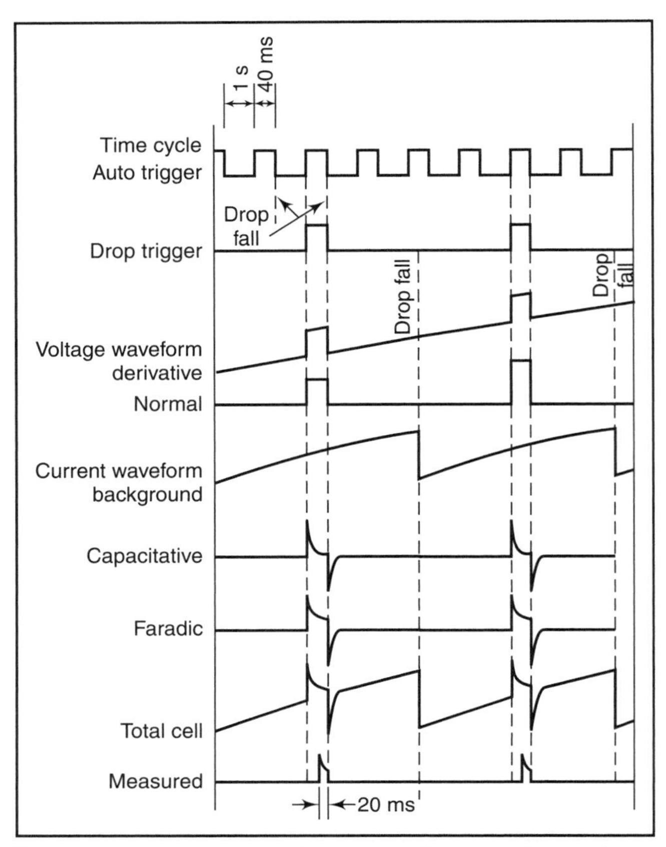

Figure 20.22 :: Typical time cycle in pulse polarography

For the polarogram to be independent of background and capacitive currents, the effect of background current is estimated instrumentally and is subtracted from the cell current. The remaining current is measured during the last 20 ms of the pulse, when the capacitive current has fallen to zero. Pulsed polarography is useful in determining trace metal pollutants in air and water.

Modern polarographic instruments are PC based which perform all typical voltametric and polarographic analysis such as dc polarography, pulse polarography and cyclic voltametry. Figure 20.23 shows a typical PC-based instrument. A range of electrodes are available to meet the needs of various applications.

20.9 Coulometers

Coulometry

Coulometry is an analytical teachnique for measuring an unknown concentration of an analyte in solution by completely converting the analyte from oxidation state to another. This technique is an absolute measurement similar to gravimetry or titration and requires no chemical standards or calibration.

Thus, coulometric methods of analysis depend upon the exact measurement of the quantity of electricity that passes through a solution during the course of an electrochemical reaction. The quantity of reactant formed between the beginning and the interruption of current at the end of the process is directly related to the net charge transferred, Q. Analytical methods based on the measurement of the quantity of electricity are designated by the generic term of coulometry, a term derived from coulomb.

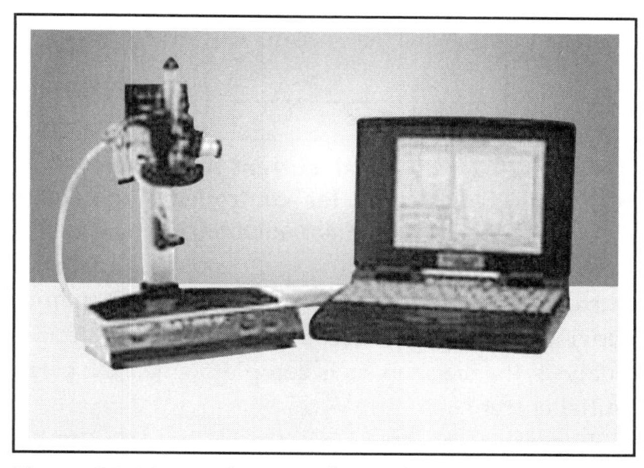

Figure 20.23 :: Polarography–voltametry system with PC support (Courtesy, M/s Topac, USA)

According to Faraday's law, the quantity of electricity involved in the electrolysis of one equivalent of substance is one Faraday or 96,494 coulombs. The weight W of a substance consumed or produced in an electrolysis involving Q coulombs is:

$$W = W_m Q / 96{,}494 n \qquad (1)$$

where W_m is the gram atomic weight or gram molecular weight of the substance being electrolysed. Here, n is the number of electrons involved in the electrode reaction. Equation (1) can also be expressed as:

$$Q = nFVC_b$$

where F equals 96,490 coulombs, V equals volume of solution in litres and C_b equals bulk concentration of the electrolysed analyte in moles/litre.

The total amount of electricity (Q) which is required to electrolyse a certain species is the current time integral:

$$Q = \int_0^t i\,dt$$

This integral is equal to the area under the i-t curve.

Coulometric methods employ the following two techniques:

- Potentiostatic coulometry or controlled potential coulometry, and
- Amperostatic coulometry or controlled current coulometry.

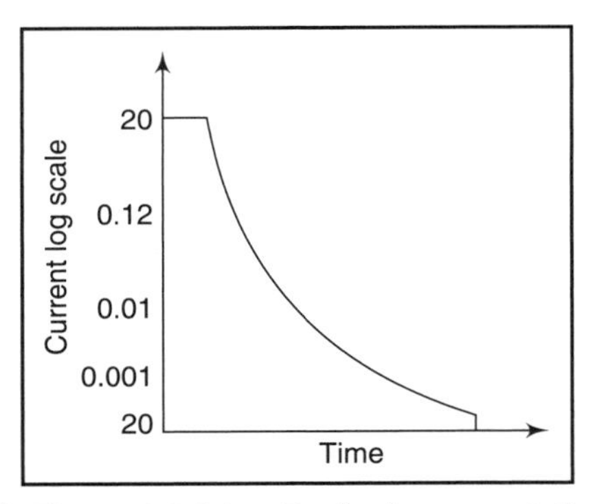

Figure 20.24 :: Typical current-time curve for controlled potential coulometry

In controlled potential coulometry, the potential of the working electrode is controlled at a constant value and the electrolysis current is measured against time. Completion of the electrolysis is indicated by decay of the current to a negligibly small value. Typical working electrodes used in this analysis are platinum, silver, silver-chloride and mercury.

Figure 20.24 shows a typical current–time curve for controlled potential coulometry. Before starting the electrolysis, the working electrode potential is chosen, which will electolyse the species of interest. If this potential is not known before hand, it can be determined from a polarogram of a standard solution of the given analyte. When the electrolysis is begun, the current increases to a high value initially. It then falls exponentially, as the analyte is consumed. Obviously, it is possible to carry out the electrolytic generation to infinite time. In practical analysis, the electrolysis is complete when the current has decayed to less than 0.1 per cent of the initial current.

Integration of the *i-t* curve may be done by graphic, mechanical, electromechanical or electronic means. The integration unit may be attached to the potentiometric recorder which draws the *i-t* curve.

The current decay in controlled potential coulometry is given by the equation:

$$i_t = i_o \times e^{-kt}$$

or

$$2.3 \log (i_o/i_1) = kt,$$

where i_o is the initial current and i_t the current at time t. When the logarithm of the current is plotted as a function of time, the intercept at $t = 0$ is i_o and the slope is $-k/2.3$.

Controlled current titration

In a controlled current titration, the current is set at a working electrode and maintained throughout the titration, with the potential at the indicating electrode measured against time. A mercury or platinum electrode is used as the working electrode, along with a reference and indicating electrode.

The success of most coulometric titrations depends upon the ability to attain 100% current efficiency. This implies that the amount of titrant produced in the electrolysis is exactly equal to that predicted by Faraday's law, judicious selection of the current density (current/unit electrode area) can mean the difference between success and failure of a coulometric titration. A value of 0.5 mA, cm^{-2}, mN^{-1} is commonly stated as near the maximum limiting current density for many substances. Coulometry involves only the fundamental quantities of current and time. It is thus free from many of the uncertainties and errors associated with standard solutions. It offers excellent precision and accuracy and is useful for analysis at very small amounts, ranging as low as a few hundredths of a microgram in volumes of 5–50 ml.

A coulometric cell is an electrochemical cell, in which the two electrodes can be a platinum pair. As electrochemical reactions can occur at either electrode, the two-electrode compartments must be separated by a suitable membrane to prevent interaction between the products. A sintered glass disc or an agar–gel plug is often used for this purpose. The electrode compartment of interest must be suitably stirred. The end-points are usually detected with a microammeter.

Coulometric experiments range from chrono-coulometry, whereby steps or sweeps of electrode potential produce current transients of less than 10^{-3} s duration to the recording of charge-time relations or total final charge values in electro-synthesis, electrode-position or controlled potential coulometry operations, which last several minutes or hours.

20.10 Amperometers

Amperometry

Amperometry is the method of determining the concentration of an electroactive substance by applying a fixed voltage across an indicator and reference electrode, and then measuring the current passing through the cell. This technique is particularly well suited to trace analysis. The current measured is generally on the diffusion-current plateau of the current–voltage curve, a region where the current is independent of the potential of the indicator electrode. The rate of diffusion and hence the current is proportional to the concentration of diffusing material in the bulk of the solution. The most common use of amperometry is in titrations, wherein the current is measured as a function of the volume of titrant added. Concentration changes during a titration are reflected in a change in the current.

The amperometric technique may be used with either single polarized electrode or two polarized electrodes. The electrodes in either case are usually small with surface areas of a few tenths of a square centimetre.

In the single electrode method, a polarized electrode, coupled with an unpolarized electrode (saturated calomel electrode), is immersed in the solution being titrated. The polarized electrode may be a dropping mercury electrode, a rotated electrode or a stationary electrode. The cell is connected to a manual type polarograph. The potential of the polarized electrode is held at a constant value during the titration and the current which flows through the system is observed. The second electrode acts as a reference electrode. Typical titration curves in amperometric titrations with one polarized electrode may take a variety of forms depending on whether the electroactive species is the titrate, titrant or a product of the titration reaction.

In the two-electrode system, two stationary platinum wire electrodes are immersed in the titration cell. The potential of both electrodes varies during the titration, but the potential difference between them is kept constant. The current through the cell is measured during titration. The equivalence point is deduced from the plot of current against volume of titrant. The applied potential difference is relatively small, 0.01 to 0.1 V and the current is generally not as large as in single indicator electrode amperometry. Two electrode system is also called dead stop end point, and is particularly applicable when a reversible oxidation-reduction system is present, either before or after the end point.

Amperometric titrations are carried out in a polarographic cell suitably modified to permit entry of a burette stirrer. H-type cells are also convenient, but a wide-mouthed 100 ml flask or beaker fitted with a suitably pierced cover is mostly employed. When oxygen is known to interfere, it may be removed in the usual way. For stirring, gas bubbling is frequently employed. Any polarographic instrument may be employed to carry out amperometric titrations.

Advantages of amperometric titrations are that, the electrode characteristics are unimportant and the method offers greater sensitivity than conductance and potentiometric titrations. The method is applicable in very dilute solutions, even down to 10^{-4} or 10^{-5} M, according to the type of electrode used.

20.11 Aquameters

Aquametry involves the determination of small amounts of water in solids, liquids and gases. This is done by titration with Karl Fischer Reagent (KFR) and the end-point is determined colorimetrically or electrometrically. The electrometric method is generally preferred, because highly coloured samples obscure the colorimetric end point.

Aquametry

Automated titration equipment have been designed, in which the titrant delivery is controlled by an automatic correcting circuit, that varies the rate of titrant addition to minimize titration time, while maintaining high precision and repeatability. A sensitivity as great as 1 ppm of water in 100 ml of sample is attainable.

Karl Fishcer Trirator

Karl Fishcer Titrator essentially comprises of two parts: The Burette Assembly consists of a piston traveling in precision bore glass tubing. The inside diameter of the tube is very accurately controlled with a tolerance of ± 0.0002 inch. The pistons are coupled with the counters which are driven from the lead screw by a toothed belt. Backlash must be held to less than 0.002 ml. A small dc motor is connected to the lead screw by means of a friction drive and gear set with a speed reduction of 200:1. Limit switches are used to automatically stop the motor at both ends of the piston stroke.

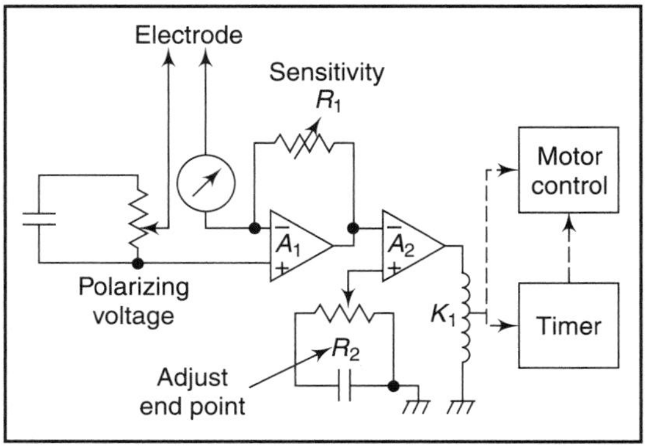

Figure 20.25 :: Block diagram of electronic circuit for Karl Fischer titrator

Figure 20.25 shows the block diagram of a typical Karl Fischer Titrator. A voltage is applied to two platinum electrodes immersed in the sample solution. The resultant current is measured by amplifier A_1 and compared to a reference voltage by amplifier A_2. When the output of A_1 exceeds the reference voltage set at R_2, relay K_1 is energized. The relay contacts turn the burette drive motor on and off. A timer circuit is used to provide selectable end-point delays from ten seconds to two minutes. The burette drive motor should be speed controlled during the course of the titration, so that time required to reach a precise end-point is minimized.

In the Karl Fischer titration, each addition of titrant causes an initial increase in electrode current, which decays rapidly at first, and then slowly, as the end-point is approached. The end-point is reached when the current stays above a set level for some pre-selected time. Figure 20.26 shows a plot of electrode current versus time for a typical titration. With the advent of microprocessors, major improvements in instrument design have occurred in the field of titrimetry.

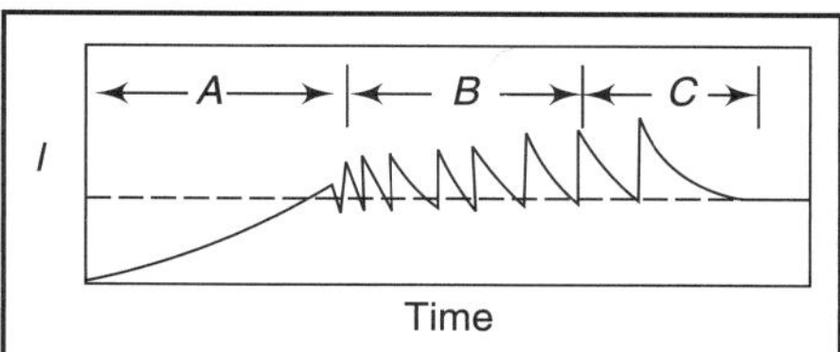

Figure 20.26 :: Electrode current versus time for a simulated titration

20.12 Microcomputer-controlled Electrochemical Instrumentation

Several electrochemical techniques have proved to be powerful tools in analytical chemistry. For example, for an overview of chemistry, cyclic voltametry is powerful, but normal and reverse pulse voltametry, rotating electrode methods, or chrono-coulometry provide simpler means for obtaining quantitative evaluation of parameters of interest. A greater insight of a sample is obtained by examining it by various types of pulse voltametry, possibly by ac voltametry in any of several forms, and perhaps by one or more stripping procedures. The objective, of course, is to find the combination of conditions that leads to optimum sensitivity, precision and flexibility in the format of electrochemical excitation and observation.

Ideally, in an electrochemical laboratory, one would have a full range of options on a single instrument that should be simple, provide common style for selecting and setting up the experiments, regardless of experimental mode and the time needed to switch between modes ought to be negligible. The only realistic method for achieving this is to place the entire repertoire under the charge of a computer. The computer must have full control of the potentiostat and cell by an automated switching network. it must provide a range of excitation waveforms and schemes for acquisition of data, processing the data for taking certain decisions and communication with the operator.

He, *et al.* (1982) constructed an apparatus based on these ideas. The functional block diagram is shown in Figure 20.27. The system is constructed around Intel SBC-80/10B single board computer, which supervises all operations. It operates in a standard Intel Multibus configuration by which it interacts with additional memory and additional custom designed boards, that carry: (i) the potentiostat, (ii) circuitry for acquisition of electrochemical responses, (iii) apparatus for switching cell and scale connections, (iv) a controller for video display, and (v) circuitry for automatic

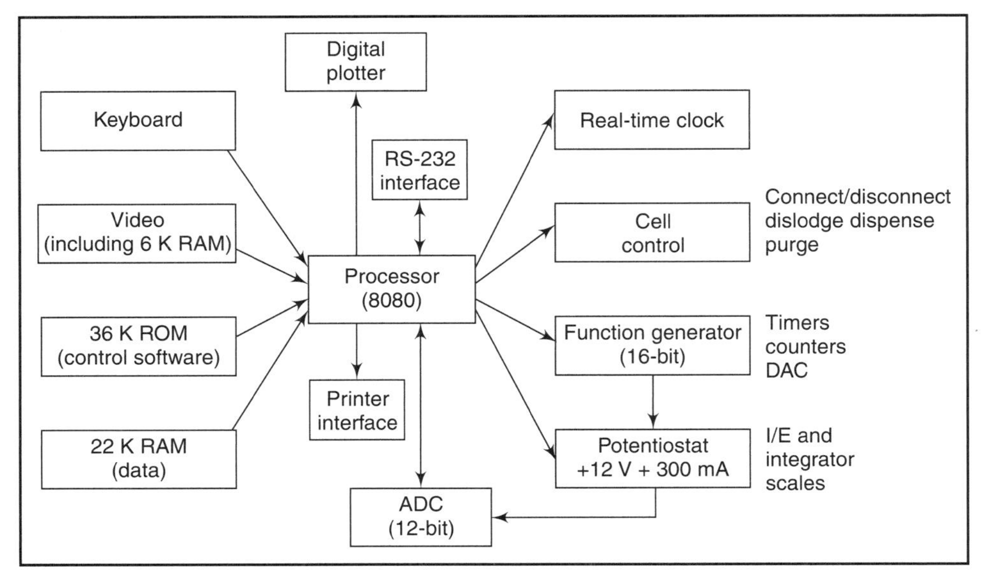

Figure 20.27 :: Block diagram of a microcomputer based electrochemical instrument

compensation of cell resistance. Table 20.1 lists out electrochemical repertoire of the instrument. Control of all functions is exercised through software resident in 36 kbyte of ROM. The remaining 28 kbytes of memory exist as random access storage for data and other temporary information.

Table 20.1 Electrochemical Techniques [After He, *et al.* (1982)]

Pulse Methods	: Normal pulse polarography and voltametry : Square wave polarography and voltametry : Differential pulse polarography and voltametry : Reverse pulse polarography and voltametry
AC Methods	: AC polarography and voltametry : Phase selective ac polarography and voltametry : Second harmonic ac polarography and voltametry
Sweep Methods	: Linear sweep voltametry : Cyclic voltametry
Stripping Methods	: Differential pulse stripping : Square wave stripping : Linear sweep stripping
Miscellaneous	: Chronocoulometry : Bulk electrolysis with coulometry

One has access to almost the complete range of electroanalytical methods, that can be carried out with a single potentiostat. The potentiostat has a standard adder design and is programmed by the output of a 16-bit digital-to-analog converter. Step and pulse measurements are possible on time scales, from a few milliseconds to several seconds; cyclic and linear sweep voltametry can be done at rates upto 50 v/s, and ac voltametry is possible at 10–250 Hz.

The instrument contains control circuitry and software that makes it fully compatible with static mercury drop electrode. In addition, it has a real-time clock and a set of interface to external devices, such as printers and control terminals. A separate repertoire of display and plotting operations allows results to be examined, smoothed or unsmoothed, as averages of multiple runs, as differences between sample and background voltamgrams, as first or second derivatives, as magnified views of data in specified potential ranges or as plots of individual sweep segments in multi-cycle voltametry.

A significant feature of the instrument is a provision for automatic compensation of cell resistance. Through software, the system first makes an actual measurement of the uncompensated resistance in the cell, then it decides the maximum degree of compensation that can be accurately obtained by positive feedback.

The interaction of the operator with the instrument takes place through an integral video monitor capable of presenting text and graphics. Commands are entered through a standard keyboard. Permanent copies of graphical data are available at high resolution from a digital plotter, or at low resolution from a dot-matrix printer.

Chapter 21

pH Meters and Ion Analyzers

21.1 What is pH ?

The concept of pH was introduced by Sorensen in 1909. He recognized that hydrogen ion concentrations, as distinct from total acidities, are frequently of importance in chemical processes. While studying enzymatic reactions, he found it convenient to define a symbol which could represent the concentration of hydrogen ions and called this symbol as pH. It is defined by the following equation:

$$pH = -\log_{10} C_H,$$

where C_H is the hydrogen ion concentration.

$$C_H = 10^{-pH} \qquad \text{(i)}$$

Pure water is known to be a weak electrolyte and it dissociates to form hydrogen ions and hydroxyl ions as follows:

$$H_2O = H^+ + OH^-$$

Assuming that activity coefficients are unity, the dissociation constant K_W of pure water is given by:

$$K_W = C_H^+ \times C_{OH}^- \qquad \text{(ii)}$$

The product of hydrogen and hydroxyl ions in water at 25°C is 1.008×10^{-14} moles^2 litres^{-2} and the concentrations of hydrogen and hydroxyl ions will of necessity be equal. Since the positive and negative electric charges in the solution must balance, each of these concentrations is given by:

$$C_{H^+} = C_{OH^-} = \sqrt{K_W}$$

$$= \sqrt{1.008 \times 10^{-14}}$$

$$= 1.004 \times 10^{-7}$$

$$C_H^+ \approx 10^{-7} \qquad \text{(iii)}$$

Therefore, pH of pure water = 7.

It is obvious that the neutral point or point at which the hydrogen and hydroxyl ions are present in equal concentrations, is located at pH 7.

The lower-case letter 'p' in pH stands for the negative common (base ten) logarithm, while the upper-case letter "H" stands for the element hydrogen. Thus, pH is a logarithmic measurement of the number of moles of hydrogen ions (H^+) per liter of solution. Incidentally, the 'p' prefix is also used with other types of chemical measurements where a logarithmic scale is desired, pCO_2 (Carbon Dioxide) and pO_2 (Oxygen) being two such examples.

The pH of an acidic solution [i.e. (H^+) > ($0H^-$)] at 25°C will be less than 7 and that of an alkaline solution greater than 7. The peculiarity of the logarithmic scale is exemplified by the fact that a ten-fold decrease in hydrogen-ion concentration corresponds to an increase of one pH unit, whereas a two-fold increase of the concentration means pH will decrease by log 2, i.e. by 0.301 units.

The dissociation constant K_W is a function of temperature and therefore, the neutral point will vary as the temperature is changed. The pH for neutral conditions decreases to 6.5 at 60°C from the value of 7 at 25°C and increases to 7.5 at 0°C. The range of the pH scale also depends upon the magnitude of K_W.

The approximate practical range of the pH scale is from –1 to 15 at room temperature, although most of the commercial instruments are designed to measure 0 to 14 pH.

There is, at present, a wide utilization of pH measurements in the chemical laboratories, industries and the clinics. This has been made possible by the discovery of the hydrogen ion function of glass membranes, which led to the development of convenient, practical glass electrodes, pH meters and controllers that allow the pH of process solutions to be adjusted automatically. The technology of pH instrumentation has been greatly developed and extremely sensitive instruments are now commercially available.

With the developments in chemical thermodynamics, it has gradually become clear that Sorensen's experiments did not, in reality, yield hydrogen ion concentrations. No doubt the numbers obtained depended in a complex manner on the activity of the electrolytes in the solution under investigation, they were not an exact measure of the hydrogen ion activity and indeed could never be made so. Sorensen's measured value of pH are not, therefore values of C_{H^+} (Hydrogen ion concentration) as he originally considered them to be. However, the methods of pH measurement have been standardized by convention, in a manner which allows a maximum of theoretical significance to be placed upon the experimental results.

21.2 Principle of pH Measurement

The measurement of hydrogen ion concentration (pH) in a test solution is made by measuring the potential developed in an electrochemical cell. The electrochemical pH cell consists of a measuring electrode and a reference electrode, both immersed in the solution under investigation. The two electrodes are connected to a measuring instrument and the emf (electromotive force) between these two electrodes is measured. The measuring electrode is pH sensitive and its potential is proportional to the pH of the solution, in which it is immersed while the reference electrode would always develop a constant electrical potential, against which the potential of the glass electrode is measured.

Nernst equation

The potential of the measuring electrode may be written by means of the Nernst equation:

$$E = E_0 + 2.3026\ RT/F \log C_H$$

$$E = E_0 - 2.3026\ RT/F\ pH_C,$$

when E_0 = standard potential
R = gas constant
T = absolute temperature
F = Faraday constant
pH_C = pH value deviation from 7.

The above equation shows that the emf developed in the electrochemical pH cell is a linear function of pH_c. Figure 21.1 shows the relationship between pH and temperature of a typical glass electrode.

Change of pH of one unit = 58.2 mV at 20°C
= 62.2 mV at 40°C

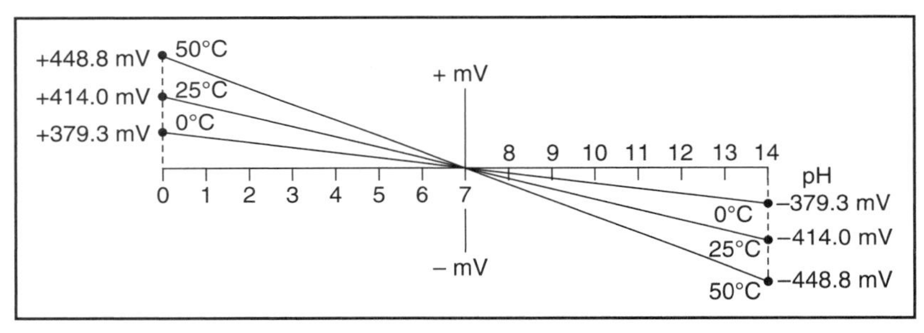

Figure 21.1 :: Relationship between pH and emf at different temperatures (Courtesy, M/s Beckman Instruments, USA)

Slope factor

The factor –2.3026 RT/F is called the slope factor and is obviously dependent upon the solution temperature. It is clear that with 1°C change in temperature, the emf changes by 0.2 mV. Further, pH measurement is essentially a measurement of millivolt signals by special methods.

Buffer solution

For measurement of pH, the electrodes are first immersed in a buffer solution of known pH. The pH meter zero reading is adjusted by the standardization control, until the pH value of the buffer is indicated by the meter. This standardization automatically compensates for the various potentials in the electrodes system. Subsequently, immersion of the electrodes in a test solution produces a potential that is proportional to the pH of the solution. This potential registers directly as pH on the scale of the pH meter. Temperature compensation knob is set at the temperature of the solution.

21.3 Electrodes for pH Measurement

21.3.1 The Hydrogen Electrode

The hydrogen electrode is the primary electrode to which all electrochemical measurements are referred. However, owing to the experimental difficulties associated with it, other electrodes are commonly employed for routine pH measurements. Nevertheless, the performance of all other electrodes is always evaluated in terms of the hydrogen electrode.

The hydrogen electrode consists of an inert but catalytically active metal surface, most frequently platinum, over which hydrogen is bubbled to achieve electrochemical equilibrium with the hydrogen ions in the solution. The following redox reaction takes place:

$$H^+ + e^- \rightleftharpoons 1/2\ H_2$$

The electrode is immersed in the solution under investigation and electrolytic hydrogen gas at 1 atm pressure is bubbled through the solution and over the electrode, in such a way that the electrode surface and the adjacent solution gets saturated with the gas at all times. Electrode life is 7-20 days before its response becomes sluggish. The potential set up at the hydrogen electrode by a given activity of hydrogen ions is governed by the Nernst equation. When the partial pressure of the hydrogen is other than 1 atm, correction would have to be applied.

Since the hydrogen electrode is essentially a redox system and, as such, is affected by the presence of oxidising and reducing agents, it is therefore subject to a number of limitations in its application.

21.3.2 Glass Electrode

The widespread application of pH measurements in the control of industrial processes, as well as in research can be largely attributed to the development of the convenient and versatile glass electrode. Its action is based on the principle, that when a thin membrane of glass is interposed between two solutions, a potential difference is observed across the glass membrane, which depends on the ions present in the solutions. Depending on the composition of the glass, the response may be to H^+ ion or it may be to other cations. The selective response of certain glass compositions to H^+ has led to the development of pH responsive glass electrodes.

In construction, the glass electrode consists of a thin walled bulb of pH-sensitive glass sealed to a stem of non-pH sensitive high resistance glass. The pH response is limited entirely to the area of the special glass membrane, thus making the response independent of the depth of immersion. The membrane normally has a thickness of the order of 0.05–0.15 mm, and the bulbs are of the order of 10 mm in diameter. Figure 21.2 shows typical construction of a glass electrode. Both surfaces of the membrane are pH sensitive.

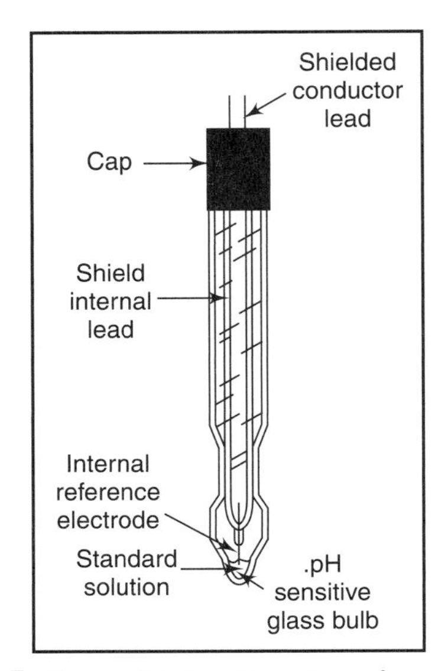

Figure 21.2 :: Construction details of Beckman glass electrode

The glass pH electrodes are constructed of special glass to create the ion-selective barrier needed to screen out hydrogen ions from all the other ions floating around in the solution. This glass is chemically doped with lithium ions, which is what makes it react electrochemically to hydrogen ions. Of course, glass is not exactly what you would call a "conductor" rather, it is an extremely good insulator. This presents a major problem if it is intended to measure voltage between the two electrodes. The circuit path from one electrode contact, through the glass barrier, through the solution, to the other electrode, and back through the other electrode's contact, is, therefore, one of extremely high resistance.

On the inside of the membrane is a system of effectively constant pH. It is composed of a silver–silver chloride or calomel electrode dipped in hydrochloric acid. Changes in electrical potential of the outer membrane surface are measured by means of an external reference electrode and its associated salt bridge. The complete pH cell is represented as follows:

Internal Reference Electrode	Internal Electrolyte	Glass Membrane	Test Solution	External Reference Electrode

The ideal pH response of a glass electrode behaving exactly in the same manner as a hydrogen electrode is given by:

$$E_2 - E_1 = 2.3026\ RT/F(\text{pH}_2 - \text{pH}_1)$$

where E_1 and E_2 are the values of the electromotive force of cell 1 in test solutions of pH equal pH_1 and pH_2, respectively.

pH response

The equation shows that ideal pH response is 54.2 mV at 0°C, 59.16 mV at 25°C and 73.04 mV at 95°C. Unfortunately, no glass electrode yet constructed has the theoretical response in all types of test solutions and over the entire pH range. The most important characteristics of a glass electrode are, low melting point, high hygroscopicity and relatively high electrical conductivity. For many years, the best pH sensitive glass available was Corning 015 or Schott 4073 glass.

pH range

The useful pH range for a glass electrode generally lies between pH 1 and pH 11. Below pH 1, acid errors may be expected requiring frequent standardization against a buffer of a pH close to the actual pH of the test solution. Above pH 11, alkaline errors are observed. The alkaline error depends upon the composition of the membrane glass. Some manufacturers like Radiometer (Denmark) supply two different types of glass electrodes, one for 0-12 pH and the other for 0-14 pH. An approximate correction can be made for alkaline errors by means of empirically determined correction for the type of membrane glass used.

Due to a large change with temperature of the electrode resistance, glass electrodes will, as a rule, perform satisfactorily only within a temperature range of about 60°C. However, special glass electrodes covering temperature to about –10°C or up to about +120°C are also available, but they need to be specified.

The useful lifetime of a glass electrode is generally several years, if it is handled with care. The bad condition of the electrode manifests itself in slow response and reduced sensitivity. Usually, a glass electrode may be considered good if it has a fast response (<1 min), gives stable response during measurements on a well stirred solution and exhibits sensitivity better than 95 per cent.

Glass membrane

A glass membrane exhibits quite a high electrical resistance and consequently the internal resistance of the cell with a glass electrode is of the order of 50-500 MΩ. The emf measurement therefore, necessitates the use of measuring circuits with high input impedance. Furthermore, the electrodes and the leads from them must be supported on holders made of a good insulating material, in order to eliminate electrical leakage across the outside surface of the glass bulk. Sometimes, the upper part of the outside of the glass electrode is rendered water repellent by the application of a silicone oil or parafin wax. The high resistance of glass electrodes renders them very susceptible to capacitive pick-up from ac mains or charged bodies. In order to minimize such effects, it is necessary to screen the electrode cable. The screen may be connected directly to earth or is grounded to the case of the instrument.

Commercial glass electrodes are available in a wide variety of sizes and shapes. They are designed to operate with samples as small as one drop; others require at least 5 ml of solution. Special microcells for the pH measurement of blood are supplied by several manufacturers.

The correct physical dimensions of the electrode are dictated by the sample size and the sample vessel. Electrodes with epoxy bodies and a protective cap in place should be used to combat harsh environmental conditions. An electrode with a flat membrane should be used for measurements performed directly on flat surfaces. In short, an electrode design literally exists for almost every possible measuring situation.

Glass electrodes have two disadvantages. These are:

- Measuring solutions containing particulate matter can damage the glass membrane.
- The glass membrane is easily broken. There are alternatives to the glass membrane, though they are used seldom due to other drawbacks, such as limited pH range or long response time. The antimony electrode is used as an alternative, mainly in solutions containing HF. A thin oxide layer formed on the surface of the antimony is sensitive to pH.

Care and maintenance of the glass electrode on a regular basis ensures:

- a faster response,
- more reliable measurements, and
- a long lifetime.

The entire glass membrane must always be clean. Rinsing the membrane with distilled water will often suffice for aqueous solutions. Rinsing the electrode with a mild detergent solution once a week will be beneficial. An alkaline hypochlorite solution can be used to clean electrode membranes subjected to solutions containing fat or proteins.

Between measurements, store the glass electrode in a pH buffer with pH< 7. High temperature measurements, compounded by constant use in strong alkaline solutions or weak solutions of hydrofluoric acid will drastically reduce the lifetime of the electrode, since the glass membrane will slowly dissolve. Dry storage is recommended if the electrode will not be used for two weeks or more. Before use, the electrode should be soaked well.

New electrodes or those that have been stored dry should be conditioned or activated before use by soaking the bulb for a period of 12–24 h in 0.1 N hydrochloric acid. After overnight soaking rinsing, soaking in a buffer of pH = 4, and again rinsing, the electrode should be ready for use. The relative fragility of glass demands reasonable care in its handling. The pH-sensitive tip of the electrode should not become scratched or cracked through contact with the sample or container. Therefore, it should not be allowed to rest on the bottom of the sample container. The tip should be dried by gentle rubbing with absorbent tissue.

Trapped air bubbles around the inner reference electrode will produce an unstable reading. Swing the electrode in an arc or tap it gently to remove the bubbles. The electrode may have to be heated gently to approximately 60°C in a water bath if the air bubbles are trapped by KCl crystals.

A sluggish response for a glass electrode, even after proper maintenance has been performed, may dictate the need for a slight etching of the outer glass layer of the membrane. The following treatment is only recommended after all other measures have been used to improve response and have failed. Soak the glass membrane portion of the glass electrode in a 20% ammonium bifluoride solution for one minute, followed by 15 seconds in 6 M hydrochloric acid. Rinse the electrode well and soak for 24 hours in a pH buffer with pH< 7. The electrode deteriorates after 5–7 etchings.

The proper functioning of the glass electrode depends on the hydration of the glass layer that takes place on the surface of the pH sensitive glass membrane during soaking and measurement in aqueous solutions. As long as the electrode is frequently rehydrated, accurate measurements in non-aqueous or partly aqueous solutions are also possible. This can be accomplished by soaking in a slightly acidic buffer. In non-aqueous solvents completely immiscible with water and before soaking, the electrode should first be rinsed with a solvent which is miscible with both water and the solvent before rinsing with water.

The electrode cable and the electrode plug must be kept clean and dry if reliable measurements are to be obtained because of the very small electrode currents which pass through the glass electrode.

A number of factors dictate the useful lifetime of the glass electrode membrane and is highly individualistic. High temperatures, frequent measurements in alkaline solutions, repeated etchings, and improper maintenance will reduce the electrode's lifetime, whereas proper maintenance will prolong the useful lifetime. The glass membrane will, however, deteriorate gradually even when stored dry. A standard glass electrode, whether a monoprobe or a combination electrode, will usually last for 12–18 months.

21.3.3 Calomel Electrode or Reference Electrode

The purpose of the reference electrode is to provide a stable, reproducible voltage to which the working (detector) electrode potential may be referenced. A reference electrode may be considered a small battery whose voltage is determined by the chemistry taking place between a solid conductor, which is usually a metal salt, and the electrolytic solution around it. Ideally, if a small current is passed through the electrode, the potential change is negligible. Also, the potential value of the reference electrode should not vary with time and should be reproducible from electrode to electrode.

The most common reference electrodes which meet these criteria are:

- Mercury/mercurous electrode (calomel), and
- Silver/silver chloride electrode (Ag/AgCl) electrode.

Therefore, in order to measure the potential changes of the pH-sensitive electrode directly, it is necessary that the pH cell be completed by means of a stable reference electrode, whose potential remains unaffected by changes in the composition of the cell solution. The reference electrode against which the potential of the glass electrode is measured is the calomel electrode. It consists of (Figure 21.3), a metallic internal element, typically of mercury–mercurous chloride (calomel) or silver–silver chloride, immersed in an electrolyte, which is usually a saturated solution of potassium chloride.

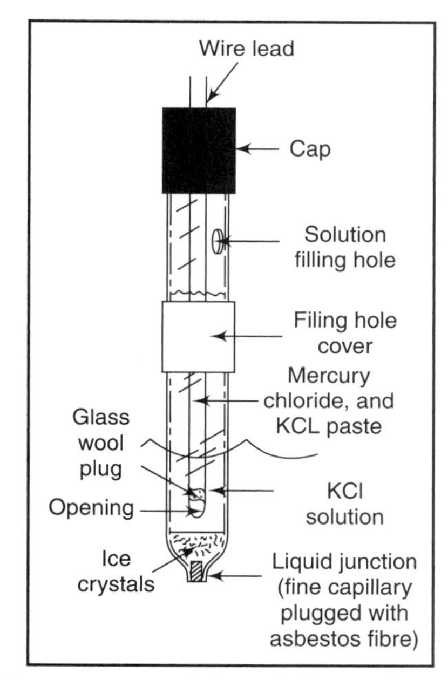

Figure 21.3 :: Calomel electrode (Courtesy, M/s Beckman Instruments, USA)

The electrolyte solution forms a conductive salt bridge between the metallic element and the sample solution, in which the measuring and reference electrodes are emplaced. For a stable electrical connection between the internal metallic element and the sample solution, a small but constant flow of electrolyte solution is maintained through a liquid junction in the tip of the outer body of the reference electrode. Depending upon the nature of the application, this junction may be formed in several ways. For example, the tip of the electrode could be formed by an embedded linen or asbestos fibre, or by a permeable composition of pressed-sintered carborundum and glass pellets.

$$1/2\ Hg_2Cl_2 + e \rightleftharpoons Hg + Cl^-$$

Since at a given temperature, the activity of the mercurous chloride is constant and that of the mercury is unity by definition, it is the chloride ion activity which is potential determining. When this is fixed, the electrode has a fixed potential at a fixed temperature. The most commonly used source of chloride ion is potassium chloride at saturated 3.8 M, 3.5 M or 0.1 M concentrations. Saturated electrodes have largely replaced electrodes with other KCl concentrations. They are used in practical pH measurements at high as well as low temperature. However, their useful life is known to be short at temperatures above 70°C.

Although calomel reference electrode is the most commonly used electrode, it is sometimes necessary to use other types of reference electrodes in special circumstances. For example:

- The Ag/AgCl electrode is recommended as a reference electrode at high temperatures (–10 to +110°C).
- The mercury/mercurous sulphate electrode with a salt bridge solution of potassium sulphate is used in test solutions that must not contain chloride ions.
- The calomel chloride with a salt-bridge solution of saturated lithium chloride (LiCl) is used for measurement is non-aqueous solutions. Li Cl is more suitable in organic solvents than is KCl.
- A calomel electrode with open liquid junction facilitates measurements in contaminated samples and samples containing solid particles.
- A calomel electrode with duplicate salt-bridge (potassium nitrate as extra salt-bridge), may be used if test solution must not contain chloride ions.

The reference electrode should preferably be stored in a small beaker containing the salt bridge solution for short-term storage. For long term storage, the electrode should be rinsed, dried, and stored with the end cap on and the rubber band covering the filling hole in place.

The solution in the salt bridge should always be on a higher level than the solution to be measured, as infiltration of the sample solution may occur in the salt bridge, that is, the direction of flow should always be from the salt bridge into the sample solution. If this cannot always be achieved, monthly changing the salt bridge solution should be the norm.

Annular ceramic junctions and porous pin junctions can occasionally become blocked due to crystallization of the salt bridge filling solution. Soaking in the salt bridge solution usually remedies the situation, but, on those instance that it doesn't, raising the temperature to the maximum allowable for the reference system will often help. A precipitate of silver chloride or sulfide may clog the porous pin. The gentle use of an abrasive paper, e.g., emery cloth, will remove most blockages. Soaking the porous pin for a few hours in acidic solution of thiourea (1 M thiourea in 0.1 M HCl) will usually do what the abrasive paper sometimes fails to do, that is, chemically clean the blockage.

Trapped air bubbles can also cause malfunctions. These bubbles can be removed by gently tapping the electrode or shaking it downward as one would do to a clinical thermometer.

The useful lifetime of a reference electrode depends on the maintenance and care given to the electrode. The electrode should never be allowed to dry out, the junction should be kept clean, and the salt bridge should always be filled to the level intended by the manufacturer. With proper maintenance, the lifetime for a reference electrode is indefinite, but usually greater than two years.

By providing good maintenance care to the electrodes, proper calibration should be able to be performed easily. If there is a continued problem, the electrodes should be replaced or examined again.

When performing a calibration with two buffers, stability should occur within approximately one minute in each case. The zero point and sensitivity should be written down after each calibration since a large deviation from one calibration to the next indicates a problem.

21.3.4 Silver/Silver Chloride Reference Electrode

The Silver/Silver chloride (Ag/AgCl) electrode features glass-body construction with a porous ceramic frit at the solution interface. The filling solution is 3M NaCl gel that has been saturated with AgCl, the gel is semi-solid and will appear cloudy with occasional particles.

Figure 21.4 shows the basic construction of Silver/Silver chloride electrode. A porous reference junction separates the filling solution in the electrode from the solution whose pH is to be measured. The filling solution's constant chloride ion concentration generates potential at a pure silver wire with silver chloride on it. The silver wire passes the signal from the solution being measured to the electrode's cable or connector. This configuration of the electrode is called 'Single Junction Reference'.

Single junction reference electrode

Double junction reference electrode

For most applications, a single junction reference electrode is adequate. However, if samples contain proteins, sulfides, heavy metals or any other material which interacts with silver ions, they may react with the gel, causing a reduction in the reference output. This reaction can lead to erroneous reference signals or to precipitation at the reference junction leading to a short service life. A double junction reference electrode design as shown in Figure 21.5 offers a barrier of protection to combat the above interactions. In this design, the inner chamber contains the usual high (3.5 M or higher) salt concentration solutions so that stable outputs are generated. The outer chamber, which contacts the sample through the porous reference junction, is filled with 0.1 M KCl. This lower ionic strength material more closely matches that of the sample and further reduces spurious potentials.

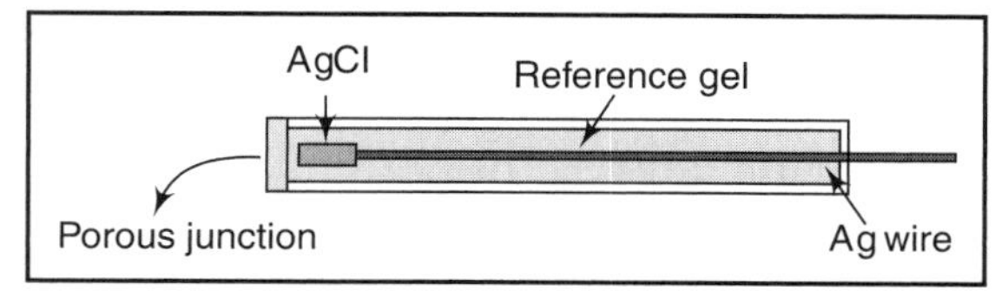

Figure 21.4 :: Silver/Silver chloride reference electrode—single junction

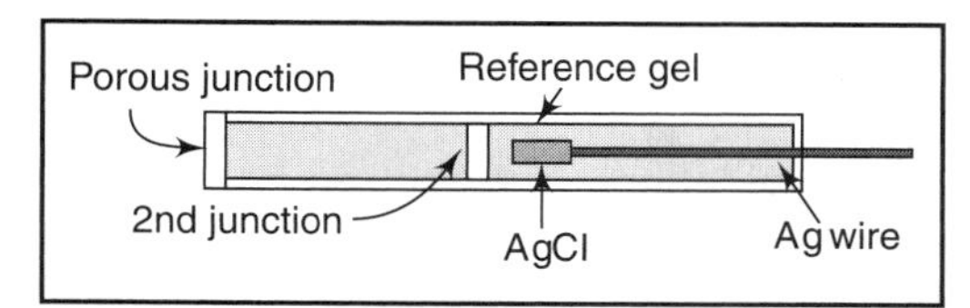

Figure 21.5 :: Silver/Silver chloride reference electrode—double junction

The Ag/AgCl reference electrodes are easily spoiled by drying. It is, therefore, advisable to keep the tips wetted at all times and store in 3M NaCl when not in use. This helps to extend its life time. They usually last for 3-6 months.

The use of a double junction electrode with a second internal junction not containing KCl or a modified internal filling solution can be used. An electrode utilizing mercurous sulfate and potassium sulfate is one example. Several other combinations can be found in Table 21.1.

Table 21.1 Potentials for Different Reference Electrodes

Type of reference electrode	Salt-bridge solution(s)	Potential vs. standard H_2 electrode	Potential vs. Sat. calomel electrode
Hg/Hg_2Cl_2	Sat. KCl	244 mV	0 mV
Ag/AgCl	Sat. KCl	200 mV	−44 mV
Hg/Hg_2SO_4	Sat. K_2SO_4	640 mV	408 mV
Calomel	Sat. LiCl	~200 mV	~ −45 mV
Hg/Hg_2Cl_2	Sat. KCl/KNO_3	244 mV	~ 0 mV

21.3.5 Combination Electrode

A combination electrode is one in which the pH-sensitive glass electrode and the reference electrode are built together (Figure 21.6). In this arrangement, the internal reference electrode in the glass electrode is identical with the external reference electrode. Both reference electrodes are the silver/silver chloride type. Also, the inner solutions in the two parts of the electrode are identical and they both are symmetrical electrode chains. The two silver/silver chloride reference electrodes are protected against light by means of ruby red glass, which absorbs light. The construction ensures that the two electrodes are at the same temperature during operation.

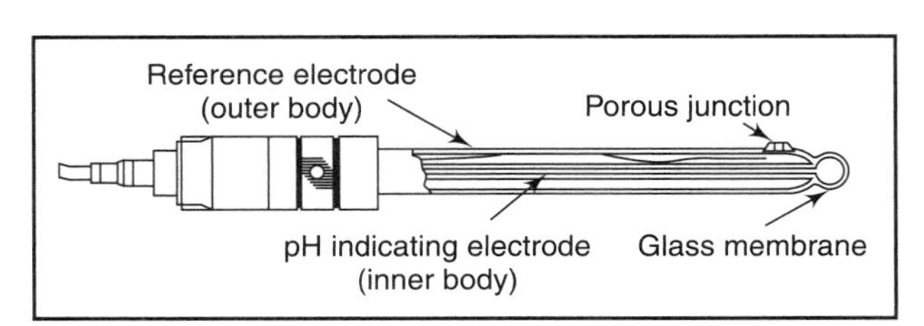

Figure 21.6 :: Combination electrode

A combination electrode can be used within the temperature range –10°C to + 100°C as they have silver/silver chloride reference electrodes. Also, the measuring error due to small temperature differences of the buffer solutions and the sample are negligible.

Combination electrodes containing the electrolyte in the gel format have been introduced, which provide the advantage of low maintenance. They are generally supplied with plastic bodies, and are more rugged than liquid filled glass electrodes. Hence, they are commonly used with portable pH meters as shown in Figure 21.7. The gel used is a non-toxic, USP grade, organic material or it may be polyacrylamide.

Gel-filled electrodes

Most gel-filled electrodes utilize a wick or diffusion style junction. These junctions allow the sample to diffuse in and the reference gel to diffuse out, causing a shift in potential. Some samples also react with the silver ion in the gel, which forms a precipitate. Both of these processes can cause clogging of the junction, as manifested in slower and less accurate response over time.

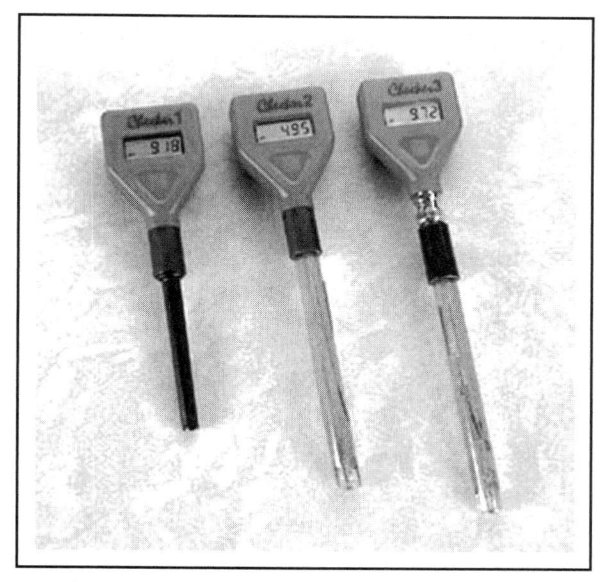

Figure 21.7 :: Portable pH meter (Courtesy, M/s Hanna Instruments, *www.hannainst.com*)

Beckman coulter gel filled combination electrodes provide a special micropore junction which overcomes the above problems. The junction consists of several small pores in the actual body of the electrode which can be wiped clean each time the electrode is rinsed. There is direct contact between the sample and the gel, thus resulting in faster readings. Additionally, this allows the electrode to be stored dry and eliminates the need for a special storage solution. The gel-filled electrodes operate in a wide temperature range (–5 to 100°C) and are available in various shapes, sizes and lengths to meet varied requirements.

The electrodes should never be stored dry. The preferred choice for storing is KCl solution of any concentration between 2.0M and 3.8M. Another good choice is buffer solution (pH4). If it is necessary to clean the probe with acid, caustic, solvent or other cleaning solution, it is best to soak the electrode in KCl solution after cleaning and prior to use or calibration. This will re-condition the bulb and reference, extending the probe life and improving calibration accuracy.

21.3.6 The Asymmetry Potential

If identical solutions are placed inside and outside the bulb of the glass electrode, it is found that in spite of the apparent symmetry of the cell thus formed, there exists an emf of a few millivolts. This potential difference is called the asymmetry potential and is thought to arise from the different states of strain on the inside and outside surfaces of the glass electrode bulb. The asymmetry potential of a glass electrode is not absolutely constant, but may drift slightly from day to day. However, it is not ordinarily subject to large and sudden fluctuations, particularly in glass electrodes, and the existence of this potential is therefore, of no practical consequence in the pH measurement. It may be regarded, for a short series of measurements, as a constant of the cell assembly. In most of the commercial pH meters, this constant potential is compensated by adjustment of the standardizing control knob or zero adjustor, when the instrument is made to read the pH for the standard solution. The standardization of the instrument is carried out immediately before the measurements are made on unknown solutions. As an extra precaution, the standardization procedure can be repeated at the end of the measurement.

21.3.7 Buffer Solutions

A buffer may be defined as a solution whose pH remains nearly constant, despite the addition of a substantial quantity of acid or base. Buffers are employed for the standardization of pH cells. The cell emf is measured for given buffers and related to the known pH values for the buffers. The pH value of an unknown solution is then derived from this calibration. Many buffer solutions have been reported in literature. There are also British standards and standards of NBS on buffers.

Buffer tablets

Buffer tablets are commercially available, which when dissolved in an appropriate quantity of fresh distilled water, give buffer solutions. These tablets may contain the buffer material admixed with substance, which aid in tabletting but do not significantly affect the pH. Buffer tablets for pH 4, 7 and 9.2 are available commercially.

Buffers should be protected from exposure to the atmosphere, where gases such as carbon dioxide, ammonia and oxygen are present, as they will tend to change the pH value of the buffer solution. Since air would enter the bottle everytime the buffer containing bottle is opened, a buffer solution would have a limited useful lifetime, when reliable pH measurements are required. Also, the pH value of a buffer solution depends upon the temperature and the actual value for the test temperature can be found on the label on the bottle.

Buffers with high precision have only a limited lifetime and a limited stability. They must be used within a short period of time, governed by how precise the measurements must be. Though alkaline buffers pose the greatest problem because of their tendency to absorb carbon dioxide from the air, a buffer solution in an opened and capped bottle, will last for a limited period of time. Even unopened buffers in thin, plastic bottles have a tendency to absorb CO_2 from the atmosphere. Thick plastic bottles, sealed in cans under a nitrogen atmosphere, have the greatest advantage for maintaining optimal solutions.

High quality buffers contain a small amount of germicide to prevent microbiological growth, since many of the buffers are excellent culture media. Buffers with other additives, including many colored dyes, could disturb the pH value or the stability of the solution. They may have an adverse effect on the liquid junction.

21.3.8 Calibration

Each electrode has slightly different characteristics. Electrodes with different nominal values are produced by different manufacturers. Also, the zero pH and the sensitivity vary with time. Calibration matches the pH meter to the characteristics of the electrodes being used, and continued calibration of the electrode on a regular basis is required in new of the continually changing characteristics during the lifetime of the electrodes. The most accurate calibration is performed using two different buffer solutions (Figure 21.8). This enables both pH^0 (zero pH) and the slope (sensitivity) to be determined.

The sensitivity is independent of temperature and is usually stated as a percentage of the theoretical value. The slope, when expressed as mV/pH, is directly dependent upon temperature. The slope at 25°C is often used as an alternative to sensitivity in per cent (100% = 59 mV/pH).

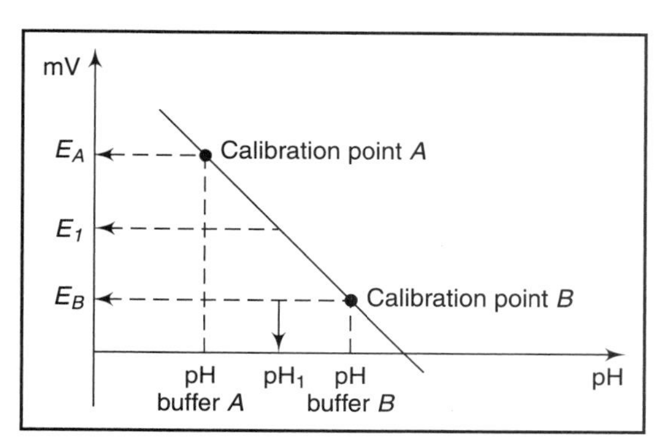

Figure 21.8 :: Calibration curve developed using two buffers

Zero pH is generally used to describe electrode characteristics, though the potential at pH 0 or pH7 at 25°C can also be given.

In modern pH measuring instruments, it is not normally necessary for the operator to construct a calibration graph and interpolate the results for unknown samples. Most pH electrodes when connected directly to a pH meter perform the calibration automatically. This determines the slope mathematically and calculates the unknown pH value for immediate display on the meter.

21.4 pH Meters

The function of the pH meter is to measure the potential difference (in mV) between the electrodes and convert it to a pH display.

In the earlier days, the pH of a solution was determined with the change in colour observed on pH paper when dipped in the solution of interest. The colour was compared with the colour chart. It is possible to read pH to an accuracy of 0.1 pH by this method.

With the advent of glass and calomel electrodes and with the development of very stable, drift-free dc amplifiers with extremely high input impendence, pH measurements with much better accuracy have become possible.

pH electrodes are basically voltage sources with a very high internal resistance. In order to avoid errors in the measurement of voltage, no current should flow from the source if possible. This means that the whole measuring circuit, from the electrode via cable, plug, socket, switching element upto the measuring amplifier itself must be very well insulated. Only high-quality insulation material such as PTFE (Teflon), polyethylene, glass, etc., must be used to get insulation resistance of upto 10^{14} ohms. It may be noted that even a small contamination by atmospheric deposits or from split liquids can influence the insulation values.

The following considerations govern the design of pH meters:

- The internal resistance of the electrode is very high. It is of the order of 1000 MΩ. Therefore, the input impedance must be at least 1000 times more than the resistance of the pH cell. Also, the measurement should be unaffected by large changes in magnitude of this resistance.
- Current should neither be drawn by the pH meter from the solution, nor any current should flow on the electrode, which might result in the polarization of the electrode. Such polarized electrodes will give rise to erroneous results.
- The meter must have a provision for compensating changes in pH readings due to changes in temperature.

In general, commercial pH meters can be broadly categorized into two main types:

- The null-detector type or the potentiometer type, and
- The direct-reading type.

In the instruments of the first type, the procedure followed is essentially that used in potentiometers. An emf equal and opposite to that of the pH cell is applied, so as to give zero reading on a galvanometer.

Direct-reading instruments

Direct-reading instruments are similar to voltmeters of the deflection type. The current signal available after the amplifier is used to operate a meter, suitably calibrated in pH units. Of the two types, the null-detector type of instrument is inherently capable of greater accuracy than the direct reading type. However, the convenience and the usefulness of the direct-reading type meters in following the changes of pH that occur during the course of a reaction have made them more popular.

The pH meters invariably make use of some amplifying device to amplify emf produced in the pH cell. As the amplifier input stage must have an extremely high input impedance, a number of commercial pH meters have been making use of special electrometer tubes. They have a very low grid current and the stray currents emanating from surface leakage emission of photoelectrons by the grid under the action of light and soft X-rays. Quite recently, electrometer tubes have been replaced by solid state devices, like field effect transistors, MOSFET's and integrated circuits having high input impedance. IC 8007 is an example of the integrated circuit having FET input stage and which is quite suitable to be used in the input stage of a pH meter. With modern semiconductor components, a pH meter with an input resistance of up to 10^{17} Ω can be built with little difficulty.

For making pH measurements, the emf of a pH cell is amplified in direct coupled amplifiers. Dc amplifiers give rise to zero drift errors, which must be eliminated for obtaining accurate results. Improvements in zero stability are possible through the use of low drift and low input offset voltage integrated circuits and highly stable power sources. Zero drift can also be reduced through the use of balanced and differential amplifiers. They are so constructed that their responses to external signals are additive, while those to internal noise or drift are subtractive. Several other methods are available, by which zero stability can be achieved to a great extent by using zero corrected dc amplifiers, contact-modulated amplifiers and vibrating capacitor modulated amplifiers. These methods are discussed in the subsequent sections.

The Nernst equation, on which pH measurements are based contains a temperature-dependent component. Therefore, arrangements are invariably made for automatic or manual compensation in changes of pH due to changes in temperature in the commercial pH meters. The instrument is calibrated at one temperature (say, 25°C) and compensation is applied by suitable adjustments of the output current of the amplifier. The current is adjusted by incorporation of a variable resistance in the output circuit, so that the calibration point may then correspond to the desired temperature. This ensures that for a given pH, the current to the meter is constant.

Automatic temperature compensation is achieved by using a resistance thermometer or thermistor in the output circuit. The thermistor is mounted on the electrode holder and is immersed in the test solution, along with the electrodes. As the temperature of the solution changes, the circuit constants are altered accordingly.

21.4.1 Null-Detector Type pH Meters

The simplest approach for measurement of pH is to use a potentiometric 'null-balance' method. Figure 21.9 shows the principle of this method in which the voltage output between a pair of pH electrodes is measured without drawing any current from the circuit under test.

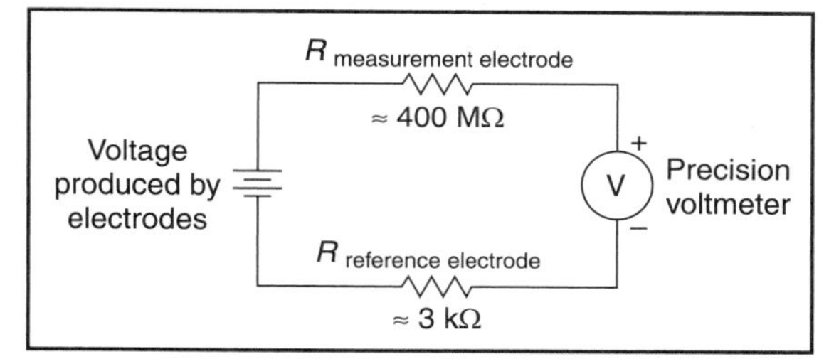

Figure 21.9 :: Principle of null detector type pH meter

In the circuit, the emf developed on the pH electrodes is shown along with series resistors of both the glass electrode and reference electrode. The precision voltage can be adjusted until the null detector shows zero. The reading on the voltmeter connected in parallel with the precision voltage would show the electrode potential representing pH of the solution. At the 'nulled' detector condition, there would be zero current in the pH electrode circuit, and therefore no voltage drop across the resistances of either electrode, giving the real electrode voltage at the voltmeter terminals. The pH value is read from the calibrated precision voltage source dial marked in pH units.

Instruments of this type are fundamentally simple in principle, as the electrode potential is determined by direct comparison with a precision voltage. A better accuracy than 0.01 pH can be achieved by using this type of instruments. They can be easily made battery operated and portable. Also, they are easy to maintain, as the electronic circuit is very simple. As these circuits require readjustment frequently, they are unsuited to long-time unattended operations in the industrial field. They are seldom seen in contemporary use.

21.4.2 Direct Reading pH Meters

The earliest type of pH meters were simple analog devices with a resolution of only one or two millivolts. The original meters were calibrated in millivolts and the corresponding pH value was read from a calibration graph. However, it was soon realized that pH electrodes are reasonably uniform and reproducible instruments with very similar slopes so that it is not necessary to have a unique calibration graph for each electrode. In this case, the meters can be calibrated directly in pH units by the manufacturer and can simply be re-calibrated each time they are used, to compensate for temperature changes or slight differences in the electrode response.

Direct-reading instruments are characterized by a simplicity of operation and speed of measurement, which are of considerable importance for making routine laboratory and industrial pH measurements. In these instruments, the pH value is indicated by the display system in either the analog or the numerical form in the digital display system without requiring any balancing process by the operator. After making an adjustment on one buffer solution only, samples may be measured in rapid succession, without having to operate any control on the instrument. This type of meter is very simple and quick to use and is perfectly adequate for many pH measurements because it requires a change of more than 5 mV in order to change the pH value by more than 0.1 pH units. Also, if employed to measure a solution in which the pH value is changing, the direct reading meter will give a continuous indication of the value, which could even be recorded for subsequent control of a process.

21.4.2.1 Chopper Amplifier Type pH Meters

In the chopper amplifiers, the direct voltage from the electrodes is chopped at the mains frequency and amplified as an alternating voltage. The output is rectified by a phase-sensitive circuit and indicated by a meter.

Basically, a chopper is a form of relay, consisting of very light moving parts, so that they may be operated repetitively at a fairly high frequency. The moving contact vibrates in response to the presence of an alternating current electromagnet and alternately makes and breaks contact with a pair of contactors connected to the ends of the centre tapped primary winding of a transformer. The secondary winding is connected to the input of the amplifier.

Figure 21.10 is the block diagram of a chopper amplifier type pH meter. The oscillating contact of the chopper, in one position, short-circuits the input of the amplifier, while in the other it is connected to the glass electrode. The reference electrode is connected back to the amplifier through a negative feedback resistor R_r. If the current in the feedback resistor is zero and the chopper contact is at position A, the glass electrode potential will be applied to the amplifier input. When the relay changes to position B, the amplifier input is brought to zero abruptly. When the contact

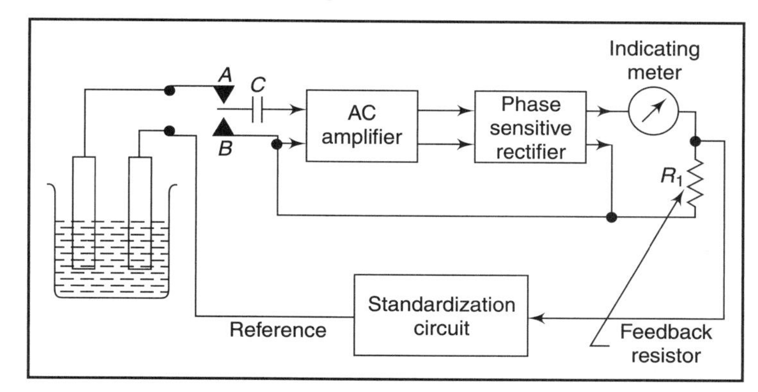

Figure 21.10 :: Chopper amplifier type pH meter

moves at a fast rate between the positions *A* and *B*, a square-wave voltage waveform is produced, which can be capacitively coupled to the input amplifier.

The alternating voltage is amplified by an ac amplifier and given to a phase-sensitive rectifier. The polarity of the dc output of the phase-sensitive rectifier reverses with the changes in the phase of the ac input by 180°C. A resistance R_1 is included in the feedback circuit and the current flowing though it produces a potential across it, which is in opposition to the electrode potential, indicating that the meter indication gives the difference voltage.

The use of chopper amplifiers for high-input resistance signals presents problems not encountered with low-input resistance signals. For example, the making and breaking of the contact to the glass electrode leads to changes in the capacity at the surfaces very abruptly, and gives rise to a sudden voltage surge, which is transmitted to the amplifier as a series of spikes on the waveform. This results in some degree of zero instability in the dc output of phase-sensitive rectifier. Also, with aging, the surface conditions change and the magnitude of the surges increase. These factors demand special care in design and construction of pH meters based on this principle. Solid state switching devices are now used in the chopper amplifiers.

21.4.2.2 Vibrating Condenser Amplifier Type pH Meter

In a chopper amplifier, the main problem relates to the presence of spikes due to transient potentials at the contactors. This disadvantage is overcome by using a vibrating condenser in place of a mechanical chopper. The dc signal is converted to an ac one by applying it to one plate of a condenser, whose capacity is changed at a constant frequency. The capacity is changed by vibrating one of its plates. If the distance between the plates is varied sinusoidally, capacity will vary in an inversely sinusoidal way. So, for a given value of charge q, the voltage V across the condenser will vary sinusoidally. The ac signal is processed in the same manner, as in a chopper amplifier. The method offers infinite input resistance, which is limited only by the insulating material used in its construction. The zero stability is much better than a direct coupled amplifier.

For a good performance, the frequency and amplitude of the vibrator should be stable and constant. Figure 21.11 shows the block diagram of a vibrating capacitor amplifier type pH meter.

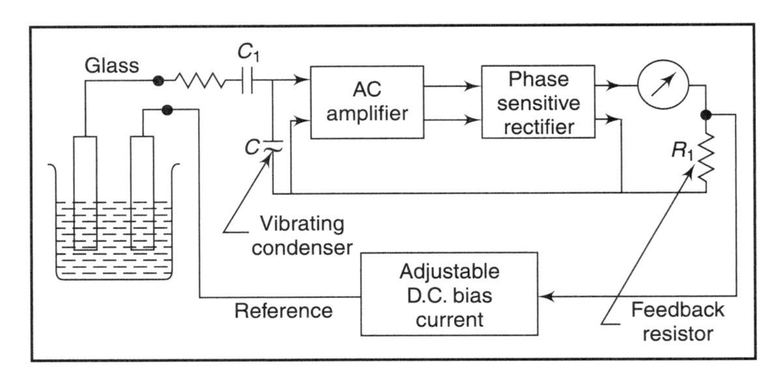

Figure 21.11 :: Vibrating capacitor amplifier type pH meter

21.4.2.3 Zero Corrected dc Amplifier Type pH Meter

The zero stability of a direct-coupled amplifier can be improved by incorporating an additional circuit for measuring the amount of drift present in the amplifier and making a corresponding correction automatically. This type of amplifier is used in the Beckman Zeromatic pH meter. In this instrument, automatic zero correction is employed at intervals of one second, without disturbing the reading. The pH meter is provided with a switch between the glass electrode and the input amplifier, which makes it possible to disconnect the electrode and short-circuit the amplifier input. This is used to measure the amplifier zero drift periodically, making an automatic correction at once. The correction is made by charging a capacitor in series with the amplifier input, to a voltage equal and opposite to the amplifier drift.

Referring to Figure 21.12, the pH measuring circuit consists of a two-stage amplifier A_1 followed by a emitter follower stage A_2 connected to the meter and feedback resistor R_f so that the whole system acts as a negative feedback amplifier. Capacitor C_M acts as a memory capacitor. The relay contacts in position 1, remain there for the measuring period of 1 s. The correcting action is applied when the contacts are moved to position 2, where they remain for 15 ms. On change of the contacts to position 2, the input voltage V to amplifier A_1 would normally decay to zero, and if no drift is present, the output would also fall to zero. If drift has occurred in A_1, the resulting output is amplified by A_1 and returned through C_2 to produce a voltage across R equal to the amount of drift. Since C_M and R are parallel in position 2, this voltage is stored on C_M and constitutes a correcting voltage when the contacts are brought back to position 1. R_3 and C_3 work as a smoothing circuit for the disturbances, which are produced during the zero correction and prevent them from reaching the meter.

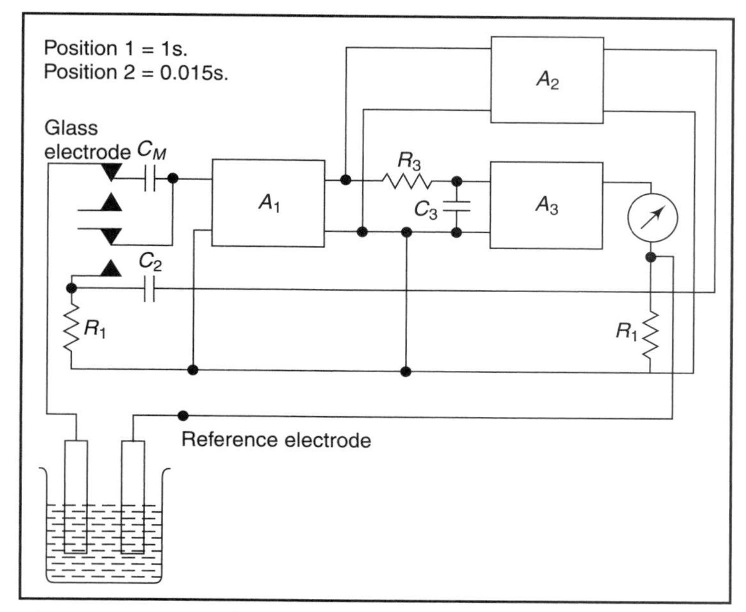

Figure 21.12 :: Schematic diagram of Beckman Zeromatic pH meter

21.4.2.4 Modern Design of pH Meters

Inherently, bipolar transistors are current amplifying devices with a low input impedance. For these reasons, conventional transistor amplifiers are seldom employed for pH measurements. Instead, field-effect transistors are used at the input stage to achieve a high input impedance. Metal-oxide silicon field-effect transistors of the insulated gate type have very small input leakage currents. A pair of these can be used to construct a pH meter. Figure 21.13 shows one such circuit. This circuit makes use of two MOSFET's in a differential amplifier configuration. This circuit is suitable to be used with combination type pH electrode. The differential input cancels the common mode errors, such as effects of temperature and supply voltage variations. The meter used is of centre zero type. Initially the input is grounded and the potentiometer R_7 is adjusted to bring the meter pointer to centre of the scale i.e. pH7. R_3 is adjusted to read 0 pH after a simulated electrical signal (for 25°C) is given to the input, corresponding to a change of seven pH units.

Before the pH of a solution is measured, the pH meter is standardized by dipping the electrode in a solution of known pH and adjusting R_6 (standardize control), until the meter indicates the known pH. After the meter is thus standardized, the probe is rinsed in distilled water, wiped dry and placed in the solution whose pH is to be measured.

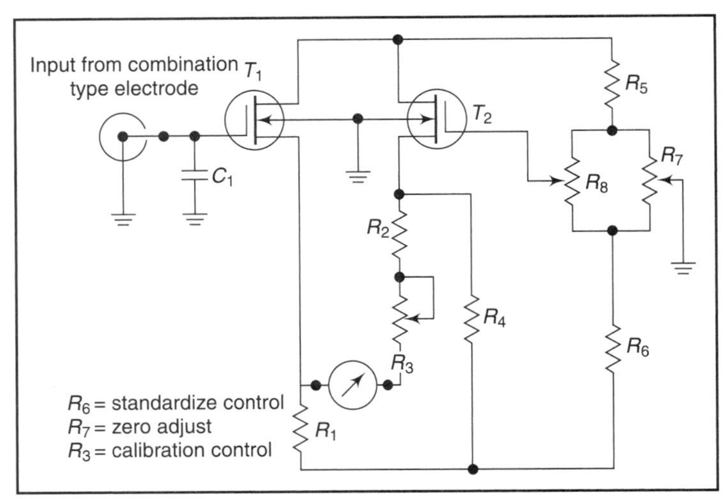

Figure 21.13 :: Circuit arrangement of a pH meter using MOSFETs at the input stage

The pH readings can be corrected if the measurements are made at temperatures other than 25°C by the following formula:

$$\text{pH error} = (T - 25)(pH_1 - pH_2)/T + 273,$$

where pH_1 is the instrument reading and pH_2 is the pH of the buffer.

Care should be taken to wire the circuit on high quality glass epoxy base printed circuit board. As MOSFET's are easily damaged by static electric charges, care should be taken to short their leads with wires during assembly and wiring of the circuit. This circuit is usually employed in pocket pH meters.

Figure 21.14 shows another circuit arrangement which uses a matched pair of field effect transistors housed in a single can. The circuit would provide input impedance greater than 10^{12}.

The emf produced by the measuring electrode is given to the gate of transistor T_1. The potential applied to the input 1 of the operational amplifier would depend upon current, which passes through the transistor and its corresponding resistance R. The potential applied to the gate of transistor T_2 is set by the buffer bias adjustment, which is fed from a zener stabilized potential supply. The potential developed across R_2 would depend on the current through T_2. Resistance R_1 and R_2 are kept equal. Therefore, the output of the operational amplifier will depend upon the difference in potential developed on the measuring electrode and the potential set up in the instrument. The current flowing through the indicator also flows through the manual or automatic temperature compensating resistor. Thus, potential applied to the reference electrode can be arranged to compensate for the change in slope of the pH/temperature relationship, i.e. the gain of the system can be changed by the negative feedback across the temperature compensator, so as to match the slope of the pH/temperature relationship.

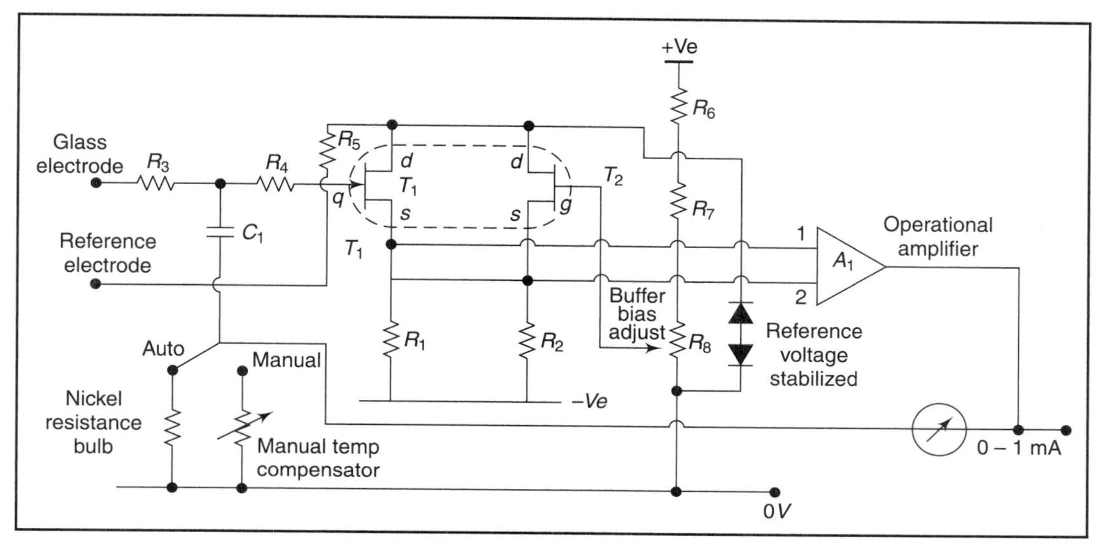

Figure 21.14 :: Direct reading type pH meter with temperature compensation arrangement

21.4.2.5 Digital pH Meters

A digital pH meter signifies a major advancement in pH measurements by offering a resolution of 0.1 or even 0.01 mV, thereby enabling much greater accuracy. This means that the stability and reproducibility of the electrode response become the main limiting factor in determining its accuracy and precision.

The amplifier and other circuits must have a small temperature coefficient, i.e., the influence of temperature variations must be under control, to attain reliable and consistent results. Operational amplifiers with FET input stage are available in the integrated form. They can be directly coupled to the electrodes to amplify signals from the pH cell.

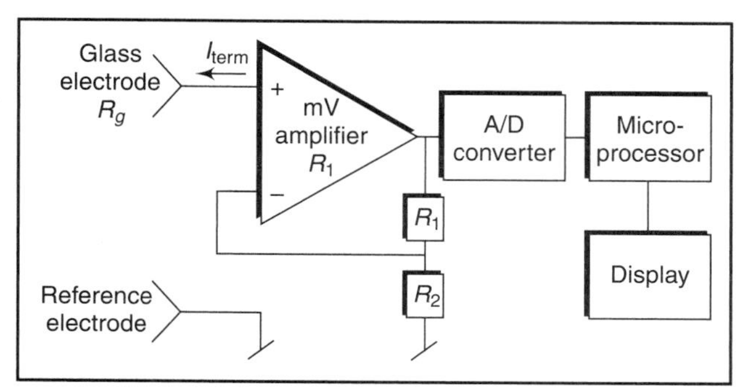

Figure 21.15 :: Block diagram of a microprocessor based digital read-out direct reading pH meter

Figure 21.15 shows a microprocessor-based digital pH meter. The amplifier works under the same conditions all the time and is directly connected to an A/D converter. The input amplifier and the converting circuit must meet certain requirements in order to obtain a correct measurement. A potential difference between the glass electrode and the reference electrode is amplified in the mV amplifier before the A/D converter feeds the signal to the microprocessor for calculation of the result.

The amplifier's input resistance, R_1, must be considerably higher than the inner resistance of the typical glass electrode, that is, higher than 10^8 Ohms. It is also important that the amplifier does not send any current through the glass electrodes as this will give an error potential and could even damage the electrode. The bias current or so-called terminal current, I_{term}, should therefore be below 10^{-12}A.

A microprocessor-based circuit then translates the converter's output and the calculated pH is displayed. A temperature sensor provides both temperature display and a temperature correction. For some microprocessor systems, automatic recognition of calibration buffers and automatic stability control of the electrode signal is possible.

Microprocessor-based instruments contain simple program to calculate the slope and intercept from the calibration data and then use these to calculate the sample concentration from the millivolt reading in the sample. The operator can simply enter the concentrations of the standards and measure the millivolts, then immerse the electrodes in the sample and read the sample concentration directly from the meter. These instruments usually have small keypads and are prone to errors in data entry.

21.4.3 Industrial pH Meters

The measurement of pH is carried out in industry to ensue either that the proper reaction to take place or the quality of the end-product, thereby contributing to the creation of better quality products at lower production costs. Measurements and control of pH normally need to be done either in a tank or on steam in a pipe. For these examples, the electrode systems could be dip-type or mounted on-line.

Quite often, the electrode system and the control unit could be as far apart from each other as 50 m. There would be a strong possibility of interference due to pick up by the cable connecting the two. Special type of shielded cables are used for this purpose. The second factor is the loss of signal strength in the connecting cable. This problem can be overcome either by compensating the loss by the amplifier circuit, or alternatively the amplifier is fitted into the electrode system and the amplified signal is transmitted through the cable to the display system kept at a convenient place. This, however, introduces problems of on-site maintenance.

There could be situations wherein the measurement is to be carried out in an explosive atmosphere. In this case, the design should be such as not to ignite the combustible gases. Normally, the amplifier part is housed in an explosive-proof chamber and it is operated preferably on batteries, thereby minimizing chances of sparking in the circuit. Industrial pH measuring instruments are fitted with contacts for operating visual or audible alarms, or activating control valves providing pH control of the process within preselected limits.

The output from the industrial pH amplifier is fed into recorders, controllers, indicators, etc. There is usually a choice of outputs and in many cases these are isolated from earth. Any leakage current from external equipment would be fed back to the circuit via the reference electrode causing an error in reading. An isolator eliminates this problem. The dc output current from the pH meter is periodically reversed by a transistor switch. The resultant square wave is then coupled through a transformer and reconverted to dc by a second synchronized switch. The transformer interrupts the dc path between the pH meter and the external equipment, so that it may be earthed.

21.4.4 Failures in pH Meters

The failures in a pH meter can be classified into three main categories: (a) defective electrodes, (b) defective input circuitry, and (c) defective electronic circuitry.

If the instrument balances at zero with the input shorted and fails at the pH measurement, it is logical to replace the electrodes with new ones. If the instrument works with new electrode, defective electrode must be replaced. Since majority of failures occur because of defective electrodes, it is advisable to have a spare set of electrodes with every instrument and use them by rotation to avoid drying up due to long storage. If the instrument fails even with new electrodes, input circuitry must be checked. The failures in input circuit could be due to: (a) poor insulation between glass electrode terminal and common terminal, and (b) excessive leakage current in the input circuit. Because of high resistance associated with the glass electrode, input circuit should have high insulation resistance and low leakage currents. These may be impaired at times by collection of dust and other vapours. In such cases, input terminals and function-selector switch should be cleaned with a solvent such as carbon tetrachloride. The capacitor connected at the amplifier input should also be checked for leakage and it should be replaced if defective. If the instrument is still defective it may be due to excessive input current. In such a case, the input stage should be checked.

If the instrument fails to balance at zero, it is most likely that electronic circuitry is defective. Before attempting the repairs, it is advisable to short-circuit the input externally and note the position of the pointer in all measuring ranges. If the pointer is within the scale, coarse buffer adjustments may be tried.

21.5 Selective Ion Electrodes

Over the past decade, the pH meter has been at the centre of a most important change in the field of analytical measurements due to the introduction of selective ion electrodes. As their name implies, these electrodes are sensitive to the activity of a particular ion in solution and quite insensitive to the other ions present. As the electrode is sensitive to only one ion, a different electrode is needed for each ion to be studied. Approximately 20 types of selective ion electrodes are presently available.

In recent years, a number of measuring electrodes, which possess selectivity toward some particular ion, have increased in popularity among many fields of science. These electrodes have been used in such applications as pollution control, biomedical research, clinical medicine, food processing, and metallurgy, to name a few. Various ion selective electrodes give direct measurement of the activity of many common anions and cations such as sodium, potassium, calcium, chloride, nitrate and sulfide, as well as fluoride and many more. There are about twenty of these types of electrodes. Some electrodes measure dissolved gases such as ammonia and, sulfide, dioxide, and carbon dioxide. In addition, several enzyme selective electrodes have been developed.

Galvanic cell

Just like the most common pH electrodes, ion-selective electrodes (ISEs), work on the basic principal of the galvanic cell. By measuring the electric potential generated across a membrane by "selected" ions, and comparing it to a reference electrode, a net charge is determined (Figure 21.16). The strength of this charge is directly proportional to the concentration of the selected ion.

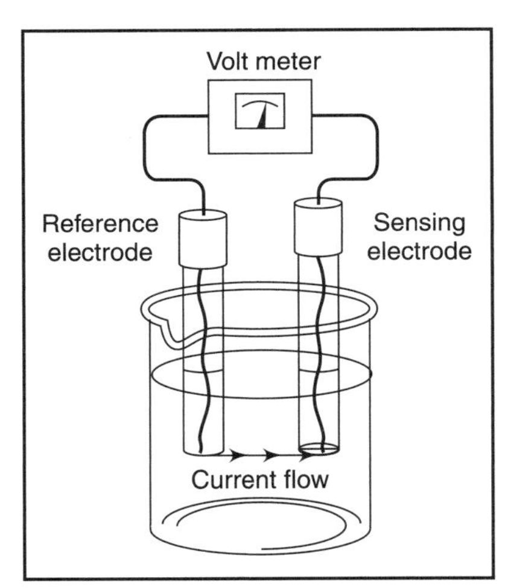

Figure 21.16 :: Principle of measurement of specific ions using ion-selective electrodes

The development of a range of ion-selective electrodes has stimulated interest in the solid-state chemistry of glass and crystals and in the specificity for metal ions of synthetic and natural organic and complexing agents. Following developments in cation-responsive glasses and precipitate-impregnated silicone rubber membranes, the real impact of ion-selective electrodes was made by the fluoride ion-selective solid-state lanthanum fluoride crystal membrane electrode. The fluoride electrode was quickly followed by a calcium ion-selective liquid ion-exchanger membrane electrode. These and a whole host of other ion-selective electrodes in various styles have been applied in fields like environmental and industrial monitoring, reaction rate studies, enzyme reactions, general analysis and non-aqueous media studies.

Ion selective electrodes come in various shapes and sizes. All of them consist of a cylindrical tube, generally made of a plastic material, between 5 and 15 mm in diameter and 5 to 10 cm in length. An ion-selective membrane is fixed at one end so that the external solution can only come into contact with the outer surface, and the other end is fitted with a low noise cable or gold plated pin for connection to the millivolt measuring device. In some cases the internal connections are completed by a liquid or gel electrolyte, in others by an all-solid-state system. Figure 21.17 shows different technologies used in ion-selective electrodes.

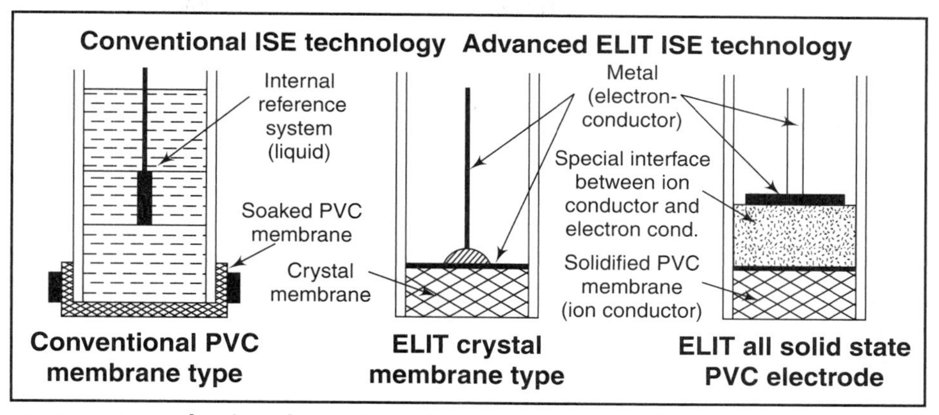

Figure 21.17 :: Ion-selective electrode technology (Courtesy, *www.nico2000.net*)

Ion-selective membranes are currently available for a number of commonly occurring ionic species. A perusal of manufacturer's catalogues reveals that the most common are:

- CATIONS: Ammonium (NH_4^+), Barium (Ba^{++}), Calcium (Ca^{++}), Cadmium (Cd^{++}),
- Copper (Cu^{++}), Lead (Pb^{++}), Mercury (Hg^{++}), Potassium (K^+), Sodium (Na^+), Silver (Ag^+).
- ANIONS: Bromide (Br^-), Carbonate (CO_3^-), Chloride (Cl^-), Cyanide (CN^-), Fluoride (F^-),
- Iodide (I^-), Nitrate (NO_3^-), Nitrite (NO_2^-), Perchlorate (ClO_4^-), Sulphide (S^-), Thiocyanate (SCN^-).

The manner in which these different membranes select and transport the particular ions is highly variable and, in many cases, highly complex. There are two main types of membrane material, the other based on a solid crystal matrix, either a single crystal or a polycrystalline compressed pellet, and the other based on a plastic or rubber film impregnated with a complex organic molecule which acts as an ion-carrier.

(i) Crystal-Membrane Electrodes, e.g. Fluoride

The fluoride electrode is a typical example of this type. Here the membrane consists of a single lanthanum fluoride crystal which has been doped with europium fluoride to reduce the bulk resistivity of the crystal. It is 100% selective for F^- ions and is only interfered with by OH^- which reacts with the lanthanum to form lanthanum hydroxide, with the consequent release of extra F^- ions.

(ii) Impregnated-PVC-Membrane Electrodes, e.g. Potassium

The potassium electrode was one of the earliest developed and simplest examples of this type. Its membrane is usually in the form of a thin disc of PVC impregnated with the macrocyclic antibiotic valinomycin. This compound has a hexagonal ring structure with an internal cavity which is almost exactly the same size as the diameter of the K^+ ion. Thus, it can form complexes with this ion and preferentially conducts it across the membrane. Unfortunately, it is not 100% selective and can also conduct small numbers of sodium and ammonium ions. Thus, these can cause errors in the potassium determination if they are present in high concentrations.

Several limitations must be kept in mind when using ion selective electrodes. First, it must be remembered that activity and not concentration is being measured. Since the activity coefficient is dependent upon the total ionic strength of the solution, the activity will deviate from the concentration to a greater extent as the solution of the concentration increases. Also, the variation in concentration of any other ionic species that happens to be in the solution will affect the total ionic strength and change the activity of the ion being measured. These effects may be swamped by adding high and constant concentrations of non-interfering ions, such as sodium chloride that are available in the Total Ionic Strength Adjustment Buffer (TISAB).

Second, the electrode responds only to free ions. The presence of any species which complexes with the ion being measured will lower its activity and therefore the electrode response. Finally, the electrode response is selective, not specific. Selectivity means interference from competing ions is possible.

The potential of the ion-selective electrode (ISE) arises when there are solutions of a given ion, X, on either side of a membrane. The type of membrane categorizes the electrode: there are glass membranes (H^+, Na^+, K^+, Ag^+ selective), crystal membranes (F^-, Ce^{3+}, Pb^{2+} selective), liquid ion-exchange membranes (NO^{3-}, K^+, Ca^{2+} selective), and so forth.

Ion-selective electrodes are classified into four major groups. These are detailed below.

- *Glass Electrodes:* The first glass ion-selective electrode to be developed was the one sensitive to hydrogen ions. Glasses containing less than 1 per cent Al_2O_3 are sensitive to hydrogen ions (H^+) but almost insensitive to other ions present. Glasses, of which the composition is Na_2O 11 per cent, Al_2O_3 18 per cent, SiO_2 71 per cent is highly selective towards sodium, even in the presence of other alkali metals. Glass electrodes have been made that are selectively sensitive to sodium, potassium, ammonium and silver.
- *Solid State Electrodes:* These electrodes use single crystals of inorganic material doped with a rare earth. Such electrodes are particularly useful for fluoride, chloride bromide and iodide ion analysis.
- *Liquid-Liquid Membrane Electrodes:* These electrodes are essentially liquid ion-exchangers, separated from the liquid sample by means of a permeable membrane. This membrane allows the liquids to come in contact with each other, but prevents their mixing. Based on this principle, cells have been developed that are selective to calcium and magnesium. These cells are used for measuring water hardeners.
- *Gas Sensing Electrodes:* These electrodes respond to the partial pressure of the gases in the sample. The most recent of these to be developed are the gas sensing electrodes for ammonia and sulphur dioxide. Ammonia or sulphur dioxide is transferred across a gas permeable membrane, until the partial pressure in the thin film of filling solution between the glass electrode membrane and the probe membrane equals that in the sample. The resultant pH

change is measured by a combination pH electrode, a potential is developed related to the partial pressure and hence the ammonia or sulphur dioxide concentration.

The applications using ion-selective electrodes are many, most of which are time-saving and simple to use. The electrodes are now used in continuous monitoring of ammonia, nitrate, fluoride, chloride, cyanide, sodium, etc. providing vital information for the power industry environmental work and process control industries.

Figure 21.18 shows examples of various types of Orion Ion-selective electrodes.

Advantages of Ion-selective Electrode

The main advantages of ion-selective electrodes are:

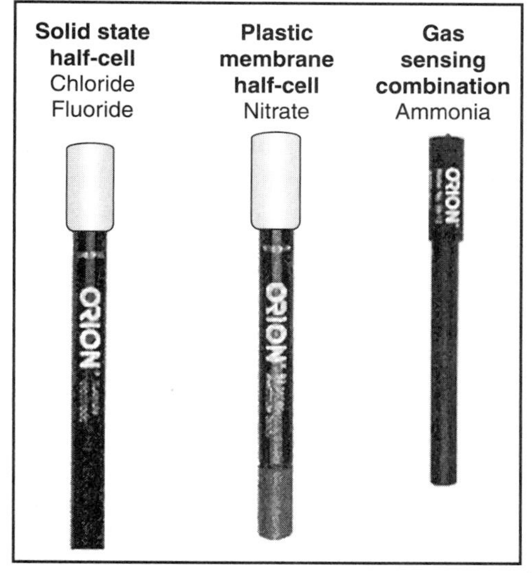

Figure 21.18 :: Different types of ion-selective electrodes (Courtesy, M/s Orion, USA)

- When compared to many other analytical techniques, ion-selective electrodes are relatively inexpensive and simple to use. They have an extremely wide range of applications and wide concentration range.
- The plastic-bodied all solid state or gel-filled models are very robust and durable and ideal for use in the field or in laboratory environments.
- When measuring ions in relatively dilute aqueous solutions and where interfering ions are not a problem, they can be used very rapidly and easily e.g. simply dipping in lakes or rivers.
- They are highly useful for the continuous monitoring of changes in concentration: e.g. in potentiometric titrations or monitoring the uptake of nutrients, or the consumption of reagents.
- They are particularly useful in biological/ medical applications because they measure the activity of the ion directly, rather than the concentration.
- With careful use, frequent calibration, and an awareness of the limitations, they can achieve accuracy and precision levels of ± 2 or 3 per cent for some ions and thus compare favourably with analytical techniques which require far more complex and expensive instrumentation.
- ISEs constitute one of the few techniques which can measure both positive and negative ions.
- They are unaffected by sample colour or turbidity.
- ISEs can be used in aqueous solutions over a wide temperature range. Crystal membranes can operate in the range 0°C to 80°C and plastic membranes from 0°C to 50°C.

The majority of pH electrodes are produced in the form of combination electrodes in which the reference system is housed with some cylindrical body as the sensor head. This arrangement minimizes the effect of any stray electrostatic fields. In contrast to pH electrodes, most of the ISE's are produced as mono-electrodes for use with separate reference systems. One reason for this is because ISE membranes have a far lower impedance than pH sensors and are less susceptible to stray electrostatic fields. Thus, it is not necessary to screen the sensor head by surrounding it with the reference system (www.nico2000.net, Chapter 5).

Problems with ISE Measurements

The main problems with ISE measurements are:

- Effect of interference from other ions in solution,
- Effect of the ionic strength of the solution, thereby reducing the measured activity relative to the true concentration at high concentration,
- Drift in electrode potential during a sequence of measurements, and
- Blocked or contaminated ISE or reference electrode by organic molecules.

These problems have led to the definition of the following parameters:

- *Selectivity Coefficient*: The ability of an ion-selective electrode to distinguish between different ions in the same solution is termed as selectivity coefficient. For example, if the primary ion for which the electrode is sensitive is '*A*' and the interfering ion is '*B*', then a selectivity coefficient of 0.01 would mean that the electrode is 100 times more sensitive to '*A*' than to '*B*'.
- *Ionic Strength*: It is a measure of the total effect of all the ions in a solution. It is the sum of the molar concentration multiplied by the square of the valency of all the ions.
- *Activity*: The effective concentration measured at the electrode head is known as the activity of the ion,
- *Activity Coefficient*: It is the ratio of the activity divided by the concentration. The activity coefficient is always less than one and becomes smaller as the ionic strength increases. Thus, the difference between the measured activity and the actual concentration becomes higher at higher concentration.

21.5.1 Ammonia Electrode

One of the most commonly used ion-selective electrodes is the ammonia electrode. The electrode uses a hydrophobic gas-permeable membrane to separate the sample solution from the electrode internal solution (Figure 21.19). Dissolved ammonia on the sample solution diffuses through the membrane until the partial pressure of ammonia is the same on both sides of the membrane. In any given sample, the partial pressure of ammonia is proportional to its concentration.

When ammonia diffuses through the membrane, it will dissolve in the internal filling solution and react reversibly with water as follows:

$$NH_3 + H_2O \rightleftharpoons NH_4^+ + OH^- \quad \text{(i)}$$

The relationship between ammonia, ammonium ion and hydroxide is given by:

$$[NH_4^+][OH^-]/[NH_3] = \text{constant} \qquad \text{(ii)}$$

The internal filling solution contains ammonium chloride at a sufficiently high level so that the ammonium ion concentration can be considered fixed. Therefore:

$$[OH^-] = NH_3 \text{ constant} \qquad \text{(iii)}$$

The potential of the electrode sensing element with respect to the internal reference element varies with changes in the hydroxide level:

$$E = E_0 - S \log [OH^-], \qquad \text{(iv)}$$

where S is the electrode slope.

Since the hydroxide $[OH^-]$ concentration is proportional to ammonia concentration [Eq. (iii)], electrode response to ammonia is given by:

$$E = E_0 - S \log [NH_3]$$

The reference potential E_0 is partly determined by the internal reference electrode, which responds to the fixed level of chloride in the internal filling solution.

When plotted on a semi-logarithmic paper, electrode potential response as a function of ammonia concentration is shown in Figure 21.20. The response is a straight line with a slope of about 58 mV per decade.

The response time of the electrode is about 1 min (95 per cent of the total mV reading) for ammonia concentrations above 6×10^{-5} M (1 ppm NH_3). Below this value, response times are usually longer. Also, there could be errors due to ammonia absorption from the air. Samples above 1 M in ammonia concentration should be diluted before measurement.

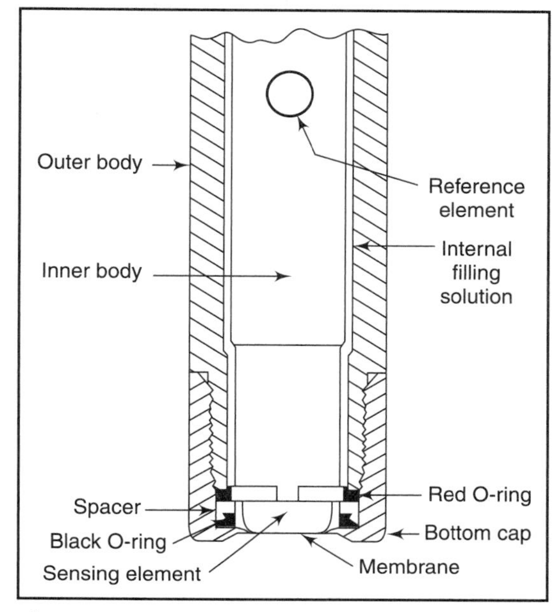

Figure 21.19 :: Construction of ammonia electrode

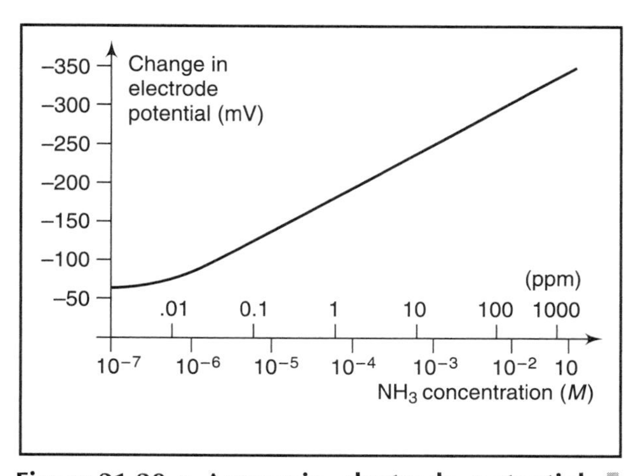

Figure 21.20 :: Ammonia electrode potential response as a function of ammonia concentration

Temperature changes cause electrode response to shift and change slope. At 10^{-3}M, a 1°C temperature change produces a 2 per cent error. Therefore, it is necessary that the samples and standards should be at the same temperature.

Water vapour is a potential electrode interference, as it can move across the membrane, changing the concentration of the internal filling solution under the membrane. Such changes manifest themselves as electrode drift. The problem is minimized if the total level of dissolved species in solution (osmotic strength) is below 1 M. Therefore, samples with osmotic strengths above 1 M should be diluted before measurement. However, dilution should not reduce the ammonia level below 10^{-5} M.

The reproducibility of results in ammonia electrodes is a ticklish problem and is limited by several factors, such as temperature fluctuations, drift and noise. With calibration every hour, electrode measurements to ± 2 per cent can be obtained. Within the operating range of the electrode, reproducibility is independent of concentration.

Electrode slope

Ion-sensitive electrodes are characterized by what is called an electrode slope. Slope is defined as the change in millivolts observed with every tenfold change in concentration. For ammonia electrode, the change in millivolts should be in the range of –54 to –60 mV/decade, when the solution temperature is between 20 and 25°C.

Ammonia electrode not only allows measurements not only of dissolved ammonia in aqueous solutions, but can also be used to measure the ammonium ion after conversion to ammonia or organic nitrogen after Kjeldahl digestion of the sample. The performance of the electrode is not affected by sample colour and turbidity. There is no interference from almost all anions, cations and dissolved species, other than volatile amines.

21.5.2 Fluoride Electrode

The fluoride ion-selective electrode is a solid state device that measures fluoride ion in water and some organic solvents. The fluoride selective electrode is a solid-state type electrode consisting of a lanthanum fluoride crystal sealed over the end of an inert plastic tube which contains an internal electrode and filling solution usually of 0.1M NaCl and 0.1M NaF. A potential arises because of the difference in fluoride activity on either side of the crystal. The crystal is an ionic conductor in which only fluoride ions are mobile.

When the membrane is in contact with a solution containing fluoride ions, an electrode potential develops across the membrane. The magnitude of this potential is dependent upon the level of fluoride ion in solution. The potential across the membrane is measured against an external reference electrode with a digital pH/millivolt meter. The potential corresponding to the level of fluoride ions in solution is described by the Nernst equation:

$$E = E° - 0.0591 \log A,$$

where E = measured electrode potential

$E°$ = reference potential (a constant)

A = fluoride level in solution

The level of fluoride, A, is the activity or 'effective concentration' of free fluoride ions in solution. The total fluoride concentration, C_{total}, may include some bound or complexed ions as well as free ions. The electrode responds only to free ions, whose concentration is:

$$C_{free} = C_{total} - C_{bound},$$

where C_{bound} is the concentration of fluoride ions in all bound or complexed forms.

The fluoride activity is related to free fluoride concentration by the activity coefficient. Ionic activity coefficients are variable and largely depend on ionic strength.

The pH range for the fluoride selective electrode is limited at the acid end by the formation of HF and at the basic end by hydroxide interference. Polyvalent cations of Si^{+4}, Fe^{+3}, and Al^{+3} also interfere by forming complexes with fluoride. To avoid problems arising from these possible interferences, TISAB (total ionic strength adjustment buffer) is added in excess and in exactly the same amount to each of the solution that will be measured. This TISAB buffer (Tissue, 2000b) renders all the solutions to about a pH of 5.0 and contains a complexing agent to complex polyvalent cations. Most importantly, however, it contains a high concentration of NaCl to provide a constant ionic strength to each solution.

The fluoride electrode also responds to metal ions that complex with fluoride (e.g. Al^{3+} and Fe^{3+}). Such ions constitute an interference if present in sufficiently large concentrations. The response of the electrode to another interfering ion can be included in the Nernst equation which is shown below for a general case:

$$E = E° + (0.0591/\mathrm{n}) \log(a_A + K_{AB} a_B \, n/z)$$

where $E°$ = a combination of several constants

a_A = activity of species A with charge n

a_B = activity of interferent B with charge z

K_{AB} = the selectivity constant of the electrode for A over B.

Selectivity constant

The selectivity constant is a measure of the extent of the interference posed by a particular ion that might be present in the sample. The selectivity of the fluoride electrode is excellent, OH^- being the main interference.

The pH range for satisfactory measurements with the fluoride electrode lies between 5 and 7 at 10^{-6} M F^- (up to pH 11 at 0.1 M F^-). In acid solutions with a pH below 5, hydrogen complexes a portion of fluoride in solution, forming undissociated acid HF and the ion H_2^-. Hydroxide ion interferes with the electrode response to fluoride when the level of hydroxide is greater than one-tenth the level of fluoride ion present.

A calibration curve is prepared for the response of a particular electrochemical cell to a series of standard fluoride solutions. The measured potentials are plotted versus the logarithm of the concentrations or activities of the fluoride ion. The result is a straight line with a slope of 59.16 mV if a Nernstian response is obtained. In neutral solutions, fluoride concentration can be measured down to 10^{-6} M (0.02 ppm) fluoride. The upper limit of detection is a saturated fluoride solution.

The electrode response curve will not only shift but will change slope with changes in temperature. A 1°C change in temperature will result in a 2% error at the millimolar level. Ideally, standards and samples should be kept at a constant temperature, preferably 25°C. The fluoride electrode has a operable temperature range of 0° to 80°C under continuous use.

Solid state electrodes have a typical life of 1-2 years in the laboratory and 1-3 months when used continuously at elevated temperature. The main advantage of solid state sensors over silver metal–silver halide electrodes of the second type is their insensitivity to redox interferences and surface poisoning.

Care and Maintenance of ISEs

When handling ISEs, care should be taken to avoid damaging the membrane surface. If the electrodes are in frequent use, then they can simply be left hanging in the electrode holder with the membrane surface open to the air but protected by a clean dry beaker. For prolonged storage in a cupboard or drawer, the membrane should be protected by covering with the rubber or plastic cap, which is normally provided with the electrode. After extensive use, the membranes may become coated with a deposit or scoured with fine scratches which may cause a slow or reduced response (low slope) or unstable readings.

Crystal membranes can be regenerated by washing with alcohol and/or gently polishing with fine emery paper to remove any deposit or discoloration, followed by thoroughly washing with de-ionised water to remove any debris. After this, they may require soaking in the concentrated standard solution for a few minutes before a stable reading can be re-established. However, that prolonged immersion of crystal membranes in aqueous solutions should be avoided because this can cause a build up of oxidation products on the membrane surface and thus inhibit performance. Conversely, PVC membranes should not even be touched, let alone polished. They can often be regenerated by prolonged soaking over several days in the standard solution, after removing any deposit with a fine jet of water, or rinsing in alcohol.

Differences between pH and other Ion-selective Electrodes

The differences between pH and other ISEs are:

- In contrast to the pH membrane, the membranes used in the ion-selective electrodes are not entirely ion-specific. They can permit the passage of some of the other ions, leading to the problem of ionic interference.
- Most ISEs have a much lower linear range and higher detection limit than the pH electrode.
- The calculation of ionic concentration is far more dependent upon a precise measurement of the potential difference than the pH because the pH depends on the order of magnitude of the concentration rather than the precise value.
- ISEs mostly work effectively over a narrow pH range.

21.6 Ion Analyzer

Ion analyzers are basically pH/mV meters (Figure 21.21), which enable the operator to calculate the concentration of specific ions from the potentials developed at the ion-sensitive electrode, when dipped in sample solution. By measuring both the electrodes potential in a standard solution and in the sample solution, it is possible to calculate the unknown solution concentration by solving the following equation:

$$C_X = C_S \times 10\Delta E/S,$$

where, C_X = concentration of the unknown solution

C_S = concentration of the standard solution

E = difference between the observed potential in the sample solution and the observed potential in the standard solution

S = electrode slope (change in electrode potential per ten-fold change in concentration)

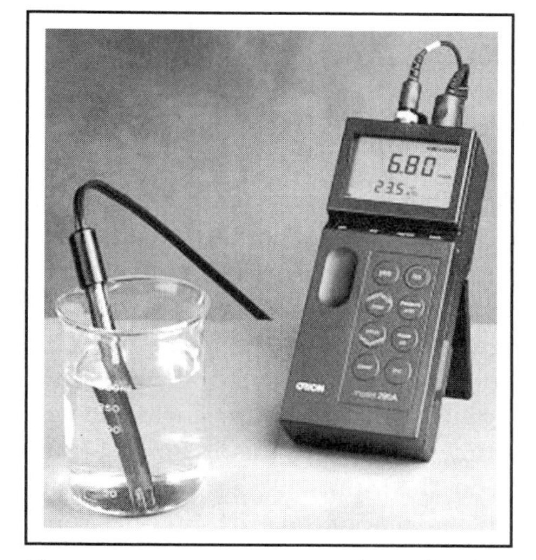

Figure 21.21 :: pH meter/ORP/ISE/T Meter Model 290A (Courtesy, M/s Orion)

Ion analyzers are mostly microprocessor-based instruments, which are programmed to calculate sample concentration from a set of input data, such as electrode potentials, standard concentration, slope and blank correction. The instruments measure relative milliovolts, pH and concentration of specific ions. The programme for direct measurement concentration is based on Nernstian electrode response:

$$E_X = E_0 + S \log (C_X + C_b),$$

where, E_x = electrode potential

E_0 = constant

C_b = blank concentration.

Blank correction

The blank correction (C_b) accounts for the finite lower limit of detection of electrodes. If a solid or liquid-membrane electrode is placed in pure water, the membrane dissolves slightly, producing an equilibrium concentration of the measured ion. This concentration is a constant background for all measurements and is represented by C_b. The typical electrode response curves are generally given by the electrode manufacturers. If the sample concentration falls in the linear response region, a blank correction may not be necessary. But, if the sample concentrations are low, and fall in the non-linear region of the response curve, blank correction must be applied.

Ion analyzers need recalibration every two to three hours. If there is no change in the sample temperature, one needs to repeat only the set standard. If temperature has changed, increase the magnitude of the slope setting by 1 mV (mono-valent) or 0.5 mV (divalent ion) per 5°C increase in temperature.

The standard calibration procedure for a specific ion meter is similar to that used to calibrate a pH-meter with pH buffers. Two standard solutions are used, which are a decade apart in concentration and approximately bracket the expected concentration range of the unknown sample solution.

Ion-selective electrodes and ion analyzers find numerous applications in agriculture, agronomy, food technology, medicine, chemistry, biology, physics, geology, oceanography, pharmocology, engineering and environmental sciences. This is because hundreds of different analyses can be made in a faster and easier manner through electrode methods.

A block diagram of the ion analyzer is shown in Figure 21.22. The first stage is the input buffer amplifier, which provides a very high input impedance and less than 1 pA input bias current. The electrode potentials are individually buffered by unity gain amplifiers with FET front ends. Figure 21.23 shows the input buffer stage. The two FET's are operated as source followers, each running at a constant drain current determined by its associated op-amp. The voltage at the + input of each op-amp is held constant, and therefore the drain current in FET's will be constant. In order to do this, the op-amp output voltage must maintain a constant V_{GS} and must therefore, follow the input voltage. The op-amps effectively serve the dual purpose of controlling the operating current of the FET and providing current gain. As with other similar circuits, the high input impedance of the buffer amplifier gets degraded by the presence of dirt, moisture or solder flux. Also, the input FET is delicate and will get destroyed by static discharge. When the inputs are not being driven by a signal, they must be grounded with shorting straps. The input amplifier is followed by a differential amplifier, before the signal is given to an A-D converter.

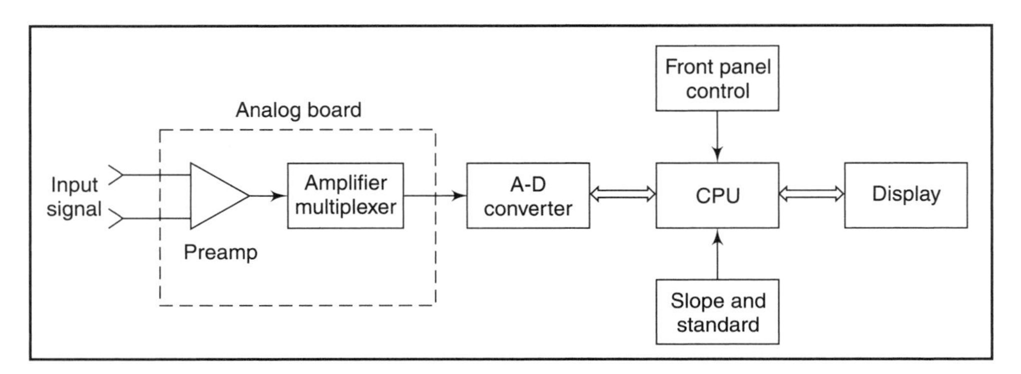

Figure 21.22 :: Block diagram of a microprocessor-based ion analyzer

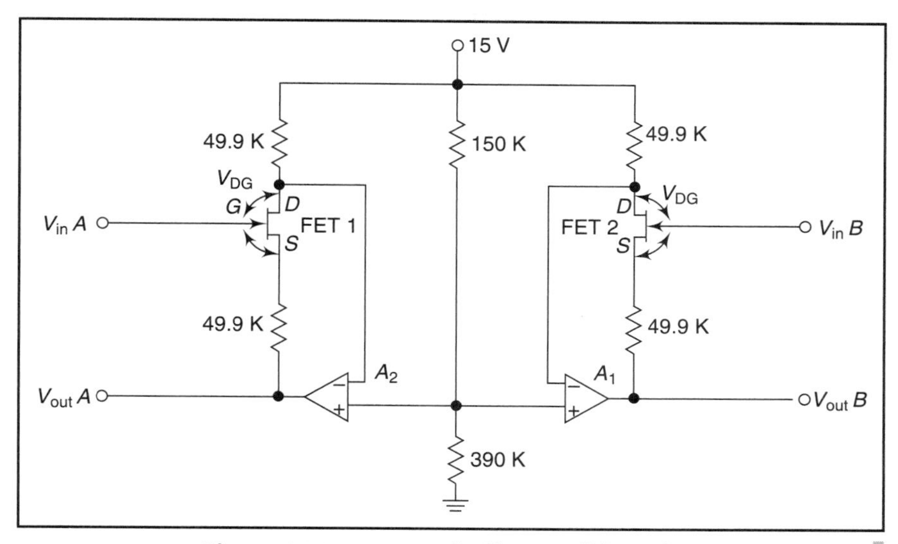

Figure 21.23 :: Input-buffer amplifier of an ion-analyzer

The A-D converter could be a dual-slope integrator type. The results of the A-D converter are held in the A-D data latch by using shift-registers and the loading function is controlled by the A-D converter. The output of the latch remain in high impedance state, until they are enabled by a signal from the control port decoder. Thus, the loading and reading of data from the A-D are independent. The microprocessor may read data from the A-D converter, regardless of the timing of the analog-to-digital conversion cycle.

The microprocessor sends and receives information through the input-output (I/O) bus. The bus is driven by only one source at a time and all other sources must be disabled, i.e, kept in a high impedance state. The bus may be driven by the CPU, A-D converter, slope switches, standard value switches and mode switches. The CPU and display receive data from the bus.

The microcomputer, consisting of a microprocessor, memory interface and read-only memory, (ROM), performs well-defined processing functions. Therefore, the program is stored in permanent read-only memory. Under program control, the microprocessor generates signals on the control port to select the path along which data will flow on the I/O bus. The CPU communicates with the memory and the memory interface through the microprocessor data bus. Through this bus, instructions and numerical constants flow from the memory outputs into the CPU. The memory interface performs the task of generating the address for each instruction stored in memory. It does this by maintaining a program counter according to commands from the CPU. The timing for the microprocessor and for all signals on the buses is generated by the CPU clock.

Grounding system

Because of the low level of signal generated and high impedance of the ion-selective electrodes, the grounding system is designed very carefully. Usually, the ion analyzing instruments have three grounds: (i) the chassis and the electrostatic shield in the power transformer are connected to earth ground through the third wire of the AC line. This provides isolation from line noise; (ii) digital ground provides the return path for all the logic signals, including the microprocessor signal and the display current; (iii) analog ground provides a reference point for electrode input signals and a return path for all analog current.

The analog and digital grounds are kept separate so that the digital signal return currents never flow through the same conductor as analog signal returns. The earth ground is not connected either digital or analog ground.

Kollman and Reddish (1983) illustrate a continuous ion-selective analyzer for the monitoring of gases detectable via ion selective electrodes.

21.6.1 PC-based pH Meter Ion-analyzers

The development of computer interfaces for connecting the sensor electrodes directly to a personal computer, without the need for a meter to measure the voltage, has revolutionized ISE data acquisition and processing.

Specially designed electronic hardware can be used with any type of electrode plugged into the interface with a standard connector. The electrode signals are processed by a low-noise amplifier and analog-to-digital converter for direct transfer into any computer. Versions are available for connecting to either the serial or parallel printer port using standard cables. The visual basic software is designed to run on any computer capable of running Windows 3.1 or later version operating system. Apart from the ease and speed of operation, the great advantage of this type of measuring system is that it virtually eliminates any possibility of operator error in recording and transferring data. It also facilitates far more sophisticated data processing, and display and archiving of results, once the raw data are securely recorded in the computer.

Data processing

Data processing software is designed to carry out most of the functions previously undertaken by the operator, and much more. The software basically carries out the following functions:

- *Hardware set-up*: This is meant to configure the system for the sensors and reference electrodes being used (e.g. temperature, pH, Redox, or any ISE) and give a full documentation of electrode types, serial numbers, operator details, date and time, etc.
- *Signal measurement*: The software interprets the signal from the interface and uses a calibrating factor to produce an accurate display of the millivolts. The operator can select to take single readings, or the average of multiple readings. A time delay after immersion of the electrodes can also be specified before taking a reading, if required.
- *Calibration*: The only operator entry required is of the concentration (or pH) of the calibration standards. For ISE measurements, calibration graphs are plotted automatically and instantaneously and can be assessed with various line-fitting techniques.
- *Sample Calculation*: The sample results are calculated instantaneously with no possibility of operator error. All results are saved and displayed in data tables where sample numbers and comments can be added. Concentration results can be reported as ppm or moles/l.

- *Display, reporting and archiving of results*: All data can be stored, printed or imported to other software packages. Continuous recording of pH measurements can be displayed graphically, in real-time, to facilitate such activities as pH titrations or process monitoring. Figure 21.24 shows a typical screen display in a PC based system.

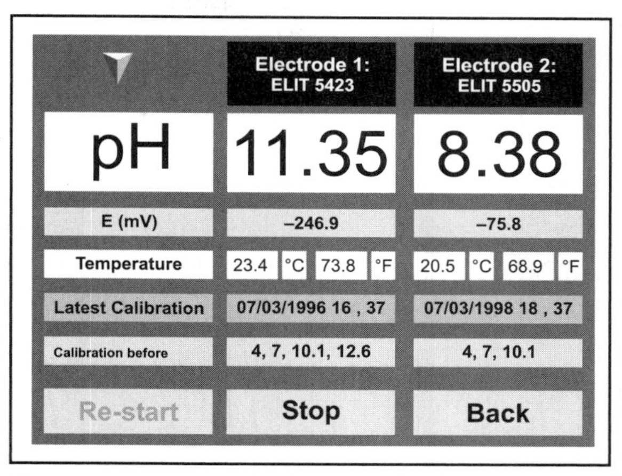

Figure 21.24 :: Typical screen display on a PC-based pH meter (Courtesy, Nico 2000 Ltd., UK)

21.7 Chemically Sensitive Semiconductor Devices

Considerable effort has recently been directed towards the development of ion-sensitive electrodes based on a modification of the metal-oxide-semiconductor field effect transistor. In these devices, the chemical sensitivity is obtained by fabricating the gate insulation of the FET out of ion-sensitive materials, usually a polymer or SiO_2. These devices are called ISFETs (ion-selective field effect transistors). A simple review, which discusses the chemistry and physics of chemically sensitive semiconductor devices, is due to Janata (1989).

In these devices, the ion-sensitive material is bonded to the FET itself. This requires the material and its method of fabrication to be compatible with the substrate (high purity silicon). This significant requirement puts a severe limitation on the use of some of the best-characterized membrane materials, including ion-sensitive glasses. Martin and Sinclair (1976) report the development of pH-sensitive electrode by means of thick-film screening techniques. This electrode retains the advantages of ion-sensitive FET transducers, but eliminates the restrictions on membrane selection and fabrication. Here, the ion-sensitive structure is physically separated from the FET. In this way, the ion-sensitive membrane can be fabricated on a compatible substrate and the FET can then be attached approximately and placed in close proximity to the ion-sensitive membrane. A hybrid electrode structure permits the incorporation of a source follower FET amplifier, directly adjacent to the pH membrane, significantly reducing response time and noise pick-up (Bergveld and Rooji, 1979).

Figure 21.25 presents the cross-section of an ISFET, which is essentially a conventional insulated gate-field effect transistor that has its metallic gate contact replaced by a chemically sensitive coating and a reference electrode. In solution, the gate region can be coated with an ion-sensitive membrane. Interaction of ions in solution with the membrane results in a change of the interfacial potential and corresponding alteration of drain current. By this technique, numerous cations and anions have been sensed (H^+, K^+, Ca^{2+}, Cl^-, I^-) and (CN^-). The ISFET has advantages

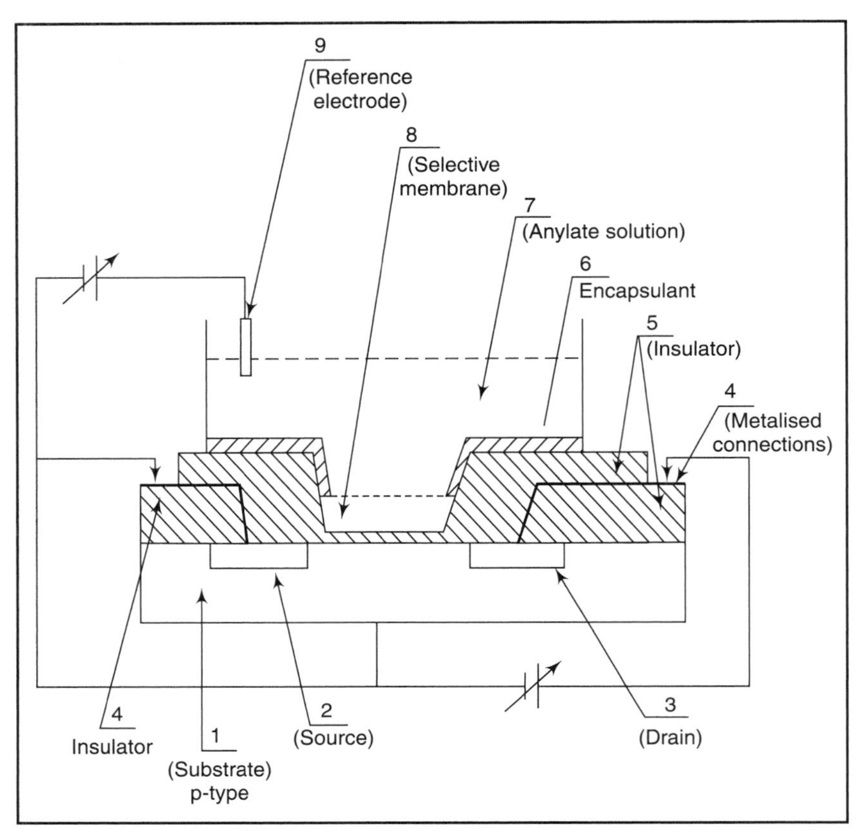

Figure 21.25 :: Constructional details of an ion-sensitive field-effect transistor (after Wohltgen, 1984)

in its small size (less than 1 mm^2) and low output impedance, which makes it ideal for *in vivo* monitoring or analysis of small sample volumes. However, problems like ion-selective coating adhesion and device encapsulation have prevented large scale use of ISFETs.

The construction and working of micro-sensor devices have become possible as a result of extraordinary advances in micro-electronics. These devices fall into two categories. The first category consists of devices that sense chemical species, whose presence modulates the transport of an electronic charge in the device. Examples of micro-sensors in this category include CHEMFETs, ion-controlled diodes, Schottky diodes, thin-film tin oxide gas sensors and chemi-resistors, etc. The second category consists of all micro-sensor devices that do not fit into the first category. These devices, which are based on unique techniques, include surface acoustic wave devices, potentiometric gas sensors and pyroelectric enthalpimetric sensors. Seitz (1984) has discussed several other chemical sensors based on fibre optics.

On the basis of this principle, a urea biosensor can be constructed by the addition of the specific bio-component to the ISFET i.e. the enzyme urease, which reacts with the urea, splitting it into carbon dioxide and ammonia, which, in turn, are in equilibrium with bicarbonate ion and ammonium ion. These two chemicals can be detected directly as gases by the appropriate gas-sensitive FET. Alternatively, bicarbonate and ammonium ion can be detected with appropriate ion-sensitive, polymeric membrane on an ISFET. Due to the use of the enzyme, the dense is referred to as an ENFET (enzyme field-effect transistor). Its constructional details are shown in Figure 21.26.

Enzyme field-effect transistor

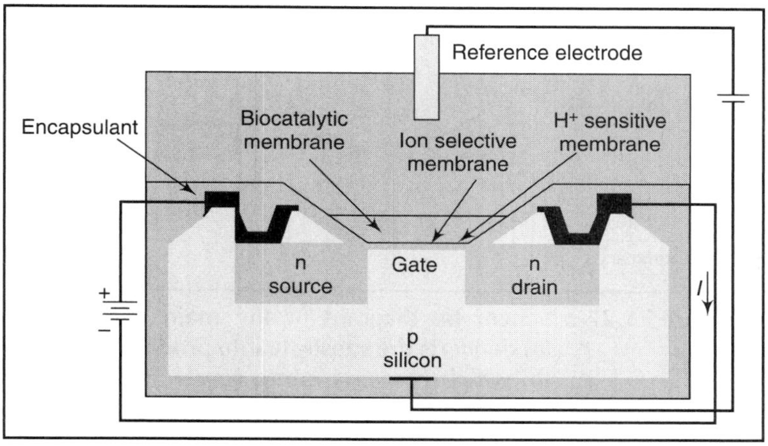

Figure 21.26 :: Schematic diagram of an ENFET (Frazer, 1995)

While time-saving is the major benefit of ISFETs, they are notorious for their sensitivity to changes in temperature, external pH, ionic strength, characteristics of the reference electrode, necessitating calibration before every measurement. pH probes based on ISFET are now commercially available as a replacement for fragile glass electrodes. These probes are virtually unbreakable and give faster and more reliable measurements. They can be stored dry and require minimal routine maintenance.

The associated electronics is embedded in the probe itself. The pH-related and temperature signals are processed in such a way that they match the glass pH electrode signal thus enabling a perfect match with pH meters facilities such as automatic temperature compensation, self-diagnostics, auto-shut-off and calibration.

21.8 Biosensors

Modification of the ion-selective electrode, in which an enzyme is immobilized on the surface of the glass electrode behind a polymer membrane, would result in bio-molecule specificity. Such type of sensors are known as biosensors, a term which is used to describe a wide variety of analytical devices based on the union between biological and physico-chemical components. A biosensor is therefore, an analytical device which converts a biological response into an electrical signal (Figure 21.27). The term 'biosensor' is often used to cover sensor devices used in order to determine the concentration of substances and other parameters of biological interest even where they do not utilize a biological system directly. A more practical definition states, "A biosensor is a self-contained integrated device that is capable of providing specific quantitative or semi-quantitative analytical information using a biological recognition element which is in direct spatial contact with a transduction element".

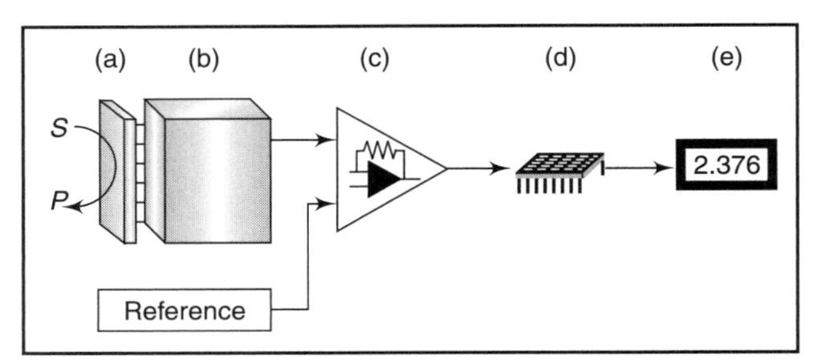

Figure 21.27 :: Schematic diagram of the main components of a biosensor. The biocatalyst (a) converts the substrate into product. The reaction is determined by the transducer (b) which converts it into an electrical signal. The output from the transducer is amplified (c), processed (d) and displayed (e) (adapted from Chaplin, 2004)

The biological response of the biosensor is determined by the biocatalytic membrane which accomplishes the conversion of the reactant to a product. Immobilized enzymes possess a number of advantageous features, which makes them particularly applicable for use in such systems. They may be re-used, which ensures that the same catalytic activity is present for a series of analyses. This is an important factor in securing reproducible results. Many enzymes are intrinsically stabilized by the immobilization process, but even where this does not occur, there is usually considerable apparent stabilization.

The biological component can consist of enzymes, antibodies, whole cells or tissue slices and is used to recognise and interact with a specific analyte (Figure 21.28). The physico-chemical component, often referred to as the transducer, converts this interaction into a signal, which can be amplified and which has a direct relationship with the concentration of the analyte. The transducer may use potentiometric, amperometric, optical, magnetic, colorimetric or conductance change properties.

Some typical examples of biosensors are:

- The heat output (or absorbed) by the reaction (calorimetric biosensors),
- Changes in the distribution of charges causing an electrical potential to be produced (potentiometric biosensors),

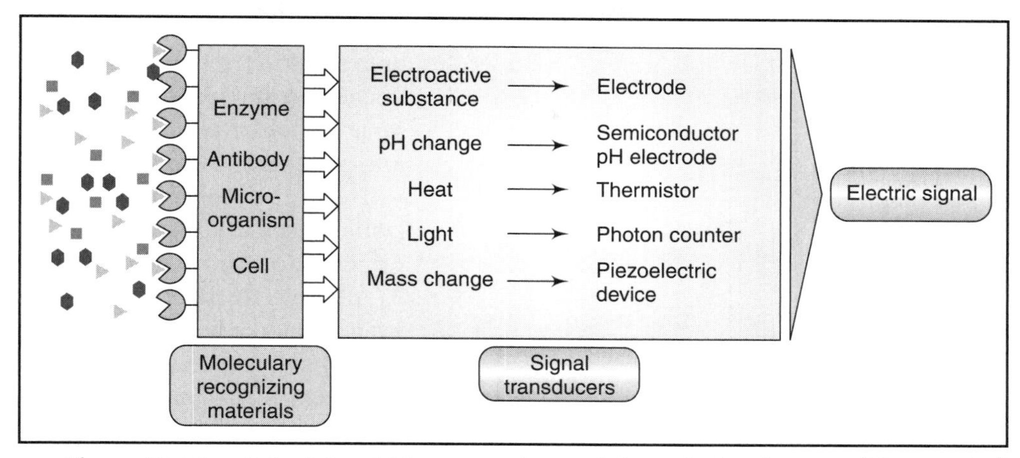

Figure 21.28 :: Principle of biosensors (*www.jaist.ac.jp/~yokoyama/biosensor*)

- Movement of electrons produced in a redox reaction (amperometric biosensors),
- Light output during the reaction or a light absorbance difference between the reactants and products (optical biosensors), or
- Effects due to the mass of the reactants or products (piezo-electric biosensors).

Urea electrode

The best known example of a biosensor is the urea electrode, which employs an ammonium ion-selective electrode coated with urease behind a Teflon membrane. Urea diffuses through the membrane and is converted to NH^{4+} by the reaction with urease and can then be determined potentiometrically at the glass membranes. Another important biosensor developed is for measuring blood glucose in diabetic patients. The biosensor combines glucose oxidase with an electrode to measure the oxygen generated by the reaction. Alternatively, glucose oxidase has been coupled to an electrode by ferrocene derivatives, to facilitate electron transfer between the two. These biosensors measure glucose concentrations between 10^{-1} and 10^{-7} *M*, but generally have a linear response in the range of 10^{-2} to 10^{-4} *M*.

The critical areas of biosensor construction are the means of coupling the biological component to the transducer and subsequent amplification system. Most of the early biosensors immobilized enzymes on selective electrodes, such as the Clark O_2 electrode, which measured one of the reaction products (e.g. O_2) of the enzyme–analyte interaction. Figure 21.29 shows the construction of this type of electrode. If the immobilized enzyme is soluble glucose oxidase between the two membranes, it becomes a glucose sensor. It works on the principle that in the presence of glucose, oxygen is electrode-consumed, providing a change in the signal from a conventional oxygen electrode. For constructing the sensor, glucose oxidase entrapped in a polyacrylamide gel was used. In general, the response times of such types of bio-electrodes were slow and subsequent work has concentrated on closer coupling of the biological component to the transducer. The present research in biosensors attempts to dispose of the coupling

Glucose sensor

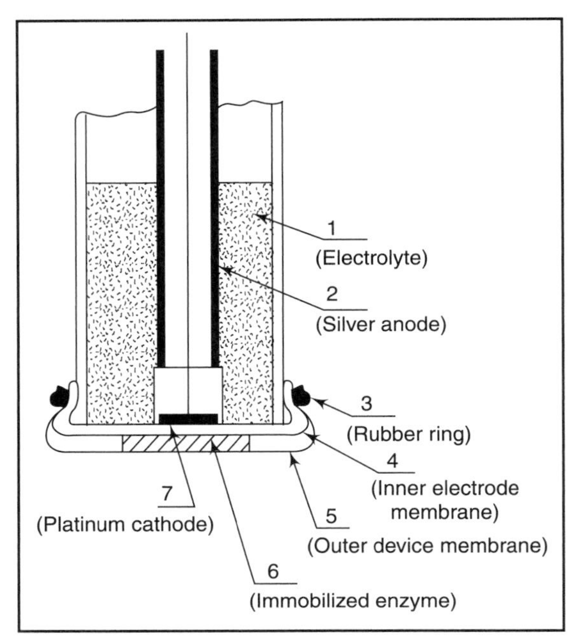

Figure 21.29 :: Constructional details of an enzyme utilizing sensor with oxygen electrode as the underlying analytical tool

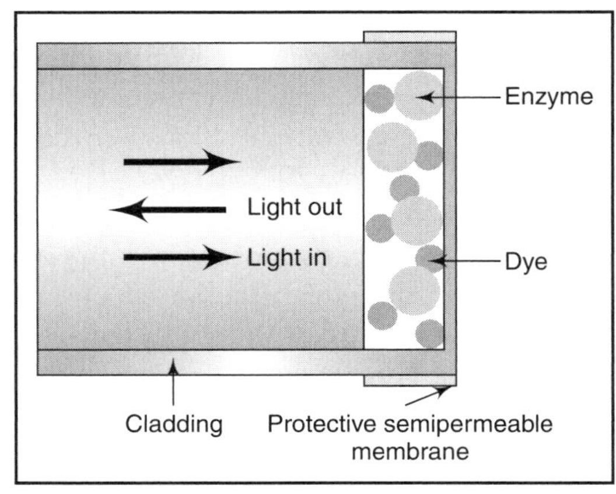

Figure 21.30 :: Optical biosensor (Frazer, 1994)

agent by direct immobilization of the enzyme onto a field-effect transistor. If such a system is developed for glucose, it will enable the biosensor to be attached to cathodes and injected into patients to measure blood glucose directly.

The transducers employed in the bulk of enzyme electrodes use either the potentiometric principle, or the amperometric principle.

After electrochemical biosensors, the second major family of biosensors are optical biosensors which have been exploited commercially. There are based on advances in fibre-optic technology, laser miniaturization and reproducible manufacture of prisms and waveguides. These can be considered as miniature traditional spectrophotometers. Figure 21.30 shows a typical arrangement of optical biosensor (Sloper and Flanagan, 1994). The fibre optic cable is coated with a biological element. Light is introduced into the sample via the fibre. Depending upon their number and type, molecules present at the surface of the fibre will absorb certain wavelengths of incoming light. The light that traverses or scattered by the biological layer is measured. The difference between incoming and outgoing intensity of light when measured and compared to a calibration or reference enables to determine the level of the analyte.

Tan, *et al.* (1999) discuss the development of a new device called nanoscopic optical biochemical sensor, with dimensions in the micro-meter to nanometer range, which offers fast response time and excellent biochemical sensitivity. The device uses the biochemical selectivity of optically sensitive dye molecules or bio-molecules, such as enzymes, antibodies, DNA molecules or living cells to recognize substances of interest. These ultrasmall sensors are expected to have increasing applications in the biomedical field at intracellular and extra cellular levels.

Biosensors thus combine the selectivity of biology with the processing power of modern micro-electronics and opto-electronics to offer powerful new analytical tools with major applications in medicine, environmental diagnostics and the food and processing industries. Biosensors can benefit bioprocesses in the following ways:

- *Specificity*: This is due to their remarkable ability of biological sensing to distinguish the analyte of interest and similar substances.
- *Speed*: They enable direct and instantaneous measurement of analytes as compared to normal lengthy analytical procedures.
- *Simplicity*: They dispense with conventional methods of analysis which requiring many steps to treat the sample as the receptor and transducer are integrated into one single sensor.
- *Continuous Monitoring Capability*: This is because in enzyme-based biosensors, the immobilized enzyme can be used for repeated assays.

Biochips and biocomputers are situated at the extreme limits of biosensor technology within the domains of bioelectronics, nanotechnology and micro-machining. It is being attempted to replace those components of computers presently made from doped inorganic semi-conducting oxide and insulating nitrides (silicon chips) by engineered protein molecules, capable of communicating via electron and light exchange. The advantage would mainly be one of size and speed. While photolithography can create sub-micron features in inorganic material and organic polymers, proteins are usually on the nanometer scale. Intense research activity is presently underway in the field of nanotechnology.

Besides the medical field, biosensors have tremendous applications in the food and beverage industries. Although several biosensors have been developed over the past few years and there are already numerous working biosensors, various problems still need to be resolved. Most complex problems awaiting solution are their limited lifetime, which restrict their commercial viability, necessitating improvements in their stability.

At present, most biosensors are based on a layer of enzyme setting on top of an electrode, i.e. such a device is an enzyme electrode. Commercially, the most successful biosensor enzyme electrodes have been those based on glucose oxidase. These are used by diabetics as a glucometer—a portable and home device to measure the concentration of glucose in the blood.

Chapter 22

Blood Gas Analyzers

Blood gas analyzers are used to measure pH, partial pressure of carbon dioxide (pCO_2) and partial pressure of oxygen (pO_2) of the body fluids with special reference to human blood. The measurements of these parameters are essential to determine the acid-base balance in the body. A sudden change in pH and pCO_2 could result in cardiac arrhythmias, ventricular hypotension and even death. This shows the importance of the maintenance of physiological neutrality in blood, and consequently the crucial role the blood gas analysers play in clinical medicine.

22.1 Acid Base Balance

The normal pH of the extra-cellular fluid lies in the range 7.35 to 7.45, indicating that the body fluid is slightly alkaline. When the pH exceeds 7.45, the body is considered to be in a state of alkalosis. A body pH below 7.35 indicates acidosis. Both acidosis or alkalosis are disease conditions widely encountered in clinical medicine. Any tendency of pH of blood to deviate towards these conditions is dealt with by the following three physiological mechanisms: (i) buffering by chemical means, (ii) respiration, (iii) excretion, into the urine by kidneys.

The blood and tissue fluids contain chemical buffers, which react with added acids and bases and minimise the resultant change in hydrogen ions. They respond to changes in carbon dioxide concentration in seconds. The respiratory system can adjust sudden changes in carbon dioxide tension back to normal levels in just a few minutes. Carbon dioxide can be removed by increased breathing and therefore, hydrogen concentration of the blood can be effectively modified. The kidney requires many hours to readjust hydrogen ion concentration by excreting highly acidic or alkaline urine to enable body conditions to return towards normal.

Arterial blood has a pH of approximately 7.40. As venous blood acquires carbon dioxide, forms carbonic acid and hydrogen ions, the venous blood pH falls to approximately 7.36. This pH drop of 0.04 units occurs when CO_2 enters the tissue capillaries. When CO_2 diffuses from the pulmonary capillaries into the alveoli, the blood pH rises 0.04 units to bring the normal arterial value of 7.40. It is quite difficult to measure the pH of fluids inside the tissue cells, but from estimates based on CO_2 and $(HCO_3)^-$ ion concentration, intracellular pH probably ranges from 7.0 to 7.2.

In order to maintain pO_2, pCO_2 and pH within normal limits, throughout the wide range of bodily activity, the rate and depth of respiration vary automatically with changes in metabolism. Control of alveolar ventilation takes place by means of chemical as well as nervous mechanisms. The three important chemical factors regulating alveolar ventilation are the arterial concentrations of CO_2, H^+ and O_2. Carbon dioxide tension in the blood stream and cerebro-spinal fluid is the major chemical factor regulating alveolar ventilation. The carotid and aortic chemoreceptors stimulate respiration when oxygen tension is abnormally low. In fact, so many organs participate in the control of respiration; it is difficult to include all aspects in the text. The readers may like to read any standard textbook on human physiology to appreciate the mechanism of respiration control and maintenance of physiological neutrality of the blood.

Table 22.1 lists out the normal range for pH, pCO_2, pO_2, total CO_2, base excess and bicarbonate, all measurements made at 37°C (Gambino, 1967).

Table 22.1 :: Typical Expected Values of Blood Gas Parameters

Plasma Parameter		Arterial or Arterialized Capillary Blood	Venous Plasma (separated at 37°C)
pH		7.37 to 7.44	7.35 to 7.45
pCO_2	men women	34 to 35 mmHg 31 to 42 mmHg	36 to 50 mmHg 34 to 50 mmHg
pO_2	resting adult resting adult over 65 years	80 to 90 mmHg 75 to 85 mmHg	25 to 40 mmHg
Biocarbonate	men women	23 to 29 mmol/l 20 to 29 mmol/l	25 to 30 mmol/l 23 to 28 mmol/l
Total CO_2 (plasma) Base Excess	men women men women	24 to 30 mmol/l 21 to 30 mmol/l –2.4 to +2.3 mmol/l –3.3 to +1.2 mmol/l	26 to 31 mmol/l 24 to 29 mmol/l 0.0 to +5.0 mmol/l –1.0 to + 3.5 mmol/l

22.2 Blood pH Measurement

When making pH measurements, the Sorenson's definition for pH of blood and plasma is adopted, which is given by:

$$pH = -\log_{10} (H^+),$$

where H^+ designates the concentration of hydrogen ions.

According to this relation, whole blood with a (H^+) of 4×10^{-8} moles/l would have a pH of 7.4. This implies that an increase in the (H^+) to 1×10^{-7} moles/l would correspond to a decrease in pH to 7.0.

Blood or plasma pH is a fairly constant number. The range of pH in health and disease is relatively narrow: in health 7.36 to 7.42, in disease from about 7.00 to 7.80. Any greater fluctuations of pH may result in acidosis or alkalosis and death may occur on either side of variation.

	7.00		7.36		7.44		7.88	
DEATH	|	ACIDOSIS	|	NORMAL	|	ALKALOSIS	|	DEATH

Electrochemical pH determination utilizes the general method of measuring the difference in potential occurring between two solutions of different pH, using the glass electrode. However, there are special problems associated with the measurement of pH of blood. Particular attention must therefore be given to the problems mentioned below:

22.2.1 Electrodes for Blood pH Measurement

Several types of electrodes have been described in literature for the measurement of blood pH. They are all of the glass electrode type but made in different shapes so that they may accept small quantities of blood and yield accurate results. The most common type is the syringe electrode, which is preferred for the convenience of taking small samples of blood anaerobically. The small 'dead space' between the electrode bulb and the inner surface of the syringe barrel is usually filled with dilute heparin solution to prevent blood coagulation. Before making measurements, the syringe should be rolled between the hands to ensure thorough mixing.

Syringe electrode

Microcapillary glass electrodes are preferred when it is required to monitor pH continuously: for example during surgery. These types of electrodes are especially useful when very small volume of the sample is to be analyzed.

Micro-electrode

Typically, a micro-electrode for clinical applications requires only 20-25 µl of capillary blood for determination of pH. The electrode is enclosed in a water jacket with circulating water at a constant temperature of 38°C. The water contains 1% NACl for shielding against static interference. The capillary is protected with a polyethylene tubing. The internal reference electrode is silver/silver chloride and the calomel reference electrode is connected to a small pool of saturated KCl, through a porous pin. An accuracy of 0.001 pH can be obtained with this electrode against a constant buffer. Figure 22.1 shows the constructional details of a typical blood pH electrode and the measurement set up used in practice.

Quite often, combination electrodes comprising both measuring and reference electrodes offer single-probe convenience for all pH measurements. Several instruments offer the ability to measure pH in small containers with as little as 250 µl of sample.

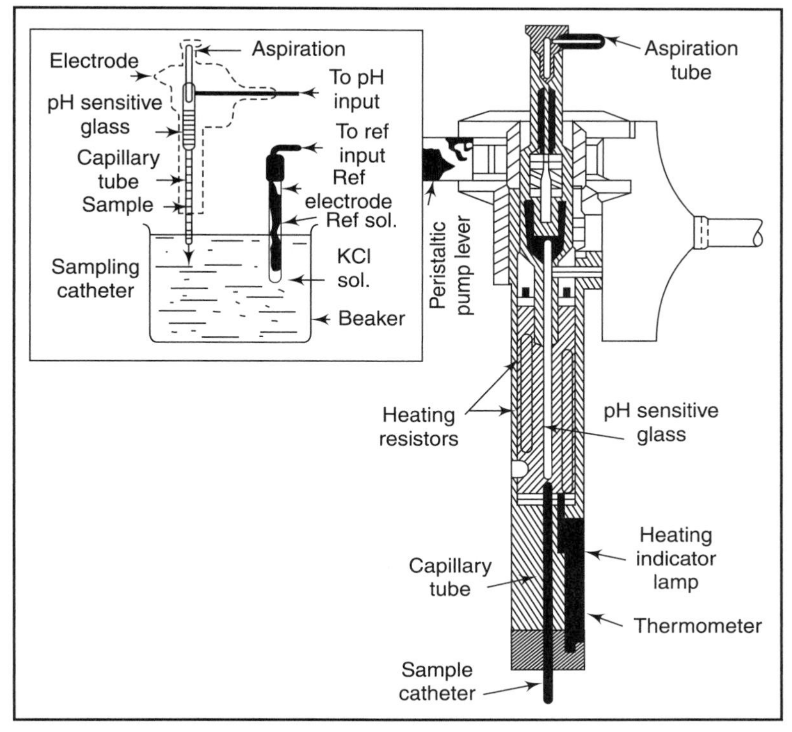

Figure 22.1 :: Constructional details of a blood pH electrode (Courtesy Corning Scientific Instruments, USA)

Ahn, *et al.* (1975) bring out the drawbacks of the conventional macro- and micro- size pH electrodes when used for biomedical applications. These are due to the relatively large size of the macro-electrode and the fragility of the micro-electrode. They constructed a miniature pH glass electrode, using Corning 015 glass as hydrogen-ion-sensitive glass. The dimensions of the electrode are 1.0 mm outside diameter and 0.25 mm wall thickness. The inner electrolyte is a solution of 0.1 N hydrochloric acid and the inner reference electrode is a silver/ silver chloride electrode. The silver/silver chloride electrode is made from a silver wire (0.127 mm in diameter and 99.9 per cent purity) by electrolytic method. A FET input operational amplifier is integrated into the pH electrode. Temperature response of the electrode was –1.51 mV/K at a pH value of 7. Evaluation of the stability of the electrode showed a 1% drift over a seven-hour operational time. The pH temperature hysteresis effects showed a 0.5 and 1.0 per cent deviations, respectively. The response time was within 4s for 99 per cent response.

Miniature pH glass electrode

22.2.2 Effect of Blood on Electrodes

Glass electrodes deteriorate if allowed to remain in contact with blood for a long time. This results in change of emf-pH slope. The poisoning effect appears to be due to protein deposition. Therefore, as a precautionary measure, in an apparatus where blood necessarily remains in contact around the electrode for long periods (more than 20 min.) the response must be checked frequently against buffer solutions. The poisoning effect can be reduced by putting the electrode in pepsin and 0.1 N HCl, followed by careful wiping with tissue paper.

The pH of blood is found to change linearly with temperature in the range 18° to 38°C. The temperature coefficient for the pH of blood is 0.0147 pH unit per degree centigrade. This necessitates the use of a highly accurately temperature controlled bath to keep the electrodes with the blood sample at 37°C ± 0.01°C. A circuit diagram for controlling temperature of the bath is shown in Figure 22.8.

Another important point to be kept in mind while making blood pH measurements is that because of the possible individual variations in the temperature coefficient of blood pH, the method of measuring at some temperature other than 37°C followed by correction is not recommended. It is advisable to keep both the glass as well as the reference electrode at the temperature of measurement.

22.2.3 Buffer Solutions

Buffer

Buffer solutions are primarily used for the: (i) creation and maintenance of a desired, stabilized pH in a solution, and (ii) standardization of electrode chains for pH measurements. A buffer is, therefore, a substance which by its presence in a solution is capable of counteracting pH changes in the solution as caused by the addition or the removal of hydrogen ions. Buffer solutions are characterized by their pH value. They are available in tablets of pH value 4.7 and 9.2.

Buffer solutions used in blood pH measurements are the following:

- 0.025 molar potassium dihydrogen phosphate with 0.025 molar disodium hydrogen phosphate. This solution has a pH value of 6.840 at 38°C and 6.881 at 20°C.
- 0.01 molar potassium dihydrogen phosphate with 0.04 molar disodium hydrogen phosphate. This buffer has a pH value of 7.416 at 38°C and 7.429 at 20°C.

These buffers should be stored at a temperature between 18° to 25°C. In order to maintain an accurate pH, the bottles containing them should be tightly closed.

22.3 Measurement of Blood pCO_2

The blood pCO_2 is the partial pressure of carbon dioxide of blood taken anaerobically. It is expressed in mmHg and is related to the percentage CO_2 as follows:

$$pC0_2 = \text{Barometric pressure} - \text{water vapour pressure} \times \frac{\%CO_2}{100}$$

At 37°C, the water vapour pressure is 47 mmHg, so at 750 mm barometric pressure, 5.7 per cent CO_2 corresponds to a pCO_2 of 40 mm.

pCO_2 electrode

All modern blood gas analyzers make use of a pCO_2 electrode of the type described by Stow *et al.* (1957). It basically consists of a pH-sensitive glass electrode having a rubber membrane stretched over it, with a thin layer of water separating the membrane from the electrode surface. The technique is based on the fact that the dissolved CO_2 changes the pH of an aqueous solution. The CO_2 from the blood sample defuses through the membrane to form H_2CO_3, which dissociates into (H^+) and (HCO_3) ions. The resultant change in pH is thus a function of the CO_2 concentration in the sample. The emf generated was found to give a linear relationship between the pH and the negative logarithm of pCO_2. Although the electrode could not provide sensitivity and stability required for clinical applications, it made way for realising a direct method for measurement of pCO_2.

The basic construction of the electrode was modified by Severinghaus and Bradley (1958) to a degree that made it suitable for routine laboratory use. In the construction worked out by them, water layer was replaced by a thin film of an aqueous sodium bicarbonate ($NaHCO_3$) solution. The rubber membrane was also replaced by a thin Teflon membrane, which is permeable to CO_2 but not to any other ions, which might alter the pH of the bicarbonate solution. The CO_2 from the blood diffuses into the bicarbonate solution. There will be a drop in pH due to CO_2 reacting with water forming carbonic acid. The pH falls by almost one pH unit for a tenfold increase in the CO_2 tension of the sample. Hence, the pH change is a linear function of the logarithm of the CO_2 tension. The optimum sensitivity in terms of pH change for a given change in CO_2 tension is obtained by using bicarbonate solution of concentration of about 0.01 mole/l. The electrode is calibrated with the known concentration of CO_2. The response time of the CO_2 electrode is of the order of 0.5 to 3 min. This electrode is twice as sensitive and drifted much less than the Stow's electrode. Figure 22.2 shows the construction of a typical pCO_2 electrode.

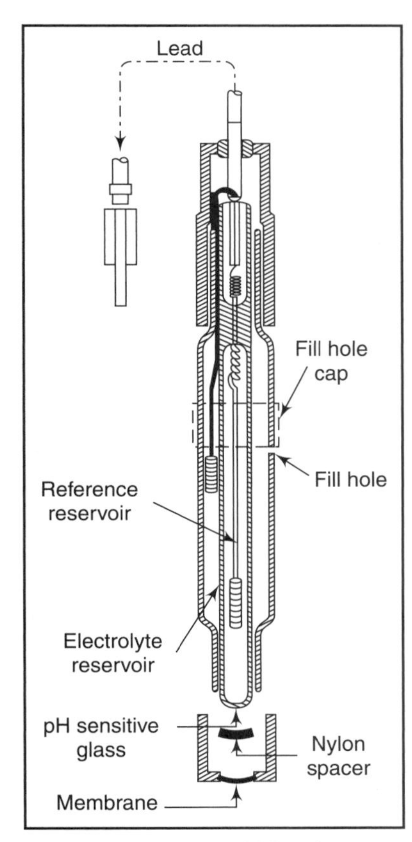

Figure 22.2 :: Parts of blood pCO_2 electrode (Courtesy, Corning Scientific Instruments, USA)

Further improvements in stability and response time were achieved by Hertz and Siesjo (1959). They used a dilute solution of $NaHCO_3$ (0.0001 N) which helped in reducing the response time but the drift introduced posed serious problems. The compromise between response time and drift was achieved by using 0.001 N solution of $NaHCO_3$. Silver/silver chloride reference electrode was replaced by a calomel cell which was made an integral part of the electrode.

Severinghaus (1962) made a further improvement upon the earlier Severinghaus-Bradley electrode in the low pCO_2 range by replacing the cellophane spacer with a very thin nylon mesh. Glass fibres or powdered glass wool were also found to be good separators. He used a membrane of 3/8 mil Teflon and glass wool for the separator. Electrodes with 95 per cent response in 20 s were constructed in this way.

Reyes and Neville (1967) constructed a pCO_2 electrode by using 0.5 mm polyethylene as a membrane and used no separator between the glass surface and this membrane. They added carbonic anhydrase to the electrolyte. The response time was found to be 6 seconds for 90 per cent of a step change from 2 per cent to 5 per cent CO_2. Use of pCO_2 electrode for measuring blood or plasma pCO_2 has been studied repeatedly and has been found to be accurate, precise and expedient, Hill and Tilsley (1973). An extension of the miniature pH electrode (Ahn, *et al.*, 1975) is the miniature pCO_2 electrode described by Lai, *et al.* (1975).

22.3.1 Performance Requirements of pH Meters Used for pCO_2 Measurement

The emf generated by pCO_2 electrode is a direct logarithmic function of pCO_2. It is observed that a tenfold change in pCO_2 causes the potential to change by 58 ± 2 mV. The pH versus log pCO_2 relationship is linear within ±0.002 pH unit from 1 to 100 per cent carbon dioxide. Since 0.01 unit pH change corresponds to a 2.5 per cent change in pCO_2 or 1 mmHg in 40 mmHg, for achieving an accuracy of 0.1 mmHg, it is desirable to read 0.001 pH unit, i.e., a resolution of 60 μV. This order of accuracy can be read only on a digital readout type pH meter or on analog meter with expanded scale. The instrument should have a very high degree of stability and a very low drift amplifier. The input impedance of the electronic circuit must be at least 10^{12} Ω.

It is essential to maintain the temperature of the electrode assembly constant within close limits. It is experimentally shown that variation in the temperature of ± 1°C produces an error of ±1.5 mmHg or about ±3 per cent at 5 mm pCO_2. The combined effects of temperature change upon the sensitivity of the pH electrode and upon the pCO_2 of the blood sample amount to a total variation in sensitivity of 8 per cent per degree centigrade.

Calculated Bicarbonate, Total CO_2 *and Base Excess*: Acid base balance determinations are based on several calculations, which are routinely used in conjunction with blood pH and gas analysis. An accurate picture of the acid base balance can be determined from the equilibrium.

$$CO_2 + H_2O \rightarrow H_2CO_3$$

$$H_2CO_3 \rightarrow H^+ + HCO_3^-$$

which, for bicarbonate has an equilibrium constant

$$K_{H_2}, C_{O_2}/HCO_3^- = \frac{[H^+][HCO_3^-]}{H_2CO_3},$$

where (H^+), (HCO_3^-) and (H_2CO_3) refer to the concentration of these substances.

Since $H_2CO_3 = 0.03\ pCO_2$

and since $pH = -\log [H^+]$

Therefore, $$pH = pK + \log \frac{[HCO_3^-]}{0.03\ pCO_2},$$

where pK equals 6.11 for normal plasma at 37°C. This formula is used in blood gas analyzers for calculating actual bicarbonate.

Total CO_2 is calculated from the relationship:

$$[HCO_3^-] + (0.03 \times pCO_2) = \text{total } CO_2 \text{ in millimoles/l}$$

Base excess is calculated using the following formula described by Siggaard-Andersen (1963):

$$\text{Base excess} = (1 - 0.0143 \times Hb)[HCO_3^-] - (9.5 + 1.63\,Hb) \times (7.4 - pH) - 24,$$

where Hb represents the patients' haemoglobin value.

Base excess

Base excess is the number of milli-equivalents of strong acid or base which would be required per litre of blood to restore it to a pH of 7.400 at 37°C with pCO_2 held at 40 torr. This is usually estimated from pH and pCO_2 measurements done at 37°C in a sample of blood using Siggaard-Andersen's alignment monogram (Siggaard-Andersen, 1963).

22.4 Blood pO_2 Measurement

The partial pressure of oxygen in blood or plasma indicates the extent of oxygen exchange between the lungs and the blood, and normally, the ability of blood to adequately perfuse the body tissues with oxygen. The partial pressure of oxygen is usually measured with polarographic electrode. There is a characteristic polarizing voltage at which any element in solution is predominantly reduced and in the case of oxygen, it is 0.6 to 0.9 V. In this voltage range, it is observed that the current flowing in the electrochemical cell is proportional to the oxygen concentration in the solution.

Oxygen electrode

Most of the modern blood gas analyzers utilize an oxygen electrode first described by Clark (1956) for measuring the partial pressure of oxyzen. This type of electrode consists of a platinum cathode, a silver/silver chloride anode in electrolyte

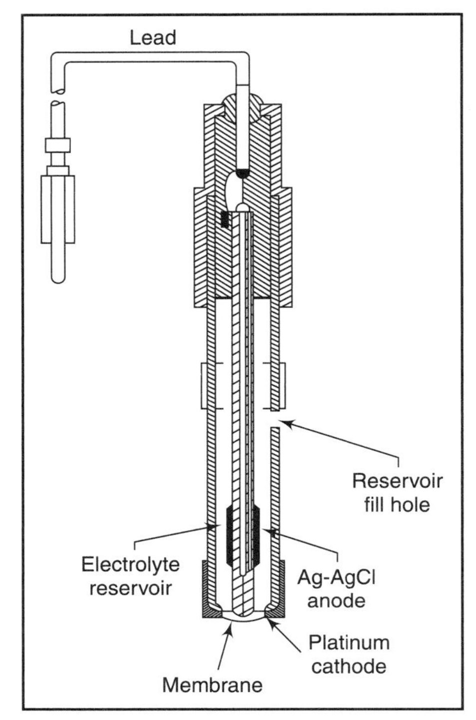

Figure 22.3 :: Constructional details of pO_2 electrode (Courtesy, Corning Scientific Instruments, USA)

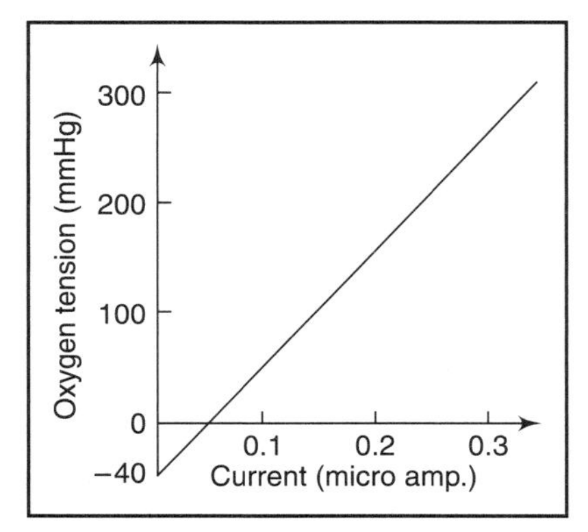

Figure 22.4 :: Calibration curve used to obtain a calibration constant of pO_2 electrode (after McConn and Robinson, 1963)

filling solution and polypropylene membrane. The electrode is of single unit construction and contains the reference electrode also in its assembly. Figure 22.3 shows the construction of a typical Clark type oxygen electrode. The entire unit is separated from the solution under measurement by the polypropylene membrane.

Oxygen from the blood diffuses across the membrane into the electrolyte filling solution and is reduced at the cathode. The circuit is completed at the anode, where silver is oxidised, and the magnitude of the resulting current indicates partial pressure of oxygen. The reactions occurring at the anode and cathode are as follows:

Cathode reaction:

$$O_2 + 2H_2O + 4e^- \rightarrow 4OH^-$$

Anode reaction:

$$4Ag \rightarrow 4Ag^+ + 4e^-$$

The Clark electrode for measuring PO_2 has been extensively studied and utilized. It is found to be of particular advantage for measuring blood samples. The principal advantages are: (i) the sample size required for the measurement can be extremely small, (ii) the current produced due to pO_2 at the electrode is linearly related to the partial pressure of oxygen, (iii) the electrode can be made small enough to measure oxygen concentration in highly localised areas, and (iv) the response time is very low, so the measurements can be made in seconds. As compared to this, it takes a very long time if the measurements are made by chemical means.

McConn and Robinson (1963) observed that zero electrode current was not given by a solution having zero oxygen tension, but occurred at a definite oxygen tension, which they called the 'electrode constant'. So, for calibrating the electrode it is necessary to know this constant for that particular electrode. They further showed that when the straight line calibration curves

electrode constant

(Figure 22.4) were extended backwards, they did not pass through the origin, but intersected the oxygen tension axis at a negative value. To obtain a true zero current (less than 10 nA), the electrolyte of the electrode is deoxygenated by bubbling nitrogen through it for about half an hour and then placing the electrode in water redistilled from alkaline pyragallol.

The platinum cathode of the oxygen electrode tends to become contaminated or dimensionally unstable with time and use. The result is usually an inability to calibrate and slope the electrode on any pO_2 range. The manufacturers usually recommend application of ammonium hydroxide on the tip of the electrode (10 per cent solution), with a gentle, rotary motion using a swab. The silver chloride gets dissolved in ammonium hydroxide. It is then flushed with distilled water.

The polarographic electrodes usually exhibit an ageing effect by showing a slow reduction in current over a period of time, even though the oxygen tension in the test solution is maintained at a constant level. Therefore, it needs frequent calibration. It is probably associated with the material depositing itself to the electrode surface. The effect due to ageing can possibly be avoided by covering the electrode with a protective film of polyethylene, but it has the undesirable effect of increasing the response time.

The measurement of current developed at the pO_2 electrode due to the partial pressure of oxygen presents special problems. The difficulty arises because of the extremely small size of the electrical signal. The sensitivity (current per torr of oxygen tension) is typically of the order of 20 pA per torr for most commercial instruments. It is further subject to constant drift and is also not independent of the sample characteristics. Measurement of oxygen electrode current is made by using high input impedance, low noise and low current amplifiers. Field effect transistors usually form the input stage of the pre-amplifiers.

Hahn (1969) used a field-effect transistor operational amplifier to measure small polarographic currents. The op-amp is connected as a trans-resistance converter, the output of which can be read directly by a digital voltmeter. Figure 22.5 shows the circuit in which the polarizing voltage is supplied by the cell B (1.3 V) and variable resistance VR_1. The standing current from the electrochemical cell is cancelled by means of VR_2, Battery B and 1 GΩ resistance. Capacitor C (100 pF) is

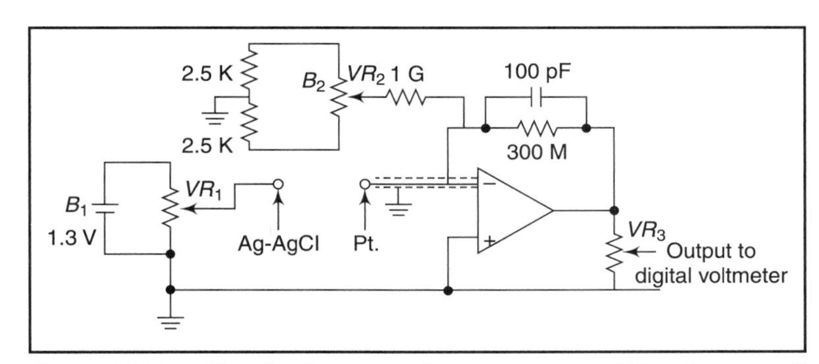

Figure 22.5 :: Circuit diagram of a current amplifier for use with pO_2 electrode

included to limit the bandwidth of the amplifier to reduce noise and to ensure good dynamic stability.

22.5 A Complete Blood Gas Analyzer

Blood gas analyzers are designed to measure pH, pCO_2 and pO_2 from a single sample of whole blood. The size of the sample may vary from 25 µl to a few hundred microlitres. The estimations take about 1 min. With built-in calculators, the instruments can also compute total CO_2, HCO_3 and Base Excess. A typical block diagram of a blood gas analyzer machine is shown in Figure 22.6. In this machine, separate sensors are used for pH, pCO_2 and pO_2. The instrument contains three separate high input impedance amplifiers designed to operate in the specific range of each measuring electrode. A separate module houses and thermostatically controls the three

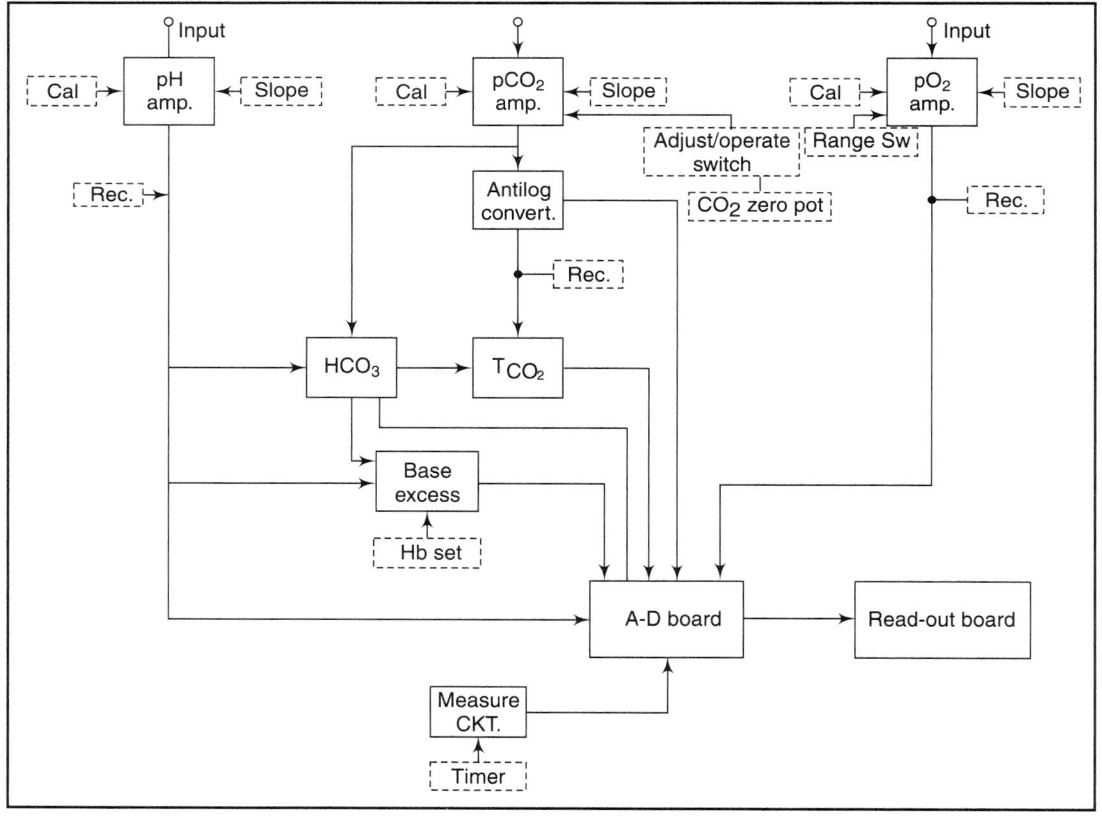

Figure 22.6 :: Block diagram of a blood gas analyzer

electrodes. It also provides thermostatic control for the humidification of the calibrating gases. A vacuum system provides aspiration and flushing service for all three electrodes. Calibrating gases are selected by a special push button control and passed through the sample chamber when required. Two gases of accurately known O_2 and CO_2 percentages are required for calibrating the analyzer in the pO_2 and pCO_2 modes. The gases required are: O_2 value of 12% Cal and 0% Slope and CO_2 value of 5 per cent Cal and 10 per cent Slope. These gases are used with precision regulators for flow and pressure control. Two standard buffers of known pH are required for calibration of the analyzer in the pH mode. The buffers that are used are 6.838 (Cal) and 7.382 (Slope). It is generally recommended that the sample chamber should control 7.382 buffer when in the standby mode.

The input signal to (HCO_3^-) calculator [Figure 22.7 (a)] comes from the outputs of pH and pCO_2 amplifiers. The outputs are suitably adjusted by multiplying each signal by a constant and are

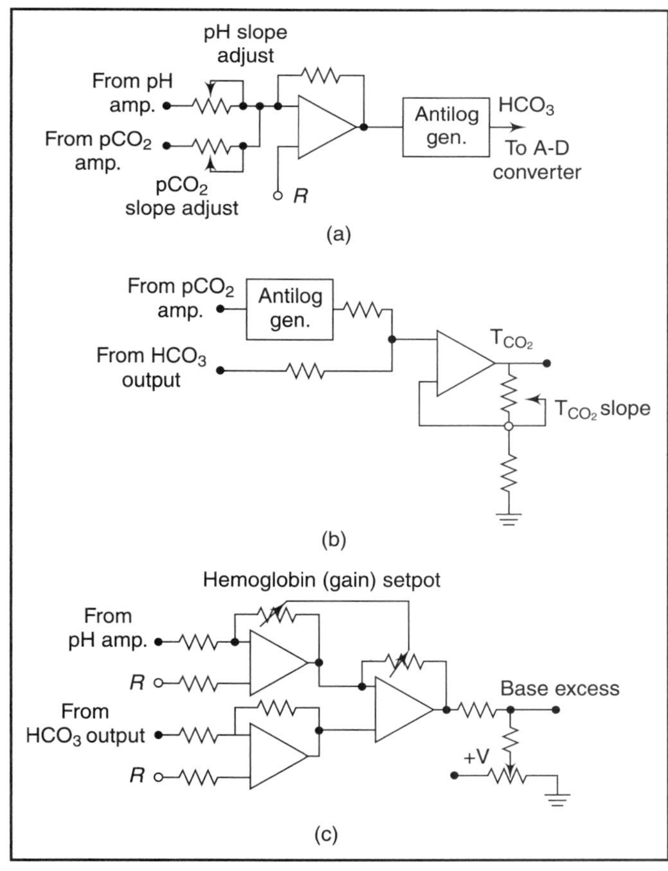

Figure 22.7 :: Circuit diagram for computation of: (a) Bicarbonate (HCO_3^-) (b) Total CO_2 (c) Base excess

given to an adder. The next stage is an antilog generator similar to the one used in pCO_2 amplifier. The output of this circuit goes to A-D converter for display. Resistance R is used to adjust zero at the output.

The total CO_2 is calculated [Figure 22.7 (b)] by summing the output signals of (HCO_3^-) calculator and the output of the pCO_2 amplifier. The facilities for adjusting slope and zero at the output are available.

The base excess calculator [Figure 22.7 (c)] consists of three stages. In the first stage, the output of the pH amplifier is inverted in an operational amplifier whose gain is controlled with a potentiometer (Haemoglobin value) placed on the front panel. The output of the HCO_3^- calculator is inverted in the second stage. The third stage is a summing amplifier A_3 whose output is given to an A-D converter.

The analog output of the selected parameter channel is given to the input of the A-D converter. The output of the A-D converter goes to a digital read-out circuit like LEDs.

The three electrodes (pH, pO_2 and pCO_2) are housed in a thermostatically controlled chamber. It also provides thermostatic control for the humidification of the calibrating gases. The thermal block and the humidifier block heat control circuits are of the same type (Figure 22.8). The temperature is set with a potentiometer for exactly 37°C. The heater circuit is controlled by a thermistor in the block, which acts as a sensor. As the heat increases, the resistance of the thermistor decreases. At 37°C, the thermistor is calibrated to have a resistance of 25 K.

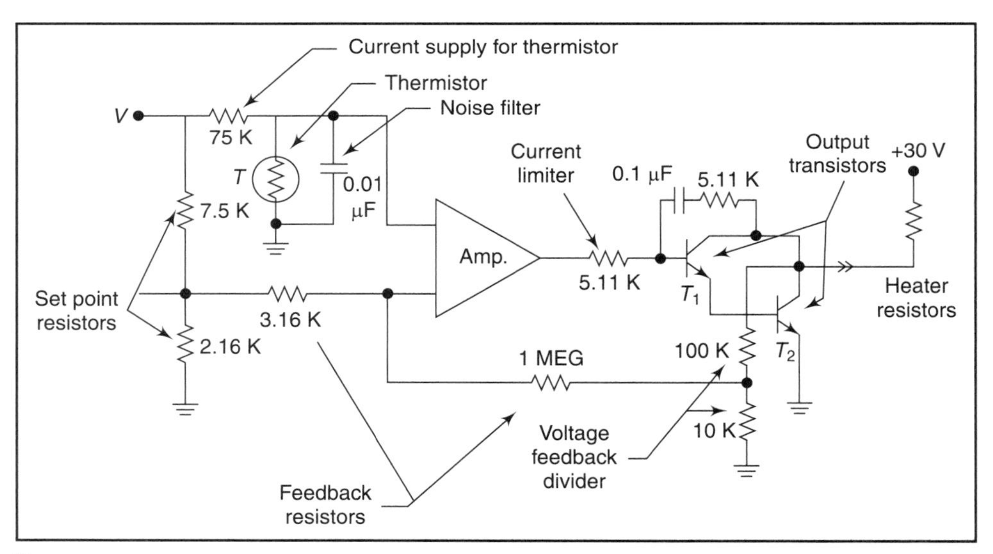

Figure 22.8 :: Temperature control circuit for thermostated chamber (Courtesy, Corning Scientific Instruments, USA)

If the temperature of the block decreases, the resistance of the thermistor will increase. The increase in resistance will cause the voltage at inverting input of op-amp to become more negative. This result in the output voltage to go to more positive, increasing the base current of transistors T_1 and T_2. The increase in base current increases the collector current, which goes directly to the heater resistor on the block. As the heater resistor heats up the block, the thermistor will decrease until it returns to 25 K.

Many of the blood gas analyzers have a provision for checking the membrane of pO_2 and pCO_2 electrodes. In the check position, a potential is applied across the membrane. Any leak in the membrane of sufficient magnitude will result in a considerable lowering of the resistance may be from 100 MΩ to 500 kΩ. The change in resistance can be used to have a change of potential to switch on a transistor, which would cause a lamp to light on the front panel of the instrument. This would indicate that a new membrane is needed.

22.5.1 Fibre Optic-based Blood Gas Sensors

For *in vivo* measurements and reliably analyzing blood gases, a small, stable, accurate and bio-compatible sensor is required which could be inserted in the blood flow of an artery through an arterial cannula and remain in place for several days. In addition, it has to be low cost so that it could be used as a disposable item. Advances in fibre optics and the development of pH and oxygen sensitive dyes have made such a sensor possible. Blood gas analyzers based on such sensors are now commercially available.

Figure 22.9 shows the schematic diagram of a fibre optic-based blood gas analyzer (Soller, 1994). The sensors are interfaced with an electro-optic monitor. The monitor supplies the excitation light , which may be from a monochromatic source such as a diode laser or a broadband source like xenon lamp whose light is filtered to provide a narrow bandwidth of excitation. Two wavelengths of light are provided, one wavelength is sensitive to changes in the

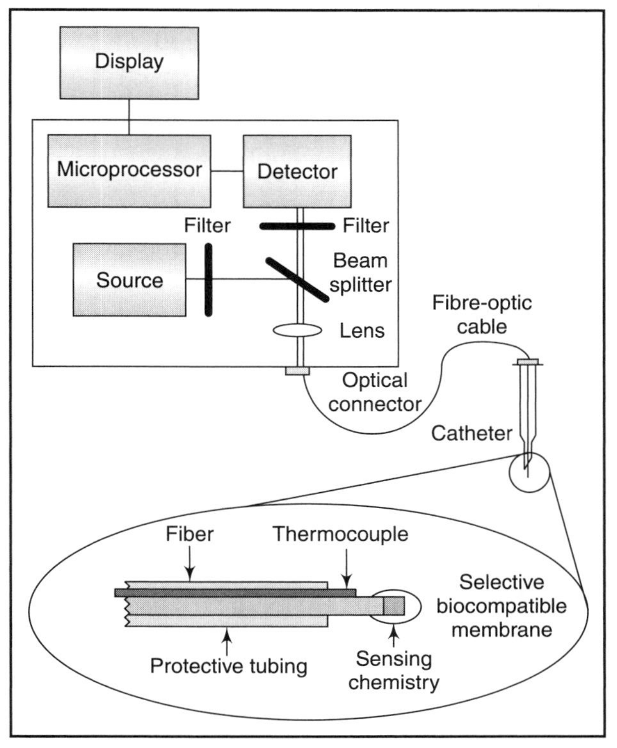

Figure 22.9 :: Block diagram of fibreoptic based blood gas sensor and monitor (after Soller, 1994)

species to be measured, while the other wavelength is unaffected by changes in the analyte concentration. The unaffected wavelength serves as a reference and is used to compensate for fluctuations in source output and detector efficiency. The light output from the monitor is coupled into a fibre-optic cable through appropriate lenses and optical connectors. The cable is sufficiently long to permit easy patient access by allowing the monitor to be placed at a distance.

Sensor assembly

Within the sensor assembly (Figure 22.10) are three optical fibres—one each for measuring blood O_2, CO_2 and pH. The optical-fibre is approximately 10 cm long and also has a thermocouple or thermistor wire running alongside the fibre to measure temperature near the sensor tip. Temperature correction is necessary for optical blood gas sensors. The solubility of the gases, O_2 and CO_2 in the sensing material is a function of temperature and the optical properties of the sensing chemistry also change as the temperature varies. The fibers and the temperature sensor are encased in a protective tubing to contain any fiber fragments in case of sensor breakage.

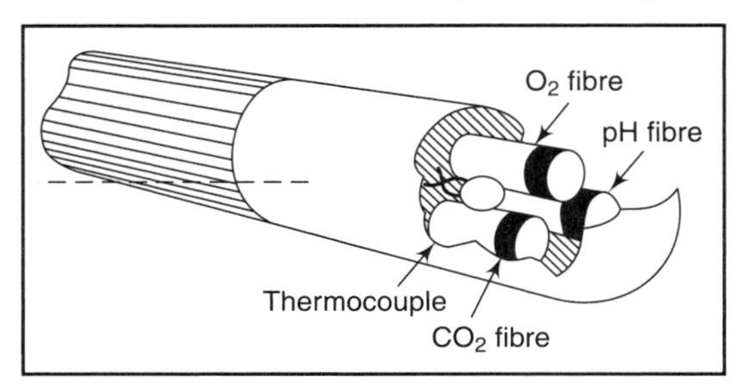

Figure 22.10 :: The sensor assembly contains three fibres that measure pH, pCO_2 and pO_2 bundled together with the thermocouple

Each fibre is as thin as human hair and coated at the tip with a specific chemical dye (Figure 22.11). When light of a known wavelength strikes the dye, the dye fluoresces, giving off light of a different wavelength. The fluorescent emission changes in intensity as a function of the concentration of the analyte (O_2, CO_2 or pH) in the blood. The emitted light travels back down the fibre to the monitor where it is converted into an electrical signal by using a solid state detector or a photomultiplier. The signal is amplified

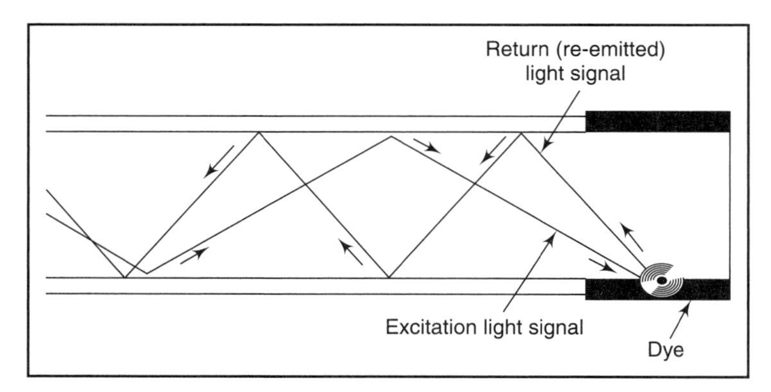

Figure 22.11 :: Within each fibre's core, excitation light reflects along, the fibre toward the fluorescent dye at the fibre's tip. The dye at the tip reacts to the excitation light and analyte concentration by fluorescing. The fluorescent signal then returns in the same fibre to the monitor, which measures the intensity of the signal

before it is given to a digitizer. Signal processing to relate the light intensity to the analyte concentration is achieved using a microprocessor and digitally displayed.

Since a detector produces noise or dark current when it is not illuminated, accurate signal measurements require that ambient light be subtracted from the total signal. Thus, a signal measurement is made with the flash lamp off. This ambient light is subtracted from the total signal by correlated double integration circuits. Another factor that affects accurate blood gas calculations is the background fluorescence of the materials in the optical block. This value is obtained by measuring the current developed by the detector when no sensor is connected. As with ambient light, the background fluorescence is subtracted from the signal measurement.

Considerable effort has gone into identifying organic molecules, which would make suitable sensors. These molecules must have high fluorescent intensity at excitation and emission wavelengths that match the available light sources and detectors. They must be photostable, i.e. their emission properties should not change as they are continually illuminated by the excitation source. Sensors based on fluorescence quenching of organic dyes such as perylene dibutyrate have been reported for measurement of pO_2. Oxygen sensors based on the phosphorescence quenching of metal—loporphyrins and terbium complexes have also been successfully tried.

It has been found from experimental studies that as the partial pressure of oxygen increases, the sensitivity decreases. The best sensitivity is achieved in the region of 30 to 150 mmHg, but drops off considerably by higher pO_2, making it difficult to resolve small changes in pO_2, when the partial pressure of oxygen is greater than 200 mmHg. Further, at a high pO_2, as the quenching increases, the light reaching the detector decreases. A compromise is thus required to be made in selecting a sensing material that provides adequate sensitivity over the required pO_2 measurement range and simultaneously offers good signal-to-noise ratio at the detector. The performance range of the sensor is normally limited to under 300 mmHg in order to achieve both good sensitivity and adequate light detection.

pH sensor

pH sensor designs are based on dye molecules whose optical properties change as the pH is varied between 6.8 and 7.8. At any pH in the range of interest, both the acid and base forms of the dye molecule are present and each form has distinct optical characteristics. pH sensors have been developed which take advantage of the fact that the excitation wavelength for fluorescence emission of some dyes is different for the acid and base forms and the ratio of emission excited at these two wavelengths can be used to calculate pH. Additionally, sensors have been developed which utilize the difference in absorption maxima for both the acid and base forms of the dye.

The commonly used pH-sensitive dye is phenol red whose absorption spectra is shown in Figure 22.12. The largest peak is observed from base form of phenol red at 560 nm and is used to measure pH because it is more sensitive to pH changes than the acid peak at 430 nm. A wavelength, which is insensitive to pH changes, is used as a reference, either a wavelength greater than 600 nm or the isobestic point at 480 nm. The relationship between pH and base form of the dye is given by the following Henderson-Hasselbalch equation:

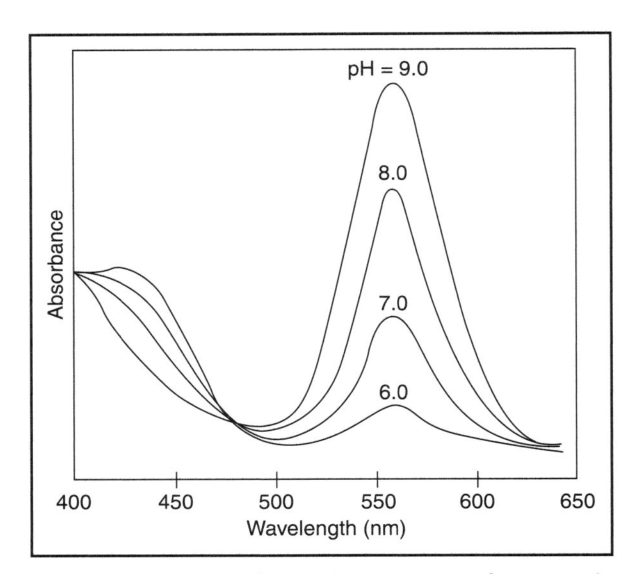

Figure 22.12 :: Absorption spectra of pH sensitive dye (Phenol red)

$$pH = pK_a - \log \frac{[HA]}{[A^-]},$$

where pH is the negative logarithm of the hydrogen ion concentration and pK_a is the negative logarithm of the equilibrium constant k_a, which describes the dissociation of the acid, HA.

One of the difficulties in designing a pH sensor is to achieve a resolution of 0.01 pH units over the range of 6.8 to 7.8. An effective way to achieve this is to optimize the pK_a of the dye material. This can be done through the proper choice of a functional group attached to the dye molecule or by immobilizing the pH-sensitive material on a polymer with the appropriate ionic characteristics.

Most fibre optic sensors designed for measuring pCO_2, use the same approach as a pCO_2 electrode. A pCO_2 sensor is fabricated by surrounding a pH sensor with a gas permeable membrane containing a bicarbonate ion (HCO^-_3) buffer. The membrane allows gaseous CO_2 and water vapour to enter the sensor, and they combine to form carbonic acid as per the following equations:

$$CO_2 + H_2O \longrightarrow H_2CO_3$$
$$H_2CO_3 \longrightarrow H^+ + HCO_3^-$$
$$HCO_3^- \longrightarrow H^+ + CO_3^-$$

The partial pressure of CO_2 can be related to the measured pH through the equilibrium constants for the above reactions and the equation is

$$pH = \log N + pK_1 - \log (K_s\, pCO_2),$$

where N = concentration of bicarbonate ion in the sensor

pK_1 = negative log of the acid dissociation constant for H_2CO_3 times the hydration constant for CO_2.

K_s = solubility coefficient for CO_2

This principle for the design of pCO_2 sensors has been implemented by using both fluorescence-based and absorption-based pH sensors (Vurek, *et al.*, 1983).

The methods for measuring pH, CO_2 and O_2 are similar, except that the wavelength of light used for different blood gas parameters varies. The optics are composed of three channels, each for measuring one of the parameters. Provision for calibration is made in the measuring system to compensate for individual physical variations between sensors and monitors. The calibration technique involves placing the sensor in a calibration solution, then bubbling precision mixtures of O_2, CO_2 and nitrogen (N_2) through the fluid. When equilibrium is reached, there are known partial pressure of pO_2 and pCO_2 in the solution. The pH is also known from the gas tensions and the chemical composition of the solution. The bubbling is repeated with a second gas mixture to provide a second calibration point. Using both calibration point the monitor can calculate the appropriate calibration factors for that sensor.

With the development of fibre optic-based blood gas sensors, routine electrode membraning and maintenance have become history. Continuous self-monitoring provides clear and immediate information of instrument performance. The keyboard based user interface provides advanced analytical performance and data processing capabilities. Along with measurement of the blood pH, pO_2 and pCO_2, some also include facilities for measuring other important ions such as Na^+, Ka^+, Ca^{++} and Cl^- in the blood. This is possible with the development of optical sensors based on fluorescence emission. All the sensors are mounted on a cassette shown in Figure 22.13. The syringe adapter shown at the right side of the cassette allows the automatic aspiration of a sample directly from a syringe. Removing the adapter allows for direct sample aspiration from a capillary or microsampler. The sensor calibration is verified by the system automatically after insertion of the cassette. The cassette is removed after sample analysis.

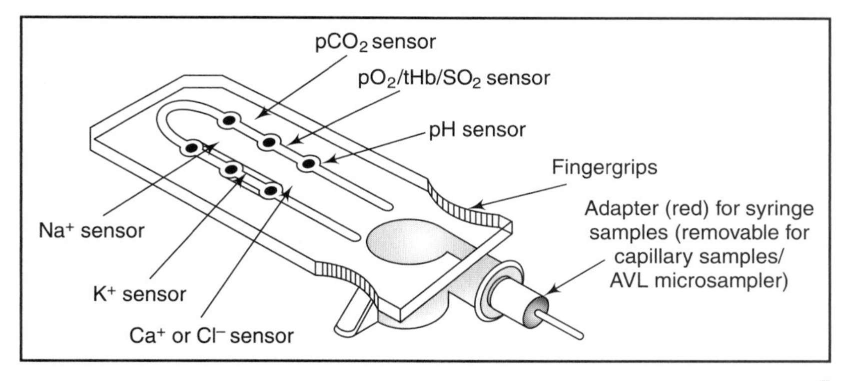

Figure 22.13 :: Sensor cassette (Courtesy, M/s AVL Medical Instruments)

The optical sensors in the cassette are designed in a way that the analytes bind with the fluorescent sensor molecule. The sensor molecule is selective for the specific analyte, i.e. the pH sensor molecule reacts only with "H^+", and the O_2 sensor only with O_2 molecules etc. The intensity of the emitted fluorescent light varies with the concentration of the ions (H^+, Na^+, K^+) or the partial pressure of the gas molecules (O_2 and CO_2) in the sample. The relationship is specific for each analyte. The corresponding calibration information for each component is encrypted in the bar code. Before the analyte can bind to the fluorescent molecule, it is made to pass through an optical isolator. The isolator prevents interference by unspecific light with the light detection system. The pO_2 sensor also enables to measure and compute total haemoglobin and oxygen saturation. The equipment works on a minimum sample size of 125 μl.

With the miniaturization of the direct reading electrodes, it is possible to combine them into a single cuvette so that a complete blood gas determination could be made on a single small sample. The introduction of the microprocessor and its use in blood gas analyzers free medical personnel from the task of monitoring the reaction in the electrode chamber and from the tedious chores of calculating and copying the results

All commercial blood gas analyzers make use of the same basic electrodes and signal conditioner circuitry. The main differences between instruments manufactured by various companies relate not to the measurements of the parameters but to the degree of automation and the technique by which the sample is presented to the electrodes.

Chapter 23

Industrial Gas Analyzers

23.1 Types of Gas Analyzers

Several physico-chemical properties have been utilized for the analysis of gases in simple or multi-component mixtures. However, most of the commercially available analyzers work on the measurement of quantities, such as infrared absorption, paramagnetism, thermal conductivity and gas density. In addition to these, many of the methods described in other chapters can also be used for the determination of gaseous components. These methods include visible and ultraviolet spectrophotometry, infrared spectrophotometry, mass spectrometry and various electrochemical methods.

All molecules, with the exception of the noble gases, consist of several atoms which exhibit a regular three-dimensional structure by chemical forces and whose valence electrons can attain defined energy states. The molecules are primarily in rotational and translational motions, relative to their surroundings, thus showing Brownian molecular motion. Also, the single atoms within a molecule can vibrate mutually and the shells of the valence electrons within the molecule can reach different states of energy. Eventually, analyzers can be designed so that the molecules of measuring gases may be made to undergo a physical or chemical reaction, which may reveal their nature and extent. Three different interactions are generally utilized, which form the basis of various gas analyzers. These are shown in Figure 23.1.

23.2 Paramagnetic Oxygen Analyzer

Oxygen has the property of being paramagnetic in nature, i.e. it does not have as strong a magnetism as that of permanent magnets, but at the same time it is attracted into a magnetic field. Nitric oxide and nitrogen dioxide are other two gases which are paramagnetic in nature. Most gases are, however, slightly diamagnetic, i.e. they are repelled out of a magnetic field.

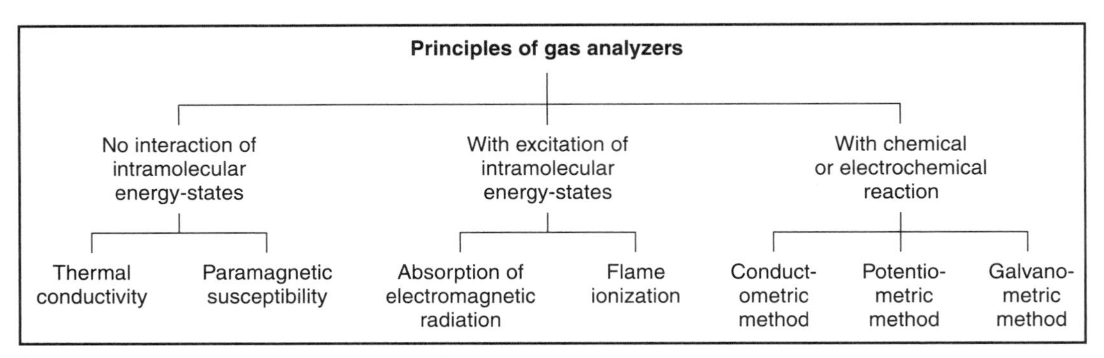

Figure 23.1 :: Principle of gas analyzers

Paramagnetism Figure 23.2 shows relative paramagnetism exhibited by different gases. The magnetic susceptibility of oxygen can be regarded as a measure of the tendency of an oxygen molecule to become temporarily magnetized when placed in a magnetic field. Such magnetization is analogous to that of a piece of soft iron in a field of this type. Similarly, diamagnetic gases are comparable to non-magnetic substances. The paramagnetic property of oxygen has been utilized in constructing oxygen analysers.

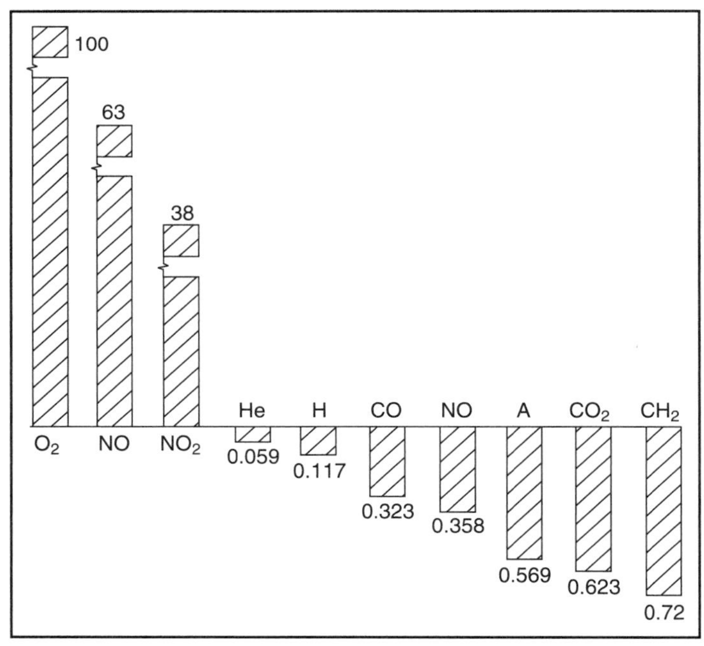

Figure 23.2 :: Relative paramagnetism of various gases

The paramagnetic oxygen analyzer was first described by Pauling, *et al.* (1946). Their simple dumb-bell type of instrument has formed the basis of more modern instruments. Figure 23.3 shows the schematic of a paramagnetic analyzer from M/s Beckman. The arrangement incorporates a small glass dumb-bell suspended from a quartz thread between the poles of a permanent magnet. The pole pieces are wedge-shaped in order to produce a non-uniform field.

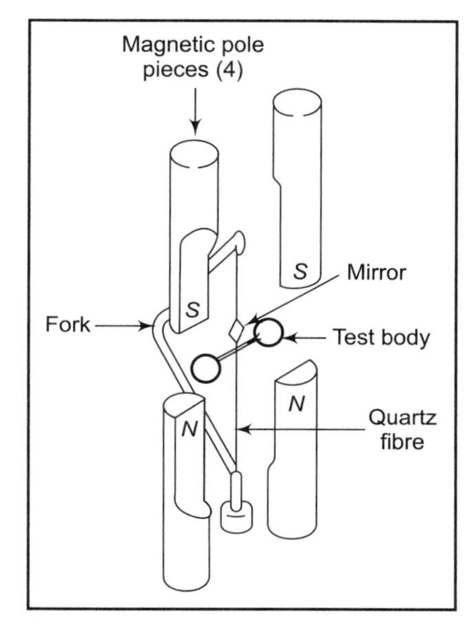

Figure 23.3 :: Arrangement of magnets in paramagnetic oxygen analyzer (Courtesy, M/s Beckman, USA)

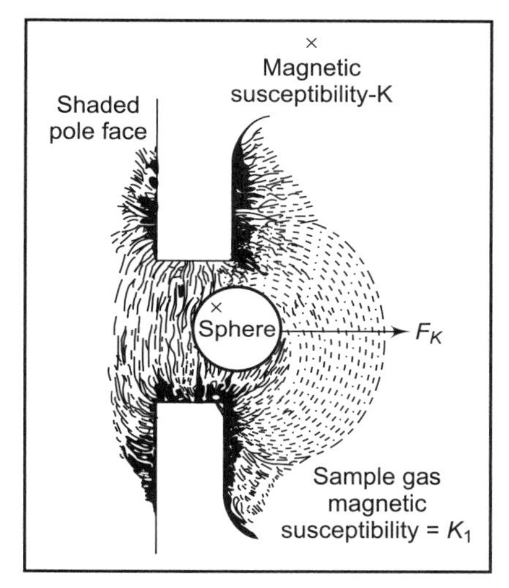

Figure 23.4 :: Sphere in non-uniform magnetic field (Courtesy, M/s Beckman, USA)

Referring to Figure 23.4, when a small sphere is suspended in a strong non-uniform magnetic field, it is subject to a force proportional to the difference between the magnetic susceptibility of this sphere and that of the surrounding gas. The magnitude of this force can be expressed as:

$$F = C(K - K_0),$$

where C = a function of the magnetic field strength and gradient,

K_0 = magnetic susceptibility of the sphere,

K = magnetic susceptibility of the surrounding gas.

The forces exerted on the two spheres of the test body are thus a measure of the magnetic susceptibility of the sample and therefore, of its oxygen content.

The magnetic forces are measured by applying to one sphere an electrostatic force that is equal and opposite to the magnetic forces. The electrostatic force is exerted by an electrostatic field established by two charged vanes mounted adjacent to the sphere (Figure 23.5). One vane is held

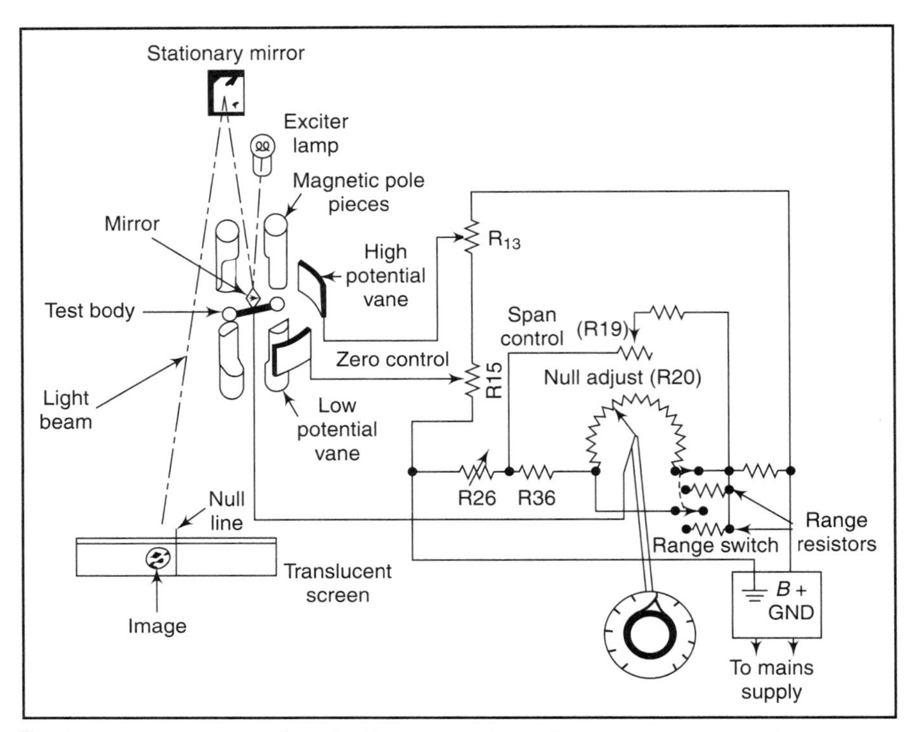

Figure 23.5 :: Functional diagram of Beckman paramagnetic oxygen analyzer

at a higher potential than the test body, the other at a lower potential. Since the glass test body must be electrically conductive, it is sputtered with an inert metal.

The test body is connected electrically to the slider of the null-adjust potentiometer R_{20}. This potentiometer is part of a voltage-dividing resistor network connected between ground and B^+. Potential to the test body can be adjusted over a large range.

An exciter lamp directs a light beam on to the small mirror attached to the test body. From the mirror, the beam is reflected to a stationary mirror and then on to a translucent screen mounted on the front panel of the instrument. The geometry of the optical system is so arranged that a very small rotation of the test body causes an appreciable deflection of the image cast by the beam.

Zero control of the instrument is provided by the ganged R_{13}–R_{15} setting, which changes the voltage present on each vane with respect to the ground, but does not change the difference in potential existing between them. This adjustment alters the electrostatic field. Rheostat R_{19} sets the upscale standardization point, i.e. it provides span or sensitivity control.

When no oxygen is present, the magnetic forces exactly balance the torque of the fibre. However, if oxygen is present in the gas sample drawn in the chamber surrounding the dumb-bell, it would displace the dumb-bell spheres and they would move away from the region of maximum magnetic flux density. The resulting rotation of the suspension turns the small mirror and deflects the beam of light over a scale of the instrument. The scale is calibrated in percentages by a volume of oxygen or the partial pressure of oxygen. Paramagnetic oxygen analyzers are capable of sampling static or flowing gas samples.

Only a few improvements have been suggested and carried out in the development of oxygen analyzers based on paramagnetism. The original quartz suspension has been replaced largely with a platinum–iridium suspension, which is more robust. Instead of measuring the deflection of the dumb-bell, a null-balance system is preferred, wherein the deflection is off-set by passing a current through a coil of wire attached to the dumb-bell. The current required is proportional to the deflecting couple and thereby to the oxygen tension of the gas. The control of current is carried out by a helical potentiometer which is duly calibrated.

Displacement of the dumb-bell results in unbalancing the output from a pair of photocells. The difference in their output signals is fed to a differential amplifier, which supplies its output current to the dumb-bell coil to null the deflection. The current is indicated on a meter. Oxygen analyzers are available with continuous readout 0–25 per cent or 0–100 per cent oxygen. The instruments are calibrated with the references as specified. The standard cell volume is 0–10 ml and the response time is about 10 s.

The recommended flow rate in the Beckman instrument is 50–250 cc/min, when the sample enters the analysis cell through a porous diffusion disc. If the sample enters directly for rapid response, the flow rate is 40–60 cc/min.

Before the gas enters the analyzer, it must be pressurized with a pump and passed through a suitable cleaning and drying system. In many cases, a small plug of glass wool is sufficient for cleaning and drying functions. The entry of moisture or particulate matter into the analyzer will change the instrument response characteristics. Therefore, the use of a suitable filter in the sample inlet line is recommended.

Any change in the temperature of the gas causes a corresponding change in its magnetic susceptibility. In order to hold this temperature of the gas in the analysis cell constant, the analyzer incorporates thermostatically-controlled heating circuit. Once the instrument reaches a temperature equilibrium, the temperature inside the analysis cell is approximately 60°C. The sample should be admitted to the instrument at a temperature between 10 and 45°C. The sample may not have time to reach temperature equilibrium before entering the analysis cell.

Calibration of the instrument consists of establishing two standardization points, i.e. a down-scale and an up-scale standardization point. These two points can be set by passing standard gases through the instrument at a fixed pressure, normally atmospheric pressure. First, a zero standard gas is admitted and the zero-control is adjusted. Then, the span gas is admitted and the span control is adjusted. Alternatively, the required practical pressures of oxygen are obtained by filling the analysis cell to the appropriate pressures with non-flowing oxygen or air. If the highest point is not greater than 21% oxygen, dry air is used to set the span point. If the point is greater than 21% oxygen, oxygen is used to set this point.

The instrument should be handled carefully, as the fine quartz fibre supporting the test body may break.

23.3 Magnetic Wind Instruments

Magnetic wind

Figure 23.6 shows a paramagnetic oxygen analyzer based on the phenomenon of magnetic wind, which rises due to the motion of paramagnetic oxygen molecules into the non-uniform magnetic field. The flowing gas is made to pass over a heated filament connected as one arm of a Wheatstone bridge circuit. The gas cools the filament and unbalances the bridge circuit. The output voltage from the bridge can be fed to a recorder, to show a continuous indication of oxygen concentration in the sample.

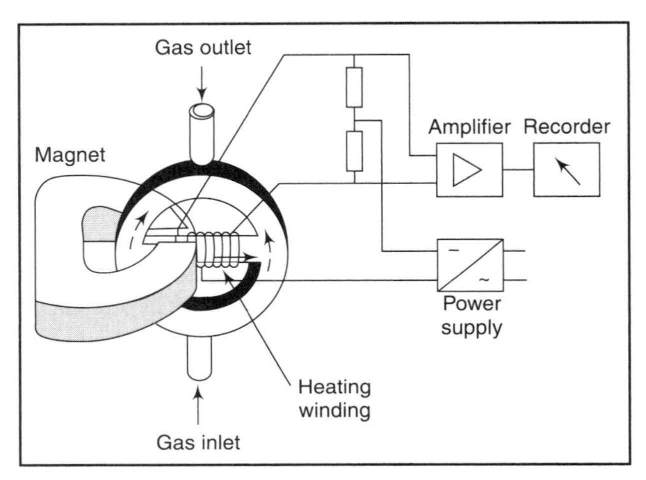

Figure 23.6 :: Principle of Hartman and Braun oxygen analyzer

The Hartman and Braun oxygen analyzer type Magnos 2T and Magnos 5T work on the principle of thermomagnetic action. This method makes use of the dependency of paramagnetic susceptibility on the temperature according to the Curie law.

The analyzer comprises a ring chamber, which consists of a metallic hollow ring arranged in vertical plane, with a gas inlet below and a gas outlet above. The gas is conducted in the upward direction. The two halves of the ring chamber are connected by a glass tube. The transverse tube has taped wire winding on its outside. The winding forms a part of a Wheatstone bridge circuit, whose one-half is placed in the field of a permanent magnet.

In case the sample gas does not contain any oxygen, there is no flow in the horizontally placed transverse tube. If, however, the sample gas contains oxygen, it is attracted into the magnetic field. This gas flow is heated and due to an increase in the temperature, the susceptibility of the paramagnetic oxygen molecules decreases. In this way, the heated gas is pushed away by the cold gas coming from the left hand side. A gas flow thus arises inside the transverse tube. This flow is called the magnetic wind, whose velocity depends upon the oxygen concentration in the sample gas. The magnetic wind cools down the left side of the heated winding more than the right side. Therefore, a change in the bridge balance takes place, which depends upon the temperature gradient. The voltage difference of the bridge is proportional to the oxygen concentration in the measuring gas.

The instrument suffers from the following sources of error:

(i) The filament temperature is affected by changes in the thermal conductivity of the carrier gas. Thus, the calibration is correct for only one gas mixture, which must be specified for each analyzer.

(ii) Hydrocarbons and other combustible gases in the sample stream react on the heated filaments causing changes in temperature and therefore their resistance, which results in extremely large errors. These hydrocarbons have to be removed by means of a cold trap, but if the percentage in the gas is high, an error will result due to the change in sample volume.

(iii) The cross tube must be horizontal, to avoid an error due to gravitational chimney flow effects.

It is important to note that magnetic wind and thermo-magnetic analyzers are synonymous. The first name is generally used in Europe and the second in the United States. In some instruments, the paramagnetic type of sensor works by passing the unknown gas between two powerful magnetic poles, wherein a thin aluminium membrane is suspended. The magnetized oxygen is pulled into the space between one of the magnetic poles and the membrane, thereby displacing the membrane in the opposite direction. The displacement of the membrane can be conveniently measured by using LVDT or strain gauge transducer. The extremely low inertia of the thin membrane in the oxygen sensor allows it to come to equilibrium in 0.25–0.5 s. This facilitates the recording of rapid changes in oxygen concentration. This is particularly useful for making measurements during each breathing cycle, while making measurements of respiratory gases.

Oxygen analyzers are employed in the areas of oxygen absorption studies on plants and tissues, respiratory studies, food processing, air pollution studies and anaesthesiology.

Figure 23.7 shows the arrangement of a rapid oxygen analyzer OXYMAT-M from M/s Siemens used for medical applications.

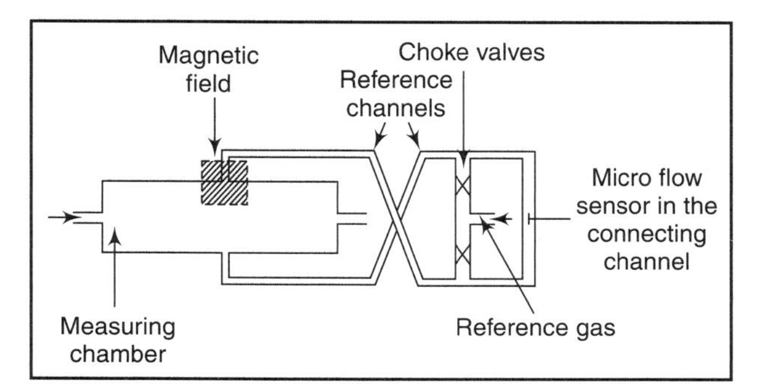

Figure 23.7 :: Schematic representation of the measuring system of Siemens OXYMAT—M oxygen analyzer for medical applications

The measuring procedure is based on the pressure difference which develops between two gases having different oxygen concentration in a magnetic field. One of the two gases, usually ambient air, serves as the reference gas, which flows in the channels, both of which open into a single measuring chamber. One channel is located within the magnetic field, and the other in a position which is virtually free from magnetic influences. The respiratory air to be analyzed is sucked directly through the measuring chamber. On account of the paramagnetic property of oxygen, it is subjected, in the inhomogeneous parts of the magnetic field, to forces which act upon the oxygen molecules in the direction of higher field strengths. If the oxygen concentrations of the two samples of gas differ, a pressure difference develops between the two points of entry of the reference gas into the measuring chamber. This pressure difference is compensated via a connecting channel, in which a micro-flow sensor converts the stream of gas into an electrical signal, which is proportional to the difference in oxygen concentration between the gas to be analyzed and the reference gas.

The instrument employs an electromagnet with changing flux intensity. As a result, an alternating pressure is created in the measuring chamber, and an alternating electric voltage of 25 Hz is developed at the micro-flow sensor, which is rectified and amplified in the electronics part of the equipment. The measurement of alternating pressure has an advantage in that it prevents any unsymmetry and drift phenomena.

23.4 The Electrochemical Methods

Analyzers based on the electrochemical methods are mostly used for the determination of the oxygen content of a gas. They utilize an electrolytic cell and can be broadly classified as the galvanic, polarographic and conductometric methods.

23.4.1 Galvanic Methods

Galvanic cell

Galvanometric methods are based on the fact that the electrical current of a galvanic cell, which is equipped with appropriate electrodes and an appropriate electrolyte, would depend upon the oxygen concentration, with it being related to the rate of oxygen uptake by such a cell. It is the reverse of electrolysis in which oxygen is evolved at the anode and hydrogen at the cathode, when electrical current is made to pass through the cell. These analyzers are used for the measurement of small oxygen concentrations.

The galvanic cell has two electrodes, one of which is made of noble metal such as silver and the other of a base metal such as lead. The oxygen contained in the sample gas is made to bubble through the electrolyte. A magnetically driven stirring system helps to ensure a quick and efficient mixing of the reaction liquid and the gas.

Analyzers based on this principle are used to measure the content of dissolved oxygen in boiler feed water. For this purpose, the boiler feed water flows through the cell and acts as an electrolyte and the cell is used for continuous monitoring.

There is a need to control the oxygen uptake at the cathode. This is generally done by having a porous carbon cathode and semi-permeable membranes. Since the electrons are supplied by the dissolving anode, the lift of the cell is limited. The cell is affected with a very high temperature gradient, approximately 4 per cent/°C. This is compensated by using a combination of NTC (negative temperature coefficient) and PTC (positive temperature coefficient) thermistors.

The speed of response and sensitivity are improved by using silver gauze instead of a smooth electrode and by reducing the volume of electrolyte. The cell could be cylindrical in shape with a central anode of porous material like lead saturated with electrolyte and dipped in a reservoir, over which the sample gas flows. The cathode is formed by gauze which surrounds the anode. This arrangement is known as the Hersch cell. Several improvements have been suggested in the basic form of the cell to increase its life and sensitivity. One common type is that in which diffusion of the oxygen through a Teflon membrane causes a current flow between two electrodes separated by a liquid or gel electrolyte.

The noble metal cathode, if used, is not attacked by the electrolyte. Therefore, the drifts and instability associated with porous cathodes are eliminated. Also the cell has no output in the absence of oxygen, and a definite zero is obtained which does not need calibrating. Current in the galvanic cell obeys Faraday's law, which is given by the following relation:

$$I = 0.263\, CFP\,(298/\mathrm{T}),$$

where I is the expected current in micro amps, when a gaseous sample containing C ppm of oxygen by volume passes through the cell at a flow rate F cm^3/min measured at P atmosphere and T°K. The expression assumes that the perfect gas laws apply. With a sample flow rate of 100 cm^3/min at one atmosphere pressure and 298°K, the theoretical sensitivity is 26.3 μA/ppm.

These instruments are generally slow in operation and the sample gas must be scrubbed to remove CO_2, SO_2, H_2S or any acidic gas, but one attraction is that they can be used to measure the dissolved oxygen in liquids.

Another type of electrochemical analyzer employs the high temperature galvanic cell. This cell is manufactured by a number of companies.

All these cells consist of a calcium stabilized zirconium oxide electrolyte, with platinum electrodes. At the operating temperature, oxygen molecules on the side of the cell exposed to a high partial pressure of oxygen (the anode) gain electrons. Simultaneously, oxygen molecules are formed by the reverse action, at the other electrode (the cathode).

For a cell operating at 850°C, the standard Nernst equation for an oxygen cell is:

$$\text{EMF (open circuit)} = 55.7 \log_{10} (P_a/P_b),$$

where P_a is the partial pressure of oxygen within the cell, and P_b is the partial pressure of oxygen outside the cell. Since this effect is specific for oxygen, the instrument output is not affected by the presence of water or CO_2. However, hydrocarbons, hydrogen and other combustible gases will burn at the operating temperature and result in an indication of less oxygen than is actually present. The response of analyzers using such cells is very fast.

The probe can be used for direct determination of oxygen in the flue, provided that the flue temperature is less than 800°C. The instrument can be calibrated only by removing the probe from the flue and inserting it into an enclosed container, which can be filled with calibration gases.

The instrument can be mounted outside the flue. The sample gas is drawn through a short sample tube to the measuring cell by an air ejector pump. This analyzer provides a means of introducing calibration gases for setting zero and span. It is important to note that both these types of instruments require a supply of clean, dry air at the reference side of the zirconium electrolyte.

23.4.2 Polarographic Cells

Polarographic cells are generally used to measure the partial pressure or percentage of oxygen from injected samples, continuous streams or in static gas monitoring. They find maximum utility in the respiratory and metabolic laboratories. Polarographic cells are based on the redox reactions, in a cell having both the electrodes of noble metals. When a potential is applied, oxygen is reduced at the cathode in the presence of KCl as the electrolyte and a current will flow. The cathode is protected by an oxygen permeable membrane, and the rate at which oxygen reaches the cathode will be controlled by diffusion through the membrane. The voltage–current curve will be a typical polarogram (Figure 23.8). A residual current flows in the cell at low voltages. The current rises with an increase in voltage, until it reaches a plateau where it is limited by the diffusion rate of oxygen through the membrane. For a given membrane and at a constant temperature, this would be proportional to the partial pressure of oxygen across the membrane.

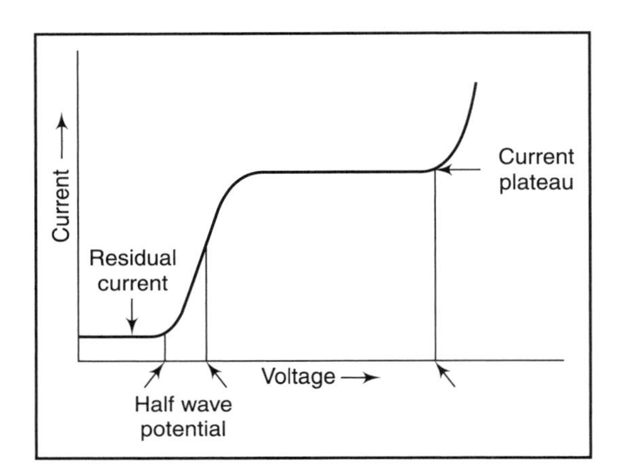

Figure 23.8 :: Response of polarographic sensor

When the voltage is applied in the plateau region, the current in the cell is proportional to the oxygen concentration.

Polarographic cells are temperature-sensitive, as the diffusion coefficient changes with temperature. The temperature coefficient is usually 2–4 per cent/°C. Therefore, temperature compensation circuits are used to overcome this problem. Polarographic oxygen cells are used mainly for portable gas detectors, wherein simplicity, low cost and light weight are important. They are preferably used for measuring oxygen in liquids, especially in water pollution and medical work.

The oxygen analyzer incorporates an oxygen sensor, which contains gold cathodes, a silver anode, potassium chloride electrolyte gel and a thin membrane. The membrane is precisely retained across the exposed face of the gold cathode, compressing the electrolyte gel beneath, into a thin film. The membrane, permeable to oxygen, prevents airborne solid or liquid contaminants from reaching the electrolyte gel. The sensor is insensitive to other common gases. A small electrical potential (750 mV) is applied across the anode and cathode.

Although the composition of the atmosphere is remarkably constant from sea level to the highest mountain, i.e. oxygen 21 per cent and nitrogen 79 per cent, there is a great difference in the partial pressure of oxygen at different altitudes. The polarographic sensor, which actually senses partial pressure, would therefore require some adjustment to read approximately the percentage oxygen at the altitude at which it is used. Humidity can also affect oxygen readings, but to a lesser degree. Water vapour in the air creates a water vapour partial pressure that lowers the oxygen partial pressure slightly. Therefore, for precision work, it is often desirable to use a drying tube on the inlet sample line. Also, care should be taken to calibrate and sample under the same flow conditions, as required for the gas to be analyzed. The range of the instrument is 0–1000 mm Hg of O_2 and the response time is 10 s for 90 per cent, 35 s for 99 per cent and 70 s for 99.9 per cent. The instrument can measure oxygen against a background of nitrogen, helium, neon, argon, etc., with no difficulty. The sensor is slightly sensitive to carbon dioxide and nitrous oxide, with a typical error of less than 0.1 per cent oxygen for 10 per cent carbon dioxide and 4 per cent oxygen for 100 per cent nitrous oxide.

23.4.3 Conductometric Method

The conductometric method is convenient and is the most widely used method for trace gas analysis. In practice, the sample gas is passed through a cell containing a liquid reagent, which can react with the gas of interest. The conductivity of the liquid is measured before and after the reaction with the gas. The difference in conductivity is proportional to the gas concentration.

In order to obtain reproducible results, the flow of gas and reagent must be kept constant. Therefore, the measuring gas must enter the analyzer at a constant velocity, which is generally adjusted by a pneumatic bypass and indicated by a capillary flow meter. A slow stream of reaction solution enters the reaction cell via a second capillary, and its flow rate is also kept constant. In some analyzers, the chemical reaction between the measuring gas and the reagent takes place quantitatively in a spiral reaction cell, where the gas is separated from the liquid. The liquid is then passed through the conductivity measuring cell, while the gas is vented from the analyzer.

In order to minimize errors due to change in ambient temperature, the measuring cell is kept in precisely controlled temperature environment.

By proper selection of the reagent, analyzers based on this principle can be made specific for various gas components. The method is especially suitable for measuring traces of H_2S, SO_2, NH_3 and H_2O in the ppb range.

23.5 INFRARED GAS ANALYZERS

For their operation, infrared gas analyzers depend upon the fact that some gases and vapours absorb specific wavelengths of infrared radiation. One of the most commonly measured gases using infrared radiation absorption method, is the carbon dioxide. The technique used for this purpose is the conventional double-beam infrared spectrometer system having a pair of matched gas cells in the two beams. One cell is filled with a reference gas, which is a non-absorbing gas like nitrogen, whereas the measuring cell contains the sample. The difference in optical absorption detected between the two cells is a measure of the absorption of the sample at the particular wavelength. Since the vibration excitation occurs only if we have hetero-atomic molecules, the infrared absorption principle is not applicable for the analysis of gases, whose molecules are formed by two identical atoms like oxygen, hydrogen and nitrogen.

Infrared analyzers are used for the determination of a large number of components, including CO, CO_2, SO_2, NH_3, H_2O, nitric oxide as well as most gaseous hydrocarbons. The selectivity is, however, restricted by the fact that the absorption bands partly overlap mutually. This can be eliminated by providing a filter cell, which is filled with the interfering component. The selectivity can also be enhanced by the negative filtering, so that it is possible to distinguish the interfering components.

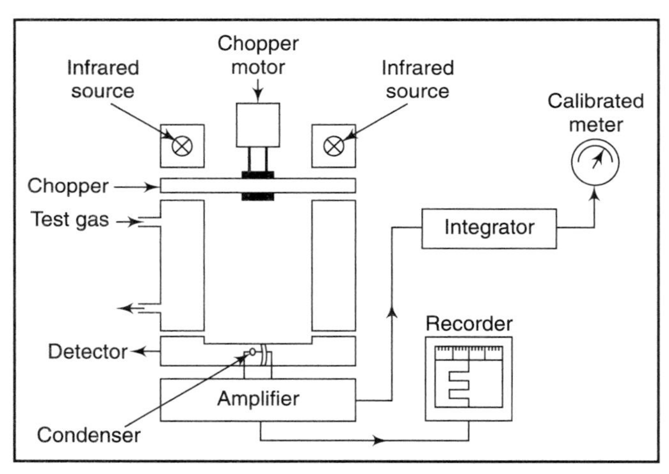

Figure 23.9 :: Infrared gas analyzer

A simple method of using the infrared technique for gas analysis is shown in Figure 23.9. This method does not require the use of a wavelength dispersing device. Two identical infrared sources in the pick-up or sensing head emit beams of radiation that are pulsed by a motor-driven chopper. The source of infrared radiation is the hot-wire spiral. The rotating chopping disc occludes each beam twice per rotation. For industrial analyzers, the chopping frequency is 2–10 Hz, whereas for medical application, it is 2–50 Hz. One beam passes through the sample cell, and the other beam through a reference cell, and both beams enter opposite ends of the detection chamber. The detection chamber is a permanently sealed unit divided into two compartments by a thin, metal diaphragm. Both compartments are charged to the same pressure, with the gas being measured. For example, the detector of a CO_2 analyzer would contain pure CO_2 while that of a halothane analyzer would contain pure halothane vapour.

When the gas being measured enters the sample cell, it absorbs infrared radiation at the same wavelength as the gas in the detection chamber. This reduces the amount of radiation reaching the gas in the sample side of the detection chamber and produces a lower pressure in that side. The diaphragm bends towards the side of lower pressure, and this movement is converted into electrical impulses. The diaphragm thus vibrates at the chopping frequency and periodically bends towards the sample half of the detector. The metal diaphragm usually forms one plate of a capacitor. The movement of the diaphragm thus results in a variable capacitor, which, in turn, forms part of the tuning circuit of a radiofrequency oscillator, used in an amplitude modulation arrangement. In an alternative arrangement, the capacitor is supplied with a constant charge and the resulting voltage changes at the chopping frequency are amplified in a three-stage tuned amplifier with high input impedance in the input stage. The signal is rectified and smoothed and the output signal is displayed on a meter or recorder.

Conventional infrared analyzers of the type described above have the following drawbacks:

- The source must be balanced either electrically or mechanically.
- Unbalanced chopper causes failure of chopper motor bearing.
- Luft detector cell using gas as detection mechanism is vibration sensitive and subject to cross-talk with other gases. Cell is difficult to maintain and expensive to replace.

In order to overcome these difficulties, a range of infrared gas analyzers employing a single source in an arrangement shown in Figure 23.10 have been developed. The solid state detector is PbSe. The chopper has a high speed of 3000 rpm and provides response times of up to 100 ms for 90 per cent reading.

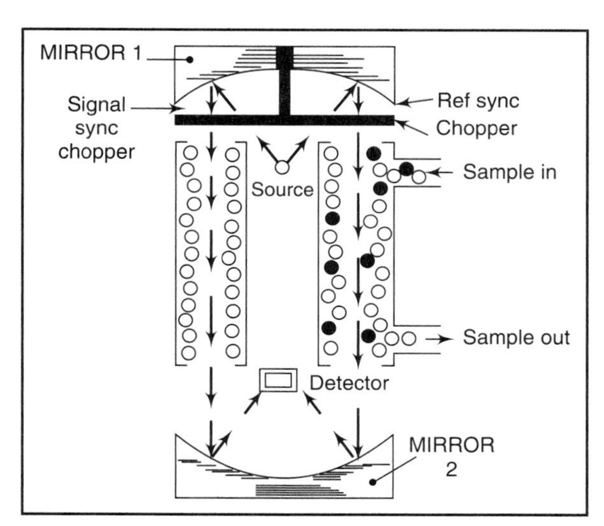

Figure 23.10 :: Improved version of infrared gas analyzer (Courtesy, M/s Infrared Industries, USA)

The infrared source operates at a temperature of about 830°C, wherein it emits infrared energy optimized for the spectral bands of interest and long life. The infrared energy source is located at the focal plane of a parabolic reflector, so that the reflected energy from the reflector is effectively collimated. The collimated energy is chopped by the coaxial chopper, which allows the energy to pass alternately through the reference and sample tubes. Since the energy is collimated, it passes through these tubes without internal reflections, so that gold foil coatings on the inside of these tubes are not necessary. The sample tube length can be selected according to the absorption strength and concentration of the sample gas. At the output end of the two tubes, a second parabolic reflector images the energy onto the detector filter assembly. The filter is a narrow bandpass interference filter, with the bandpass characteristics matched with the absorption spectra of the gas of interest.

Infrared gas analyzers are particularly useful for measuring carbon dioxide in respired air in the medical field. In these instruments, two types of samples are employed—a micro-catheter cell and a breathe-through cell. The micro-catheter cell is used with a vacuum pump to draw off small volumes from the nasal cavity or trachea. Its typical volume is 0.1 ml and it is particularly useful when larger volumes could cause patient distress. The breathe-through cell accepts the entire tidal volume of breath with no vacuum assistance. It can be connected directly into the circuit of an anaesthesia machine. These instruments have a typical response time of 0.1 s and a sensitivity range of 0 to 12 per cent CO_2. The calibration of the CO_2 analyzer is done in the following way:

- In order to establish zero calibration, an inert gas is sampled and the meter adjusted to zero. Zero calibration is generally made with room air.
- The upper end of the meter scale is calibrated with a cylinder of calibration gas. The frequency of calibration primarily depends upon the accuracy desired.

Infrared sources applied in the early approaches of infrared analyzers had significant power and used opto-acoustic detectors, characterized by a relatively small sensitivity. The method of modulation of radiation emitted by such sources was to apply mechanical choppers. Thermal or photon detectors equipped with interference filters are more often used in contemporary analyzers. They have greater sensitivity and can operate with sources of lower power and small thermal mass. The radiation sources are now modulated directly by using pulsed circuits, thus offering a high reliability as no moving parts are involved (Puton, *et al.*, 2002).

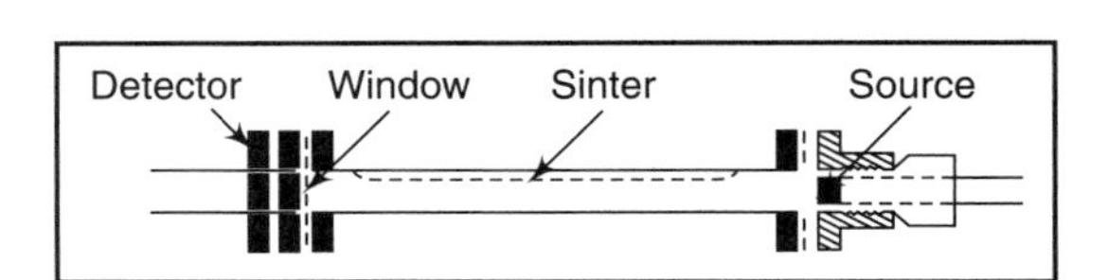

Figure 23.11 :: Infrared gas analyzer for detection of hydrocarbons (after Jones, *et al.*, 1971)

A schematic diagram of a simple infrared gas analyzer developed especially for detection of hydrocarbons is shown in Figure 23.11. This analyzer employs a partially selective source and a partially selective detector, so chosen that their combined characteristics limit the sensitivity of the combination to a narrow spectral region, with the absorption band centred at 3.4 μm. The indium arsenide photovoltaic detector operating at ambient temperature has maximum detectivity at about 3.4 μm, a rapid decrease in sensitivity at longer wavelengths and no response to radiation of wavelength greater than 4 μm. It is used in conjunction with a source of radiation consisting simply of a bead, (about 3 mm diameter), of borosilicate glass encapsulating a platinum-rhodium heating coil. The optimum temperature of operation was found to be in the region 400–500°C by these workers.

The instrument is very simple in construction. It consists of a single absorption cell, the inner wall of which is silvered, with the source at one end and the detector at the other. The gas is made to diffuse into the cell through a fine stainless steel sinter. The change in the dc signal from the detector is amplified and can be displayed either on a suitably calibrated meter or chart recorder. No chopper arrangement is necessary.

A similar system comprising a heated quartz source and an indium antimonide detector, which would cover the spectral region from 4 to 6 μm, could be used for the measurement of high concentrations of carbon monoxide and carbon dioxide. To determine either gas unambiguously, a filter cell would be required; this would be filled with carbon dioxide to eliminate the unwanted radiation when detecting carbon monoxide and vice-versa.

For process applications, the multiple internal reflections technique is used (Frant, 1980). A block diagram of the equipment based on this technique is shown in Figure 23.12. The scheme is

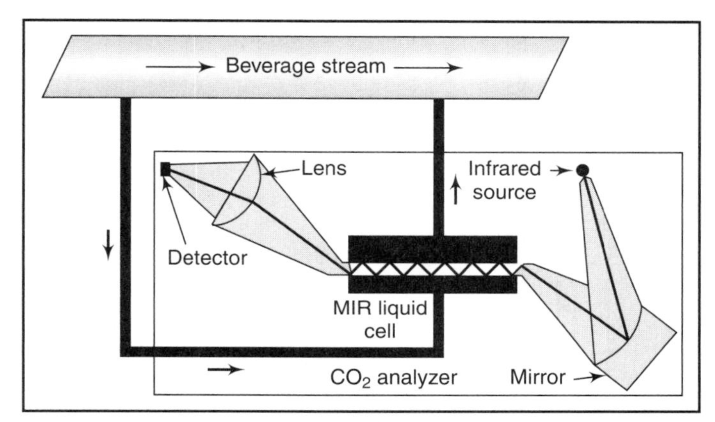

Figure 23.12 :: Infrared gas analyzer for process applications (after Frant, 1980)

basically applicable for measuring CO_2 in the flowing beverage. A portion of the main beverage stream is brought through a small stainless steel pipe to the analyzer and then returned to the main stream. The measurements are made by using two IR sensors, one measuring sugar and the other measuring the CO_2 content.

23.6 Thermal Conductivity Analyzers

Thermal conductivity

The thermal conductivity of a gas is defined as the quantity of heat (in calories) transferred in unit time (seconds) in a gas between two surfaces 1 cm^2 in area, and 1 cm apart, when the temperature difference between the surfaces is 1°C. The ability to conduct heat is possessed by all gases, but in varying degrees. This difference in thermal conductivity can be employed to quantitatively determine the composition of complex gas mixtures. Changes in the composition of a gas stream may give rise to a significant alteration in the thermal conductivity of the stream. This can be conveniently detected from the rise or fall in temperature of a heated filament, placed in the path of the gas stream. The changes in temperature can be detected by using either platinum filament (hot wire) or thermistors.

Figure 23.13 shows the relative thermal conductivity of a series of gases of interest for analysis. A gas analysis based on the thermal conductivity procedure presupposes binary gas mixtures or such gas mixtures, respectively, which include a measuring component, whose thermal

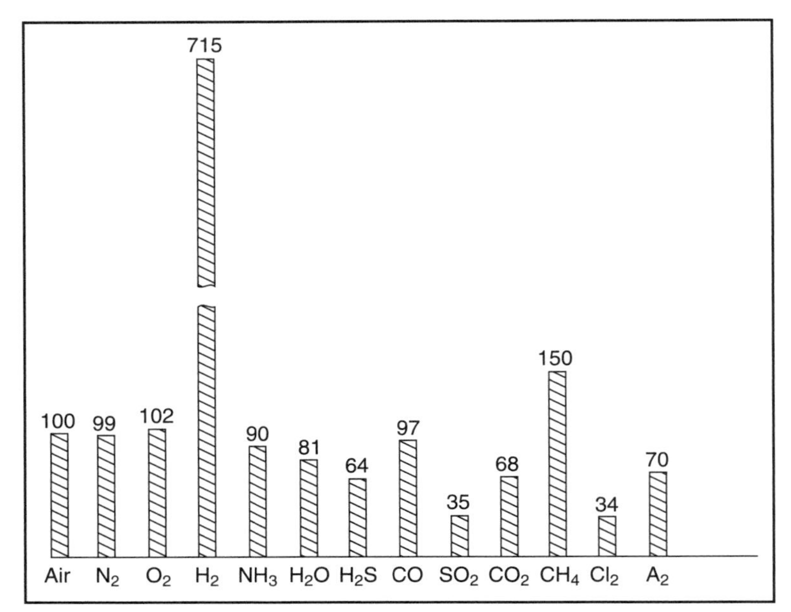

Figure 23.13 :: Relative thermal conductivity of different gases

conductivity differs sufficiently from the thermal conductivity to the carrier gas. Typical examples of application are the measurement of hydrogen in blast furnace gases, the determination of argon in oxygen in the process of air decomposition and of sulphur dioxide in roasting gases in the production of sulphuric acid.

In a typical hot-wire cell thermal conductivity analyzer, four platinum filaments (Figure 23.14) are employed as heat-sensing elements. They are arranged in a constant current bridge circuit and each of them is placed in a separate cavity in a brass or stainless steel block. The block acts as a heat sink. The material used for construction of filaments must have a high temperature-coefficient of resistance. The materials generally used for the purpose include tungsten, Kovar (alloy of Co, Ni and Fe) or platinum.

Two filaments connected in opposite arms of the Wheatstone bridge act as reference arms, whereas the other two filaments are connected in the gas stream, which act as measuring arms. The use of a four-cell arrangement serves to compensate for temperature and power supply variations.

Initially, reference gas is made to flow through all the cells and the bridge is balanced precisely with the help of potentiometer *D*. When the gas stream passes through the measuring pair of filaments, the wires are cooled and there is a corresponding change in the resistance of the filaments. The higher the thermal conductivity of the gas, the lower would be the resistance of the wire and vice versa. Consequently, the greater the difference in thermal conductivities of the

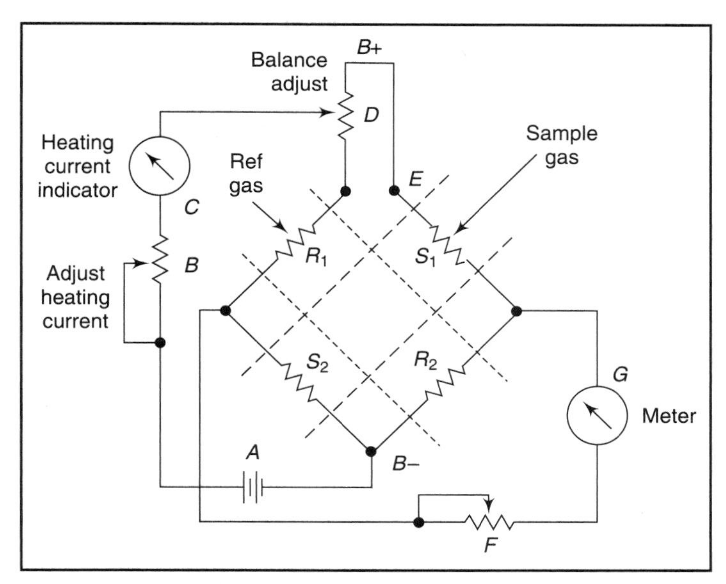

Figure 23.14 :: Schematic diagram of a hot wire thermal conductivity analyzer

reference and sample gas, the greater would be the unbalance of the Wheatstone bridge. The unbalance current can be measured on an indicating meter, or on a strip chart recorder.

Thermistors

Thermistors can also be used as heat-sensing elements arranged in a similar manner, as hot-wire elements in a Wheatstone bridge configuration. Thermistors possess the advantage of being extremely sensitive to relatively minute changes in temperature, and have a high negative temperature coefficient. When used in the gas analyzers, they are encapsulated in glass. Thermistors which are fairly fast in response are also available. The circuit arrangement is shown in Figure 23.15.

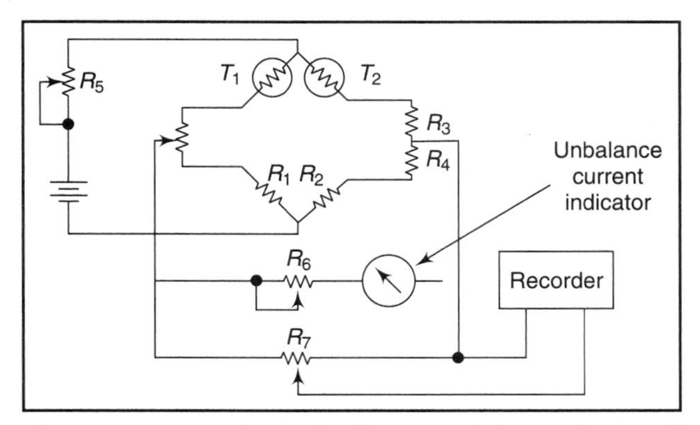

Figure 23.15 :: Schematic diagram of thermal conductivity analyzer using thermistors

Thermal conductivity gas analysis is inherently non-specific. Therefore, the simplest analysis occurs with binary gas mixtures. A thermal conductivity analyzer can be used in respiratory physiology studies to follow CO_2 concentration changes in the individual breaths of a patient. A high speed of response necessary for this purpose can be obtained by reducing the pressure of the gas surrounding the filaments, to a few millimetres of mercury absolute. The variations in the proportions of oxygen and nitrogen in the sample stream will have little effect, since they both have almost the same thermal conductivity. The effect of changes in water vapour content can be minimized by arranging to saturate the gas fed to both the sample and reference filaments.

An analysis of a multi-component mixture is possible, if all components but one have almost the same thermal conductivity, so that the mixture can be treated as a binary mixture. Similarly, analysis is also possible if all components of the mixture other than the one being measured vary in the same ratio from each other.

23.7 Analyzers Based on Gas Density

It is known that the density of an ideal gas has a direct linear relation with the molecular weight of that gas. Fortunately, all real gases behave as ideal gases at room temperature and normal atmospheric pressure. Instruments based on the principle of gas-density balance are commercially available.

Figure 23.16 illustrates the principle of operation of a gas-density balance. The reference gas enters the balance at *A*, where it splits itself into two streams and leaves the balance at *D*. Two detectors (B_1 and B_2), which may be either hot wires or thermistors and mounted in the path of the two streams, are connected as two arms of a Wheatstone bridge. When the reference gas flows such that the flow is balanced, the two detectors are equally cooled and the recorder would indicate a zero base line.

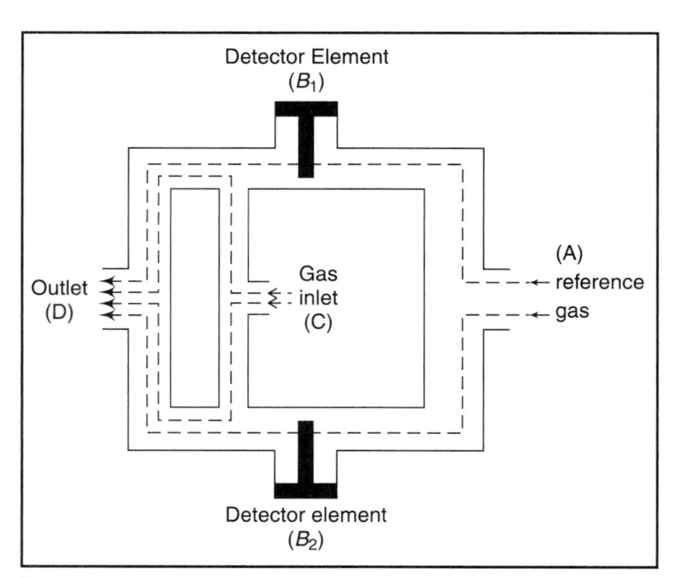

Figure 23.16 :: Principle of gas analyzer based on gas density

The sample gas enters the balance at *C* and it also splits into two streams. It mixes with reference gas in the horizontal conduits and leaves at *D*. If the sample gas has the same density as the reference gas, there will be no unbalance of reference streams or of the detector elements. If the sample carries a gas having a higher density than the reference gas, it will cause a net downward flow, partially obstructing the flow in the lower path, like *A*-B_2-*D*. This would result in raising the temperature of the detector element B_2.

This, in turn, increases the flow in the path A-B_1-D and causes more cooling of the element B_1. This temperature differential causes an unbalance in the bridge, the unbalance being linearly proportional to the gas-density difference between the reference and the sample gas. If the detectors used are hot-wires, it may require some factor of amplification, before the signal can be given to a recorder. The use of thermistor generally eliminates the requirement of amplification. The effective sample volume is typically 5 ml.

23.8 Method Based on Ionization of Gases

The spectral regions for maximum radiation absorption for different gases are of different wavelengths. For example:

N_2	less than 900 Å (far ultraviolet)
O_2	1450 Å (ultraviolet)
CO_2	2.73, 4.25 and 14.93 μ
Water vapour	2.6, 20 and 52 μ (infrared)

Neither nitrogen nor oxygen analyses are routinely done by using these absorption bands. However, with sufficient electrical excitation and at suitable pressures, gases emit radiation in different ways (like spark, arc, glow discharge in different parts of the radiation spectrum). The measurement of the emitted radiation can help in the determination of the unknown concentration of a gas in a mixture. This technique has been utilized for the measurement of nitrogen gas, particularly in respiratory gases.

The measuring technique utilized for measuring nitrogen is essentially that of a photospectrometer, wherein a gas sample is ionized, selectively filtered, and detected with a photocell, which provides an appropriate electrical output signal. The presence of nitrogen is detected by the emission of a characteristic purple colour, when discharge takes place in a low pressure chamber containing the gas sample. Nitrogen meters are usually employed in the medical field for the measurement of nitrogen concentration, to follow a breath-by-breath variation in respiratory gases and other nitrogen gas analysis applications.

Nitrogen meters

The instrument generally operates in two parts. The sampling head contains the ionizing chamber, filter and the detector. The other part contains the power supply, amplifier and the display system. The ionizing chamber or the discharge tube is maintained at an absolute pressure of a few torr. A rotary oil vacuum pump draws a sample and feeds it to the discharge. The voltage required for striking the discharge in the presence of nitrogen is of the order of 1500 V dc. This voltage is generated by using a dc-dc convener, or by rectifying the output of a high voltage transformer.

The light output from the discharge tube is interrupted by means of a rotating slotted disc (Figure 23.17), so that a chopped output is obtained. This light is then passed through optical filters to the wavelength corresponding to the purple colour. The intensity of light is measured

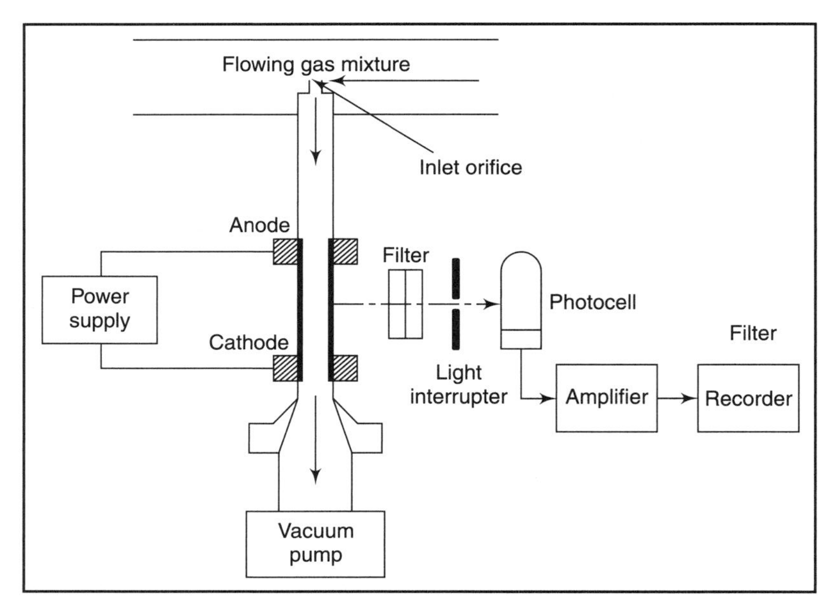

Figure 23.17 :: Schematic diagram of a nitrogen meter (Courtesy, M/s Med Science, USA)

with a photocell and an amplifier specifically tuned to the chopping frequency. The light intensity is proportional to the nitrogen concentration.

Modern nitrogen analyzers measure and digitally display the concentration of nitrogen. The sampling rate in these instruments is adjusted with the help of a needle valve, which is normally set at 3 ml/min. The vacuum system provides 600–1200 μ of Hg. The instrument is calibrated for water saturated mixtures of nitrogen and oxygen, as a reading error of up to 2 per cent can be expected with dry gases. Compensation for this error can be simply made by adjusting the sampling head needle valve, if it is desired to monitor dry gases.

Chapter 24

Environmental Pollution Monitoring Instruments

Awareness of and concern about the deteriorating environment are increasing the world over. It is necessary to monitor changes taking place in the quality of the environment, for initiating efforts to control it. A large number of instruments are required to accomplish the environmental pollution monitoring programme. The representative sampling of pollutant concentrations at the point of discharge or in the environment requires an understanding of the pollutant characteristics as well as of specific objectives, in considerations like monitoring and control. Only then can appropriate measurement techniques be identified and employed. Pollution monitoring is thus a very complex task, which involves the systematic collection and evaluation of physical, chemical, biological and related information pertaining to environmental quality and effluent discharges.

24.1 Air Pollution Monitoring Instruments

Rapid industrialization and steadily increasing vehicular traffic on the roads have problem of air pollution. At places, gaseous and dust pollutants in the atmosphere reach such a magnitude, that suitable measures to limit the emission of pollutants become imperative. One of the basic measures to environmental quality management, lies in the ability to continually monitor environmental characteristics and provide reliable, accurate and automatically recorded data for timely interpolation. Pollutants are required to be monitored, not only in the environment but also as they are discharged from the multitude of stacks, exhaust pipes from industrial establishments and vehicles. Today, the pollutants that must be measured are many in number and they are emitted from a still larger number of sources.

Gaseous and dust pollutants

The most reliable and useful information on the degree of pollutant concentrations is obtained by continuous sampling using officially accepted methods. The data about the peak concentrations obtained from such sampling is indispensable, when monitoring the air for potentially hazardous pollutants. For some specific purposes, however, samples collected intermittently may be useful, as a basis towards a more meaningful monitoring programme. Regardless of whether one is involved with continuous monitoring or monitoring on an intermittent basis, the analysis of the air we breathe has become a necessity for the survival of mankind.

24.1.1 Representation of Concentrations of Gases

Gas concentrations in the atmosphere are generally represented as parts per million by volume, i.e. ppm/v or simply ppm or parts per hundred million (pphm), i.e. parts per billion (ppb). On the other hand, toxicological data is generally represented on a gravimetric basis, e.g. micrograms per cubic metre or milligrams per litre.

Conversion from volumetric to gravimetric concentration can be obtained by applying gas laws, the general equation for this being:

$$\mu g/m^3 = (ppm) \times PM/RT \times 10^3,$$

where P = total pressure (atm)
M = molecular weight of gas of interest
R = gas constant = 0.0821 1-atm/(mole) (°K)
T = absolute temperature, °K.

24.1.2 Types and Concentration of Various Gas Pollutants

Many types of pollutants are present in the atmosphere. However, for many reasons, it is not simple to make definite statements about pollutant concentrations. This is because of the extreme variability of the pollutant concentration themselves, which vary constantly with air turbulence and the strength of emissions. The major gas pollutants are carbon monoxide, sulphur oxides, hydrocarbons, nitrogen oxides and particulates.

Carbon Monoxide

Carbon monoxide is colourless, odorless, poisonous gas that has an affinity for haemoglobin, which is 210 times that of oxygen. By combining with haemoglobin in the blood, it inhibits the delivery of oxygen to the body's tissues, thereby causing asphyxia or shortness of breath.

Carbon monoxide is an especially hazardous pollutant with which even healthy individuals get affected. It is a by-product of combustion processes, in which incomplete oxidation of fossil fuels takes place. It is basically associated with automotive exhaust and deep mining operations. Industrial processes also contribute to carbon monoxide pollution levels. The average concentrations of this gas are found to be much below 200 ppm.

Hydrocarbons

Hydrocarbons enter the atmosphere from a wide variety of sources like petroleum-refining processes, incomplete combustion of organic fuels and evaporation of fuels and solvents. Gasoline is the major source of their emission from internal combustion engines, since they exhaust unburned and partially burnt hydrocarbons. These are also important, because of their reaction in the atmosphere in the formation of ozone with nitrogen oxides and sunlight, and also to form photochemical smog. Methane constitutes the major component of the total hydrocarbon emission.

Sulphur Oxides

Sulphur dioxide is the most common and the most abundant gaseous pollutant. The major health effects associated with its high exposure include breathing and respiratory problems. Sulphur dioxide damages trees, plants and agricultural crops. It can also accelerate corrosion of natural and man-made materials. It thus, damages ecosystems and a major precursor in the formation of acid rain. It is emitted into the atmosphere from heating and from industrial plants using high-sulphur coal and other sulphur-containing fossils. It is also associated with the most serious urban pollution disasters. Sulphur trioxide, which is estimated to be one-hundredth the concentration of sulphur dioxide, is formed in a secondary reaction and becomes sulphuric acid aerosol. It is reported that SO_2 is present at an average level of 0.024 ppm in urban areas.

Nitrogen Oxides

Nitrogen oxides are a class of pollutants formed when fuel is burned at a very high temperature (above 1200°F), such as in automobiles and power plants. For pollution purposes, the main pollutants are nitric oxide (NO), nitrogen dioxide (NO_2) and other oxides of nitrogen. Nitrogen dioxide (NO_2), for example, is a highly toxic, reddish brown gas that is formed through the oxidation of nitric oxide emitted primarily from the combustion of fuels. It basically causes an odorous, brown haze that irritates the eyes and particularly, affecting sufferers of respiratory disease.

The emissions of nitrogen oxides are chiefly from the products of fuel combustion in furnaces and engines. It has been observed that the distribution of nitrogen oxides closely follows the population concentrations. Its level ranges from 0.5 to 0.12 ppm on an annual average basis.

Oxidants (Ozone)

The presence of oxidants in air can have a significant effect on the ambient air quality. The major component of total oxidants is ozone that has damaging effect on plants, animals and material if present in higher concentrations.

Other pollutants which may be present are hydrogen sulphide, ammonia, halides (chlorine, fluorine and bromine) and carbon dioxide.

24.1.3 Instrumental Techniques and Measurement Range

Measurement techniques for various gas pollutants are given in Table 24.1.

Table 24.1 :: Measurement Techniques with Full Scale Measurement Range for Various Gas Pollutants

Gas	Full-scale range	Measurement technique
Carbon monoxide	0–50 ppm 0–200 ppm	infrared absorption Gas chromatography (flame lonization detector)
Hydrocarbons	0–80 ppm	Ultraviolet absorption Gas chromatography (flame ionization detector) Mass spectrometry
Sulphur oxides	0–2 ppm	Ultraviolet absorption Infrared absorption Gas chromatography (Flame photometric detector) Colorimetric method Conductimetric method Coulometric method Electrochemical transducers
Nitrogen oxides	0–1 ppm	Colorimetric method Coulometric method Chemiluminescence method Electrochemical transducers infrared spectroscopy Lasers
Oxidant	0–500 ppb	Chemiluminescence method Coulometric method Colorimetric method Ultraviolet absorption method

24.2 Carbon Monoxide

24.2.1 Non-dispersive Infrared Analyzer

Non-dispersive infrared analysis depends upon the characteristic energy of absorption of CO molecule at wavelength of 4.6 μ. Infrared energy is also absorbed by other gases like CO_2, H_2O,

SO_2 and NO_2. The differentiation of CO from such type of gases depends on the difference in energy absorbed, as infrared radiation is passed through a sample cell containing CO and a reference cell containing a fixed quantity of nitrogen, CO and water vapour.

Figure 24.1 shows a block diagram of the instrument. The reference cell contains a fixed quantity of the gases, whereas air sample is made to flow at about 150 ml/min through the sample cell.

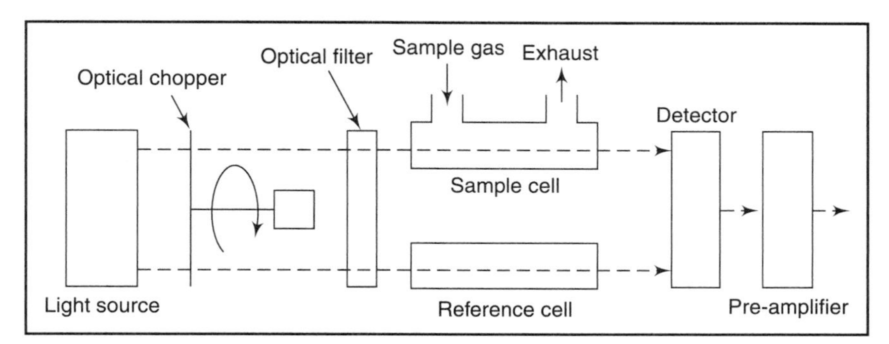

Figure 24.1 :: Block diagram of measuring system for carbon monoxide

CO analyzers based on infrared absorption would give greater sensitivity with larger cell path lengths. Instruments with 1 m cell length would measure from 1 to 50 ppm. However, some of the latest instruments are capable of measuring 1 to 25 ppm, even with cell paths of 10 cm. The response time of such instruments varies from 1 to 5 min. They are calibrated by passing a known part per million concentration of CO in nitrogen into the sample cell. The zero is set by using pure nitrogen as a sample. The effects of interfering gases like CO_2 and water vapour can be further minimized by placing optical filters ahead of the sample cell, so that IR radiation window is limited to a range, where radiation absorption by these does not significantly take place.

Non-dispersive infrared absorptiometry has the following advantages:

- The effect of the flow rate is small.
- The response speed is high.
- High sensitivity measurements are possible.
- The effect of interfering components is small.
- The equipments are easy to maintain.

24.2.2 Gas Chromatography

When an air sample containing CO is passed through a stripper column, the heavy hydrocarbons are retained and CO and methane are passed into a chromatographic column and then into a catalytic reducing chamber. The methane would pass through the reducing chamber unaffected, while CO is reduced to methane. By using hydrogen flame ionization detector, both methane peaks can be detected. The first peak is due to methane, while the second peak would correspond to CO. The accuracy is about ±2 per cent. Peak heights of CO and CH_4 would give sensitivity of about 50 ppb.

24.3 Sulphur Dioxide

24.3.1 Colorimetry

In this method, a known volume of air is passed through an aqueous solution, which contains reagents that absorb SO_2 and produce a coloured substance. The amount of coloured substance is proportional to the component of interest (SO_2). This is determined by measuring a solution's optical absorbance spectrophotometrically. Within limits, the absorbance is linearily proportional to the concentration of the coloured species, in accordance with the Beer's Law.

The most widely used method involving colorimetry for determination of SO_2 is that of West–Gaeke method, which is applicable to concentrations in the range of 0.005–5 ppm. The air to be analyzed is passed through an aqueous solution of 0.1 M sodium tetrachloromercurate. The sulphur dioxide reacts with the mercuric salt to form a sulphatomercuric compound. The solution is further treated with acid-bleached para-rosaniline and formaldehyde, which would result in a red-purple colour. The amount of colour is determined photometrically.

Sulphur dioxide may also be determined by measuring the decolorization of an aqueous starch–iodine solution; which normally has a deep-blue colour. When SO_2 is passed through the solution containing starch–iodine solution, the iodine is reduced by the sulphur dioxide, which in turn results in decrease of the intensity of starch-iodine colour.

The advantages of colorimetry are its simplicity, high sensitivity and good specificity. Colour intensity is sensitive to temperature, purity of reagents, pH, development time and age of solutions. The method is free from interference from H_2SO_4, SO_3, NH_3, etc.

24.3.2 Conductivitimetry

Analyzers based on measuring the change in conductivity of a solution, when a sample of air containing sulphur dioxide is bubbled through it, are the oldest and most commonly utilized instruments for ambient air monitoring. The solution consists of sulphuric acid and hydrogen peroxide. The change in electrical conductivity takes place due to the formation of sulphuric acid by oxidation of the sulphur dioxide.

$$H_2O_2 + SO_2 \rightarrow H_2SO_4 \rightarrow H^+ + HSO_4^-$$

These instruments are characterized by fast response and high sensitivity. However, their performance gets affected due to interference by non-SO_2 gases, that produce or remove ions in solution. Therefore, these analyzers are employed only if interfering gases are not in high concentrations, or if they can be effectively removed from the air sample.

Conductivity cell

Figure 24.2 describes the construction of a conductivity cell for the continuous measurement of sulphur dioxide in air. The cell *C* is made of glass, 1 cm inside diameter with a jet *J*, orifice 0.5 mm diameter, located 1 cm above the reagent surface.

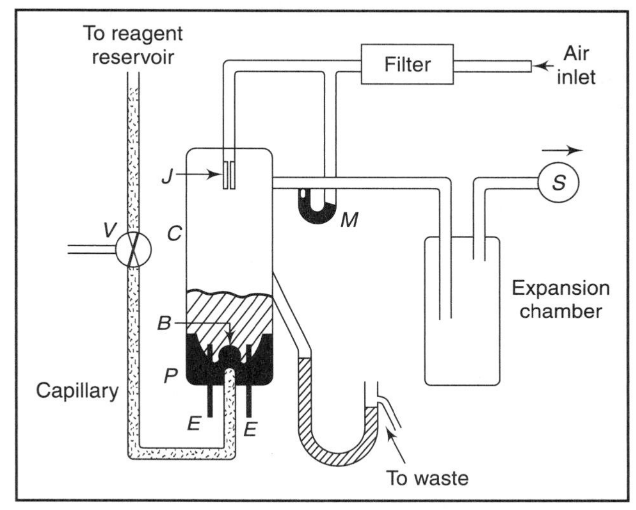

Figure 24.2 :: Conductivity method for the measurement of SO_2 in air (after Killick, 1969)

Two electrodes *E* made of 18 SWG stainless steel wire are inserted through a perspex cap *P*. The cap is sealed to the base of the cell. Reagent enters the cell from a central tube inserted in the perspex cap. A small glass bead *B* in the cell acts as a non-return valve on the entry of the feed tube and prevents sulphuric acid diffusing from the cell. The end of the jet is made from a piece of capillary tube. A filter is placed before the jet to prevent blocking due to solid materials. The flow of air through the jet is maintained at approximately 200 ml/min. The cell is of small size (1.5 ml) and its capacity to absorb SO_2 is thus limited. It is therefore operated intermittently, the electrolyte being discharged and replaced at regular intervals of 15 min. The resulting output is recorded as a saw-tooth waveform.

To measure the conductivity of the cell, 5 V alternating current is applied across the electrodes. Alternating current avoids polarization. When normal urban concentrations of SO_2 are measured, the current through the cell increases from its zero value of 20 μA to up to 40 μA at the end of 15 min. sampling period. In conditions of heavy pollution, it may go to 2 mA. Because the current is recorded every 15 minutes, the concentration of sulphur dioxide at any instant is proportional to the slope of the saw-tooth at that instant. The calibration is carried out by using known concentrations of sulphur dioxide in air.

24.3.3 Gas Chromatography

There are two major problems when the gas chromatographic technique is employed for measuring pollutants in air. These are: (a) Most of the pollutants are extremely reactive materials. Therefore, they may not pass through the column and appear at the detector. This necessitates the use of special column and support materials. (b) The amount of pollutants of interest is generally so small that when a reasonably sized gas sample is injected into the carrier gas stream, it is not detected by ordinary GC detectors. Hence, extremely sensitive and specific detectors must be employed.

These detectors measure the emissions from sulphur compounds introduced into hydrogen-rich flame. A narrow band optical filter selects the 394 nm emission band. These detectors are not susceptible to interference from non-sulphur pollutants. However, they are subject to interference from other sulphur compounds. This type of interference can be minimized by using selective filters.

24.3.4 Coulometry

The ability of sulphur dioxide to reduce iodine can be employed to monitor SO_2 by using iodine coulometry in an aqueous solution. The mass of I_2 reacted per unit time during any given interval of time would indicate concentration of SO_2 in air.

A coulometric arrangement would require two electrodes made of platinum, which act as anode and cathode. These electrodes maintain a trace of I_2 in equilibrium with potassium bromide to maintain conductance. The shift in the anode–cathode potential is sensed by a third reference electrode, when SO_2 from sample air is bubbled through the solution in a detector cell. The coulometric method provides a detection limit of 0.01 ppm. The response is not instantaneous and it may take about 4 minutes for the 90 per cent of the signal to appear for any concentration of SO_2.

24.3.5 Flame-Photometric Detector

Flame photometers, in which sample air is introduced into a hydrogen-rich air flame by using a narrow band interference filter that shields the photomultiplier tube detector from all but 394 nm emission energy of flame-excited sulphur atoms, offer another method for monitoring combined sulphur. Research has continued to increase the specificity of the method to actual SO_2 rather than total sulphur measurements. Suitable filters have been developed to remove interferences from hydrogen sulphide and some other organic sulphur. It is expected that improvements in selective filtration will result in a flame photometer for monitoring of ambient SO_2 within acceptable accuracy levels. Sensitivity to total sulphur by this method is 0.01 ppm.

24.3.6 Ultraviolet Fluorescence Method

This method is based on the principle that SO_2 molecules absorb ultraviolet (UV) light and became excited at a particular wavelength. These molecules then decay to a lower energy state emitting UV light at a different wavelength. The intensity of fluorescence is proportional to the SO_2 concentration.

Figure 24.3 shows a block diagram of a fluorescence SO_2 analyzer. It consists of a:

- *Hydrocarbon Scrubber*, which removes hydrocarbons contained in ambient air, which are also excited by the light source and consequently emit fluorescence. The SO_2 molecule passes through the hydrocarbon unaffected.
- *Fluorescence Chamber*, in which the fluorescence of SO_2 is emitted.
- *UV Light*, which is an emitter of ultraviolet light by an electric discharge
- *Photoelectric Detector*, which is located adjacent to the fluorescence chamber and converts fluorescence to electrical signal. The technique offers a highly sensitive technique of monitoring SO_2.

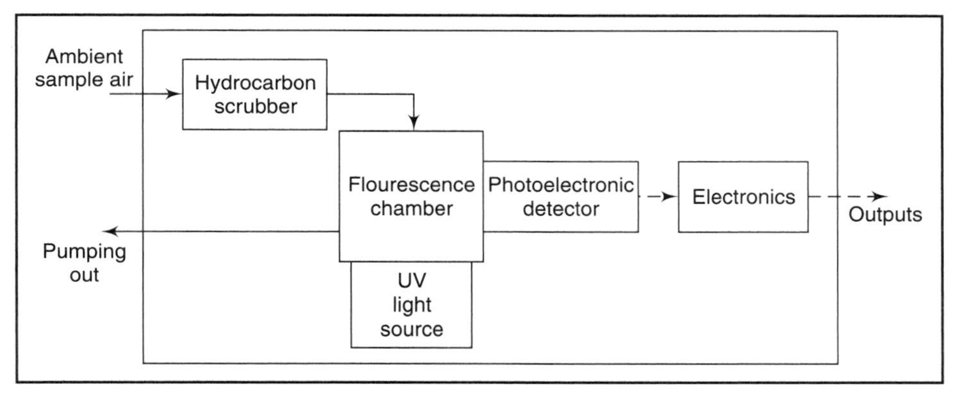

Figure 24.3 :: Fluorescence SO_2 analyzer

24.4 Nitrogen Oxides

24.4.1 Colorimetry

The colorimetric method can be used for the determination of nitrogen dioxide by using Saltzman method. This method is based on a reaction, in which a pink coloured dye complex is formed when the air containing NO_2 is passed in an absorbing solution consisting of the sulphanilic acid and diamine dissolved in the acetic acid medium. This method is sensitive in the parts per million range.

Nitric oxide may be determined by the same procedure, by first passing the sample through an acid permanganate bubbler, which oxidizes nitric oxide to nitrogen dioxide.

The method used for monitoring NO_x in stack effluents in concentration range of 5 ppm to several thousand ppm, consists in passing the sample into an evacuated flask containing a solution of H_2O_2 in sulphuric acid. The oxides of nitrogen are converted into nitric acid, and the nitrate ions react with phenol-disulphonic acid to produce a yellow colour, which is measured colorimetrically.

24.4.2 Chemiluminescence

The phenomenon of emission of radiation from chemi-excited species is known as chemiluminescence. It results due to the formation of new chemical bonds. The species in the excited state possess higher energy levels than the ground state and usually have a very short life. This phenomenon is very useful for measurement of air pollutants, particularly NO and NO_2. Instruments based on the measurement of chemiluminescent emission, based on the following reaction have been developed.

$$NO + O_3 \rightarrow NO_2 + O_2$$

$$NO_2 \rightarrow NO_2 + h\upsilon \quad (\lambda_{max} = 6300 \text{ Å})$$

Since NO_2 reacts only slowly with ozone and the reaction which produces NO_3 is not accompanied by chemiluminescence, it is necessary to reduce NO_2 to NO before admission into the reactor.

$$NO_2 \xrightarrow{\text{energy}} NO + 1/2O_2$$

Nitric oxide and ozone containing gas stream are mixed in a vessel at a sub-atmospheric pressure of about 2 mm of Hg. Light emission is measured with a photomultiplier. With the use of high gain, low dark current photomultiplier tubes, extremely low levels of radiation can be measured. The response of the instruments based on chemiluminescence is linear from I ppb to 1000 ppm of NO. The technique is extremely useful for measurement of NO in automotive exhaust gases.

24.4.3 Use of CO Laser

Figure 24.4 is a block diagram of an apparatus for detecting nitric oxide in 0.25 ppm concentration. The apparatus consists of a CO laser, which emits radiation that is absorbed by the NO in the mixture, the amount of absorption being proportional to the concentration of NO present. The wavelength match between laser and NO is made exact, and hence the absorption is enhanced by placing the NO in a magnetic field of a few kG intensity. The field shifts the absorption wavelength of the NO into coincidence with the fixed laser wavelength.

The CO laser used is a dc-excited continuous working laser, which operates at a single wavelength of 5.307 μ and at liquid nitrogen temperature. A diffraction grating is used at one end of the cavity as a line selector. The laser yields 5–30 mW of single line power.

The absorption cell is made of Pyrex and is of 15 mm diameter and 90 cm length. It is evacuated to a pressure of 10^{-6}–10^{-5} torr. In order to produce a modulating audio frequency magnetic field along the axis of the cell, insulated wire is closely wound around the outside of the cell over about half its length. The coil is excited with a current of 1A in the frequency range of 5–150 kHz. This produces a varying magnetic field of about 50G peak-to-peak intensity. The dc magnetic field is produced by a solenoid, which produces a field up to 2.5 kG.

The detector is a liquid nitrogen cooled Ge–Au element. The signal is amplified in a lock-in amplifier, before it is given to the recorder. The signal amplitude varies linearly with the concentration of NO in the sample.

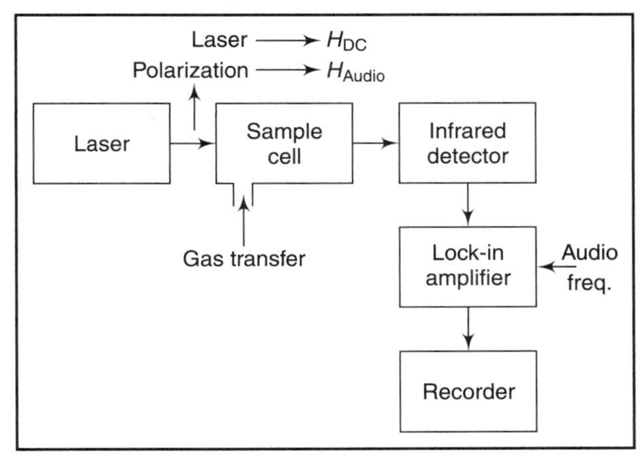

Figure 24.4 :: Measurement of nitric oxide using CO laser (after Bonczyk, 1975)

24.4.4 Laser Opto-acoustic Spectroscopy

Opto-acoustic detectors, in conjunction with thermal IR sources, have been widely used in gas detection and measurement systems. These have been developed for application in air pollution measurement. This technique has been used to measure trace amounts of nitrogen oxides in the stratosphere. The opto-acoustic detector involves the absorption of an amplitude modulated beam of IR by a gas which results in the generation of sound. The energy absorbed by the gas molecules from the IR beam excites the molecules to the rotational-vibrational energy levels above the ground state. The main path for decay of these excited states is collisional de-excitation, which results in the transfer of absorbed energy into heat and raises the gas temperature. The temperature rise causes a corresponding pressure rise in the gas. When the beam intensity is modulated, the gas temperature and pressure change accordingly. Thus periodic pressure variations in the gas result in the generation of sound.

Opto-acoustic detector

A block diagram of a laser opto-acoustic detector is shown in Figure 24.5. The arrangement makes use of a CO_2 laser that is tuned by rotating a diffraction grating at one end of the laser

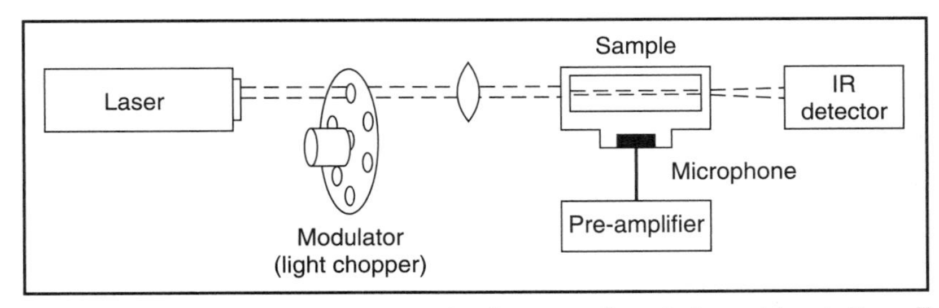

Figure 24.5 :: Laser opto-acoustic detector for nitric oxides (after Kreuzer, 1978)

cavity. It is tuneable to 64 different emission lines in the range from 927 to 1085 cm^{-1}. The laser beam is brought to a focus at the chopping wheel and then refocused into the detector. The chopping frequency is selected to optimize the signal-to-noise ratio. The microphone is commercially available, model 4144 manufactured by M/s Bruel and Kjaer Instruments Inc. The acoustic signal is amplified in a pre-amplifier and displayed. The detector is highly sensitive and sample amounts as small as 20 pg may be detected.

24.5 Hydrocarbons

24.5.1 Flame Ionization Detector

Organic compounds easily pyrolyse when introduced into an air-hydrogen flame. The pyrolysis produces ions that can be collected by having a cylindrical grid surrounding the flame. The detector response would be in proportion to the number of carbon atoms in the chain. For example, propane would roughly give three times the intensity of response as compared to methane. The ions collected on the positively charged grid are amplified in a high input impedance amplifier, whose output is given to a chart recorder. The variation in the ion intensity resulting from flame ionization of any organic compound in the air sample is recorded on the chart. Generally, 0–20 ppm range is adequate for atmospheric sampling.

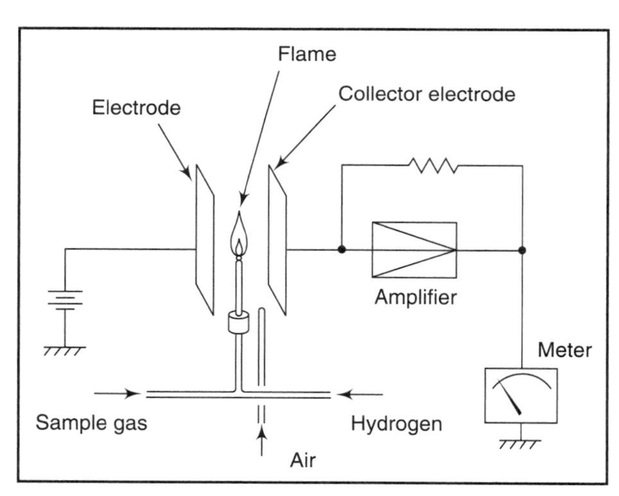

Figure 24.6 :: Schematic diagram of a measuring system for hydrocarbons based on flame ionization detection principle

Figure 24.6 shows a block diagram of the apparatus used for measuring hydrocarbons based on flame ionization detection principle. The sample gas containing hydrocarbon controlled at a constant flow rate is mixed with hydrogen for fuel use. The mixed gas is burned at the end of a very fine nozzle. Two electrodes are placed on either side of the flame and an appropriate electric field is applied to them. Ionic current will flow between the collector electrode and the other electrode. The current is then amplified and displayed on a meter-analog or digital.

It may be noted that the flame ionization detection method generally shows responses in proportion to carbon numbers. However, if carbon coexists with oxygen, the response may differ depending on the kind of the hydrocarbon. The analyzer, in general, provides good linearity across a broad range of concentrations, high sensitivity and fast response.

24.5.2 Gas Chromatography

The gas chromatographic technique has been applied for the detection and measurement of hydrocarbons, carbon monoxide and carbon dioxide in air. They are generally detected with flame ionization detectors. Flame ionization detectors are not sensitive to carbon monoxide and carbon dioxide. These are first converted to methane by hydrogenation, and then measured with this detector.

24.5.3 Use of Lasers

Fixed frequency, infrared emitting lasers have demonstrated their potential in atmospheric monitoring and research. This is based on the absorption of CO_2 laser frequencies by some gases. For example, a tunable dual-line CO_2 laser can be used for atmospheric spectroscopy and pollution monitoring. This arrangement facilitates measurement of 0.5 and 2 ppb densities of ethylene in a calm and turbulent atmosphere, respectively. Ethylene is of considerable interest as a pollutant to the petrochemical industry and for automobile exhausts.

24.6 Ozone

24.6.1 Colorimetry

Total oxidants are usually measured by liberation of iodine from potassium iodide under buffered conditions. The gas is passed through a solution of neutral, buffered potassium iodide (1 per cent KI buffered at pH 6.8). The iodide is oxidized to iodine in the presence of oxidizing agent and gives the solution a pale-yellow colour, whose intensity is measured colorimetrically.

24.6.2 Chemiluminescence

The chemiluminescence method is based on the emission of light from solid organic dye samples, due to the passage of ozone over the surface. Very low zone concentrations can be detected. However, periodic calibrations are necessary for both the instrument and the ozone source used.

Other chemiluminescence reactions due to ozone have also been used to determine ozone concentrations. A portable ozone meter, which uses the reaction between O_3 and ethylene, has been constructed. This instrument depends upon the reaction which occurs when air and ethylene are drawn into a Pyrex container and mixed directly in front of a Pyrex window. A photomultiplier tube is mounted behind this window, which produces an electrical signal due to photons of chemiluminescence energy and is proportional to the total ozone present in the sample. The concentrations as low as 0.001 ppm are detectable.

Ray, *et al.* (1986) describe an apparatus using the fast chemiluminescence method for the measurement of ambient ozone. The design of the instrument is shown in Figure 24.7. Sample gas is drawn through the detector cell at 1–7 L/min by a diaphragm pump. Fluid and air are separated at the reservoir, and the dye solution is circulated. Sample gas flows across a fibre pad of either paper or glass mat that is saturated with the organic dye dissolved in an alcohol solvent.

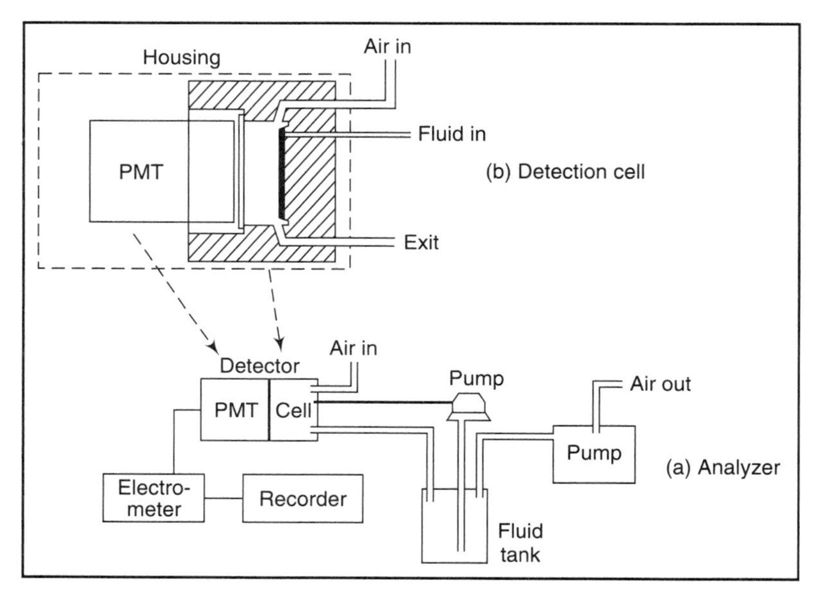

Figure 24.7 :: Chemiluminescent ozone analyzer and design of the ozone detection cell

The air and dye solution enters at the top of the detector cell through separate holes. Both air and fluid exit at the bottom together. The detection cell also houses a photo-multiplier tube that responds to the chemiluminescence, as the ozone reacts with the dye. The photocurrent is amplified by an electrometer and the signal output is recorded on a strip-chart recorder. The dye solution is pumped into the cell at 1 ml/min. Response of ozone is from 1 to 10 nA/ppb, depending on dye type and solvent.

24.6.3 Absorptiometry

The optical technique measures the absorption of light by ozone in the ultraviolet part of the spectrum. This method is the standard procedure for total ozone determination in the atmosphere. This method has an advantage in that the absorption is specific to ozone. However, a long optical path is necessary for appreciable light extinction to occur under expected O_3 concentrations. Also several absorptions at different wavelengths must be taken, since comparison with ozone-free air is clearly impossible and data reduction is somewhat complicated.

24.6.4 Conductivitimetry

A wet chemical method which uses the oxidizing properties of O_3 can be employed to construct a sensitive automatic meter for continuous sampling of contaminating oxidants in the atmosphere. The ozone containing air is bubbled into a potassium iodide solution and the resulting iodine determined by measuring the current through the cell. The current is related to ambient O_3 levels by previous calibration with a known ozone source. This technique has been employed to construct an air-ozone meter, which measures and records instantaneous ozone concentrations.

The accuracy of this method has been improved by determining the concentration of iodine and hence ozone by titrating the solution with sodium thiosulphate. The apparatus (Figure 24.8) consists of an hermetically sealed glass jar containing 150 cm^3 of buffered 10 per cent KI solution

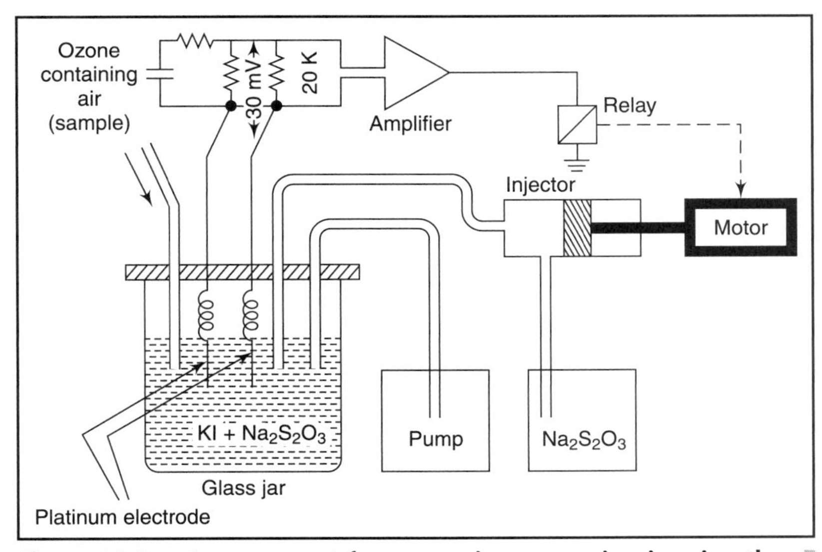

Figure 24.8 :: Arrangement for measuring ozone in air using the oxidizing properties of ozone (after Steinberger and Goldwater, 1972)

and about 0.5 cm^3 sodium thiosulphate of known concentration. Two spiral platinum electrodes dip into the solution and a bias voltage of 30 mV is applied across them. The air above the solution is evacuated, whereas the outside is let in through a Tygon tube which is inert to ozone.

When ozone enters the solution, the following reaction takes place:

$$O_3 + 2I + H_2O \rightarrow I_2 + O_2 + 2OH^-$$

The iodine then reacts with the thiosulphate

$$I_2 + 2(S2O_3)^- \rightarrow 2(I^-) + S_4O_6^-$$

Reaction continues so long as there is thiosulphate in the solution. When all the thiosulphate has been reacted, free iodine appears and reacts at the electrodes.

The electrical resistance is high as long as there is an excess of thiosuiphate. The resistance decreases when it is used up. This change is used to control the operation of the instrument. The voltage drop across 20 kΩ resistance, which is in series with the electrode, is used to operate the recorder as well as a relay, which controls a motorised injector, which injects 0.5 cm^3 thiosulphate in each operation. The recorder serves mainly to indicate as to when the injection was made, and thus the average ozone concentration between any two injections can be calculated. Since the pumping rate is known, knowledge of the time intervals gives the total volume of air sampled.

24.7 Automated Wet-Chemical Air Analysis

Wet chemical methods are the accepted means for obtaining data in a majority of environmental studies. Specific chemistries are recommended for most of the common pollutants in the atmosphere. Basically, an automated wet-chemical analysis system would be as shown in Figure 24.9.

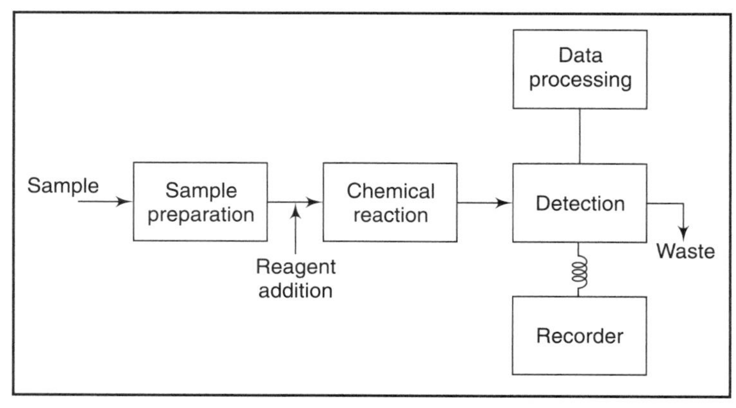

Figure 24.9 :: Basic automated air analysis system

In this scheme, the sample is automatically introduced and prepared. Reagents are then added in proper quantities and sequence. A chemical reaction will take place. The presence of a particular constituent is then detected, displayed and recorded.

In any air analysis system, the equipment must be capable of obtaining a proper sample. This is essential, in order to obtain correct results. In automated wet-chemical methods, the accuracy of the sampling system is dependent upon maintaining a constant ratio between the amounts of sample ratio and absorbing solution. The rate of flow of the absorbing solution is maintained constant by a proportionating pump.

Sampling System: Figure 24.10 shows the principle of operation of one type of sampling system schematically. The air sample control system utilizes the fluid dynamic principle, whereby a constant pressure drop is maintained across a fixed orifice, thus resulting in constant rate of flow. In order to avoid dependence upon critically dimensioned orifices, a precision needle valve is used to establish the constant orifice. Once set, the orifice is maintained constant by a 0.45 μ filter, which is placed in front of the needle valve. The relatively small flow rates combined with the large surface area of the filter, afford long filter life with negligible change of filter pressure drop. This arrangement permits the flow rates to be adjusted between 300 and 800 ml/min, depending upon the chemical parameter being analyzed. A carbon vane pump equipped with a vacuum relief valve provides the necessary vacuum for air to flow through the system.

Thus, with the pressure losses of the needle valve constant, the orifice area maintained constant by a protective filter, and the pressure vacuum downstream of the orifice held constant by the vacuum relief valve, a constant flow system is established. This arrangement provides constant air flow, within better than ± 5 per cent per week of continuous operation. Air is then made to enter a bubbler through a glass-fibre plug. A series of samples are thus collected for fixed durations at a pre-determined flow rate. Automatic program control the duration of sampling and the interval between samples, and switch automatically from one collector to the other.

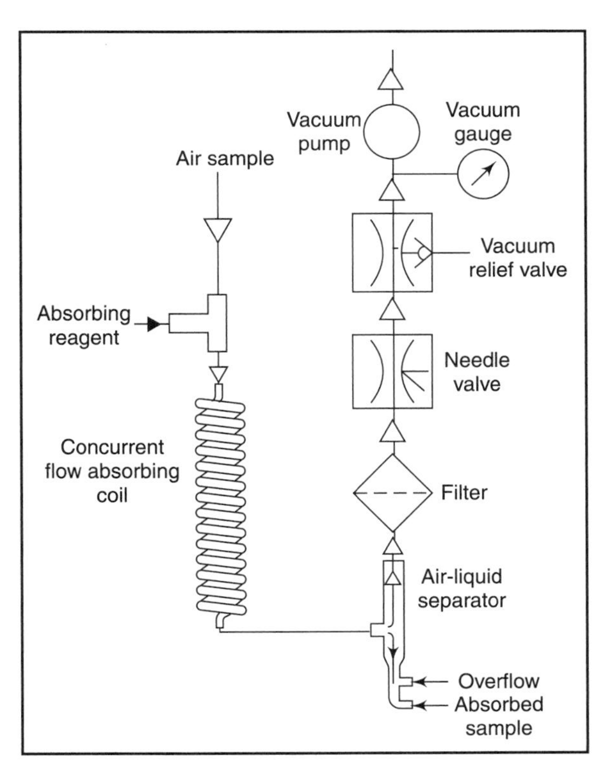

Figure 24.10 :: Sampling system (after Zehnder and Belew, 1970)

Methodologies: Wet chemical methods have been used consistently by regulatory agencies.

They have the advantage over other methodologies, in that they are specific, sensitive, accurate and reproducible.
The measurement of total oxidants and SO_2 are two perfect examples of how automated wet-chemical analysis fulfils the requirements outlined above.

24.7.1 Total Oxidants

Figure 24.11 shows the flow diagram for measurement of total oxidants in air. It also incorporates an ozone generator, to enable dynamic calibration and stabilization of the system.

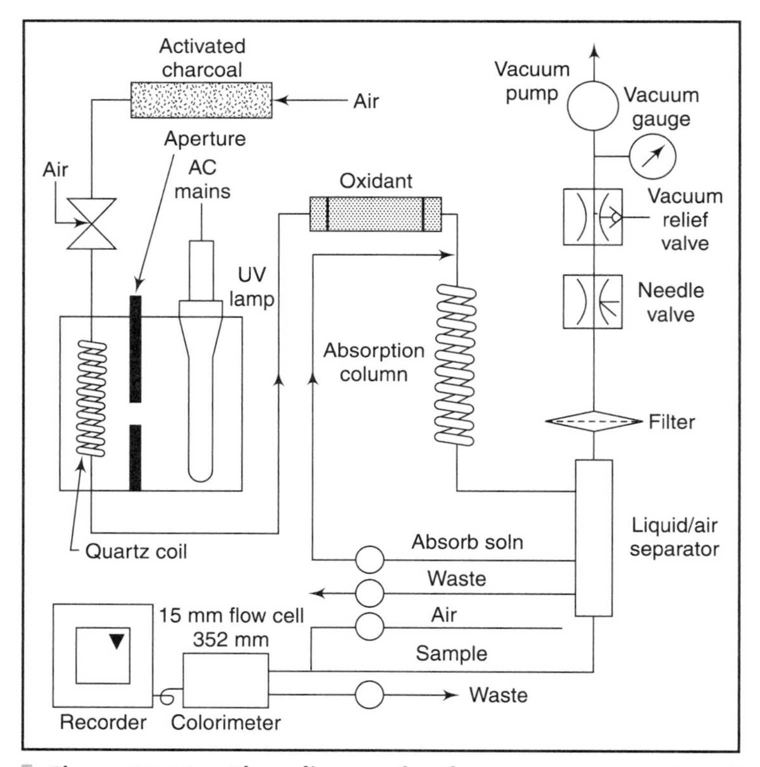

Figure 24.11 :: Flow diagram for the measurement of total oxidants in air (Technicon Air Monitor IV)

The measurement system makes use of the well-known neutral-buffered potassium iodide reagent method. First, the sample air is purified by means of an activated charcoal filter. Upon leaving the filter, the purified air passes through a quartz helix, where part of the oxygen is converted to ozone by a constant current UV source. The sample then passes through a chromium trioxide-firebrick column, which eliminates negative interference of SO_2. It then enters the absorption column, where the ozone reacts with a solution of 1 % potassium iodide. After separating the spent air, the liberated iodine is measured colorimetrically at 352 nm. The photoelectric colorimeter comprises of a dual-beam optical assembly, having usual provisions for filters, flowcells and phototubes. Phototubes range from 340 to 900 nm. Interchangeable flow cells are provided for 15 and 50 mm.

Ozone in fixed amounts is generated in the quartz helix. The amount generated is dependent upon the intensity of the incident UV radiation and is adjustable by altering the aperture between the quartz helix and the UV lamp. By inserting the proper size apertures, known amounts of ozone, e.g. 0.05 ppm, 0.10 ppm and 0.20 ppm are obtained

In an alternative method of monitoring total oxidants, the oxidation tube is first removed and the ozone generator aperture adjusted to produce a constant concentration of approximately 0.10 ppm of ozone, to be read as a baseline set at the mid-scale of the recorder. Depending upon their oxidative or reductive conditions prevailing in the ambient air, the instrument will read either above or below the mid-scale baseline, thus enabling the user to obtain quantitative measurement, both of oxidising as well as reducing substances. This mode of operation is advantageous in areas where, for example, excessive emissions of SO_2 and/or H_2S occur periodically.

24.7.2 Sulphur Dioxide

The West–Gaeke procedure for the estimation of SO_2 in air can be automated by having a scheme as shown in Figure 24.12.

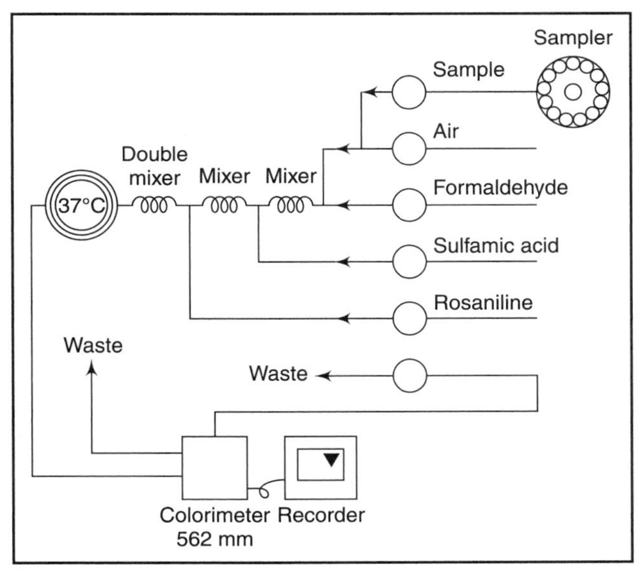

Figure 24.12 :: Automated wet analysis system for SO_2 based on West-Gaeke procedure

Here, the specific absorbant, sodium tetrachloromercurate, is used to form a stable, non-volatile complex with sulphur dioxide. Its subsequent reaction with acid-bleached prosaniline and formaldehyde gives a sensitive, specific and temperature-independent colour reaction. It is measured with colorimeter at a wave length of 560 nm. It is found that this method is not prone to interference by sulphur trioxide, sulphuric acid, ammonia or carbon monoxide, which can be present in the atmosphere. Nitrogen dioxide, known to interfere at equivalent concentrations to that of SO_2, is eliminated by adding sulphamic acid. Where any influence of heavy metals is known to be present, it is removed by complexing with EDTA.

Air Monitor IV from M/s Technicon provides for the totally automated, simultaneous measurement of sulphur dioxide, nitrogen dioxide, oxides of nitrogen, total oxidants, aldehydes and hydrogen sulphide, with a built-in feature of flexibility to change parameters, as the need arises for a system in a permanent location, as well as, in air mobile laboratories.

24.8 Water Pollution Monitoring Instruments

Pollution of water sources can occur from a very large number of compounds, with varying degrees of pollutants. Their access to water resources can adversely affect the quality of water and may render it unsuitable for drinking purposes, industrial use, production of fish and aquatic foods, irrigation and even transport through waterways. Sometimes, the constituents of industrial waste waters may present health hazards, even when present in minute amounts. The problem of water pollution is thus a serious one for society.

The quality of water can be determined by measuring the quantity of specific pollutants present in it. The measurements generally include elemental analyses, physical determinations, microbiological and bacteriological examinations, radiochemical analyses, etc. Water analyses were earlier carried out either at the field site or in the laboratory, on samples were gathered at the site. It is obvious that just taste, smell and visual inspection are inadequate to determine the chemical, physical and biological characteristics of water, and therefore, instrumental methods have been adopted to perform this work. Instruments such as the pH meter, colorimeter, conductivity meter and turbidity meter are routinely used in water analysis laboratories. Modern instruments, such as the atomic absorption spectrophotometer, gas chromatograph, thin layer chromatograph and polarograph are increasingly used for detecting trace metals, pesticides and toxic organics. These instruments are preferred over wet chemical methods due to their speed of estimation, greater precision and accuracy of measurements. With these instruments, it is possible to estimate several pollutants at the microgram and nanogram levels.

24.8.1 Types of Pollutants and Techniques

Different types of investigations are required for monitoring pollutants, depending upon the purpose of the monitoring programme. The following standard parameters are determined for the assessment of normal pollution characteristics of waste waters: alkalinity, biological oxygen demand (B.O.D.), ammonia, nitrates, kjeldahl nitrogen, total phosporous and total dissolved suspended and volatile solids. Several other parameters are monitored in case of industrial effluents and similarly in the case of water used for drinking or domestic purposes.

24.8.1.1 Conductivity

Conductivity is the measure of dissolved ionized solids in water. Conductivity is temperature-dependent and is therefore standardized at 25°C. It is non-specific and measures, to some degree, all the ions present in water. A sudden increase in conductivity is an indication of pollution by strong acids, bases or other highly ionized substances. Conductivity was expressed in micromhos (μ mhos). However, it is now expressed as microSiemen. High conductivity values can be expressed as milliSiemens. For ultrapure water, the conductivity is 0.055 μ mhos/cm at 25°C. Conductivity, in simple terms, is a measurement of the ability of a solution to conduct an electric current. An instrument measures conductivity by placing two plates of conductive material with known area and distance apart in a water sample. Then a voltage is applied and the resulting current is measured. Using Ohm's Law,

$V = 1/R$

and the conductivity $G = 1/R$, then

G is determined as $G = 1/R = 1/V$

Typically an ac signal is used as a voltage source to prevent ionization of the electrodes.

An alternative technique is to use the non-contacting type of cell, as shown in Figure 24.13. It uses a magnetic field to sense conductivity. A transmitting coil generates a magnetic alternating field that induces an electric voltage in a liquid. The ions present in the liquid enable a current flow that increases with increasing ion concentration. The ionic concentration is then proportional to the conductivity. The current in the liquid generates a magnetic alternating field in the receiving coil. The resulting current induced in the receiving coil is measured and used to determine the conductivity value of the solution. This type of cell offers complete galvanic separation of measurement from medium, reduced maintenance and resistance to chemical attack and there is no polarization.

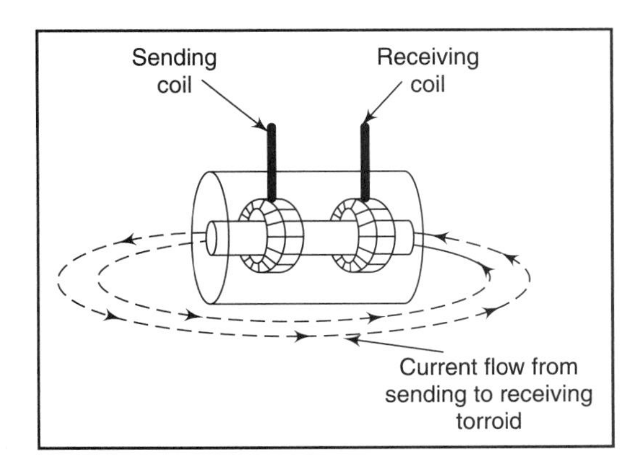

Figure 24.13 :: Non-contact type conductivity meter

24.8.1.2 Dissolved Oxygen

The term 'dissolved oxygen' is used to describe the amount of oxygen dissolved in a unit volume of water. It is a measure of the ability of water to sustain aquatic life. The solubility of oxygen in water decreases with increase in temperature. Also, almost all other types of waste consume oxygen from the receiving stream. Dissolved oxygen (DO) is expressed in mg per litre (ppm). The dissolved oxygen can be measured by a special sensor kept in an electrochemical cell by the amperometric method. The cell comprises a sensing electrode, a reference electrode and a supporting electrolyte, a semi-permeable membrane, which serves dual function. It separates the water sample from the electrolyte, and at the same time, permits only the dissolved oxygen to diffuse from the water sample through the membrane into the supporting electrolyte. The dissolved gas may subsequently react at the sensing electrode and thus cause a current flow. The range for this measurement is 0–20 ppm, with an accuracy of ± 1 ppm. The biological oxygen demand of water could be determined from dissolved oxygen measurements.

Although dissolved oxygen is usually displayed as mg/l or ppm, DO sensors do not measure the actual amount of oxygen in water. Instead, it is the pressure of oxygen in water which is measured. Oxygen pressure is dependent upon both salinity and temperature.

Two techniques are used to measure DO. These are: (i) galvanic, and (ii) polarographic. Both probes make use of an electrode system where the DO reacts with the cathode to produce a current.

Galvanic DO sensors

In galvanic DO sensors, the two electrodes, an anode and cathode, are immersed in electrolyte. An oxygen permeable membrane separates the anode and cathode from the water being measured. Oxygen diffuses through the membrane and interacts with the probe internals to produce the electrical current. This is shown in Figure 24.14.

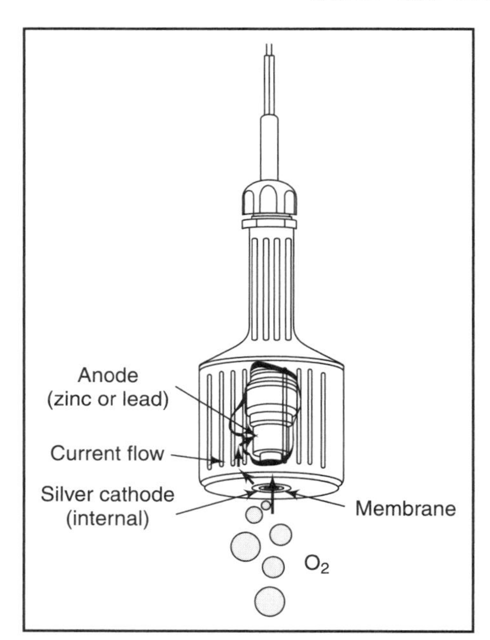

Figure 24.14 :: Membrane type DO (dissolved oxygen) probe

This current is passed through a thermistor and therefore, the actual output from the sensor is in millivolts. The thermistor connects for membrane permeability errors due to temperature change. Similarly, atmospheric pressure affects the saturation of oxygen. DO probes must be calibrated for the barometric pressure when reading in mg/l.

The polarographic technique makes use of the Clark-type cell, which is described in Chapter 22.

24.8.1.3 pH

pH is the logarithm of active hydrogen ion concentration in moles per litre. The pH of a neutral solution is 7. Values lower than 7 are considered acidic and higher than 7 are basic. Many important chemical and biological reactions are strongly affected by pH. In turn, chemical reactions and biological process (e.g: photosynthesis and respiration) can affect pH. Natural water shows only slight fluctuations in pH. Rapid changes in pH are generally due to pollution, such as dumps of acids and alkalies. The pH of the water sample is measured with the combined glass-calomel electrode immersed in the water sample. Instrument manufacturers guidelines should be followed for calibrating the pH meter.

24.8.1.4 Oxidation-reduction Potential (ORP)

ORP is an electrochemical potential developed by oxidizing or reducing materials in water. Industrial wastes usually contain strong oxidizing and reducing agents. Oxidants and reductants present in neutral waters tend to balance out to give zero ORP. A platinum electrode is used in conjunction with the calomel electrode for oxidation-reduction potential measurements.

24.8.1.5 Temperature

Temperature affects the solubility of oxygen and other chemical contents of water. The changes in water temperature are normally slow. They are seasonal unless pollution is experienced. The temperature of water can be measured by using a forward-biased silicon diode as a sensor. The diode forms part of a Wheatstone bridge, and the unbalanced voltage is amplified and given to the meter for display. The meter is calibrated in terms of temperature in °C. The range is 0–50°C with a reading accuracy of ± 0.5°C.

24.8.1.6 Turbidity

Turbidity is a commonly accepted criterion of water quality in water treatment for industrial or potable purposes. It is an expression of the optical property of a sample, which causes light to be scattered and absorbed rather than transmitted in a straight line through the sample. It is a measure of the undissolved solids in water including silt, clay, algae, rust, bacteria and other micro-organisms. Turbidity is expressed in Jackson Turbidity units.

The size, shape and refractive index of the particles affect the transparency of water. It is, therefore, not possible to correlate turbidity and the quantity of these particles by weight. Instruments used for measuring turbidity work on the principle of measuring the intensity of scattered light at an angle from a strong light beam. The two methods commonly used are:

(i) *Turbidimetric* analysis is based upon the measurement of the diminution in power of a collimated beam due to suspension of particles.

(ii) *Nephelometry* is based upon the measurement of light scattered by a suspension. This is usually done by measuring the scattered radiation at right angles to the collimated beam.

The choice between nephelometric and turbidimetric measurement depends upon the fraction of the light scattered. When scattering is extensive, owing to the presence of many particles, the turbidimetric measurement is more satisfactory. If the suspension is less dense, and the diminution in power of the incident beam is small, the nephelometric method provides a more satisfactory method. In dilute suspensions, the attenuation of a parallel beam of radiation by scattering is given by:

$$\rho = P_o e^{-jb}$$

where P_o and P are the power of the beam before and after passing through the length b of the turbid medium. The quantity J is called the turbidity coefficient and is linearly related to the concentration C of the scattering particles. So:

$$\log_{10} P_o/P = kbc$$

where $k = 2.3\ J/c$.

The relationship between $\log_{10} P_o/P$ and C is established with standard samples, the solvent being used as a reference to determine P_o. The resulting calibration curve is used to determine the concentration of samples from turbidimetric measurements.

Turbidimetric measurements are carried out with a filter photometer and a typical optical path as shown in Figure 24.15. Highly monochrometric light is not required. It can be done by directing a light beam from a light-emitting diode into the water sample and measuring the light that scatters off the suspended particles present in the water. The nephelometric measurement method is used in most commercially available instruments with a sensor range of 0–1000 nephelometric turbidity units (NTU) and an accuracy of ±5 per cent of 2 NTU, whichever is greater.

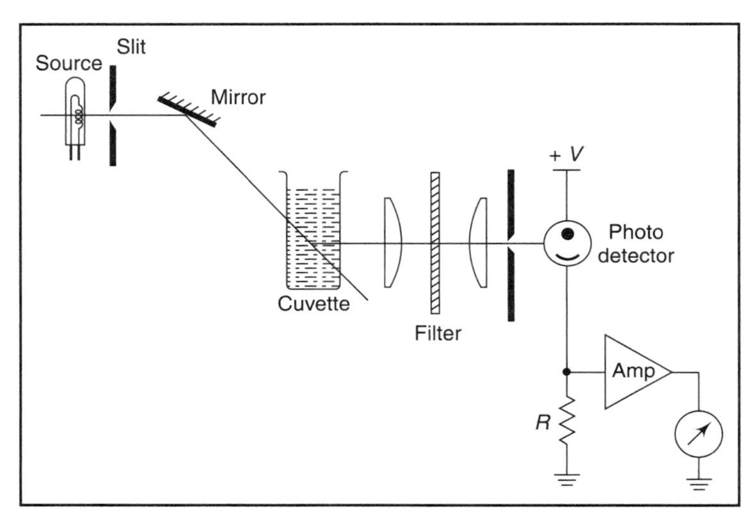

Figure 24.15 :: Optical arrangement for measurement of turbidity of water

Turbidity readings are affected by suspended sediment particle size, entrained air bubbles, floating debris, and other particles in the water that may collect on or near the optic sensor during data collection. Modern instruments are microprocessor based in which the software provides for temperature compensation.

The field calibration of the turbidity sensor is done by using Formazin or other approved primary standards and following the manufacturers calibration instructions.

The concentration of a variety of ions can be determined by this technique, by the use of suitable precipitating agents, so as to form a stable colloidal suspension. Surface-active agents such as gelatin are frequently added to the sample to prevent coagulation of the colloid. A computer-controlled laser nephelometer has been described by Koyuncu (1986) to determine the concentrations of liquid solutions.

Figure 24.16 shows the block diagram of the laser-based nephelometer. It uses a 5 mW He–Ne laser as a light source. The laser has the advantage that it gives a coherent monochromatic beam (λ = 632.8 nm) of high intensity. D_1 and D_2 are the two diaphragms which provide a Gaussian intensity distribution across the beam profile. The sample cell is placed in front of the beam at an angle. A lens L focuses the scattered light on a photodiode. The main exit beam is then directed into an optically absorbent medium to avoid secondary reflections. The output of the photodiode is proportional to the incident intensity of the scattered light. This, in turn, is directly related to the concentration of the liquid in the sample cell. The electronic circuitry eliminates any high frequency noise effects and produces compatible analog voltage levels for a 12-bit analog-to-digital converter. A/D output is interfaced via buffers to a personal computer. Suitable software enables to display the liquid concentrations.

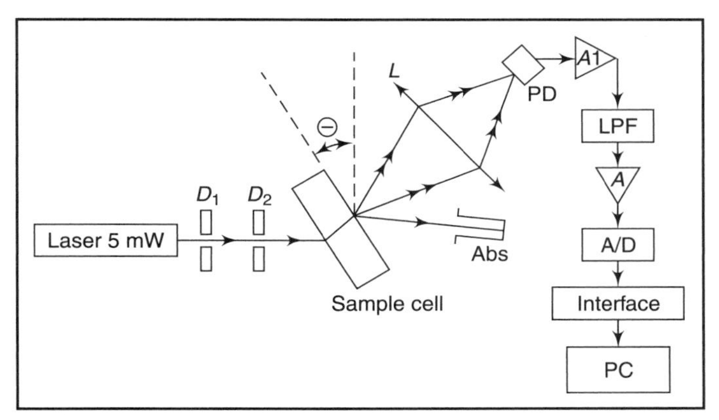

Figure 24.16 :: Block diagram of the nephelometer system (redrawn after Koyuncu, 1986)

The pollution of water is also determined to a large extent by the presence of anions, such as halides, nitrate, sulphate, cyanide, carbonate, etc., and cations such as lead, zinc, chromium, arsenic, copper etc. Inorganic anions are generally estimated by spectrophotometric and electrochemical techniques. With the availability of ion-selective electrodes, the task of estimating both anions and cations has become easy and at the same time, accurate and reliable. Ion-selective electrodes for anions such as fluoride, chloride, carbonate, cyanide, etc., and for cations such as sodium and potassium are now available. Polarography is useful in the analysis of metal ions in water, because of its high sensitivity, its ability to analyse mixtures and to tolerate large quantities of dissolved solids. Bacteriological and radioactive measurements are limited only to very special cases.

Table 24.2 gives the performance requirements for typical automatic water pollution monitoring instruments.

Table 24.2 :: Performance Requirements of Water Pollution Monitoring Instruments

Parameter	Transducer	Range	Full-scale accuracy
Conductivity	Potentiometric	0–60,000 mohs	1%
pH	Glass and calomel electrode	2–12	1%
Dissolved oxygen	Polarographic	0–24 mg/l	1%
Temperature	Silicon diode thermistor or thermocouple	0–50°C	0.5°C
Chloride	Ion-selective electrode	0–240 mg/l 0–2400 mg/l	5%
Turbidity	Optical	0–120 JTu 0–1200 JTu 0–2400 JTu	2%

The instrument must have automatic temperature compensations and should be stable for at least up to two weeks. It is not always possible to fix up the water sample from the site thoroughly and expect the results not to vary in the laboratory measurements. Therefore, portable instruments have been designed, which can be conveniently taken to the site and their results can be obtained almost immediately.

Chapter 25

Electronic Devices and Circuits

Analytical instruments are used to obtain qualitative and quantitative information about the composition of a sample of matter. For doing so, the chemical or physical quantity to be measured undergoes a series of transformations from the input to the output stage. Modern analytical instruments, depend to a large extent, upon the use of various electronic devices, components and circuits to carry out these transformations. In order to understand the principle of working of various analytical instruments, a good level of understanding is required about the various electronic devices and circuits which form the basic building constituents of these instruments.

25.1 Electronic Components

An electronic component is any device that handles electricity. These components come in many different shapes and sizes and perform different electrical functions depending upon the purpose for which they are used. Accordingly, electronic equipment make use of a variety of components.

25.1.1 Active vs Passive Components

There are broadly two types (Figure 25.1) of components: passive components and active components.

Passive Components: A passive device is one that contributes no power gain (amplification) to a circuit or system. It has no control action and does not require any input other than a signal to perform its function. They always have a gain less than one, thus, they cannot oscillate or amplify a signal. A combination of passive components can multiply a signal by values less than one; they can shift the phase of a signal, they can reject a signal because it is not made up of the correct frequencies, they can control complex circuits, but they cannot multiply by more than one because they basically lack gain. Passive devices include resistors, capacitors and inductors.

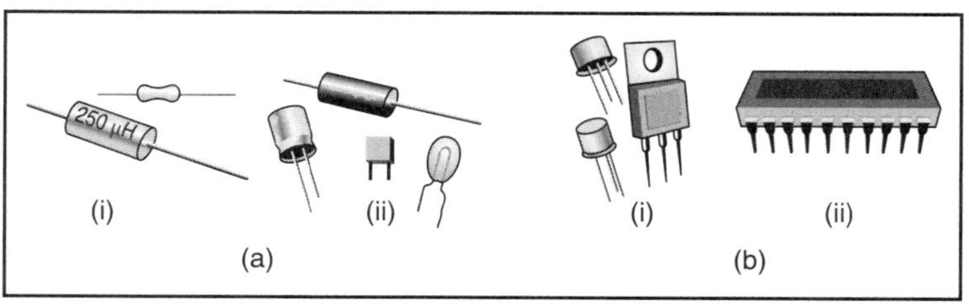

Figure 25.1 :: Types of components: (a) passive components: (i) resistors (ii) capacitors (b) active components: (i) transistors (ii) integrated circuits

Active Components: Active components are devices that are capable of controlling voltages or currents and can create a switching action in the circuit. They can amplify or interpret a signal. They include diodes, transistors and integrated circuits. They are usually semi-conductor devices.

25.1.2 Discrete vs Integrated Circuits

When a component is packaged with one or two functional elements, it is known as a 'discrete' component. For example, a resistor used to limit the current passing through it functions as a discrete component. On the other hand, an integrated circuit is a combination of several interconnected discrete components packaged in a single case to perform multiple functions. A typical example of an integrated circuit is that of a microprocessor which can be used for a variety of applications.

25.2 Passive Components

Resistance: Resistance is the opposition to the flow of current offered by a conductor, device or circuit. It is related to current as follows:

Resistance = voltage/current (Ohm's Law)

The resistance is expressed in ohms (abbreviated Ω). The value of resistance of a metal is temperature-dependent and increases with the rise in temperature. The most commonly encountered resistors in electronic circuitry are carbon resistors. Their values are either printed in numbers or put in the form of colour coded bands around the body. Each number from 0 to 9 has been assigned a colour according to the following table and Figure 25.2.

Colour coded bands

The first band closest to the end of the resistor represents the first digit of the resistance value, the second band gives the second digit and the third band gives the number of zeros to be added to the first two digits to get the total value of the resistor. If there is a fourth band, it indicates either ±5 per cent or a ±10 per cent tolerance. If it is absent, the tolerance is ±20 per cent.

Black	0	Blue	6
Brown	1	Violet	7
Red	2	Grey	8
Orange	3	White	9
Yellow	4	Gold	± 5% tolerance
Green	5	Silver	± 10% tolerance

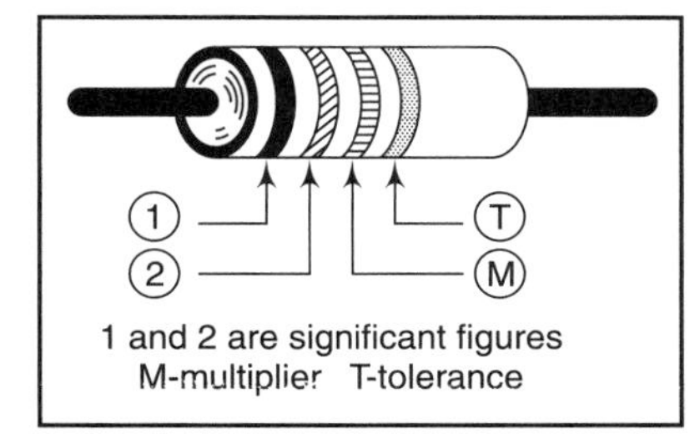

Figure 25.2 :: Colour code of resistors

High stability resistors are made by depositing a layer of cracked carbon on a ceramic rod and completely insulating the resistive element, so that it can withstand arduous operating conditions. Still superior in stability and for low-noise applications are metal oxide film resistors.

Variable resistance

Some resistors are made to give an adjustable or variable resistance. They are three-terminal devices (Figure 25.3), in which the central terminal is connected to the moveable contact or wiper. A variable resistance is obtained between the wiper arm and any one end of the resistor track. Such a device is called a rheostat.

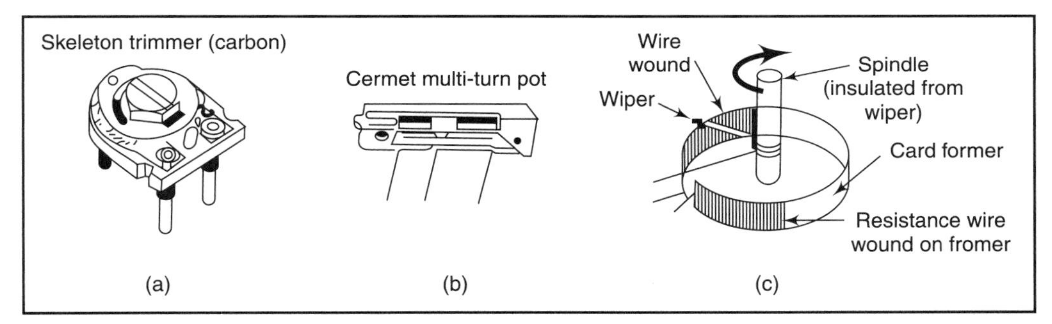

Figure 25.3 :: Types of variable resistors (a) carbon composition (b) multi-turn cermet (c) wire wound

If several resistances (R_1, R_2, R_3, R_4) are connected in series, then the total value (R_T) is given by:

$$R_T = R_1 + R_2 + R_3 + R_4.$$

When these resistances are connected in parallel, the value R_T is given by

$$\frac{1}{R_T} = \frac{1}{R_1} + \frac{1}{R_2} + \frac{1}{R_3} + \frac{1}{R_4}$$

Capacitor: A capacitor (also called a condenser) consists of two conductors separated by a dielectric or an insulator. The dielectric can be paper, mica, ceramic or plastic film or foil. High value capacitors are usually of electrolytic type. They are made of a metal foil with a surface that has an anodic formation of metal oxide film. The anodised foil is in an electrolytic solution. The oxide

Electrolytic capacitors

film is the dielectric between the metal and the solution. High value of capacity of electrolytic capacitors in a small space is due to the presence of a very thin dielectric layer. An electrical charge can be placed on the plates by connecting a voltage source to the capacitor. Figure 25.4 shows various types of capacitors used in electronic circuits.

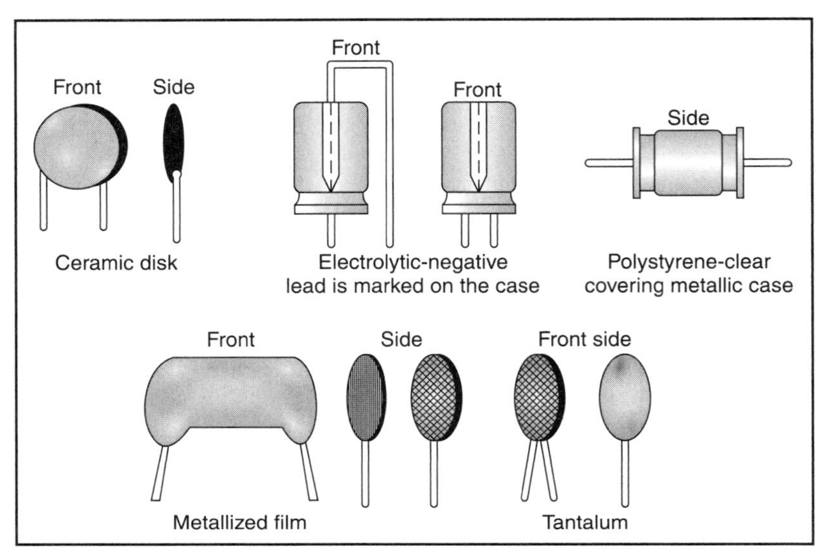

Figure 25.4 :: Various types of capacitors

Capacitance

Capacitance is measured in farads. A capacitor has a capacitance of one farad when one coulomb charges it to one volt. The farad is too large a unit. Usual sub-units used are microfarad (μF) and the picofarad (10^{-12} F).

Variable capacitance

Trimmer capacitor is a variable capacitor. Spacing between the metal plates which are separated by a dielectric can be adjusted to give a variable capacitance.

Inductance: Inductance is the characteristic of a device which resists a change in the current through the device. Inductors work on the principle that when a current flows in a foil of wire, magnetic field is produced, which collapses when the current is stopped. The collapsing magnetic field produces an electromotive force, which tries to maintain the current. When the coil current is switched, the induced emf would be produced in such a direction, as to oppose the build-up of the current.

$$e = -L\, di/dt$$

Henry

The unit of inductance is henry. An inductance of one henry will induce a counter emf of one volt, when the current through it is changing at the rate of one ampere per second. Inductances of several henries are used in power supplies as

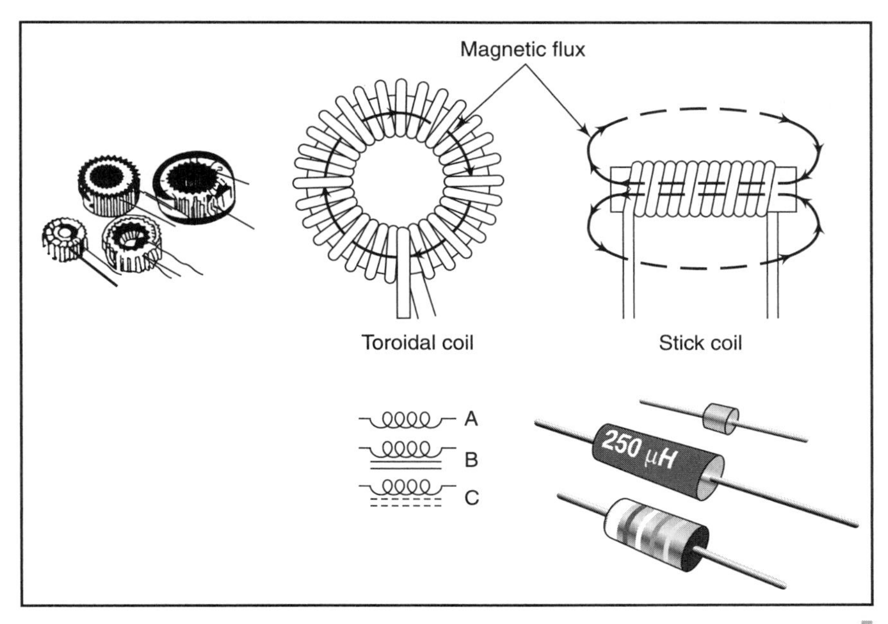

Figure 25.5 :: Shapes of commonly used inductors

smoothing chokes, whereas smaller values (in the milli or microhenry ranges) are used in audio and radiofrequency circuits. Figure 25.5 shows the shapes of commonly used inductors.

25.3 Semiconductor Devices

Unlike other electron devices, which depend for their functioning upon the flow of electrons through a vacuum or a gas, semiconductor devices make use of the flow of current in a solid. These solids, called semiconductors, have poorer conductivity than a conductor, but better conductivity than an insulator. These elements belong to the fourth group of the periodic table. The materials most often used in semiconductor devices are germanium and silicon. The atoms of these elements have a crystal structure and the outer or valence electrons of individual atoms are tightly bound to the electrons of adjacent atoms in electron pair bonds or the covalent bonds. In order to obtain free electrons from such structures, very small amounts of other elements having different atomic structure can be added as impurities. These impurities can thus modify and control the basic electrical characteristics of the semiconductor materials.

Electron pair

When the impurity atom has one more valence electron than the semiconductor atom, this extra electron cannot be bound in the electron pair and is held loosely by the atom. This free electron requires very little excitation to break away from the atom and the material becomes a

N-type material

better conductor. The presence of this extra free electron makes the material to have a negative charge and the resulting material is called N-type material.

A different effect is produced when the impurity added is the atom having one less valence electron than the semiconductor atom. This means that one of the bonds in the resulting crystal structure cannot he completed, because the impurity atom lacks one valence electron. A vacancy or hole thus exists in the lattice structure, which encourages the flow of electrons in the semiconductor material, consequently increasing its conductivity. The vacancy or hole in the crystal structure is considered to have a positive electrical charge, because it represents the absence of an electron. Semiconductor material which contains these holes or positive charges is called P-type material.

P-type material

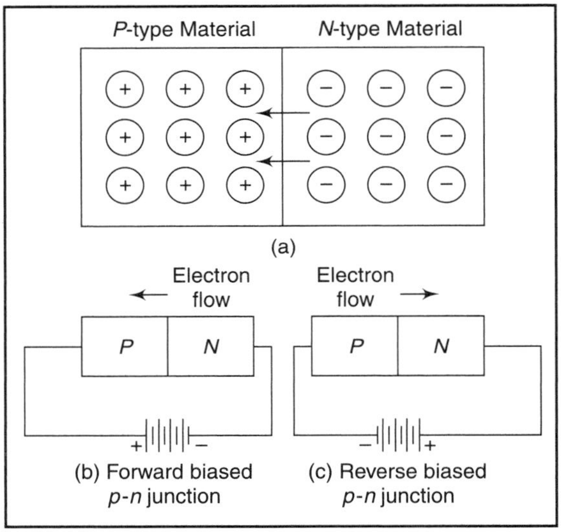

Semiconductor devices are considerably small in size and are light in weight. They have no filaments or heaters for electron emission and therefore require no heating power or warm-up time. They are solid in construction, extremely rugged and are free from microphonics and can therefore withstand severe environmental conditions. Because of these advantages, semiconductor devices have almost displaced vacuum tubes from many applications, in which it was formerly pre-eminent. The emphasis in the following text will therefore be on circuits making use of semiconductor devices in discrete and integrated forms.

P-type Material
N-type Material
(a)
Electron flow
Electron flow
P
N
P
N
(b) Forward biased *p-n* junction
(c) Reverse biased *p-n* junction

Figure 25.6 :: (a) P-N junction (b) Forward biased P-N junction (c) Reverse biased P-N junction

25.3.1 P–N Junction

When N-type and P-type materials are joined together (Figure 25.6 (a)), a very important phenomenon takes place at the junction of the two materials. The free electrons from the N-type material diffuse across the junction and recombine with holes in the P-type material. This is called diffusion current.

Diffusion current

Let us consider the condition when an external battery is connected across a P–N junction (Figure 25.6 (b)) with its positive terminal connected to the P-type material and the negative terminal to the N-type material. Under these circumstances, the junction is said to be forward-biased, as the polarity of the external battery facilitates the movement of the charge carriers across the junction and the junction offers a very low resistance to the current flow. The P-N junction under this condition is called forward-biased.

If the polarity of the externally applied battery is reversed [Figure 25.6 (c)], the charge carriers, under the influence of the external battery, move towards the adjacent sides of the junction and form a barrier layer, with the result that the electron flow in the external circuit is very low. The P-N junction under this condition is said to be reverse-biased.

25.3.2 Semiconductor Diode

The simplest type of semiconductor device is the diode, which is basically a P-N junction (Figure 25.7). The N-type material which serves as the negative electrode is called cathode, and the P-type material which serves as the positive electrode is referred to as the anode. The arrow symbol shows the direction of the flow of conventional current, which is opposite to the flow of electrons.

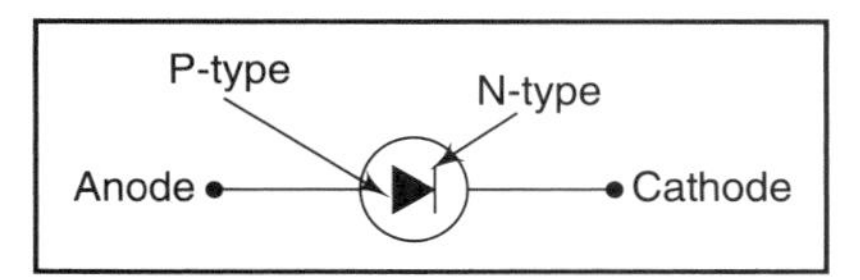

Figure 25.7 :: Semiconductor diode

The semiconductor diode conducts current more easily in one direction than in the other, so that it is an effective rectifying device. The semiconductor diodes are available in a wide range of current capabilities, suitable practically for almost all applications.

Conventional diodes normally show a low value of forward resistance and a very high value of reverse resistance. The variation in resistance is due to the non-linear voltage/current characteristics of the diode. Figure 25.8 the shows voltage–current diagram for a typical semiconductor diode.

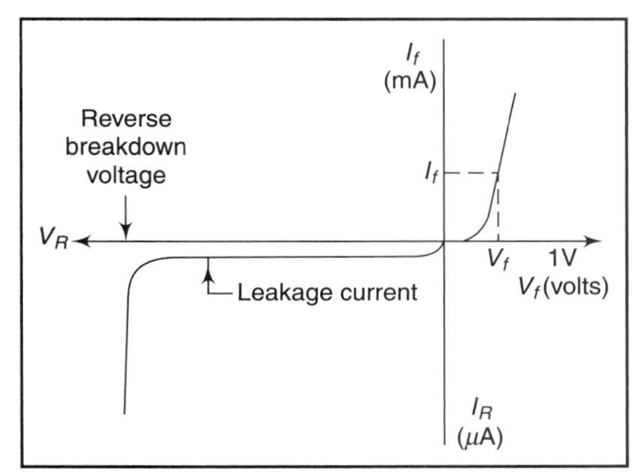

Figure 25.8 :: Voltage current characteristics of semiconductor diodes

Signal Diodes are general purpose diodes, which find applications involving low currents and a wide range of voltages, sometimes extending upto 50 kV.

Switching Diodes change their state from conducting to a one non-conducting and vice-versa in a very short time when the voltage is reversed. *Rectifiers* are similar to signal diodes, but are more suitable for high currents.

Low and medium power diodes are usually available in axial packages where as high power diodes are available in a large variety of packages of a vast range of shapes and sizes. Very high power diodes have a thread for mounting on to a PCB or a heat sink. Figure 25.9 shows various shapes of diodes, which are commercially available. Diode arrays or networks, containing up to 48 devices are also available in packages similar to integrated circuits.

A single diode when used for rectification gives half wave rectification. When four diodes are combined, full wave rectification can be obtained. Devices containing four diodes in one package

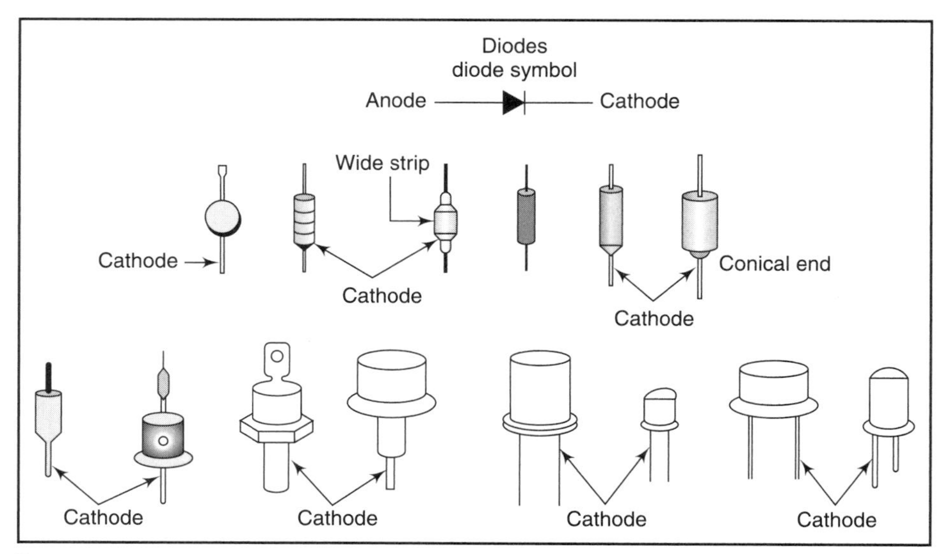

Figure 25.9 :: Various shapes of diodes

are called diode bridges. Diode bridges with large current capacities require a heat sink. Typically, they are screwed to a piece of metal or the chassis of equipment in which they are used. The heat sink allows the device to radiate excessive heat.

Zener Diode: A silicon diode has a very low reverse current, say 1 μA at an ambient temperature of 25°C. However, at some specific value of reverse voltage, a very rapid increase occurs in reverse current. This potential is called breakdown avalanche or the Zener voltage and may be as low as 1 volt or as high as several hundred volts, depending upon their construction.

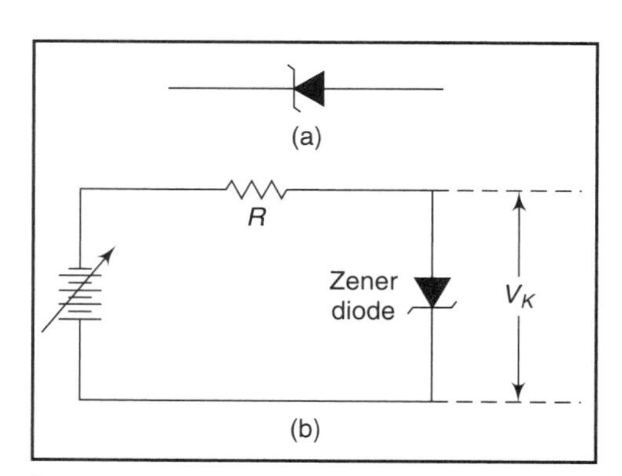

Figure 25.10 :: Zener diode (a) Symbol (b) Used as a constant voltage source

A Zener diode has very high resistance at bias potentials below the Zener voltage. This resistance could be several megohms. At Zener voltage, the Zener diode suddenly shows a very low resistance, say between 5 and 100 Ω.

A Zener diode behaves as a constant voltage source in the Zener region of operation, as its internal resistance is very low. The current through the Zener diode (Figure 25.10) is then limited only by the series resistance *R*. The value of series resistance is such that the maximum rated power rating of the Zener diode is not exceeded.

In order to aid in distinguishing the zener diode from a general purpose diode, the former is usually labelled with its specified breakdown voltage. Since this voltage is required in the circuit design, the value is generally indicated on the diode. For example, some common values are 6.8 V, 7.2 V, 9.6 V etc.

Varactor Diode: A varactor diode is a silicon diode that works as a variable capacitor in response to a range of reverse voltage values. Varactors are available with nominal capacitance values ranging from 1 to 500 pF, and with maximum rated operating voltages extending from 10 to 100 volts. They mostly find applications in automatic frequency control circuits. In a typical case, a varactor shows 10 pF capacitance at reverse voltage of 5 volts and 5 pF at 30 volts. Figure 25.11 shows different shapes of varactor diodes.

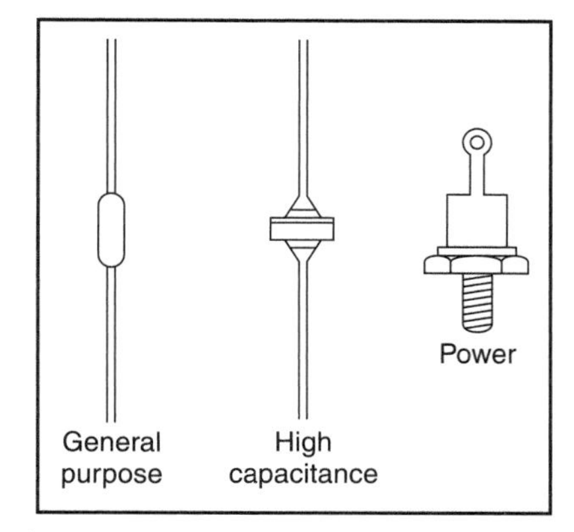

Figure 25.11 :: Varactor diode—different types

Varistor: A varistor is a semiconductor device that has a voltage-dependent non-linear resistance which drops as the applied voltage is increased. A forward biased germanium diode shows such type of characteristics and is often used in varistor applications, such as in bias-stabilization circuits.

Light Emitting Diodes (LEDs): A light emitting diode is basically a P-N junction that emits light when forward-biased. LEDs are available in various types (Figure 25.12) and mounted with various coloured lenses like red, yellow and green. They are mostly used in displays employing seven segments that are individually energized to form alphanumeric characters.

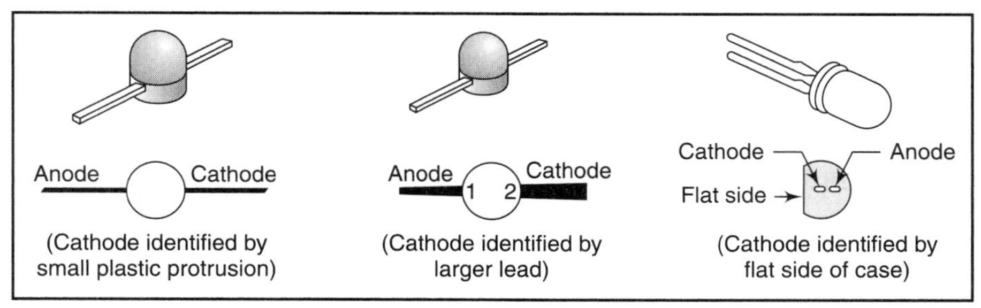

Figure 25.12 :: Light emitting diodes

LED displays are encountered in test equipment, calculators and digital thermometers whereas LED arrays are used for specific applications such as light sources, punched tape readers, position readers, etc.

Electrically, LEDs behave like ordinary diodes except that their forward voltage drop is higher. For example, the typical values are; IR (infrared): 1.2 V, Red: 1.85 V, Yellow: 2 V, Green: 2.15 V. Further, the actual voltages may vary depending upon the actual technology used in the LED.

Photodiode: A photodiode is a solid state device, similar to a conventional diode, except that when light falls on it (P-N junction), it causes the device to conduct. It is practically an open circuit in darkness, but conducts a substantial amount of current when exposed to light.

25.4 Transistors

25.4.1 Bipolar Transistors

The most commonly used semiconductor device is the transistor having characteristic to control voltage and current gain in electronic circuit. These properties enable the transistor to provide amplification, rectification, switching, detection and almost any desired function in the circuit. It is the basic device of all solid state electronics, both as a single component or as an element of integrated circuit.

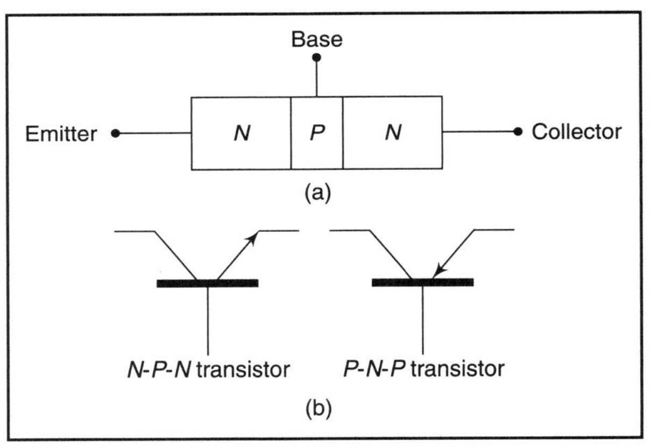

Figure 25.13 :: (a) Functional diagram of a transistor, and (b) Symbols used for transistors

A transistor is a three-terminal device. The terminals are called base (*B*), collector (*C*) and emitter (*E*). Basically, it is made up of two diodes: a base–emitter diode and a base–collector diode. In normal amplifier operation, the base–emitter diode is forward-biased and the base–collector diode is reverse-biased.

N-P-N transistor

P-N-P transistor

Depending upon the direction of current flow, different symbols (Figure 25.13) are used for N-P-N and P-N-P transistors. The first two letters of the N-P-N and P-N-P designations indicate the respective polarities of the voltages applied to the emitter and the collector in normal operation. In an N-P-N transistor, the emitter is made negative with respect to both the collector and the base, and the collector is made positive with respect to both the emitter and the base. In a P-N-P transistor, the emitter is made positive with respect to both the collector and the base, and the collector is made negative with respect to both emitter and the base.

Transistor packages

More than 500 transistor packages are listed in the component manufacturers' catalogues. However, only about 100 types are in common use. Metallic packages (TO-3, TO-5 and TO-18) have been in use for a long time. However, they have been mostly replaced in low and medium power applications by cheap plastic packages due to their low cost. For high power applications, however, metallic packages, both stud or bolt type are still common, although flat type packages are being replaced by plastic versions, with a metallic tab to improve heat dissipation. Figure 25.14 shows commonly used transistor packages and their terminals.

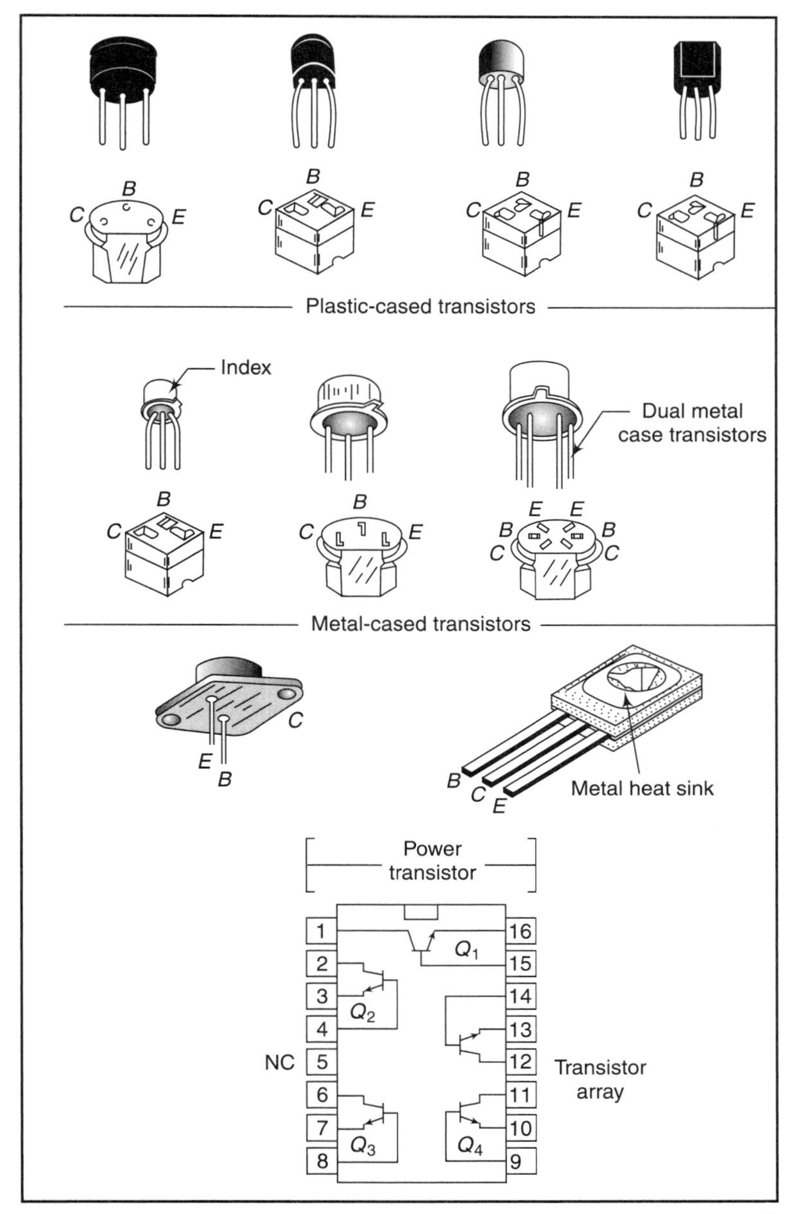

Figure 25.14 :: Transistor symbols and terminals

Figure 25.15 indicates the bias polarity required to forward-bias the base-emitter junction. The arrow head distinguishes between the emitter and the collector and shows the direction of a 'conventional' current flow. Electron flow is opposite from the direction of the arrow points. The

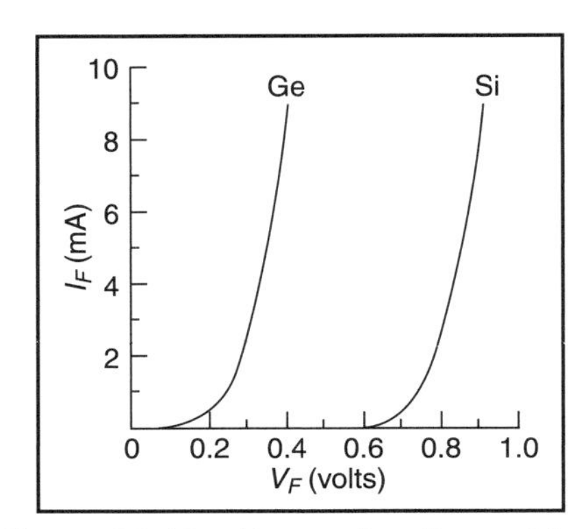

Figure 25.15 :: Forward voltage of base-emitter junction in Ge and Si transistors

Figure also compares the biasing required to cause conduction and cut-off in N-P-N and P-N-P transistors. If the transistor's base-emitter junction is forward-biased, the transistor conducts. However, if the base-emitter junction is reverse-biased, the transistor is cut-off.

The voltage drop across a forward-biased emitter-base junction varies with the transistor's collector current. Figure 25.16 shows the relationship between voltage and current for base-emitter junction in silicon transistors.

A common problem in transistors is the leakage, which can shunt signals or change bias voltages, thereby upsetting circuit operation. This problem is

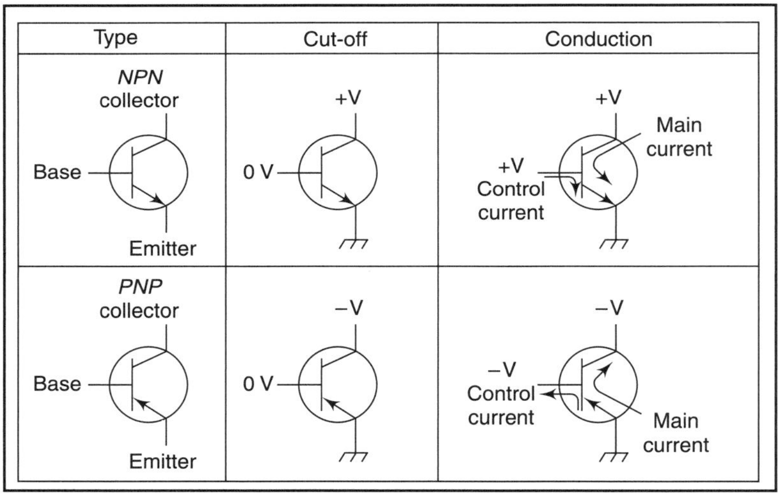

Type	Cut-off	Conduction
NPN collector, Base, Emitter	0 V, +V	+V, +V Control current, Main current
PNP collector, Base, Emitter	0 V, −V	−V, −V Control current, Main current

Figure 25.16 :: Three basic biasing arrangements in transistors

Leakage current

particularly serious in direct-coupled or high frequency stages. Leakage current is the reverse current that flows in a junction of a transistor when specified voltage is applied across it, the third terminal being left open. For example, I_{CEO} is the dc collector current that flows when a specified voltage is applied from collector to emitter, the base being left open (unconnected). The polarity of the applied voltage is such that the collector-base junction is reverse-based. Obviously, in a transistor, six leakage paths are present (with the third electrode open), as shown in Figure 25.17. Leakage current increases with temperature and doubles about every 10°C.

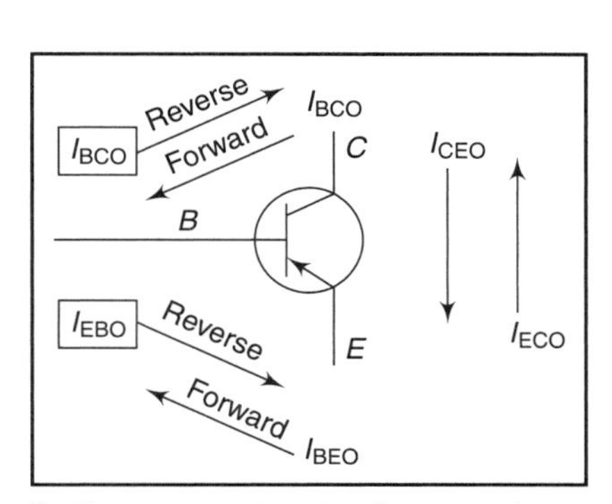

Figure 25.17 :: Leakage paths in a transistor

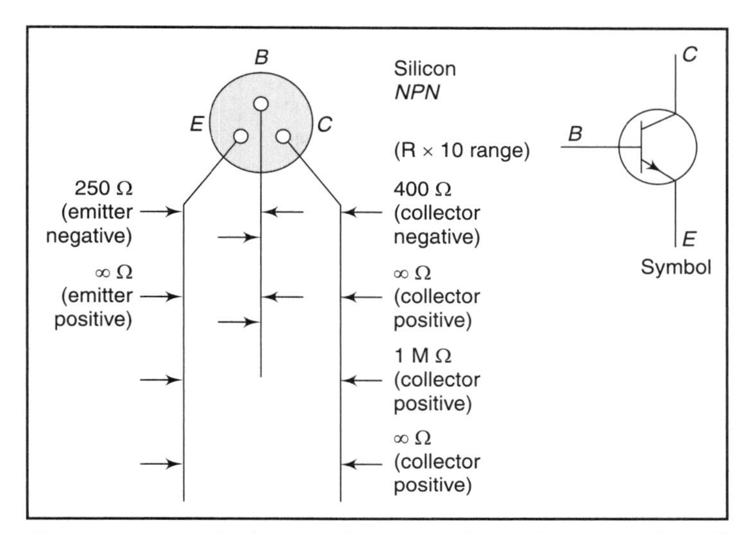

Figure 25.18 :: Bipolar transistor junction resistance values

Figure 25.18 shows typical bipolar transistor junction resistance readings. The polarity of the ohmmeter to be applied on the various transistor leads is also indicated in the Figure.

Power Transistors: The junctions of the power transistors have comparatively larger areas than small signal transistors and have the following characteristics:

- Forward resistance values are generally lower than those for small signal silicon transistors.
- Similarly, they have lower reverse resistance values. The test results with an ohmmeter on a silicon power transistor are shown in Figure 25.19.

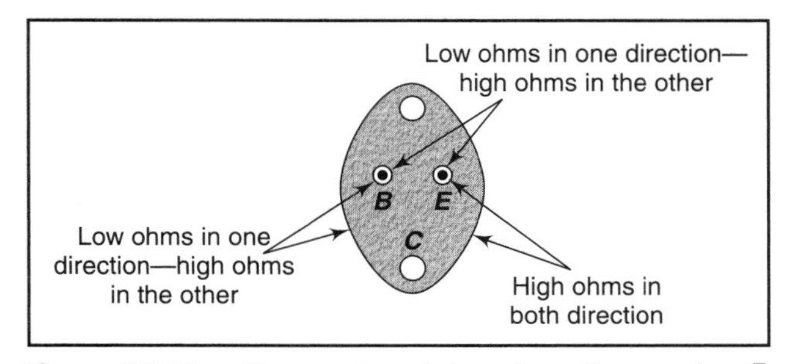

Figure 25.19 :: Power transistor: junction resistance values

Power transistors are usually mounted on the heat sinks or heat radiators (Figure 25.20). They are sometimes mounted on the chassis using silicone grease to increase heat transfer.

Darlington Transistors: A Darlington is a special type of configuration usually consisting of two transistors fabricated on the same chip or at least mounted in the same package. Darlington pairs are often used as amplifiers in input circuits to provide a high input impedance. Darlingtons are used where drive is limited and high gain typically over 1000 is needed. In this configuration,

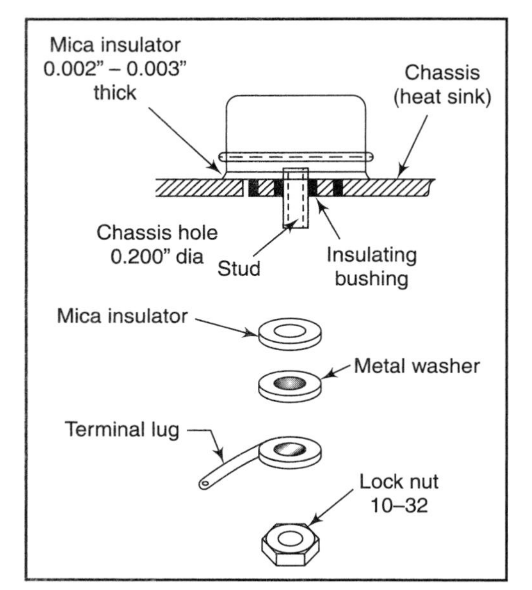

Figure 25.20 :: Power transistor: mounting arrangement

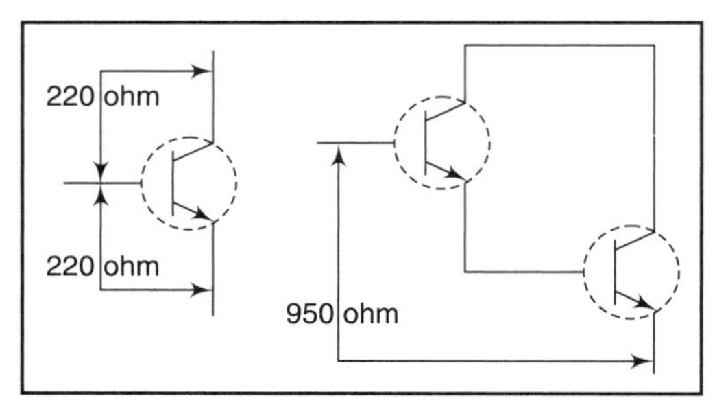

Figure 25.21 :: Darlington pair: forward resistance values

(Figure 25.21) the emitter base junctions are connected in series and the collector terminals are connected in parallel. A Darlington configuration behaves like a single transistor where the current gains (h_{fe}) of the individual transistors it is composed of are multiplied together and the base-emitter voltage drops of the individual transistors are added together.

25.4.2 Field-Effect Transistor

The field-effect transistor is basically a three-terminal semiconductor device. It is a voltage-operated device and is biased by a voltage. No input current normally flows and hence its input resistance is virtually infinite.

Field-effect transistors are of two types: the junction FET (JFET) and the metal-oxide semiconductor FET (MOSFET). The MOSFET is also called insulated-gate FET (IGFET).

In the JFET, a thin conducting channel (Figure 25.22) of finite conductance is established between two P-N junctions. The current from the source to the drain, for a given voltage, depends on the dimensions of the channel. If the P-N junctions are reverse biased by applying a voltage to the gate, a depletion region containing no mobile carriers is formed, and the width of the conducting channel is reduced. Thus, the magnitude of current between source and the drain can be controlled by the reverse bias applied to the gate electrode. This provides a means of controlling the amplification in the FET.

Depending upon the type of material of the channel, the FET may be N-channel or P-channel.

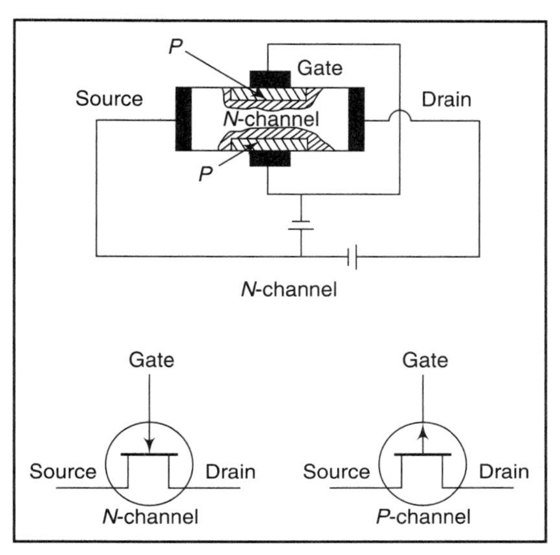

Figure 25.22 :: Field-effect transistor

25.4.3 MOSFET

In the metal-oxide semiconductor FET (MOSFET), a thin layer of silicon oxide insulates the gate contact from the channel. There are two types of MOSFET: depletion (normally on at zero bias) and enhancement (normally off at zero bias). Figure 25.23 shows depletion type *N* and *P* channels and enhancement type *N* and *P* channels. Also there is a fourth lead. This is the substrate or body of the MOSFET and is usually connected to the source or ground. The substrate is not always shown in the illustration of a MOSFET and for all intents and purposes, can be disregarded in considering the FET operation. We can identify depletion MOSFET by the solid channel and enhancement MOSFET by the broken channel.

The P-channel enhancement type transistor is mostly used. The MOSFET exhibits an extremely high input resistance, which may be in the range of 10^{12} – 10^{15} Ω. Unlike the junction FET, the MOS maintains a high input resistance without regard to the magnitude or polarity of the input gate voltage.

The thickness of gate insulating material in a MOSFET is typically below 2×10^{-5} cm. Consequently, the gate to substrate voltage of the order of 50 V or so will cause breakdown of the insulation and the MOSFET will be ruined. Any static charge on a person's finger or some tool can destroy a MOSFET. To prevent destruction, most MOSFETS are supplied by the manufacturers with their leads twisted, so that the gate is shorted to the substrate. Shorting leads should be clipped to the gate when soldering or installing MOSFET.

As in the bipolar transistor, the MOSFET is capable of both switching and amplification. The MOSFET gain is measured in terms of a voltage ratio while the bipolar gain is represented as a current ratio. While the bipolar transistors have a much higher speed than the MOS transistors, the MOS transistors offer a packing density roughly four times, as compared to bipolar transistors due to the difference in their fabrication techniques.

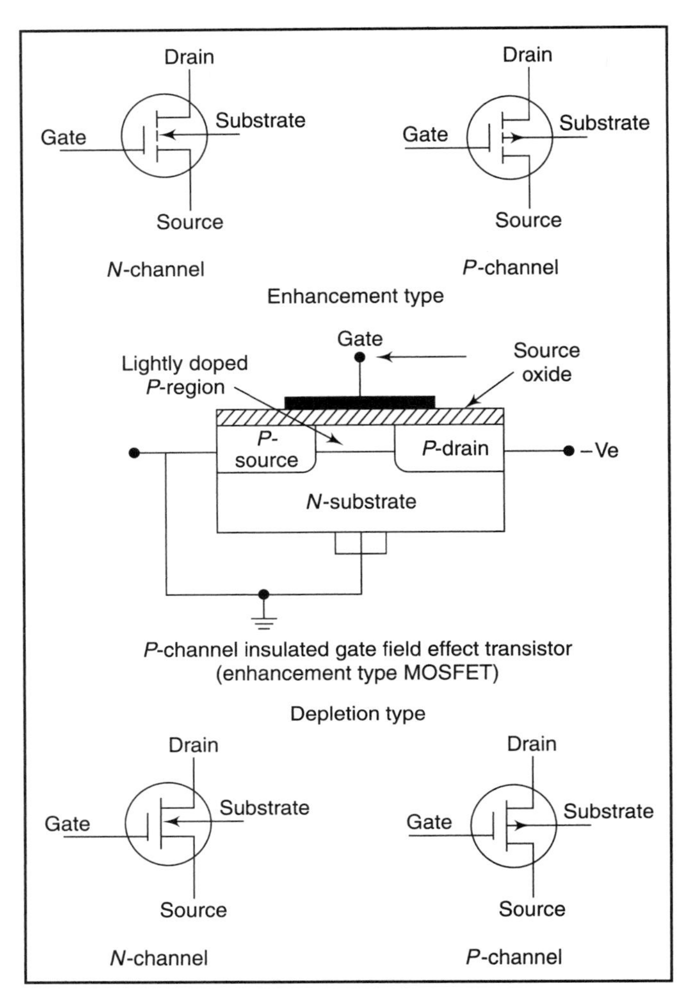

Figure 25.23 :: MOSFET (metal-oxide insulated gate field-effect transistor)

25.5 Integrated Circuits

Integrated circuits eliminate the use of individual electronic parts such as resistors, capacitors and transistors, etc., as the building blocks of electronic circuits. In their place, we have tiny chips of semiconductor material, whose functions are not those of single components but of scores of transistors, resistors and capacitors and other electronic elements, which are all interconnected to perform the function of a complete and complex circuit. So, instead of usual semiconductor devices, resistances and capacitances, complete multistage amplifiers, complex flip-flops and dozens of other functional circuits have become the basic components of electronic equipment.

Integrated circuits are placed in a protective housing and connections to the outside world are provided through pins. The common packages are the TO-5 metal can, the ceramic flat pack and the dual-in-line ceramic or plastic flat packs known as dual-in-line packages (DIPs). The popular packages have 14, 16, 18, 24 or 40 connecting pins. Figure 25.24 shows standard IC packages.

Common packages

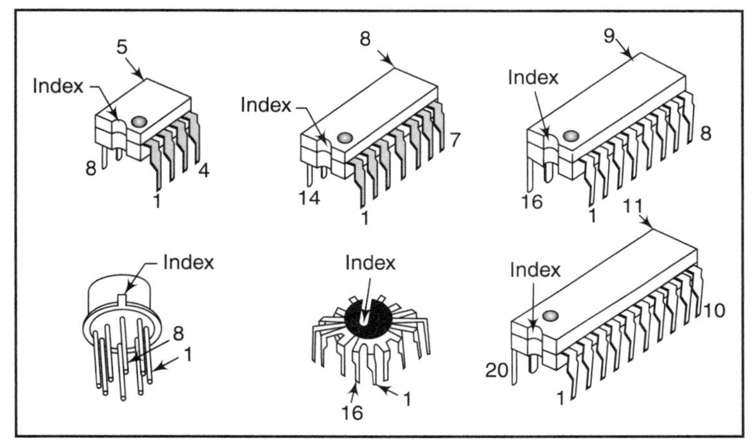

Figure 25.24 :: Commonly available IC packages

Circuits requiring close matching and tracking of components can be more advantageously adapted for integrated circuits. Also, it is less expensive to fabricate integrated transistors than passive elements, such as resistors and capacitors. Large values of resistances and capacitors are to be avoided, as they are not economical. Also, the circuit design has to be such that it should be non-critical as to component tolerances. For example, 20% tolerances of resistances and capacitances shall have much lower cost than 10% tolerance values. Another severe limitation of integrated circuit technology for linear applications is the non-availability of inductances as integrative circuit elements. So, either the inductances are usually eliminated while designing a particular integrated circuit, or hybrid circuit would have to be used. In digital electronics, standard circuits are available for virtually every application.

Linear integrated circuits are characterized by an output that is proportional to its inputs. Such circuits are designed as dc amplifiers, audio amplifiers, rf amplifiers, if amplifiers, power amplifiers, differential amplifiers etc.

An important class of linear integrated circuits is operational amplifiers (op-amps). These amplifiers were originally utilized in analog computers to perform various mathematical operations such as addition, subtraction, integration and differentiation. Op-amps are now used to perform precise circuit functions, some of which will be discussed in this chapter.

Operational amplifiers

25.6 Operational Amplifiers

An operational amplifier is a high gain dc amplifier originally intended to be used for mathematical operations. But the versatility of this amplifier, owing to the provision for external selectable feedback, has made it possible for adaption to many applications in the field of analytical instruments. It is used for the construction of ac and dc amplifiers, active filters, phase inverters, multivibrators, comparators, etc. The operational amplifiers are available in the integrated form and thus simplify the design of equipment by offering a high quality amplifier in one package and result in considerable size reduction. They are popularly known as op-amps.

Block diagram of a typical operational amplifier is shown in Figure 25.25. The input stage is a conventional differential amplifier with a constant current source placed in the emitters of the two transistors. It is desirable to have a high gain in this section, so that any imbalances or imperfections in the succeeding stages have little or no effect on the output signal.

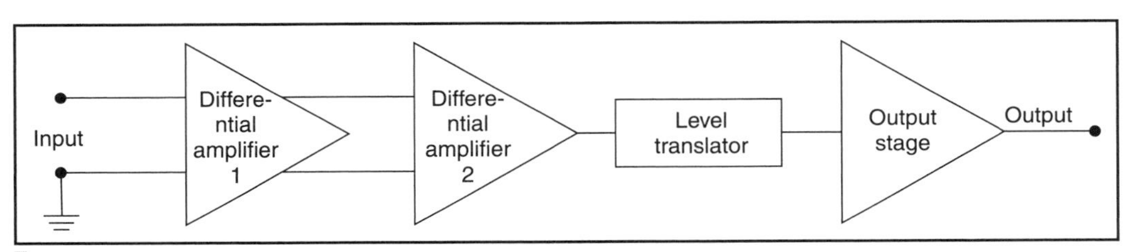

Figure 25.25 :: Block diagram of an operational amplifier

The signal output from the first stage is fed differentially to the second stage differential amplifier. Because common mode rejection capability is not as stringent in this stage, this stage does not require a constant current source in the emitters. Normally, the second stage is needed only to provide some additional gain. Its input resistance should be high enough to prevent excessive loading of the first stage. Therefore, an emitter follower circuit is often employed.

The output of the second stage is normally taken to be single ended. Here, the amplified signal is associated with a certain amount of dc voltage at its output. Some means of level translation is thus necessary between the second and the final stages. By eliminating the dc level at the final stage, the output voltage will vary about a zero reference level, thus preventing any undesired dc current in the load and also enhancing the permissible output voltage swing.

25.6.1 Symbolic Representation

The op-amp is symbolically represented as a triangle (Figure 25.26) on its side. In digital circuit symbols, the inverter is represented as a triangle, but the op-amp symbol is much larger. The triangle indicates the direction of signal flow. It is associated with three horizontal lines, two of which (*A* and *B*) indicate signal input and the third (*C*), the output signal connections.

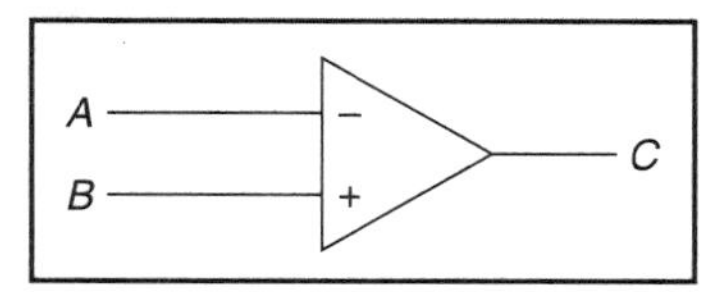

Figure 25.26 :: Symbolic representation of an op-amp

The signal input terminals are described by minus (–) and plus (+) signs inside the triangle. The (–) input is called the 'Inverting input', because the output voltage is 180° out of phase with the voltage to this input. On the other hand, the (+) input is called the non-inverting input, because the output voltage is in phase with a voltage applied to this terminal. The names inverting and non-inverting terminals have been given to indicate the phase of output signal, in comparison to the voltage applied at the inputs. Figure 25.27 shows the operation of the op-amp as inverting and non-inverting amplifier.

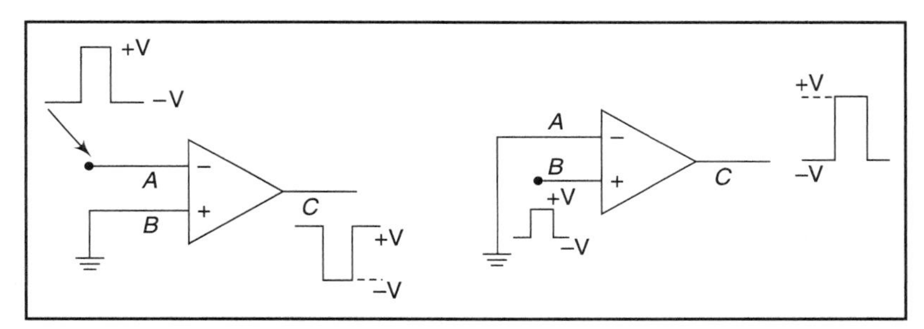

Figure 25.27 :: (A) Inverting operation of an op-amp (B) Non-inverting operation of an op-amp

25.6.2 Power Supply Requirements for Op-Amps

Op-amps need to be powered by dc power supply, like any other transistor amplifier. The power supply should be of proper voltage regulation and filtering for correct operation of the op-amp.

The power supply leads on the op-amp are usually marked as +V and –V, to which the positive and negative supply voltages should respectively be connected, with reference to ground. The positive and negative supply voltages are usually symmetrical, i.e. the two voltages are equal but opposite in sign. The most commonly used voltage to power op-amps are +15 V and –15 V. However, this is not always the case. Therefore, it is advisable to consult manufacturer's data manuals on the op-amp of interest to determine the power supply requirements.

It may be noted that usually the power supply leads are not shown on the op-amps in the circuit schematics. It is assumed that the reader is aware that dc voltage is necessary for operation of the op-amp.

25.6.3 Output Voltage Swing

Just as standard transistor amplifiers are limited in their output voltage swing, so also are the op-amps. The limitations on the output voltage swing are generally dependent upon the magnitude of the positive and negative power supply voltages of the op-amp. Usually, the output of an op-amp can go to a voltage value no more positive than the positive power supply and no more negative than the negative power supply can provide. When the output of an op-amp goes to +V and –V, it is said to be in positive or negative voltage saturation.

25.6.4 Output Current

Op-amps are designed to provide a limited current in the output, which is usually a few milliamperes, less than 10 mA in most standard op-amps. If more current is drawn, the output signal begins to change, because of the current limiting provisions built into the op-amp output circuit, which limit their own output current to a safe operating region.

Some op-amps are designed to deliver larger currents, of the order of amperes, at the output pin, but they can only be characterized as special devices and not standard op-amps.

25.6.5 Characteristics of Op-Amps

An ideal op-amp would have the following characteristics:

1. Infinite open-loop voltage gain (A),
2. Infinite bandwidth,
3. Infinite input impedance,
4. Zero output impedance,
5. Zero offset (voltage at output when input is zero), and
6. Maximum output voltage equals +V and minimum –V.

By carefully examining these characteristics, the following implications are obvious:

- The voltage across the input terminals of an amplifier with infinite gain must be zero or negligibly small (input voltage = output voltage/gain of amplifier)
- No (zero) current can flow between the input terminals of the amplifier, because of the infinite input impedance (input current = input voltage/input impedance). If one input terminal is at ground reference potential, then the other must be at ground potential. The terminal is thus called a virtual ground and the input is called virtual short-circuit (in virtual short-circuit, no potential can exist across it, and no current can flow through it)
- With the infinite voltage gain, we can expect to get a very very large voltage output from a very small voltage input. In fact, with a small voltage across the input terminals, the amplifier output is driven into positive or negative saturation very easily. An op-amp is said to be operating properly when the output is in the linear region of operation, unless the op-amp circuit is designed to operate to perform non-linear function.

25.6.6 Performance Characteristics of Op-Amps

The performance characteristics of an operational amplifier are defined as:

- *Gain*: The voltage gain of an amplifier consisting of n similar stages is equal to the gain of a single stage raised to the power n.
- *Linearity:* It defines the relationship between input and output. It is generally desirable that a plot of output signal against input signal level shows a linear relationship that passes through zero on both axes.
- *Offset*: If the plot between input and output does not pass through the origin, the amplifier is said to exhibit offset.
- *Noise:* Noise is a spurious signal that is superimposed on the amplified signal. The noise signal generally has components at all frequencies within the bandpass of the amplifier.
- *Drift*: Very low frequency noise is termed as drift.
- *Reliability*: The ability of an amplifier to provide a reproducible output for a given input over an extended period of time is called reliability.
- *Stability*: It refers to the ability of the amplifier to avoid uncontrolled oscillation.

It is obvious that in order to provide accurate performance, an amplifier must be linear, reliable and free from offset. However, in practice, the components used in amplifiers are sensitive to environmental changes, especially temperature, and are affected by aging. To achieve desirable standards of accuracy, the design of the instrument is made in such a way that is not critically dependent as a whole, on individual amplifier behaviour.

25.6.7 Typical Op-Amp Circuits

Typically, an operational amplifier has two input terminals and one output terminal. It is basically a differential amplifier and is normally used with external feedback networks that determine the function performed. Several examples of the use of operational amplifiers appear at different places in the text. Some of the basic application circuits are given below:

25.6.7.1 Basic Inverting Circuits

Many circuits require a signal to be inverted as well as amplified. Op-amps are easily configured for such negative gain. In the circuit of Figure 25.28, the input signal is applied to the inverting (negative) terminal and the non-inverting (positive) terminal is grounded. The input voltage E_1 is applied in series with input resistance R_1. The feedback resistance R_2 is connected between the input and the output terminals. At point 1, the input current is equal to the feedback current.

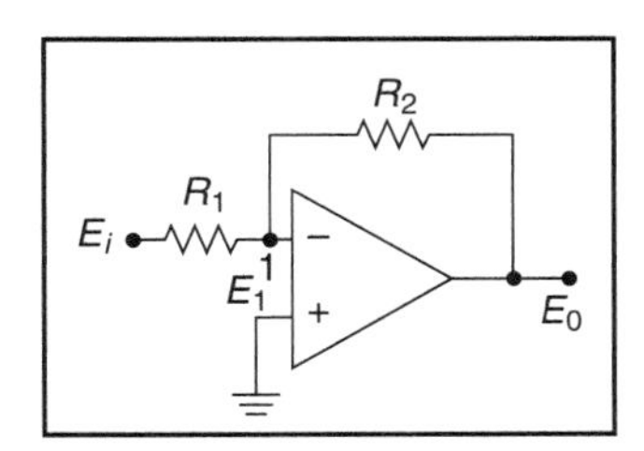

Figure 25.28 :: Basic inverting circuit

Therefore, we can write:

$$\frac{E_1 - E_i}{R_1} + \frac{E_0 \sim E_i}{R_2} = 0$$

If the voltage gain of the amplifier is A, then $E_0/E_i = A$

$$E_i = E_0/A$$

$$E_1/R_1 - \frac{E_0/A}{1} \cdot \frac{1/R_1}{2} + E_0/R_2 - E_0/A \cdot 1/R_2 = 0$$

$$E_0 = \frac{E_1(R_2/R_1)}{1-[1/A][1+R_2/R_1]}$$

Since $A \gg (1 + R_2/R_1)$

$$E_0 = -R2/R_1 \cdot E_1$$

Hence $E_0/E_1 = -R_2/R_1$

This shows that the gain of the circuit is independent of the voltage gain of the amplifier. Also, the gain is unaffected by changes in temperature, device parameters or frequency, as the gain depends primarily upon the components in the external feedback circuit.

25.6.7.2 Summing Circuit

A useful and practical extension to the simple inverter circuit is obtained by providing a number of inputs, so that the voltage signals can be added to each other. The various input signals can be summed up in different proportions by suitably adjusting the values of the input resistors. Referring to Figure 25.29,

$$E_0 = R_F[E_1/R_1 + E_2/R_2 + E_3/R_3]$$

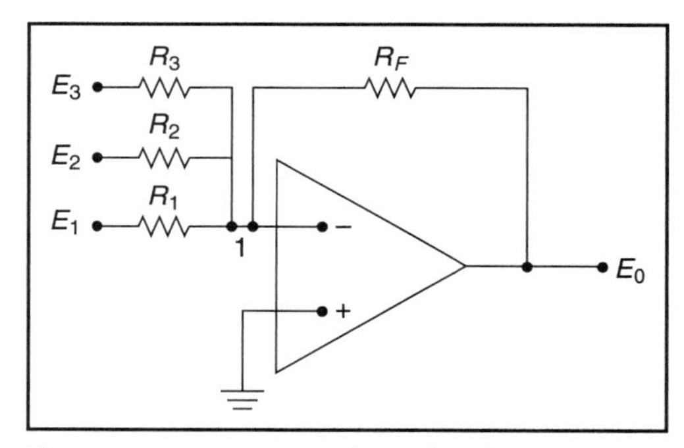

Figure 25.29 :: Summing circuit

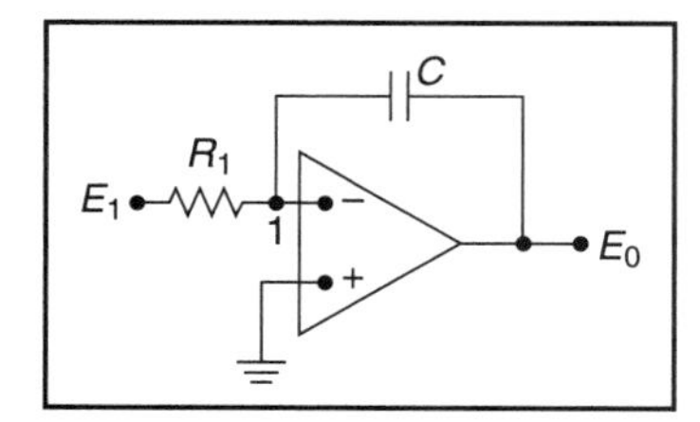

Figure 25.30 :: Integrating circuit

25.6.7.3 Integrating Circuit

Figure 25.30 shows a simple running integrator (no reset or hold logic) that can be used within a stable feedback loop. Here the feedback path is provided by a capacitor. Hence, sum of the currents at point 1 is given by:

$$E_1/R_1 + C\,dE_0/dt = 0$$

Integrating with respect to time

$$E_0 = -1/R_1C\int E_1\,dt$$

Integrating circuits are required for modelling dynamic systems and solving differential equations.

25.6.7.4 Differentiating Circuit

The circuit shown in Figure 25.31 is arranged to generate an output voltage which is proportional to the differential with respect to time of the input voltage. The circuit can be analyzed by considering sum of the currents at point 1:

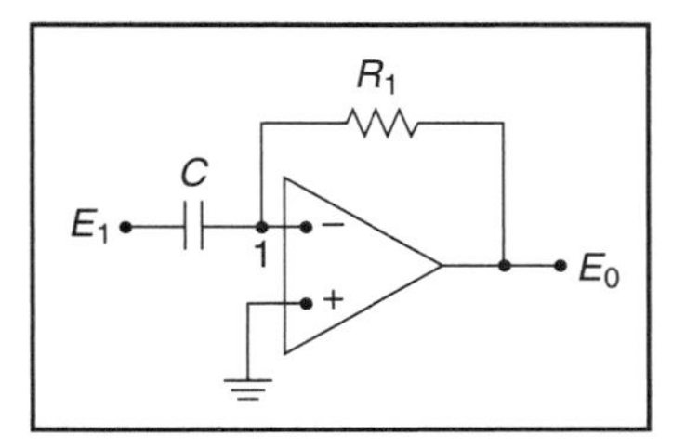

Figure 25.31 :: Differentiating circuit

$$C\cdot\frac{dE_1}{dt}+\frac{E_0}{R}=0$$

$$E_0=-RC\cdot\frac{dE_1}{dt}$$

This shows that the output voltage is proportional to the derivative of the input. Differentiating circuits are susceptible to noise and instability and are, to be used with care. They are usually followed by a filter to limit the effective bandwidth.

25.6.7.5 Voltage Follower

If the output of the operational amplifier is connected back to the input at the inverting terminal and, while the input is given at the non-inverting terminal, the circuit functions as a voltage follower (Figure 25.32). In this case, the voltage gain of the amplifier is almost unity and the output voltage follows changes in the input voltage. The circuit provides a very high input impedance and low output impedance. Therefore, this configuration is ideal for isolating and driving other circuits.

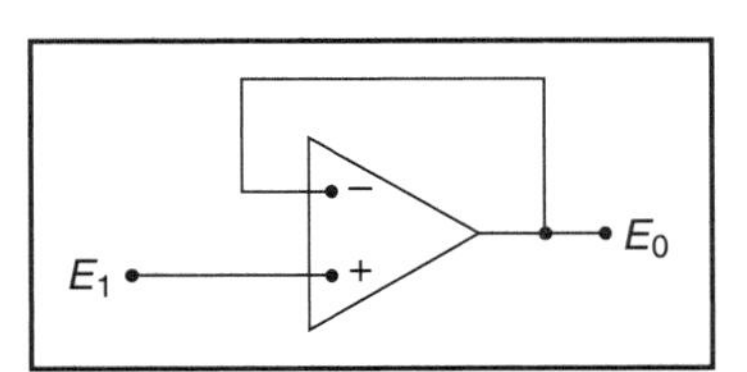

Figure 25.32 :: Voltage followers

25.6.7.6 Unity Gain Inverter

A unity gain voltage inverter is formed by using identical resistances in the input and feedback paths (Figure 25.33). Inverters are used wherever sign changes are necessary. It can also be used simply to lower the impedance level or raise the power level of a signal. The circuit is capable of functioning well over a wide range of signal levels, frequencies and impedances.

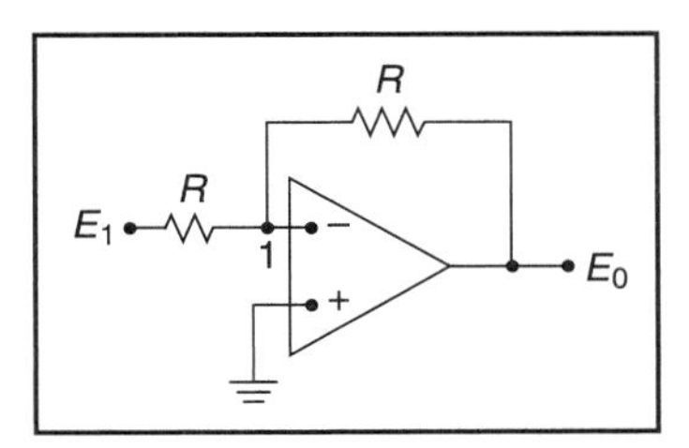

Figure 25.33 :: Unity gain voltage inverter

25.6.7.7 Integrator

The integrator is shown in Figure 25.34 and performs the mathematical operation of integration. This circuit is essentially a low-pass filter with a frequency response decreasing at 6 dB per octave. An amplitude-frequency plot is shown in Figure 25.35.

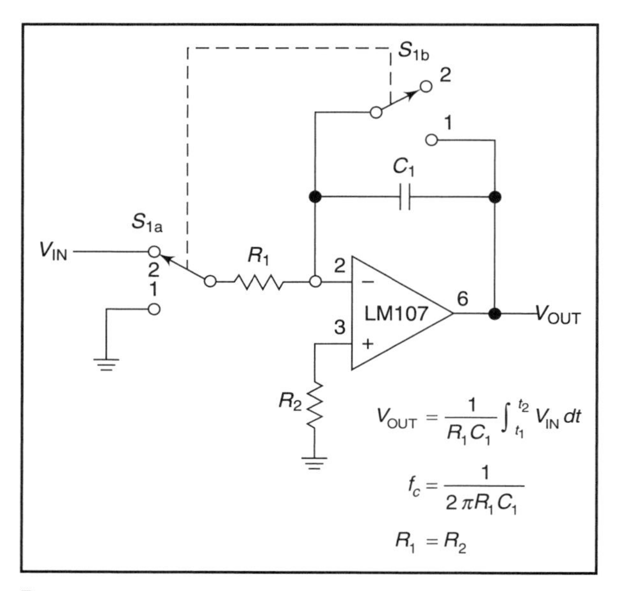

Figure 25.34 :: Integrator

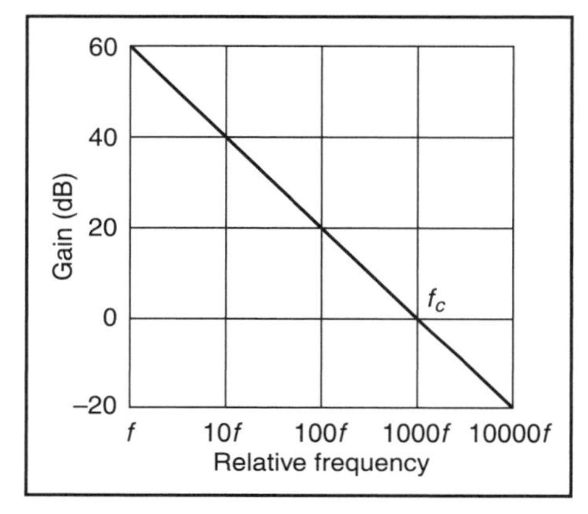

Figure 25.35 :: Integrator frequency response

The circuit must be provided with an external method of establishing the initial conditions. This is shown in the figure as S_1. When S_1 is in position 1, the amplifier is connected in unity-gain and capacitor C_1 is discharged, setting an initial condition of zero volts. When S_1 is in position 2, the amplifier is connected as an integrator and its output will change in accordance with a constant times the time integral of the input voltage.

For integration operation, the amplifier used should generally be stabilized for unity–gain operation and R_2 must equal R_1 for minimum error due to bias current.

25.6.7.8 Current to Voltage Converter

Some transducers produce an output voltage, others an output current. Current transducers include photodiodes, some temperature sensors and a variety of biological probes. Often the currents preduced are very small of the order of nanoamps or less. Such currents require amplification before they can be used in a system and the first stage of such amplification is usually a current-to-voltage converter.

Current may be measured in two ways with an operational amplifier. The current may be converted into a voltage with a resistor and then amplified or the current may be injected directly into a summing node. Converting into voltage is undesirable for two reasons: first, an impedance is inserted into the measuring line thereby causing an error; second, the amplifier offset voltage is also amplified with a subsequent loss of accuracy. The use of a current-to-voltage converter, using an operational amplifier and a feedback resistor prevents the occurrence of both these problems.

The current-to-voltage converter is shown in Figure 25.36. The input current is fed directly into the summing node and the amplifier output voltage changes to extract the same current from the summing node through R_1. The scale factor of this circuit is R_1 volts per amp. The only conversion error in this circuit is I_{bias} which is summed algebraically with I_{IN}.

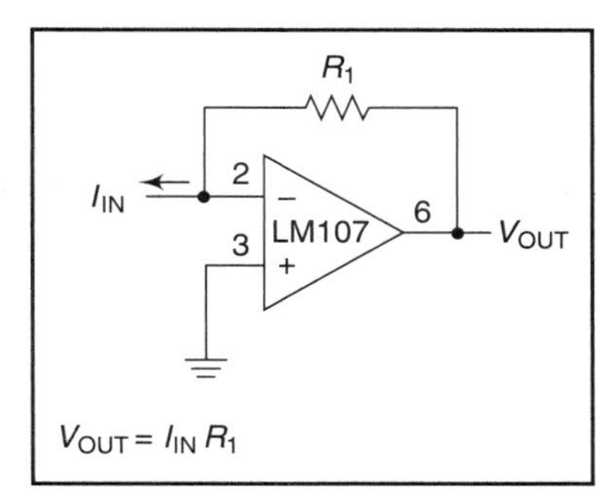

Figure 25.36 :: Current-to-voltage converter

25.6.7.9 Photocell Amplifier

Amplifiers for photoconductive, photodiode and photovoltaic cells are shown in Figure 25.37 respectively.

All photo-generators display some voltage dependence on both speed and linearity. It is obvious that the current through a photoconductive cell will not display strict proportionality to incident light if the cell terminal voltage is allowed to vary with cell conductance. Somewhat less obvious is the fact that photodiode leakage and photovoltaic cell internal losses are also functions of terminal voltage. The current-to-voltage converter neatly sidesteps gross linearity problems by fixing a constant terminal voltage, zero in the case of photovoltaic cells and a fixed bias voltage in the case of photoconductors or photodiodes.

A photodetector is optimized by operating into a fixed low load impedance. Currently available photovoltaic detectors show response times in the microsecond range at zero load impedance and photoconductors, even though slow, are materially faster at low load resistances.

The feedback resistance, R_1, is dependent on cell sensitivity and should be chosen for either maximum dynamic range or for a desired scale factor. R_2 is elective: in the case of photovoltaic cells or of photodiodes, it is not required in the case of photoconductive cells, it should be chosen to minimize bias current error over the operating range.

25.6.7.10 Log Converter

A number of instrumentation applications can benefit from the use of logarithmic or exponential signal processing techniques. The principle used for making log converters is based on the relationship between collector current and emitter base voltage in a transistor which is precisely logarithmic from currents below one picoamp to currents above one milliamp. Using a matched pair of transistors and integrated circuit operational amplifiers, a linear to logarithmic converter can be constructed with a dynamic range in excess of five decades.

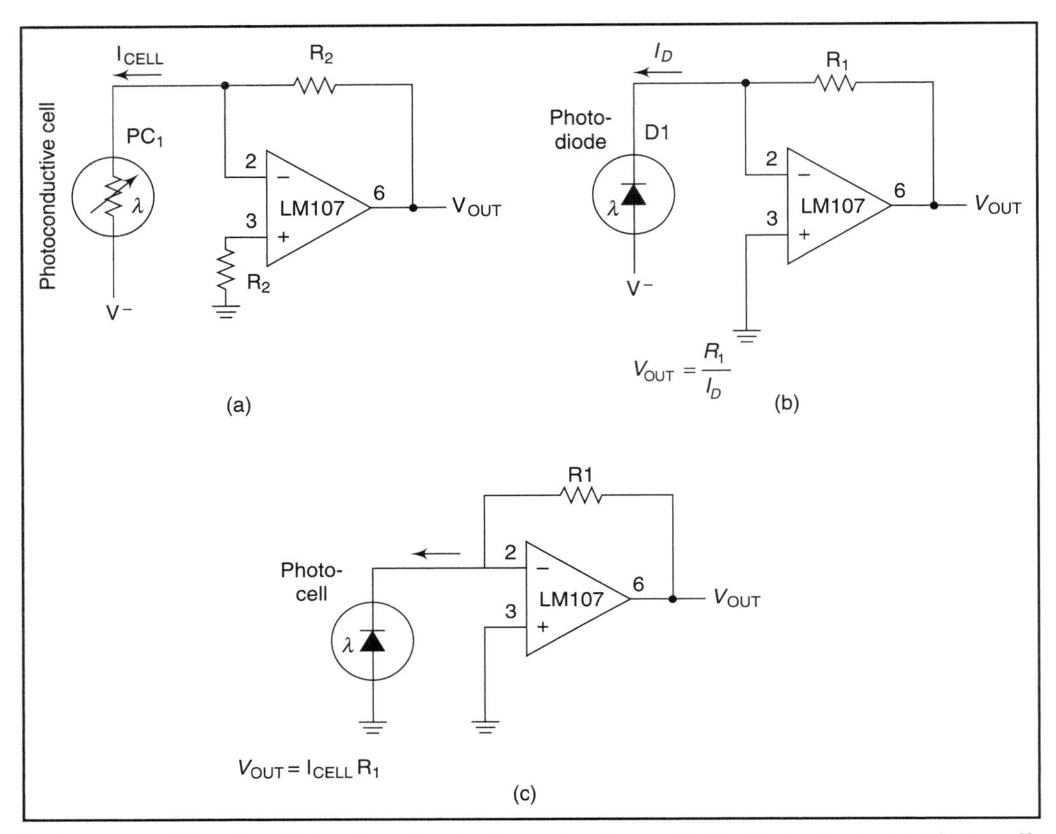

Figure 25.37 :: Amplifiers for: (a) photoconductive cell (b) Photodiode (c) Photocell

The circuit in Figure 25.38 generates a logarithmic output voltage for a linear input current. Transistor Q_1 is used as the non-linear feedback element around an operational amplifier A_1. Negative feedback is applied to the emitter of Q_1 through divider, R_1 and R_2, and the emitter base junction of Q_2. This forces the collector current of Q_1 to be exactly equal to the current through the input resistor. Transistor Q_2 is used as the feedback element of an A_2 operational amplifier. Negative feedback forces the collector current of Q_2 to equal the current through R_3. For the values shown, this current is 10 μA. Since the collector current of Q_2 remains constant, the emitter base voltage also remains constant. Therefore, only the V_{BE} of Q_1 varies with a change of input current. However, the output voltage is a function of the difference in emitter base voltages of Q_1 and Q_2.

$$E_{OUT} = \frac{R_1 + R_2}{R_2} (V_{BE2} - V_{BE1})$$

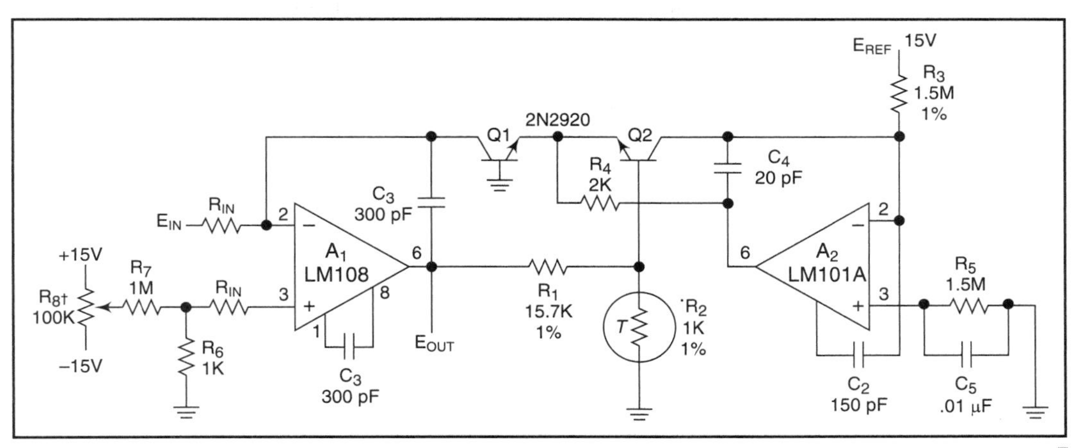

Figure 25.38 :: Log converter

For matched transistors operating at different collector currents, the emitter base differential is given by

$$\Delta V_{\text{BE}} = \frac{kT}{q} \log_e \frac{I_{\text{C1}}}{I_{\text{C2}}},$$

where k is Boltzmann's constant, T is temperature in degrees Kelvin and q is the charge of an electron. Combining these two equations and writing the expression for the output voltage gives:

$$E_{\text{OUT}} = \frac{-kT}{q}\left(\frac{R_1 + R_2}{R_2}\right)\log_e\left(\frac{E_{\text{IN}}\, R_3}{E_{\text{REF}}\, R_{\text{IN}}}\right)$$

For $E_{\text{IN}} \geq 0$. This shows that the output is proportional to the logarithm of the input voltage.

25.7 Sources of Noise In Electronic Circuits

Noise enters the measurement system from two sources: external interference and the inherent noise of the circuit itself.

External interference

External interference is noise which originates from sources not related to the actual circuit. Such noise sources include ground and power-supply noise created by other circuitry in a system, stray electromagnetic pick-up of line frequency energy (and the harmonics thereof) and radio and radar transmissions, contact arcing in mechanical switches and relays, and transients due to switching in reactive circuits. External interference can often be eliminated once the interfering source is identified and appropriate action taken.

Inherent noise

The second type of noise is the inherent noise of the circuit itself and it appears as a result of the fundamental, intrinsic properties of the system. Unlike external interference, it cannot be totally eliminated since it is caused by components in the actual circuit such as resistors and sources within the electronic circuitry. The best that can be accomplished is to minimize the noise in a specific bandwidth of interest.

Fundamental noise

Since the *fundamental noise* arises from the discontinuous nature of matter and energy, it ultimately limits the sensitivity of every instrument measurement. The main types of fundamental noise associated with solid state electronic devices are thermal, shot and flicker noise.

25.7.1 Thermal Noise or Johnson Noise

Thermal excitation of the electrons in conductors causes random movement of charge. In a resistance, this random current causes a noise voltage, known as Johnson Noise, whose amplitude is given by the formula:

$$E_N \text{ (rms)} = \sqrt{4kTRB},$$

where k = Boltzmann's Constant (1.38×10^{-23} J/°K)
T = Temperature °K
R = Resistance in Ohms
B = Bandwidth in Hertz

At room temperature (25°C) this may be simplified to:

$$E_N \text{ (rms)} \approx 1/8\sqrt{RB} \quad \text{or} \quad e_n \approx 4\sqrt{R}$$

E_N = Total Noise in μV rms
R = Resistance in kilohms
B = Bandwidth in Kilohertz

e_n = Spectral Density in nV/$\sqrt{\text{Hz}}$
R = Resistance in kilohms

Johnson noise

White Noise

Johnson noise is a fundamental property of resistances and is always considered when designing low noise circuitry. It is also called 'White Noise' because it is independent of the absolute values of frequencies. Johnson noise can be reduced by three ways: by reducing the temperature, the resistance itself or the working bandwidth.

Reducing the temperature is generally impractical. Since the function is a square root, cooling a resistor from room temperature (25°C/298K) to liquid nitrogen temperature (–196°C/77K) will only reduce the noise by 42 per cent. Reducing the resistance itself or the working bandwidth are generally more useful. Again, reducing the bandwidth is recommended only if the frequencies important to the measurement are not excluded. As a reference point it is useful to remember that at room temperature a 1K resistor has 4nV/$\sqrt{\text{Hz}}$ white noise. This is equivalent to 128nV rms noise in a 1kHz bandwidth.

25.7.2 Shot Noise

This kind of noise originates from the movement of charge carriers as they cross p-n junctions in semiconductor devices. There are statistical variations in the rate of electron flow and these manifest themselves as a noise current, given by the formula:

$$I_N = 5.7 \times 10^{-4} \sqrt{I_j B},$$

Where I_N is junction current in picoamps
I_J is junction current in picoamps
B is bandwidth of Interest (in Hertz).

Since the electronic charge is extremely small, the noise current is also very small and is only significant when the bandwidth is very large or when the noise current is an appreciable fraction of the total current. Since noise current is proportional to the square root of the current, the second case occurs only at very low currents. Thus shot noise is important at high frequencies and in amplifiers with very low bias currents but rarely elsewhere.

Like thermal noise, shot noise is proportional to the square root of the measurement bandwidth, and is also termed as 'white noise.' Thus shot noise is also minimized by reducing the bandwidth. The magnitude of shot noise is usually small and therefore can often be ignored

25.7.3 Flicker Noise

Flicker noise is the dominant noise at low frequencies. It has a power spectral density which is inversely proportional to frequency (hence the term '1/f Noise'). The noise voltage spectral density is therefore inversely proportional to the square root of the frequency and is represented by the following equation:

$$V_{av} = \sqrt{KI^2/F}$$

where K is a constant depending upon factors such as resistor materials and geometry, I is the dc current, and f is the frequency. Flicker noise predominates in measurements from 0 Hz (dc) up to about 300 Hz; it is due primarily to the contribution of the 1/f term. Flicker noise in amplifier systems is commonly referred to as 'Drift'. In sensitive measurements, flicker noise may be eliminated by avoiding the use of low frequencies (including dc).

Noise testing is slow and time-consuming and therefore expensive. So it is normal for manufacturers to quote typical specifications rather than absolute maxima.

25.7.4 Environmental Noise

Low-level signals which are usually encountered in analytical instruments are quite sensitive to external contaminations. This is especially troublesome when the signal source impedance is very high. The spurious signal or noise is an unwanted signal caused by the stray capacitance, inductance or resistance, which accidentally couple various parts of the circuit or its surroundings. It can produce errors in measurements and completely obscure useful data. The ratio of the wanted-signal to the unwanted or noise signal is expressed as the signal-to-noise ratio.

The most common and omnipresent stray signals are those derived from the 50 Hz line voltage and they are readily identified on the recordings. They are caused by numerous reasons, but are picked up more if the instrument has poor connections.

A major consideration in combating stray signals in all low-level measurements and recording systems is the need for proper grounding of the circuit. Its primary function is to assure that electronic enclosures and chassis are maintained at zero potential. Modern laboratories have a third copper conductor in all electrical circuits, which is non-current carrying and is connected to the electric power ground or the cold water mains pipeline. This will usually provide a satisfactory system ground.

System ground

In cases where it is not practical to connect the signal source to the system ground, it is imperative for a second low impedance grounding point to be established. This is called the signal ground. It is generally undesirable to connect the signal ground to the system ground. Moreover, the signal circuit should be grounded but only at one point.

Signal ground

Interference is sometimes caused, when the ground current is returned by more than one path. Two separate grounds are seldom, if ever, at the same absolute voltage. Their potential difference creates an unwanted current in series with one of the signal leads, and causes a noise signal to be combined with useful signal. In order to prevent noise pick-up from electrostatic fields, low level signal conductors must be surrounded by an effective shield. A woven metal braid under an outside layer of insulation is adequate for many applications. However, for transmission of microvolt-level signals, very low leakage capacitance is essential. Specially designed cables having lapped foil shields plus a low resistance drain wire in place of the braided wire shield are used for this purpose. This design reduces leakage capacity from about 0.1 μμf per foot to 0.01 μμf per foot for a typical cable.

The signal cable shield is grounded at the signal source. This prevents signal-cable capacity from shunting the amplifier's impedance to ground. It also preserves the high common mode rejection of the amplifier. The shield is connected to the low side of the signal source.

To sum up:

- Every low-level recording or display system should have a stable system ground and a good signal ground.
- The signal cable shield should not be attached to more than one ground, and this ground should be at the signal source.
- More than one accidental or intentional ground on either the signal circuit or the signal cable shield will produce excessive electrical noise in any low-level circuit. Therefore, the signal circuit should be grounded at only one point and never at more than one point
- Always ground a floating signal circuit and the signal cable shield only at the signal source.

Electromagnetic radiation that adversely affects circuit performance is generally termed EMI, or electromagnetic interference. Normally, electronic equipments are well shielded to avoid electromagnetic interference. *Shielding* is the use of conductive materials to reduce radiated EMI by reflection and/or absorption. Shielding can be applied to different areas of the electronic package, from equipment enclosures to individual circuit boards or devices. However, some interference may still get into the system, for one reason or other. A very common effect is that the mains and supply cables are infected with very high frequency pulse type noise due to electromechanical switches, commutators, motors etc. This pulse type noise is carried into the casing by mains cable and disturbs the whole system, influencing both the power supply and signal conductors. These problems can be solved by providing EMI filter at mains supply side. Robinson (1990) explains the types of shielding required to protect electronic equipment against electromagnetic and radiofrequency interference.

Shielding

25.8 Sources of Noise in Low–Level Measurements

25.8.1 Electrostatic and Electromagnetic Coupling to AC Signals

The distributed capacitance between signal conductors and from signal conductors to ground provides a low impedance ac path, resulting in signal contamination from external sources like power lines and transformers. Similarly, the alternating magnetic flux from the adjacent power line wires induces a voltage in the signal loop which is proportional to the rate of change of the disturbing current, the magnitude of the disturbing current and the areas enclosed by the signal loop, is inversely proportional to the distance from the disturbing wire to the signal circuit. Unequal distances of the two signal carrying conductors from the disturbing current wire result in unequal mutual inductances which cause the magnetic field to produce a noise voltage across the amplifier input terminals.

Low-level signals are sensitive to external contamination especially in the case of high source impedance. Referring to Figure 25.39, it is obvious that the currents generated by various noise signals will flow through the signal source impedance Z and result in an unwanted addition to the transducer signal. This may include electromagnetic noise pick up, electrostatic pick up and the unwanted current generated by a ground loop between two separate grounds on the same signal circuit. The magnitude of these unwanted signals will be directly proportional to signal source impedance as shown by the relationship:

Amplifier input signal = $E + IZ$,

where E = normal signal amplitude
Z = impedance of signal source
I = current generated by noise.

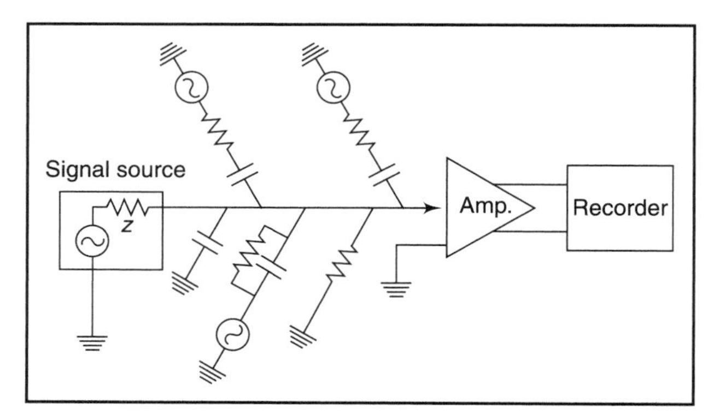

Figure 25.39 :: Currents produced by various forms of noise flow through the signal source impedance and become an unwanted addition to the useful signal. The noise amplitude is directly proportional to signal source impedance (Courtesy, Gould Inc., USA)

It is obvious that as the signal source impedance approaches zero, so will the noise input to the amplifier. In fact low-source impedance effectively shunts out the noise.

In order to prevent noise pick-up from electrostatic fields, low-level signal conductors are surrounded by an effective shield. This usually is a woven metal braid around the signal pair, which is placed under an outside layer of insulation. A more effective shielding is provided by a special type of signal cable, which has lapped foil shields, plus a low resistance drain wire instead of the conventional braided wire shield.

The easiest and generally the best way to protect a signal cable against external electromagnetic disturbances is to twist the circuit conductors closely together to electrically cancel the effect of an external magnetic field. The shorter the lay of the twist, the greater the noise rejection. Thus, electromagnetic coupling is reduced by shielding, wire twisting and proper grounding which provide a balanced signal pair with satisfactory noise rejection characteristics.

25.8.2 Proper Grounding (Common Impedance Coupling)

Grounding

Placing more than one ground on a signal circuit produces a ground loop which may generate so much noise that it may completely obscure the useful signal. The term 'grounding' means a low-impedance metallic connection to a properly designed ground grid, located in the earth. Stable grounding is necessary to attain effective shielding of low level circuits to provide a stable reference for making voltage measurements and to establish a solid base for the rejection of unwanted common mode signals. There are generally two grounding systems-a system ground and a signal ground. All low-level measurements and recording systems should be provided with a stable system ground to assure that electronic enclosures and chassis operating in an electromagnetic environment are maintained at zero potential. In most instances the third copper conductor in all electrical circuits, which is firmly tied to both electric power ground, the building ground and the water system, will provide a satisfactory system ground. The signal ground, on the other hand, is necessary to ensure a low noise signal reference to ground. This ground should be a low-impedance path to wet earth to minimize the introduction of spurious voltages into the signal circuitry. It is important to note that a signal circuit should be grounded at one point only.

Two separate grounds seldom lie at the same absolute voltage. If we connect more than one ground to the same signal circuit, an unwanted current will flow in the ground loop thus created. This current combines itself with the useful signal (Figure 25.40). Also, there is a second ground

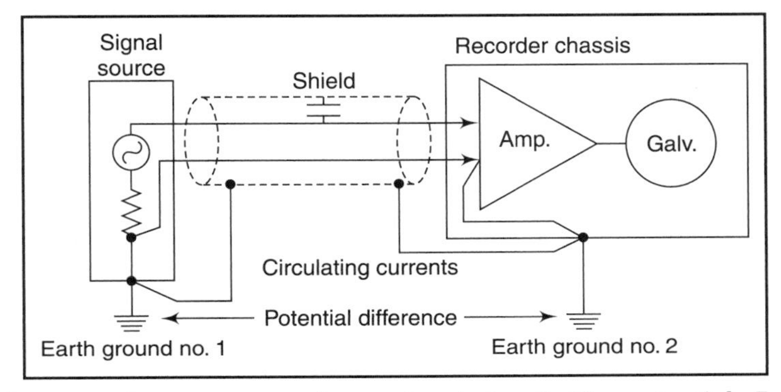

Figure 25.40 :: Ground loop created by more than one ground on a signal circuit. The potential difference between earth ground No.1 and earth ground No.2 causes current to circulate in the signal cable shield and also in the lower signal conductor, producing two separate ground loops.

loop through the signal cable-shield from the signal source to the amplifier. The current in the shield is coupled to the signal pair through the distributed capacitance in the signal cable. This current then flows through the output impedance of the signal source and back to ground, thus adding a second source of noise to the useful signal. Either one of these ground loops generates a noise signal that is larger than a typical millivolt useful signal. Ground loops are eliminated by floating lower input terminal of the amplifier. The amplifier enclosure is still solidly grounded to earth-ground No. 2 but this will not create a ground loop, since the amplifier enclosure is insulated from the signal circuit. The ground-loop through the signal cable is removed by grounding the shield only at the signal source which is the proper configuration for minimum noise pick-up (Figure 25.41).

25.9 Noise Reduction Techniques

25.9.1 Hardware Techniques

In a measurement system, it is often possible to ignore noise if the magnitude of the desired signal is relatively large as compared to the undesirable (noise) signal. However, many situations require the extraction of useful signal from the noisy background. Noise reduction has traditionally been carried out with electronic hardware circuits. This is mostly achieved with frequency as the most commonly used property. It was mentioned in the previous section that 'white noise' can be

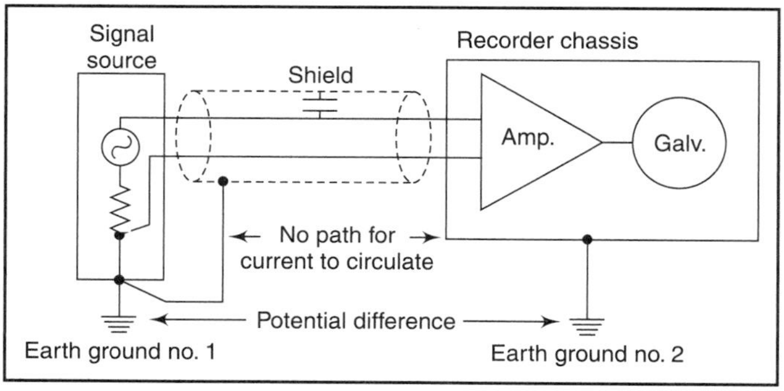

Figure 25.41 :: Eliminating multiple grounds. The ground loop in the lower signal lead has been broken by removing the jumper wire to earth ground No. 2. The ground loop in the cable shield has been broken by removing its connection to earth ground No. 2. (Courtesy, Gould Inc., USA)

reduced by narrowing the range of measured frequencies i.e. limiting the frequency range but ensuring the proper signal fidelity. The methods commonly used for frequency selection are filters, integrators and modulators. Modulations are used in chopper amplifiers. The modulation/dc modulation techniques are used to process a signal in a region of minimum noise and to discriminate between signal and noise on the basis of the signals unique modulation configuration relative to the random pattern of noise. Commonly used noise reduction techniques are as follows:

25.9.1.1 Differential Amplifier

Amplifiers designed for use in the input stage (pre-amplifiers) are mostly of the differential type, that is, they have three input terminals out of which one is arranged at the reference potential while the other two are live terminals. The differential amplifier is employed when it is necessary to measure the voltage difference between two points, both of them varying in amplitude at different rates and in different patterns.

The differential amplifier is an excellent device for use in recording systems. Its excellence lies in its ability to reject common-mode interference signals, which are invariably picked up along with the useful signals. Also, as a direct coupled amplifier, it has good stability and versatility. High stability is achieved because it can be insensitive to temperature changes which is often the source of excessive drift in other configurations. It is versatile in that it may be adapted for a large number of applications, e.g., applications requiring floating inputs and outputs or for applications wherein the grounded inputs and/or outputs are desirable.

The working of a differential amplifier can be explained with the help of Figure 25.42(a), where the two transistors with their respective collector resistances, form a bridge circuit. If the two resistors and the characteristics of the two transistors are identical, the bridge is perfectly balanced and the potential difference across the output terminals is zero.

Let us now apply a signal at the input terminals 1 and 2 of this circuit. The signal is to be such that at each input terminal, it is equal in amplitude but opposite in phase with reference to the ground. This signal is known as the differential mode signal. Because of this signal, if the collector current of T_1 increases, the collector current of T_2 will decrease by the same amount, and the collector voltage of T_1 will decrease while that of T_2 increases. This results in a difference voltage between the two output terminals that is proportional to the gain of the transistors. On the other hand, if the signal applied to each input terminal is equal in amplitude and is in the same phase (called the common-mode input signal), the change in current flow through both transistors will be identical, the bridge will remain balanced, and the voltage between the output terminals will remain zero. Thus, the circuit provides high gain for differential mode signals and no output at all for common-mode signals.

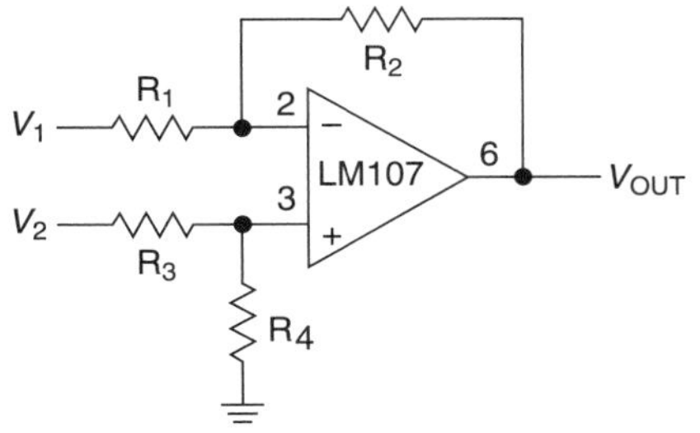

$$V_{out} = \left(\frac{R_1 + R_2}{R_3 + R_4}\right)\frac{R_4}{R_1}V_2 - \frac{R_2}{R_1}V_1$$

For $R_1 = R_3$ and $R_2 = R_4$

$$V_{out} = \frac{R_2}{R_1}(V_2 - V_1)$$

Figure 25.42 :: (a) Differential amplifier

Differential mode signal

Common-mode input signal

Common mode rejection ratio

The ability of the amplifier to reject these common voltages on its two input leads is known as common mode rejection and is specified as the ratio of common mode input to differential input to elicit the same response. It is abbreviated as CMRR (Common mode rejection ratio). CMRR is an important specification referred to the differential amplifier and is normally expressed as decibels. CMRR of the pre-amplifiers should be as high as possible so that only the wanted signals find a way through the amplifier and all unwanted signals get rejected in the pre-amplifier stage. A high rejection ratio is usually achieved by the use of a matched pair of transistors in the input stage of the pre-amplifier and a large 'tail' resistance in the long-tailed pair to provide maximum negative feedback for inphase signals. The technique of long-tailing (a technique used to current drive an active device) improves the CMRR in differential amplifiers without upsetting the gain for the desired signal. Very high CMRR can be achieved with the use of an active long-tail. Also, a high input impedance is very necessary in order to obtain a high CMRR.

The design of a good differential amplifier essentially implies the use of closely matched components which has been best achieved in the integrated circuit form. High gain integrated dc amplifiers, with differential input connections and a provision for external feedback have been given the name 'operational amplifiers' because of their ability to perform mathematical operations. These amplifiers are applied for the construction of ac or dc amplifiers, active filters, phase inverters, multivibrators and comparators, etc., by suitable feedback arrangement, and therefore find a large number of applications in analytical instrumentation field.

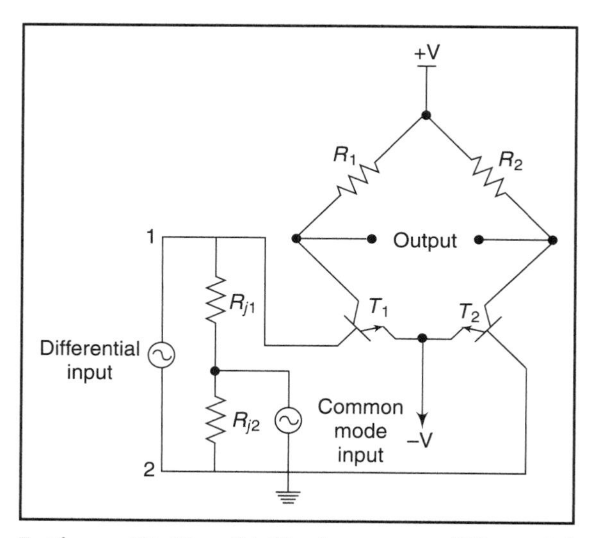

Figure 25.42 :: (b) Single op-amp differential amplifier

Figure 25.42 (b) shows a single op-amp in a differential configuration. The common mode rejection for most op-amps is typically between 60 dB and 90 dB. This may not be sufficient to reject common mode noise generally encountered in low signal level measurements. Also, the input impedance is not very high to handle signals from high impedance sources. One method to increase the input impedance of the op-amp is to use field effect transistors (FET) in the input differential stage. A more common approach is to use an instrumentation amplifier in the pre-amplifier stage.

25.9.1.2 Instrumentation Amplifier

The differential amplifier is well-suited for most of the applications in analytical measurements. However, it has the following limitations:

- The amplifier has a limited input impedance and therefore, draws some current from the signal source and loads them to some extent.
- The CMRR of the amplifier may not exceed 60 dB in most of the cases, which is usually inadequate in modern analytical instrumentation systems.

These limitations have been overcome with the availability of an improved version of the differential amplifier, whose configuration is shown in Figure 25.43. An instrumentation ampli-

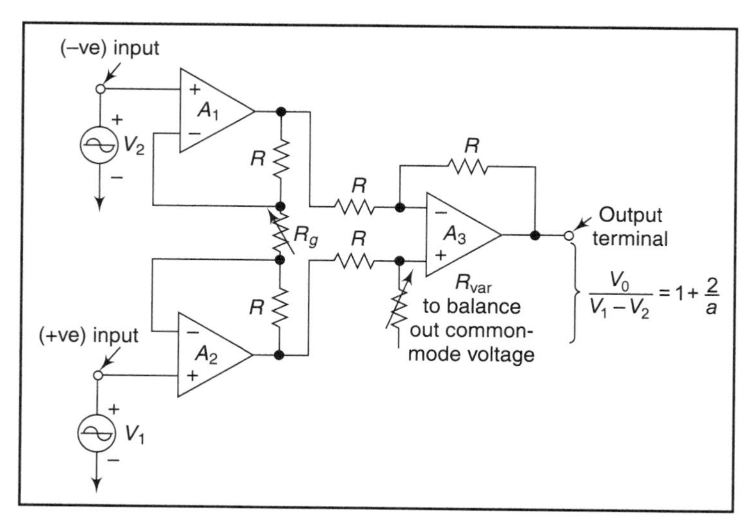

Figure 25.43 :: Instrumentation amplifier

fier is a precision differential voltage gain device that is optimized for operation in an environment hostile to precision measurement. It basically consists of three op-amps and seven resistors. Basically, connecting a buffered amplifier to a basic differential amplifier makes an instrumentation amplifier.

In Figure 25.43, op-amp A_3 and its four equal resistors R form a differential amplifier with a gain of 1. Only A_3 resistors have to be matched. The variable resistance R_{var} is varied to balance out any common mode voltage. Another resistor R_g, is used to set the gain using the formula:

$$\frac{V_0}{V_1 - V_2} = 1 + \frac{2}{a},$$

where $a = R_g/R$.

V_1 is applied to the + ve input terminal and V_2 to the –ve input terminal. V_0 is proportional to the difference between the two input voltages.

The important characteristics of the instrumentation amplifier are:

- Voltage gain from differential input ($V_1 - V_2$) to single ended output, is set by one resistor.
- The input resistance of both inputs is very high and does not change as the gain is varied.
- V_0 does not depend upon common mode voltage, but only on their difference.

If the inputs are prone to high voltage spikes or fast swings, which the op-amps cannot cope with, they may be protected by using back-to-back connected diodes at their inputs. However, this reduces the input impedance value substantially and also limits the bandwidth.

The instrumentation amplifier offers the following advantages for its applications:

- Extremely high input impedance,
- Low bias and offset currents,
- Less performance deterioration if source impedance changes,
- Possibility of independent reference levels for source and amplifier,
- Very high CMRR,
- High slew rate, and
- Low power consumption.

Good quality instrumentation amplifiers have become available in single IC form such as μA725, ICL7605, LH0036, etc.

25.9.1.3 Filters

In order to achieve reduction of noise, it is a common practice to narrow the range of measured frequencies by the use of electronic filters.

A filter is a circuit which amplifies some of the frequencies applied to its input and attenuates others. There are four common types of filters:

- *High Pass:* which only amplifies frequencies above a certain value,
- *Low Pass:* which only amplifies frequencies below a certain value,
- *Band Pass:* which only amplifies frequencies within a certain band, and
- *Band stop:* which amplifies all frequencies except in those in a certain band.

Passive filters

Active filters

Digital filters

Filters may be produced by using many different methods. These include passive filters which use only passive components such as resistors, capacitors and inductors; active filters which use amplifiers in addition to resistors, capacitors and inductors in order to obtain performance impossible with passive filters. Operational amplifiers are frequently used as the gain blocks in active filters; digital filters which use analog-digital converters to convert a signal to digital form and then use high-speed digital computing techniques to filter it.

Low-pass filter

The simple low-pass filter is shown in Figure 25.44. This circuit has a 6 dB per octave roll-off after a closed-loop 2 dB point defined by f_c. Gain below this corner frequency is defined by the ratio of $R3$ to $R1$. The circuit may be considered as an ac integrator at frequencies well above f_c, however, the time domain response is that of a single RC rather than an integral.

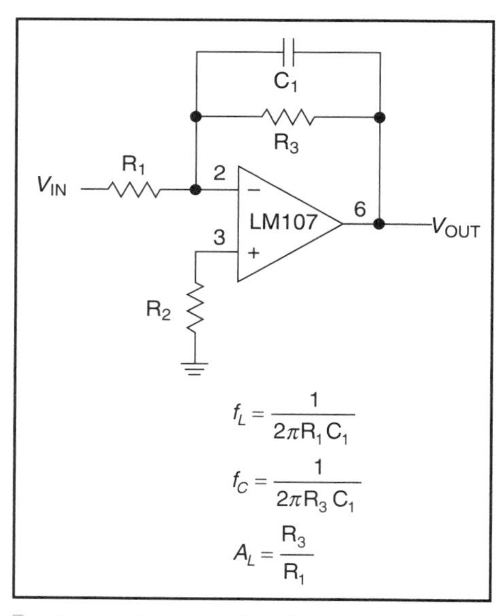

Figure 25.44 :: Simple low pass filter

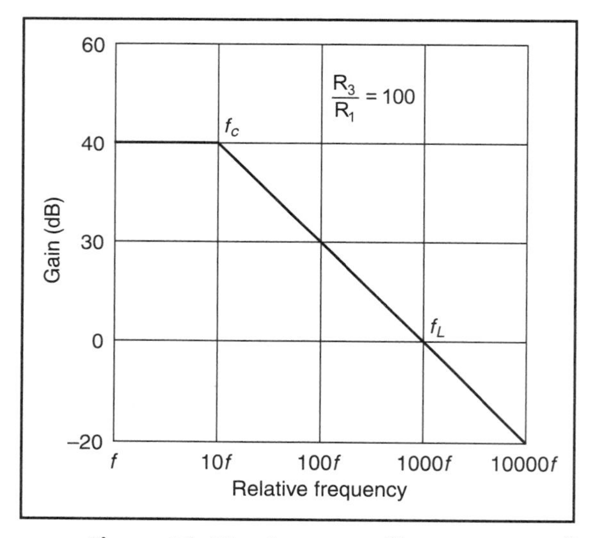

Figure 25.45 :: Low pass filter response

A gain frequency plot of circuit response is shown in Figure 25.45 to illustrate the difference between this circuit and the true integrator.

25.9.1.4 Modulators (Carrier Amplifiers)

If the signal and noise cannot be separated by simple filtering, modulation/demodulation techniques can be used to process a signal in a region of minimum noise. The signal and noise are discriminated on the basis of the unique modulation configuration of the signal relative to the random pattern of the noise. An amplifier based on this technique is called 'carrier amplifier'. This type of amplifier enables one to obtain zero frequency response of the dc amplifier and the inherent stability of the capacitance coupled amplifier. The carrier amplifier consists of an oscillator and a capacitance coupled amplifier. The oscillator is used to energize the transducer with an alternating carrier voltage. The transducers, which require ac excitation, are those whose impedance is not purely resistive. An example of this can be of a capacitance-based pressure transducer whose impedance is mainly capacitative with a small resistive component. The frequency of the excitation voltage is usually around 2.5 kHz. The transducer shall change the amplitude of the carrier voltage in relation to the changes in the chemical variable being measured. The output of the transducer therefore, would be an amplitude modulated (AM) signal (Figure 25.46). The modulated ac signal can then be fed to a multistage capacitance coupled amplifier. The first stage produces amplification of the AM signal. The second stage is so constructed that it can respond only to signal frequency of the carrier. It can be further amplified in the following stage. After amplification, the signal is demodulated in a phase-sensitive demodulator circuit. This helps to extract amplified signal voltage after the filter circuit. The voltage produced by the demodulator can then be applied to the driver stage of the display system.

Phase-sensitive demodulator

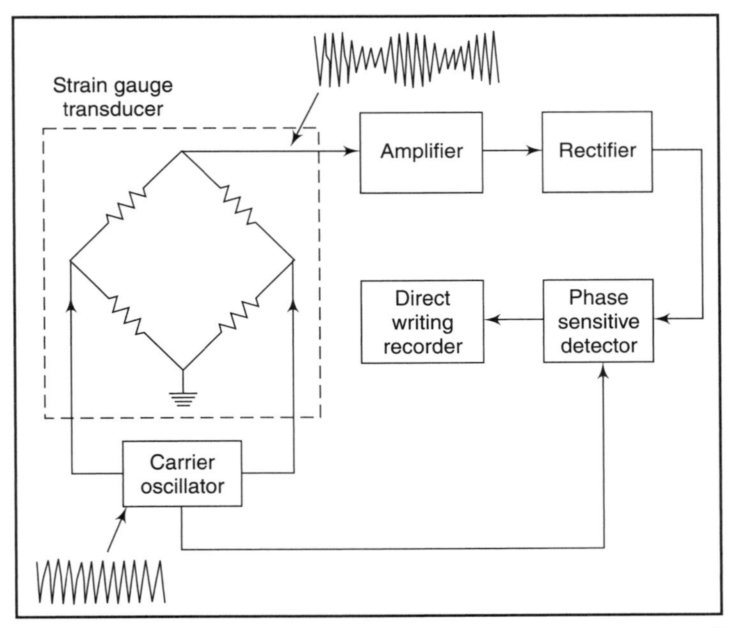

Figure 25.46 :: Carrier amplifier

25.9.1.5 Chopper Amplifier

The chopper amplifier is a useful device in the field of analytical instrumentation, as it provides a good solution to the problem of achieving adequate low frequency response, while preventing the drift problem inherent in direct coupled amplifiers.

This type of amplifier makes use of a chopping device, which converts a slowly varying direct current to an alternating form, with amplitude proportional to the input direct current and with phase dependent on the polarity of the original signal. The alternating voltage is then amplified by a conventional ac amplifier, whose output is rectified back to get amplified direct current. A chopper amplifier is an excellent device for signals of narrow-band width and reduces the drift problem to zero.

Mechanical chopper

There are two types of choppers, viz. mechanical chopper and transistor chopper. A mechanical chopper is simply an electronic switch driven by an alternating current. It is so designed that, the flux saturates a magnetic circuit, such that switching operation occurs only near the cycle zero points. For mechanical choppers, 50 Hz mains frequency is usually used as the chopping frequency. Choppers which operate at higher frequencies, say about 400 Hz are also available.

Transistor chopper

The use of a transistor as a chopper increases the possible rate of switching, and can therefore be useful for signals of wider bandwidths. Several typical transistor chopper circuit configurations are available in literature. The action of the transistor as chopper is based on its low saturation resistance of ON mode and high resistance for OFF mode. The transistor can be used as a signal chopper at rates conveniently up to 100 kHz. Greater stability against drift can be achieved in the chopper stabilized dc amplifier, shown in the Figure 25.47.

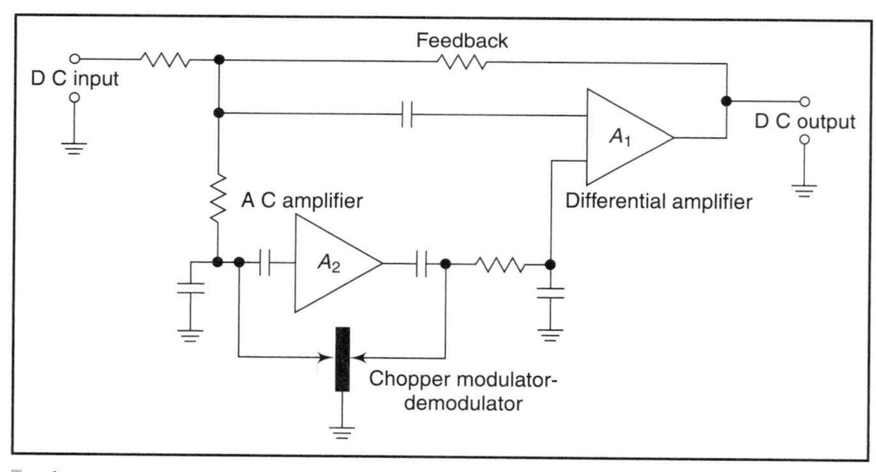

Figure 25.47 :: Chopper amplifier

Since drift is a low frequency phenomenon, the signal can be applied simultaneously to a chopper modulated ac amplifier of the narrow band-width and through an ac coupling to a differential output dc amplifier of wide band-width. The low frequency components are given high gain over the restricted band-width and the drift output is fed back for drift stabilization. For higher frequency components, the gain is provided by the dc amplifier. In this arrangement, medium and higher frequencies pass through the dc amplifier with the gain, say as A_1. The drift and very low frequency signals pass through both amplifiers in series with the gain A_1/A_2. The drift factor is reduced by A_1/A_2 and it is possible to obtain drift figures of only a few microvolts.

Chopper amplifiers find applications in analytical instrumentation in the amplification of small dc signals of a few microvolts. Such order of amplitudes are obtainable in atomic absorption spectrophotometers, pH meters, etc. The frequency response of a chopper amplifier depends upon the value of the chopping frequency. The input impedance can be made high by using a sub-miniature electrometer tube or insulated gate field effect transistor, at the input stage.

25.9.1.6 Lock-in Amplifier

A lock-in amplifier is a useful version of the carrier technique designed for measurement of low-level signals buried in noise. This type of amplifier, due to its extremely narrow-width output band in which the signal is carried, reduces wideband noise and increases the signal-to noise ratio. Thus, the difference between carrier amplifier and lock-in amplifier is that the former is a general purpose instrument amplifier while the latter is designed to measure signals in a noisy background.

In principle, the lock-in-amplifier works by synchronizing on a single frequency, called the reference frequency. This frequency is made to contain the signal of interest. The signal is modulated by the reference frequency in such a way that all the desired data is at the single reference frequency whereas the inevitable noise, being broadband, is at all frequencies. This permits the signal to be recovered from its noisy background.

Lock-in amplifier primarily work on the principle of phase-sensitive detection to improve the signal-to-noise ratio in continuous wave experiments. A typical example of application of a lock in amplifier is shown in Figure 25.48 wherein the analytical signal (the fluorescence) is modulated by chopping the optical excitation source (the laser) at the reference frequency.

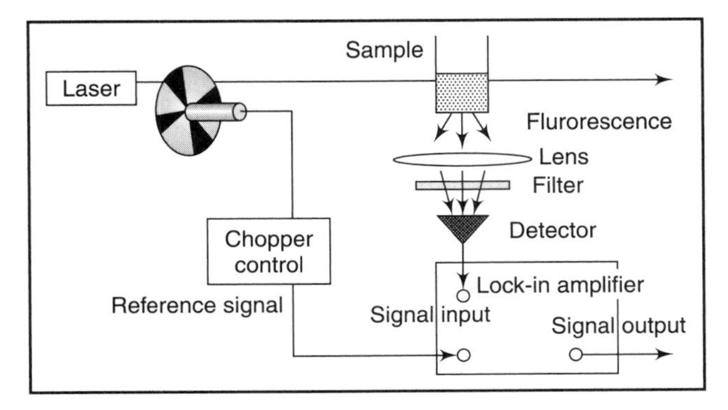

Figure 25.48 :: Lock-in amplifier

25.9.1.7 Box Car Integrator

Integration of dc signals over a limited time period provides another method for reducing white noise. The signal with its coherent characteristics gets added with respect to the integration time, while the random noise adds as the square root of the integration time. Thus, the signal/noise ratio is enhanced as the square root of integration time. Although a simple RC circuit can be used to integrate signals, using an operational amplifier with a capacitor in the feedback circuit offers a more stable and precise integrating circuit.

An improvement over the simple integrator is the box car integrator in which the same portion of the signal is periodically sampled for a fixed period of time and then averaging the sample using a low pass RC filter. The scheme is shown in Figure 25.49. the integrator is triggerable and gated and provides an S/N enhancement for the portion of the signal that is sampled, which is equal to the square root of the number of pulses integrated. The technique is particularly useful in applications requiring pulsed signal detection.

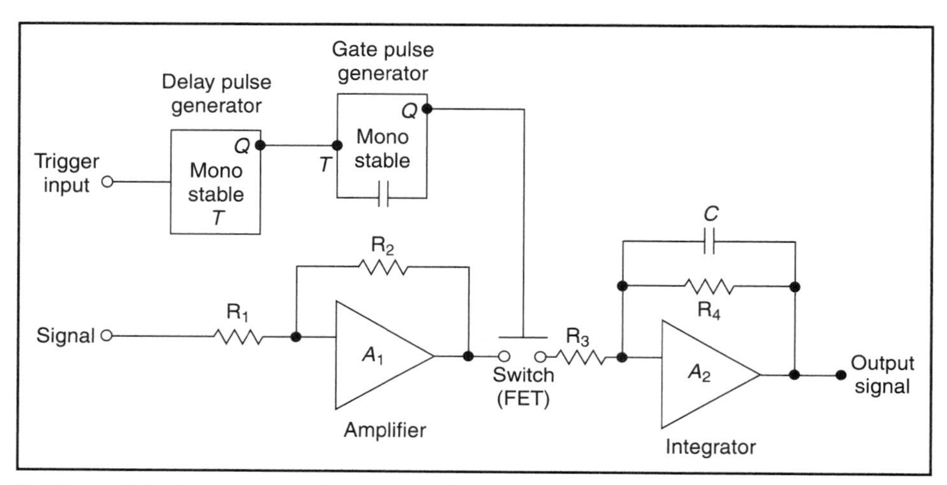

Figure 25.49 :: Box car integrator (redrawn after Willard, *et al.*, 1988)

25.9.2 Software Techniques

25.9.2.1 Fourier Transform

In analytical instrumentation, sensors/transducers pick up signals from sources of chemical information. Signal processing employs sophisticated mathematical analysis tools and algorithm to extract information buried in these signals received from various sensors and transducers. Signal processing algorithms attempt to capture signal features and components that are of analytical value. Since most signals of analytical interests are time varying, there is a need to capture transient phenomena when studying the behaviour of such signals.

Fourier transform

The fourier transform technique depends upon the basic principle, that any wave function could be represented as a series of sine and cosine functions with different frequencies. This is illustrated in Figure 3.17. The determination of the sine and cosine components of a given wave function is (d) known as Fourier transformation (FT).

Time-domain

Frequency domain

As illustrated in section 3.5 (Chapter 3), there are two different presentations of the same experimental data, known as domains. These are the time-domain in which the data are recorded as a series of measurements at successive time intervals and the frequency domain in which the data are represented by the amplitude of its sine and cosine components at different frequencies. For example, for recording and display purposes, the signals are represented in the time domain i.e. the signal is represented by means of its value (*Y*-axis) on the time axis (*X*-axis). In the frequency domain, any signal may be described as consisting of sine-waves and having different amplitudes and phases (Y-axis) as a function of frequency (X-axis). The transformation between the two representations is given by the Fourier Transform (FT).

The basic motivation for developing the frequency analysis tools is to provide a mathematical and pictorial representation for the frequency components that are contained in any given signal. The term spectrum is used when referring to the frequency content of a signal. The process of obtaining the spectrum of a given signal using the basic mathematical tools is known as frequency or spectral analysis. Most time-varying signals of practical interest can be decomposed into a sum of sinusoidal signal components. For the class of periodic signals, such a decomposition is called a Fourier series. For the class of finite energy signals, the decomposition is called the Fourier transform.

Spectrum

Referring to Figure 25.50 a system's response to a varying input signal $s(t)$, with a frequency spectrum $S(f)$, can be described essentially interchangeably, by the response $r(t)$ in the time domain (as a time history) or $R(f)$ in the frequency domain (as a frequency spectrum).

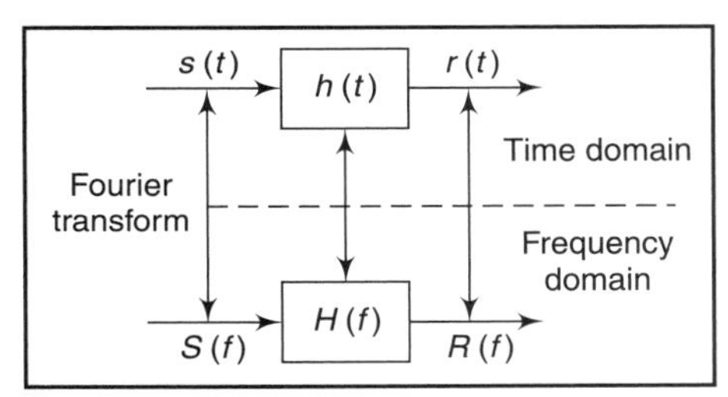

Figure 25.50 :: A system transfer function, h(t) and H(f) characterizes its response r(t) in the time domain and R(f) in the frequency domain, to receive stimuli, S(t) and S(f). The two domains are related by the Fourier Transforms.

The key system characteristics operate on the signal to produce the response to the stimulus. In the frequency domain, the operation can be expressed as a simple product, the ratio of response to stimulus is called the transfer function, $H(f)$:

$$H(f) = R(f)/S(f)$$

The time domain and frequency domains are closely connected via the Fourier transform.

25.9.2.2 Fast Fourier Transform (FFT)

Using the Fourier transform can become tedious and time-consuming even when one is using computers and especially when a large number of factors have to be considered. The introduction of fast Fourier transform (FFT) algorithm, which expands the signal into sine and cosine functions, has been of tremendous help. The frequency spectrum is computed for each discrete segment of the signal. The output of the FFT algorithm is a set of coefficients, two for each frequency component in the signal's spectrum. One coefficient (A) is multiplied by the cosine, or amplitude portion of the component. The other (B) is multiplied by the sine, or phase portion, of the component. Each component in the FFT series can then be represented as:

$$A \cos(\omega t) + i\, B \sin(\phi),$$

where

ω = Angular frequency of the component
A = An FFT coefficient
B = An FFT coefficient
ϕ = The phase angle of the component
i = The imaginary number, $\sqrt{-1}$.

The number of frequency components and pairs of FFT coefficients required to represent a given waveform is a function of the highest frequency to be resolved and the sample rate used.

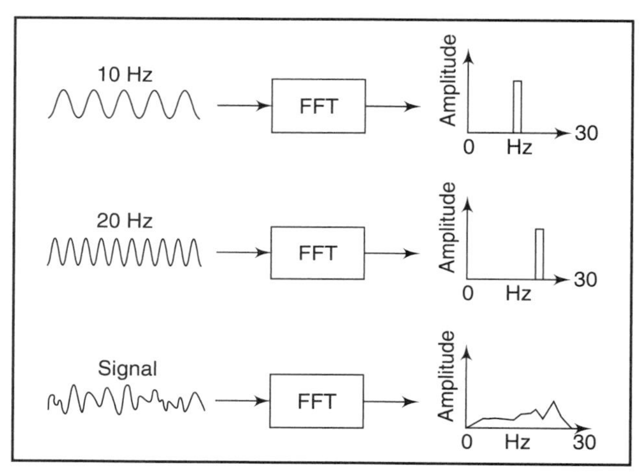

Figure 25.51 :: Time varying signal decomposed into basic frequency components

Figure 25.51 illustrates the decomposition of a typical time varying signal into its basic frequency components and then displays them as a frequency spectrum. The diagram shows frequency components along with the amplitude present in each frequency. Once the frequency spectrum of a given segment of the signal has been calculated, a number of techniques are available to display the information.

The FFT method assumes the signal to be stationary and is therefore insensitive to its varying features. However, most signals are non-stationary and have highly complex time frequency characteristics. The stationary condition for the non-stationary signal can be satisfied by dividing the signals into blocks of short segments, in which the signal segment can be assumed to be stationary. This method is called the Short-Time Fourier transform (STFT). However, the problem with this method is the length of the desired segment. Choosing a short analysis window may cause poor frequency resolution. On the other hand, a long analysis window may improve the frequency resolution but compromises the assumption of stationarity within the window.

The Fourier theorem states that any waveform can be duplicated by the superposition of a series of sine and cosine waves. As an example, the Fourier expansion of sine waves provides an approximation of a square wave as shown in Figure 3.17. As more terms are added the superposition of sine waves better matches a square wave.

The Fourier transform uses the above concept ot convert between two different descriptions of a physical system.

$$f(t) = 1/2\pi F(\omega)e^{-i\omega t}\, d\omega$$

$$F(\omega) = f(t)^{i\omega t}\, dt$$

In these equations ω is angular frequency ($2\pi \times$ frequency), t is time, and:

$$Ei\omega t = \cos(\omega t) + I \sin(\omega t),$$

where $I = \sqrt{-1}$

The $F(\omega)$ function gives the frequencies at which the signal is non-zero and the $f(t)$ function gives the times at which the signal is non-zero. Both of these functions are suitable descriptions of a waveform or physical system

Given a function in time, $f(t)$, we can transform it into an equivalent function in frequency, $F(\omega)$. We can look at the second expression in detail to understand what is happening.

$$F(\omega) = \int f(t)^{j\omega t}\, dt$$

In order to do the transform, we multiply $f(t)$ times $[\cos(\omega t) + I \sin(\omega t)]$. We do this at all times between 00 and −00.

25.9.2.3 DFT and FFT

Experimental data usually consists of discrete data points rather than a continuous function as used in the above equations. The Discrete Fourier-Transform (DFT) is an algorithm for doing the transform with discrete data. The DFT is an order N^2 calculation, meaning that the number of multiplications is equal to the square of the number of data points. This algorithm has been supplanted by Fast Fourier-Transform (FFT) algorithms, which reduce redundancies and take much less computer time. The order of this calculation is N/logN.

Sampling frequency

A key experimental parameter is the sampling frequency, which must be at least twice as large as the highest frequency component that is present in the data. This sampling rate is called the Nyquist critical frequency. Sampling at less than the Nyquist frequency results in aliasing of the result. Aliasing can be prevented by filtering out all frequencies above the Nyquist frequency so that they do not create artifacts in the transformed spectrum.

Fast Fourier-Transform (FFT) algorithms are used in NMR, infrared, Raman, and mass spectrometry instrumentation. Their advantages are:

- Multiplex (or Fellgett) advantage
- Throughput (or Jaquinto) advantage
- Rapid signal averaging of data to greatly improve the signal-to-noise ratio of the data

The multiplex advantage arises from recording all signal frequencies simultaneously. The throughput advantage arises because no physical slit is necessary to obtain resolution in the resulting spectra.

25.9.2.4 Wavelet Transform

An emerging method used to analyze non-stationary signals is the wavelet transform. The wavelet method acts as a mathematical microscope in which we can observe different parts of the signal by just adjusting the focus. In practice, it is not necessary for the wavelet transform to have continuous frequency (scale) parameters to allow fast numerical implementations; the scale can be varied as we move along the sequence. Therefore, the wavelet transform has very good time resolution at the high frequencies, and good frequency resolution at the low frequencies. In analytical instrumental field, wavelet transform have been widely used in many research areas including spatial filtering, edge detection, feature extraction, data compression, pattern recognition, speech recognition, image compression and texture analysis.

Wavelets constitute a relatively new signal processing method. A wavelet transform is almost always implemented as a bank of filters that decompose a signal into multiple signal bands. It separates and retains the signal features in one or a few of these bands. Thus, one of the biggest advantages of using the wavelet transform is that signal features can be easily extracted. In many cases, a wavelet transform outperforms the conventional FFT when it comes to feature extraction and noise reduction.

Another method of signal analysis is that of adaptive filter which can continuously adjust itself to optimally perform under the changing circumstances. This is achieved by correcting the signal according to the specific application. The correction may be enhancement or some reshaping, for which a correction algorithm is required. This can be best implemented digitally. Most adaptive filters, therefore, are implemented by means of computers or special digital signal processing chips.

25.9.2.5 Digital Signal Processing

When we pass a signal through a device that performs an operation, as in filtering, we say we have processed the signal. The type of operation performed may be linear or non-linear. Such operations are usually referred to as *signal processing*.

The operations can be performed with a physical device or software. For example a digital computer can be programmed to perform digital filtering. In case of digital hardware operations (logic circuits), we have a physical device that performs a specified operation. In contrast, in digital processing of signals on a digital computer, the operations are performed on a signal consist of number of mathematical operations as specified by the software.

Most of the signals encountered in analytical instrumentation are analog in nature, i.e. the signals are functions of a continuous variable such as time or space. Such signals may be processed by analog systems such as filters or frequency analyzers or frequency multipliers. Until about two decades ago, most signal processing was performed using specialized analog processors. As digital systems became available and digital processing algorithms were developed, the digital processing became more popular. Initially, digital processing was performed on general purpose microprocessors. However, for more sophisticated signal analysis, these devices were quite slow and not found suitable for real time applications. Specialised designs of microprocessors have resulted in the development of digital signal processors, which although perform a fairly limited number of functions, but do so at very high speeds.

Digital signal processor

A digital signal processor (popularly known as DSP) requires an interface (Figure 25.52) between the analog signal and digital processor, which is commonly provided by an analog-to-digital converter. Once the signal is digitized, the DSP can be programmed to perform the desired operations on the input signal. The programming facility provides the flexibility to change the signal processing operations through a change in the software, whereas hardwired machines are difficult to configure. Hence programmable signal processors are common in practice. On the other hand when the signal processing operations are well defined, as in some applications, a hardwired implementation of the operations can be optimized so that it results in cheaper and faster signal processors. In cases when digital output from processor is to be given to user in analog form, a D/A converter is required.

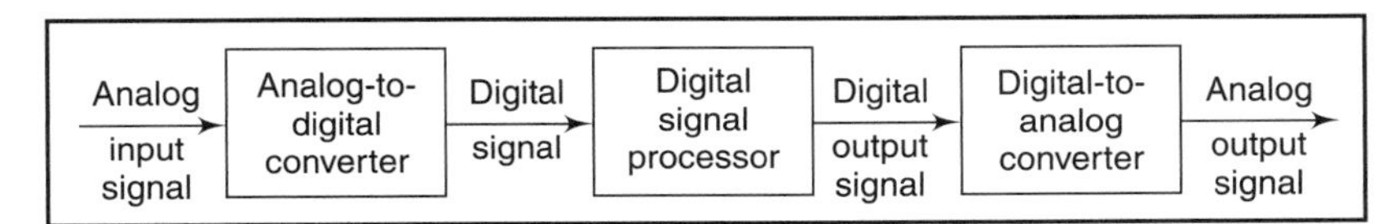

Figure 25.52 :: Basic elements of a digital signal processor (DSP) system

DSPs are available as single-chip devices and are commercially available. The most widely used DSP family is the TMS320 from Texas Instruments. Another range of processors is available from Motorola as DSP56001. As regards the comparison of their speeds, the 16-bit Motorola 68,000 microprocessor can handle 2,70,000 multiplications per second while the DSP56001 is capable of 10,000,000 multiplications per second, thus helping increase the speed 37 times. The flexibility to reconfigure the DSP operations allows them to be used in most of the modern analytical instruments for signal processing applications like transformation to the frequency domain, averaging and in a variety of filtering techniques.

Chapter 26

Digital Circuits

26.1 Digital Circuits

In analytical instrumentation, most circuitry is concerned with the amplification and processing of signals, which are available in an analog form. However, the introduction of compact digital computers have made possible the digital manipulation of the analog signals after they have been converted into a digital form. To do this, an analog-to-digital converter is employed, which basically samples the analog signal at a predetermined rate to get their digital equivalents. Using combinations of pulse circuits such as logic gates and flip-flops, it is possible to carry out arithmetical manipulations upon the series of digital values. Logic circuits are used to test whether a predetermined set of conditions has been obeyed and can route the path of signals accordingly. Flip-flops are used to store information in the form of noughts and ones, corresponding to a transistor being cut-off, or switched on, respectively. They can be cascaded to get an electronic counter. Measuring frequency and voltage are two important applications of electronic counters. The availability of integrated circuit logic modules at low costs has led to an universal use of digital techniques. These modules are used in large quantities in digital computers, because of their small size, low power consumption and high reliability. The small size of MOS elements has enabled to have large-scale integration (LSI), in which thousands of elements are created on a single chip. The digital IC's are available in a big range as simple logic gates, memory units and even as complete data processing units like microprocessors.

Logic circuits

Flip-flops

There does not seem to be any area in modern analytical instrumentation wherein digital circuits are not or will not be used in some form. The basic reason for this is that digital circuits operate from defined voltage levels, and this reduces any uncertainty about the resulting output and behaviour of a circuit. Many circuits operate with voltages that can only be on or off, e.g. a light can be on or off, a motor can be running or stopped or a valve can be open or shut. All these are digital operations and would need digital circuit elements for their operation and control.

26.1.1 Binary Number System

The binary number system is used practically exclusively in digital computers. This is because of its most remarkable feature of simplicity, as just two digits, namely 0 and 1, are used. The binary system has a base of 2 and any number can be expressed in the binary number with powers of 2. For example, the number 10 in the binary number is 1010.

Binary number

$$10 = 1.2^3 + 0.2^2 + 1.2^1 + 0.2^0$$
$$= 8 + 0 + 2 + 0$$
$$10_{10} = 1010_2$$

The subscripts 2 and 10 are used to indicate the base in which the particular number is expressed. The binary system is also a positional value system, in which each digit has its value expressed as powers of 2.

Bits

Byte

Nibbles

Word

The zeroes and ones in the binary notation are commonly called 'bits'. This is an abbreviated form of binary digits. Most frequently, we deal with 8-bit combinations. An 8-bit unit is known as 'byte'. Thus a byte represents numbers in the range 0 to 255. Four-bit units are often referred to as 'nibbles'. Thus, a byte consists of two nibbles. 16-bit unit is generally known as word. A word, thus consists of two bytes and four nibbles.

Binary numbers consisting of many digits are often tedious to handle by the human operators. The digits of binary numbers are therefore bunched together in groups of three or four, octal or hexadecimal notation respectively. Their relationship is shown in Table 26.1

Table 26.1 Binary Numbers and Their Equivalents

Decimal	Binary	Octal	Hexadecimal
1	1	1	1
2	10	2	2
3	11	3	3
7	111	7	7
8	1000	10	8
9	1001	11	9
10	1010	12	A
11	1011	13	B
12	1100	14	C
13	1101	15	D
14	1110	16	E
15	1111	17	F
16	10000	20	10
17	10001	21	11
27	11011	33	1B

In order to provide a shorthand notation for the system of logic based on a single valued function with two discrete possible states, Boolean algebra is used. This type of algebra, based on logical statements that are either true or false, is a very useful tool in the design and troubleshooting of digital logic circuits. The validity of a Boolean statement can be verified by drawing a truth table.

Boolean algebra

26.1.2 Truth Tables

Truth tables provide a tabular means of presenting the output side of logic devices for any set of inputs. Truth tables contain one column for each of the inputs and a column for the output. In basic truth tables, the column notations are usually H or L (for high and low) or, for binary notation, 1 or 0. For example in a logic circuit, the truth table can be represented as:

Input states		Output states
A	B	
0	0	0
0	1	0
1	0	0
1	1	1

26.1.3 Logic Circuits

Logic circuits are decision-making elements in electronic instruments. They are the basic building blocks of the circuits that control data flow and processing of signals. In most systems which use logic, the output function represents a voltage level, either high or low:

There are several ways to represent two state 'yes' and 'no' decisions. Some of these are given below:

Yes	**No**
Open	Closed
1	0
Positive	Negative
True	False
High	Low
On	Off

26.1.4 Logic Convention

In digital circuits, 0 and 1 are represented by two different voltage levels, often called HIGH and LOW. The logic convention usually employed to relate these two entities is as follows (Figure 26.1):

- In the positive logic convention, logic 1 is assigned to the most positive (HIGH) level of the voltage and logic '0' to the least positive (LOW) level.
- In the negative logic convention, logic 1 is assigned to the most negative (LOW) level and logic 0 to the least negative (HIGH) level.

This convention is important to understand the interpretation of digital data. For example, suppose 1001 (binary) data is presented on a set of binary coded decimal output lines. In positive logic, this would mean 1001 (binary) data is presented on a set of binary coded decimal output lines. In positive logic, this would mean 1001 (binary) = 9 (decimal), while in negative logic, the same would mean 0110 (binary) = 6 (decimal).

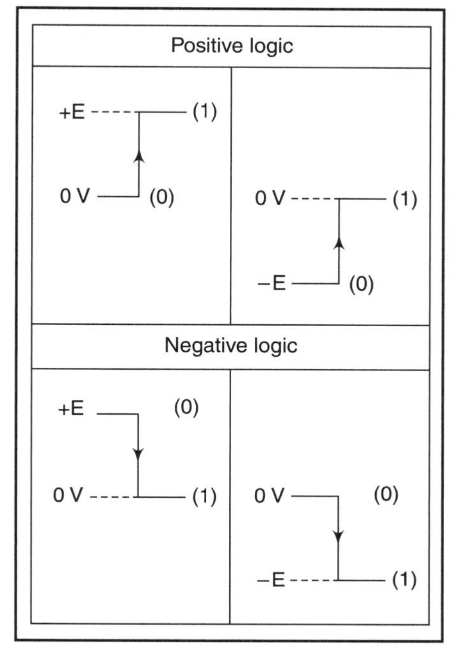

Figure 26.1 :: Logic convention usually employed to represent two levels, high and low in digital circuits

26.2 Types of Logic Circuits

26.2.1 The AND Gate

When the presence of two or more factors is necessary to produce a desired result, the AND gate is employed. This implies that the output of the AND gate will stand at its defined '1' state if and only if, all of the inputs stand at their defined '1' states.

Figure 26.2 shows the functioning of an AND gate, which is like that of a set of switches in series. Only when they are all closed simultaneously, can there be an output, i.e. when the power is applied, both switches (*A* and *B*) must be closed before the lamp *X* will light. The logical notation is expressed in Boolean terms as $AB = X$ (A and B equal X). The graphical representation and the truth table of the AND gate is shown in Figure 26.3.

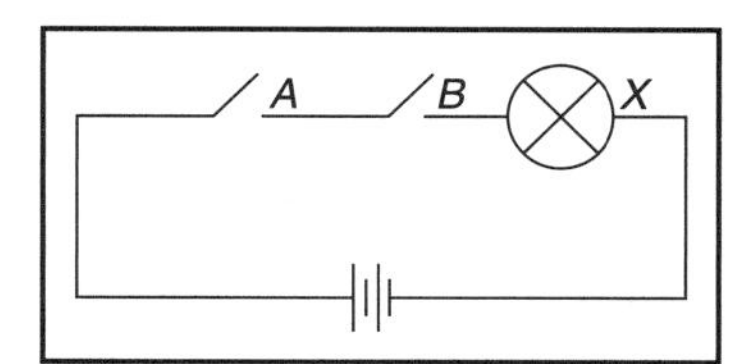

Figure 26.2 :: AND gate equivalent circuit

The AND gate is used primarily as a control element with one input regulating the traffic through others. If a word is to be allowed to pass through the gate, a '1' at the control input will open the gate. The '0's in the word are maintained in the right position at the output, because the gate is closed whenever there is at least one '0' input.

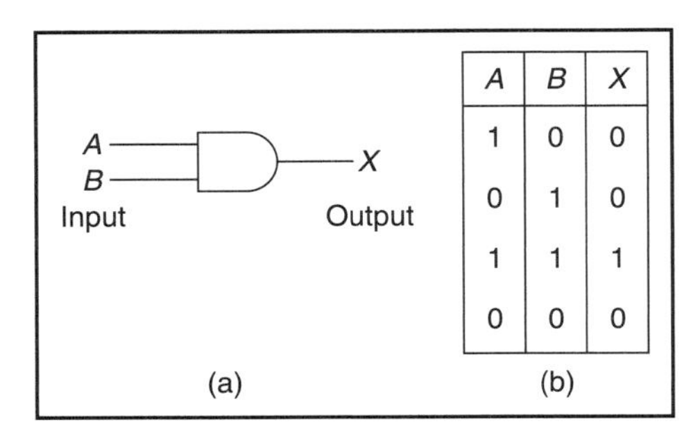

A	B	X
1	0	0
0	1	0
1	1	1
0	0	0

Figure 26.3 :: AND gate (a) Graphical representation (b) Truth table

26.2.2 The OR Gate

The OR gate provides the means of achieving a desired result with a choice of two or more inputs. This means that the output of the OR gate will stand at its defined '1' state if and only if one or more of its inputs stand at their defined '1' state.

As shown in Figure 26.4, the functioning of the OR gate is similar to a set of switches connected in parallel; If either or both of the switches (A and B) is/are closed, power will be applied to the lamp X causing it to glow. The logical notation is expressed in Boolean terms as $A + B = X$ (A or B equals X). The truth table and graphical representation of the OR gate is shown in Figure 26.5.

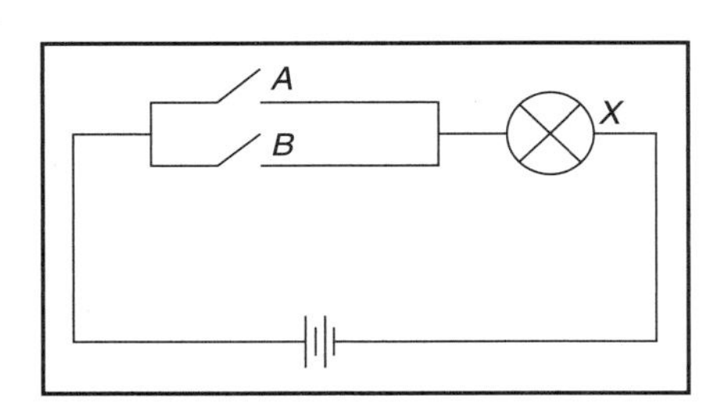

Figure 26.4 :: OR gate equivalent circuit

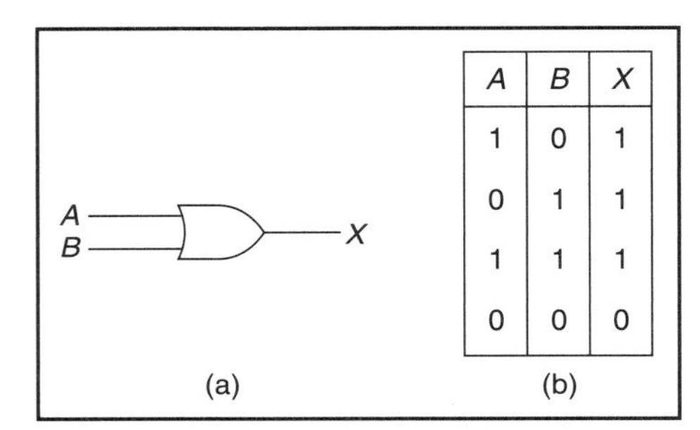

A	B	X
1	0	1
0	1	1
1	1	1
0	0	0

Figure 26.5 :: OR gate (a) Graphical representation (b) Truth table

The OR gate differs from the AND gate, in that a '1' at one input or the other input will give a '1' out-put: hence the OR gate. However, two '0' inputs give a '0' output and two '1' inputs give a '1' output.

The OR gate is designed to prevent interaction or feedback between inputs.

26.2.3 The INVERTOR (NOT) Gate

An INVERTOR is used if it is necessary to change the state of information before it is used. Therefore, the output of an INVERTOR is always the complement of the input, i.e. '0' becoming '1' and '1' becoming '0'. Figure 26.6 shows the symbol and truth table of an INVERTOR circuit.

The INVERTOR is never used by itself, but always in conjunction with another logic element. It is then represented by a small circle directly connected to the other logic element.

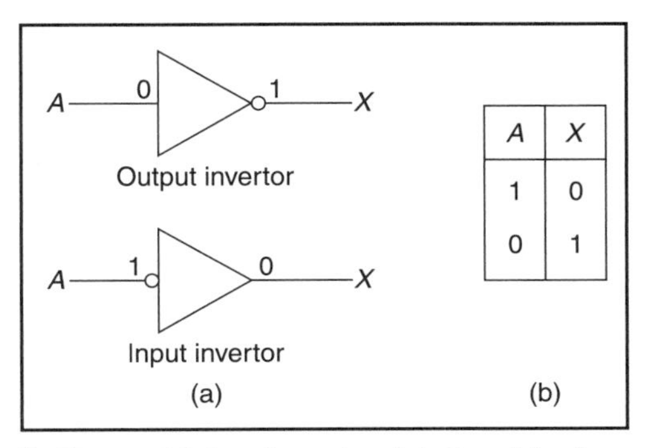

A	X
1	0
0	1

Figure 26.6 :: Invertor (a) Graphical representation (b) Truth table

26.2.4 The NAND (NOT-AND) Gate

When an AND gate has an invertor at the output, the combined circuit is called a NAND gate. This is in effect the opposite of the AND gate. When all inputs are '1' the output is '0'. Typical NAND gate circuitry is shown in Figure 26.7.

The functioning of this gate is equivalent to a number of switches in series, in parallel with a lamp (Figure 26.8). At least one switch must be open in order to make the lamp 'ON'.

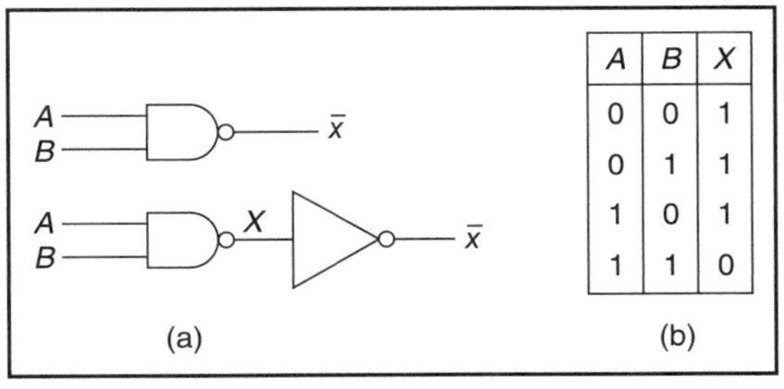

A	B	X
0	0	1
0	1	1
1	0	1
1	1	0

Figure 26.7 :: (a) Typical NAND gate configuration (b) Truth table of NAND gate

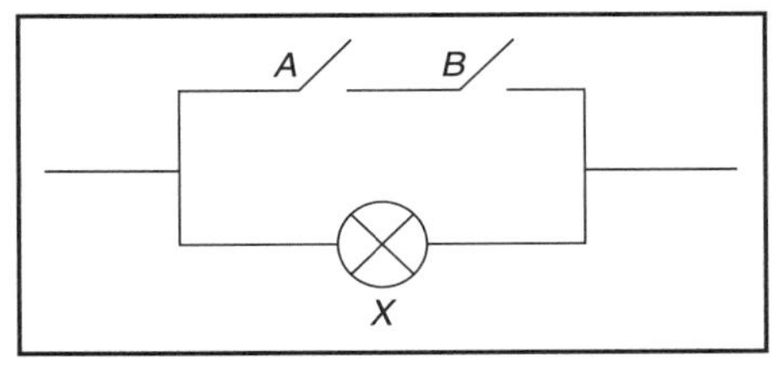

Figure 26.8 :: Equivalent circuit of NAND gate

26.2.5 The NOR Gate

The NOR gate is an OR gate with an invertor in its output circuit, which produces a NOT OR. Thus, when neither one input NOR the other is a '1', the output is '1'; hence the name NOR gate. The symbol and truth table of NOR gate is shown in Figure 26.9.

The functioning of NOR gate is similar to that of a number of switches in parallel with a lamp. All switches must be open if the lamp is to glow (Figure 26.10).

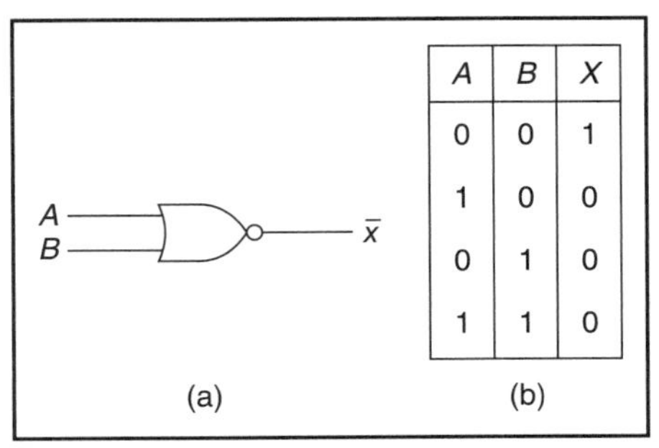

A	B	X
0	0	1
1	0	0
0	1	0
1	1	0

Figure 26.9 :: NOR gate (a) Symbol (b) Truth table

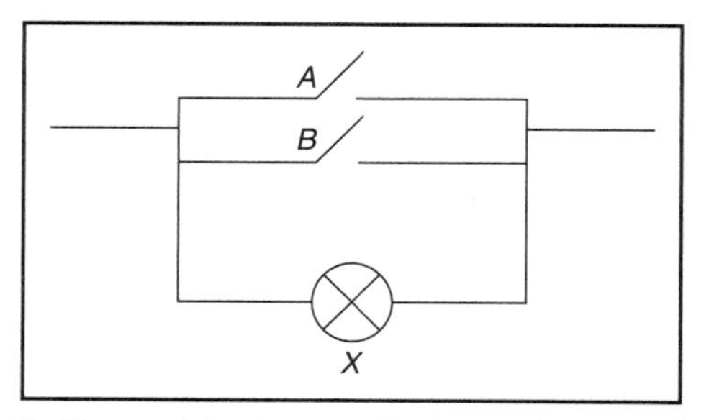

Figure 26.10 :: Equivalent circuit of NOR gate

26.2.6 The EXCLUSIVE-OR (EX-OR) Gate

The EX-OR gate may be regarded as a combination of AND and OR gates. It produces a 1 output only when the two inputs are at different logic levels.

A (two-input) EX-OR gate as shown in Figure 26.11 can be regarded as the combination of two AND gates and an OR gate.

An EX-OR gate has always two inputs, and its output expression is:

$$X = A + B$$

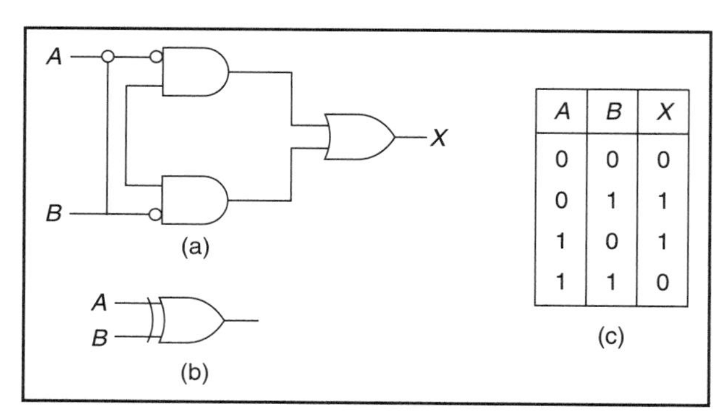

A	B	X
0	0	0
0	1	1
1	0	1
1	1	0

Figure 26.11 :: EX-OR gate (a) Configuration (b) Symbol (c) Truth table

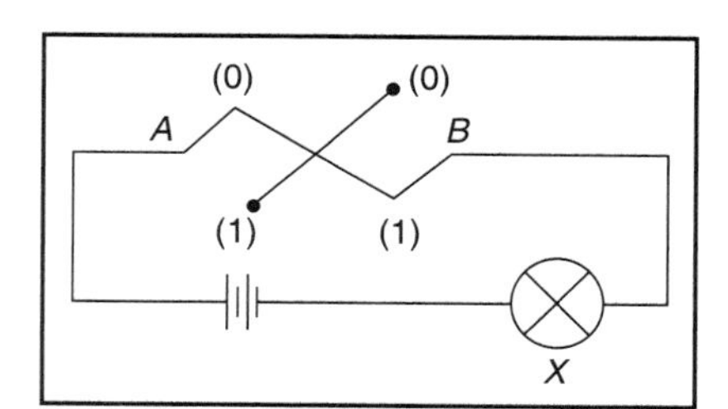

Figure 26.12 :: EX-OR equivalent circuit

The equivalent circuit of an EX-OR gate is given in Figure 26.12.

26.2.7 The INHIBIT Gate

The INHIBIT gate is an OR gate with an inhibiting input. In this gate, the output will stand at its '1' state if, and only if, the inhibit input stands at its defined '0' state, AND one or more of the normal OR inputs stand at their defined '1' state.

The symbol and truth table of the INHIBIT gate is shown in Figure 26.13 The equivalent circuit of an INHIBIT gate is shown in Figure 26.14.

This gate is very useful for controlling inputs (*A* and *B*) by means of the inhibiting signal *C*. When the inhibiting signal is present ($C = 1$), the output is always OFF ($X = 0$), but when the inhibiting signal is absent ($C = 0$), the signals *A* and *B* can pass to the output *X*.

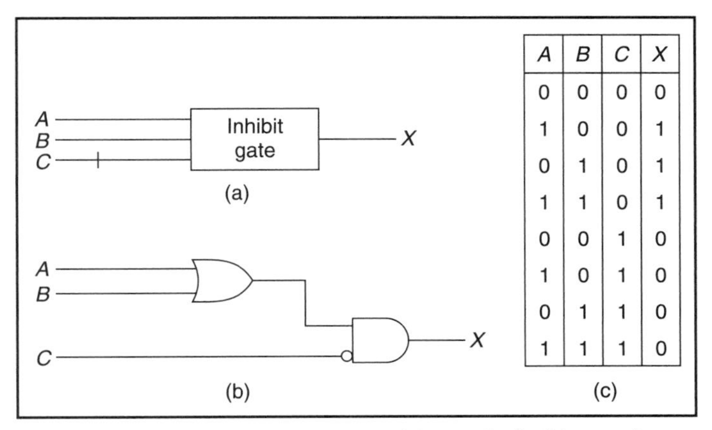

A	B	C	X
0	0	0	0
1	0	0	1
0	1	0	1
1	1	0	1
0	0	1	0
1	0	1	0
0	1	1	0
1	1	1	0

Figure 26.13 :: INHIBIT gate: (a) Symbol, (b) Configuration (C) INHIBIT gate truth table

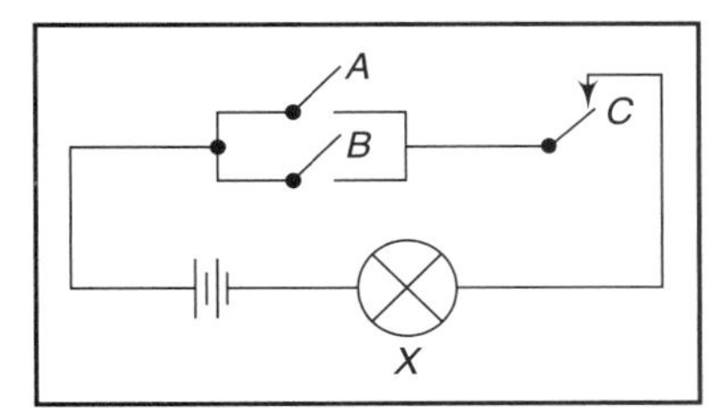

Figure 26.14 :: Equivalent circuit of INHIBIT gate

26.3 Logic Families

The nature of the basic logic elements depends upon the properties of the electrical components used to realize them. In the early days of digital techniques, when diodes were largely used in the circuitry, it was natural to take the AND or OR gates as the basic elements. Later, when transistors came to the fore, it became natural to take basic logic circuits on the NAND and NOR gates. This is because the output signal of the transistor is opposite to the sign of its input. The most popular and most widely used circuits in modern digital equipment are the transistor–transistor logic and complementary metal–oxide semiconductor logic families. The logic circuits have become increasingly complex and bulky. The introduction of integrated circuits and in particular the development of planar technology have solved the problem of bulk along with the possibilities of obtaining several functions on one chip.

26.3.1 Transistor-Transistor Logic (TTL)

The most popular and widely employed logic family is the transistor-transistor family. The various logic gates are manufactured in the integrated circuit form by most manufacturers of semiconductors.

The basic function of the TTL gate is the NAND function. This is shown in Figure 26.15. The input transistor has a number of emitters equal to the desired fan-in of the circuit. The multi-emitter input has the advantage, that less space is required on the semiconductor chip for a given number of inputs.

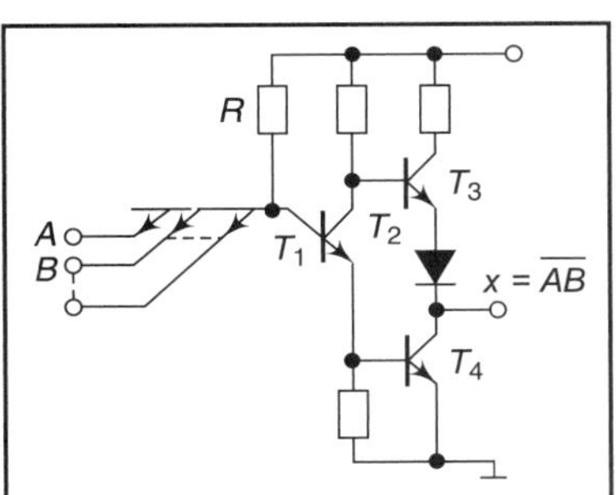

Figure 26.15 :: Basic transistor–transistor logic (TTL) NAND gate

If one or more inputs are at logic '0', current will flow through input resistance *R*. Consequently, the collector of the input transistor will be LOW. Only when all inputs are HIGH will the collector be HIGH too. The input circuit in fact gives normal AND operation. The next stage acts as a kind of phase splitter for driving the totempole' output. When T_2 is cut-off, T_4 remains cut-off too. However, if T_3 is on, T_4 will be on and T_3 off, resulting in a LOW output, which is a NAND function. The diode in the output chain ensures that T_3 is cut-off when T_4 goes on.

The popularity of TTL family rests on its good fan-in and fan-out capability, high speed (particularly Schottky TTL version), easy interconnection to other digital circuits and relatively low cost. The main characteristics of TTL logic are: propagation delay 10 ns, flip-flop rate 20 MHz, fan-out 10, noise margin 0.4 V, dissipation per gate 10 mW. The standard TTL gates are marketed as 74 series which can operate up to 70°C. However, 54 series are operatable up to a temperature of 125°C. Most IC packages contain more than one gate. For example IC 7400 is a quad 2-input NAND gates whereas 7420 is a dual 4-input NAND gates. There are various types of TTL families mostly differing only in speed and power dissipation.

The propagation time of 10 ns for most TTL circuits is too high for certain applications. This can be reduced by keeping out the transistor off saturation region by clamping the base collector junction to a voltage below the saturation level. This is achieved in Schottky TTL logic family. Schottky TTL logic gates are available in the integrated form as 74S/54S series. A low power Schottky TTL series is also commercially available as 54LS/74LS. The gates in the family are faster than standard TTL and consumes much less power. The internal power dissipation is minimized by increasing resistance values.

26.3.2 Emitter-Coupled Logic (ECL)

This (ECL) family provides another means of achieving higher speed of the gate. This differs completely from the other types of logic families, in that the transistors when conducting are not saturated, with the result that logic swings are reduced. For example, if the ECL gate is operated from 5 V, the logic '0' is represented by 0.9 V and logic '1' by 1.75 V.

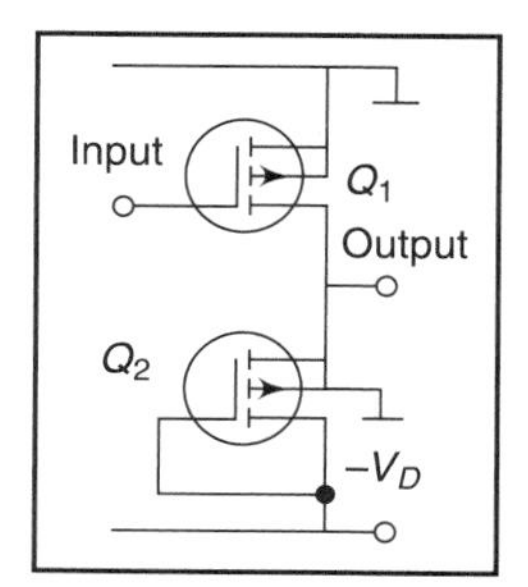

Figure 26.16 :: MOS invertor logic gate configuration

26.3.3 CMOS Logic Families

The complementary metal-oxide semiconductor (CMOS) logic families offer significant advantages over bipolar transistor-based logic circuits, particularly because they feature very low power dissipation and good noise immunity.

The function of a MOS logic gate can be explained with the MOS invertor shown in Figure 26.16. When the input is '0', the driver MOST Q_1 (metal oxide semiconductor transistor) is OFF and Q_2 (lower) pulls the output to the –V (minus the threshold voltage) level, i.e. logic '1'. The threshold voltage for MOSTs is defined as the gate voltage at which conduction between source and drain first starts. When the input is '1', Q_1 is conducting and thus has a low impedance. The output voltage thus drops to logic '0'. When two or more driver MOSTs are connected in parallel to the inverter, a NOR gate is obtained. Replacing the driver MOST Q_1 in series gives, in negative logic, a NAND gate.

The main disadvantage associated with MOS gates is their limited switching speeds as compared to bipolar circuits. This is due to the very high impedance of the MOST devices, because of which stray capacitances cannot be charged quickly. The typical switching times are of the order of 1 μs. One way of improving switching times is to make use of CMOS (complementary MOS) techniques, in which both p-channel and n-channel MOSTs are employed. Figure 26.17 shows a typical configuration of CMOS inverter. With this arrangement, only one device is ON (low impedance) at a time, the other device being OFF (high impedance), resulting in a lower power consumption. With a '0' at the input, the n-channel MOST will be OFF and the p channel will be ON, which gives an output at logic '1'. With a '1' at the input,the conditions are reverse, with an output voltage at logic '0'. The advantage of the circuit is that, when the impedance of one device is decreasing, the impedance of the other is increasing, giving a push-pull effect, which narrows the transition region, giving sharper transfer characteristics and thus increasing the speed.

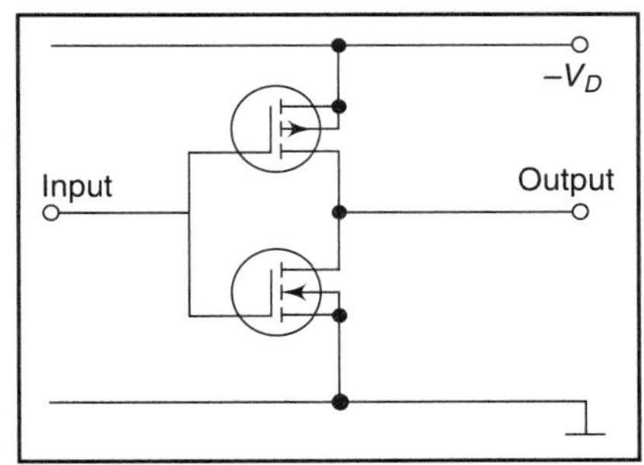

Figure 26.17 :: Circuit configuration of CMOS invertor

A number of p-channel gates connected in parallel are used to construct NOR gates. Similarly, a NOR gate is made by using a number of such devices in series.

A great advantage of the MOS technique is that it facilitates high density packing of a large number of devices. The technique is most suitable for the construction of large-scale integrated circuits rather than simple gates and flip-flops. Commercially available CMOS gates are available as 4000 series. For example, Quad 2-input AND gate in CMOS comes as 4081(7408 TTL) while Quad 2-input NOR gate as 4001(7402 TTL).

26.3.4 Characteristics of Integrated Circuit Logic Gates

The various families of logic gates are associated with different characteristics. This means that one of them may prove to be best suited for a particular application. The important characteristics are as follows:

- *Speed of Operation:* The speed of operation of a logic gate is the time required by it to pass from one state to another. This is generally expressed in terms of the propagation delay. The propagation delay of a gate takes place on account of the switching time of a transistor and rise time of the switching input voltage. The rate at which a flip-flop can switch from one state to the other is called its clock rate.

- *Threshold Value:* The threshold voltage of a gate circuit is defined as the input voltage at which the gate just switches from one state to the other. For TTL logic family, the threshold voltage is 1.4 V (Figure 26.18). However, the maximum input voltage that will definitely give logic '0' is 0.8 V, where as the minimum input voltage giving a definite logic '1' is 2.0 V.

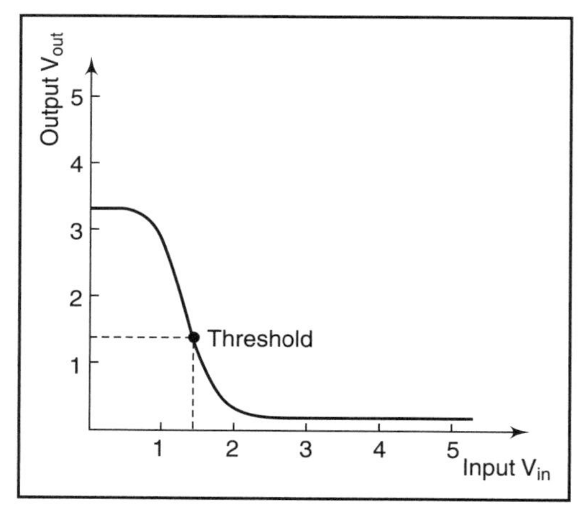

Figure 26.18 :: Positive logic threshold level in TTL gate

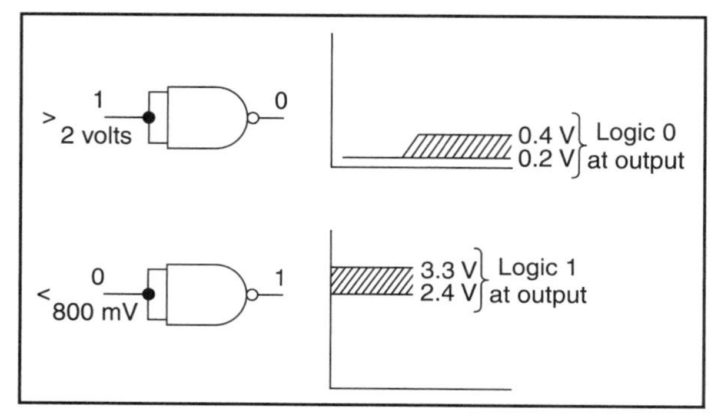

Figure 26.19 :: Logic levels at gate outputs

For correct operation of a gate, specific voltage levels must be applied. For example, for a TTL gate, logic '1' has a typical voltage of 3.3 V and a minimum value of 2.4 V (Figure 26.19). On the other hand, logic '0' is typically 0.2 to 0.4 V.

- *Noise Margin:* In order to avoid errors in a logic system due to parasitic voltages like spikes, logic devices should have a wide voltage swing between the two states, i.e., logic devices should have wide noise margin. Figure 26.20 shows noise margins in high and low states.
- *Power Dissipation:* Power dissipation generally implies the power required for operation of the logic device. As circuit complexity increases, the power dissipation per gate must decrease, so as to limit the amount of heat which may be dissipated in the semiconductor junctions.
- *Fan-in and Fan-out:* The fan-in of a gate is the number of inputs which can be connected to the gate without seriously affecting its performance. Similarly, the fan-out is defined as the maximum number of circuits that can be connected to its output terminals without the output falling outside the specified logic levels.

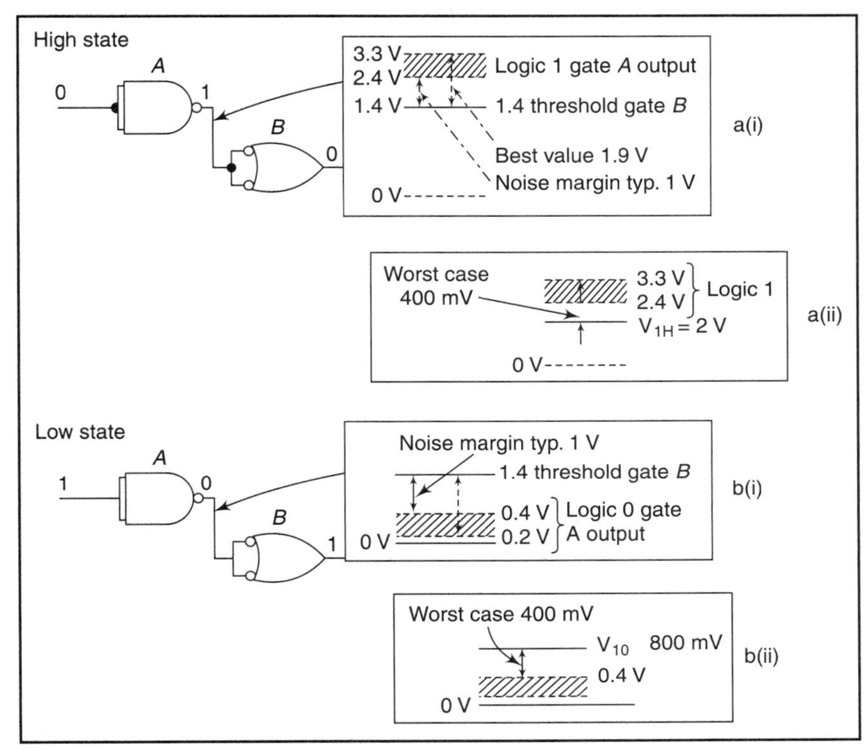

Figure 26.20 :: Noise margins in TTL gate
(a) High state (i) Typical 1 V (ii) Worst case 400 mV
(b) Low state (i) Typical 1 V (ii) Worst case 400 mV

26.4 Categories of IC's Based on Packing Density

- *SSI (Small scale integration)* means integration levels typically having about 12 equivalent gates on chip. Available in 14 or 16 pin DIP or Flat packs.
- *MSI (Medium scale integration)* means integration typically between 12 and 100 equivalent gates per IC package. Available in 24 pin DIP or Flat pack or 28 pin ceramic chip carrier package.
- *LSI (Large scale integration)* implies integration typically upto 1000 equivalent gates per IC package. Includes memories and some microprocessor circuits.
- *VLSI (Very large scale integration)* means integration levels with extra high number of gates, say upto 100000 gates per chip. For example, a RAM may have more than 4000 gates in a single chip, and thus comes under the category of VLSI device.

26.5 Typical Digital Integrated Circuits

26.5.1 Flip-Flops

A flip-flop is a two-state electronic device, which can be either turned on or turned off when commanded to do so. It is a bistable logic element with one or more inputs and two complementary outputs. A flip-flop essentially remains in its last state, until a specific input signal causes it to change state. Because of the ability of the flip-flop to store bits of information (1 or 0), it has memory characteristics. It is thus one of the most important basic building blocks in digital circuitry.

Nearly all flip-flops have two output levels [Figure 26.21(a)] designated as Q and $\bar{Q}$ on which the true state (Q) and the complement ($\bar{Q}$) of the stored function is available. The input terminals may receive either discrete level or pulse signals depending on the circuit. A flip-flop is known by several other names; the most common being bistable multivibrator, binary or latch.

The truth table [Figure 26.21(b)] indicates the two possible output conditions for a flip-flop and the corresponding definition for the state. There are many forms of flip-flops, each of which has its specific features. However, all those various forms of flip-flops contain essentially the same bistable element, the RS flip-flop.

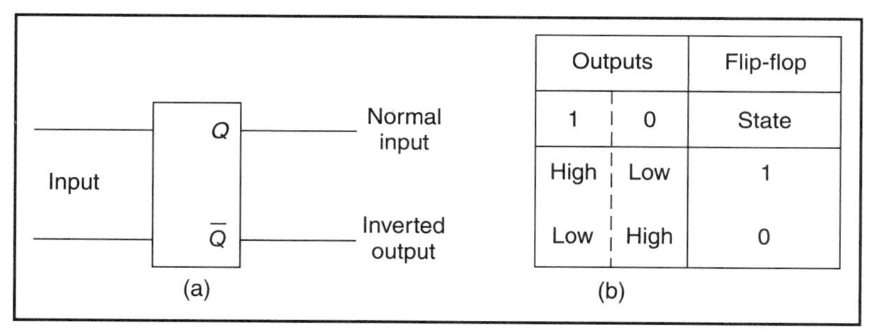

Figure 26.21 :: (a) Basic flip-flop configuration (b) Truth table of a basic flip-flop circuit

Reset-set Flip-flop or the R-S Flip-flop: A basic flip-flop can be constructed by using two NAND gates (Figure 26.22) cross-coupled to the inputs. When power is applied, opposite states will appear on the outputs of gates A and B. If the Q output of gate—A is '1', this '1' will be applied to the input of gate B whose output ($\bar{Q}$) will then become '0'. When this '0' is applied to the input of gate A, a '1' will remain on the Q output of gate A. Thus, the gates are latched into a stable state. Since output Q is high, the flip-flop is in the high or '1' state. Any additional pulse at the SET input will have no effect on the output. However, when a pulse is applied to the RESET input, the output reverses or flips. Further, pulses at the RESET input have no effect on the outputs. Switching the inputs again causes the output to flop back to their original condition.

The flip-flop is like a toggle switch, either in one position or the other; and once the change-over has been made, repeating the action has no further effect. The condition is stable, either way. The R-S flip-flop is commonly used when there are no possibilities of simultaneous set and reset inputs. If both the inputs are simultaneously enabled with a low pulse, both outputs will go high for the duration of the pulse. Removing the enabling pulses will make the flip-flop to go over to an indeterminate state, i.e. it could latch in either the '1' or '0' condition. The stable state to which it will finally race depends upon the relative time delay of the two NOR gates used in the circuit. Though R-S flip-flop in itself has limited applications, it is the basic building block of flip-flop chains in integrated circuit form. They normally utilise a clock input to synchronise the changes from one state to another.

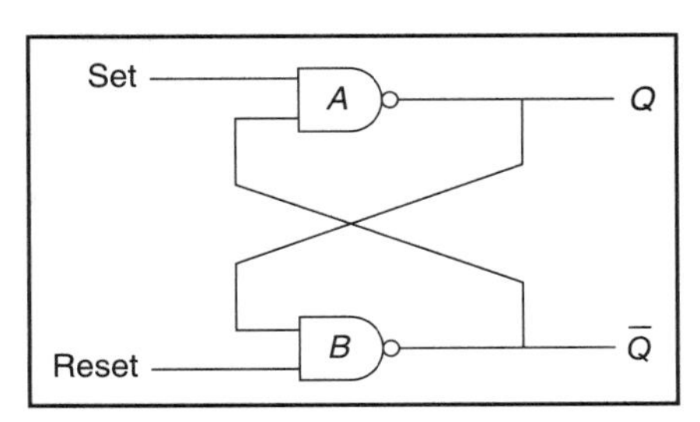

Figure 26.22 :: R-S flip-flop configuration

Clocked Flip-Flops: The problem of races can be avoided by using a synchronous or a clocked flip-flop, in which the inputs can only be applied in coincidence with a clock signal. It is convenient to enable or condition the flip-flop by applying the appropriate levels to the inputs first and then arrange for the flip-flop to change state on receipt of a pulse from another source. The pulse is called clock, which may be from an oscillator circuit. Thus, the clock signal is a signal that initiates action at regular spaced intervals. Figure 26.23 shows a typical clock signal. The operations in the system take place at the time when the transition occurs from 1 to 0 (falling edge) or 0 to 1 (rising edge).

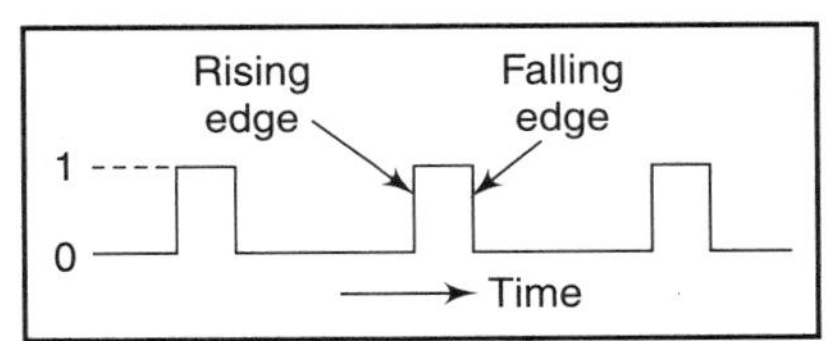

Figure 26.23 :: Typical clock signal

Clock signal

Clocked flip-flops are designed to change states at the appropriate clock transition and rest between successive clock pulses. Figure 26.24 shows the circuit configuration of a clocked R-S flip-flop. The flip-flop section (gates *C* and *D)* is identical to the R-S flip-flop. In addition, there is a circuit arrangement which applies clock pulses to either the SET or RESET input terminal of the flip-flop. If the SET input line is enabled with a high signal, gate *A* will be enabled when a high pulse is presented at the clock input. When gate *A* is enabled, it will provide a low SET signal to the R-S flip-flop, which will go to the '1' condition.

If the SET input line has a low level and the RESET line is enabled with a high level, gate *B* will be enabled when a high clock pulse occurs. When gate *B is* enabled, a low RESET signal will be presented to the flip-flop and it will go to the '0' condition. It may be noted that the flip-flop will only change state when a clock pulse is applied.

D Flip-Flop: One way of avoiding the intermediate state found in the operation of the simple R-S flip-flop is to provide only one input, which can be either high or low. This input is called the D input or data input and the flip-flop thus constructed is called D flip-flop. Figure 26.25 shows the

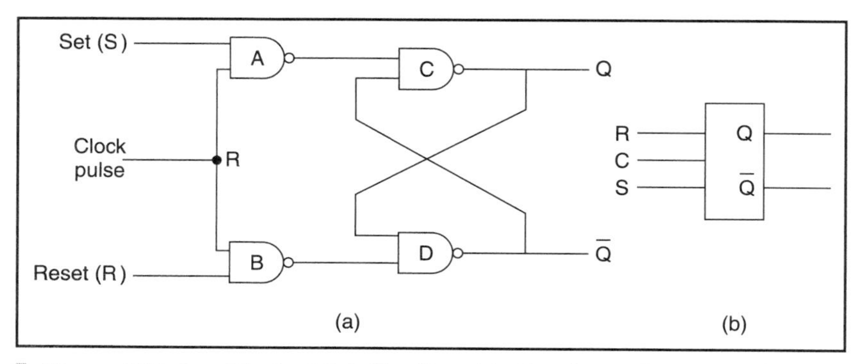

Figure 26.24 :: Clocked R-S flip-flop: (a) Circuit configuration (b) Symbol

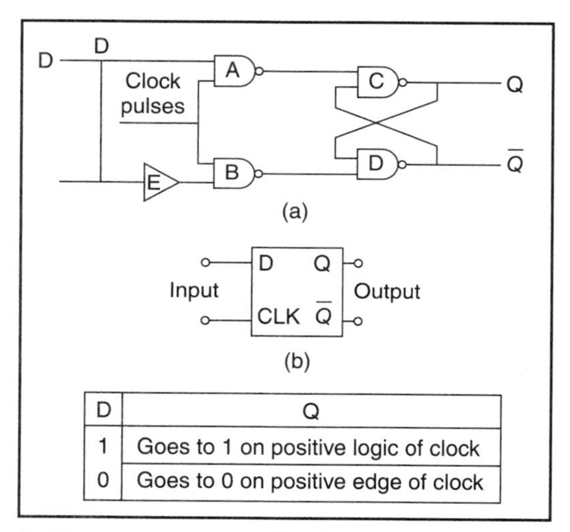

D	Q
1	Goes to 1 on positive logic of clock
0	Goes to 0 on positive edge of clock

Figure 26.25 :: D flip-flop: (a) Circuit configuration (b) Representative symbol (c) Truth table for D flip-flop

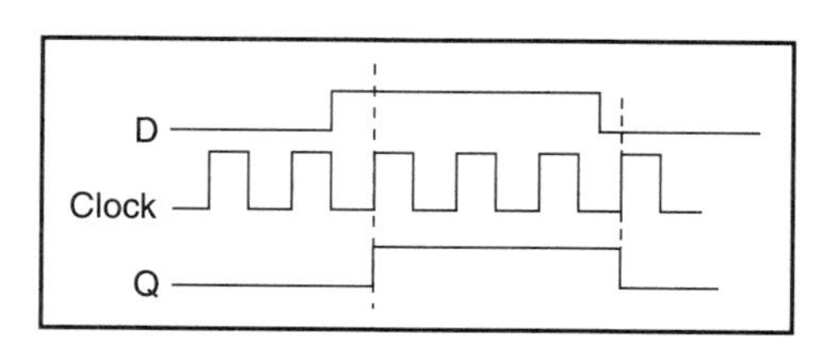

Figure 26.26 :: Positive edge triggering in a D flip-flop

circuit arrangement of a D-type flip-flop. A '1' or a '0' applied to this input is passed directly to one of the inputs of the flip-flop proper and inverted to the other input.

From Figure 26.26, it is obvious that whatever information is present at the D input prior to and during the clock pulse is propagated to the Q output when the clock pulse is applied, while the inverse of that information appears at the $\bar{Q}$ output. The flip-flop is thus set in the '1' state if the D input is made '1' and in the '0' state if the D input is made '0'. This type of circuit is known as edge-triggered D flip-flop and is one of the most commonly used in computers. In D-type latch, which is similar to the D flip-flop, except that it can change states during the HIGH portion of the clock signal, i.e. as long as clock is HIGH, Q will follow the D input even if it changes when the clock goes LOW, Q will store (or latch) the last value it had and the D input has no further effect.

J-K Flip-Flop: One of the most useful members of the flip-flop family is the J-K flip flop. A unique feature of the J-K flip-flop is that it has no ambiguous state. It is the most widely used flip-flop type in logic circuitry and is the ideal memory element to use.

A J-K flip-flop often has more than one J input and more than one K input. In this case, one J input and one K input are generally connected together for use as input for clock pulses. This kind of circuit arrangement is shown in Figure 26.27. The operation of the circuit is controlled as follows:

- A clock pulse will not cause any changes in the state of the flip-flop if neither the *J* nor the *K* input is activated.
- If both the *J* and *K* inputs are activated, the flip-flop will change state when the next clock pulse is received.
- The *J* and clock pulse inputs used together set the flip-flop in the set position, while the *K* and clock pulse inputs reset it.

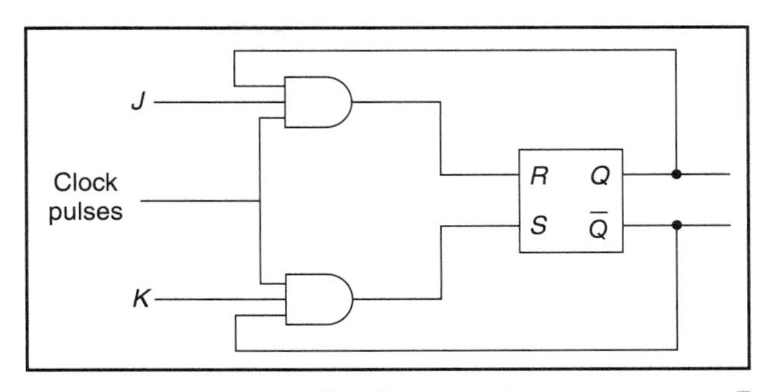

Figure 26.27 :: J-K flip-flop circuit arrangement

It may be observed that propagation delay time prevents the *J-K* flip-flop from racing (toggling more than once during a positive clock edge). This is because the outputs change after the positive clock edge has struck. By then, new *Q* and $\overline{Q}$ signals are too late to coincide with positive spikes driving the AND gates. Commercially available *J-K* flips-flops also give the facility of a synchronously setting the output to 1 (preset) or 0 (clear). Integrated circuit DM 74733 has two *J-K* flip-flops with clear facility. Dual *J-K* with separate preset and clear facilities are available in DM 74766 and with common preset and clear in DM 7478.

Master-slave flip-flop

It may be noted that the output of *J-K* flip-flop becomes stable only when the clock pulse goes to zero. To avoid this problem, a special form of the '*J-K* flip-flop known as the '*J-K* master-slave flip-flop' is used. As the name implies, it comprises two flip-flops - a master and a slave - being triggered at positive and negative edge of the clock respectively. This is advantageous, because a sequence of such master-slave flip-flops could then be triggered simultaneously by a clock pulse derived from the same source, without ambiguity. Figure 26.28 shows master-slave *J-K* flip-flop.

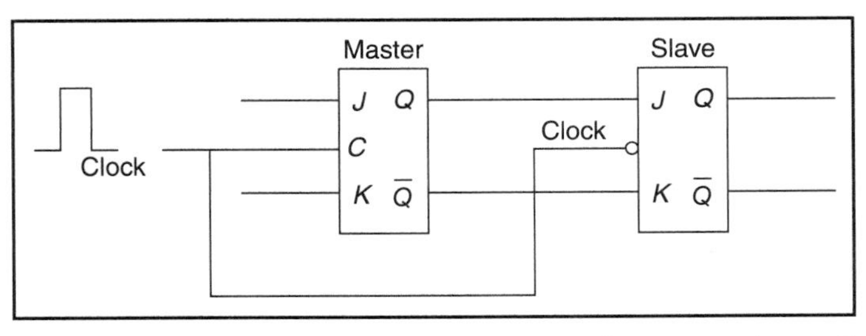

Figure 26.28 :: Circuit configuration of master-slave J-K flip-flop

T flip-flop

If *J* and *C* inputs of the *J-K* flip-flop are connected together it becomes a toggle or a *T* flip-flop, whose output toggles whenever a '1' is applied at the input.

Flip-flops are the basic blocks, using which, it is possible to build many other useful sequential circuits, some of which are described in the subsequent section.

26.5.2 Counters

A counter is a sequential circuit consisting of a series of flip-flops, which go through a sequence of states on the application of pulses at its input. Counters are constructed out of T or J-K flip-flops.

In Figure 26.29 it may be seen that three *J-K* flip-flops are combined to give a three-bit counter. One output gives the information at each stage, the other is to carry to the next stage. Let us assume that all outputs at Q are initially cleared to '0'. A pulse applied to the input of the first stage switches its counting output *Q* (*A*) to '1', while the other output $\bar{Q}$ (*A*) becomes '0' and thus has no effect on the second stage. The circuit has now counted the first pulse (binary 1). On the receipt of the second pulse, the count output *Q* (*A*) of the first stage switches back to '0', while the carry output $\bar{Q}$ (*A*) flips to '1'. This switches the second stage (*B*) to count '1'. The circuit has now counted two, which in binary notation is 10. The third pulse switches stage one to count '1', *Q* (*A*), but the second stage is not switched. The count is 11, binary notation for three. The counting continues in this fashion until the circuit output is one-one-one, which as we know is 7 in binary notation. The eighth count resets the circuit to zero. The pulse diagram of a 3-bit binary counter is shown in Figure 26.29 (a). Figure 26.29 (b) gives the truth table of this counter.

Decimal counters

Decimal counters are binary counters which employ four flip-flops and are so constructed that they count up to 9 (instead of *15)* and reset to 0000 on receipt of the tenth input pulse. By cascading more flip-flops to the chain, we can build a ripple counter of any length. Eight flip-flops give an 8-bit ripple counter, 12 flip-flops result in a 12-bit ripple counter and so on.

Ripple counters

The binary counters as shown in Figure 26.29 are also called ripple counters, because the carry moves through the flip-flops like a ripple on water. Because of this, the most significant flip-flop cannot change states until three propagation delay times have elapsed. Because the carry has to propagate from the least to the most significant flip-flop, ripple counters are too slow for some application. This problem is overcome by using *synchronous counters,* in which all flip-flops clock simultaneously. Because of this, the correct binary word appears after one propagation delay time rather than three, thus offering high speed of operation. For an 8-bit counter, the maximum binary number is 1111 1111, which is equivalent to decimal 255. Similarly, for 16-bit counter, the binary number and its decimal equivalent are:

Synchronous counters

1111 1111 1111 1111 = 65,535

In order to design a counter, one has to choose the kind of flip-flop to be used and decide the number of flip-flops required. These counters are capable of counting only upwards from zero to some maximum count and then reset to zero. Counters have also been designed to count in either direction, which are called up/down counters.

up/down counters

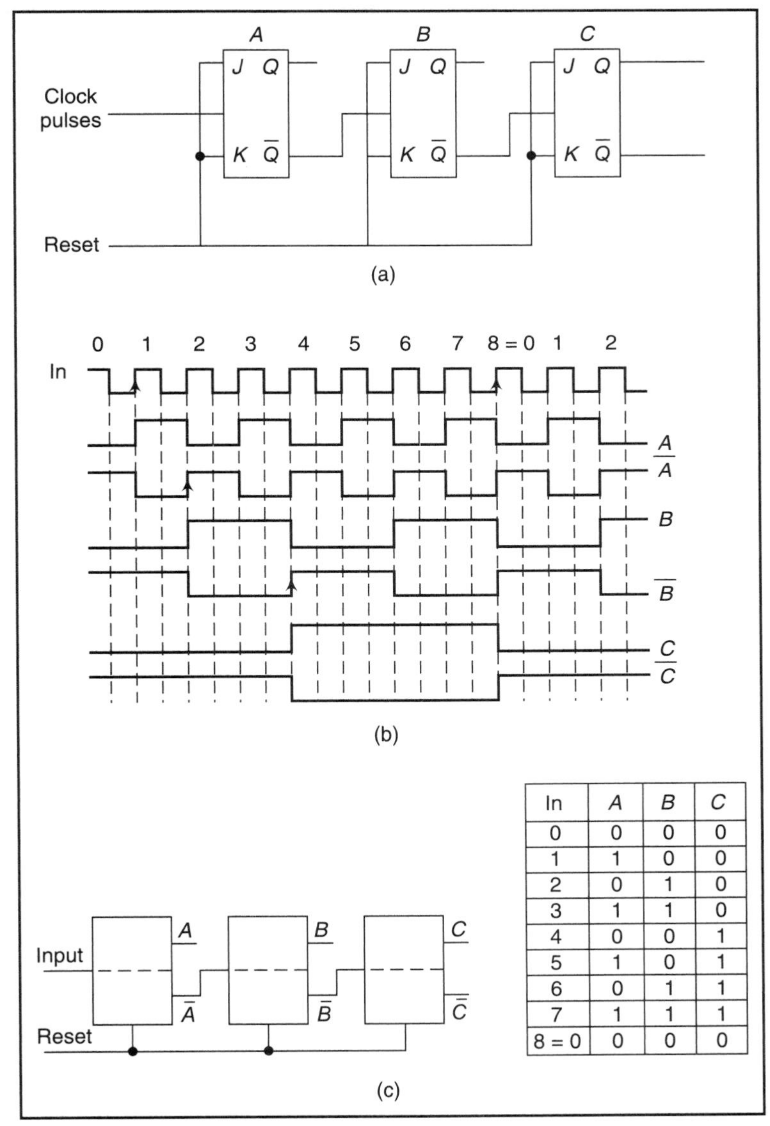

In	A	B	C
0	0	0	0
1	1	0	0
2	0	1	0
3	1	1	0
4	0	0	1
5	1	0	1
6	0	1	1
7	1	1	1
8 = 0	0	0	0

Figure 26.29 :: (a) Logic diagram of a 3-bit counter (b) Timing diagram of a 3-bit counter (c) Truth table of a 3-bit counter

These counters have an up/down input, which is used to control the counting direction, i.e. one type of logic level causes the counter to count up from zero and the other logic level applied to it causes the counter to count down, say from the 1111 to 0000 (in 4-bit counters). In some up/down counters, two separate clock inputs are employed, one for counting up and the other for counting downwards. Up/down counters fall in the class of controlled counters, i.e. where the counting sequence can be changed by changing the value of a control signal.

A wide variety of counters are commercially available as standard integrated circuit packages. They preclude the need for constructing counters using individual flip-flops. Among the popular counters are DM 7492, which is a divide by 12 counter, and a binary counter. DM 74191, DM 74192 are synchronous up-down counters.

26.5.3 Registers

A register is a group of memory elements employed to store binary information. Registers have an important place in digital computers, because the very operation of computers is based on transferring binary information from one register to another and carrying on certain operations before it is again transferred. The simplest register is a flip-flop.

In digital circuitry, a register usually consists of parallel latches. It can represent a number in the range from 0 to $2^n - 1$, where n is the number of latches in parallel. The register works under the control of a clock, which signals when the register should record the input, the data at the last latch always appears at the output. Registers are internal to the microprocessor and are very important because of the rather lengthy process involved in accessing data in memory. For example, intermediate results can be temporarily kept in registers, rather than returning them to main memory repeatedly. Hence, the number of programmable registers in the microprocessor (central processing unit) is very important.

Shift Registers: A shift register is a group of serially connected flip-flops that is used for the temporary storage of information. They can also be used for shifting of the information stored in the register either one position to the right or left with each clock pulse. This is accomplished by gating the outputs of the flip-flops to the appropriate inputs for performing either a left or right shift. This shift direction is controlled by a mode input.

A shift register can be built using series connected R-S, J-K or D-type flip-flops. They are so connected that the output of each flip-flop becomes the input to the next flip-flop. As the register is clocked, the data is shifted one position to the left or right for each clock pulse. They are, however, constructed using a number of integrated flip-flops of TTL or CMOS families. The capacity of an integrated shift register ranges from 4 bits in the TTL family to 2048 bits in the CMOS family. CMOS shift registers are usually only of the serial-in-serial-out type, because of the many stages involved. There are not enough pins for parallel-input or output connections.

Figure 26.30 shows a 4-bit shift register employing D flip-flops. This is a serial input shift register. Initially, a clear pulse (logic 0) is applied to the RESET which sets the outputs at Q_A' Q_B' Q_C and Q_D to 0. Next, the first data bit (D_1) is applied to the SERIAL INPUT. A pulse will appear at Q_A at the leading edge of D_1. When the next data bit D_2 is applied to the input, $Q_A = D_2$ and $Q_B = D_1$. Continuing this process after four clock pulses, $Q_A = D_4$, $Q_B = D_3$, $Q_C = D_2$ and $Q_D = D_1$.

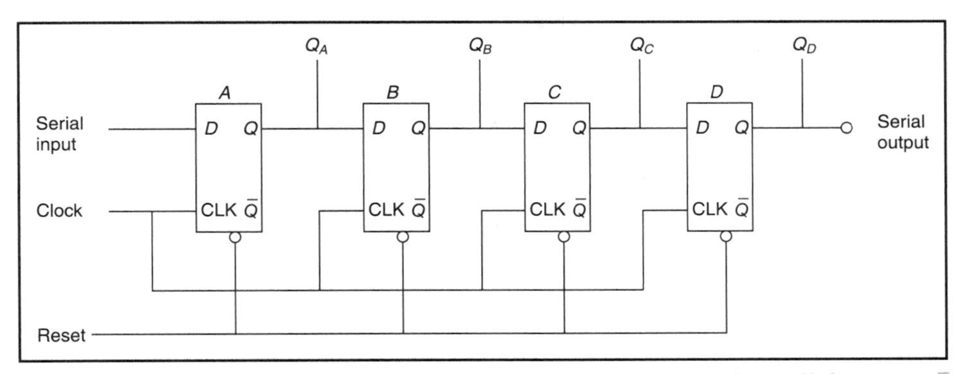

Figure 26.30 :: 4-bit shift-register with serial entry and serial parallel output

The circuit explained above pertains to serial-to-parallel converter. This arrangement is not practical in shift registers with a large number of bits. In such cases, serial output registers are employed.

Methods of Using Shift Registers: Shift registers can be used in any one of the following four ways:

- Serial-in/parallel-out,
- Parallel-in/serial-out,
- Serial-in/serial-out, and
- Parallel-in/parallel-out.

With a serial-in/parallel-out shift register [Figure 26.31(a)] data are fed in the serial form, and when the complete word is stored, the bits are read off simultaneously from the output of each stage.

Figure 26.31(b) shows parallel-in/serial-out shift register in which the data is stored, after clearing all the stages, in each flip-flop. The data are then read out serially, i.e. one bit at a time under the control of the clock.

The serial-in/serial-out shift register [Figure 26.31(c)] merely acts as a temporary delay circuit. The data are read out in the same order in which it has been stored.

Similarly parallel-in/parallel-out shift registers [Figure 26.31(d)] act as a temporary storage.

26.5.4 Multiplexer

A multiplexer is a logic circuit which accepts several data inputs and outputs only one of them at a time. In essence, it behaves as a multi-position switch which operates under the control of SELECT or ADDRESS inputs. Figure 26.32 shows the representation of a digital multiplexer.

Multiplexers are available in the form of integrated circuits. For example, 7415 lA is an eight input multiplexer which gives out complementary outputs.

Other popular multiplexers are 74153, which is dual 4:1 multiplexer and 74150, which is 16:1 multiplexer.

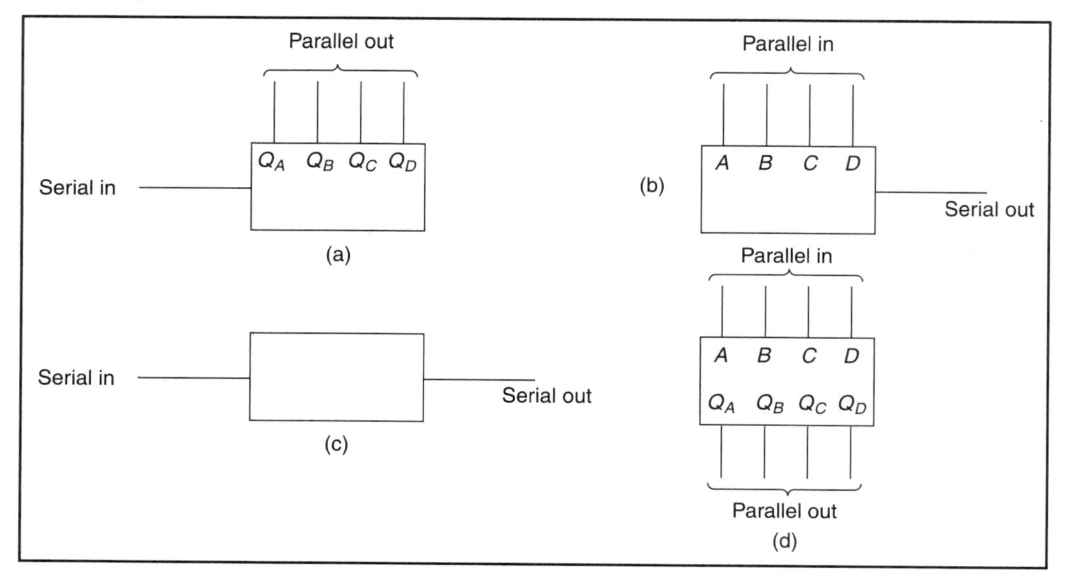

Figure 26.31 :: Methods of using a shift-register: (a) Serial-in/parallel-out (b) Parallel-in/serial-out (c) Serial-in/serial-out (d) Parallel-in/parallel-out

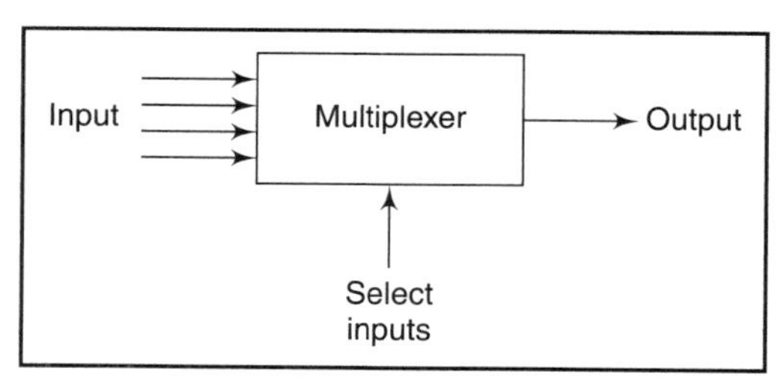

Figure 26.32 :: Symbolic representation of a multiplexer

26.5.5 Demultiplexer

A demultiplexer performs the reverse operation of multiplexer. It receives a single input and distributes it over several outputs. The SELECT input code determines to which output the data input will be transmitted. Figure 26.33 shows a schematic diagram of a demultiplexer.

A typical example of demultiplexer is 74155, which is a dual 1:4 demultiplexer. It converts line inputs to 4-line outputs.

26.5.6 Encoders

In digital circuits including microcomputers, the data are handled in the binary form, whereas the most common language of communication is decimal numbers and alphabetic characters. Therefore, there is a need to devise interface circuits between digital systems and human operators. Several binary codes have been developed, which carry out the function of code conversion. The process of generating binary codes is known as encoding.

Some of the commonly used codes are decimal-to-BCD encoder, octal-to-binary encoder and hexadecimal-to-binary encoder. 74147 IC is a priority encoder which can be used for decimal-to-BCD conversion. The block diagram of 74147 is given in Figure 26.34. Similarly 74148 IC provides octal-to-binary encoding.

Hexadecimal code is commonly used in microcomputers, especially when long binary words are handled. Hexadecimal-to-binary encoder can be realized by using two 74148 octal-to-binary encoders and a data selector.

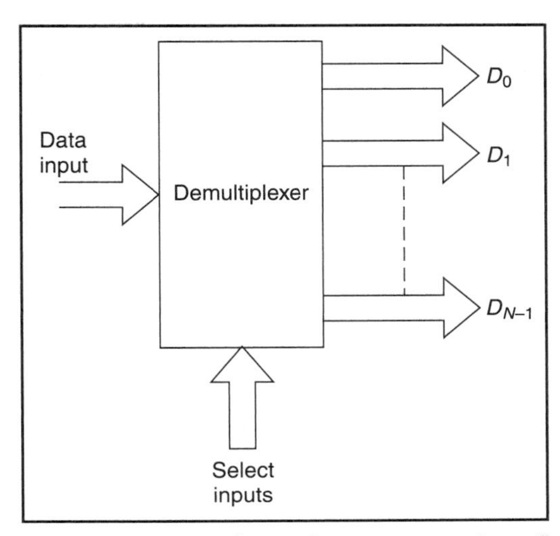

Figure 26.33 :: Schematic representation of a demultiplexer

26.5.7 Decoders

A decoder is a logic circuit which converts the input bit binary code into an appropriate output signal,

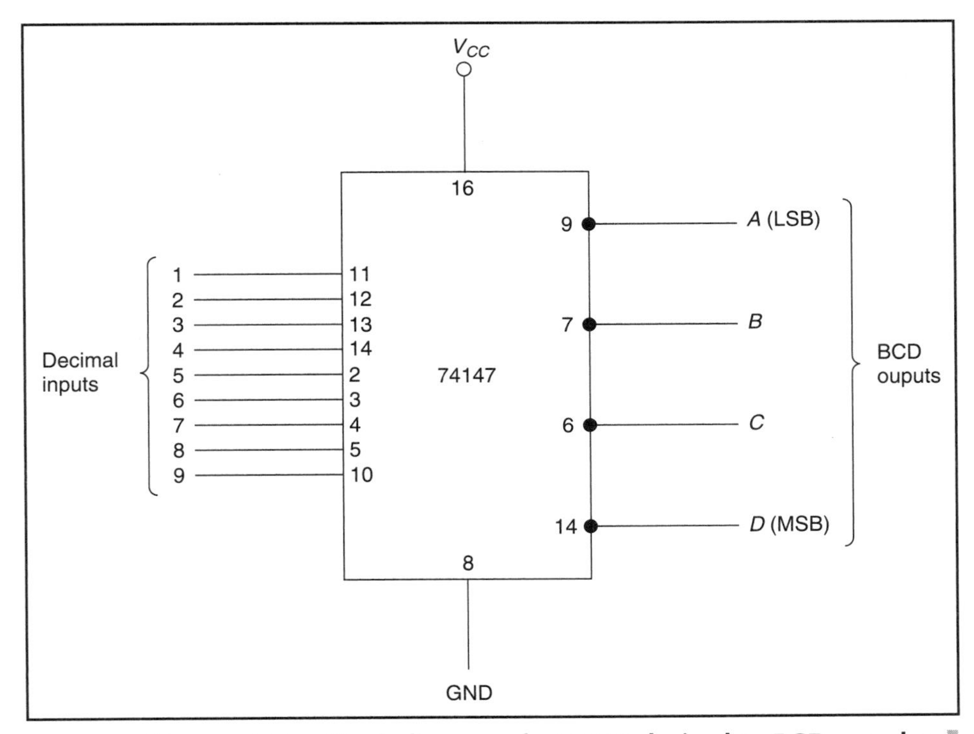

Figure 26.34 :: Block diagram of a 74147 decimal-to-BCD encoder

to identify as to which of the possible 2^n combinations is present. The most commonly used decoder is BCD-to decimal decoder, which provides decoding from 4 line-to-10 line decoding function. The input is 4-bit binary information, out of which only 10 BCD input codes are used. Correspondingly, there are 10 output pins.

In digital display systems, the digits are displayed on seven-segment light emitting diodes (LEDs). Therefore, it is more convenient if the BCD code is decoded into seven-segments. Thus, BCD-to-seven-segment decoder is the most popular display device used in digital systems. Figure 26.35 shows the block diagram of BCD-to-seven segment decoder along with seven-segment LED (Figure 26.36) display unit. The decoder circuit has four input lines for BCD data and seven output lines to drive a seven-segment display, *a* through *g* outputs of decoder are to be connected to *a* through *g* inputs of the display, respectively. The outputs of the decoder can be active-low or active-high and the seven segments of the display may be seven cathodes, with a common anode or seven anodes with a common cathode. The decoders normally include drivers in the chip itself. The typical examples of BCD-to-seven segment decoder driver ICs are 7446A and 7447A (active-low, open collector) and 7449 (active-high, open collector).

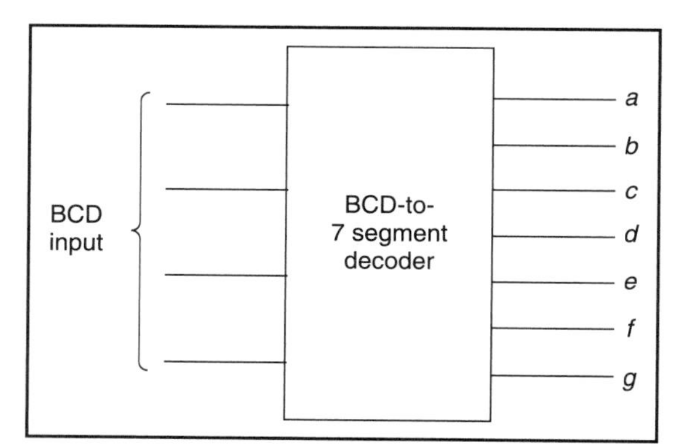

Figure 26.35 :: Block diagram of BCD-to-7 segment decoder

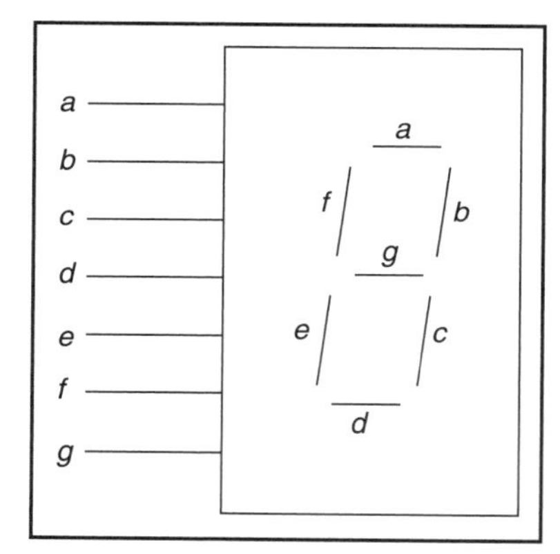

Figure 26.36 :: 7-segment display arrangement

26.5.8 Tri-state Logic

Tristate gates are designed so as to give output in three distinct states. Besides the normal two states of logic '1' (high) and logic '0' (low), a third state having a very high output impedance is available in tristate gates.

Figure 26.37 shows two possible arrangements of tristate buffers. When the control signal is high, the switch is off and no signals flow through the device. When the control line is low, the switch is on, and the input signals are passed through to the output.

Tristate buffers are usually used at each outlet of data and address bus. This is one of the first ICs a signal encounters as it comes into the computer.

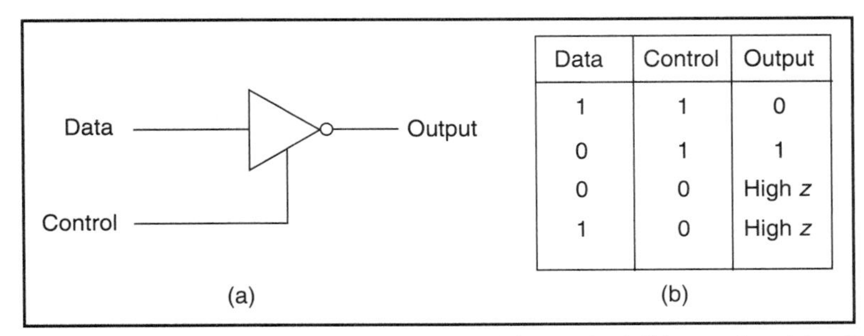

Data	Control	Output
1	1	0
0	1	1
0	0	High *z*
1	0	High *z*

Figure 26.37 :: TTL tristate invertor: (a) Symbol (b) Truth table

26.6 Semiconductor Memories

A digital memory is an array of binary storage elements arranged in a manner that it can be externally accessed. The memory array is organized as a set of memory words. Each word consists of a number of single bit storage elements called memory cells. The word length of a memory word is typically one, four, or eight memory cells. Therefore 1-bit, 4-bits or 8-bits (byte) of information can be stored by the memory word respectively. The memory capacity is the product of the number of memory words and the number of memory cells in each word. It is measured in bits and frequently expressed in kilobits where 1 kilobit = 2^{10} = 1024.

Memory words

26.6.1 Random Access Memory

Random access memory (RAM) is used in a microprocessing system to store variable information. The CPU under programme control can read or change the contents of a RAM location as desired. RAMs are a generic category that encompass all memory devices in which the contents of any address can be accessed at random in essentially the same time as any other address.

There are two types of RAMs: static and dynamic. Both dynamic and static MOS random access memories are popular; the dynamic ones for their high circuit density per chip and low fabrication costs and the static RAMs for single power supply operation, lack of refresh requirements and low power dissipation. Bipolar RAMs are used for very high speed scratch-pad memories.

Dynamic RAM: In the dynamic RAM, information is stored as electrical charge on the gate capacitance of MOS transistors. Since these capacitors are not perfect, the charge will leak away and the information is likely to be lost with time if the charge is not periodically refreshed. This can be done in several ways and depends upon the type of device in use.

Static RAM: Static RAM does not need to be refreshed, as the memory cells are bistable and similar in design to conventional flip-flops. In general, a static RAM consumes more power than its dynamic counterpart. However, it requires less support circuitry. Also, there are no problems of synchronizing the memory refresh cycles with normal CPU read and write operations.

When information is stored in the memory, it is written into the memory. When information is retrieved from a semiconductor memory, it is read from the memory. These are the only two functions that are done to static memories. Writing information into a memory is done in a write cycle. Reading information from a memory is done in a read cycle. The term cycle means a fixed period of time required to perform the function of writing into or reading from a memory. In fact, the electrical data or information is stored as a level of dc voltage. One dc voltage level corresponds to a '1' stored in the memory. A different dc voltage level corresponds to a '0' being stored in the memory.

Cycle

In a semiconductor memory (Figure 26.38), data are entered on an input pin on the physical device labelled 'data in'. Data being read from a memory are read on a device output pin labelled 'data out'. Therefore, a one-bit memory device will have the following four major physical connections: power input (V_{CC}), data input (D_1), data output (D_0) and read cycle or write cycle (R/W). If we want 16 bits of information storage, four address pins will be required ($2^4 = 16$) in the memory device (Figure 26.39). Under those conditions, no particular sequence will be needed to read or write information in the memory.

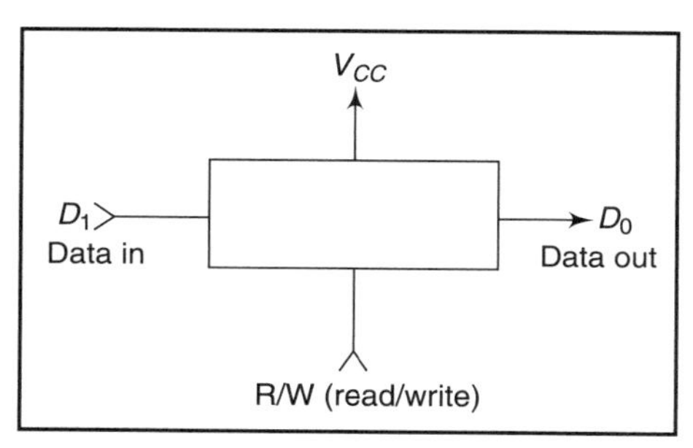

Figure 26.38 :: Block diagram of one-bit memory

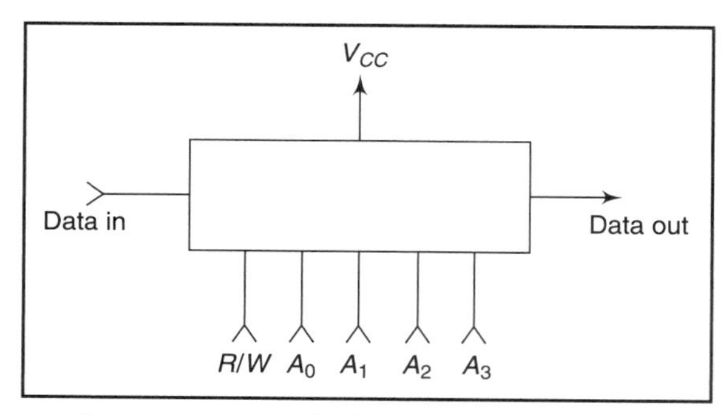

Figure 26.39 :: Block diagram of a 16-bit memory

Most semiconductor random access memory is volatile, i.e. the stored information is lost when the power supply is removed. This problem is avoided by using a battery-maintained power supply.

26.6.2 Read-Only Memory

In a microprocessor-based system, ROMs are normally used to hold the program of instructions and data constants such as look-up tables. Unlike the RAM, the ROM sis nonvolatile, i.e., the contents of the memory are not lost when the power supply is removed. Data stored in these chips is either unchangeable or requires a special operation to change. This means that removing the power source from the chip will not cause it to loose any data.

There are five basic ROM types. They are:

- ROM,
- PROM,
- EPROM,
- EEPROM, and
- Flash Memory.

(a) *Mask programmed ROMs,* which are programmed by the manufacturer to the user's requirements. ROM chips contain a grid of columns and rows and use a diode to connect the lines if the value is a (1). If the value is (O) then the lines are not connected at all. This type of ROM is only used if fairly large number of units are required, because the cost of preparation of creating the bit pattern on the chip is quite high. The contents of these ROMs cannot be altered after manufacture. Once the chip is made, the actual chip can cost very little money. They use very little power, are extremely reliable and, in the case of most small electronic devices, contain all the necessary programming to control the device.

(b) Programmable Read Only Memory (PROM) is programmed by the user. Selectively fusing (open-circuiting) the metal or polysilicon links in each memory cell sets that cell to a fixed state. The process is irreversible. In one form of PROM, the information is stored as a charge in a MOSFET cell. Blank PROM chips can be coded with anyone with their programmer.

PROM chips have a grid of columns and rows just as ordinary ROMs do. The difference is that every intersection of a column and row in a PROM chip has a fuse connecting them. A charge sent through a column will pass through the fuse in a cell to a grounded row indicating a value of "1". Since all the cells have fuse, the initial, or blank, state of a PROM chip is all "1"s. To change the value of a cell, to "0", the programmer is used, which sends a specific amount of current to the cell. The higher voltage breaks the connection between the column and the row by burning out the fuse. The process is known as burning the PROM.

The contents of a PROM can be erased by flooding the chip with ultraviolet radiation. Following this process, a fresh pattern can be entered. PROMs are used in the microprocessor based systems during the system development phase and on the production system when the total production run is not high enough to justify the use of mask-programmed ROMs.

PROMs can only be programmed once. They are more fragile than ROMs. A jolt of static electricity can easily cause damage to the chip. But blank PROMs are inexpensive and are great for prototyping the data for a ROM before committing to the costly ROM fabrication process.

(c) *Erasable Programmable Read Only Memories (EPROM)* provide the facility of re-writing the chips several times. EPROMs are configured using an EPROM programmer that provides voltage at a specified levels, depending upon the type of EPROM used.

For erasing the chips of its previous contents, an EPROM requires a special tool that emits a certain frequency of ultra-violet (UV) light. Because the UV light will not penetrate most plastics or glasses, each EPROM chip has a quartz window on top of the chip. The EPROM is kept very close to the eraser's light source, within an inch or two, to work properly. An EPROM eraser is not selective, it will erase the entire EPROM. The EPROM must be removed from the device it is in and placed under the UV light of the EPROM eraser for several minutes. An EPROM that is left under UV light too long can become over-erased. In such a case, the chip can not be programmed.

(d) *Electrically Erasable Programmable Read Only Memories (EEPROM) or Read-Mostly Memories (RMM)* are designed such that the contents of these memories can be altered electrically. However, this is a fairly slow process. It often requires voltages and circuit techniques that are not commonly found in normal logic circuitry.

EEPROMs remove the following drawbacks of EPROMs:

- The chip does not have to be removed to be rewritten
- The entire chip does not have to be completely erased to change a specific portion of it
- Changing the contents does not require additional dedicated equipment

(e) *Flash Memory* is a type of EEPROM that uses *in-circuit wiring* to erase by applying an electrical field to the entire chip or predetermined sections of it called *blocks*. Flash memory works much faster than traditional EEPROM because it writes data in chunks, usually 512 bytes in size, instead of a byte at a time.

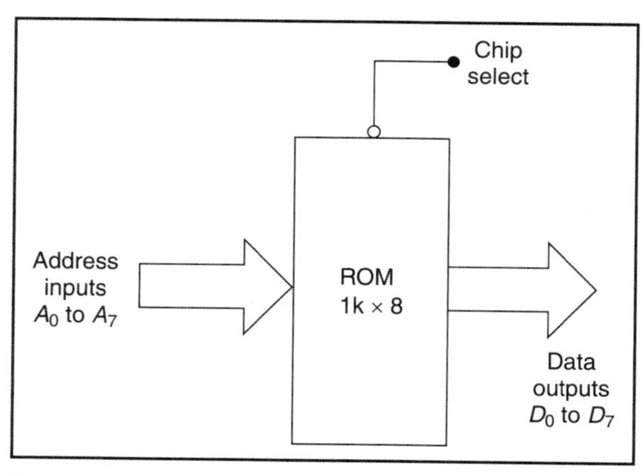

Figure 26.40 :: Typical symbol of ROM

Figure 26.40 shows a typical symbol of ROM, for storing 1024 8-bit words. This is also called a 1 K × 8 ROM where 1 K represents 1024. Similarly, a 2048 × 8 can be written as a 2 K × 8 and so on. Since 1 K ROM stores 1024 different words, it needs 10 address inputs (2^{10} = 1024). The word size is 8 bits, so there are eight output lines. The memory chip is enabled or disabled through the chip select (CS) input. ROMs do not provide for data input or read/write control because they do not normally have write operation. Some ROMs do provide for special input facilities for initially writing the data into the ROM which is generally shown on the symbol.

The Intel 2716 is a 16,384 (2 K × 8) bit ultraviolet erasable and electrically programmable read only memory. Figure 26.41 shows pin configuration and block diagram of the 2716 EPROM.

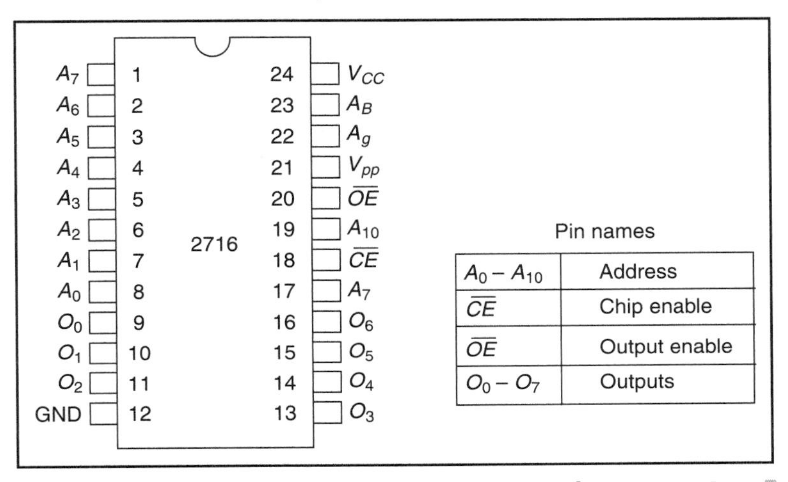

Figure 26.41 :: Pin configuration of Intel 2716 EPROM

The 2716 has a standby mode which reduces the active power dissipation by 75%. The 2716 is placed in the standby mode by applying a TTL high signal to the $\overline{CE}$ input. When in standby mode, the outputs are in a high impedance state, independent of the $\overline{OE}$ input.

The 2716 provides two line control functions which allow for the low power dissipation and assurance that output bus contention will not occur. To use these control lines, $\overline{CE}$ (pin 18 is decoded and used as the primary device selecting function while $\overline{OE}$ (pin 20) be made a common connection to all devices in the array and connected to the READ line from the system control bus. Thus, the output pins are only active when data are desired from a particular memory device.

26.7 Microprocessor

The microprocessor, in essence, consists of basic circuit elements such as transistors, resistors and diodes which when combined form the basic logical elements, namely AND, OR and INVERTORS. In principle, the complete operation of the microprocessor could be described by a combination of these devices. More complex circuit elements such as flip-flop, counters, registers and arithmetic logic unit are formed from these gates and go to make the complete microprocessor. Microprocessor is a single integrated circuit with 40 or even 64 are even higher connection pins.

Microprocessors are usually classified on the basis of their word length. The word length of a microprocessor defines the basic resolution and memory addressing capability. For example, an 8-bit microprocessor will perform all calculations on binary numbers with 8 digits. 8 binary digits give a decimal number between 0–255.

The arithmetic and logic unit (ALU) and control section of a computer have traditionally been constructed together, and form the core of the microprocessor (Figure 26.42). The ALU is that portion of the hardware which performs the arithmetic and logical operations on the data. It essentially contains an adder which is capable of combining the contents of the registers in accordance with the logic of the binary arithmetic. This enables the processor to perform the arithmetic manipulations on the data, it obtains from memory and from its other inputs. Using only the basic adder, a programmer can achieve complete arithmetic operations.

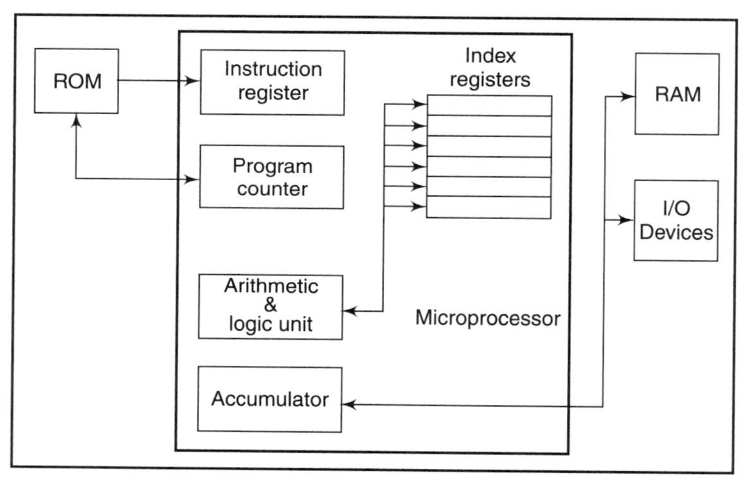

Figure 26.42 :: Microprocessor—basic parts

The control unit constitutes the brain of the microprocessor. It co-ordinates all the parts of the microprocessor so that the events take place in a logical sequence and at the right time. Numerically encoded instructions, which are stored in the memory, are sent to the control unit to direct it to carry out certain basic algorithmic steps. The control unit is also capable of responding to external signals, such as an interrupt or wait request.

The microprocessor contains registers which are temporary storage devices. Some registers, such as the program counter and instruction register have dedicated uses, whereas most other registers, such as accumulator are more for general-purpose use.

The operation of the microprocessor and synchronization of various activities under its control is maintained by a crystal controlled clock or oscillator, which is usually at a fixed frequency, generally greater than 5 MHz.

The most powerful asset of the microprocessor is its enormous speed of operation. This is possible because the microprocessor can store all the necessary instructions and data, until required, in the memory. Two types of memories are used in the microprocessor, namely, read-only memory (ROM) and Random access memory (RAM). The ROM generally stores the programs, whereas the RAM stores the data. The microprocessor can rapidly access any data stored in memory, but often the memory is not large enough to store the entire data bank required for a particular application. This problem can be solved using external storage equipment, such as floppy disk or hard disk system. A microprocessor also requires input/output ports, through which it can communicate its results with the outside world, like a display or peripheral device or provide control signals that may direct another system.

Most practical microprocessors use 40 pins, and even to keep to that number requires that the eight data pins be used for both reading and writing. Figure 26.43 shows a representation of a typical CPU. As already mentioned above, the function of the CPU is to manipulate data in accordance with the instructions stored in the memory. For this, the CPU transfers data and internal state information via an 8-bit bidirectional 3-state data bus Memory and peripheral

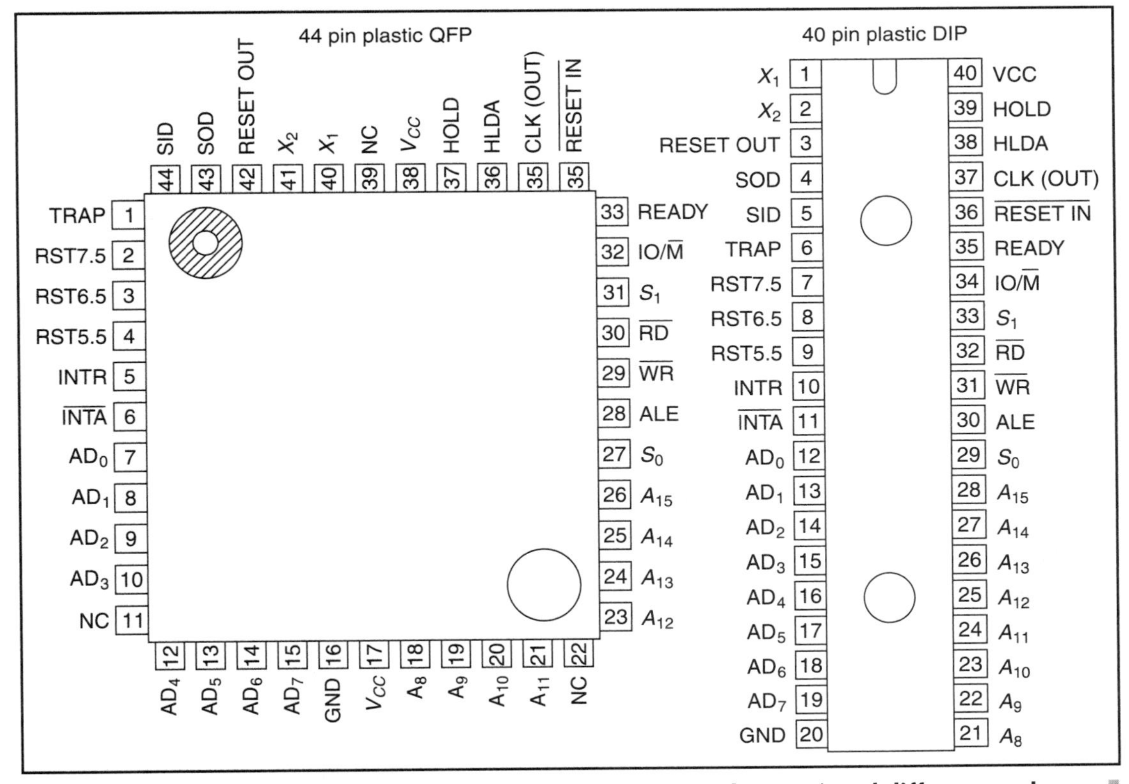

Figure 26.43 :: A typical microprocessor chip configuration (Intel 80C85) and different packages

device addresses are transmitted over a separate 16-bit 3-state address bus. Timing and control outputs are given out for synchronization and control purposes, whereas control inputs like reset, hold and ready and interrupt are used to perform specific functions.

Data Lines: Data-in lines are used to transfer instructions or data from memory or data from input device to the CPU. On the other hand, data-out lines are used to transfer results from CPU to memory or output devices. The number of lines in data-in or data-out is related to the word length of the computer. Normally, a n-bit processor has n data-in lines and n data-out lines. In a 8-bit CPU, one bus of eight lines can be used to carry all the data signals, whether for reading or writing. To ensure that this operation is correctly carried out, a read/write control is provided in the microprocessor package. A bus which carries signals in both directions is called a bidirectional bus and the data bus is the only one which is bidirectional.

The CPU's internal data bus is isolated from the external data bus by an 8-bit bidirectional three-state buffer. In the output mode, the internal bus content is loaded into an 8-bit latch that, in turn, drives the data bus output buffers. The output buffers are switched-off during input or non-transfer operations.

During the output mode, data from the external data bus is transferred to the internal bus. The internal bus is precharged at the beginning of each internal state, except for the transfer state.

Address Lines: These are employed to address the memory to fetch instructions or data from it. These are also used to address and connect I/O devices to the CPU. The number of address lines determine the size of the memory a particular CPU can handle, or maximum number of I/O devices that can be connected to the CPU.

A typical microprocessor has 16 pins as address outputs, which are labelled as A_0 to A_{15}. They are used to select particular locations in the memory. For example, the first signal on the address lines will comprise of 16 logic '0', address '0'. As the microprocessor starts operating, A_0 will change to 1 and when the read/write control signal is used for reading, a byte will be fetched from that particular memory location corresponding to this address. After this, the address will step up at the end of first instruction and call for the next instruction.

The 16 address lines can carry up to 2^{16} (65,536) bits of information, which is equivalent to 64 k memory. Microprocessors in general will, however, need much smaller memory for small machines. Thus, there will always be unused address lines, which can be connected to video display unit or output signals to other circuits, under the control of the program.

Multiplexing

Multiplexing is a useful technique employed with microprocessors for obtaining additional address and data lines For example, by multiplexing, it is possible to handle 16-bit data signals, through an eight-line data bus. Similarly, the address lines and data lines are also multiplexed and we have address and data available on the same physical lines at different instants of time. However, the result of any time-multiplexing is slowing down the system.

As microprocessor systems are based on the binary numbering system, it is necessary to use multiple connections generally 8,16 or 32 between each of the integrated circuits (chips). These interconnections are usually referred to as buses. There are three buses in a microprocessor system.

The assembly language of a microprocessor enables to extract the greatest run-time performance because it provides for direct manipulation of the architecture of the processor. However, it is also the most difficult language for writing programs, so it falls far from the optimal language line.

The C language which is used to develop modern versions of the Unix operating system provides a significant improvement over assembly language for implementing most applications, it is the language of choice for real time programming. It is an excellent compromise between a low level assembly language and a high level language. C is standardized and structured. C programs are based on functions that can be evolved independently of one another and put together to implement an application. These functions are to software as black boxes are to hardware. C programs are transportable. By design, a program developed in C on one type of processor can be relatively easily transported to another.

Historically, the first microprocessor was introduced by Intel in November 1971. It was 4004, the 4-bit CPU having about 2200 transistors on it. Shortly afterwards, Intel introduced an 8-bit microprocessor chip, the 8008. It had more computing power and flexibility than 4004, had 45 instructions with an average execution time of 30 μs. It was better suited for applications of data handling and control. The 8008 remained the sole 8-bit microprocessor for two years till Intel announced an upgraded version, the 8080. Simultaneously, National Semiconductor and Rockwell among others, announced their own entries.

In mid-1974, Motorola came out with 6800 8-bit microprocessor. Unlike other microprocessors that required multiple power supplies (for example, 8080 required three), the 6800 was the first +5 V single power supply microprocessor. 1974 saw the introduction of the 1802, the first CMOS processor by RCA. Two major introductions in 1976 were the Intel 8085 and the Zilog Z80. The 8085 offered additional features to the 8080 and required a single +5V power supply. The Z80 contained improvements in architecture over the 8080 and incorporated all the 8080 instructions in its instruction set.

The era of 16-bit microprocessors began in 1974 with the introduction of the PACE chip by National Semiconductor. The Texas Instruments TMS 9900 was introduced two years later. Subsequently, the Intel 8086 became commercially available in 1978, the Zilog Z800 in 1979 and the Motorola MC 68000 in 1980. Several higher performance versions of the original chips are now available.

Microprocessors with 32-bit internal paths and 16-bit external paths have been in existence since 1980. However, the era of true 32-bit microprocessors began in 1981 with the commercial introduction of the Intel iAPX 432, followed by Motorola 68020, Zilog Z80000 and Intel iAPX 386. All these chips reflect a sustained and conscious attempt towards the integration of an enormously large numbers of transistors.

26.8 Microcontrollers

A microcontroller contains a CPU, clock circuitry, ROM, RAM and I/O circuitry on a single integrated circuit package. The microcontroller is therefore, a self-contained device, which does not require a host of associated support chips for its operation as conventional microprocessors do (Simpson, 1980). It offers several advantages over conventional multichip systems. There is a cost and space advantage as extra chip costs and printed circuit board and connectors required to support multichip systems are eliminated. The other advantages include cheaper maintenance, decreased hardware design effort and decreased board density, which is relevant in portable analytical instruments.

Microcontrollers have traditionally been characterized by low-cost high volume products requiring a relatively simple and cheap computer controller. The design optimization parameters require careful consideration of architectural tradeoffs, memory design factors, instruction size, memory addressing techniques and other design constraints with respect to area and performance. Microcontrollers functionality, however, has been tremendously increased in the recent years. Today, one gets microcontrollers, which are stand alone for applications in data acquisition system and control. They have analog-to-digital converters on chip, which enable them direct use in instrumentation. Another type of microcontroller has an on-chip communication controller, which is designed for applications requiring local intelligence at remote nodes and communication capability among these distributed nodes. Advanced versions of the microcontrollers in 16-bit configuration have been introduced for high performance requirements particularly in applications where good arithmetical capabilities are required.

26.9 Data Converters

In the world of analytical equipment, most electric circuitry is concerned with acquisition, amplification and processing of signals which are available in an analog form. However, the availability of low cost and highly programmable microprocessors and compact digital computers have made possible the digital manipulation of the analog signals after they have been converted into a digital form. On the other hand, the display systems are often of analog type, requiring the digital output to be converted it into an analog form. The class of devices for converting analog signals to digital form and digital signals to analog form are called Data Converters. They are obviously of two types

Analog-to-Digital Converter (ADC) is a device which has an analog signal as its input and a digital representation of that input as its output.

Digital-to-Analog Converter (DAC) is a device which receives a digital input and outputs an analog quantity.

Figure 26.44 shows the principle of data converters. It may be seen that output of a DAC is not purely analog, but it is quantized and made up of discrete steps. This is because the input to DAC is digital. There can be only a finite number of input states i.e. there can be only a finite number of output states. The size of the discrete steps depends on the resolution and conversion rate of the system.

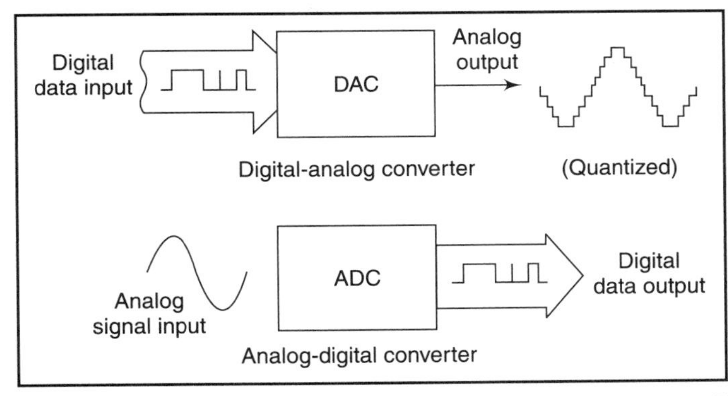

Figure 26.44 :: Principle of data converter

Resolution

Resolution of an ADC is defined as the smallest analog change that can be discriminated. From a digital point of view, resolution is the number of discrete steps into which the full analog input signal range of the converter is divided. It is usually expressed as a number of bits (binary) or a number of digits. For example, in a 12-bit ADC, if the input voltage range is ±5V, the smallest analog change that can be resolved is 10/4096 (2^{12}) = 2.44 mV dc.

Conversion Time is the time elapsed between application of a convert command and the availability of data at its output.

The working of a data converter is usually checked by using an oscilloscope and observing the analog/digital waveform at the input/output terminals.

26.9.1 A/D Converters

A/D conversion can be divided into two basic groups: open loop types and feedback types. The open loop converter generates a digital code directly upon application of an input voltage and is generally an asynchronous operation. The feedback type, on the other hand, generates a sequence of digital codes, reconverts (D/A) each one in turn back to an analog value and compares it to the input. The resulting digital output will be the closest value of the reconstructed analog voltage compared to the real analog voltage. The feedback method is more popular. Under this technique, we shall consider only successive approximation A/D converter and dual slope A/D converter.

26.9.1.1 Successive Approximation A/D Converters

The most direct approach to analog-to-digital conversion is that of using a D/A converter in digital feedback loop, as shown in Figure 26.45. The loop is closed at the analog comparator, which compares the current output of the D/A converter against the input current developed by the analog input voltage. This comparison is made one bit at a time and the method is, therefore, known as successive approximation technique.

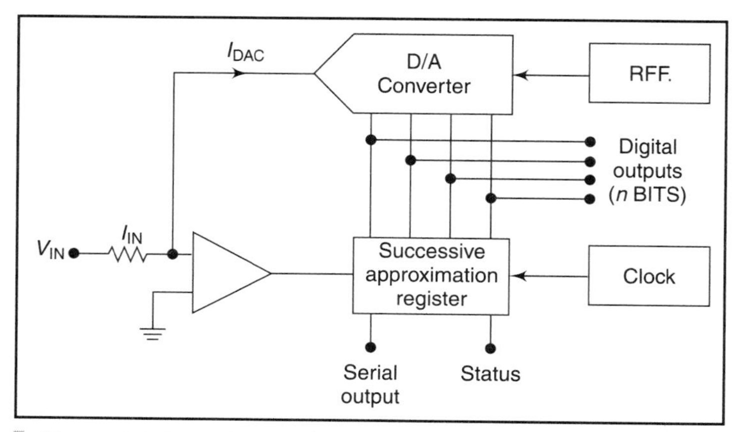

Figure 26.45 :: Successive approximation A-D converter

Most significant bit

A clock circuit steps the converter through n comparison steps, where n is the resolution of the converter in bits. In the first clock period, the D/A converter's MSB (most significant bit) output which is one-half of full scale, is compared against the input. If it is smaller than the input, the next largest bit is turned on. If the MSB current is larger than the input, in the next clock period it is turned off when the next bit current is turned on. The D/A converter output at any given time is the cumulative total of all the previous bit currents which have been left on.

Least significant bit

The comparison process is continued one bit at a time from the MSB down to LSB (least significant bit). After the last clock period, the output of the successive approximation register contains the digital word represent ing the analog input. The converter also puts an end to conversion or status pulse, indicating that conversion is complete. In addition to the parallel data output on n digital lines, there is also a useful serial output from most converters derived from the comparison process.

This class of A/D converters represents an excellent compromise between circuit complexity, speed and ability to produce high accuracy codes. This type of converter can have resolution of from 8 to 16 bits. This is the commonest of A/D converters and there are numerous areas of application.

For this type of conversion, the digital output corresponds to some previous value of the analog input during the conversion. Thus, the aperture is equal to the total conversion time, which vary with these devices from 1 to 220 μs.

26.9.1.2 Dual Slope A/D Converter

The dual slope technique operates on a voltage-to-time principle. The basic idea involved is to integrate the unknown input and the known reference and compare their slopes. Figure 26.46 shows a block diagram of the dual-slope A/D converter.

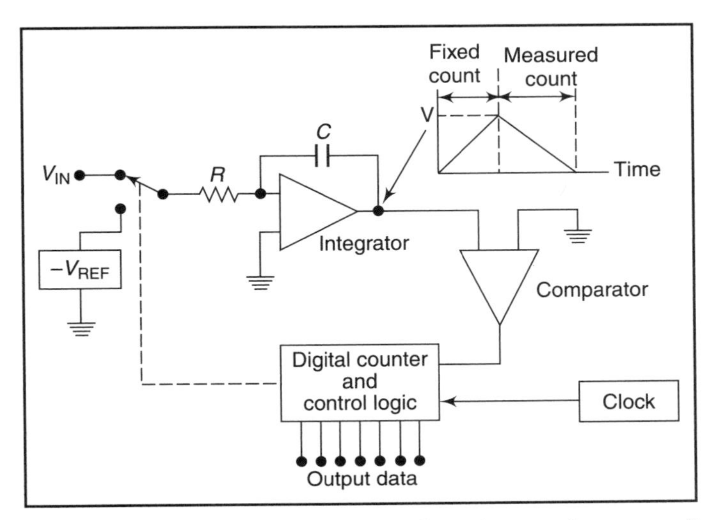

Figure 26.46 :: Principle of dual slope integration method of analog to digital conversion

The conversion cycle begins by switching the operational integrator to the input voltage, which is then integrated for a fixed time period. After this time, the integrator is switched to a reference voltage of opposite polarity and the integrator output integrates back to zero for a period of time, which is measured by the counter. The resultant count is then the digital value of the input voltage.

The advantage of the dual slope technique is that, the accuracy and stability are dependent only on the reference, and not on other components in the circuit. This assumes, of course, that the operational integrator is linear. Some devices also incorporate automatic zeroing circuitry to reduce the effect of offset drift with time and temperature. The technique offers simplicity, accuracy and noise immunity due to integration of the input signal. The chief drawback is relatively slow conversion time.

26.9.1.3 Charge Balancing A/D Converter

Another integrating conversion technique which is quite popular is the charge-balancing A/D converter, illustrated in Figure 26.47. In this arrangement, an operational integrator is enclosed in a digital feedback loop consisting of a comparator, pulse timer circuit and a switched reference. A positive input voltage causes the integrator output voltage to cross zero volts, which is detected by the comparator and triggers the pulse timer circuit. The output pulse from the timer switches a negative reference current to the integrator output to increase in the positive direction. Therefore, every time the integrator output crosses zero, another pulse is generated and integrated.

A state of equilibrium exists when the average current developed by the pulses just equals the input current. Since each current pulse is a fixed amount of charge, the name charge-balancing is given to this technique.

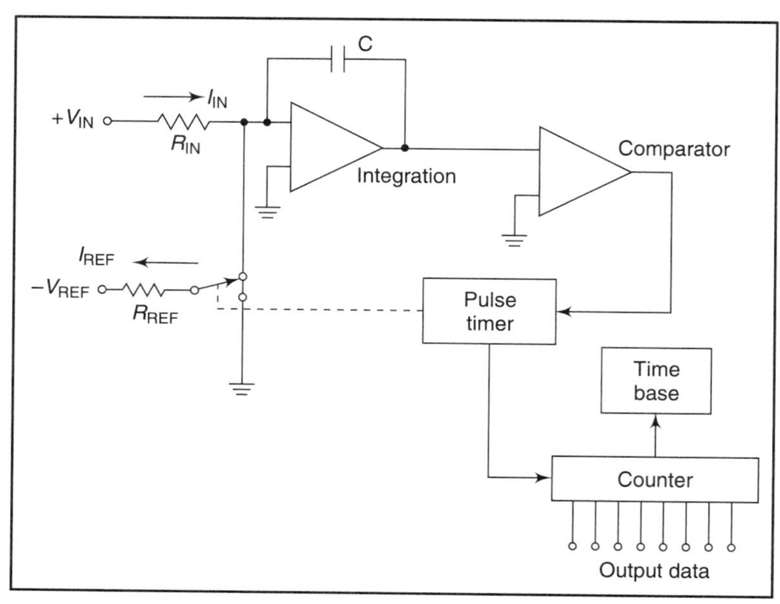

Figure 26.47 :: Charge-balancing A-D converter

Figure 26.48 shows a typical A/D converter for microprocessor-based applications. The digital output lines D_7 through Do come from tristate latches, which are part of the converter. A HIGH on the ENABLE input will enable these outputs, so that the digital representation of the analog input is present on these lines. A LOW on the ENABLE input puts these output lines in their high-z state. Normally, the ENABLE input is pulsed HIGH only after the BUSY output has indicated that the conversion is complete. If the ENABLE input is made HIGH during the conversion time interval, the output lines will indicate the results of the previous A/D conversion.

A typical example of an integrating analog-to-digital converter is that of ICL 7109 from Intersil. It is monolothic, CMOS, 12-bit A/D converter and is not intended for very fast applications. The data sheet specifications are quoted at 7.5 conversions per second, corresponding to an internal clock frequency of about 61.5 kHz, or a clock period of 16.3 μs. The separate tristate byte wide outputs of the ICL 7109 make it ideal for interfacing to 8-bit microprocessor buses.

26.9.2 Key Parameters in A/D Converters and their Selection

The choice of an A/D converter and associated front end signal-conditioning hardware demands a careful analysis. Choosing an A/D converter subsystem would usually depend upon the following considerations (Lockhart, 1990).

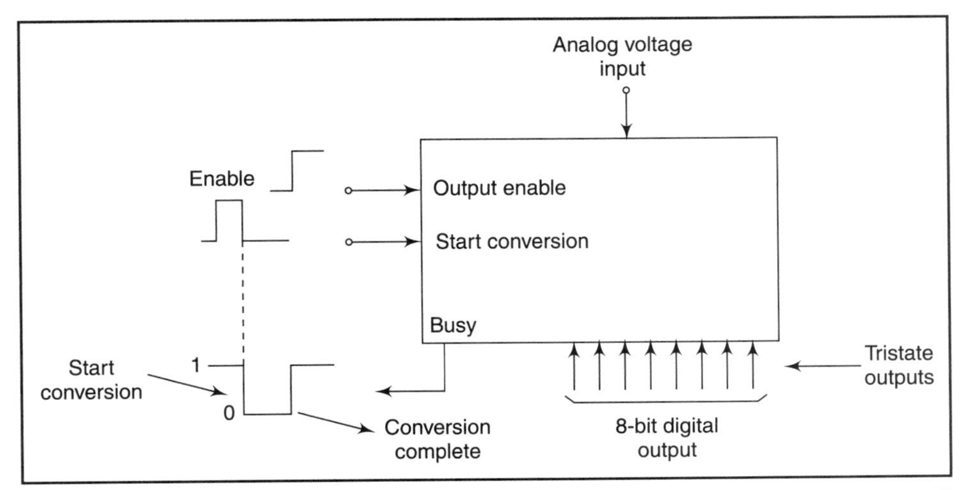

Figure 26.48 :: A-D converter for μP-based applications

26.9.2.1 Resolution

Resolution is defined as the smallest analog change that can be discriminated. From a digital point of view, resolution is the number of discrete steps into which the full analog input signal range of the converter is divided. It is usually expressed as a number of bits (binary) or as number of digits (BCD).

Successive approximation A/D converters are 12-bit devices, and so divide a voltage range into 4096 steps. For an A/D converter whose analog input voltage range is ±5V dc, the smallest analog change that can be discriminated is 10 ÷ 4096 = 2.44 mV dc. Integrating converters are frequently 16-bit devices and therefore have a wide dynamic range; for example they can discriminate the same analog voltage change of 2.44 mV at a full-scale input voltage 16 times as large. The resolution can be improved by using an amplifier at the input, and the effective resolution is increased by the magnitude of the amplifier gain.

26.9.2.2 Sampling Rate

The sampling rate on a particular experiment depends on how much information is required from the analog signal (such as intensity as a function of time), and on what kind of signal processing is planned for the collected data. For example, if the peak is 1s long and it is decided that the system should collect at 10 points over the length of that peak, the minimum data collection rate would be 10 points/s or 10 Hz if the peak is 0.1s long, the sampling rate would have to be at least 100 Hz.

Signal processing usually requires a higher sampling rate. For example, according to the Nyquist frequency criterion, the FF (fast Fourier Transform) analysis requires that the sampling rate be more than twice the highest frequency component of the input. It is important to know as to whether the computer memory can handle the sample throughput rate. For example, if the throughput is 10,000 Hz (points per sec.) and lasts only for a burst of one second, the computer should have enough main memory to handle the 10,000 samples of raw data. But if this rate is sustained for l0s, the computer may not have enough main memory to handle 100,000 samples. This problem of handling volumes of data that exceed the capacity of main memory is tackled by periodically transferring the current block of data to mass storage disc units.

Throughput in data collection alone and with real-time processing can be substantially improved, by upgrading the computer to a more powerful model having a larger main memory. Another approach is to use parallel computers. One machine is dedicated to collecting and storing data on a shared mass storage device and the second computer devotes all its capabilities to processing the data and controlling the experiment.

26.9.2.3 Speed of an A/D Converter or Conversion Rate

The speed of an A/D converter is generally expressed as its conversion time, i.e. the time elapsed between application of a convert command and the availability of data at its outputs. A/D converters can operate in three different ways.

Wait Loops: Here a program starts the converter and then waits for the Status line to show an end of conversion. The computer cannot perform other tasks during this operation. If the program is written in assembler or machine code, each pass through the loop might take 20 μs. Also, it may take about 30 μs to store the digital values in memory, increment a pointer to memory, preparing for the storage of the next data point and decrement a counter used by the program to detect if all the desired points have been taken. All these may take about 30 μs. Therefore, even if the conversion time of the A/D converter were infinitely fast, the A/D converter cannot take data faster than 20 kHz.

Interrupts: The A/D converter is started and the CPU continues executing other programs. When the conversion is over, it demands a program interrupt that temporarily stops execution of the current program. Next operation automatically attends to the service routine for the A/D converter. These operations may take 200–300 μs. Even if the A/D converter is infinitely fast, data collection would be limited to 3–5 kHz.

Direct Memory Access (DMA): In DMA, the A/D converter takes control of the computer's bus during transfer of a data word into memory. The transfer is completed in a few hundred nanoseconds, thereby offering a great stealing advantage. The actual transfer takes place by stealing time from the CPU.

26.9.2.4 Single Ended vs Differential Mode

Laboratory instruments are often connected to a microcomputer through a suitable interface. There are two choices of wiring between the instrument and interface.

Single-ended: In this, the signal (V_0) received by the interface (Figure 26.49) is the difference between the potential at an input terminal and ground. Though only one input terminal is needed per channel, the differences in ground potential at the instrument and interface result in common mode noise pickup, in which the received signal:

$V_0 = V_s + V_n$, where V_n is the ground loop noise component.

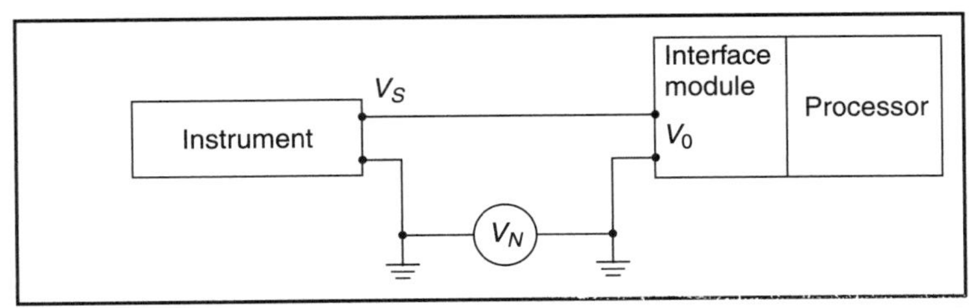

Figure 26.49 :: Single-ended wiring arrangement—here the signal to be processed is between the input terminal of the interface module and the ground. Common mode signal pick up limits the data collection rate and resolution.

Differential Wiring: In this, the signal of interest (V_0) is the instantaneous potential difference between the input signal ($V_1 + V_n$) and the return signal ($V_2 + V_n$), so that common mode noise is not present in V_0 (Figure 26.50). In addition, electrostatic noise may be eliminated by grounded cable shielding and magnetically induced noise eliminated by using twisted pair. Any reasonable noise condition that cannot be handled with shielded, twisted differential wiring can usually be suppressed satisfactorily with signal averaging and other digital filtering.

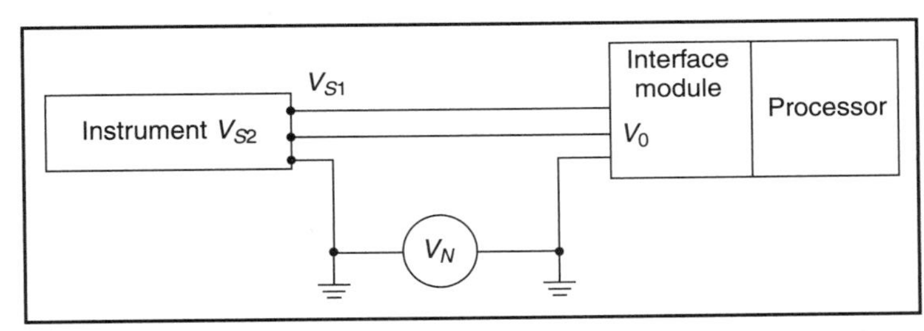

Figure 26.50 :: Differential wiring arrangement, in which the common mode signals are effectively cancelled

Many A/D converter subsystems give the user the option of choosing single-ended or differential input. Obviously, the latter will cut the available number of multiplexed channels into half.

When an extreme noise is encountered, it may be necessary to transmit the signal from a source as a variable current or frequency, and the information converted back to a potential at the A/D converter end. Voltage-to-frequency converters can generate a pulse train proportional to voltage at the source end of a transmission cable, and a frequency-to-voltage converter regenerates a voltage at the A/D converter end. Like FM radio, frequency transmission is quite noise-immune.

26.10 Data Acquisition Systems for Analytical Instruments

Most analytical signals are in the form of analog voltages. Data acquisition is simply the gathering of information about such signals present in a system or process. It is a core tool to the understanding, control and management of such systems or processes. Parameter information such as pH, temperature or pressure is gathered by transducers that convert the information into electrical signals. Depending upon the number of parameters to be measured, the signal from the transducers are transferred by wire, optical fibre or wireless link to an instrument which conditions, amplifies, measures, scales, processes, displays and stores the sensor signal. This constitutes the Data Acquisition system. Today's data acquisition systems make use of powerful microprocessors and computers which do this job more accurately, more flexibly, with more sensors, more complex data processing, and an elaborate presentation of the final information.

For interfacing analog signals to microprocessors/microcomputers, use is made of some kind of data acquisition system. The function of this system is to acquire and digitize data, often from hostile clinical environments, without any degradation in the resolution or accuracy of the signal. The functional block diagram of a data acquisition system with multiple inputs is show in Figure 26.51. Since software costs generally far exceed the hardware costs, the analog/digital interface structure must permit software effective transfers of data and command and status signals to avail of the full capability of the microprocessor.

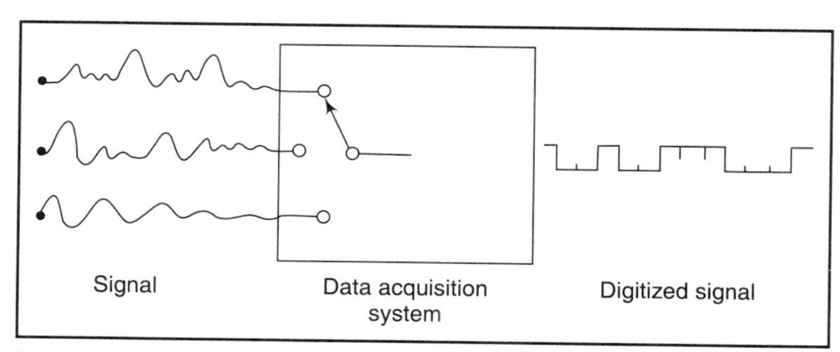

Figure 26.51 :: Basic data acquisition system

The analog interface system, in general, handles signals in the form of voltages. The physical parameters such as temperature, flow, pressure, etc. are converted to voltages by means of transducers. The choice and selection of appropriate transducers is very important, since the data can only be as accurate as the transducer.

Figure 26.52 shows a block diagram of a universal interface circuit for connecting analog signals to microprocessors. It basically comprises a multiplexer, instrumentation (buffer) amplifier, a sample-and-hold circuit, analog-to-digital converter (ADC), tristate drivers and control

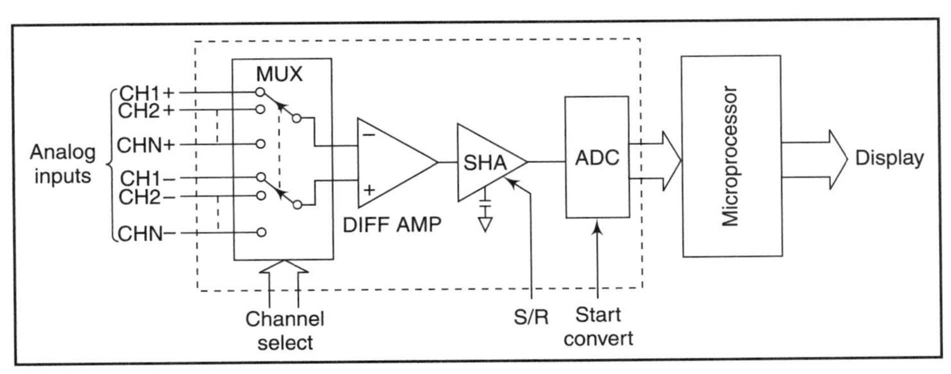

Figure 26.52 :: Interfacing analog signals to microprocessors

logic. These components operate under the control of interface logic that automatically maintains the correct order of events.

Multiplexer: The function of the multiplexer is to select under address control, an analog input channel and connected to the buffer amplifier. The number of channels is usually 8 or 16. Depending on its input configuration, the multiplexer will handle either single ended or differential signals.

The address logic of most multiplexers can perform both random and sequential channel selection. For real time systems, the random mode permits the multiplexer to select any channel when the program responds to a peripheral service request. Sequential channel selection, as the name imploys involves addressing each channel in order.

Buffer Amplifier: The buffer amplifier conditions the selected input signal to a suitable level for application to the A/D converter. Driven by the multiplexer, the buffer amplifier, which is usually an instrumentation amplifier, provides impedance buffering, signal gain and common mode rejection. It has a high input impedance, 100 Mohms are more to reduce the effects of any signal distortion caused by the multiplexer. The high input impedance also minimizes errors due to the finite on-resistance of the multiplexer channel switches.

In order to improve the system sensitivity, the amplifier boosts the input signal. If it is required to have analog signals of differing ranges, connected to the multiplexer input, then a programmable gain amplifier would be preferable where the gain would be set in accordance with the multiplexer selection address. The use of programmable gain amplifiers removes the necessity to standardize on the analog input ranges.

Sample and Hold Circuit: The A/D converter requires a finite time for the conversion process, during which time the analog signal will still be hanging according to its frequency components. It is therefore necessary to sample the amplitude of the input signal, and hold this value on the input to the A/D converter during the conversion process. The sample and hold circuit freezes its output on receipt of a command from the control circuit, thereby providing an essentially constant voltage to the A/D converter throughout the conversion cycle.

The sample hold is essentially important in systems having resolution of 12-bits or greater, or in applications in which real time inputs are changing rapidly during a conversion of the sampled value. On the other hand, a sample hold may not be required in applications where input variation is low compared to the conversion time.

A/D Converter: The A/D converter carries out the process of the analog-to-digital conversion. It is a member of the family of action/status devices which have two control lines—the start conversion or action input line and the end of conversion or status output line.

An A/D converter is a single chip integrated circuit having a single input connection for the analog signal and multiple pins for digital output. It may have 8,12,16 or even more output pins, with each representing an output bit. The higher the number of bits, the higher the precision of conversion. Each step represents a change in the analog signal: 8-bits gives 256 steps, 12-bits provides 4096 steps and we get 32768 steps with 16 output bits.

Tri-state Drivers: The tri-state drivers provide the necessary isolation of the A/D converter output data from the microprocessor data bus and are available as 8-line units. Thus, for the 10 or 12-bit converters, two drivers would be required which would be enabled by two different read addresses derived from the address decoder.

Some A/D converters have in-built tri-state drivers. However, because of their limited drive capability, they can be used only on lightly loaded buses. For heavily loaded systems, as in microcomputers, the built-in drivers are permanently enabled and separate tristate drivers employed for the data bus isolation.

Control Logic: The control logic provides the necessary interface between the microprocessor system but and the elements of the acquisition unit in providing the necessary timing control. It is to ensure that the correct analog signal is selected, sampled at the correct time, initiate the A/D conversion process and signals to the microprocessors on completion of conversion.

Output Interface: Digital output signals often have to be converted into analog form so that they can be used and acted upon by external circuits, e.g., oscilloscope, chart recorder, etc. Therefore, digital-to-analog (D/A) converters are used for converting a signal in a digital format into an analog form. The output of the D/A converter is either current or voltage when presented with a binary signal at the input.

The input coding for the D/A converter is similar to the output coding of the A/D converter, while full-scale outputs are jumper-selectable for 0 to 1, ± 5 and ± 10 V. D/A converters generally deliver the standard 4 to 20 mA output and loading can range from 50 Ω to 4 kΩ. The important parameters which govern the choice of an A/D converter or D/A converter are resolution, measurement frequency, input characteristics, offset error, noise, microprocessor compatibility and linearity, etc.

Data acquisition can be divided into two broad classifications—real-time data acquisition and data logging. Real-time data acquisition is when data acquired from sensors is used either immediately or within a short period of time, such as when controlling a process. Data logging on the other hand is when data acquired from sensors is stored for later use. In reality, there is a continuum of devices between real time data acquisition and data logging that share the attributes of both of these classifications.

Real-time data acquisition

Data logging

A data logger, on the other hand, is an electronic instrument which connects to real world devices for the purpose of collecting information. For example, data logger can be pictured as a black box recorder in aeroplanes, which record voice and the plane status data information.

Building a data acquisition system with a hardy cross-breed of both analog design and digital design no longer requires elaborate design efforts involving design of individual circuit blocks. The design problem of data acquisition systems have been greatly simplified with the introduction of single chip systems. Included on-chip is an 8-bit analog-to-digital computer with bus-oriented outputs, a 16-channel multiplexer, provisions for external signal conditioning and logic control for the systems interface.

Chapter 27

Computer-Based Analytical Instruments

27.1 Computers in Analytical Laboratories

It is well-recognized that computers have much to offer in the analytical field, in both routine applications as well as in research. This can be in the way of the automatic calculation of the calibration curve, automatic drift correction, processing the data from the interfaced instruments into directly readable signals and increased productivity. The primary benefit, however, is in the reduction of the time and labour of the technician, which is usually required for conversion of raw results into proper functional scales. This reduces the associated errors, which inadvertently creep in with manual methods of computation. Computers, in fact, have become basic laboratory tools that aid the chemist in carrying out the required analyses in a more efficient manner, particularly their applications as components of analytical instrument systems (Dessy, 1984). New methods of measurement have also been developed, which are possible only through the use of computerized instrumentation and high speed data processing techniques.

Computer

A computer is a combination of devices, which serves to make mathematical calculations and thus to find solutions to various types of problems, which are otherwise difficult and time-consuming. This is possible because the computers do two things with extraordinary speed, reliability and patience (Friedman, 1984). They 'crunch' numbers and shuffle data. When we say a computer can 'crunch numbers', we mean that it will allow you to feed in seemingly endless numerical data and mathematically process the data until it arrives at the solution to your problem.

When we say that a computer can shuffle data, it is meant that a computer will accept any kind of information—alpha (letter and punctuation), numeric (numerals) and alphanumeric, store it, relate it to other data, extract specific data, re-arrange the data in the desired order and display it to the user in any preferred order. Depending upon the kind of data, the sophistication of the computer processing power and the way in which the information is displayed to the user, the software that shuffles data is called a database. In other words, all data handling software is referred to as the database.

Database

27.2 Digital Computer

There are two main classes of electronic computers, viz. analog and digital. In general, digital computers have a much higher degree of accuracy than analog computers and they have almost replaced the latter.

A digital computer, as the name implies, is based for its operations on the handling of digits and numbers. The operation is basically arithmetic and covers addition, subtraction, multiplication and division. Digital computers also incorporate additional facilities to perform calculus type of solutions.

All of us are accustomed to work with the decimal system of dealing with numbers. But in digital computers, the system used for solutions of problems is the binary system, which means that we take 2 as the base instead of 10 in the decimal system. Any number when expressed in binary system contains a series of only ones and zeros. Representation of numbers in this way very much facilitates the transmission and storage of numbers by two simple signals for one and zero. Such signals may merely be positive pulses transmitted at regular intervals. The occurrence of the pulses at any time may mean a one and its absence at another time interval, a zero. This time generated pulses are employed to perform the desired computing operations.

The block diagram (Figure 27.1) shows the two major functional parts of a typical digital computer. These are delineated below.

The central processor includes:

- Internal storage or memory
- Arithmetic unit
- Control unit

The peripheral unit includes:

- Input devices
- Output devices
- Auxiliary memory or back-up store.

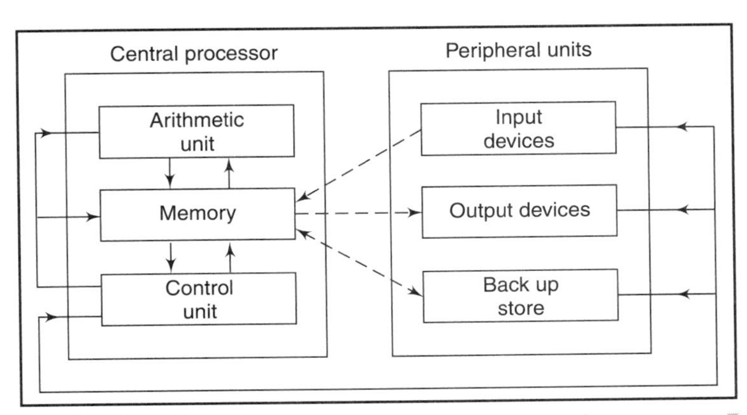

Figure 27.1 :: Block diagram of a digital computer

In order to arrive at the solution of any problem, all essential information (data) is fed into the input. This information consists of the numerical data involved in the problem and the coded instructions of what should be done with the data. The combined data are passed on to the memory unit in the central processor, wherein each part of the information is stored in separate locations in the form of magnetic storage or in solid state memory.

The memory unit stores the information received from the input and from other functional parts, and delivers this information at a designated time during the computing process. The control unit gets instructions from numerical codes, which are stored in the memory and interprets these codes in the appropriate sequence and directs the operations of the entire machine. The function of the control unit can be illustrated by considering a simple problem of summation. The control unit sends a signal to the memory unit to pass the stored numbers to the arithmetic unit, where they are temporarily stored in the registers. As a next step, it signals the arithmetic unit to add these numbers and send the output back to the memory unit from where it is passed to the output under instructions from the control unit. In practice, however, the problems are not so simple and they involve a series of arithmetic operations to be carried out before the results can be passed to the output. If the usual memory unit is not sufficient to accommodate all the data involved in a certain problem, it can be supplemented with an auxiliary or a back-up storage unit.

The function of an input device is to present the basic symbols, 0, 1, 2, ..., 9; A, B, C, ..., Z; +, - ,; %, %, x, etc., in a coded form, so that they can be held within the computer's internal storage. This is achieved by using a keyboard which is similar to that of a typewriter.

The solution of a problem by a digital computer is ultimately transferred to the output section. The output can be in the form of a display on the cathode ray tube (monitor). A graph plotter incorporated in the system can also be operated with programmed instructions so that graphs can be drawn from a pattern of discrete digits held in internal storage. The output data may also be obtained in the printed form from printers.

27.2.1 Off-Line/On-Line Computers

An off-line computer accepts the experimental data from input devices such as floppy disks, compact magnetic tape or punched paper tapes. It does not take data directly from the analytical instrument, nor does it process in real-time.

The off-line configuration is generally implemented with large computers in situations requiring complex calculations and manipulation of sizable amounts of data. The computer running in an off-line configuration is not required to respond to the instantaneous needs of a specific instrument. On the other hand, an on-line computer takes data directly from the instrument or process, and analyses it to take decisions based on it.

The instrument and the computer, in this case, are linked through an electronic interface to perform tasks such as acquisition and processing of data as well as instrument control functions. The computer responds instantaneously to the data acquired, carries out compilations, provides output information rapidly enough to improve the dynamic operation of the process and the instrument. Examples of the need for rapid execution of complex mathematical transformations functions are found in Fourier transform of nuclear magnetic resonance and infrared spectroscopy, which would not have been possible without the help of on-line computers.

It is thus obvious that with the off-line computer, only the past history of an experiment can be studied, whereas with the on-line computer, it is possible to determine results in real-time to investigate new and unusual occurrences during the course of an experiment.

27.2.2 Types of Digital Computers

Computers are generally classified as super, mainframes, minis and micros. Although these categories are now well-established, understood and talked about by everyone, the manner in which computers are actually defined for these groups continues to change. This is because computer technology changes so rapidly that such definitions quickly become obsolete. For example, what is called a minicomputer today might have been considered a mainframe computer ten years ago. This implies that almost any criteria accepted today quickly become invalid or obsolete.

The generally accepted basis of classification of computers is according to word size, memory capacity and processing speed. It may be remembered that the greater the word size, memory capacity and processing speed, the larger is the computer. For example, a minicomputer has a greater word size, memory capacity and processing speed than a microcomputer. Similarly, a supercomputer would process data faster than a mainframe, which is faster than a mini, which is again faster than a micro. Let us first explain something about word size, processing speed and memory capacity.

Word Size: A word' is a group of bits that the computer treats as a single unit of information. The number of bits in a word varies. Small computers may use words of four or eight bits, large computers may use words of 32, 64 or more bits. Figure 27.2 illustrates words consisting of 4, 8 and 16-bits, which shows that the larger the word size, the larger the computer category that the computer fits into.

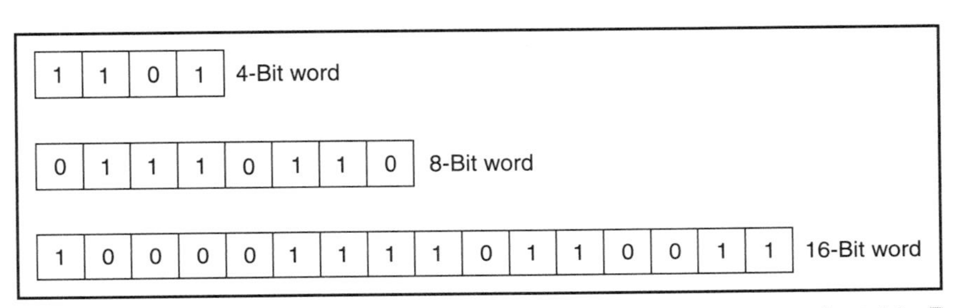

Figure 27.2 :: Word sizes—4-, 8- and 16-bit

Memory Capacity: Memory capacity is determined by the number of memory locations available in a system. Memory size is generally measured in thousands of words or kilowords. One kiloword (k) is actually equivalent to 1024 words. For example, if a computer has 64 k memory, it means a capacity of 65, 536 locations (64 × 1024), if the measurement unit is bits. If the memory is expressed in bytes, the same 64 would represent 524, 288 (64 × 1024 × 8) memory locations. The memory of large computers is expressed in megabytes.

Processing Speed: Processing speed implies the number of instructions (spelled out in computer program) executed per unit time. Obviously, supercomputers can process data faster than mainframes, which can process faster than a mini, which can process faster than micros.

Microcomputers are more likely to be used as components of electronic systems, whereas mini, mainframe and supers are used as general purpose computers, i.e. they are likely to be used for a wide range of data processing activities, many times for off-line applications. Larger (mini, mainframe, super) computers generally have far more extensive software support than microcomputers.

It should be noted that the distinction between the various categories is not sharp, and the differences are only matters of degree rather than kind.

27.3 Microcomputers

The microcomputer is the most fascinating development of the last century. It is a digital computer which uses a microprocessor as its central processing unit. Because of its low cost, small size and tremendous capabilities, the microcomputer's range of applications is presently so large that they are limited only by the imagination of the user. Microcomputers were originally designed to replace hard-wired logic and control circuits in relatively simple machines such as microwave ovens, sewing machines, cash registers, washing machines, etc. However, the present generations are capable of data processing tasks in highly sophisticated instrumentation systems. The microcomputer applications include those for industrial control (robots), hobby and personal computers, word processing, office automation, scientific instruments and processes involving data acquisition and analysis.

It may be noted that personal computers (PCs) are general purpose microcomputers. They are now used with analytical instruments, many times as attachments, for data processing and control.

27.3.1 Dedicated Microcomputers

The general purpose microcomputer dedicated to a specific function in a laboratory has been largely responsible for a revolution in the methodology, economy, quantity and quality of experiments performed and the data analyzed. In combination with an analytical instrument, the microcomputer becomes an element in a total analytical instrumentation system and is capable of acquiring and analyzing data, and controlling the instrument and experiments based on the data received. This has facilitated the designing of considerably more sophisticated instrumentation systems that were not possible earlier by having microcomputers as internal components.

Laboratory systems built around dedicated computers are already in use in a variety of analytical fields such as mass spectroscopy, gas chromatography, nuclear magnetic resonance and clinical chemistry. Programmed data processing systems and associated software are available as black box devices for applications.

Dedicated computers are of low initial cost and provide much more computer capability for a particular problem than when it is interfaced to a timed-sharing system. In addition, a dedicated computer system does not need programming staff. (Sumpson and systems, 1980)

Microcontrollers are essentially microcomputers having CPU and memory (ROM and RAM) on the same chip and are controlled by programs stored in the ROM. These programs (firmware) are placed in the computer by the manufacturer and cannot be altered by the instrument user. In case use of microcomputers as integral components of instruments, the configuration is termed as in-line, which is conceptually shown as Figure 27.3.

Firmware

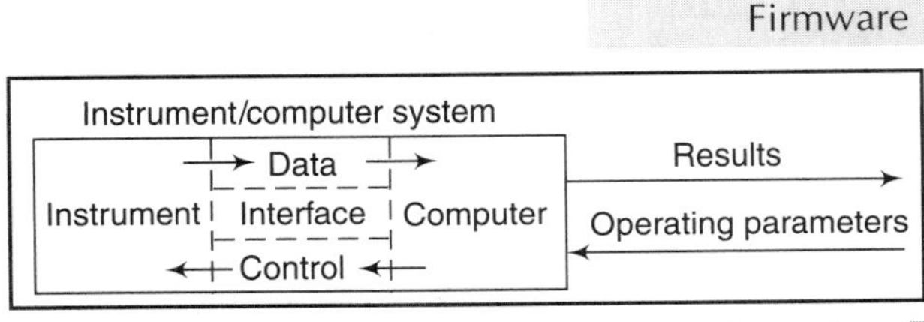

Figure 27.3 :: In-line computer configuration

In-line microcomputers offer several advantages in terms of increased reliability (decreased maintenance) due to substitution of discrete components, improved accuracy of results with automatic periodic instrument calibration, higher precision of results using digital signal processing and ease of communication with other devices being external to the instrument.

In-line microcomputers have become standard components in present-day analytical instrumentation. Therefore, their presence in instrument design is established by references to microcomputers in almost every chapter of this book.

27.4 Components of a Microcomputer

A block diagram of a microcomputer system is shown in Figure 27.4. It consists of all or some of the following sub-systems:

- Microcomputer unit (CPU, memory, I/O),
- Video display unit,
- Keyboard,
- Printer,
- Plotter,
- Controllers,

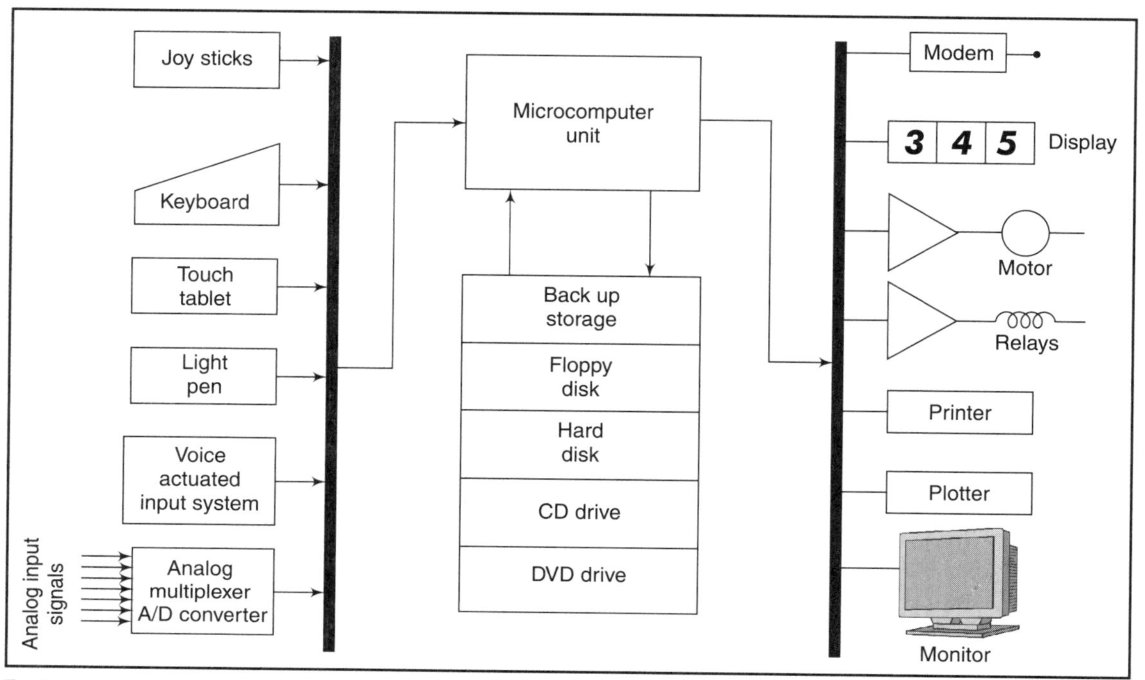

Figure 27.4 :: Various sub-systems of a microcomputer system

- Bulk storage devices,
- Floppy disk and hard disk unit, and
- Modems.

27.4.1 Microcomputer Unit

A simplified block diagram of a microcomputer is shown in Figure 27.5. It consists of the following components:

- *CPU Module:* This contains the central processing (CPU) system, timing and interface circuitry to memory and I/O devices.
- *Memory:* This contains read-only-memory (ROM or PROM) for program storage and random access memory (RAM) for data storage.
- *Input/output (I/O) Ports:* These contain circuitry that allows communications with devices outside of the system. For example, keyboard, cassette tape, video display unit.

There are three buses that interconnect the blocks. These are the:

- *Data Bus:* A bidirectional path on which data can flow between the CPU and memory or input/output. It carries the actual data being manipulated

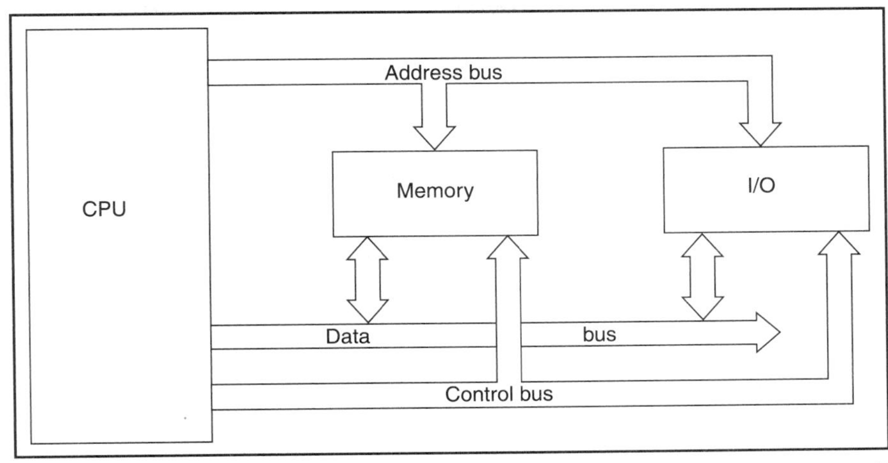

Figure 27.5 :: Building blocks of a microcomputer

- *Address Bus:* A unidirectional group of lines that identify a particular memory location or input/output device.
- *Control Bus:* It carries all the control and timing signals. It is a unidirectional set of signals that indicate the type of activity in current process. The types of activities could be memory read, memory write, input/output read, input/ output write and interrupt acknowledge.

The central processing unit is the most important component of any computer system—big or small. In a microcomputer, it is the microprocessor chip which acts as the central processing unit. The microprocessor can be termed as the heart of the microcomputer, because its capabilities determine the capabilities of the microcomputers. In such a system the central processing unit (CPU) requests instructions prepared by the programmer, calls for data and makes decisions related to the instructions. On the basis of the data, the processor determines the appropriate actions to be performed by the other parts of the system. Since there are many peripherals associated with the given system, the microprocessor must be capable of selecting a particular device. It identifies each device by means of a unique address code. A typical microprocessor has 16 binary address lines providing 65, 536 addressing codes. Data to and from the processor is carried across a bi-directional 8- or 16-bit wide data bus. Many processors also provide a serial data path. Several microprocessors use a multiplexed address/data bus on which both the address and data are transmitted on the same signal paths. In this case, the first portion of the bus cycle transmits the address while data transfer takes place later in the cycle. This architecture is popular for microprocessors with an 8-bit data bus.

Another important link in the system is the set of input-output (I/O) interfaces. These interfaces include all the information channels between the system and the real world. There are digital ports through which programs and control commands may be loaded and from which digital data may be transmitted to peripherals such as the keyboard, printers and floppy drive, etc.

The memory stores the data to be manipulated by the CPU, as well as the program that directs that manipulation. A program is a group of logically related instructions. The CPU reads each instruction from memory in a logically determined sequence, and uses it to initiate processing actions. The CPU can rapidly access any data stored in memory. Also, the CPU can address one or more output ports, that are added to receive information from external equipment, and input the data contained there.

Program

The computer also requires one or more output ports that permit the CPU to communicate the result of its processing to the attached equipment. Like input ports, output ports are addressable, and both the ports permit the processor to communicate with the outside world.

The CPU unifies the system by controlling the functions performed by the other components. The CPU must be able to fetch instructions from memory, decode their binary contents and execute them. It must also be able to reference memory and I/O ports, as required in the execution of instructions. In addition, the CPU should be able to recognize and respond to certain external control signal, such as INTERRUPT requests.

Hardware alone does not make a microcomputer. Before any microcomputer, and for that matter any computer, can be put to work, it must be given a set of instructions (program). It is the program that states the procedure that the computer has to follow in solving the program at hand. By changing the program, the same hardware can perform many different functions.

27.4.2 The Video Display Unit

The most obvious part of a microcomputer is the video display unit, which facilitates the man-machine communication. It provides a means of establishing a dialogue with the microcomputer system by entering data, whereas the microcomputer system displays appropriate messages on a video display screen. The operator responds by typing a reply at a keyboard. Anything typed is displayed on the screen to confirm the correctness of the data which the microcomputer has accepted. A shift key provided on the keyboard facilitates movement between the upper and lower case character displays.

Cursor

All video displays use what is called a cursor. A cursor is a small spotter square or arrow of light sitting where the next character will be displayed. When text is displayed on the screen, it can be selectively modified by moving the cursor to the error spot and correcting the character.

A graphic video display allows the display of pictures in addition to characters. The pictures are created out of straight lines, circles, plots, solid shapes and block forms. In less expensive systems, graphic display units display black and white, but not grey pictures. There are also colour graphic displays which provide the added capability of specifying the colour in which a character or graphic segment will appear on the screen.

One of the important applications of microcomputers with video monitors is word processing. For this, data in the text form are typed into the computer and edited from the display on the video monitor. Finally, the finished text is automatically transferred to a hard copy printer. The text can be manipulated to add, delete and re-arrange words and paragraphs under the program control and with the help of the video monitor.

27.4.3 Keyboard

Although a large number of devices are available for inputting data into microcomputers, the keyboard is the most commonly used. The keyboard's suitability for alphanumeric data entry as well as its long-standing association with the inputting function will continue to support its pre-eminence for a long time to come.

In order to represent the input and output data in a format that is easily understood by the analyst, the alphanumeric characters and punctuation makes have to be coded in binary form. The most commonly used code is the ASCII (American National Standard Code for Information Interchange). Here, each alphanumeric character is represented by 7 bits, which gives 128 possible combinations. Thus, all the 52 letters (upper and lower case), 10 numerals, and 22 additional special symbols can easily be represented.

Keyboards are provided with appropriate electronics for coupling in the microcomputer system. Most keyboards include buffered outputs to prevent false triggering arising from switch bounce, i.e. the generation of more than one actuation signal from a single key closure. The microcomputer examines each key closure several times and rejects signals whose duration is shorter than a set value as noise.

Switch bounce

The key arrangements could be of two types: typewriter and data entry. The latter, a calculator-style alphabetic arrangement plus numeric keypad is more convenient to operate for non-typists. It permits greater speed and lower error rates. The typing layout, on the other hand, is more apt for users who touch-type long text messages.

Most of the keyboards employ the QWERTY layout (also called the Sholes layout, after its inventor Christopher Sholes). QWERTY takes its name from the first six letters on the keyboard top row of letters.

27.4.4 Controllers

A keyboard is not the only item used to control computers. Other popular controllers include joysticks, game paddles, light pens, a mouse, a digitizer (graphic pad) and track balls.

Joysticks

Joysticks are currently the most popular controllers. They are used to move the cursor or player in a desired direction by manipulating the stick in one of several directions, usually numbering eight. Joysticks are basically potentiometer-based and are coupled to the computer with an analog-to-digital converter.

Game paddle

Trackballs

Game paddle is a variable potentiometer, which gives an analog signal to the computer for achieving action in a straight line on the screen. This is intended for triggering the fire action on the game computers. Trackballs are used for games in which rapid changes in direction are required. They consist of a billiard ball in a stable mounting, with an optical sensor that can track small changes in direction.

Mouse

A *mouse* is another popular device used in most applications. Essentially, it is an upside-own trackball in which the mounting is moved across the computer desk, by holding a hidden ball. The computer mouse comes in one to three buttons, whose use is determined by the program.

Light pen

A *light pen* is a tool that allows the co-ordinates of a point on the screen to be entered into a computer. It is based on the principle of sending a pulse to the screen control circuit at the precise moment when it sweeps the spot just in front of the light pen. A light pen is a very good controller for artwork.

Touch-tablet

A *touch-tablet* is another kind of graphical input device, which is entirely independent of any display and provides the co-ordinates of the pen position. An interesting feature of the tablet is that a finger can act as the pen, by using high frequency alternating current (ac) and capacitive coupling with a uniform resistive sheet instead of dc with direct pen coupling.

Digitizer

The *digitizer* or the *graphics pad* consists of vertical and horizontal rows of wires placed closely. When a pen is touched to the cover of the matrix, one pair of wires touch each other, thereby informing the computer about the location of the pen.

The above controllers are usually available as plug-into connector options.

27.4.5 Storage Systems for Microcomputers

Microcomputers are becoming more and more powerful and are being employed in applications requiring large memories. The semiconductor memories are still quite expensive and it is economical to use other storage devices to serve as back-ups for storing bulk information. Therefore, two levels of memories are used in the microcomputers. These are:

(a) The memory unit which communicates directly with the CPU: This is called the main memory.
(b) Back-up storage or auxiliary memory to store program and data not currently needed by the processor: The information in this case is transferred to the main memory on a demand basis. The most commonly used auxiliary memory device in microcomputers are floppy disks, compact discs (CD) and hard disks.

The speeds of these two levels of memory are matched by using a buffer memory.

The magnetic disk medium is, at present, the best computer peripheral for large capacity data storage equipped with fast program access and retrieval. The disk systems are of three types, which are the:

- Floppy disk system,
- Hard disk system, and
- Compact disc.

Floppy Disk: A floppy disk system comprises the following three elements (Figure 27.6):

- Floppy diskette on which data are stored and played back,
- Mechanical disk drive unit, and
- Electronic controller that interfaces the disk drive to the computer.

Floppy disk

The floppy disk, generally 3.5" in diameter, consists of a flexible, thin plastic disk coated on both sides with a layer of magnetic oxide similar to that used on audio recording tapes. Data are written on the diskette by using a form of frequency modulation (FM). Double density recording is obtained by using a modified FM (MFM) or modified-modified FM (M^2FM).

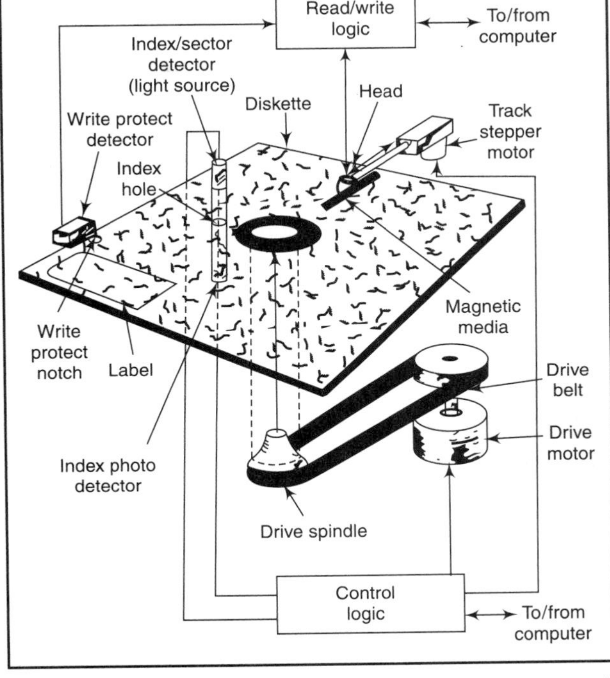

Figure 27.6 :: Elements of a floppy disk drive

Hard Disk

Hard Disk: Hard disk (Winchester) systems are high density storage units and are faster than floppy disk systems. Hard disks with enormous storage capacity, of 100 GB, are also available.

Compact Disc

Compact Disc: Another powerful medium for storing digital data is the compact disc (or CD). It is a non-volatile optical data storage medium and was originaly invented for digital audio. The CD-ROM (abbreviation for Compact Disc—Read-Only Memory) reading devices are frequently included as a component in personal computers. A CD is a flat, plastic disc with digital information encoded on it in a spiral from the centre to the limit, or the outside edge. CDs can either be burned or pressed. Small amounts of data are burned, while larger amounts are pressed. The break-even point seems to be somewhere between 100 and 500 copies.

Digital information is encoded at near-microscopic size in CDs, thereby allowing the storage of a large amount of information. CDs record data as tiny pits (or non-pits) pressed into the lower surface of the plastic disc; a semiconductor laser beam in the player reads these. Most CDs cannot be written with a laser, but CD-ROM discs have coloured dyes that can be burned (written to) once, and CD-RW (re-writable) discs contain phase-change material that can be written and over-written several times. Most CD-ROM drives can read CD-ROM discs; modern drives carrying the MultiRead mark can read CD-RW discs.

Compact discs are available in a range of sizes but the most commonly available CD is 120 mm in diameter. A 120 mm disc can store about 74 minutes of music or about 650 megabytes of data. Discs that can store about 700 megabytes (80 minutes of music) have become more common. For drives installed in computers, all current CD-ROM and DVD-ROM drives can read and write CD-ROM and CD-RW discs. CD-ROM drives may connect to an IDE (ATA) interface, a SCSI interface or a proprietary interface, such as the Panasonic CD interface. Most CD-ROM drives can also play audio CDs.

27.4.6 Printers

Video terminals are indispensable in microcomputer systems, but many of these systems also require hard copy which requires specifying a printer. For the same reason, some systems can operate adequately with a printer terminal alone. The choice of a particular type of printer depends upon its speed, quality of print, ruggedness and multiple-copy capability, etc. Basically, all printers can be characterized as either impact or non-impact types.

The impact type printers employ mechanical forces to transfer character images to paper, whereas the non-impact type use the thermal, electrical, chemical or optical technique to transfer such images to specially treated paper or film. The technique of data generation leads to another classification of printers. A serial printer produces one character at a time, while a line printer generates hard copy by the line, in a parallel format.

Printers are also classified on the basis of the character formation technique. A printer can output either fully formed or dot matrix characters. A dot matrix is an array of dots, of which five to seven are usually placed horizontally and seven to nine, vertically. Impact printers can produce either type of characters, whereas the non-impact type generally employ the dot matrix technique. A more commonly used printer is the laser printer which has almost replaced all other types of printers.

27.4.7 Plotters

Plotters are used to present information in the graphical form. Data generation on plotters occurs in either off-line, on-line or remote time sharing modes. In the off-line mode, a magnetic disc supplies the plotting instructions and plotting control comes from the host computer. Hence, the plotting system speed remains consistent with that of the host. The plotting system usually incorporates a controller that can communicate with the host computer or disc, and provides the necessary drive signals for the plotter. By far, the largest numbers of plotters are connected to computers in on-line or remote/time sharing modes. This is because of the expanded use of microprocessor-based plotter controllers, which unload the host computer from some of its time-consuming tasks. The microprocessor can also generate lines and characters, beside handling communication protocol. This results in a distributed processing system, in which the plotter system operates at on-line computer speeds.

Plotters can be classified into three types: flat bed, drum and electrostatic. If plot quantity takes priority over quality, the plotting area assumes importance and governs the selection of a plotter. Usually, the flat bed plotter area ranges from about 11 × 17 to over 60 × 100 inches. Drum and electrostatic plotters use paper measuring 8.5 to 72 inches wide and up to hundreds of feet long.

In a plotter, software commands control the generation of lines or dots that combine to produce the desired graphic information. Application programmes which specifically tailor system operation are available from system suppliers. They are designed to generate such information as flow charts, contour maps, three-dimensional perspective drawings and symbols for electronic schematics, logic diagrams and floor layouts. The software for flat bed and drum plotters specifies individual line segments that can be connected to produce an infinite variety of graphic data. Electrostatic plotter software, on the other hand, formats data to a dot at each desired location along the raster line. In other words, an electrostatic plotter performs the same function in software that the flat bed and drum do by controlling their pens.

27.4.8 Modems

The transfer of relatively small amounts of data from one computer to another can be done via a physical storage medium such as a magnetic disc or compact disc. But transferring larger amounts of data necessitates wiring of the computers together to enable direct exchange. This can be done by making use of telephone lines, but they are in analog form while computers are digital. The translation of data from one signal form to another for transmission over telephone lines is achieved by using a modem.

The term modem combines two functions: modulation and demodulation. It consists of a transmitter that converts digital (base-band) signals from the computer into analog (pass-band) signals suitable for transmission over telephone lines. The modem receiver demodulates the received pass-band signals into base-band signal. The working of the modem is transparent to the computing task and the operator is not concerned with it. The computer and remote terminal operate as if they are connected directly. Modems operating on telephone lines are termed as voice band or voice grade modems.

Each application makes different demands on a modem. The parameters for selecting a modem are the data transmission rate, communication modes, synchronization modulation techniques, types of lines and industry standards. The most important parameter is the data transmission rate and the modems can be categorized as: low speed (bit rate up to 600 bps), medium speed (1200 to 2400 bps) and high speed (4800 bps and up). Data transmission rates are specified as bit rate or baud rate. The bit rate is the actual transfer rate of transmitted data (bits of data per unit time), whereas the baud rate is the unit of signalling. In order to achieve higher data transmission rates, special modulation techniques are used, which enable the transmission of multiple numbers of bits per baud. High speed modems require two, three or four bits per baud to implement bit rates of 2400, 4800 or 9600 bps. In low speed modems, the bit and baud rates are the same.

Bit rate

Baud rate

The applications determine whether one or two-way communication is needed, whether two locations are required to be connected point-to-point, or whether a multiple location network is required.

27.5 Computer Software

In computer science, computer software refers to the information processed by the computer system. It means a program or a set of programs which is needed to get the computer to perform its intended function. In general, the term software is used to refer to all instructions, routines and programs, and their associated techniques. Since the programs are changeable, they are called software. This is in contrast to hardware which is the term used for computer equipment and the logic circuits of physical components. The programs stored in the read-only memory (ROM) are often referred to as Firmware to indicate that they are somewhere between hard logic and soft programs.

Software is often divided into the following two major categories:

- System software, and
- Application software.

27.5.1 System Software

System software provides the basic system-specific functions of the computer. It is responsible for controlling, integrating, and managing the individual hardware components of a computer system so that other software and the users of the system see it as a functional unit without having to be concerned with the low-level details such as transferring data from memory to disk, or rendering text onto a display. Generally, system software consists of an operating system and some fundamental utilities such as disk formatters, file managers, display managers, text editors, user authentication (login) and management tools, and networking and device control software.

Operating Systems

Operating Systems: An operating system is essentially an agent between the hardware, the software and the user (Figure 27.7). Every general purpose computer must have an operating system to run other program. Operating systems perform the basic tasks, such as recognizing input from the keyboard, sending output to the display screen, keeping track of files and directories on the disk, and controlling peripheral devices such as disk drives and printers. It is a set of programs that collectively manage all the resources available to the computer, including the CPU, the peripheral devices and the software. For achieving maximum speed, most successful operating systems are written in machine language.

The operating system ensures that different program and users running at the same time do not interfere with each other. The operating system is also responsible for security, and ensures that unauthorized users do not access the system. Operating systems can be classified as follows:

- *Multi-user:* Allows two or more users to run programs at the same time; some operating systems permit hundreds or even thousands of concurrent users.
- *Multi-processing*: Supports running a program on more than one CPU.
- *Multi-tasking:* Allows more than one program to run concurrently.
- *Multi-threading*: Allows different parts of a single program to run concurrently.
- *Real time*: Responds to input instantly; general-purpose operating systems, such as DOS and UNIX, are not real-time.

Operating systems provide a software platform on top of which other programs, called application programs, can run. The application programs must be written to run on top of a

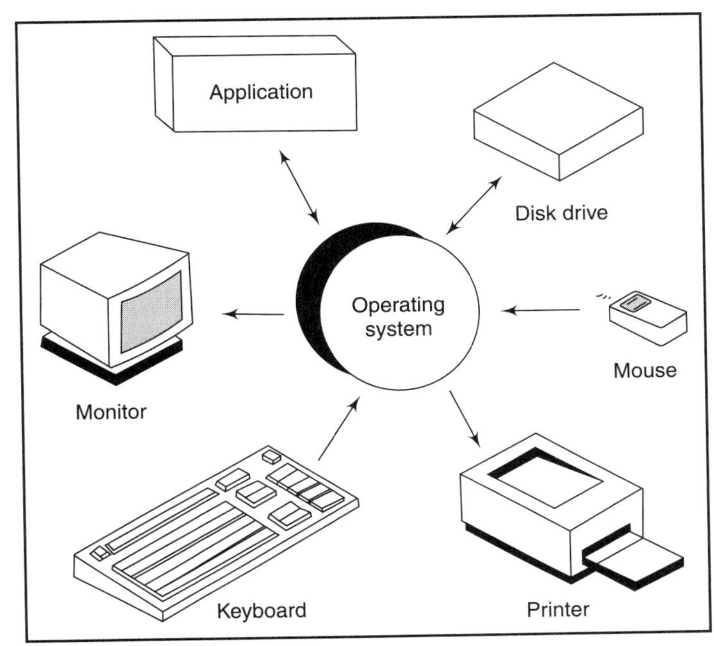

Figure 27.7 :: Function of operating system in computers

particular operating system. One's choice of operating system, therefore, determines, to a great extent, the applications one can run. For PCs, the most popular operating systems are DOS, OS/2, and Windows, but others such as Linux, are also available.

A user, normally interacts with the operating system through a set of commands. For example, the DOS operating system contains commands such as COPY and RENAME for copying files and changing the names of files, respectively. The commands are accepted and executed by a part of the operating system called the command processor or command line interpreter. Graphical user interfaces allow the user to enter commands by pointing and clicking at objects that appear on the screen.

At present, the major operating systems that are widely use on personal computers have been consolidated into two families: the Microsoft Windows family and the UNIX-style family, which includes various definitions of UNIX, Linux and Mac OS X. The most popular operating systems are listed below.

- ***Windows XP***: Windows XP is a server, workstation, and desktop operating systems made by Microsoft that run on the Intel/Cyrix/AMD Pentium.
- ***Windows 2000***: Windows 2000 Professional, Windows 2000 Server, and Windows 2000 Advanced Server are server and workstation operating systems made by Microsoft that run on the Intel/Cyrix/AMD Pentium.

- *Windows NT:* Windows NT, Windows NT Server, and Windows NT Server Enterprise Edition are server and workstation operating systems made by Microsoft that run on Intel/Cyrix/AMD Pentium, Intel 80x86, and DEC Alpha.
- *Windows ME:* Windows ME (Millennium Edition) is a desktop operating system made by Microsoft that runs on Intel/Cyrix/AMD Pentium.
- *Windows 98:* Windows 98 is a low quality desktop operating system made by Microsoft that runs on Intel/Cyrix/AMD Pentium and Intel 486DX.
- *Windows 95:* Windows 95 is an obsolete desktop operating system made by Microsoft that runs on Intel/Cyrix/AMD Pentium and Intel 80486.
- *Windows 3.11:* Windows 3.11 is an obsolete desktop operating system made by Microsoft that runs on Intel 80486, 80386, and 80286.
- *MS-DOS:* MS-DOS is a text-based desktop operating system made by Microsoft that runs on Intel 80 × 86.
- *PC-DOS:* PC-DOS-2000 is a text-based desktop operating system made by IBM as an update of the older MS-DOS. It runs on Intel 80 × 86. Only low end servers can run on this operating system.
- *MXS:* Macintosh OS X Server (MXS) is a server operating system based on Rhapsody. MXS is made by Apple Computers and runs on Motorola/IBM Power PC.
- *Mac OS X:* Macintosh OS X (ten) is a desktop operating system based on Rhapsody. Macintosh OS X is made by Apple Computers and will run on Motorola/IBM Power PC.
- *Classic Macintosh:* Macintosh OS 9, OS 8, OS 7, and OS 6 are desktop operating systems made by Apple Computers that run on Motorola/IBM Power PC and Motorola 680 × 0.
- *OS/2:* OS/2 is a high performance desktop and high-end operating system made by IBM that runs on Intel/Cyrix/AMD Pentium and Intel 80 × 86.
- *Linux:* Linux is a free version of UNIX that runs on Intel/Cyrix/AMD Pentium, Intel 80 × 86, Motorola/IBM Power PC, Motorola 680 × 0, Sun SPARC, SGI MIPS, DEC Alpha, HP PA-RISC, DEC VAX, ARM, API 1000+, and CL-PS7110.
- *Solaris:* Solaris is a UNIX-based operating system made by Sun Computers that runs on Sun SPARC and Intel/Cyrix/AMD Pentium.
- *SunOS:* SunOS is an older text-based UNIX that runs on Sun SPARC. Solaris is an enhancement of SunOS that includes a graphic user interface.
- *HP-UX:* HP-UX is a UNIX-based operating system made by Hewlett Packard that runs on HP PA RISC.
- *IRIX:* IRIX is a UNIX-based operating system made by SGI that runs on SGI MIPS.
- *Rhapsody:* Rhapsody is a UNIX-based operating system that includes capabilities from the NeXT and Macintosh operating systems. Rhapsody is made by Apple Computers and will run on Intel/Cyrix/AMD Pentium and Motorola/IBM PowerPC and will have a Run Time Library for Windows.
- *AIX:* AIX is IBM's version of UNIX.
- *MVS:* MVS is a mainframe operating system made by IBM.

- ***NetWare:*** NetWare "is a dedicated network operating System" that runs on Intel/Cyrix/AMD Pentium, Intel 80486, and Intel 80386.[e16]
- ***SCO OpenServer:*** SCO OpenServer is a UNIX-based operating system that runs on Intel/Cyrix/AMD Pentium.
- ***SCO UnixWare:*** SCO UnixWare is a UNIX-based operating system that runs on Intel/Cyrix/AMD Pentium.
- ***BSDi:*** BSDi Internet SuperServer (also known as BSD/OS) is a commercial BSD UNIX for Pentiums and other processors.

UNIX is widely used in academic institutions and back-end implementations, while Windows is popular among home users as well as businesses for front-end use. In general, Windows is the most widely used operating system with various studies placing Microsoft's market share anywhere from 90 to 98 per cent. Linux is widely used in web servers, and is making inroads into home and business environments. Mac OS X, which has incorporated the major parts of UNIX, and its predecessors, are popular with multimedia designers. Mainframe computers and embedded systems use a variety of different operating systems, many of them with no direct connection to Windows or UNIX.

Utilities: Utilities are the programs which are used in housekeeping functions and I/0 manipulation. They usually cover those features which are omitted from an operating system or programming language. Programs such as those used for developing graphics and sounds, programs that repair damaged data diskettes and programs that write programs are some of the commonly known utilities.

27.5.2 Application Software

Application software is used to accomplish specific tasks other than just running the computer system. Application software may consist of a:

- Single program, such as an image viewer;
- Small collection of program (often called a software package) that work embedded systems, the application software and the operating system closely together to accomplish a task, such as a spreadsheet or text processing system;
- Larger collection (often called a software suite) of related but independent program and packages that have a common user interface or shared data format, such as Microsoft Office, which consists of closely integrated word processor, spreadsheet, database management system, etc.; or
- Software system, such as a database management system, which is a collection of fundamental programmes that may provide some service to a variety of other independent applications.

In the above types of embedded systems, the application software and the operating system software may be indistinguishable, such as in the case of the software used to control a microwave oven, a VCR or DVD player.

27.5.3 Software Creation

Software is created with programming languages and related utilities. Therefore, it is of special importance for the users to understand the computer language(s) used by his or her equipment. A programming language or computer language is a standardized communication technique for expressing instructions to a computer. It is a set of syntactic and semantic rules used to define computer program. A language enables a programmer to precisely specify the data which a computer will act upon, how these data will be stored/transmitted, and precisely the actions to be taken under various circumstances.

Machine language

Assembly language

The most basic computer language is the machine language, which is a rudimentary method of communicating directly to the computer through a system of binary digits. Programming with a machine language or assembly language (based on use of mnemonic to each instruction) is a long and tedious process. That is why various high-level languages have been developed. These languages allow the user to work in a language that is closer to English, thereby improving efficiency and simplifying communication.

High level languages

High level languages make computer program less dependent upon particular machines or environments (Figure 27.8). This is because programming languages are converted into a specific machine code for a particular machine rather than being executed directly by the machine. The concept that easier-to-understand languages could lead to less error-prone and more rapid development was the basis for Fortran II way back in 1958, and the foundation for the high-level languages of today. Since the 1950s, however, numerous high-level languages have cropped up. Users have heard and worked with most of these. Fortran is still used in engineering, and general programming problems are often solved with C, C++, Java, and Perl, which are all considered to be high-level language.

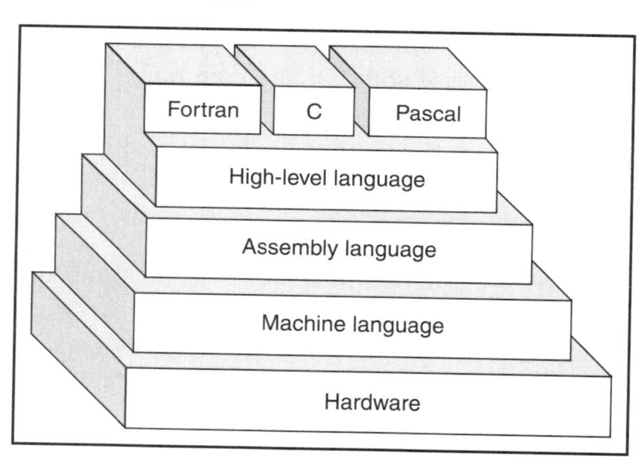

Figure 27.8 :: Combination of hardware–software vis-à-vis languages

Two mechanisms are used to translate a program written in a programming language into the specific machine code of the computer being used. The translation mechanism used is the one that translates the program text as a whole and then runs the internal format. This mechanism is defined as a compilation.

Compiler

The compiler is therefore a programme which takes the human-readable programme text (called source code) as data input and supplies object code as output. The resulting object code may be a machine code which will be executed directly by the computer's CPU, or it may be a code matching the specification of a virtual machine.

If the program code is translated at runtime, with each translated step being executed immediately, the translation mechanism is defined as an interpreter. Interpreted programmes run usually more slowly than compiled programmes, but have more flexibility because they are able to interact with the execution environment. Although the definition may not be identical, these typical fall into the category of scripting programming languages.

Interpreter

Most languages can be either compiled or interpreted, but they are better suited for one or the other. In some programming systems, program are compiled in multiple stages, into a variety of intermediate representations. Typically, the later stages of compilation are closer to the machine code than the earlier stages. The popular high level languages include BASIC, Pascal and FORTAN etc.

BASIC

BASIC (Beginners All-purpose Symbolic Instruction Code) is the most widely used and familiar language. It is easy to learn, because of its English-like statements and simple syntax. The main application areas of BASIC are for commercial, business and educational purposes. Virtually all microcomputer systems offer it, generally in ROM.

Pascal

Pascal is another high level language, which is popular in the academic field as it has several features that help in structured programming. However, Pascal has not become popular in home and business markets. UCSD (University of California at San Diego) Pascal is available on some microcomputers.

FORTRAN

Cobol

FORTRAN (Formula Translation) is most commonly used for mathematics, engineering and science problems. Cobol (Common Business-oriented Language) is most often used in business and data processing applications. Other languages used for microcomputers include LOGO, FORTH, PILOT, COMAL, LISP and OCCAM.

Widespread developments in electronics technology in terms of high density integrated circuit chips, have led to the evolution of more and more reliable and usable computers. This was paralleled by the development of a variety of standardized computer languages to run on them. The explosive growth of the Internet became possible with the large-scale availability of personal computers and the increased use of computer programming through more accessible languages such as visual Basic and Python.

Choosing a language is a purely subjective matter. The choice of a language is, however, usually based on the user's experience in the language and on how one which is well-established the language is (Owen, 1984).

Software can be recorded on the printed page or on a disk. Reading the data off a printed page and feeding it manually to the computer via a keyboard is the slowest process. Feeding the data into the computer's memory from a pre-recorded floppy disk or compact disk (CD) is a much more efficient and presently the most widely used method.

27.6 Connecting Laboratory Instruments to Computers

For real-time processing of data, the instrument has to be connected to a computer for which a variety of interfacing possibilities are available. The purpose of an interface is to accept experimental data as presented by the instrument and to transmit that data to the computer in a form it can deal with. The interface configuration depends upon the following factors (Liscouski, 1982):

- Characteristics of the data to be collected, which would, in turn, depend upon the needs of the experiment, and
- Capabilities of the microcomputer, i.e. whether it can accept the data rates and perform the real-time processing required.

The interface between the instrument and the computer consists of hardware and software resident in the computer or in the form of embedded intelligence in the instrument itself. The proportions of interfacing functions that are provided by hardware and software depend upon the type of interface (Dessy, 1986b).

27.6.1 Types of Interfaces

There are four common types of interfaces for obtaining information from an instrument into a computer. These are:

- *Analog I/0,* which requires the computer to read directly an analog signal such as voltage. Here the user determines where to pick up the instrument signal and how to get it to the computer.
- *Digital or parallel I/O,* in which the two voltage levels are used to represent logical 0 and 1
- *Serial ASCII,* in which the instrument generates a bit stream of alphanumeric characters as eight data bits (the seven-bit ASCII code is not generally used for instrument applications).
- *IEEE-488 standard interface,* which defines the 16-line bus, linking the instrument and computer, the shape, size and number of pins in the connections, all voltages, the handshaking procedure including addressing; and the method of data transfer.

Irrespective of the type of interface selected, application software needs to be written for the computer to handle the data and to process it. Depending upon the requirements of the experiment, the development of software may be a trivial or a major task. Many data reduction and analysis sub-routines are available in commercial packages for the most popular computers.

For data collection from instruments under external control, all three of the communication protocols—RS 232 C, binary and IEEE 488—can be employed. The fundamental control element is a binary output representing an ON/OFF request. The TTL signals generated by the basic parallel output boards are given to solid-state relays to allow the user to control 8-16 solid state switches with TTL inputs. These switches are optically isolated and can control 110-220V ac upto 5A.

The control of external devices is an inherent part of the IEEE 488 standard, whereas commercial modules are available that interpret a pre-selected set of ASCII characters transmitted over a serial line as the command to activate or de-activate a target device.

A timing element is usually needed to facilitate the external control of a device. It is common to incorporate clock-timer modules into the computer's option slots to provide controllable real-time functions. These devices have programmable counters, in which externally provided pulses cause them to count down towards zero. When a time-out occurs, a system interrupt is generated, indicating that a real-time period has elapsed and some service action is required. Although it is possible to program complex situations with one clock, multiple overlapping asynchronous events are best handled with multiple real-time clocks.

27.6.2 Analog Interfaces

Instruments which lack a digital output require an analog interface to convert voltage or the other analog output of an instrument into a binary form that can be handled by the digital computer.

Figure 27.9 shows the functions of an analog interface, which converts the analog dc signals from a primary measuring instrument into a digital signal. A pre-amplifier may be necessary to obtain satisfactory resolution from the A/D converter. Analog-to-digital converters are characterized by their resolution, dynamic range and maximum throughput, i.e. total samples per second which can be handled. Since software costs generally far exceed hardware costs, the analog-digital interface structure must permit software-effective transfers of data command and status signals to realize the full capability of the host microcomputer. At the same time, it must acquire and digitize data, often from hostile industrial environments, with no degradation in the resolution or accuracy of the signal.

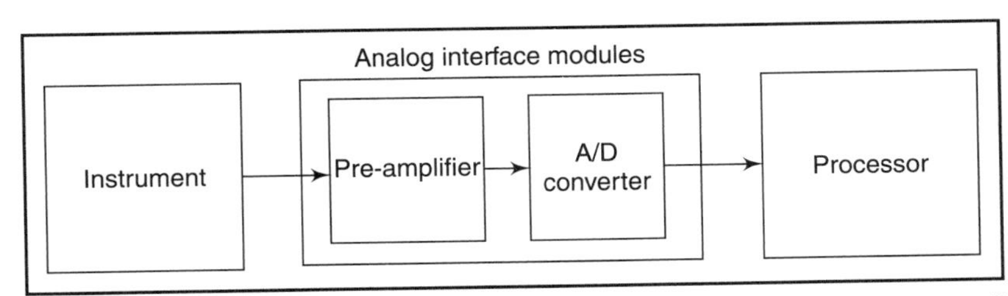

Figure 27.9 :: An analog interface. It converts the dc signal from the instrument into binary numbers. A pre-amplifier may be necessary to obtain a better resolution from the A-D converter

27.6.3 Digital I/O Interfaces

Digital I/O was the first type of computer interface developed for the discrete signal outputs of instruments like pH meters, balances and spectrophotometers. In such interfaces, the data are handled in digital form (1 and 0). BCD (binary-coded decimal) was the first such method and is still the most commonly used method of digital data notation. Each digit may be transmitted individually and successively (bit parallel, digital serial). Alternatively, all the digits for a datum may be transmitted at one time (bit parallel, digital parallel). In the former, only four bits of data need to be collected at one time, while in the latter format, the bit string length is determined by the dynamic range and resolution of the internal BCD converter.

If the BCD instrument output uses TTL, it may be connected directly to a parallel port of a computer. This parallel port must have a sufficient number of input bits to handle the data string. In addition, it must have at least one output bit that can control the update line of the data latch associated with BCD output devices. Figure 27.10 shows a typical parallel BCD interface. The A/D converter in these systems feeds BCD output to a set of data flip-flops used to latch the information. An update line prevents the system from changing the latch contents, while the computer is reading the data.

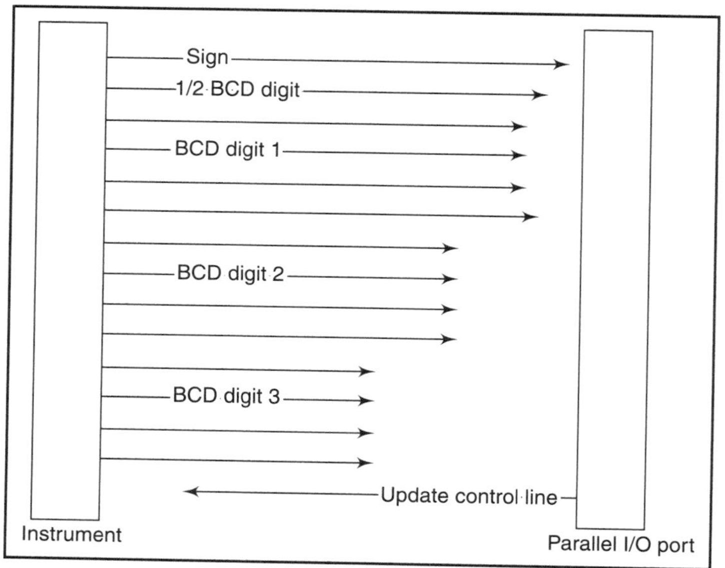

Figure 27.10 :: Parallel BCD interface

It may be noted that incoming data are in BCD and most computer programs (and languages) implement straight binary arithmetic. A BCD-to-binary conversion is generally required.

Data transfers through digital I/O interfaces are customarily asynchronous. They take place whenever the instrument and the computer are prepared to accept it. Co-ordination of the instrument and computer is facilitated by handshaking software on the computer. In the handshaking procedure, the instrument sends the computer a signal, asking whether it is ready to receive data, receives an acknowledgement in the affirmative from the computer on a separate REPLY line, and then sends the data. A representative data word architecture used in transmitting experimental data from an instrument to a computer through a digital I/O interface is shown in Figure 27.11. The software, indicates what the bit positions represent in the data word. The 1 and 0 levels are represented in both input and output voltages by 3.5 V dc and 0 V dc, respectively.

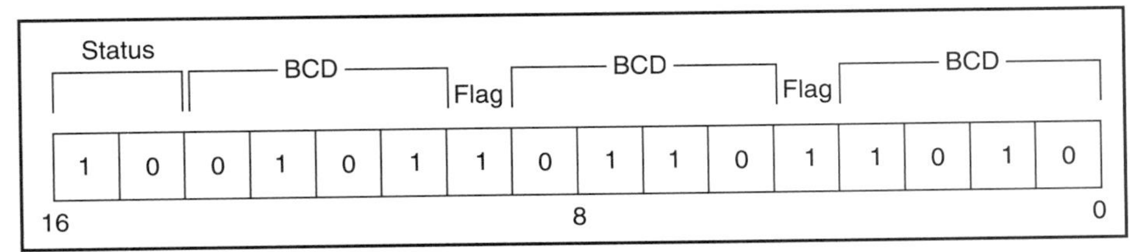

Figure 27.11 :: Digital I/O interface for transmitting experimental data from an instrument to a computer

27.6.4 Serial ASCII Interface

A serial ASCII instrument generates a bit stream of alphanumeric characters as defined by the ASCII Code by American Standard Committee for Information Interchange. This code is a 7- or 8-bit code consisting of 1's and 0's that represent letters, numbers and control characters. The 7-bits can encode 128 possible values. The 26 uppercase and 26 lower-case letters, 10 Arabic digits and a variety of grammatical symbols can be expressed in the code. Computer keyboards generate most of these codes by using combinations of the regular alphanumeric (Seven bit ASCII code is not generally used for instrument applications). Thus data and control information are sent along the two data channels in ASCII code.

There are two types of arrangements for transmitting bits between the instruments and interface modules: the 20 mA serial ASCII interface and RS 232 C interface.

20 mA Serial Interface: Figure 27.12 shows the 20 mA transmission interface, which is based on a two-wire closed current loop between the instrument and computer. The instrument generates an

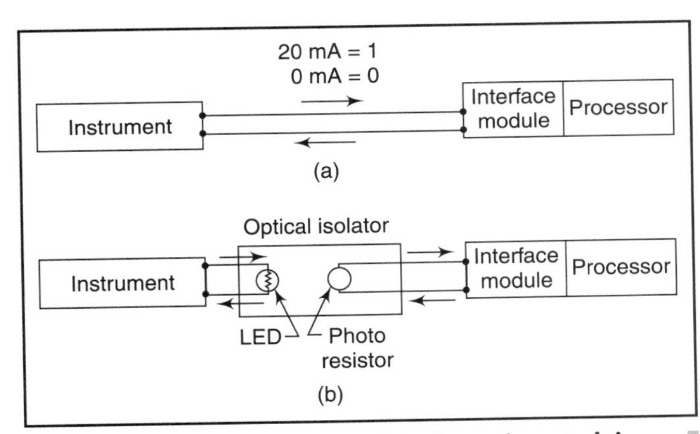

Figure 27.12 :: (a) 20 mA serial ASCII interface (b) use of an optical isolator inserted in a 20-mA loop to avoid two active devices transmitting signals over the same loop. For two-way communications between instrument and computer, two independent 20 mA loops with isolators are used

instantaneous current of 20 mA to represent a '1' bit and zero current to represent a '0' bit. Obviously, there cannot be two active devices on the same 20 mA loop. This is done by inserting an optical isolator between the instrument and interface.

The 20 mA bit transmission interface permits long distance communication between the instrument and the computer, wherein the distance could be as much as 1000 ft. Its main disadvantage is that, there are no extra lines for handshaking or modem control, and therefore 20 mA transmission is limited to local wiring only. In the absence of handshaking lines, the synchronization of the instrument and computer is provided by software on both sides.

RS-232 C Interface: The most prevalent digital connection between instruments and computers is the RS-232C standard recommended by EIA (Electronic Industry Association), in which data are represented by the voltage level rather than current as in a 20 mA loop. The '1' state is represented by any voltage that is more negative than 3V dc to ground and the '0' state by any voltage more positive than + 3V dc to ground (*Dessy, 1986a*).

Figure 27.13 shows a typical RS 232C interconnect, which makes use of the pins on the 25-pin output connector on the instrument. Pin '2' is the transmit pin and pin '3' is the receive pin, whereas pin '7' is the ground connection. In addition to these, there are a number of specifically assigned control and handshake pins that may or may not be utilized by the instrument.

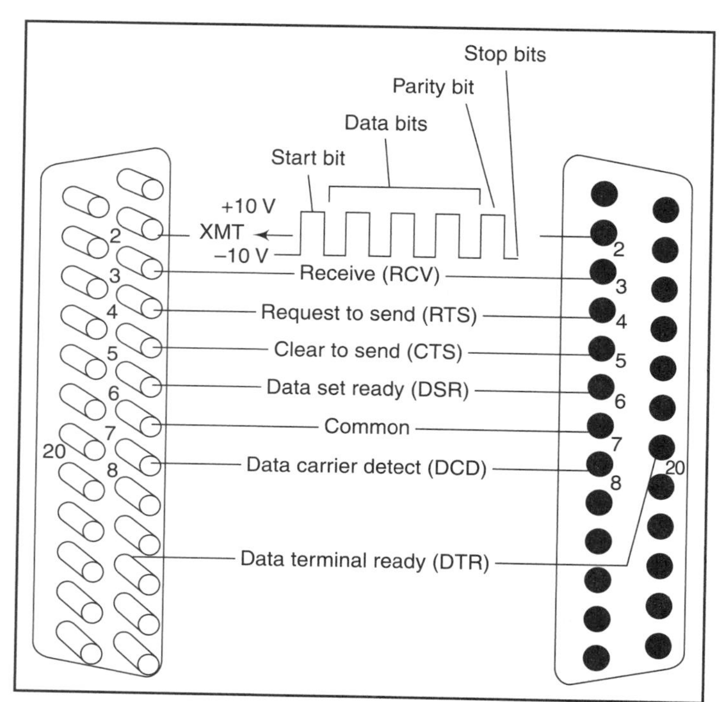

Figure 27.13 :: RS-232C interface—use of 25 pin connectors

In order to co-ordinate exchange, the data bit string is encapsulated between the START and STOP bits. This is the only synchronizing element in the communication link. After the START bit, the computer samples the incoming data stream at a rate determined by its own internal clock independent of the instruments clock. The communication is essentially asynchronous. A parity bit is also transmitted to enable the computer to ascertain whether the character received is valid.

In order to co-ordinate information exchange, the RS 232C protocol provides hardware 'handshake', which are shown in Figure 27.14. Referring to a modem (modulator-demodulator), one end is called the data terminal and the other the data set. According to convention, pin 20 (data terminal ready) and pin 6 (data set ready) are used to communicate that the units are physically present, power has been applied and they are operational. Pin 4 (request to send) is used to demand the attention of the computer while pin 5 (clear to send) acknowledges this request. Hardware handshake lines can be replaced by software handshakes involving the exchange of control characters along the pin 2 and pin 3 data lines.

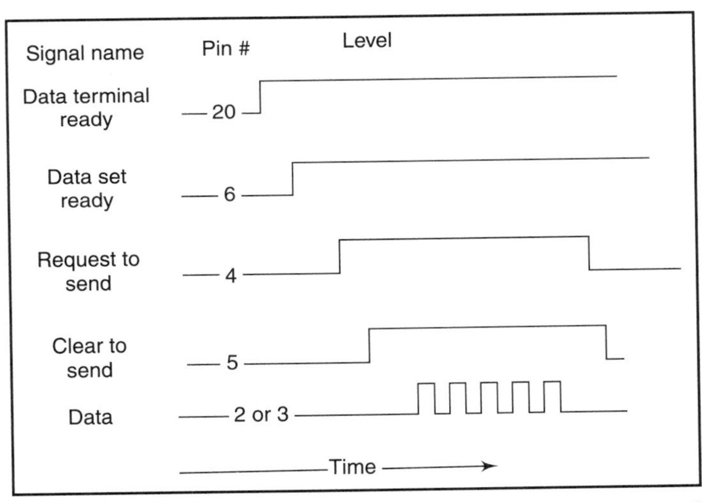

Figure 27.14 :: RS-232C protocol. The figure shows relative timing of the handshake signals

The computer must expect bits to arrive at the same rate at which the instrument places them on the line. This is expressed in bits per second, often referred to as the baud rate. The common values of the baud rate are 150, 300, 600, 1200, 2400, 4800, 9600 and 19,200 baud. At 9600 baud, it takes about 1 ms for a character to transit the wire. Directly connected RS 232C protocols usually support communication at 2400 baud for a kilometer. For longer distances, modems can be used. These are frequency shift keying (FSK) modems that encode data at speeds of up to 1200 baud. Modems operating at 9600 bits/s, which are based on the quadrature amplitude modulation technique, are also available.

The RS 232C circuit's noise susceptibility rises with the increasing baud rate and distance. RS 232C interface lines are therefore usually limited to a short distance between the instrument and the computer—20 meters is the maximum range for some instruments. Nevertheless, RS 232C transmission is far more popular than other interfacing techniques.

27.6.5 IEEE-488 Standard Interface (HPIB, GPIB)

The IEEE (Institute of Electric and Electronic Engineers) 488 standard defines the interface between the instrument and the computer in a digital data acquisition system. It basically defines the 16-line bus (physical cable) linking the instrument and the computer (called the controller); the shape, size, and number of pins in the connectors; all voltages, the handshaking procedure and the method of data transfer. In essence, the standard provides for a cable of eight data lines, eight parallel control and handshake lines and eight ground lines that interconnect a family of instruments (Figure 27.15). A 24-pin ribbon connector is used, with either tristate or open-collector TTL negative logic levels for the signal lines. Upto 15 instruments can be multi-dropped along IEEE-488 lines, which can reach 20 meters lengths and communication speeds of 500 k bytes/s (4000 k bits/s). Interface cards and software support programs for this popular interface are available from a number of vendors (Harman, 1986).

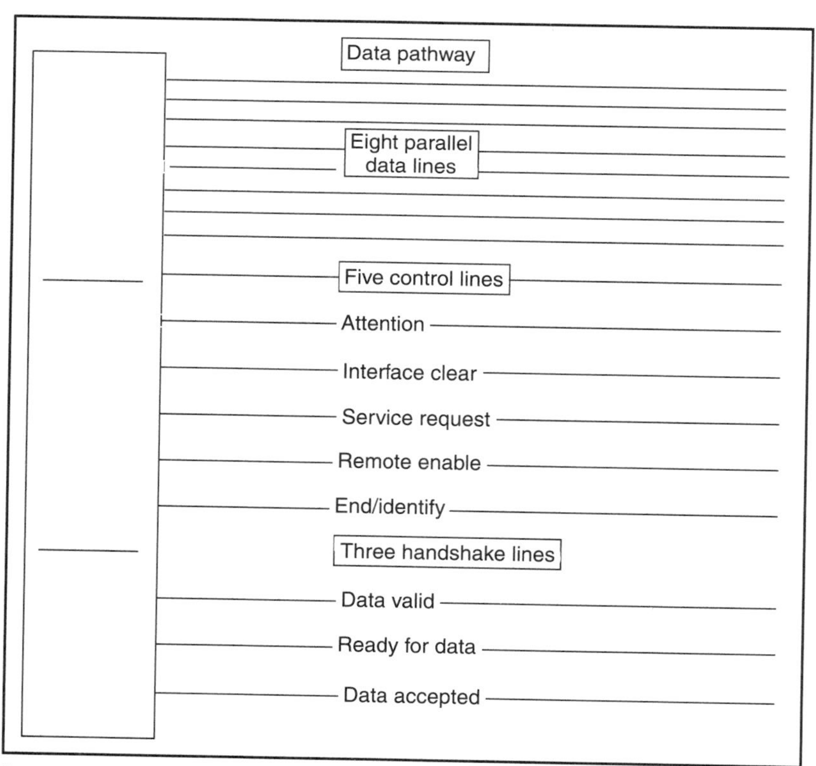

Figure 27.15 :: IEEE-488 standard

The IEEE-defined communications software in the instruments and the computers interface modules provides for the exchange of messages on the bus between what are called 'Talkers' and 'Listeners'. As the bus controller, the computer observes activity on the bus and issues commands, designating which instruments talk and which listen at any given time. Each instrument on the bus is assigned an address between zero and 30 that the controller uses to identify the instrument when instructing it to talk or listen. Each instrument's address can be set with switches located on the instrument itself.

The IEEE-488 standard describes a number of message routines through which instruments communicate over the bus. The contents of each message string and the vocabulary of meaningful characters are unique to the particular instrument and this information is provided by its user's guide.

All communications using GPIB, including commands and data, use a hardware handshake for every byte. All devices connected to the bus participate in that handshake. As a consequence, every device on the bus can influence the ongoing communication or cause severe communication problems such as bus hanging or data corruption. The reason for this can be firmware error or hardware failure in one of the participating devices such as a printer. Powering an idle GPIB device 'on' or 'off' during on-going communication also can cause such problems (Winter and Huber, 2000)

The IEEE-488 specification clearly defines the three elements of an active system that can most readily be standardized, namely, the mechanical interface, electrical interface, and data transfer. It precisely defines connector types, the number of connecting tunes, tune driver and receiver circuit parameters, and the exact protocol of message transfer between devices (Jensen, 1981).

The outstanding advantages of the standard IEEE-488 bus include byte serial, bit parallel digital data handling, synchronized communication among devices at varying data, rates and hardware interchangeability and interconnections (Williams, 1980). However, it requires highly complex logic protocol and time-consuming design analysis. This has been overcome with the availability of LSI (long scale integrated) chip Intel 8292 controller chip which has inbuilt IEEE-standard 488 logic controls.

27.6.6 LAN Communication Using TCP/IP

Most computer users have come to recognize the benefits of networking, both as a solution to system growth and expansion problems and as a flexible method of computer resource integration (Barrett, 1986). The success of networking, however, depends upon the ability to equip a broad range of equipment with a common interface. Ethernet has established itself as an industry standard networking medium (Wilson and Martin, 1986). Ethernet defines the electrical specifications for the physical connection of equipment to the network, but it does not specify any communications protocol standards for traffic wishing to use it. Of the various protocols available on Ethernet, only TCP/IP is truly vendor-independent.

Local area network communication using the transmission control protocol/Internet protocol (TCP/IP) thus enables devices to exchange information over a network. TCP/IP is often referred to as the 'Language of the Internet'. This technique involves the breaking of information into pieces or packets. The packets are specifically structured to allow error detection and correction by using redundancy mechanisms like checksums. In principle, checksums are a running total of all transmitted bytes attached to the packet and are used by the recipient to back-calculate and compare with the original checksum provided by the sender. If a mismatch is detected, a re-transmission is requested. This technique guarantees error-free data transport, and enables implementation of 'device checks' and system checks (Arneh, 1994).

Checksums

Communication in a TCP/IP environment is unaffected by the addition or removal of idle devices in the network and supports the safety procedures of analytical laboratories which necessitate the turning off of instruments not currently in use.

References

Ahn, B.K., A.O. Wist, E.C. Lia and W.H. Ko, (1975), "Development of a Miniature pH Glass Electrode with Field-effect Transistor Amplifier for Biomedical Applications", *Medical and Biological Engineering* Vol. 17, No. 4, p. 25.

Alegret, S., (2003), *Integrated Analytical Systems*, Elsevier, Amsterdam, London.

Alpert, N.L., (1974), Digital Flame Photometers—Part I, Instrument Series: Report #19, *LAB World*, May, p. CR-25.

Anderson, M.R., (1996a), Linear Sweep Voltametry, *www.elchem.kaist.ac.kr./vt/chem-ed/echem/linsweep.html*.

Anderson, M.R., (1996b), Normal Pulse Polarography, *www.elchem.kaist.ac.kr./vt/chem-ed/echem/npp.html*.

Anderson, M.R., (1997), Differential Pulse Polarography *www.elchem.kaist.ac.kr/vt/chem-ed/echem/dpp.html*.

Arneh, M.F., (1994), *Understanding Basic Network Concepts Inside TCP/IP*, New Riders Publishing, Indianapolis, p. 51.

Arnott, W.P., Hans Moosmuller and C. Fred Rogers, (1999), "Photoacoustic Spectrometer for Measuring Light Absorption by Aerosol: Instrument Description", *Atmospheric Environment*, Vol. 33, p. 2842.

Barrett, B., (1986), "Equipping Databases with an Ethernet Interface", *New Electronics*, 1 April, p. 48.

Baudean, J.E, L.S. Ettre, K.J. Hartigan, H. Hoberecht, E.W. March, R. Pigluicci, J.E. Purcell and B. Welton, (1977), "Improved Gas chromatographic Instrumentation", *American Laboratory*, 28 February 4 March.

Beaty, R.D., (1978), *Concepts, Instrumentation and Techniques in Atomic Absorption Spectrophotometry*, Perkin Elmer Corporation, USA.

Beaty, R.D. and Jack D. Kerber, (1993), *Concepts, Instrumentation and Techniques in Atomic Absorption Spectrophotometry*, Perkin Elmer Corporation, USA.

Bennett, W.H., (1950) Radio frequency mass spectrometer *Jr. Applied Physics*, Vol. 21, p. 143.

Bergveld, P. and N.F. de Rooji, (1979), "From Conventional Membrane Electrodes to Ion-sensitive Field-effect Transistors", *Med. And Biol Eng. Comput.*, Vol. 17, p. 647.

Blank, R.E. and Th. Wakefield II, (1979), "Double-beam Photoacoustic Spectrometer for Use in the Ultraviolent, Visible, and Near-Infrared Spectral Regions", *Analytical Chemistry*, Vol. 51, No.1, January, p. 50.

Bonczyk, P.A., (1975), Apparatus for local sensing of nitric oxide as a pollutant *Review of Scientific Instruments*, Vol. 46, p. 456.

Browne, J.P., (1986), "Modular Systems for Ion Chromatography", *Laboratory Practice*, January, p. 43.

Browner, F.R. and W.A. Boorn, (1984), "Sample Introduction Techniques for Atomic Spectroscopy", *Analytical Chemistry*, Vol. 56, No. 7, June, p. 875A.

Campana, A.M.G., W.R.G. Baeyens, and Yining Zhao (1997), "Chemiluminescence Detection in Capillary Electrophoresis", *Analytical Chemistry*, February 1, p. 83A.

Chalmers, J., (1983), "Trends in Fourier Transform Infrared Spectroscopy", *Laboratory Practice*, 1983.

Chaplin, M., (2004), What Are Biosensors? *www.lsbu.ac.uk/biology/enztech/biosensors.html*.

Chasteen, T.G., (2000), Gas Chromatographic Injectors, *www.chem.vt.edu/chem-ed/sep/gc/gcinject.html*.

Chopra, I., (1984), "Using a Scanning Densitometer", *Laboratory Practice*, November, p. 19.

Chopra, I., (1986), Derivative Spectroscopy: Theory and Practice", *Laboratory Practice*, October, p. 50.

Chu, L.D. and K.S. Zilora, (1986), "The PC Connection Part III", *Analytical Chemistry*, Vol. 58, No. 8, July, p. 919A.

Clark, L.C. Jr., (1956), "Monitor and Control of Blood and Tissue Oxygen Tensions" *Trans. Am. Soc. Artif. Internal Organs*, 2, 41.

Cooley, J.W. and J.W. Tukey (1965). An algorithm for the machine calculation of complex Fourier series, Math. Comp. Vol. 19, p. 297.

Cotter, R.J., (1999), "The New Time-of-Flight Mass Spectrometry", *Analytical Chemistry, News and Features*, July 1, p. 445A.

Covey, Th. R., D.E., Lee, P.A. Bruins and D.J. Henion, (1986), "Liquid Chromatography/Mass Spectrometry", *Analytical Chemistry*, Vol. 58, No. 14, December, p. 1451A.

Cox, M.F., G.N. Coleman and T.W. McCreary (1980), "Double Beam-in-Time Photoacoustic Spectrometer", *Analytical Chemistry*, Vol. 52, No. 9, August, p. 1420.

Davies, A.M.C, J. Franklin, Wright, K.M. M, Ring, S. and P.S. Belton, (1985), "Fourier Transforms Spectra", *Laboratory Practice*, July, p. 50.

De Graff, B.A., Mandy Hennip, Jalie M. Jones, Carl Salter and Stephanie A. Schaertel, (2002), "An Inexpensive Laser–Raman Spectrometer Based on CCD Detection", *Chemical Educator*, No. 7, p. 15.

Denton, M.B. and D.B. Swartz, (1974), "An improved ultrasonic nebulizer system for the generation of high density aerosol dispersions". *Review of Scientific Instruments*. Vol. 45, p. 81.

Dessy, E.R., (1984), "Managing the Electronic Laboratory, Part I", *Analytical Chemistry*, Vol. 56, No. 6, May, 1984, p. 725A.
Dessy, E.R., (1986a) "The PC Connection, Part I", *Analytical Chemistry*, Vol. 58, No. 6, 1986, p. 678A.
Dessy, E.R., (1986b). "The PC Connection", *Analytical Chemistry* Vol. 58, No. 7, June, p. 793A.
Diamond, D., (1998), *Principles of Chemical and Biological Sensors*, Wiley, New York.
Donovan, W.J., (1984), "Emerging Techniques, Scanning Calorimetry of Complex Biological Structures", *TIBS*, August, p. 340.
Ebdon, L., (1984), "Analytical Atomic Spectroscopy", *Laboratory Practice, Spectroscopy Supplement*, May, p. 10.
English, T.H., (1970) "An apparatus for cyclic-chronopotentiometry in non-aqueous solutions", *Jr. Phys. E. (Sc. Instruments)* Vol. 3, p. 69.
Ettre, L.S., (1978), "Selective Detection in Column Chromatography", *Journal of Chromatographic Science*, Vol. 16, September, p. 396.
Ettre, L.S., (1979), *The Nomenclature of Chromatography*, The Perkin-Elmer Corporation, Norwalk, Conn, USA, March 13, p. 235.
Eurachem, Guide 25 (1993), "Guidance Document No.1/WELAC Guidance Document No.WGD2: Accreditation for Chemical Laboratories: Guidance on the Interpretation of the EN 45000 Series of Standards and ISO/IEC."
Evans, C.A., Jr., (1972), "Secondary Ion Mass Analysis", *Analytical Chemistry*, Vol. 44, p. 67A.
Faithful, N.T., (1983), "A Rapid-response Recording System for Electrothermal Atomic Absorption Spectrophotometry", *Laboratory Practice*, Vol. 32, No. 7, July, p. 78.
Fell, F.A., (1985), "Diode Array Detection in HPLC", *Laboratory Practice*, March, p. 11.
Figeys, D. and D. Pinto, (2000) Lab-on-a-chip: a revolution in biological and medical sciences, *Analytical Chemistry* Vol. 72, p. 330A.
Floyd, A.K, A.V., Fassel, K.R, Winge, M.J. Katzenberger, and D.P.A Silva, (1980), "Inductively Coupled Plasma-atomic Emission Spectroscopy: A Computer-controlled, Scanning Monochromator System for the Rapid Sequential Determination of the Elements", *Analytical Chemistry*, Vol. 52, No. 3, March, p. 431.
Frant, S.M., (1980), "Process Infrared Measurements", *Analytical Chemistry*, Vol. 52, No. 12, October, p. 1331A.
Fraser, D.M., (1995), "Optical Biosensors-A Bright Future", *Medical Device Technology* Vol. 6, No. 5, p. 28.
Fraser, D.M., (1995), "Biosensors in Critical Care", *Medical Device Technology*, Vol. 6, No. 3, p. 36.
Frel, W.R, H. Jansen, and Th. A. U. Brinkman, (1985), "Post-column Reaction Detectors for HPLC", *Analytical Chemistry*, Vol. 57, No. 14, December, p. 1529A.
Friedman, H., (1984), "Working with Databases", *Radio Electronics*, April, p. 80.
Gagliardi, G., S. Viciani, and M. Inguscio, (2002), "Generation of Tunable Far-infrared Radiation with a Quantum Cascade Laser", *Optics Letters*, Vol. 27, No. 7, April, p. 52.
Galan, L. de, (1986), "New Directions in Optical Atomic Spectrometer", *Analytical Chemistry*, Vol. 58, No. 6, May.

Gambino, S.R., (1967), "Blood pH, pCO_2, Oxygen Saturation and pO_2, ASCP (American Society for Clinical Pathology) Commission on Continuing Education". Chicago, IL, USA.

George, R. and A. Adam, (1984), "A Challenge to Electrochemistry", *Laboratory Practice*, April, p. 41.

Gill, P.S., (1984), "Thermal Analysis Developments in Instrumentation and Applications", *Am. Lab.*, Vol. 16, No. 1, p. 39.

Glish, L.G. and D E. Goeringer, (1984), "Tandem Quadrupole/Time-of-Flight Instrument for Mass Spectrometry/Mass Spectrometry", *Analytical Chemistry*, Vol. 56, p. 2291.

Golay, M.J.E., (1947) "A pneumatic infra-red detector" *Rev. Sc. Instruments* Vol. 18, p. 347.

Gore, W.G. and G.W. Smith, (1974), An ultralow noise preamplifier and bias supply for photoconductive infrared detectors. *Jr. Phy. E (Sc. Instr.)* Vol. 7, No. 8 p. 644.

Grant, H., (1985), "NMR Spectroscopy for the Biologist", *Laboratory Practice*, September, p. 11.

Greenhalgh, B., (1983), "Key Features in Densitometry", *Laboratory Practice*, July, p. 44.

Hacke, D., (1985), "Titrimetry and Voltametry—A Marriage of Convenience", *Laboratory Practice*, December, p. 33.

Hahn. C.E.W, (1969), "The Measurement of Microcathode Currents by Means of a Field-effect Transistor Operational Amplifier System with Digital Display", *J. of Sc. Inst.* (*Physics E*), Vol. 2, p. 48.

Harman, A., (1986), "Bus Structures Find Their Application Niche", *New Electronics*, 1 April, p. 45.

Harris, T.D., (1982), "High-Sensitivity Spectrophotometry", *Analytical Chemistry*, 1982, Vol. 54, p. 741A.

Harrison, W.W., K.R. Hess, R.K. Marcus, and F.L. King, (1986), "Glow Discharge Mass Spectrometery", *Analytical Chemistry*, Vol. 58, No. 2, February, p. 341A.

Hawkes, E.G., (1984), "Developments in NMR", *Laboratory Practice*, *Spectroscopy Supplement*, May, p. 19.

He, P., P.J. Avery and R.L. Faulkner, (1982), "Cybernetic Control of an Electrochemical Repertoire", *Analytical Chemistry*, p. 1313A.

Hemingway, D.J., (1986), "Equipment and Applications for Gamma Spectrometry", *Laboratory Practice*, June, p. 15.

Hercules, M.D and H.S. Hercules, (1984a), "Analytical Chemistry of Surfaces, Part I, General Aspects", *Jour. of Chem. Edu.*, Vol. 61, No. 5, May, p. 402.

Hercules, M.D and H.S. Hercules, (1984b), "Analytical Chemistry of Surfaces, Part II", *Journal of Chemical Education* No.6, June, p. 483.

Hercules, M.D and H.S. Hercules, (1984c), "Analytical Chemistry of Surfaces, Part-III, Ion Spectroscopy", *Journal of Chem. Edu.*, Vol. 61, No. 7, July, p. 592.

Hertz, C.H. and B. Siesjo, (1959), "A Rapid and Sensitive Electrode for Continuous Measurement of pCO_2 in Liquids and Tissues *Acta Physiologica Scandinavica*, Vol. 47, p. 115.

Hibara, A., M. Tokeshi, K. Uchiyama, H. Hisamoto and T. Kitamori, (2001), *Anal. Lett.* 34, 1421.

Hill, D.W. and Tilsley, (1973), "A Comparative Study of the Performance of Five Commercial Blood Gas and pH Analyzers", *Br. gr. Anaesth*, Vol. 45, p. 467.

Hofmann, P., (2003), "Conjugated Surface Analysis System," *Electron Spectroscopy for Chemical Analysis*, *www.phys.du.dk/~Philip/q1_04/surflec/node24.html*.

Hogg, A.M., (1969), "A mass spectrometer variable leak inlet system for reference compounds used in high resolution mass measurement", *Jr. Phys. E. (Sc. Instruments)* 2, 289.

Hopkins, G.W. and Alfred Schwartz, (1980), "An Optical System for Full Spectrum Measurements", *Hewlett Packard Journal*, February, p. 17.

Houk, S.R., (1986), "Mass Spectrometry of Inductively Coupled Plasmas", *Analytical Chemistry*, Vol. 58, No. 1, January, p. 97A.

Howard, J., (1986), "Raman Spectroscopy—An Overview", *Laboratory Practice*, November, p. 13.

Human, M., (1985), *Basic UV/Visible spectrophotometry*, LKB Biochrom Ltd., UK.

Iwata, T., S. Endo and H. Okahana, (1986), "Computer-enhanced Fluorescence Spectroscopy", *International Laboratory*, October, 1986, p. 32.

Janata, J., (1989), *Principles of Chemical Sensors*, Plenum, New York.

Jarvis, K.F., A. L. Gray, and R.S. Houk (1992), *Handbook of Inductively Coupled Plasma–Mass Spectrometry*, Chapman and Hall, New York.

Jensen, M., (1981), "IEEE-488 Bus Compatible Digital Voltmeters—1", *New Electronics*, May, p. 38.

Jones, T.A, J.G. Firth, and A. Jones, (1971) A simple infra-red gas analyzer *Journal of Physics E (Scientific Instruments)*, Vol. 4, p. 792.

Katon, J.E, G.E. Pacey, and J.F. O'Keefe, (1986), "Vibrational Molecular Microspectroscopy", *Analytical Chemistry*, Vol. 58, No. 3, March, p. 465A.

Kerker, M. D. Wang, and H. Chew (1980), Surface Enhanced Raman Scattering, (SERs) by molecules adsorbed at spherical particles, *Applied optics*, Vol. 19 p. 3373.

Khetarpal, I. and Richard A. Mathies (1999), "Capillary Array Electrophoresis: DNA Sequencing", *Analytical Chemistry*, January 1, p. 31A.

Killick, C.M. (1969). An improved conductivity cell for the continuous measurement of sulphur dioxide in the atmosphere *Journal of Physics E (Scientific Instruments)* Vol. 2, p. 1017.

Kirkbright, F.G., (1984), "Photoacoustic Spectroscopy and Thermal Wave Imaging", *Laboratory Practice*, June, p. 11.

Knipe, A.C., D. Mclean, and R.L. Tranter, (1974), "A fast response conductivity amplifier for chemical kinetics", *Jr. Phys. E. (Sc. Instruments)* Vol. 7, No. 7 p. 586.

Knud, L. Knudsen and Robert W. Widmayer, (1980), "Light Detection and Measurement in a High Performance Spectrophotometer", *Hewlett Packard Journal*, February, p. 20.

Kollman M.T. and F.J. Reddish, (1983), "A Continuous Ion-selective Analyzer", *Laboratory Practice*, Vol. 32, No. 4, April, p. 80.

Koyuncu, B., (1986), "A Computer-controlled Laser Nephelometer", *Laboratory Practice*, November, Vol. 35, No. 11, p. 72.

Kreuzer, L.B., (1978) "Photoacoustic spectroscopy with light scattering samples", *Anal. Chem.* Vol. 50, p. 597A.

Lai, N.C., C.C. Lie, E.G. Brown, MR. Neumann and W.H. Ko, (1975), "Development of a Miniature pCO_2 Electrode for Biomedical Applications", *Med. And Biol. Eng.*, Vol. 13, p. 876.

Liebhafsky, H.A., (1972), *X-ray, Electrons and Analytical Chemistry*, Wiley-Inter Science, New York.

Liscouski, G.J., (1982), "Selecting Instrument Interfaces for Real-time Data Acquisition", *Analytical Chemistry*, Vol. 54, No. 7, p. 849A.

Lockhart, R.W., (1990), "Analog-to-digital Conversion", *Medical Electronics*, June, page 121.

Loon, V.C., (1980), "Direct Trace Elemental Analysis of Solids by Atomic (Absorption, Fluorescence, and Emission) Spectrometry", *Analytical Chemistry*, Vol. 52, No. 8, July, p. 955A.

Manning, R.L., (1969), *Introduction of Spectroscopy*, Pye Unicam Ltd., UK.

Manz, A. and H. Becker, (1998), *Microsystem Technology in Chemistry and Life Sciences*, Springer, Berlin.

McConn, R. and J. B. Robinson, (1963), "Notes on the Oxygen Electrode", *British Journal of Anesthesia*, Vol. 35, p. 679.

Miller, N.J., (1983), "Flow Injection Analysis—Flexible and Convenient Automation", *Laboratory Practice*, November, p. 86.

Miller, N.J., (1984), "UV/Visible, Fluorescence and Phosphorescence", *Laboratory Practice, Spectroscopy Supplement*, May, p. 6.

Moore, M.L. (2000). *www.homepage.gallaudet.edu/michael.Moore/AAS*

Morris D. M., (1986), "Thermal Effects in Chemical Analysis", *Analytic Chemistry*, Vol. 58, No. 7, June, p. 811A.

Morris, M.D., (1992), "Applied Laser Spectroscopy" in D.L. Andrews (ed.) VCH Publishers Inc., New York, Chapter 6.

Morrisson, A., (1984), "Computers Transform Infrared", *Laboratory Practice, Spectroscopy Supplement*, May, p. 5.

Muller, F., Popp, and F. Kuhnemann (2003), "Transportable, Highly Sensitive Photoacoustic Spectrometer Based on a Continuous—wave Dual-cavity Optical Parametric Oscillator, *Optics Express*, Vol. 11, No. 22, p. 2820.

Nier, A.O., (1940), "A mass spectrometer for routime isotope abundance measurements", *Rev. Sc. Instr.*, Vol. 11, p. 212.

Olesik, J.W., (1996), "Fundamental Research in ICP OES and ICP–MS", *Analytical Chemistry*, 1 August, Vol. 68, p. 469A.

Ottley, T., (1986), "The Use of Pyrolysis Mass Spectrometry", *Laboratory Practice*, October, p. 53.

Pauling, L.,R. Wood and C.O. Sturdevant, (1946), The paramagnetic oxygen meter *Science*, Vol. 103, p. 338.

Poole, C.F. and S.K. Poole, (1994), "Instrumental Thin Layer Chromatography", *Anal. Chem.*, Vol. 66, p. 27A.

Puton, J., K. Jasek, *et al.*, (2002), "Optimization of a Pulsed IR Source for NDIR Gas Analyzers", *Opto. Electronics Review*, Vol. 10, No. 2, p. 97.

Qlesik, John W., (1996), "Fundamental Research in ICP—OES and ICP-MS," *Analytical Chemistry*, 1 August, Vol. 68 p. 469A–474A.

Ray, D.J., H.D. Stedman, and J.G. Wendel, (1986), "Fast Chemiluminescent Method for Measurement of Ambient Ozone", *Analytical Chemistry*, Vol. 58, No. 3, p. 598.

Reyes, R.J. and G.R. Neville, (1967), "An Electrochemical Technic for Measuring Carbon Dioxide Content of Blood", *USAF School Aerospace Med. Tech. Report*, SAM-TR-67-23.

Robinson, M. (1990). "Shielding against EMI, Electronic Production", March 1990, p. 43.

Rosencwaig, A. and A. Gersho, (1976), "Theory of photoacoustic effect with solids", J. *Applied Physics*, Vol. 47, No. 1, p. 64.

Sacks, R., H. Smith and M. Nowak, (1998), "High Speed Gas Chromatography", *Analytical Chemistry News & Features*, January 1, p. 29A.

Sato, K., H. Akihide, T. Manabu, H. Hideaki, and K. Takehiko, (2003), "Integration of Chemical and Biochemical Analysis Systems into a Glass Microchip", *Analytical Sciences*, Vol. 19, January, p. 15.

Schrenker, H., (1975), Flow control in high-pressure liquid chromatography, *Hewlett Packard Journal*, Vol. 27, No. 2 p. 17.

Seitz, W.R., (1984), "Chemical Sensors based on Fibre Optics", *Analytical Chemistry* Vol. 56, No. 1, p. 16A.

Severinghaus, J.W and A.F. Bradley, (1958), "Electrodes for Blood pO_2 and pCO_2 Determination", *Journal Applied Physiology*, Vol. 13, p. 515.

Severinghaus, J.W., (1962), "Electrodes for Blood Gas, pCO_2, pO_2 and Blood pH", *Acta Anaesthesiology Scandanavica*, b. (suppl XI), Vol. 207, No. 18, p. 45.

Sharpe, M.R., (1984), "Stray Light in UV-VIS Spectrophotometers", *Analytical Chemistry*, Vol. 56, No. 2, February, p. 339A.

Siddiqui, A.S. and D. Stewart, (1974), "A broadband wavelength calibrator for use with low resolution for infrared monochromators", *Jr. Phys. E.* (*Sc. Instr.*), Vol. 7, p. 318.

Siggaard-Andersen, O., (1963), "Blood Acid Base Alignment Nomogram, Scales for pH, pCO_2 Base Excess of Whole Blood of Different Hamoglobin Concentration, Plasma Bicarbonate and Plasma Total CO_2" *Scand. J. Clin. Lab. Invest.*, Vol. 15, p. 211

Simpson, N. and J. Systems, (1980), "Single-chip or Single-board? The Designer's Choice", *Electronic Design*, Vol. 15, March, p. 37.

Slavin, W., (1982), "Atomic Absorption Spectroscopy, The Present and Future", reprinted from *Analytical Chemistry*, Vol. 54, 1982, p. 685A.

Sloper, A.N. and M.T. Flangan, (1994), "Metal Phosphate Planer Waveguides for Biosensors", *Applied Optics*, Vol. 33, No. 19, p. 4230.

Snook, R.D., (1986), "Ultra–violet Fourier Transform Spectrometry", *Laboratory Practice*, October, p. 60.

Soller, B.R., (1994), "Design of Intravascular Fibre Optic Blood Gas Sensors", *IEEE Engineering in Medicine and Biology*, Vol. 13, No. 3, p. 327.

Steinberger, E.H. and F. Goldwater, (1972), "A sensitive automatic meter for contenious sampling of containing oxidents in the atmoshere", *J. Phys. E.* (*Sc. Instruments*), Vol. 5, p. 373.

Stockwell, B.P., (1985), "Commentary: The Changing Role of Computers in the Laboratory", *Journal of Clinical Laboratory Automation*, Vol. 7, No. 2, April–June, p. 57.

Stow, R.W., R.F. Baer and B.F. Randall, (1957), "Rapid Measurement of the Tension of Carbon Dioxide in Blood", *Archives Physical Medicine and Rehabilitation*, Vol. 38, p. 646.

Strobel, A.H. (1984a) "Choosing the Right Instrument: The Modular Approach Part I", Analytical Chemistry Vol. 61, No. 2, February, p. A53.

Strobel, A.H. (1984b) "Choosing the Right Instrument: The Modular Approach Part II", *Analytical Chemistry* Vol. 61, No. 3, March, p. A89.

Tan, W., R. Kopelman, S.L.R. Basker and M.T. Miller, (1999), "Ultrasmall Optical Sensors for Cellular Measurements", *Analytical Chemistry*, September 1, 1999, p. 606A.

Tissue, B.M., (1996), "Electromagnetic Radiation", *www.chem.vt.edu/chem-ed/light/em-rad.html.*

Tissue, B.M., (2000a), "Introduction to Mass Spectrometry", *www.chem.vt.edu/chem-ed/ms/ms-intro.html.*

Tissue, B.M., (2000b), "Ion-selective Electrode", *www.chem.vt.edu/chem.ed/echem/ise.html.*

Tissue, B.M., (2004a), "Accuracy and Precision", *(www.chem.vt.edu/chem-ed/data/acc-prec.html.*

Tissue, B.M., (2004b), "Molecular Fluorescence Spectroscopy", *www.chem.vt.edu*/chem-ed/spec/molac/mol-fluo. html.

Tiwari, M.K, B. Gowrishankar, V.K. Raghuvanshi, R.V. Nandedkar and K.J.S. Sawhney (2002) "Development of a Total Reflection X-ray Fluorescence Spectrometer for Ultra-trace Element Analysis Bulletin Material Science."

Tiwari, M.K., (2002), "Development of a Total Reflection X-ray Fluorescence Spectrometer for Ultra-trace Element Analysis", *Bulletin Material Science* Vol. 25, No. 5, p. 435.

Turner, B., (1986), "An Evaporative Analyzer for HPLC Detection", *Laboratory Practice*, June, p. 11.

Vurek, G.G., et al., (1983), "A Fibre Optic pCO_2 Sensor", *Ann. Biomed. Eng.* Vol. 11, p. 499.

West, A. M., (1984), "Lasers in Spectroscopy, *Laboratory Practice*", *Spectroscopy Supplement*, May 1984, p. 13.

Wetsel, G.C., (1980) Photoacoustic effect in piezoelectriceramics *J. Opt. Soc. Am.*, Vol. 70, p. 471.

Wheeler, A.J. and A.R. Ganji, (1996), *Introduction to Engineering Experimentation*, Prentice-Hall. New Jersey, USA.

Willard, H., L. Merritt, J.A. Dean, and F.A. Settla Jr., (1988), *Instrumental Methods of Analysis*, seventh Ed., Wadsworth Publishing Co.

Williams, A.T.R., (1981), *An Introduction to Phosphorescence Spectroscopy*, A Perkin Elmer Publication.

Williams, M.R., (1980), "LSI Chips Ease Standard 488 Bus Interfacing", *Computer Design Magazine*, October 1979 issue, AR-113.

Wilson, J., (1985), Very Stable Power Supply for a Deuterium Lamp, *Laboratory Practice*, February 1985, p. 80.

Wilson, R. and J. Martin, (1986), "Chip Set Offers Complete Ethernet Interfacing Capability", *New Electronics*, 1 April, p. 50.

Windisch, G., (2003), FELMI-ZFE, Technische Universitaet Graz, Austria.

Winter, W. and L. Huber, (2000), "Implementing 21 CFR, Part-II in Analytical Laboratories Part-5, The importance of Instrument Control and Data Acquisition)" *Agilent Publication Number* 5988-0946 *EN.*

Worley, J. and S. Kvech, (2005), ICP—MS Instrumentation, *www.seals.manchester.ac.uk/—/Analytical Geochemistry Unit/Equipment and facilities/ICP-MS instrumentation.*

Wright J.C and Wirth, M.J. (1980b), "Principles of Lasers", *Analytical Chemistry*, Vol. 52, No. 9, August, p. 1087A.

Wright, J.C and M.J. Wirth, (1980a), "Lasers and Spectroscopy", *Analytical Chemistry*, Vol. 52, No. 9, August, p. 988A.

Wybrow, B.R.A., (1985), "A Microcomputer-based Injection System for Use in the Chromatographic Analysis of Gases", *Laboratory Practice*, Vol. 34, No. 4, April, p. 90.

Xu, Y., (1996), "Tutorial: Capillary Electrophoresis", *The Chemical Educator*, Vol. 1, No. 2, p. 1.

Yeung, E.S. and R.E. Synovec, (1986), "Detectors for Liquid Chromatography", *Analytical Chemistry*, Vol. 58, No. 12, October, p. 1237A.

Yeung, E.S. and M.J. Sepaniak, (1980), "Laser Fluorometric Detection in Liquid Chromatography", *Analytical Chemistry*, Vol. 52, No. 13, November, p. 1465A.

Yost, R.W., L.S. Ettre, and R.D. Conlon, (1980), *Practical Liquid Chromatography*, Perkin Elmer Corp., USA.

Zander, T.A., (1986), "Atomic Emission Sources for Solution Spectrochemistry", *Analytical Chemistry*, Vol. 58, No. 11, September, p. 1139A.

Zehnder, H. and W. Belew, (1970), "The Air Monitor IV, A New Approach in Air Monitoring", Presented at the Technicon International Congress, New York.

Index